Springer Series in Materials Science

Volume 293

The Springer Series in Materials Science covers the complete spectrum of materials research and technology, including fundamental principles, physical properties, materials theory and design. Recognizing the increasing importance of materials science in future device technologies, the book titles in this series reflect the state-of-the-art in understanding and controlling the structure and properties of all important classes of materials.

More information about this series at http://www.springer.com/series/856

Masataka Higashiwaki · Shizuo Fujita
Editors

Gallium Oxide

Materials Properties, Crystal Growth, and Devices

Editors
Masataka Higashiwaki
National Institute of Information
and Communications Technology
Koganei, Tokyo, Japan

Shizuo Fujita
Graduate School of Engineering
Kyoto University
Katsura, Kyoto, Japan

ISSN 0933-033X ISSN 2196-2812 (electronic)
Springer Series in Materials Science
ISBN 978-3-030-37152-4 ISBN 978-3-030-37153-1 (eBook)
https://doi.org/10.1007/978-3-030-37153-1

This Springer imprint is published by the registered company Springer Nature Switzerland AG
The registered company address is: Gewerbestrasse 11, 6330 Cham, Switzerland

Preface

Over the past 60 years, the history of semiconductor research and development has been driven by the exploration of new materials with different bandgap energy, starting from Ge and Si through III-V (GaAs, InGaAs, InP, and so on) to wide bandgap materials such as SiC and GaN. This was simply because the bandgap determines most of the fundamental optical and electrical properties of a semiconductor and thus leads to specific applications that capitalize on the strengths of each material. In the last decade, gallium oxide (Ga_2O_3) has become well recognized as a new wide bandgap semiconductor, whose material properties and device process technologies are being intensively investigated and developed these days. The renaissance in Ga_2O_3 research also stimulated a new semiconductor field called "ultrawide-bandgap semiconductors," which correspond to materials categorized by bandgap energy larger than the 3.4 eV value of SiC and GaN.

Ga_2O_3 technologies are rapidly evolving owing to active research and development worldwide. Our editing of this book was motivated by the desire to draw more attention to Ga_2O_3 from researchers, scientists, and eager students in the semiconductor field. This book provides extensive information about Ga_2O_3, covering physical properties recently elucidated to state-of-the art growth and device process technologies. Valuable knowledge was presented by about ninety leading researchers in this field.

Following an introduction in Chap. 1, this book is organized into four parts.

Part I

Chapters 2–4 introduce Ga_2O_3 melt bulk growth technologies: Czochralski, Bridgman, floating-zone, and edge-defined film-fed growth (EFG) methods. Wafer manufacturing from EFG Ga_2O_3 bulk crystals is also discussed in Chap. 4.

Part II

Chapters 5–16 treat various epitaxial growth technologies, such as molecular beam epitaxy, metalorganic chemical vapor deposition, halide vapor phase epitaxy, mist chemical vapor deposition, pulsed laser deposition, and low-pressure chemical vapor deposition.

Part III

Chapters 17–30 focus on material properties. Chapters 17 and 18 deal with the physical properties of Ga_2O_3 and its alloys from first principles calculations. Structural, electrical, optical, phonon, thermal, and scintillation properties of Ga_2O_3 and related materials are then discussed in Chaps. 19–30.

Part IV

Electrical and optical Ga_2O_3 devices are introduced in Chaps. 31–39, covering Ga_2O_3 field-effect transistors, diodes, ultraviolet detectors, and image sensors.

At the end, a personal recollection by Prof. Debdeep Jena (Cornell University, USA) of his Ga_2O_3 research is also included in Chap. 40 as a special contribution. He described why he was motivated to work in the new semiconductor research field of Ga_2O_3. This is especially useful for young and new audiences to the Ga_2O_3 field, who are the main targets of this book.

Ga_2O_3 is a unique material with many attractive properties. We hope that this book will familiarize the readers with the properties and possibilities of Ga_2O_3 and that the readers will be inspired to get involved in this emerging semiconductor field.

Last but not least, we would like to express our sincere thanks to all the contributors of the chapters in this book for their tremendous efforts in preparing the manuscripts.

Tokyo, Japan — Masataka Higashiwaki
Kyoto, Japan — Shizuo Fujita

Contents

Contributors

Fikadu Alema Agnitron Technology, Inc., Chanhassen, MN, USA

Aaron R. Arehart Department of Electrical and Computer Engineering, The Ohio State University, Columbus, OH, USA

Hagyoul Bae School of Electrical and Computer Engineering and Birck Nanotechnology Center, Purdue University, West Lafayette, IN, USA

Oliver Bierwagen Paul-Drude-Institut für Festkörperelektronik, Leibniz Institut im Forschungsverbund Berlin e.V, Berlin, Germany

Kelson Chabak Air Force Research Laboratory, Sensors Directorate, Wright-Patterson Air Force Base, OH, USA

Vanya Darakchieva Department of Physics, Chemistry and Biology (IFM), Linköpings Universitet, Linköping, Sweden

Robert F. Davis Department of Materials Science & Engineering, Carnegie Mellon University, Pittsburgh, PA, USA

Randy Elhassani University of Florida, Gainesville, FL, USA

Shizuo Fujita Kyoto University, Kyoto, Japan

Zbigniew Galazka Leibniz-Institut für Kristallzüchtung, Berlin, Germany

Krishnendu Ghosh Electrical Engineering Department, University at Buffalo, Buffalo, NY, USA;
Mechanical Engineering Department, University of Michigan, Ann Arbor, MI, USA

Ken Goto Department of Applied Chemistry, Tokyo University of Agriculture and Technology, Tokyo, Japan;
Novel Crystal Technology, Inc., Saitama, Japan

Andrew Green Air Force Research Laboratory, Sensors Directorate, Wright-Patterson Air Force Base, OH, USA

Marius Grundmann Felix Bloch Institute for Solid State Physics, Leipzig, Germany

Eric Heller Air Force Research Laboratory, Materials and Manufacturing Directorate, Wright-Patterson Air Force Base, OH, USA

Masataka Higashiwaki National Institute of Information and Communications Technology, Koganei, Tokyo, Japan

Keigo Hoshikawa Faculty of Engineering, Shinshu University, Nagano, Japan

Zongyang Hu School of Electrical and Computer Engineering, Cornell University, Ithaca, NY, USA

Debdeep Jena Departments of Electrical and Computer Engineering and Materials Science and Engineering, Cornell University, Ithaca, NY, USA

Gregg Jessen Air Force Research Laboratory, Sensors Directorate, Wright-Patterson Air Force Base, OH, USA

Makoto Kasu Department of Electrical and Electronic Engineering, Saga University, Saga, Japan

Noriaki Kawaguchi Nara Institute of Science and Technology, Nara, Japan

Sean Knight Department of Electrical and Computer Engineering, University of Nebraska - Lincoln, Lincoln, NE, USA

Keita Konishi Department of Applied Chemistry, Tokyo University of Agriculture and Technology, Tokyo, Japan

Rafał Korlacki Department of Electrical and Computer Engineering, University of Nebraska - Lincoln, Lincoln, NE, USA

Kimiyoshi Koshi Novel Crystal Technology, Inc., Sayama, Saitama, Japan

Sriram Krishnamoorthy Electrical and Computer Engineering, The University of Utah, Salt Lake City, UT, USA

Yoshinao Kumagai Institute of Global Innovation Research, Tokyo University of Agriculture and Technology, Koganei, Tokyo, Japan;
Department of Applied Chemistry, Tokyo University of Agriculture and Technology, Tokyo, Japan

Avinash Kumar Electrical Engineering Department, University at Buffalo, Buffalo, NY, USA

Akito Kuramata Novel Crystal Technology, Inc., Sayama, Saitama, Japan

Kevin D. Leedy Air Force Research Laboratory, Sensors Directorate, Wright-Patterson Air Force Base, OH, USA

Wenshen Li School of Electrical and Computer Engineering, Cornell University, Ithaca, NY, USA

Zeyu Liu Department of Aerospace and Mechanical Engineering, University of Notre Dame, Notre Dame, IN, USA

Tengfei Luo Department of Aerospace and Mechanical Engineering, Department of Chemical and Biomolecular Engineering and Center for Sustainable Energy of Notre Dame (ND Energy), University of Notre Dame, Notre Dame, IN, USA

Akhil Mauze Materials Department, University of California Santa Barbara, Santa Barbara, CA, USA

Piero Mazzolini Paul-Drude-Institut für Festkörperelektronik, Leibniz Institut im Forschungsverbund Berlin e.V, Berlin, Germany

Keitada Mineo NHK Science and Technology Research Laboratories, Tokyo, Japan

Alyssa Mock Electronics Science and Technology Division, U.S. Naval Research Laboratory, Washington, DC, USA

Bo Monemar Institute of Global Innovation Research, Tokyo University of Agriculture and Technology, Tokyo, Japan;
Department of Physics, Chemistry and Biology (IFM), Linköping University, Linköping, Sweden

Neil Moser Air Force Research Laboratory, Sensors Directorate, Wright-Patterson Air Force Base, OH, USA

Shin Mou Air Force Research Laboratory, Materials and Manufacturing Directorate, Wright-Patterson Air Force Base, OH, USA

Hisashi Murakami Institute of Global Innovation Research, Tokyo University of Agriculture and Technology, Tokyo, Japan;
Department of Applied Chemistry, Tokyo University of Agriculture and Technology, Tokyo, Japan

Adam T. Neal Air Force Research Laboratory, Materials and Manufacturing Directorate, Wright-Patterson Air Force Base, OH, USA

Hiroyuki Nishinaka Kyoto Institute of Technology, Kyoto, Japan

Jinhyun Noh School of Electrical and Computer Engineering and Birck Nanotechnology Center, Purdue University, West Lafayette, IN, USA

Go Okada Kanazawa Institute of Technology, Ishikawa, Japan

Takeyoshi Onuma Department of Applied Physics, School of Advanced Engineering and Department of Electrical Engineering and Electronics, Kogakuin University, Hachioji, Tokyo, Japan

Takayoshi Oshima Department of Electrical and Electronic Engineering, Saga University, Saga, Japan

Yuichi Oshima Optical Single Crystals Group, National Institute for Materials Science, Tsukuba, Ibaraki, Japan

Andrei Osinsky Agnitron Technology, Inc., Chanhassen, MN, USA

S. J. Pearton University of Florida, Gainesville, FL, USA

Hartwin Peelaers Department of Physics and Astronomy, University of Kansas, Lawrence, KS, USA

Lisa M. Porter Department of Materials Science & Engineering, Carnegie Mellon University, Pittsburgh, PA, USA

Siddharth Rajan Electrical and Computer Engineering, The Ohio State University, Columbus, OH, USA

Fan Ren University of Florida, Gainesville, FL, USA

Steven A. Ringel Department of Electrical and Computer Engineering, The Ohio State University, Columbus, OH, USA

Kohei Sasaki Novel Crystal Technology, Inc., Sayama, Saitama, Japan

Peter Schlupp Felix Bloch Institute for Solid State Physics, Leipzig, Germany

Mathias Schubert Department of Electrical and Computer Engineering, University of Nebraska - Lincoln, Lincoln, NE, USA;
Leibniz-Institut für Polymerforschung, Dresden, Germany;
Department of Physics, Chemistry and Biology (IFM), Linköpings Universitet, Linköping, Sweden

David I. Shahin Department of Materials Science and Engineering, University of Maryland, College Park, MD, USA

Mengwei Si School of Electrical and Computer Engineering and Birck Nanotechnology Center, Purdue University, West Lafayette, IN, USA

Uttam Singisetti Electrical Engineering Department, University at Buffalo, Buffalo, NY, USA

James Speck Materials Department, University of California Santa Barbara, Santa Barbara, CA, USA

Daniel Splith Felix Bloch Institute for Solid State Physics, Leipzig, Germany

Marko J. Tadjer Electronics Science and Technology Division, Power Electronics Branch, U.S. Naval Research Laboratory, Washington, DC, USA

Filip Tuomisto Department of Physics, University of Helsinki, Helsinki, Finland

Osamu Ueda Meiji University, Kawasaki, Kanagawa, Japan

Chris G. Van de Walle Materials Department, University of California, Santa Barbara, CA, USA

Joel B. Varley Materials Science Division, Lawrence Livermore National Laboratory, Livermore, CA, USA

Patrick Vogt Paul-Drude-Institut für Festkörperelektronik, Leibniz Institut im Forschungsverbund Berlin e.V, Berlin, Germany

Holger von Wenckstern Felix Bloch Institute for Solid State Physics, Leipzig, Germany

Günther Wagner Leibniz-Institut für Kristallzüchtung, Berlin, Germany

Shinya Watanabe Novel Crystal Technology, Inc., Sayama, Saitama, Japan

Virginia D. Wheeler Electronics Science and Technology Division, Power Electronics Branch, U.S. Naval Research Laboratory, Washington, DC, USA

Man Hoi Wong National Institute of Information and Communications Technology, Koganei, Tokyo, Japan

Minghan Xian University of Florida, Gainesville, FL, USA

Huili Grace Xing School of Electrical and Computer Engineering, Cornell University, Ithaca, NY, USA;
Department of Materials Science and Engineering, Cornell University, Ithaca, NY, USA;
Kavli Institute at Cornell for Nanoscale Science, Cornell University, Ithaca, NY, USA

Hirotaka Yamaguchi Advanced Power-electronics Research Center, National Institute of Advanced Industrial Science and Technology, Tsukuba, Ibaraki, Japan

Shigenobu Yamakoshi Novel Crystal Technology, Inc., Sayama, Saitama, Japan; Tamura Corporation, Sayama, Saitama, Japan

Yu Yamaoka Novel Crystal Technology, Inc., Sayama, Saitama, Japan

Takayuki Yanagida Nara Institute of Science and Technology, Nara, Japan

Jiancheng Yang University of Florida, Gainesville, FL, USA

Yao Yao Department of Materials Science & Engineering, Carnegie Mellon University, Pittsburgh, PA, USA

Peide D. Ye School of Electrical and Computer Engineering and Birck Nanotechnology Center, Purdue University, West Lafayette, IN, USA

Yuewei Zhang Materials Department, University of California, Santa Barbara, CA, USA

Hongping Zhao 205 Dreese Laboratory, Department of Electrical and Computer Engineering, Department of Materials Science and Engineering, The Ohio State University, Columbus, OH, USA

Hong Zhou School of Electrical and Computer Engineering and Birck Nanotechnology Center, Purdue University, West Lafayette, IN, USA

Chapter 1
Introduction

Masataka Higashiwaki

Abstract This introductory chapter provides current and comprehensive information about gallium oxide, covering crystal structures, basic physical properties, bulk melt growth and thin-film epitaxial growth methods, and representative electrical and optical devices.

1.1 Introduction

Silicon carbide (SiC) and gallium nitride (GaN) have been two leading wide-bandgap (WBG) semiconductors that have attracted much attention due to their excellent material properties based on their bandgaps. Recently, ultrawide-bandgap (UWBG) semiconductors, which are categorized as materials with a bandgap energy (E_g) larger than the 3.4-eV value of SiC and GaN, became recognized as an exciting and challenging new research field [1]. The UWBG semiconductors include high-Al composition AlGaN, diamond, boron nitride, and others. Gallium oxide (Ga_2O_3) is a representative UWBG semiconductor with an E_g of about 4.5 eV [2]. From the viewpoint of electronic device applications, Ga_2O_3 is expected to have superior material properties compared with those of SiC and GaN, which are typified by a prospective extremely large breakdown electric field (E_{br}) of >6 MV/cm [3].

Owing to the unique material properties and high crystal quality of Ga_2O_3, various applications can be envisioned. According to the current social trend that high-efficiency switching devices tend to become more and more important for our sustainable energy economy, unipolar Ga_2O_3 power switching devices are particularly promising and in need. Other potential applications of Ga_2O_3 transistors and diodes include harsh environment electronics that require device operation under extreme conditions such as high-temperature, radiation-rich, and chemically corrosive environments. The E_g of Ga_2O_3 lies in the deep ultraviolet (DUV) range and is thus also attractive for optoelectronic device applications. A variety of Ga_2O_3 DUV detec-

M. Higashiwaki (✉)
National Institute of Information and Communications Technology,
Koganei, Tokyo 184-8795, Japan
e-mail: mhigashi@nict.go.jp

M. Higashiwaki and S. Fujita (eds.), *Gallium Oxide*, Springer Series
in Materials Science 293, https://doi.org/10.1007/978-3-030-37153-1_1

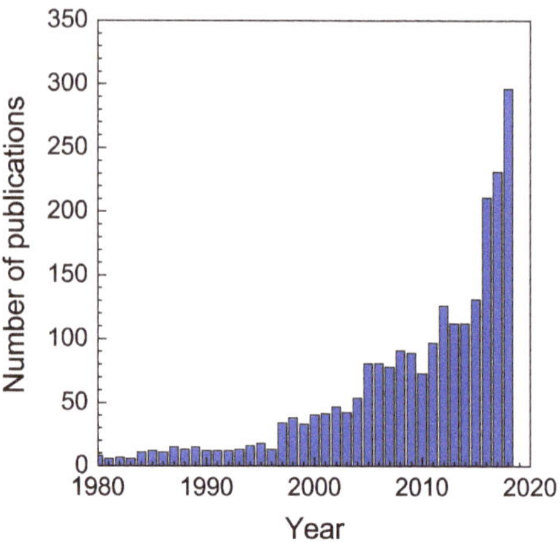

Fig. 1.1 Number of publications on Ga_2O_3 from 1975 to 2018. The papers are searched with the criterion of containing "Ga_2O_3" in the title (Data: Web of Science as of July 4, 2019)

tors and sensors have been demonstrated. Details of the device developments and applications are given in the subsequent sections.

Ga_2O_3 is not a novel material and has had a long history of more than 60 years. Five types of Ga_2O_3 polymorphs were first reported in 1952 [4], and the E_g of β-Ga_2O_3 was deduced from the optical properties of bulk single crystals in 1965 [5]. However, Ga_2O_3 had been largely ignored by a majority of the semiconductor researchers and engineers, which resulted in it falling behind SiC and GaN. The achievement of the first single-crystal Ga_2O_3 field-effect transistors (FETs) in 2011 is widely credited with changing the situation and invigorating Ga_2O_3 research [6]. As shown in Fig. 1.1, the number of publications on Ga_2O_3 significantly increases in recent years due to increased recognition among semiconductor researchers that Ga_2O_3 has unique and attractive material properties for various optoelectronic applications.

1.2 Material Properties

1.2.1 Crystal Structures of Ga_2O_3

Five polymorphs denoted by α, β, γ, δ, and ε have been confirmed for Ga_2O_3 single crystals [4]. Figure 1.2 depicts schematics of the α–ε crystal polymorphs. β-gallia structure is the thermodynamically stable form that belongs to the C2/m space group with lattice constants of a = 12.2 Å, b = 3.0 Å, and c = 5.8 Å, as shown in Fig. 1.3 [7]. It is a monoclinic structure with an angle between the a and c axes of about 104°. The unit cell of β-Ga_2O_3 contains two inequivalent Ga sites and three inequivalent O sites. Ga(I) and Ga(II) are tetrahedrally and octahedrally coordinated with O, respectively. O(I) and O(II) have threefold Ga coordination, while O(III) has fourfold. Being the only Ga_2O_3 crystal structure that can be grown from the melt, β-Ga_2O_3 has been extensively studied due to its stability and ease of production.

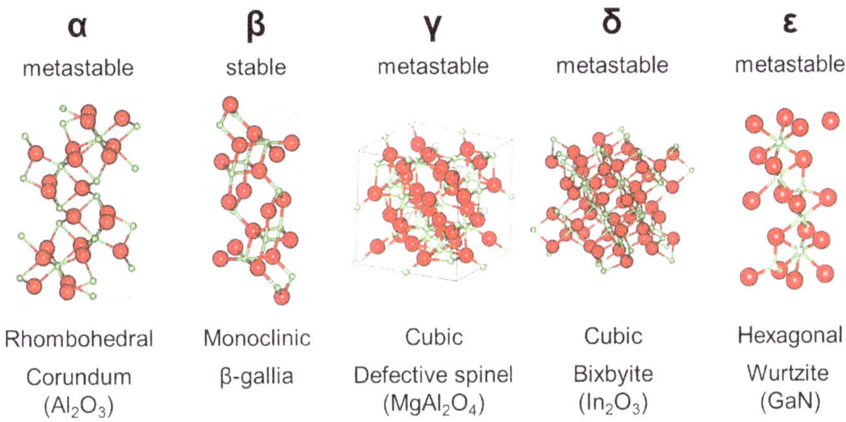

Fig. 1.2 Atomic unit cells of crystal polymorphs of Ga_2O_3 from α to ε. Courtesy of Dr. Takayoshi Oshima (Saga University, Japan)

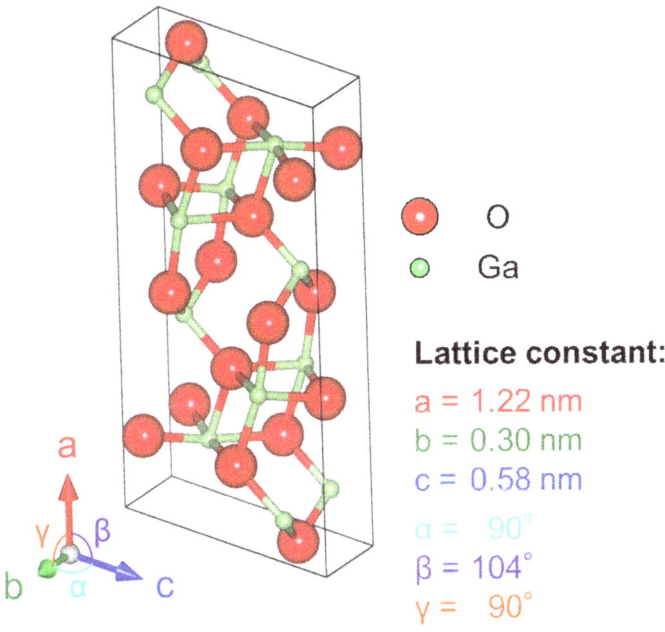

Fig. 1.3 Atomic unit cell of β-Ga_2O_3

Corundum α-Ga_2O_3 is the second most studied polymorph and thus a representative of the four metastable phases (α-, γ-, δ-, and ε-Ga_2O_3). α-Ga_2O_3 thin films are often formed in low-temperature heteroepitaxial growth on sapphire substrates. The E_g of α-Ga_2O_3 was estimated to be 5.2 eV from optical absorption [8]. Heteroepitaxial growth of defective-spinel γ-Ga_2O_3 and hexagonal ε-Ga_2O_3 thin films have also been reported. It is predicted that the hexagonal ε-Ga_2O_3 has a large spontaneous polarization; therefore, high-density two-dimensional electron gas may form at an $(AlGa)_2O_3/Ga_2O_3$ interface without any intentional doping, as is the case with an AlGaN/GaN interface. In addition to the five polymorphs confirmed in the early stages [4], an orthorhombic κ-phase polymorph, which has a very similar crystal structure to the hexagonal ε-Ga_2O_3 and is expected to exhibit a large spontaneous electrical polarization along the *c*-axis in the same manner, was discovered quite recently [9]. Formation of the metastable crystalline phases is strongly dependent on the lattice structure of the substrate and the growth temperature. High-temperature growth or annealing processes typically lead to the formation of β-Ga_2O_3 even for heteroepitaxially grown metastable phase crystals.

1.2.2 Physical Properties of β-Ga_2O_3

Some of the basic material parameters of β-Ga_2O_3 have not been fully understood, even though the β phase is the most extensively studied. The reported E_g values of β-Ga_2O_3 vary greatly between 4.4 and 4.9 eV [2, 5, 10, 11]. Onuma et al. reported that the energies of the absorption edge of β-Ga_2O_3 bulk single crystals can be divided into six ranges within 4.4–5.3 eV according to the selection rules of optical transition [2]. From their investigation, the direct and indirect E_g of β-Ga_2O_3 were estimated to be 4.48 and 4.43 eV, respectively.

n-type doping technology has been mostly established. Si, Ge, and Sn atoms are known to be shallow donors with small activation energies in Ga_2O_3. The activation energies of doped Si donors in Ga_2O_3 thin films grown by halide vapor phase epitaxy (HVPE) were estimated to be 45.6, 31.2, and 16.7 meV for electron densities (n) of 9.6×10^{15}, 1.9×10^{17}, and 1.2×10^{18} cm^{-3}, respectively [12]. Therefore, Si donors are almost 100% activated at room temperature (RT); as a result, the RT n is proportional to the doping density and can be controlled in the wide range of 10^{15} to 10^{19} cm^{-3}. The longitudinal optical (LO) phonon energies of β-Ga_2O_3 were characterized by infrared spectroscopic ellipsometry to be divided into three ranges of 35–48, 70–73, and 88–99 meV [13]. As a result, the RT electron mobility (μ) in Ga_2O_3 is limited up to 200 cm^2/V s by LO phonon scattering [14], even though the theoretically predicted and experimentally estimated electron effective mass (m^* = 0.23–0.34 m_0, m_0: free electron mass) is comparable to those of other WBG semiconductors [5, 15–17]. On the other hand, the saturation electron velocity is estimated from theoretical calculation to be larger than 1×10^7 cm/s [18], which is a sufficiently good value for many of high-frequency applications.

Three main drawbacks of Ga_2O_3 often pointed out in discussions are (i) a lack of hole-conductive *p*-type material, (ii) poor heat dissipation capacity, and (iii) absence of light emission by band-to-band transition.

Contrary to the ease of *n*-type doping, there has been no report of successful *p*-type doping with effective hole conduction in Ga_2O_3. There are three factors that make it almost impossible to realize hole-conductive *p*-type Ga_2O_3. First, it is difficult to find acceptor species with small activation energy. There are some candidates for deep acceptors in Ga_2O_3 such as Mg, Zn, Be, and N; however, theoretical studies have predicted that all these impurities would exhibit activation energies of larger than 1 eV [19–21]. This is because the valence bands of oxide semiconductors are generally composed of O 2*p* orbitals. O ions have a large electronegativity, and the O 2*p* orbitals are too stable to form shallow acceptor states. Second, the valence band maximum of Ga_2O_3 is theoretically characterized to have small dispersion and a very large effective mass, causing free holes to become almost localized with a small μ [5, 22, 23]. The O 2*p* orbitals that make up the valence band are considered to be nonbonding states since the overlaps between O ions are very weak, making it difficult to form effective conduction paths for mobile holes. Note that these two difficulties are general characteristics of oxide materials. Furthermore, it has been theoretically predicted specifically for Ga_2O_3 that free holes are localized in the bulk with a large self-trapping energy due to local lattice distortions, which result in the formation of small polarons that undoubtedly prohibit effective hole conduction [21, 22]. As discussed above, it can be considered that the realization of hole-conductive *p*-Ga_2O_3 hardly seems possible. However, even if only deep acceptors are available, it is possible to form a large energy barrier by forming a *p-n* junction.

Heat generation under high-power operation cannot be avoided even in high-efficiency power devices. The performance of many semiconductor devices is limited by their heat dissipation capacity since the resistance of the drift layer increases with rising operation temperature due to the decrease in electron μ. Therefore, low thermal conductivity is a serious potential weakness of Ga_2O_3, and thermal management will be one of the most important and challenging research and development (R&D) topics of Ga_2O_3 device technologies, even though most Ga_2O_3 devices are expected to be operational at high temperatures due to their extremely large E_g and high material stability. The experimental thermal conductivities of β-Ga_2O_3 fall in the range of 0.1–0.3 W/cm K at RT and are one order of magnitude smaller than those of SiC and GaN [24–26]. One of the effective ways to improve heat dissipation from Ga_2O_3 devices is direct bonding to a foreign substrate with good thermal conductivity. The research group of this chapter's author has developed a surface-activated bonding process to bond single-crystal Ga_2O_3 and polycrystalline SiC wafers [27]. The bonded substrate demonstrated a large mechanical bonding strength of over 10 MPa, together with a small specific electrical resistance of 2×10^{-4} $\Omega\,cm^2$ and a negligibly small thermal resistance at the bonding interface. Packaging solutions are also necessary for cooling devices from both front and back sides.

1.3 Bulk Melt Growth

The superiority of Ga_2O_3 devices for mass production stems from the availability of affordable native substrates fabricated from melt-grown bulk single crystals. β-Ga_2O_3 bulk single crystals have been synthesized by a variety of melt growth methods such as Verneuil [28], floating zone (FZ) [29–31], Czochralski (CZ) [32, 33], edge-defined film-fed growth (EFG) [31, 34], and Bridgman [35]. In contrast, SiC and GaN bulk crystals are produced by energy-intensive and cost-prohibitive methods such as sublimation, vapor phase epitaxy, and high-pressure synthesis. n-type Ga_2O_3 bulk crystals have been grown by shallow donor doping with Si or Sn, and semi-insulating bulk crystals with a resistivity of over 10^{12} Ω cm can be produced with Fe or Mg doping. Common wafer orientations are (100), (010), (001), and $(\bar{2}01)$. The principal planes of Ga_2O_3, namely (100), (010), and (001), are often used for homoepitaxial thin-film growth. The Ga_2O_3 $(\bar{2}01)$ crystal surface is almost equivalent to the c-face of corundum Al_2O_3 and is therefore suitable for heteroepitaxial growth of III-nitride materials (GaN, AlN, InN, and their alloys) [36].

1.4 Epitaxial Growth

High-quality epitaxial thin films are necessary for the design and fabrication of various types of optical and electrical devices. Avoiding undesirable impurities and minimizing structural imperfections such as dislocations, stacking faults, and point defects in epitaxial films lead directly to improvements in material properties. From this perspective, homoepitaxial growth on native substrates is considered to be ideal. As is the case with other compound semiconductors, there have been several epitaxial growth methods developed and used for Ga_2O_3, such as molecular beam epitaxy (MBE), HVPE, metalorganic chemical vapor deposition (MOCVD), and pulsed laser deposition (PLD). Furthermore, there is one unique CVD method called mist-CVD, which has been used for Ga_2O_3 homoepitaxial and heteroepitaxial thin-film growths. Ga_2O_3 is a relatively new material, whose epitaxial growth technologies are active R&D areas as of the writing of this chapter. The material quality of Ga_2O_3 epitaxial thin films has been improving under the fast-evolving growth technologies. In the early stages of development, epitaxy of Ga_2O_3 was mostly explored by MBE [37–40]. Afterward, HVPE demonstrated high-speed homoepitaxial growth of high-quality thin films with well-controlled electrical conductivity [12, 41, 42]. MOCVD is one of the most popular thin-film growth techniques for various compound semiconductors, because it produces high-quality films and is often used for mass production; therefore, similar prospects of technological success can be expected for Ga_2O_3 [43, 44]. PLD is suitable for rapidly surveying the properties of alloys and dopants [45–48]. Mist-CVD, which is a vapor deposition process using mist particles that contain source precursors, is a simple and cost-effective technique developed primarily for oxide semiconductors [8, 49]. Details of each growth technique are given in subsequent chapters contributed by specialists.

1.5 Electrical Devices

1.5.1 Advantages and Disadvantages of Ga_2O_3 Power Devices

Two types of devices currently utilized as key components for power conversion are switching transistors, which regulate the "on" and "off" states of current conduction; and rectifying diodes, which control the direction of current flow. The total energy loss in power switching devices consist of conduction and switching losses, where the former corresponds to resistive dissipation, and the latter is governed by capacitive components. Major requirements for efficient power devices include (i) low on-resistance (R_{on}), (ii) low off-state leakage current, (iii) fast switching recovery time, (iv) large breakdown voltage (V_{br}), and (v) high-temperature operation. Since the E_{br} of Ga_2O_3 is expected to be much larger than those of SiC and GaN, the drift layer resistance in Ga_2O_3 unipolar devices can be lower than those of SiC and GaN devices at a given V_{br} by reducing the layer thickness and/or increasing the donor doping concentration. The larger E_{br} enhances the V_{br} of Ga_2O_3 devices while maintaining acceptable R_{on}. Unipolar devices have strong advantages over bipolar devices for high-speed and low-loss switching due to fast drift transport of majority carriers. When bipolar devices are switched between the on and off states, charging and discharging of the depletion region by diffusion of minority carriers cause additional switching loss, because these processes correspond to intermediate states that are neither "on" nor "off". With an E_g larger than those of SiC and GaN, Ga_2O_3 promises power transistors and diodes with excellent characteristics including high V_{br}, high-power capacity, and high efficiency.

Table 1.1 compares the important material properties of β-Ga_2O_3 with those of major semiconductors. The E_{br} of Ga_2O_3 is estimated to be 6–8 MV/cm, which is a few times larger than those of SiC and GaN. This very large E_{br} is the most attractive attribute of Ga_2O_3 for power devices, because Baliga's figure of merit, which is the basic parameter to gauge how suitable a material is for power devices, is proportional to the cube of E_{br} but only linearly proportional to μ [50, 51]. From the Baliga's figure of merit, the R_{on} of Ga_2O_3 devices can be expected to be several times lower than those of SiC and GaN devices at the same V_{br} under ideal conditions. Based on the unique material properties of Ga_2O_3, unipolar Ga_2O_3 devices are considered to have large potential to outperform Si, SiC, and GaN devices for high-voltage and/or high-power applications.

1.5.2 Schottky Barrier Diodes (SBDs)

SBDs have a relatively simple structure. Their device characteristics are directly related to the material properties represented by Baliga's figure of merit. Ga_2O_3 SBDs can offer fast switching operation with a small forward voltage drop, similarly to SBDs fabricated with other WBG materials. To date, vertical Ga_2O_3 SBDs fabricated on HVPE-grown n^--Ga_2O_3 drift layers on n^+-Ga_2O_3 (001) substrates have shown promising performance exemplified by a large V_{br} of over 1 kV and a relatively small R_{on} of less than 10 $m\Omega\,cm^2$ [52, 53]. However, those characteristics are still far from the theoretical unipolar limit estimated based on the material properties of Ga_2O_3. At present, the breakdown of Ga_2O_3 SBDs is mostly limited by catastrophic damage since edge termination techniques are still immature.

Table 1.1 Comparison of material properties between β-Ga_2O_3 and major semiconductors

	Si	4H-SiC	GaN	β-Ga_2O_3
E_g (eV)	1.1	3.3	3.4	4.5
Relative dielectric constant ε	11.8	9.7	9.0	10–12[a]
Room-temperature mobility μ (cm^2/V s)	1,400	1,000	1,200	200[b]
Breakdown electric field E_{br} (MV/cm)	0.3	2.5	3.3	>6
Saturation electron velocity v_{sat} ($\times 10^7$ cm/s)	1.0	2.0	2.5	1.5[c]
Baliga's FOM ($\varepsilon\mu E_{br}^3$)	1	340	870	>1,500
Johnson's FOM ($E_{br}^2 v_{sat}^2$)	1	280	760	>1,200
Thermal conductivity (W/cm K)	1.5	2.7	2.1	0.27[d]

[a]A. Fiedler, R. Schewski, Z. Galazka, and K. Irmscher, ECS J. Solid State Sci. Technol. **8**, Q3083 (2019)
[b]N. Ma, N. Tanen, A. Verma, Z. Guo, T. Luo, H. Xing, and D. Jena, Appl. Phys. Lett. **109**, 212101 (2016)
[c]K. Ghosh and U. Singisetti, J. Appl. Phys. **122**, 035702 (2017)
[d]Z. Guo, A. Verma, X. Wu, F. Sun, A. Hickman, T. Masui, A. Kuramata, M. Higashiwaki, D. Jena, and T. Luo, Appl. Phys. Lett. **106**, 111909 (2015)

1.5.3 FETs

Ga_2O_3 FET technology is probably most suitable for high-power switching applications since FET switches can block high voltage in the off-state and conduct large current in the on-state. For general high-voltage and/or high-power applications, vertical Ga_2O_3 transistors similar to those currently utilized for Si and SiC are strongly favored and will need to be developed, because they allow for far superior field termination and current drive. Furthermore, lateral Ga_2O_3 FETs have reasonably high potential for RF applications such as generating, switching, and amplifying RF signals. Harsh environment electronics that require device operation under extremely severe conditions such as radiation-rich, wide-temperature range, chemically corrosive, and high-humidity environment are other important application fields for Ga_2O_3 FETs, because UWBG oxide semiconductors are expected to be sufficiently robust to withstand such conditions.

Transistor action for single-crystal Ga_2O_3 devices was first demonstrated by fabricating simple metal-semiconductor FETs (MESFETs) [6]. Following the MES-

FETs, there have been significant challenges of developing Ga_2O_3 metal-oxide-semiconductor field-effect transistors (MOSFETs) [54–57]. Lateral Ga_2O_3 MOSFET technology has already made remarkable progress and reached several important milestones. $(AlGa)_2O_3/Ga_2O_3$ modulation-doped FETs (MODFETs), which are analogous to AlGaAs/GaAs MODFETs, have also been demonstrated [58, 59]. Nano-membrane Ga_2O_3 FETs delivered practical limits of device performance [60]. Most of the lateral Ga_2O_3 FETs reported to date have been depletion-mode devices that operate in the normally-on mode. There have also been some reports on lateral normally-off Ga_2O_3 MOSFETs [61–64]. However, most of them were enhancement-mode devices employing an unintentionally doped or an n^--Ga_2O_3 channel and thus had a small turn-on threshold gate voltage. The challenges to develop vertical Ga_2O_3 transistors have just begun [65, 66]. All the Ga_2O_3 transistors developed so far are unipolar devices, because it is almost impossible to realize hole-conductive p-Ga_2O_3, as discussed in the preceding section.

1.6 DUV Photodetectors

Ga_2O_3 is considered to be suitable for DUV optical device applications from the viewpoint of its very large E_g of over 4.5 eV. However, there has been no report on direct conduction-to-valence band transition in Ga_2O_3 since this phenomenon may be precluded by the self-trapping effect of holes [21, 22]. Therefore, developments of Ga_2O_3 optical devices are limited to detectors. Ga_2O_3-based DUV photodetectors fit in the UV-C (200–280 nm) spectral bands and can be solar-blind, because sunlight contains no spectrum in the UV-C range. Solar-blind photodetectors are of interest for various applications in flame detection. With increasing availability of high-quality Ga_2O_3 bulk single crystals, epitaxial thin films, and nanowires, various types of DUV detectors have been fabricated and characterized [67–70].

1.7 Summary

Ga_2O_3 is a unique material that exhibits many excellent and attractive physical properties based on its extremely large E_g and has emerged as a leading material among UWBG semiconductors. Fundamentals and technologies of Ga_2O_3 materials and devices are briefly introduced in this chapter. Various melt and epitaxial growth techniques for Ga_2O_3 bulk single crystals and thin films have been investigated. Many types of devices such as SBDs, FETs, and DUV detectors have also been demonstrated. Comprehensive developments of Ga_2O_3-related technologies have just begun, and much work still remains to be done in this new semiconductor research field. Details of cutting-edge Ga_2O_3 technologies and material physics will be discussed in the following chapters.

References

1. J.Y. Tsao, S. Chowdhury, M.A. Hollis, D. Jena, N.M. Johnson, K.A. Jones, R.J. Kaplar, S. Rajan, C.G. Van de Walle, E. Bellotti, C.L. Chua, R. Collazo, M.E. Coltrin, J.A. Cooper, K.R. Evans, S. Graham, T.A. Grotjohn, E.R. Heller, M. Higashiwaki, M.S. Islam, P.W. Juodawlkis, M.A. Khan, A.D. Koehler, J.H. Leach, U.K. Mishra, R.J. Nemanich, R.C.N. Pilawa-Podgurski, J.B. Shealy, Z. Sitar, M.J. Tadjer, A.F. Witulski, M. Wraback, J.A. Simmons, Adv. Electron. Mater. **4**, 1600501 (2018)
2. T. Onuma, S. Saito, K. Sasaki, T. Masui, T. Yamaguchi, T. Honda, M. Higashiwaki, Jpn. J. Appl. Phys. **54**, 112601 (2015)
3. M. Higashiwaki, G.H. Jessen, Appl. Phys. Lett. **112**, 060401 (2018)
4. R. Roy, V.G. Hill, E.F. Osborn, J. Am. Chem. Soc. **74**, 719 (1952)
5. H.H. Tippins, Phys. Rev. A **140**, A316 (1965)
6. M. Higashiwaki, K. Sasaki, A. Kuramata, T. Masui, S. Yamakoshi, Appl. Phys. Lett. **100**, 013504 (2012)
7. S. Geller, J. Chem. Phys. **33**, 676 (1960)
8. D. Shinohara, S. Fujita, Jpn. J. Appl. Phys. **47**, 7311 (2008)
9. M. Kneiß, A. Hassa, D. Splith, C. Sturm, H. von Wenckstern, T. Schultz, N. Koch, M. Lorenz, M. Grundmann, APL Mater. **7**, 022516 (2019)
10. M. Orita, H. Ohta, M. Hirano, H. Hosono, Appl. Phys. Lett. **77**, 4166 (2000)
11. H. He, R. Orlando, M.A. Blanco, R. Pandey, E. Amzallag, I. Baraille, M. Rérat, Phys. Rev. B **74**, 195123 (2006)
12. K. Goto, K. Konishi, H. Murakami, Y. Kumagai, B. Monemar, M. Higashiwaki, A. Kuramata, S. Yamakoshi, Thin Solid Films **666**, 182 (2018)
13. T. Onuma, S. Saito, K. Sasaki, K. Goto, T. Masui, T. Yamaguchi, T. Honda, A. Kuramata, M. Higashiwaki, Appl. Phys. Lett. **108**, 101904 (2016)
14. N. Ma, N. Tanen, A. Verma, Z. Guo, T. Luo, H. Xing, D. Jena, Appl. Phys. Lett. **109**, 212101 (2016)
15. J.B. Varley, J.R. Weber, A. Janotti, C.G. Van de Walle, Appl. Phys. Lett. **97**, 142106 (2010)
16. A. Mock, R. Korlacki, C. Briley, V. Darakchieva, B. Monemar, Y. Kumagai, K. Goto, M. Higashiwaki, M. Schubert, Phys. Rev. B **96**, 245205 (2017)
17. Y. Zhang, A. Neal, Z. Xia, C. Joishi, J.M. Johnson, Y. Zheng, S. Bajaj, M. Brenner, D. Dorsey, K. Chabak, G. Jessen, J. Hwang, S. Mou, J.P. Heremans, S. Rajan, Appl. Phys. Lett. **112**, 173502 (2018)
18. K. Ghosh, U. Singisetti, J. Appl. Phys. **122**, 035702 (2017)
19. J.L. Lyons, Semicond. Sci. Technol. **33**, 05LT02 (2018)
20. H. Peelaers, J.L. Lyons, J.B. Varley, C.G. Van de Walle, APL Mater. **7**, 022519 (2019)
21. T. Gake, Y. Kumagai, F. Oba, Phys. Rev. Mater. **3**, 044603 (2019)
22. J.B. Varley, A. Janotti, C. Franchini, C.G. Van de Walle, Phys. Rev. B **85**, 081109(R) (2012)
23. H. Peelaers, C.G. Van de Walle, Phys. Status Solidi B **252**, 828 (2015)
24. M. Handwerg, R. Mitdank, Z. Galazka, S.F. Fischer, Semicond. Sci. Technol. **30**, 024006 (2015)
25. M.D. Santia, N. Tandon, J.D. Albrecht, Appl. Phys. Lett. **107**, 041907 (2015)
26. Z. Guo, A. Verma, X. Wu, F. Sun, A. Hickman, T. Masui, A. Kuramata, M. Higashiwaki, D. Jena, T. Luo, Appl. Phys. Lett. **106**, 111909 (2015)
27. C.-H. Lin, N. Hatta, K. Konishi, S. Watanabe, A. Kuramata, K. Yagi, M. Higashiwaki, Appl. Phys. Lett. **114**, 032103 (2019)
28. A.B. Chase, J. Am. Ceram. Soc. **47**, 470 (1964)
29. E.G. Víllora, K. Shimamura, Y. Yoshikawa, K. Aoki, N. Ichinose, J. Cryst. Growth **270**, 420 (2004)
30. S. Ohira, M. Yoshioka, T. Sugawara, K. Nakajima, T. Shishido, Thin Solid Films **496**, 53 (2006)
31. A. Kuramata, K. Koshi, S. Watanabe, Y. Yamaoka, T. Masui, S. Yamakoshi, Jpn. J. Appl. Phys. **55**, 1202A2 (2016)
32. Y. Tomm, P. Reiche, D. Klimm, T. Fukuda, J. Cryst. Growth **220**, 510 (2000)

33. Z. Galazka, R. Uecker, D. Klimm, K. Irmscher, M. Naumann, M. Pietsch, A. Kwasniewski, R. Bertram, S. Ganschow, M. Bickermann, ECS J. Solid State Sci. Technol. **6**, Q3007 (2017)
34. H. Aida, K. Nishiguchi, H. Takeda, N. Aota, K. Sunakawa, Y. Yaguchi, Jpn. J. Appl. Phys. **47**, 8506 (2008)
35. K. Hoshikawa, E. Ohba, T. Kobayashi, J. Yanagisawa, C. Miyagawa, Y. Nakamura, J. Cryst. Growth **447**, 36 (2016)
36. M.M. Muhammed, M. Peres, Y. Yamashita, Y. Morishima, S. Sato, N. Franco, K. Lorenz, A. Kuramata, I.S. Roqan, Appl. Phys. Lett. **105**, 042112 (2014)
37. T. Oshima, T. Okuno, S. Fujita, Jpn. J. Appl. Phys. **46**, 7217 (2007)
38. K. Sasaki, A. Kuramata, T. Masui, E.G. Víllora, K. Shimamura, S. Yamakoshi, Appl. Phys. Express **5**, 035502 (2012)
39. S.W. Kaun, F. Wu, J.S. Speck, J. Vac. Sci. Technol., A **33**, 041508 (2015)
40. P. Vogt, O. Bierwagen, Appl. Phys. Lett. **108**, 072101 (2016)
41. H. Murakami, K. Nomura, K. Goto, K. Sasaki, K. Kawara, Q.T. Thieu, R. Togashi, Y. Kumagai, M. Higashiwaki, A. Kuramata, S. Yamakoshi, B. Monemar, A. Koukitu, Appl. Phys. Express **8**, 015503 (2015)
42. Y. Oshima, E.G. Víllora, K. Shimamura, Appl. Phys. Express **8**, 055501 (2015)
43. G. Wagner, M. Baldini, D. Gogova, M. Schmidbauer, R. Schewski, M. Albrecht, Z. Galazka, D. Klimm, R. Fornari, Phys. Status Solidi A **211**, 27 (2014)
44. M.J. Tadjer, M.A. Mastro, N.A. Mahadik, M. Currie, V.D. Wheeler, J.A. Freitas, J.D. Greenlee, J.K. Hite, K.D. Hobart, C.R. Eddy, F.J. Kub, J. Electron. Mater. **45**, 2031 (2016)
45. R. Schmidt-Grund, C. Kranert, T. Böntgen, H. von Wenckstern, H. Krauß, M. Grundmann, J. Appl. Phys. **116**, 053510 (2014)
46. F. Zhang, K. Saito, T. Tanaka, M. Nishio, M. Arita, Q. Guo, Appl. Phys. Lett. **105**, 162107 (2014)
47. R. Wakabayashi, T. Oshima, M. Hattori, K. Sasaki, T. Masui, A. Kuramata, S. Yamakoshi, K. Yoshimatsu, A. Ohtomo, J. Cryst. Growth **424**, 77 (2015)
48. K.D. Leedy, K.D. Chabak, V. Vasilyev, D.C. Look, K. Mahalingam, J.L. Brown, A.J. Green, C.T. Bowers, A. Crespo, D.B. Thomson, G.H. Jessen, APL Mater. **6**, 101102 (2018)
49. T. Kawaharamura, G.T. Dang, M. Furuta, Jpn. J. Appl. Phys. **51**, 040207 (2012)
50. B.J. Baliga, J. Appl. Phys. **53**, 1759 (1982)
51. B.J. Baliga, IEEE Electron Device Lett. **10**, 455 (1989)
52. K. Konishi, K. Goto, H. Murakami, Y. Kumagai, A. Kuramata, S. Yamakoshi, M. Higashiwaki, Appl. Phys. Lett. **110**, 103506 (2017)
53. W. Li, Z. Hu, K. Nomoto, R. Jinno, Z. Zhang, T.Q. Tu, K. Sasaki, A. Kuramata, D. Jena, H.G. Xing, in *Technical Digest of the IEEE International Electron Devices Meeting* (2018)
54. M. Higashiwaki, K. Sasaki, T. Kamimura, M.H. Wong, D. Krishnamurthy, A. Kuramata, T. Masui, S. Yamakoshi, Appl. Phys. Lett. **103**, 123511 (2013)
55. M. Higashiwaki, K. Sasaki, M.H. Wong, T. Kamimura, D. Krishnamurthy, A. Kuramata, T. Masui, S. Yamakoshi, in *Technical Digest of the IEEE International Electron Devices Meeting* (2013)
56. A.J. Green, K.D. Chabak, E.R. Heller, R.C. Fitch, M. Baldini, A. Fiedler, K. Irmscher, G. Wagner, Z. Galazka, S.E. Tetlak, A. Crespo, K. Leedy, G.H. Jessen, IEEE Electron Device Lett. **37**, 902 (2016)
57. M.H. Wong, K. Sasaki, A. Kuramata, S. Yamakoshi, M. Higashiwaki, IEEE Electron Device Lett. **37**, 212 (2016)
58. E. Ahmadi, O.S. Koksaldi, X. Zheng, T. Mates, Y. Oshima, U.K. Mishra, J.S. Speck, Appl. Phys. Express **10**, 071101 (2017)
59. S. Krishnamoorthy, Z. Xia, C. Joishi, Y. Zhang, J. McGlone, J. Johnson, M. Brenner, A.R. Arehart, J. Hwang, S. Lodha, S. Rajan, Appl. Phys. Lett. **111**, 023502 (2017)
60. H. Zhou, K. Maize, G. Qiu, A. Shakouri, P.D. Ye, Appl. Phys. Lett. **111**, 092102 (2017)
61. K.D. Chabak, N. Moser, A.J. Green, D.E. Walker, S.E. Tetlak, E. Heller, A. Crespo, R. Fitch, J.P. McCandless, K. Leedy, M. Baldini, G. Wagner, Z. Galazka, X. Li, G. Jessen, Appl. Phys. Lett. **109**, 213501 (2016)

62. M.H. Wong, Y. Nakata, A. Kuramata, S. Yamakoshi, M. Higashiwaki, Appl. Phys. Express **10**, 041101 (2017)
63. K.D. Chabak, J.P. McCandless, N.A. Moser, A.J. Green, K. Mahalingam, A. Crespo, N. Hendricks, B.M. Howe, S.E. Tetlak, K. Leedy, R.C. Fitch, D. Wakimoto, K. Sasaki, A. Kuramata, G.H. Jessen, IEEE Electron Device Lett. **39**, 67 (2018)
64. T. Kamimura, Y. Nakata, M.H. Wong, M. Higashiwaki, IEEE Electron Device Lett. **40**, 1064 (2019)
65. Z. Hu, K. Nomoto, W. Li, N. Tanen, K. Sasaki, A. Kuramata, T. Nakamura, D. Jena, H.G. Xing, IEEE Electron Device Lett. **39**, 869 (2018)
66. M.H. Wong, K. Goto, H. Murakami, Y. Kumagai, M. Higashiwaki, IEEE Electron Device Lett. **40**, 431 (2019)
67. P. Feng, J.Y. Zhang, Q.H. Li, T.H. Wang, Appl. Phys. Lett. **88**, 153107 (2006)
68. T. Oshima, T. Okuno, N. Arai, N. Suzuki, S. Ohira, S. Fujita, Appl. Phys. Express **1**, 011202 (2008)
69. R. Suzuki, S. Nakagomi, Y. Kokubun, N. Arai, S. Ohira, Appl. Phys. Lett. **94**, 222102 (2009)
70. S. Nakagomi, T. Momo, S. Takahashi, Y. Kokubun, Appl. Phys. Lett. **103**, 072105 (2013)

Part I
Bulk Growth

Chapter 2
Czochralski Method

Zbigniew Galazka

Abstract The Czochralski method is one of the leading research and industrial crystal growth technologies that enables to obtain large diameter single crystals of high structural quality at low production costs per volume unit. A possibility of obtaining bulk β-Ga_2O_3 single crystals by the Czochralski method expands the diversity of growth technologies for this compound towards large volumes and high quality suitable for epitaxial growth of layers and device fabrication. Ga_2O_3 is, however, thermally unstable at high temperatures and tends to decompose that has a high impact on the growth process, size, and structural quality of obtained crystals. Additionally, the growth process is also affected by electrical/optical properties of a growing β-Ga_2O_3 crystal. Ga_2O_3 thermodynamics combined with new technical solutions allowed to obtain 2-inch diameter cylindrical single crystals of β-Ga_2O_3 of high structural quality with further scale-up capabilities. Czochralski-grown bulk β-Ga_2O_3 single crystals can be easily doped with a diversity of elements to tune their electrical and optical properties. The bulk β-Ga_2O_3 single crystals can be obtained either as electrical insulators or semiconductors both with a high transparency in the UV and visible spectral regions.

2.1 Introduction

The method was invented in 1916 by Jan Czochralski in Allgemeine Elektricitäts-Gesellschaft (AEG), Berlin, where he worked with metals and alloys. The invention was made during measuring of crystallization velocity of metals (Sn, Zn, Pb) using an apparatus of his design and construction, as described in his paper in 1918 entitled "Ein neues Verfahren zur Messung der Kristallisationsgeschwindigkeit der Metalle" [1]. The apparatus consisted of a support with a crucible containing a molten metal at its base, a pulley, and a silk thread with a glass carrier passing over the pulley and trough two guiding slots. The glass carrier was coated with a metal film for

Z. Galazka (✉)
Leibniz-Institut für Kristallzüchtung, Max-Born-Str. 2, 12489 Berlin, Germany
e-mail: zbigniew.galazka@ikz-berlin.de

M. Higashiwaki and S. Fujita (eds.), *Gallium Oxide*, Springer Series in Materials Science 293, https://doi.org/10.1007/978-3-030-37153-1_2

better wetting. The carrier was dipped into the slightly overheated melt and pulled upwards once the melt was cooled down to the crystallization temperature. The pulling was aided by a clock mechanism and the velocity was measured with a pointer and scale. The carrier initially pulled up a small amount of the metal melt which crystallized once passing through the liquid–solid interface. Differences in pulling and solidification rates led to an increase or decrease of the crystal diameter being grown.

Following a comprehensive and detailed research study on the Czochralski method by Uecker [2], the method was used for growing crystalline metal fibres in Berlin just after its invention and was subjected to a number of modifications. Gomperz [3] (Germany, 1922) introduced a die and a gas coolant for a growing crystal, Grüneisen and Goens [4] (Germany, 1923) introduced oriented crystal seeds, Hoyem and Tyndall [5] (USA, 1929) introduced a seed necking, and Walther [6] (USA, 1937) introduced a crystal rotation and control of the shape and diameter of the growing crystal, as well as re-designed a furnace for relatively high melting point materials (NaCl) that produced large diameter (2 cm) oriented crystals. These developments combined with the fundamental concept of Jan Czochralski to pull a crystal up from the melt have established key aspects of the method already before World War II.

After World War II, the Czochralski method was used to grow single crystals of elemental semiconductor Ge by Teal and Little [7] (USA, 1950). The invention of a point-contact field-effect transistor by Bardeen, Brattain, and Shockley in 1947 lead to a rapid spread of the Czochralski method for growing large diameter (6 inch) Ge and Si single crystals in 50-ties of the twentieth century and their industrialization, wherein introduction by Dash [8] a neck procedure in 1958 enabled obtaining of high quality, dislocation-free Si crystals. Two decades later Hoshikawa et al. [9] (Japan, 1980) applied a magnetic field to damp the melt convection of conductive melts. Another modification to the Czochralski method was a continuous melt feeding developed by Petrov and Zemskov [10] (Russia, 1957). The Czochralski method also spread on III–V compound semiconductors that were utilized for the first time for growing InSb and AlSb crystals by Gremmelmaier and Madelung [11] (Germany, 1953). To grow volatile compound semiconductors (GaAs, InAs, InP), Gremmelmaier [12] (Germany, 1956) introduced a hot-wall modification, while Metz et al. [13] (USA, 1962) introduced the Liquid-Encapsulated Czochralski technique. The use of the Czochralski method for growing semiconducting crystals proceeded in parallel in USA, Europe, Russia, and Japan.

A discovery of a solid-state laser based on the ruby crystal increased a demand for high structural quality oxide crystals that could be achieved with the use of the Czochralski method. First oxide crystals ($CaWO_4$) by the Czochralski method were obtained in 1960 by Nassau and Van Uitert [14] (USA). Soon after other oxides were grown by the Czochralski method in USA, Europe, and later in Japan. Introducing to the Czochralski method electronic load cells for crystal weighting by Bardsley et al. [15] (UK, 1972) and for crucible weighting by Kyle and Zydzik [16] (USA, 1973) enabled fully an automatic diameter control of the growing crystal. Main events in development of the Czochralski method are listed in Table 2.1.

Table 2.1 Main, chronological events in development of the Czochralski method

Year	Main events in development of the Czochralski method	Inventor	References
1916	Invention of pulling a crystal from the melt	Czochralski	[1]
1922	Die and gas coolant	Gomperz	[3]
1923	Oriented seeds	Grüneisen and Goens	[4]
1929	Seed necking	Hoyem and Tyndall	[5]
1937	Crystal rotation, crystal shape control, high-temperature growth furnace	Walther	[6]
1950	First elemental semiconductors (Ge)	Teal and Little	[7]
1956	Hot wall	Gremmelmaier	[12]
1957	Melt feeding	Petrov and Zemskov	[10]
1959	Neck procedure for Si crystals	Dash	[8]
1960	First oxide crystal ($CaWO_4$)	Nassau and Van Uitert	[14]
1962	Liquid encapsulation Czochralski method	Metz et al.	[13]
1972	Automatic diameter control	Bardsley et al.	[15]
1980	Magnetic field	Hoshikawa et al.	[9]

Nowadays, the Czochralski method is used worldwide to grow different type of single crystals: metals, elemental semiconductors, compound semiconductors, halides, and oxides. As large as 450 mm diameter semiconductor and halide crystals, and 200 mm diameter oxide crystals (sapphire, $Bi_4Ge_3O_{12}$) are grown nowadays by the Czochralski method. Large crystal diameter, high structural quality, and low production costs per volume unit make the Czochralski method one of the most frequently used techniques at research and industrial levels.

2.2 Czochralski Growth Furnace for Oxides

A crystal growth station for growing oxide crystals by the Czochralski method consists of a water-cooled growth chamber accommodating a growth furnace and an inductive heater (RF coil) surrounding the growth furnace that is powered by an RF generator outside the growth chamber, a puller with a pulling shaft that is coupled with a load cell and pulling/rotation motors, and a computer with a PID (proportional, integral, differential) controller for an automatic crystal diameter control, that is coupled with the motors, load cell, and generator.

A heart of the growth station is the growth furnace that defines a growth environment, an example of which is shown in Fig. 2.1. The furnace consists of a metal

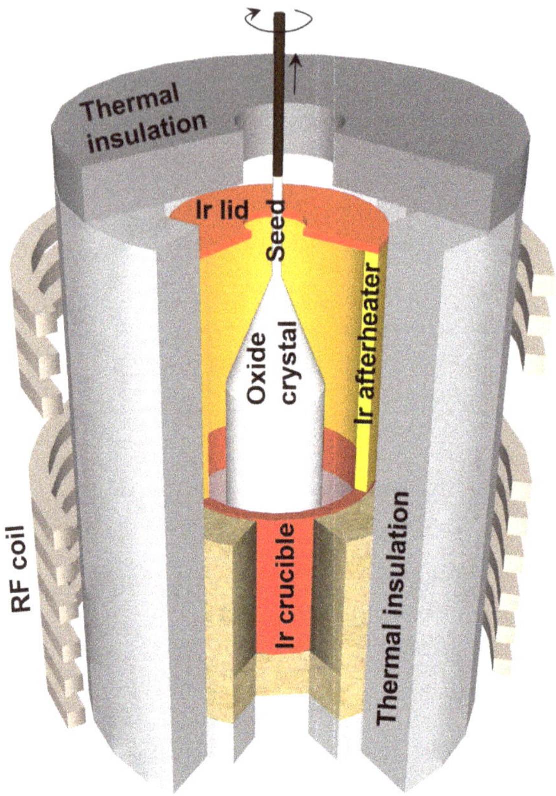

Fig. 2.1 Schematic view of a growth furnace of a general purpose for high melting point oxide crystals grown by the Czochralski method

crucible containing the melt, and a thermal insulation surrounding the crucible from all sides with a free space accommodating a growing crystal. The thermal insulation has a central opening at the top part for the pulling shaft, and a viewing window on the side (not shown) for optical observation of melting of the starting material, seeding, and crystal growth. To minimize temperature gradients in the pulling zone above the metal crucible a metal after heater (active or passive) can be utilized, possible with other metal shielding elements, such as lids. For high melting point oxides in the temperature range of about 1500–2200 °C Ir crucibles are most frequently used, at lower temperatures—Pt, while at higher ones—Mo, W, and Rh. Below about 1900 °C also alloys are sometimes utilized, such as Ir–Pt and Pt–Rh.

The thermal insulation in terms of the material type and geometry depends on the melting point of an oxide to be grown and the size of the crystal. It may comprise alumina, zirconia, magnesia, and quartz (outer part of the furnace). For high melting point oxides, to which β-Ga_2O_3 belongs, typically Ir crucibles are used with a typical wall thickness of 1.5–2 mm.

After melting and stabilizing the oxide starting material, an oriented crystal seed attached to the pulling shaft is dipped into the melt, which is then slowly pulled up while rotating. The seed diameter (usually a few mm) is next expanded to a pre-defined cylinder diameter which is typically grown at a constant rate of 1–5 mm/h. The rotation rate depends on the crystal diameter, crystal thermal properties and its transparency to infrared radiation, and on the growth rate. The crystal rotation rate values are usually between 5 and 30 rpm. Once the cylinder length achieves its final length (usually after crystallization of 40–75 wt% of the melt), it is separated from the melt and slowly cooled down to room temperature.

2.3 β-Ga_2O_3—Early Attempts with the Czochralski Method

The first mention of an application of the Czochralski method to grow β-Ga_2O_3 crystals was given in 1983 by Vasil'tsiv and Zakarko [17], which was summarized in one sentence: "*We used single crystals of the β-modification of gallium oxide grown by Czochralski's method from the melt in an iridium crucible*" (from English version published by Plenum Publishing Corporation). This work focuses on luminescence properties of Cr-doped β-Ga_2O_3 with no information about growth experiments and obtained crystals.

The next work on growth of β-Ga_2O_3 crystals by the Czochralski method was published in 2000 by Tomm et al. [18]. Here, a Ga_2O_3 decomposition issue was recognized as a limiting factor and a growth atmosphere consisting of 90 vol% Ar and 10 vol% CO_2 as an oxygen source was proposed. At such growth atmosphere small diameter (1 cm) β-Ga_2O_3 crystals with an irregular shape could be obtained.

An intensive development of the growth of β-Ga_2O_3 single crystals by the Czochralski method was started a decade later by Galazka et al. [19–23]. In-depth experimental and theoretical study of Ga_2O_3 thermodynamics, combined with a diversity of comprehensive growth experiments, led to understanding of the underlying physical phenomena involved in the growth of β-Ga_2O_3 crystals by the Czochralski method and consequently to solutions of the associated problems and limitations. This resulted in 2-inch diameter cylindrical single crystals with the weight up to 1 kg.

2.4 Thermodynamics of Ga_2O_3

Ga_2O_3 is thermally unstable at high temperatures and tends to decompose when grown from the melt (melting point, MP, of Ga_2O_3 is 1793 °C [24]), therefore thermodynamics constitutes a solid background for crystal growth from the melt.

The decomposition of Ga_2O_3 proceeds through a number of reactions (some of them defined by 2.1–2.3) that produce different volatile species: O_2, Ga_2O, GaO, and Ga, with Ga_2O having the highest partial pressure among other Ga-containing species. In addition to volatile species that evaporate, the decomposition of Ga_2O_3 also leads to a formation of metallic Ga in the melt (2.4).

$$Ga_2O_3(s) \rightarrow 2GaO(g) + \frac{1}{2}O_2(g) \tag{2.1}$$

$$2GaO(g) \rightarrow Ga_2O(g) + \frac{1}{2}O_2(g) \tag{2.2}$$

$$GaO(g) \rightarrow Ga(g) + \frac{1}{2}O_2(g) \tag{2.3}$$

$$Ga_2O_3(s, l) \rightarrow 2Ga(l, g) + 1\frac{1}{2}O_2(g) \tag{2.4}$$

where s, g, l refer to solid, gas, and liquid phase, respectively.

Figure 2.2 depicts stability fields of Ga_2O_3 as a function of oxygen partial pressure $p(O_2)$ and temperature. Around MP, the formation of the Ga_2O_3 phase (liquid and solid) would require $p(O_2) > 10^{-4.5}$ atm. At lower $p(O_2)$ values only gas phase will be stable, while at yet lower $p(O_2)$ values Ga_2O_3 will be reduced to metallic Ga. In practice, for obtaining bulk crystals from the melt of high structural quality, $p(O_2)$ should be much higher than the minimum value required for the formation of the Ga_2O_3 phase.

However, the Czochralski method imposes farther restrictions for $p(O_2)$ when growing β-Ga_2O_3 single crystals that arise from contradicting requirements for Ga_2O_3 stability (high $p(O_2)$) and Ir crucibles being used at high temperatures (low $p(O_2)$) that tends to oxidize. In practice, values of $p(O_2)$ should bring the partial

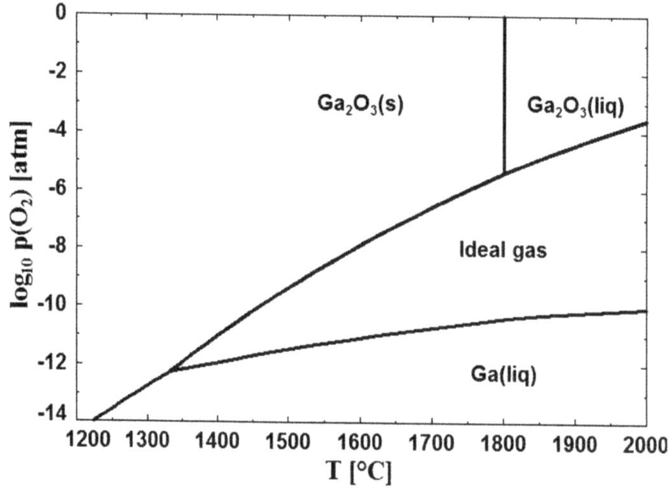

Fig. 2.2 Stability fields of the Ga–O system (s and l stand for solid and liquid, respectively)

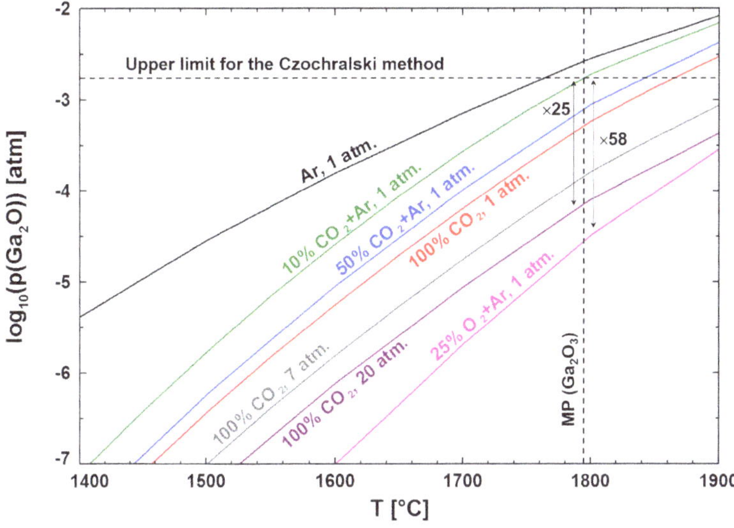

Fig. 2.3 Partial pressure of Ga_2O versus temperature at different oxygen sources and pressures

pressure of Ga_2O, $p(Ga_2O)$ to the level below 2×10^{-3} atm., which seems to be a limit for the Czochralski growth of β-Ga_2O_3, as indicated by the horizontal dashed line in Fig. 2.3. It corresponds to a minimum $p(O_2)$ of about 2×10^{-3} atm. Such conditions cannot be met by a neutral atmosphere, such as Ar (Fig. 2.3) or N_2. The minimum level of $p(O_2)$ that brings the $p(Ga_2O)$ just below the upper limit can be achieved by a mixture of Ar +10% CO_2. $p(Ga_2O)$ can further be decreased by increasing the CO_2 concentration in the (1–x)Ar (or N_2) + $x CO_2$ and applying an external overpressure (e.g. 7 or 20 atm. in Fig. 2.3). However, CO_2-based growth atmospheres enable only small diameter (about 1 cm) crystals. To increase the crystal size and structural perfection, higher oxygen concentrations in the growth atmosphere must be used, typically above 10 vol%. Such high oxygen concentrations can be achieved by a direct use of O_2 in a mixture of (1–y)Ar (or N_2) + $y O_2$ (e.g. 75 vol% Ar + 25 vol% O_2 shown in Fig. 2.3).

In particular, CO_2-containing atmospheres provide good results, as CO_2 is an effective oxygen source the concentration of which increases with temperature through the decomposition of CO_2:

$$CO_2(g) \xrightarrow{T} CO(g) + \frac{1}{2} O_2(g) \tag{2.5}$$

and keep Ir crucible intact from oxidation, which proceeds at low and moderate temperatures. Already Ar + 10% CO_2 enabled to obtain small diameter crystals, however, much better results were obtained with CO_2 content up to pure CO_2, which at atmospheric pressure provides $p(O_2) = 7 \times 10^{-3}$ atm. at MP. When using, e.g., directly Ar + 2% O_2 Ir crucibles oxidized to some extent, what was not the

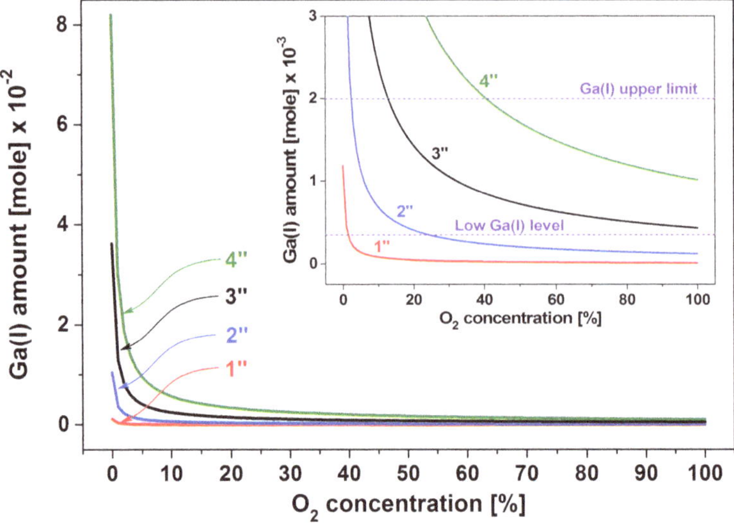

Fig. 2.4 Amount of liquid Ga in the Ga_2O_3 melt versus O_2 concentration during growth of 1, 2, 3, and 4 inch diameter crystals by the Czochralski method. For the calculations, the crucible aspect ratio was set to 1 and the crystal diameter to 0.5 of the crucible diameter. The inset shows a portion of the graph with low concentration range of metallic Ga(l) in the melt Reprinted (with some modifications) from Galazka et al. [22] according to ECS's policy

case for pure CO_2. Yet better results were obtained when using CO_2 under overpressure of 7 atm. that provided $p(O_2) = 2.5 \times 10^{-2}$ atm. at MP with substantially no crucible oxidation. CO_2 under 20 atm. provides $p(O_2) = 2.5 \times 10^{-2}$ atm. at MP and brings $p(Ga_2O)$ down about 25 times as compared with the $p(Ga_2O)$ at the upper limit. In the case of 25 vol% O_2 at atmospheric pressure, $p(Ga_2O)$ will be decreased 58 times.

CO_2-containing atmospheres, preferably under overpressure, were found effective for a decrease of $p(Ga_2O)$ that works well for crystal diameters $\leq$ 1 inch. However, larger diameter (> 1 inch) crystals could hardly be obtained at low oxygen concentrations or low $p(O_2)$. To grow larger crystals larger melt volumes are required that generate higher concentration of metallic Ga(l) (2.4), which is partly in the gas phase and evaporates, but partly remains in the melt in the liquid phase. Metallic Ga floating in the melt and on the melt surface prevents a proper seeding and degradates the crystal quality, if obtained at all. Furthermore, metallic Ga easily forms eutectic with Ir and may damage the crucible even after one growth run. To minimize the formation of metallic Ga much higher oxygen concentrations in the growth atmosphere are required. This is summarized in Fig. 2.4 that shows the amount of metallic Ga(l) in the melt depending on the oxygen concentration in the growth atmosphere during growth of 1, 2, 3, and 4 inch diameter crystals, that require 2.3, 18, 63, and 150 mol of the melt (with the crucible aspect ratio of 1).

Now it is clear that at small-scale growth (crystal diameter $\leq$ 1 inch) the amount of metallic Ga(l) is negligible small even at low oxygen concentrations (< 2 vol%) and will have only a minor impact on the crystal growth stability and crystal quality. Indeed, only small Ga-Ir eutectic signs were occasionally observed which allowed for a long-term usage of the same Ir crucible. However, the growth of 2-inch diameter crystals and larger produce so high metallic Ga(l) concentrations at low oxygen concentrations that make the growth almost impossible. To keep metallic Ga(l) concentration at low and save level (shown in the inset of Fig. 2.4 as a "Low Ga(l) level"), as in the case of 1-inch diameter crystals grown at 1–2 vol% O_2, would require about 20 and 100 vol% O_2 for 2- and 3-inch diameter crystals, respectively. In the case of 4-inch diameter crystals such low level of metallic Ga(l) concentration will never be reached, however, such large crystals still can be obtained, if Ga concentration does not pass the upper limit (inset of Fig. 2.4). Lower melt volumes for the large crystal diameters can be obtained by the use of crucibles with the aspect ratio (height to diameter) below unity.

The requirement for very high oxygen concentration to grow large diameter crystals contradicts the abovementioned use of Ir crucibles. Nonetheless, the long-lasting problem of Ir oxidation has successfully been solved and applied by Galazka et al. [21, 22]. For a better understanding of the problem solution, first we need to take a look at Ir oxidation. Ir oxidizes in such a way that at low and moderate temperatures (up to about 1200 °C) IrO_2 is formed, that is in the solid phase and remains on the crucible. This is a very efficient process and IrO_2 may contaminate the starting material (and the melt) in the crucible as well as a growing crystal. However, at higher temperatures IrO_2 is not formed any longer, but instead, IrO_3 forms, which is in the gas phase and evaporates. By a proper furnace design, the gaseous IrO_3 can be evacuated far from the melt and the growing crystal, minimizing in this way the possible contamination of the crystals with Ir. The activities of IrO_2(s) and IrO_3(g) at three different oxygen concentrations in the growth atmosphere (1.5, 25, and 100 vol%) are presented in Fig. 2.5. It can be easily concluded, that using even pure oxygen, the $p(IrO_3)$ at MP is at the level of 2×10^{-3} bar, comparable to the $p(Ga_2O)$ when growing small diameter crystals at low $p(O_2)$, so enabling the use of the Czochralski method. A suitable Ir crucible protection can significantly decrease Ir losses by evaporation.

On the other hand, Ga_2O_3 is substantially stable up to about 1200 °C even at a neutral atmosphere, as concluded from the thermal analysis and annealing experiments. Therefore, combining thermodynamics of Ir–O and Ga_2O_3 systems, we can construct a scheme of an oxygen supply to the growth furnace, which is shown in Fig. 2.6. During heating-up of a growth furnace, at low and moderate temperatures (up to about 1200 °C) solid IrO_2 is efficiently formed, while Ga_2O_3 is stable; therefore, in this temperature range (between room and transition temperatures, RT and TP) only a small O_2 concentration (< 2 vol%) can be used or even no oxygen at all. However, above the TP Ga_2O_3 quickly becomes unstable and higher O_2 concentration is needed. At the same time, IrO_2 is not formed any longer, while the $p(IrO_3)$ increases. Thus, at high temperatures (> TP) we continuously increase the O_2 concentration with increasing temperature up to a pre-defined value at MP

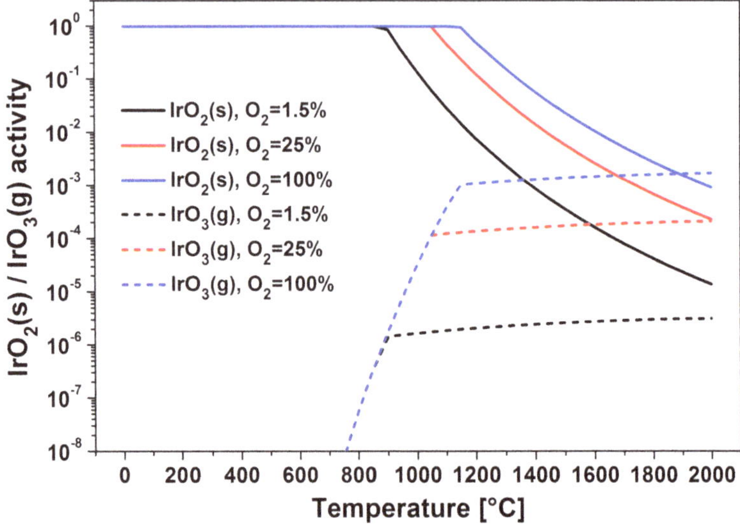

Fig. 2.5 Activities of IrO_2 (solid lines) and IrO_3 (dashed lines) versus temperature at three different oxygen concentrations in the growth atmosphere: 1.5, 25, and 100 vol%. In the case of the gas phase (IrO_3), the activity is equivalent to partial pressure in atm

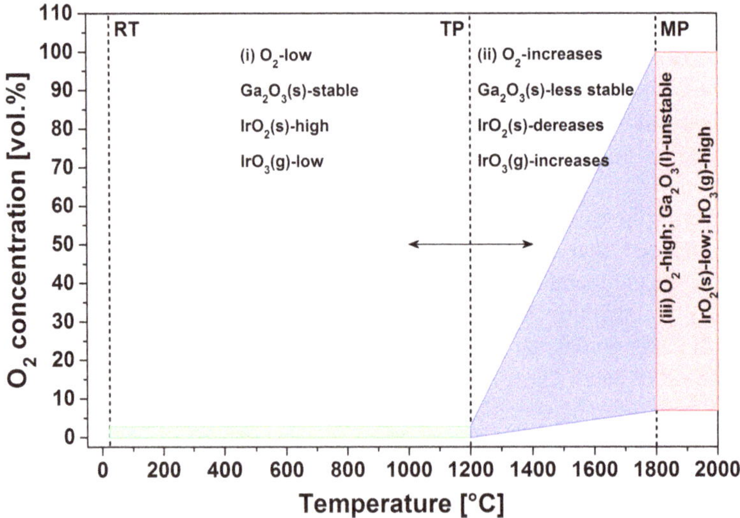

Fig. 2.6 Scheme of oxygen supply to a growth atmosphere allowing for using high oxygen concentration in the growth atmosphere and growing large diameter crystals from the melt. RT, TP, and MP are room temperature, transition point, and melting point, respectively. Reprinted from: Galazka et al. [22], according to ECS's policy

depending on the crystal size to be grown (or melt volume) and evacuate gaseous IrO_3. During crystal growth, we keep the final, high value of the O_2 concentration, which can be further increased or decreased to modify (to some extent) the electrical properties (at very high oxygen concentrations the crystals will be electrically insulating). During cooling down we can apply a reverse scheme of the oxygen supply as compared with the scheme used during heating-up of the growth furnace.

This technique is of a general purpose and can be used for any melt growth method involving a noble metal crucible. It can also be used for other materials that are thermally unstable at high temperatures, what was demonstrated for other transparent semiconducting oxides.

2.5 Crystal Growth

β-Ga_2O_3 single crystals are grown by the Czochralski method with a use of growth stations with an automatic diameter control utilizing the PID controller. The crucibles are made of high purity iridium (3N) of either 40 or 100 mm in diameter for growing 20 mm and 2-inch diameter crystals, respectively, defining the crystal to crucible ratio of about 0.5. Large crucibles had the aspect ratio <1. Above the crucible an active after heater is located that ensures low-temperature gradients in the space accommodating a growing crystal. The growth rate is typically between 1 and 2 mm/h, while the crystal rotation of 8–15 and 4–8 rpm for 20 mm and 2-inch diameter crystals, respectively. The length of the crystals is up to 70 and 100 mm for both crystal diameters, with the crystallization ratio not exceeding 50%. In each case to growth, direction is along <010> that is parallel to both cleavage planes {100} and {001}.

20 mm diameter β-Ga_2O_3 crystals obtained at low O_2 concentrations, with the use of the growth atmosphere $(1-x)$Ar + $x CO_2$ vol% with $x = 0.1$–1, or Ar + 2 vol % O_2 under atmospheric pressure, are transparent with a bit rough surface as the result of sublimation on the crystal surface of needle-shaped crystals from the gas phase, (decomposition and oxidation of Ga_2O at lower temperatures: $Ga_2O + O_2 \rightarrow \beta\text{-}Ga_2O_3$), by decomposition of the crystal surface itself, as well as by micro-cleaving of the crystal surface due to thermal stress. The degree of the surface roughness is also a function of the temperature gradients and growth time. The crystal surface becomes more smooth and shiny with higher O_2 concentrations obtained from the atmosphere $(1-y)$Ar + $y O_2$ vol% with $y = 0.02$–0.35. Relatively high O_2 concentrations (about 2.5–5 vol%) can also be obtained from the CO_2 atmosphere under a total overpressure of 7–20 bar. Examples of 20 mm diameter β-Ga_2O_3 crystals obtained by the Czochralski method at different oxygen concentrations in the growth atmosphere are shown in Fig. 2.7.

The colouration of undoped crystals varies from substantially colourless (low free electron concentration $< 10^{17}$ cm^{-3}), light blue to dark blue (high free electron concentration 10^{17}–10^{18} cm^{-3}), and yellowish (electrical insulators). The crystal colouration may also change depending on intentional doping, e.g. Sn and Si doping

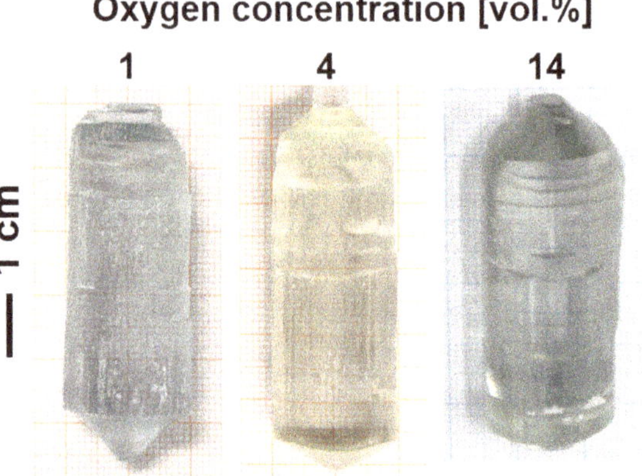

Fig. 2.7 20 mm diameter β-Ga_2O_3 crystals obtained by the Czochralski method at low (~1 vol%); high (4 vol%); and very high (14 vol%) oxygen concentration in the growth atmosphere. Different crystal colourations arise from different free electron concentrations

make the crystals dark blue (very high free electron concentration of $>10^{18}$ cm^{-3}), Mg and Ce—yellowish, Al—colourless (when grown at low O_2 concentrations), Cr—green, C (from CO_2 growth atmosphere under overpressure)—grey, but it also may change depending on O_2 concentration in the growth atmosphere.

Low O_2 concentrations (≤ 2 vol%) that produced high-quality 20 mm diameter crystals were not sufficient to obtain 2-inch diameter crystals of high structural quality, due to the formation of an excessive amount of metallic Ga in the melt. Such crystals were either polycrystalline or single crystalline with twins. In such cases, high O_2 concentrations were used during growth, typically in the range of 8–35 vol% with the use of the oxygen supply scheme shown in Fig. 2.6. The total Ir loses of unprotected crucibles is about 5–8 wt% at O_2 concentrations of 20–35% and 50–80 h growth time. Heating-up and cooling-down times for 2-inch diameter crystals were 10–15 and 20–30 h, respectively.

The 2-inch diameter crystals obtained at high O_2 concentrations and low-temperature gradients are transparent with a smooth and shiny surface due to a significant reduction of the Ga_2O_3 decomposition and evaporation. Despite of using high O_2 concentration and Ir parts, there are substantially no Ir particles on the crystal surface. This is the result of the specific scheme of the oxygen supply to the growth furnace, and a furnace design enabling an efficient evacuation of gaseous IrO_3 far from the melt and the growing crystal. Examples of Czochralski-grown 2-inch diameter β-Ga_2O_3 crystals (insulating and semiconducting) are shown in Fig. 2.8a.

The Czochralski-grown bulk crystals enable fabrication of differently oriented wafers for homoepitaxial growth of β-Ga_2O_3 layers. They could be either semi-conducting or electrically insulating. The easiest orientations for wafer fabrication are (100) and (001) being parallel to both easy cleavage planes {100} and {001}, respectively, as exemplary shown in Fig. 2.8b, including a misorientation. The most difficult orientation for wafer fabrication is (010) which is perpendicular to both easy cleavage planes, especially when a large wafer diameter (2 inch) and small thickness (0.5 mm) are in quest.

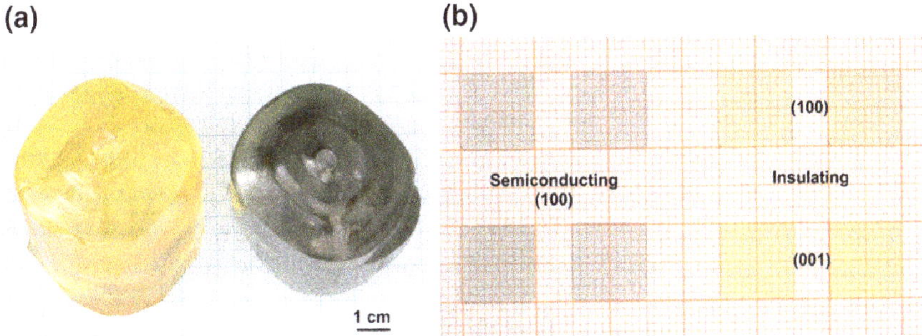

Fig. 2.8 **a** 2-inch diameter β-Ga_2O_3 crystals obtained by the Czochralski method at high oxygen concentration (8–35 vol%). The crystals are electrically insulating by doping with Mg (left) and semiconducting (right); **b** semiconducting and electrically insulating 10 × 10 mm^2 wafers prepared from 2-inch diameter crystals

2.5.1 Impact of the Free Carrier Absorption on the Growth Stability

A growing crystal is a medium for heat transfer to remove the latent heat of crystallization from the growth interface. Due to high melting point of Ga_2O_3 radiative heat transport dominates over conduction. The radiative heat transport proceeds in the near infrared (NIR) region with the maximum intensity at about

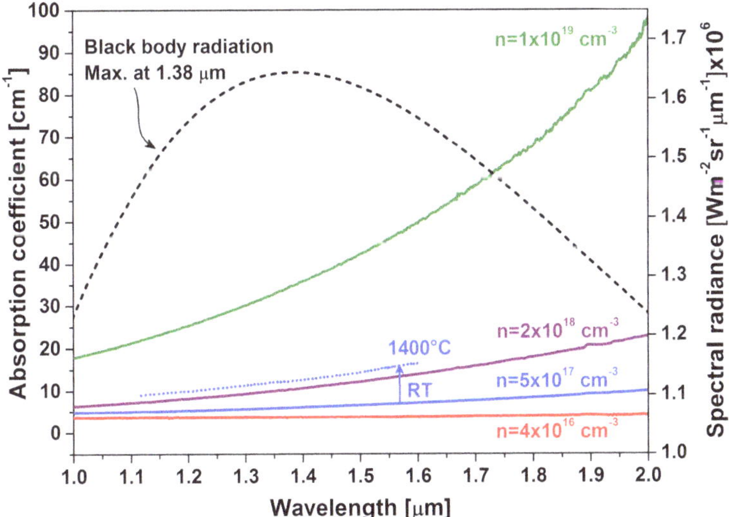

Fig. 2.9 Influence of the free electron concentration on the absorption coefficient of β-Ga_2O_3 single crystals versus wavelength at RT. The dotted line represents the absorption coefficient measured at 1400 °C for the same sample having free electron concentration at the level of 5×10^{17} cm^{-3} at RT. The dashed line represents the radiation spectrum of a black body at the melting point of Ga_2O_3. Reprinted from: Galazka et al. [20], Copyright (2014) with permission from Elsevier

1.38 μm. This is the spectral region, where the free carriers present in Ga_2O_3 (as it is a semiconductor) absorb the radiation, wherein the absorption increases with the free carrier concentration [20]. Figure 2.9 presents the black body radiation at MP of Ga_2O_3 and absorption coefficients of Czochralski-grown β-Ga_2O_3 crystals having different free electron concentration ranging from 4×10^{16} to 10^{19} cm^{-3}. These absorption coefficients were measured at RT, and the absorption increases with temperature, for example, at 1400 °C (dot line in Fig. 2.9), which is still far below the MP, the absorption coefficient increases with respect to that at RT by factor of about 2.

Problems with heat removal from the growth interface and its accumulation in the crystal due to free carrier absorption causes flattening and consequently convex-to-concave interface inversion once a crystal achieves a certain length, which decreases with the free carrier concentration. The concave interface is highly unstable and the most heat tries to dissipate through a triple point (solid crystal-melt-gas atmosphere) at the crystal periphery (meniscus). Such situation promotes a lateral growth over the melt surface (foot formation) that continuous to a certain point as it approaches the crucible wall (the hottest part of the growth furnace) and then changes the direction towards the central part of the crucible. Because the growing crystal is pulled up and rotates, this phenomenon will lead to a corkscrew (or spiral) formation. Such spiral structure enables a better heat dissipation as the crystal has a larger surface area exposed vertically to lower temperatures. The structural quality of the spiral part is at least as good as the cylindrical part of the crystal or even better; however, it provides less usable volume.

When considering 2 cm diameter crystals, the spiral substantially does not occur in electrically insulating crystals or at low free electron concentration, below 10^{17} cm^{-3}, as the interface is convex towards the melt (Fig. 2.10). The cylinder length may reach even 7 cm, corresponding to the crystallization fraction of about 50%. At higher free carrier concentration ($1–5 \times 10^{17}$ cm^{-3}) still a long cylinder can be grown, but the interface is only slightly convex towards the melt. At high free electron concentration (at the level of 10^{18} cm^{-3}), the cylinder length is of about 4–5 cm, while at very high free electron concentration of about 2–3 cm maximum. To obtain larger volume of highly conductive Ga_2O_3 the Vertical Gradient Freeze (VGF)/Bridgman methods were successfully utilized by Galazka et al. [21, 22].

In the case of 2-inch diameter crystals, the situation is more critical due to a larger optical thickness of the growing crystal. The cylinder length of 2-inch diameter crystals with the free carrier concentration of mid-10^{17} cm^{-3} approaches 2 inch, while in the case of electrically insulating crystals the cylinder length approaches 4 inch. However, it also depends on the temperature gradients. Larger temperature gradients enable longer cylinders; however, the crystal quality deteriorates due to twin formation. Electrically insulating crystals can be obtained in a different way: (i) by doping with Mg acting as a compensating acceptor, (ii) by doping with Al expanding the band gap, (iii) by using a very high oxygen concentration in the growth atmosphere that creates compensating gallium vacancies, and (iv) by a combination of any of (i)–(iii).

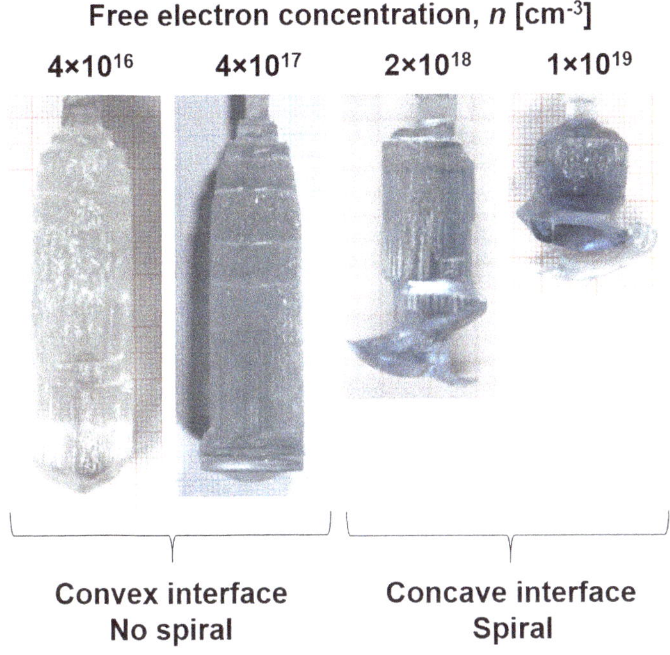

Fig. 2.10 β-Ga_2O_3 single crystals (2 cm diameter) showing an impact of the free carrier absorption on the growth stability: semiconducting with $n = 10^{16}$ cm^{-3} (undoped); semiconducting with $n = 10^{18}$ cm^{-3} (undoped); and semiconducting with $n = 10^{19}$ cm^{-3} (doped with Si)

2.6 Structural Quality

Structural quality of the obtained β-Ga_2O_3 crystals depends on many factors, in particular on (i) O_2 concentration in the growth atmosphere; (ii) temperature gradients in the growth furnace, (iii) seed quality, (iv) operating parameters; and (v) purity of the materials (starting material, crucible, insulation). Too low O_2 concentration in the growth atmosphere in relation to the melt volume may produce a polycrystal or a single crystal with a high density of twins. High-temperature gradients enable to obtain a single crystal with an extended length (better heat removal from the interface), but this is accompanied by a higher density of twins and also low angle grain boundaries, so the usable volume might be limited. High-temperature gradients during growth and cooling down may lead to an excessive thermal stress in the crystal that may result in crack formation, in particular along the {100} cleavage plane. If there are any twins in the seed, they can easily extend into the growing crystal, while propagating dislocations can be limited by a seed necking. High growth and crystal rotation rates may induce the spiral formation at early stage of growth (higher amount of the latent heat to be dissipated and enlarged forced convection promoting the interface inversion). Also, high

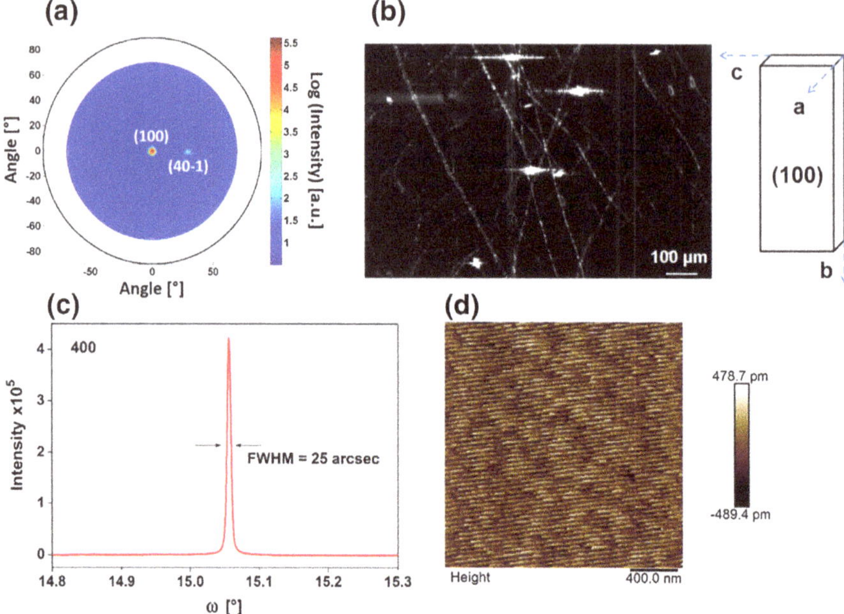

Fig. 2.11 Structural quality of Czochralski-grown β-Ga_2O_3 single crystals: **a** X-ray stereographic projection of an (100) oriented wafer, reprinted from: Baldini et al. [25], Copyright (2018) with permission from Elsevier; **b** dislocations revealed in the crystal volume by the laser scattering tomography, reprinted from: Galazka et al. [20], Copyright (2014) with permission from Elsevier; **c** FWHM of the rocking curve of an (100)-oriented wafer; and **d** roughness of an (100) wafer revealed by the atomic force microscopy, figure courtesy of Raimund Grüneberg and Günter Wagner, IKZ, Berlin, Germany

growth rates may induce higher dislocation density and other structural defects. Impurities may introduce point defects and induce growth instabilities and/or growth kinetics, and consequently decrease the crystal quality. Additionally, impurities at very low segregation coefficient may lead to a constitutional supercooling when high concentrations and/or high growth rates are used. Moreover, some of the impurities may affect heat transfer through the growing crystal (e.g. through enhancement of the free electron concentration or when they have absorption bands themselves in the NIR region) leading to growth instabilities. However, some of the impurities may have a positive, thermodynamically stabilizing effect (decrease decomposition rate of Ga_2O_3) that improves the crystal quality by lower density of point defect and more stable growth process.

Typically, β-Ga_2O_3 single crystals of 2 cm and 2 inch diameter obtained by the Czochralski method at optimized growth conditions and operating parameters are crack-free (Figs. 2.7 and 2.8) and twin-free, that is well visible in the X-ray texture shown in Fig. 2.11a. The dislocation density, as revealed by the laser scattering tomography, is below 5×10^3 cm^{-2} (Fig. 2.11b) [20]. A projection of the dislocations in the crystal volume revealed their propagation substantially parallel to the cleavage plane {100}. The full width at half maximum (FWHM) of the rocking curve is typically below 50, often below 30 arcsec [20, 22, 25], as shown in Fig. 2.11c. Chemo-mechanical polishing (CMP) of (100)- and (001)-oriented wafers prepared from the Czochralski-grown crystals produced a very smooth surface with the RMS roughness well below 0.5 nm, such as 0.15 nm shown in Fig. 2.11d. High structural quality of the β-Ga_2O_3 single crystals and fabricated wafers enable epitaxial growth of high-quality layer, as well as a preparation of a diversity of electronic and optoelectronic devices.

2.6.1 Intentional Doping and Residual Impurities

β-Ga_2O_3 single crystals grown by the Czochralski method can be easily doped with a number of elements to tune electrical and optical properties as well as to minimize the thermal decomposition and stabilize the growth process. The following elements were used for intentional doping of β-Ga_2O_3 crystals grown by the Czochralski method [20, 22, 23, 26]: Cu^{1+}, Mg^{2+}, Cr^{3+}, Al^{3+}, Ce^{3+}, Sn^{4+}, and Si^{4+}, which are added to a Ga_2O_3 starting material in the form of a corresponding dopant oxide powder. Examples of β-Ga_2O_3 single crystals doped with Mg, Ce, Al, Cr, and Si are shown in Fig. 2.12. Also other dopants, such as Li^{1+}, Ni^{2+}, Co^{2+}, and Ge^{4+},

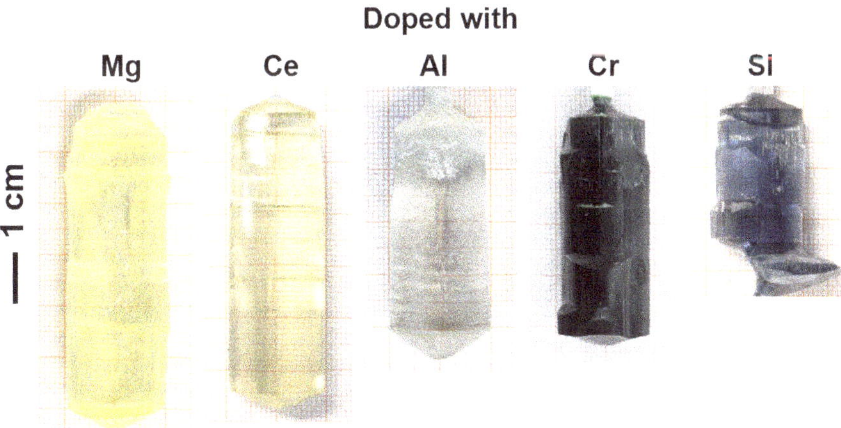

Fig. 2.12 Czochralski-grown β-Ga_2O_3 crystals doped with Mg, Ce, Al, Cr, and Si

as well as double and triple doping were studied in Czochralski growth experiments of Ga_2O_3 single crystals by Galazka et al. [27, 28].

Divalent ions aimed to compensate the electrical conductivity and, indeed, Mg fulfil this expectancy very well, while Cu has too high vapour pressure at MP of Ga_2O_3, therefore it is impractical. Tetravalent ions Sn and Si are shallow donors in β-Ga_2O_3 and increase its electrical conductivity. Cr showed electroluminescence and could potentially constitute an active element in the β-Ga_2O_3 host for solid-state lasers. Al expands the bandgap down to the UVC band, while Ce enhances scintillation properties under gamma radiation.

The incorporation of a dopant into the β-Ga_2O_3 crystal lattice strongly depends on the ionic radii of the dopant, valency, its thermal stability, coordination number, solubility in β-Ga_2O_3, $p(O_2)$ in the growth atmosphere, and on operating parameters (growth and rotation rates). The dopant concentration that is incorporated into a growing crystal depends on initial dopant concentration in the melt, solidified melt fraction, and the effective segregation coefficient of the dopant, and it is generally expressed by the Scheil-Gulliver relation:

$$C_S = C_0 k_{\text{eff}} (1 - g)^{k_{\text{eff}} - 1} \tag{2.6}$$

where C_S is the dopant concentration in the crystal, C_0 is the initial dopant concentration in the melt, k_{eff} is the effective segregation coefficient, and g is the crystallization fraction of the melt.

By measuring the dopant concentration at different locations of an obtained crystal and fitting the measured values with the Scheil-Gulliver relation, it is possible to determine the effective segregation coefficient for given growth conditions and operating parameters. The effective segregation coefficients of elements used for doping Czochralski-grown β-Ga_2O_3 crystals, along with their ionic radii in tetrahedral (IV) and octahedral (VI) sites in the β-Ga_2O_3 crystal lattice, are collected in Table 2.2. Depending on a dopant type there is some deviation from the segregation equation due to a high vapour pressures of Ga_2O, SnO and Cr-, and Cu-containing species (evaporation), which are $p(O_2)$ dependent, as well as due to oxidation states of Cr and Ce, or incorporation in structural defects (e.g. Ce).

Table 2.2 Ionic radii of dopants at tetrahedral (IV) and octahedral (VI) sites (Ga^{3+} sites in the crystal lattice), and segregation coefficients (k_{eff}) at corresponding oxygen concentrations in the growth atmosphere

		Ga^{3+}	Mg^{2+}	Cr^{3+}	Al^{3+}	Ce^{3+}	Si^{4+}	Sn^{4+}
Ionic radii (Å)	IV	0.47	0.57	–	0.53	–	0.26	0.69
	VI	0.62	0.72	0.62	0.68	0.87	0.40	0.83
O_2 (vol%)		–	1–35	4–24	1.5–21	1.5–35	1–2	2–17
k_{eff}		–	0.4	3.1–1.5	1.1	< 0.1	≈0.1	< 0.1[a]

[a]Affected by a strong dopant evaporation

Among dopants shown in Table 2.2 Cr incorporates into the growing crystal most easy and its concentration in the crystal decreases from its top to the bottom part ($k_{eff} > 1$), while Al incorporates substantially uniformly ($k_{eff} \cong 1$). Modifiers of the electrical conductivity (Mg, Sn, and Si) are built-in into the crystal structure with the fraction below about 40% ($k_{eff} < 1$); therefore, their concentrations increase from the top to the bottom part thereof. This is not a problem for Mg, which compensates the electrical conductivity even at small amount (>6 wt. ppm); however, Sn and Si create a dopant gradient and consequently inhomogeneous free carrier concentration along the crystal if their concentrations are small. An increasing dopant concentration with the crystal length ($k_{eff} < 1$) may lead to a constitutional supercooling that deteriorates the crystal quality. However, this happens at high dopant concentration in the melt (usually > 1 mol%) and large crystallization fraction (>0.7), what is not the case during growth of β-Ga_2O_3 crystals by the Czochralski method. To compensate the electrical conductivity 0.1–0.2 mol% of MgO in the melt was found to be sufficient. The same relates to SiO_2, which already in the amount of 0.1–0.2 mol% in the melt induce the free electron concentration at the level of 10^{19} cm^{-3}, that is about the maximum value that can be obtained in bulk crystals grown from the melt. Higher doping levels of SiO_2 up to 2 mol% do not increase the free electron concentration, but instead, they deteriorate the crystal quality by forming twins and grains. However, initial doping level of SnO_2 in the starting material must be higher, 1–2 mol% due to high fugacity and consequently losses of SnO at MP of Ga_2O_3.

The obtained crystals always contain some residual impurities coming mainly from a Ga_2O_3 starting material, an Ir crucible, and a thermal insulation. Their concentrations vary between crystals depending on material supplier, insulation type, and a re-use number. The main impurities are Al and Zr typically below 10 wt. ppm. Also other impurities, such as Mg, Ni, Co, Fe, Ti, Si, and Mo at concentration below 3–5 wt. ppm, are occasionally found in different bulk crystals.

2.7 Effect of Annealing

The Czochralski-grown β-Ga_2O_3 crystals can be obtained either as electrical insulators or as *n*-type semiconductors, both normal and degenerate. The electrical properties can be controlled by growth conditions, in particular by O_2 concentration in the growth atmosphere, and/or by intentional doping (e.g. Mg, Sn, Si). In this way, the free electron concentration can be controlled within three orders of magnitude (10^{16}–10^{19} cm^{-3}) with the Hall electron mobility up to 152 $cm^2V^{-1}s^{-1}$ [20]. Electrically insulating state is obtained at high O_2 concentration conditions and/or by doping with Mg. All crystals show a very steep absorption edge at about 260 nm and a good transparency in the visible/NIR spectral regions at low free electron concentrations.

Further modification of the electrical properties could be achieved by post-growth heat treatments at different atmospheres, temperatures, and times.

Annealing of the undoped crystals in the O_2-containing atmospheres at or above about 1000 °C for at least several hours results in lowering the free electron concentration to the level of about 1–2 × 10^{17} cm^{-3}, with a slight improvement of the Hall mobility. After such annealing, a thin, electrically insulating surface layer is formed [20]. Annealing in a hydrogen-containing atmosphere up to 900 °C for 10 h does not affect the free electron concentration (if at the level of about 10^{18} cm^{-3}) but significantly lowers the Hall mobility. However, annealing in the presence of hydrogen crystal samples previously annealed in the presence of O_2 ($\geq$ 1000 °C) increases the free electron concentration and improves the Hall mobility [26]. Annealing in the oxygen- and hydrogen-containing atmospheres is reversible. On the other hand, annealing Sn-doped crystals in the O_2-containing atmosphere at or above about 1000 °C for at least several hours fully or almost fully (depending on sample thickness) converts highly conducting crystals into electrical insulators [26], in contrast to undoped crystals. Modification of the electrical properties by annealing is also reflected in optical properties of the crystals, such as transmittance (e.g. through free carrier absorption) and cathodoluminescence.

In terms of thermal stability of the Czochralski-grown β-Ga_2O_3 single crystals, understood as a noticeable decomposition (i.e. detectable mass loss), they are stable up to about 1200–1300 and 600 °C in atmospheres containing O_2 and H_2, respectively, and 10 h annealing time [26]. However, a surface deterioration may start already at lower temperatures.

2.8 Summary

The Czochralski method was found to be a very efficient growth technique to obtain bulk β-Ga_2O_3 single crystals from the melt. Thermal instability of Ga_2O_3 required understanding its thermodynamics both in the gas and liquid phase. Although stabilization of the Ga_2O_3 decomposition from the point of view of evaporation can be achieved at relatively low O_2 concentration in the growth atmosphere (<2 vol%), the formation of metallic Ga in the liquid phase requires much higher O_2 concentrations with an increase of melt volumes (to obtain larger crystals). This contradicts the requirement for Ir crucibles that tend to oxidize. A combination of Ga_2O_3 and Ir thermodynamics led to a design of a new O_2 supply scheme to a growth furnace that allows to use a high O_2 concentration along with Ir crucibles, including pure O_2. Such technique enabled to obtain 2-inch diameter β-Ga_2O_3 single crystals with the weight up to 1 kg. Additionally to problems arising from Ga_2O_3 thermodynamics, the growth of bulk β-Ga_2O_3 single crystals by the Czochralski method is also affected by the free carrier absorption that results in shorter crystals with high free electron concentrations. The Czochralski method enables intentional doping to tune electrical, optical, and scintillation properties of bulk β-Ga_2O_3 crystals. Sn and Si are used to increase the electrical conductivity, Mg to compensate the electrical conductivity, Cr to introduce absorption bands in the visible spectrum, Al to expand the bandgap, and Ce to enhance the scintillation properties. Al and Ce were also found to have a thermodynamically stabilizing effect on Ga_2O_3 decomposition. In addition to intentional doping, the material's properties can also be tuned by $p(O_2)$ in the growth atmosphere and post-growth heat treatment. The obtained crystals can be either semiconductors or electrical insulators both with a very good transparency in the UV and visible spectral regions. High structural quality of the obtained bulk β-Ga_2O_3 single crystals enables to prepare differently oriented wafers for homoepitaxial growth of β-Ga_2O_3 layers and device fabrication.

Acknowledgements I would like to express my gratitude to Dr. Detlef Klimm, Dr. Steffen Ganschow, Dr. Klaus Irmscher for helpful discussions. This work was partly performed in the framework of GraFOx, a Leibniz-Science Campus partially funded by the Leibniz Association, Germany.

References

1. J. Czochralski, Z. Phys. Chem. **92**, 219 (1918)
2. R. Uecker, J. Cryst. Growth **401**, 7 (2014)
3. E.V. Gomperz, Z. Phys. **8**, 184 (1922)
4. E. Grüneisen, E. Goens, Phys. Z. **24**, 506 (1923)
5. A.G. Hoyem, E.P.T. Tyndall, Phys. Rev. **33**, 81 (1929)
6. H. Walther, Rev. Sci. Instrum. **8**, 406 (1937)
7. J.B. Little, G.K. Teal, Phys. Rev. **78**, 647 (1950)
8. W.C. Dash, J. Appl. Phys. **30**, 459 (1959)
9. K. Hoshikawa, H. Konda, H. Hirata, H. Nakanishi, Jpn. J. Appl. Phys. **19**, L33 (1980)
10. D.A. Petrov, V.S. Zemskov, Rost Kristallov **1**, 262 (1957)
11. R. Gremmelmaier, O. Madelung, Z. Naturforsch. **8A**, 333 (1953)
12. R. Gremmelmaier, Z. Naturforsch. **11A**, 511 (1956)
13. E.A.P. Metz, R.C. Miller, R. Mazelsky, J. Appl. Phys. **33**, 2016 (1962)
14. K. Nassau, L.G. Van Uitert, J. Appl. Phys. **31**, 1508 (1960)
15. W. Bardsley, G.W. Green, C.H. Holliday, D.T.J. Hurle, J. Cryst. Growth **16**, 277 (1972)
16. T.R. Kyle, G. Zydzik, Mater. Res. Bull. **8**, 443 (1973)
17. V.I. Vasil'tsiv, Y. Zakarko, Zh. Prikl. Spektrosk. **39**, 423 (1983)
18. Y. Tomm, P. Reiche, D. Klimm, T. Fukuda, J. Cryst. Growth **220**, 510 (2000)
19. Z. Galazka, R. Uecker, K. Irmscher, M. Albrecht, D. Klimm, M. Pietsch, M. Brutzam, R. Bertram, S. Ganschow, R. Fornari, Cryst. Res. Technol. **45**, 1229 (2010)
20. Z. Galazka, K. Irmscher, R. Uecker, R. Bertram, M. Pietsch, A. Kwasniewski, M. Naumann, T. Schulz, R. Schewski, D. Klimm, M. Bickermann, J. Cryst. Growth **404**, 184 (2014)
21. Z. Galazka, R. Uecker, D. Klimm, M. Bickermann, EP patent 3242965B1, 2019
22. Z. Galazka R. Uecker, D. Klimm, K. Irmscher, M. Naumann, M. Pietsch, A. Kwasniewski, R. Bertram, S. Ganschow, M. Bickermann, ECS J. Solid State Sci. Technol. **6**, Q3007 (2017)
23. Z. Galazka, S. Ganschow, A. Fiedler, R. Bertram, D. Klimm, K. Irmscher, R. Schewski, M. Pietsch, M. Albrecht, M. Bickermann, J. Cryst. Growth **486**, 82 (2018)
24. K. Hoshikawa, E. Ohba, T. Kobayashi, J. Yanagisawa, C. Miyagawa, Y. Nakamura, J. Cryst. Growth **447**, 36 (2016)
25. M. Baldini, Z. Galazka, G. Wagner, Mat. Sci. Semicon. Proc. **78**, 132 (2018)

26. Z. Galazka, Semicond. Sci. Tech. **33**, 113001 (2018)
27. Z. Galazka, K. Irmscher, R. Schewski, I. M. Hanke, M. Pietsch, S. Ganschow, D. Klimm, A. Dittmar, A. Fiedler, T. Schroeder, M. Bickermann, J. Cryst. Growth **529**, 125297 (2020)
28. Z. Galazka, R. Schewski, K. Irmscher, W. Drozdowski, M. E. Witkowski, M. Makowski, A. J. Wojtowicz, I. M. Hanke, M. Pietsch, T. Schulz, D. Klimm, S. Ganschow, A. Dittmar, A. Fiedler, T. Schroeder, M. Bickermann, J. Alloy. Compd. in print, 152842 (2019)

Chapter 3
Vertical Bridgman Growth Method

Keigo Hoshikawa

Abstract The vertical Bridgman (VB) technique developed for β-Ga_2O_3 crystals will be introduced including specific details on the VB crucible material determined by the measurement of the melting temperature of β-Ga_2O_3 and the VB growth processes of β-Ga_2O_3 in ambient air. The characteristic features of the crystallinity and the tentative electric characteristics of β-Ga_2O_3 crystals grown by the VB technique will also be introduced.

3.1 Introduction

The study of growth techniques aimed at larger size and improved quality for β-Ga_2O_3 crystals has recently intensified as β-Ga_2O_3 is a wide-bandgap semiconductor suitable for next-generation power devices due to its high breakdown voltage and high current density [1–4]. These crystals are expected to be successor materials to silicon carbide (SiC) and gallium nitride (GaN) [5, 6]. Table 3.1 shows a comparison of characteristic features of the growth techniques for crystals of the three primary power device materials: the currently used semiconductor silicon, the representative next-generation wide-bandgap semiconductor SiC, and the topical wide-bandgap semiconductor β-Ga_2O_3.

Silicon crystals are currently grown by the Czochralski (CZ) [7, 8] or the floating zone (FZ) method [7, 9]. In such melt growth processes, it should be possible to make the growth rate very high since the molten silicon changes directly to bulk solid silicon at the growth interface. Also, a low melting temperature [8] is useful for simplified furnace construction and reduced electric power consumption, advantages that can yield large crystals.

In contrast, SiC crystals are currently grown by the physical vapor transport (PVT) method [10], or the top-seeded solution growth (TSSG) method [11–13] which is currently under development. The PVT method is classed as vapor-phase growth

K. Hoshikawa (✉)
Faculty of Engineering, Shinshu University, Nagano 380-8553, Japan
e-mail: khoshi1@shinshu-u.ac.jp

M. Higashiwaki and S. Fujita (eds.), *Gallium Oxide*, Springer Series
in Materials Science 293, https://doi.org/10.1007/978-3-030-37153-1_3

Table 3.1 Comparison of crystal growth technologies for Si, 4H-SiC, and β-Ga_2O_3

	Si		4H-SiC		β-Ga_2O_3		
Bandgap eV	1.1		3.3		4.8		
Growth method	CZ	FZ	PVT	TSSG	FZ	CZ	EFG
Melting temperature °C	1420		(2000)		1790		
Growth rate mm/h	30–300		0.1–1.0		1–10		
Size enlargement	Easy		Difficult		?		

and is generally recognized as a low growth rate process due to this growth mechanism. The TSSG method is a typical solution growth, in which growth rate is limited by the rate-determining process of the amount of carbon solute that can be supplied to the growth interface from the Si solution. This makes it very difficult to achieve the same growth rate as that obtained by such direct changes as that from molten Si to solid Si. The growth rates of PVT-SiC and TSSG-SiC are 0.1–1.0 mm/h, rates drastically smaller than the 30–300 mm/h of the melt growth process of Si crystal. Furthermore, the temperature required in growth furnaces for both PVT-SiC and TSSG-SiC is higher than that in the CZ-Si method. This makes the scale-up of the growth furnace problematic, and the growth of large crystals very difficult.

β-Ga_2O_3 crystals are grown by a melt growth process similar to that used for Si crystal growth, in which the molten raw material is solidified to form bulk crystals. Therefore, we could expect to obtain the same high growth rate and large crystal size as with CZ-Si growth. As for the crystal growth atmosphere, which is strongly related to the growth furnace construction and the process control of single crystal growth, it is essential to rigorously protect the interior of a growth furnace holding molten Si, which shows thermochemically high activity with even the slightest amount of oxygen. In contrast, the presence of oxygen inside a growth furnace has little impact on the crystal growth of β-Ga_2O_3 because the crystal itself is already an oxide. The crucible for CZ-Si growth is made entirely of high-purity silica glass because of the very high thermochemical activity of molten silicon. However, we have a choice of many crucible materials for β-Ga_2O_3 growth because the molten raw materials are thermochemically stable in a furnace atmosphere with a relatively wide range of oxygen partial pressure. We might conclude from the above considerations of crystal growth technology that semiconductor β-Ga_2O_3 material, with its very wide bandgap, might be more attractive than the silicon materials, despite the fact that the development of crystal technology has reached a mature stage.

In this chapter, the growth technique developed for β-Ga_2O_3 crystals by the authors [14, 15] will be introduced, including specifics on the VB crucible material determined by the measurement of the melting temperature of β-Ga_2O_3, the VB growth technique of β-Ga_2O_3 in ambient air [14], the characteristic features of the β-Ga_2O_3 crystals thus grown [15], and the tentative electrical characteristics of β-Ga_2O_3 crystals grown using this VB technique.

3.2 VB Growth Method and Crucible Materials

3.2.1 VB Method Using Pt–Rh Alloy Crucible

β-Ga_2O_3 bulk crystals have until now been grown by optical floating zone (OFZ) [16–18], Czochralski (CZ) [19–21] and edge-defined film-fed growth (EFG) [22, 23] methods. The OFZ method has been widely used when growing β-Ga_2O_3 crystals for experimental use because the method requires no crucible and the crystals are easily grown in ambient air. However, crystals grown by OFZ are dimensionally limited due to the limited size of the light spot. Also, defects will always be present at high density as the melt zone occupies a small volume and may not cover the entire crystal diameter. The CZ and EFG methods have been expected to become the industrial techniques to produce large, high-quality β-Ga_2O_3 bulk crystals. In these methods, however, Ir crucibles and dies are used in an atmosphere with reduced oxygen partial pressure.

It is well known that in a high-temperature furnace with more than several percent oxygen partial pressure, Ir metal easily forms oxides such as IrO, IrO_2, and IrO_3 and these oxides may be lost by evaporation, which will occur faster at higher oxygen partial pressure. Therefore, a weakly oxidizing atmosphere containing Ar with a few percent of oxygen, or a mixture of Ar and CO_2, is generally used in a β-Ga_2O_3 crystal growth furnace when using an Ir crucible. However, in the weakly oxidizing atmosphere, a decomposition reaction such as Reaction (3.1) [19, 20] must be presumed to occur in a growth furnace at the high temperature required to melt β-Ga_2O_3.

$$Ga_2O_3(s) \rightarrow Ga_2O(g) + O_2(g) \tag{3.1}$$

Reaction (3.1) shows that the raw material Ga_2O_3 (*solid:* s) decomposes to Ga_2O (*gas*: g) and O_2 (g), both with high vapor pressures. Additionally, the molten Ga_2O_3 (*liquid:* l) decomposes at higher temperatures to Ga liquid (l) and O_2 (g), as in Reaction (3.2).

$$Ga_2O_3(l) \rightarrow 2Ga(l) + 3/2O_2(g) \tag{3.2}$$

The O_2 (g) generated by Reactions (3.1) and (3.2) reacts with the Ir (s) of the crucible holding the molten Ga_2O_3 and forms IrO_2 (g) by Reaction (3.3), which will then be lost by evaporation.

$$\mathrm{Ir(s)} + \mathrm{O_2(g)} \rightarrow \mathrm{IrO_2(g)} \tag{3.3}$$

Furthermore, the liquid Ga generated by Reaction (3.2) reacts with the Ir (s) and forms an Ir–Ga alloy (s and/or l) by Reaction (3.4).

$$\mathrm{Ir(s)} + \mathrm{Ga}(l) \rightarrow \mathrm{Ir{-}Ga\,alloy}(s, l) \tag{3.4}$$

This means that the Ir crucible will lose weight and/or be penetrated by melting, due to local decreases in its melting temperature [24, 25].

These considerations indicate that the suppression of Ir oxidation and the suppression of Ga_2O_3 decomposition are in conflict due to these high-temperature chemical reactions during the growth of β-Ga_2O_3 by the CZ and EFG methods, in which Ir crucibles must be used. It has been increasingly acknowledged that the high-temperature chemical stabilities of both the molten β-Ga_2O_3 and the Ir crucible during growth processes are significant challenges in the growth of large-sized and high-quality β-Ga_2O_3 crystals [25].

The authors have come to the conclusion that β-Ga_2O_3 crystals must be grown in ambient air or even with a higher oxygen partial pressure in order to suppress Reactions (3.1) and (3.2). In that case, the crucible materials used must be Pt–Rh alloys strongly resistant to an oxidizing atmosphere in a high-temperature furnace. The melting temperature of the Pt–Rh alloy must be higher than 1850 °C, as the melting temperature of β-Ga_2O_3 is near 1800 °C, and the Rh content must be higher than 10 wt% in the alloy. Finally, the VB growth method is the only choice because the temperature difference between the maximum temperature of the crucible needed to keep the raw materials molten, and the melting temperature of β-Ga_2O_3, must be smaller than that in the other growth methods of CZ and EFG.

3.2.2 *Measurement of β-Ga_2O_3 Melting Temperature and Determination of Pt–Rh Alloy Crucible Composition*

A schematic diagram of the VB growth furnace used is shown in Fig. 3.1. An RF induction heating furnace was constructed, with a work coil, heat shields, a heater, a crucible, a thermocouple, and a crucible shaft with a rotation and translation mechanism. The radio frequency of the work coil was 15 kHz. The heat shields were constructed from zirconia and alumina ceramics. The heater was made of Pt–Rh (70–30%) alloy with a melting temperature higher than 1900 °C.

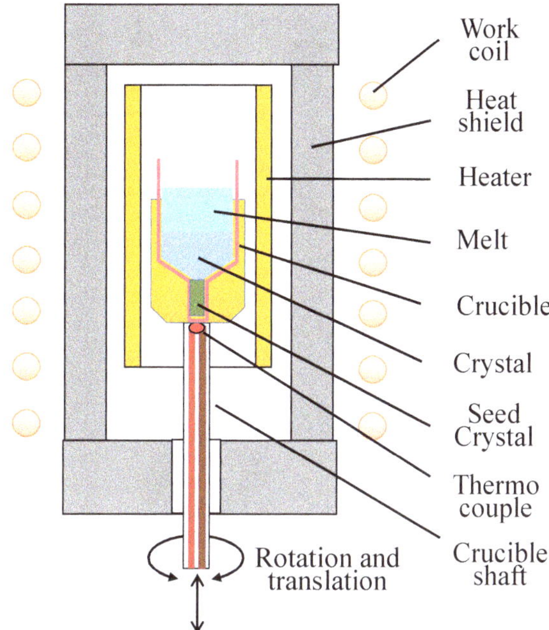

Fig. 3.1 RF heating VB furnace

Table 3.2 Reported values of the melting temperature of β-Ga_2O_3, showing large variation

• 1725 °C: Villora et al. (2014). [a]
• 1740 °C: Roy et al. (1952). [b]
• 1794 °C: Zinkevich et al. (2004). [c]
• 1795 °C: Schneider et al. (1963). [d] • 1807 °C: SGTE Database (1996). [e]
• 1820 °C: Galazka et al. (2010). [f]

[a]E.G. Villora et al., Proc. of SPIE 8987 (2014) 89871U
[b]R. Roy et al., J. Am. Chem. Soc. 74(3) (1952) 719–722
[c]M. Zinkevich et al., J. Am. Ceram. Soc. 87(4) (2004) 683–691
[d]S.J. Schneider et al., J. Res. Nat. Bur. Stand. A 67A (1963) 19–25
[e]Thermodynamic properties of individual substances, edited by L. V. Gurvich, et al., CRC Press (Boca Raton, Ann Arbor, London, and Tokyo), Vol. 3 Part 1 (1981) pp. 220
[f]Z. Galazka et al., Cryst. Res. Technol. 45(12) (2010) 1229–1236

It is necessary to know the precise melting temperature of β-Ga_2O_3 in order to design the Pt–Rh composition of the crucible. The reported values, however, differ significantly among the sources cited in Table 3.2. Thus, it was essential to measure the precise melting temperature of β-Ga_2O_3 in an air atmosphere in the same furnace used for growing the crystal.

The method used to measure the melting temperature of β-Ga_2O_3 is shown in Fig. 3.2. The 0.2 mm-thick platinum–rhodium alloy container, charged with several

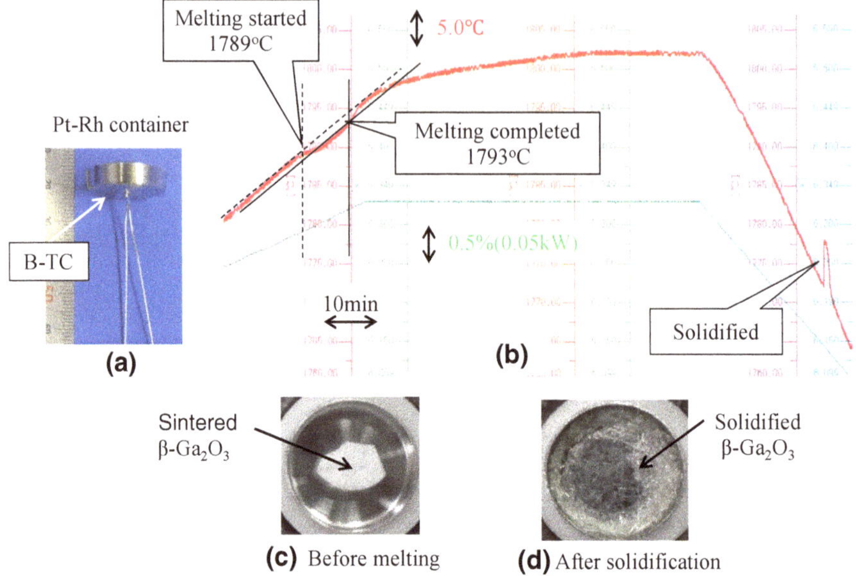

Fig. 3.2 Measuring the melting temperature of β-Ga_2O_3 [14]

grams of sintered β-Ga_2O_3, was set in the heating zone with a uniform temperature distribution in the furnace as shown in Fig. 3.1. The heating power was slowly increased at a constant rate. The temperature change at the container bottom was measured, determined by a B-type thermocouple welded to the container bottom as shown in Fig. 3.2a. A melting onset temperature of 1789 °C and a melting completion temperature of 1793 °C were observed, as shown in Fig. 3.2b. The solidification of completely melted β-Ga_2O_3 was also observed during the heating power reduction stage, as shown in Fig. 3.2b. It was confirmed that the sintered β-Ga_2O_3 in the container shown in Fig. 3.2c had changed to the solidified β-Ga_2O_3 shown in Fig. 3.2d. This experiment showed that the melting temperature of β-Ga_2O_3 was about 1793 °C in ambient air [14]. The crucible material in the VB growth furnace must have and hold a melting temperature higher by 50 °C than the β-Ga_2O_3 melting temperature. We could design and construct the crucibles with platinum-based, 10–20% rhodium alloys for the task of β-Ga_2O_3 crystal growth by the VB method in ambient air.

3.3 Growth and Characterization of VB-β-Ga_2O_3 Crystals

3.3.1 Directional Solidification of β-Ga_2O_3 Single Crystals in the VB Furnace

The crucible shown in Fig. 3.1 was made of a Pt–Rh (80–20%) alloy, inner diameter 25 mm, and 50 mm long with a conical part and a thin seed well. It was very important to measure the temperature precisely because we could not observe events inside the crucible during either the seeding or the growth process. For temperature measurement, we used a B-type thermocouple, which can be used to measure temperatures up to 1820 °C in ambient air.

Figure 3.3 shows schematics of the temperature distribution in the furnace (Fig. 3.3a), the preparation of seed and raw material (Fig. 3.3b) and the crystal growth processes (Fig. 3.3c). In the preparation, the seed crystal and the sintered β-Ga_2O_3 raw material are charged in the crucible. In the crystal growth process, melting of the raw material and the VB growth are carried out by translation of the crucible upward [Fig. 3.3(c-1)] and downward [Fig. 3.3(c-2)] each at its appropriate speed, with the temperature distribution in the furnace shown in Fig. 3.3a. In the present directional solidification, the VB growth process was started by downward translation of the crucible, and then shifting to the VGF process by decreasing the heating power during the remaining half process. The crucible translation speed in the VB process and melting temperature moving rate in the VGF process were controlled to be 1–5 mm/h; but the resultant growth rates could not be determined. The crucible was rotated at several rpm to keep the in-plane temperature distribution uniform.

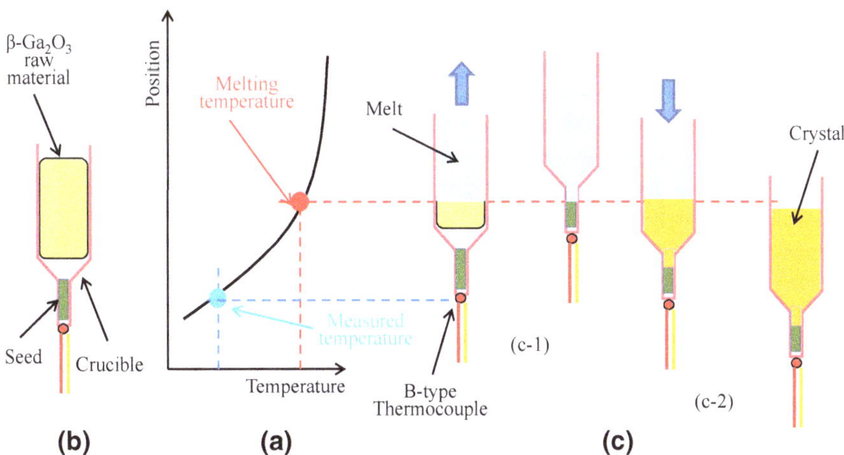

Fig. 3.3 **a** Schematics of a temperature distribution in the furnace, **b** preparation of seed crystal and raw material, and **c** crystal growth process

3.3.2 VB-β-Ga_2O_3 Crystals and Their Characteristics

β-Ga_2O_3 crystals that were directionally solidified in the VB furnace as shown in Fig. 3.1 are shown in Fig. 3.4. The single crystal 25 mm in diameter and 25 mm in length as shown in Fig. 3.4a was grown from [001] seed in the conical crucible with a thin seed well shown in Fig. 3.4a' and the crystal with 25 mm in diameter and 30 mm in length as shown in Fig. 3.4b was grown without seeding in the full-diameter crucible shown in Fig. 3.4b'. The pale brownish yellow color of these crystals may be due to Rh contamination originating from the Pt–Rh alloy crucible.

The crystals grown in the crucibles could easily be released by destructively peeling the crucible from the crystals (stripping), producing crystal surfaces that were very smooth and shiny. It is known that in the VB method, releasing the crystal without serious damage to it is usually very difficult due to its adhesion to the crucible. In the present case, it was found that β-Ga_2O_3 crystals did not adhere to the Pt–Rh alloy crucibles with either vertical or conical walls. It is concluded that single crystals of β-Ga_2O_3 can be grown not only in full-diameter crucibles but also in conical crucibles [14]. It should be concluded from the present VB-β-Ga_2O_3 growth that the seeding process enables the growth of single crystals with arbitrary growth directions from the seeds, as we grew crystals with [100], [010], and [001] growth directions. It was also found that single crystals with a growth direction normal to the (100) plane could be grown with very high probability without seeding.

Three representative crystals grown without seeding are shown in Fig. 3.5. Crystal A (Fig. 3.5a) is a polycrystalline ingot; crystal B (Fig. 3.5b) is a crystal ingot that was initially grown as a polycrystal, which abruptly changed to a single crystal with the growth direction normal to the (100) plane; and crystal C (Fig. 3.5c) is a crystal ingot grown as a single crystal from the starting edge to the

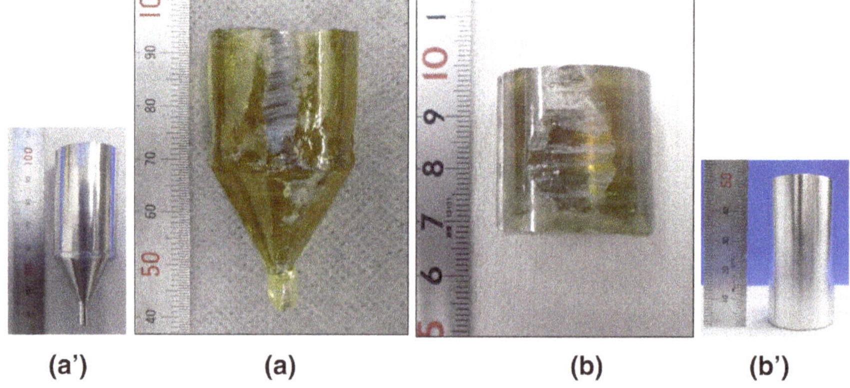

Fig. 3.4 β-Ga_2O_3 crystals grown by the VB method. a crystal grown from a [001] seed in thin seed crucible (**a'**) and b crystal grown without a seed in full-diameter crucible (**b'**)

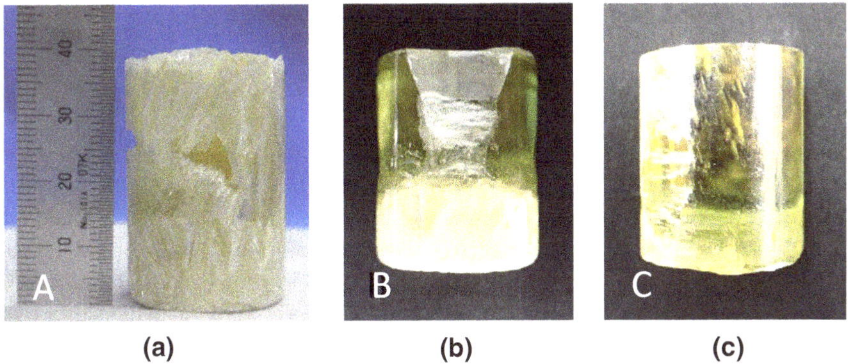

Fig. 3.5 Representative three crystals grown by VB method without seeding [14]

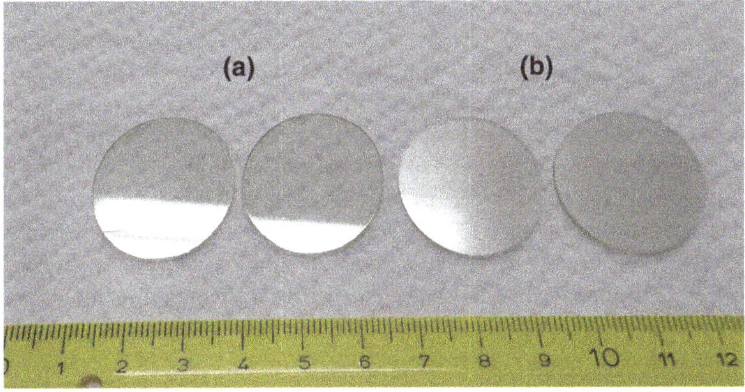

Fig. 3.6 (100) polished wafers (**a**) and as-sliced wafers (**b**) from VB-β-Ga_2O_3 crystals

terminal edge of the ingot with the growth direction normal to the (100) plane. It is well known that in the process of single crystallization by the VB method without a seed, a polycrystal with many growth directions nucleates at the starting edge, and the multiple growth directions gradually decrease in number due to growth rate anisotropy, finally yielding a single crystal. In the present β-Ga_2O_3 single crystal growth, however, a very different process was observed, as the polycrystal changed abruptly to a single crystal as shown in crystal B, or a single crystal was grown from the starting edge, as initiated by a single crystal seed, as shown in single crystal C. It can be concluded from these observations that β-Ga_2O_3 single crystals are easily grown in a specific growth direction with very strong growth rate anisotropy [14].

Figure 3.6 shows the (100) polished wafers (Fig. 3.6a) and as-sliced wafers (Fig. 3.6b) from a VB-β-Ga_2O_3 single crystal grown without seeding. It was found in both slicing and polishing the wafers that particular attention must be paid to the working direction's relation to the crystal structure of β-Ga_2O_3 single crystals because of their strong tendency to produce cleavage and cracking failures.

3.3.3 Crystal Defects and Electrical Properties

3.3.3.1 Structural Defects

Figure 3.7 shows a crossed polarizer image (Fig. 3.7a), transmitted X-ray topographic image (Fig. 3.7b), and reflected X-ray topographic image (Fig. 3.7c) of a wafer, mirror-polished on both sides, that was cut from a VB-β-Ga_2O_3 crystal ingot directionally solidified without a seed. No low-angle grain boundary (LAGB) was observed in Fig. 3.7a. The entire wafer was a single crystal with some residual internal stress at the wafer periphery (Fig. 3.7b). Figure 3.7c demonstrates the macroscopic homogeneity of the entire wafer surface as shown by the uniform reflected X-ray image (except for several localized high-stress areas as indicated by black portions). Though a strong diffraction loss was recognized at the wafer periphery in the transmitted X-ray topographic image in Fig. 3.7b, no reflection loss was observed near the wafer periphery in the reflected X-ray topographic image in Fig. 3.7c. The reason why the wafer surface is characterized as uniform by reflected X-ray topography is thought to be that this method only detects stress in the surface layer, which was rendered stress-free by the mirror polishing process. It might be considered that the uniform macroscopic reflected X-ray image in Fig. 3.7c suggests a homogeneous distribution of dislocations in both high- and low-density cases. Therefore, optical microscopy was used to investigate surface defects and their density distribution after KOH solution etching. Etch pit densities after KOH etching were measured in sample areas at nine positions, each measuring 850 μm × 1280 μm, to determine the pit density distribution in the wafer plane, and at five positions along dotted lines AA' and BB' to examine the distribution across the wafer, as marked in red in Fig. 3.7c.

Figure 3.8 shows photographs of typical etch pit images observed on the front (Fig. 3.8a) and the back (Fig. 3.8b) surfaces. From the estimation of their shape and distribution, we considered that the etch pits shown in Fig. 3.8 were likely to be dislocation pits. Figure 3.9 shows the distributions of dislocation pit density measured along lines A-A' and B-B' in Fig. 3.7c. The result shows uniform distributions without any notable patterns such as dislocations concentrated along the wafer periphery and is consistent with the result of the reflected X-ray topographic image as shown in Fig. 3.7c. These results are also consistent with the fact that the β-Ga_2O_3 crystal ingots grown in the crucibles by directional solidification did not adhere to the Pt–Rh alloy crucibles and that the surfaces of crystal ingots were very smooth and shiny, having been released from the crucibles without serious damage.

3.3.3.2 Geometric Shape and Formation Mechanism of Line-shaped Defects

Structural defects we refer to as "line-shaped defects" were detected in the present β-Ga_2O_3 crystals, with lengths from 20 to 150 μm extending in the [010] direction,

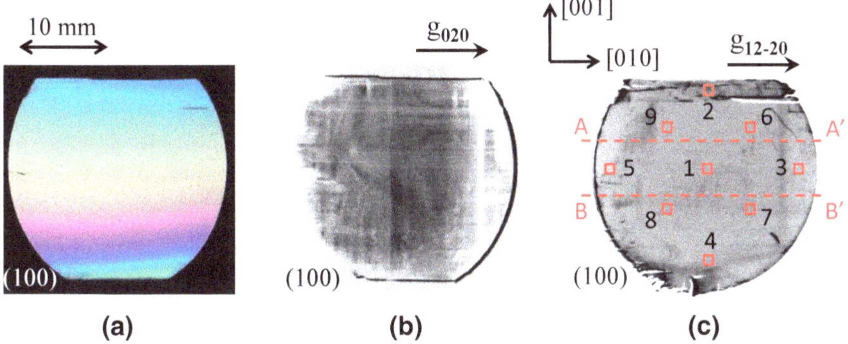

Fig. 3.7 **a** Crossed polarizer image, **b** transmitted X-ray topographic image, and c reflected X-ray topographic image of a (100) wafer [15]

and are shown in Fig. 3.10 [15]. Figure 3.10a, b, respectively, shows the defects observed on the front and back surfaces after KOH etching. Figure 3.10c shows the line-shaped defects observed in the wafers' inner regions without etching. Figure 3.10d shows a schematic of the cross section normal to the (001) plane. The line-shaped defect distribution shown in Fig. 3.10d was based on the observations shown in Fig. 3.10a, b, c. The reason why the line-shaped defects in Fig. 3.10a, b show a tapered shape is because the wafers were cut with a tilt of several degrees from the (100) plane toward the [010] direction to avoid cleavage on the (100) plane. These defects were detected by optical microscopy, and it was confirmed that they were distributed inside the wafer as shown in Fig. 3.10c. In Fig. 3.10d, two line-shaped defects revealed on the wafer surface by KOH etching (as in Fig. 3.10a, b) are indicated with solid lines (indented), and those in the bulk crystal (as in Fig. 3.10c) with dotted lines. The dislocation pit densities and the line-shaped defect densities measured in the nine regions on the front surface of the wafer in Fig. 3.7c are shown in Table 3.3. Average densities of the dislocation pits and line-shaped defects were 2.3×10^{3} cm^{-2} and 4.6×10^{2} cm^{-2}, respectively.

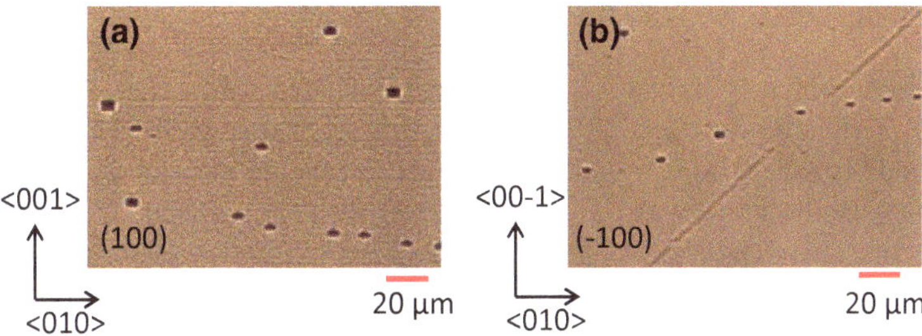

Fig. 3.8 Dislocation etch pit observation on front surface (**a**) and back surface (**b**) [15]

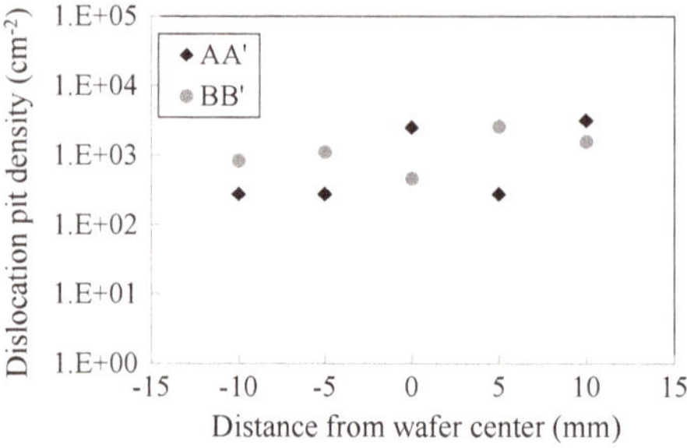

Fig. 3.9 Radial distribution of dislocation pit density of wafer A [15]

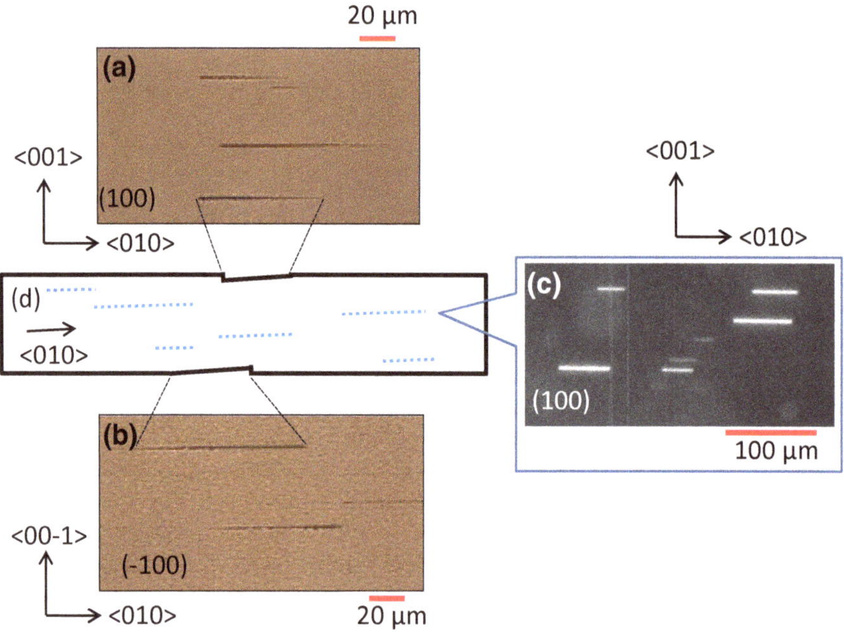

Fig. 3.10 Observations and schematics of line-shaped defects [15]

Defects similar to the line-shaped defects discussed above were reported in β-Ga_2O_3 crystals grown using the EFG method by Kuramata et al. [26], Nakai et al. [27], and Hanada et al. [28], so the shape and formation mechanism described by them may be discussed with reference to these reports in [27] and [28]. Figure 3.11 shows the cross-sectional TEM image of a rodlike defect viewed from $[\bar{1}0\bar{2}]$ direction firstly reported by Nakai et al. (Figure 3.11a) [27] and SEM image of three grooves in an array along the [001] direction reported by Hanada et al. (Figure 3.11b) [28]. Nakai et al. described defects observed by TEM as nanopipes,

Table 3.3 Dislocation pit density and line-shaped defect density of the wafer in Fig. 3.7c [15]

Position	Dislocation pit density ($\times 10^3$ cm^{-2})	Line-shaped defect density ($\times 10^2$ cm^{-2})
1	2.5	<1
2	0.4	15
3	3.1	1
4	9.0	3
5	0.3	< 1
6	0.6	15
7	2.9	5
8	1.7	2
9	< 0.1	<1
Average	2.3	4.6

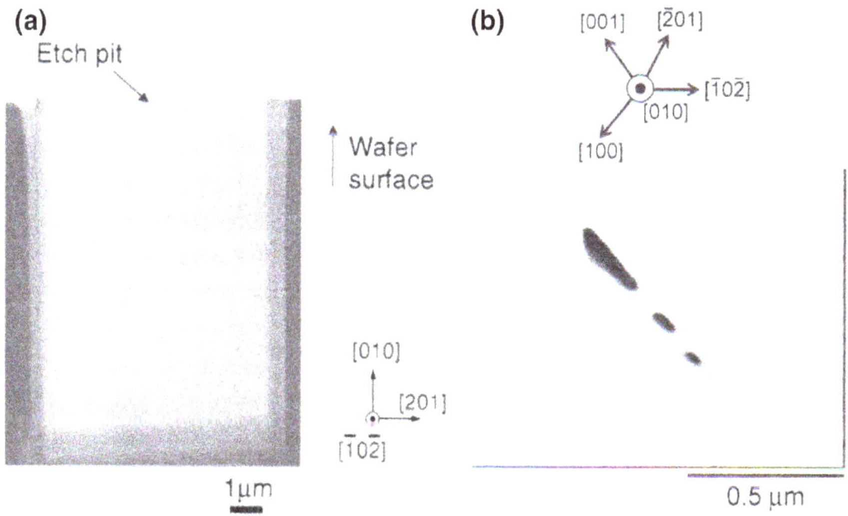

Fig. 3.11 a Cross-sectional TEM image of a rodlike defect viewed from [$\bar{1}0\bar{2}$] direction and b SEM image of three grooves in an array along the [001] direction Adapted from [27] and [28] with permission of the publisher

or rodlike defects of 0.1 μm diameter and at least 15 μm in length along [010]. Hanada et al. concluded that the defects observed by SEM were grooves with a typical length and width of 50–1200 nm in the [001] direction and less than 40 nm in the [100] direction, respectively [28].

Table 3.4 summarizes their geometric size in [15, 27, 28]. Considering that the same kind of defect was observed on different' planes and directions in different wafers with different observation methods using very different scales, no significant difference is seen between the results in [15, 27, 28]. It can be considered that the

Table 3.4 Summary of line-shaped defect reports

	Ohba et al. [15]	Nakai et al. [27]	Hanada et al. [28]
Crystal growth method	VB	EFG	EFG
Observation method	Optical microscope	TEM	SEM
Observation plane and direction	(100)	(010) and $[\bar{1}0\bar{2}]$	(010)
Defect size in <100> direction (μm)	–	0.1	<0.04
Defect size in <010> direction (μm)	20-150	>15	<0.03
Defect size in <001> direction (μm)	–	0.1	0.05–1.2
Naming of defect	• Line-shaped defect	• Nanopipe • Hollow pipe • Rodlike defect	• Nanometer-sized groove

defect length along the [010] direction is a more accurate value in [15] because it was observed on the (100) plane. It can also be considered that the defect lengths along the [001] and the [100] directions are more accurate in [28] because they were observed in the (010) plane. It can be inferred that the diameter of a nanopipe was 0.1 μm in [27]. So it could be presumed that the defect size along the [010] direction was 20–150 μm, that along the [001] direction was 0.05–1.2 μm, and that along the [100] direction was 0.1–0.04 μm.

Based on these considerations, the geometric shape of a line-shaped defect is shown in Fig. 3.12a. Nakai et al. [27] called the rodlike defects "nanopipes" and showed them to be hollow. It is speculated that the inside of the line-shaped defects is a cavity that might be thought of as a negative crystal [29, 30], formed from a

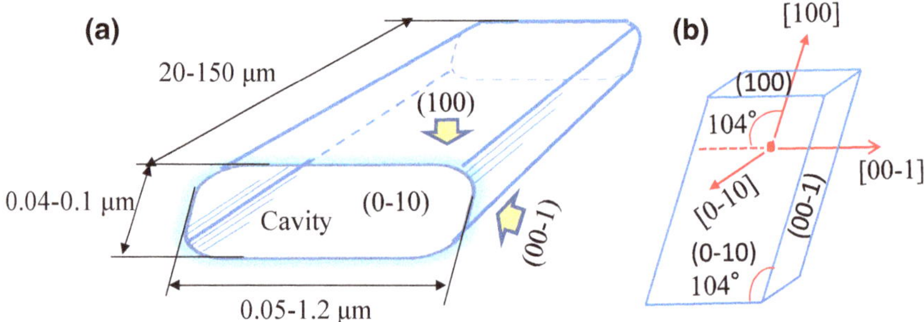

Fig. 3.12 Geometric shape of (**a**) a line-shaped defect and (**b**) β-Ga_2O_3 crystal unit cell structure

micro-void introduced into the crystal from the melt through the growth interface. This is why the defects were not sensitively detected by X-ray diffraction; however, they were sensitively detected as optical inhomogeneities inside wafers. Such micro-voids could be formed by gas generation due to the decomposition of the Ga_2O_3 melt at high temperature and/or an impurity gas due to the accumulation of impurities by segregation and following constitutional super-cooling. Similar phenomena have been reported in sapphire growth [31, 32]. It is generally known that micro-voids are transshaped spheres, or hollow due to surface tension on their inner wall [29, 30]. However, in the present case it is considered that the very strong growth rate anisotropy [33] of β-Ga_2O_3 resulted in the characteristic geometric shape shown in Fig. 3.12a. This was discussed as one of the growth characteristics of β-Ga_2O_3 crystals in [14], where it was concluded that the crystal grows in the direction normal to the (100) plane by (100) faceted growth, with a very high growth rate in the [010] direction within the (100) facet plane. Based on this, it was explained that the length of each edge of a line-shaped defect is strongly related to the directional growth rate anisotropy; i.e., the extended length in the [010] direction is due to the very fast growth rate in that direction, and similarly for the middle length in the [001] direction and the short length in the [100] direction. Additionally, it can be considered that the (010) cross section of a cavity was a cornerless rhomboid surrounded by (100) and (001) facets, as depicted in Fig. 3.12a. Finally, it was concluded that a line-shaped defect formed from a micro-void became a parallelepiped cavity extended in the [010] direction, with a cornerless rhomboid cross section (Fig. 3.12a). It was found that the three edge lengths of a defect were inversely related to the three edge lengths of the β-Ga_2O_3 crystal unit cell structure [34–36] as shown Fig. 3.12b. It has been speculated in the above discussions that the formation of line-shaped defects was strongly dependent on the growth rate anisotropy of the β-Ga_2O_3 crystal itself.

3.3.3.3 Electrical Properties

The results of impurity analysis of several VB-β-Ga_2O_3 ingots are shown in Table 3.5. Rh, as the impurity with the highest concentration, might well originate from the Pt–Rh alloy crucible, and the second commonest impurity, Si, might originate from impurities in the raw materials and/or from the ceramic heat shields

Table 3.5 Results of impurity analysis

Element	Result of analysis (wt.ppm)
Na	0.1–1.0
Al	0.1–1.0
Si	1.0–10
Fe	0.1–1.0
Rh	2.0–20
Pt	0.5–5.0

used in furnace construction. Reducing these impurities may be a future requirement to control the crystals' electrical characteristics.

The electrical properties determined by Hall effect measurements on samples from un-doped, Si-doped, and Sn-doped VB-β-Ga_2O_3 crystals are shown in Table 3.6. The carrier concentration in un-doped crystal is low, and the crystal shows a semi-insulating property. The carrier concentrations in Si- and Sn-doped crystals are very high, and these crystals are *n*-type semiconductors. The relationship between the carrier concentration and the resistivity of Si- and Sn-doped crystals is shown in Fig. 3.13. These relationships in the present VB-β-Ga_2O_3 crystals are the same as those in Si-doped MBE films reported by Onuma et al. [37].

The residual impurity concentrations shown in Table 3.5 and the electrical properties of this study's VB-β-Ga_2O_3 crystals shown in Table 3.6 and Fig. 3.13 are tentative results. It should be noted that lower levels of impurity contamination and the resulting improvement in electrical properties must be studied in relationship to the required device properties, including the device performance targets.

Table 3.6 Electric properties measured by Hall effect measurement

	n/cm^3	μ cm^2/V · s	ρ Ω cm
Un-doped	3.1×10^8	33	8.8×10^8
Si-doped	3.0×10^{18}	80	2.6×10^{-2}
Sn-doped	2.2×10^{18}	72	3.9×10^{-2}

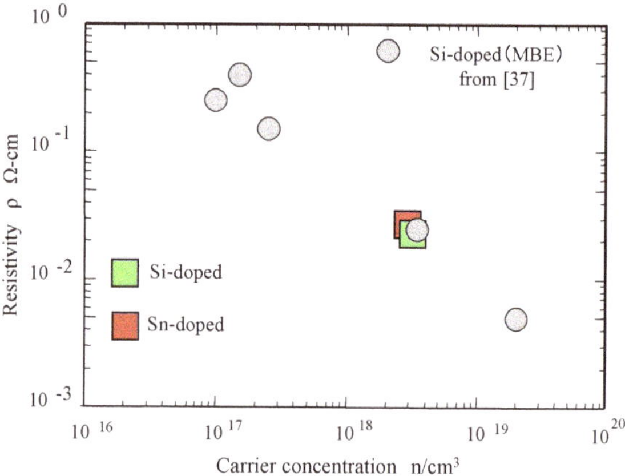

Fig. 3.13 Characteristics of carrier concentration vs resistivity of Sn- and Si-doped crystals

3.4 Summary

β-Ga_2O_3 crystals, currently attracting attention as a next-generation wide-bandgap semiconductor for power devices, are grown from a melt, as are Si crystals. However, in terms of the growth rate of large crystals, which is directly related to the production cost, β-Ga_2O_3 crystals do not have any advantage over Si crystals either for their practical capabilities or for their higher melting temperature and lower thermal conductivity when compared to Si crystals. On the other hand, β-Ga_2O_3 crystals might offer a significant advantage in that they can be grown in a high-oxygen partial pressure atmosphere, even in ambient air, because they are already oxides. Only a small oxygen partial pressure is acceptable in the growth of Si crystals due to their high affinity for oxygen at high temperatures.

In this chapter, the growth of β-Ga_2O_3 crystals by the VB method was proposed and carried out using Pt–Rh alloy crucibles in ambient air, with the results summarized below:

(1) The melting temperature of β-Ga_2O_3 in ambient air was determined to be about 1793 °C and crucibles were fabricated from platinum-based, 10–20% rhodium alloys and used for the task of β-Ga_2O_3 crystal growth by the VB method in ambient air.
(2) β-Ga_2O_3 crystals 25 mm in diameter and about 30 mm in length were successfully grown by directional solidification in the VB furnace in a conical crucible with a thin seed and a full-diameter crucible without seed.
(3) Dislocations and line-shaped defects were the characteristic defects in the β-Ga_2O_3 crystals, and the average densities of dislocation pits and line-shaped defects were 2.3×10^3 cm^{-2} and 4.6×10^2 cm^{-2}, respectively.
(4) The Rh impurity originating from the Pt–Rh alloy crucible was detected as the highest concentration impurity at 3–30 ppm, and the Si impurity originating from the raw materials and/or from the ceramic shields was the second commonest impurity at 1–10 ppm.
(5) As for tentative electrical properties, the carrier concentration in un-doped crystal is low, showing a semi-insulating property, and the carrier concentrations in Si- and Sn-doped crystals are very high, yielding *n*-type semiconductors.

References

1. M. Higashiwaki, K. Sasaki, A. Kuramata, T. Masui, S. Yamakoshi, Appl. Phys. Lett. **100**, 013504 (2012)
2. M. Higashiwaki, K. Sasaki, A. Kuramata, T. Masui, S. Yamakoshi, Phys. Status Solidi A, (1), 21 (2014)
3. T. Oishi, Y. Koga, K. Harada, M. Kasu, Appl. Phys. Express **8**, 031101 (2015)

4. M. Baldini, Z. Galazka, G. Wagner, Mater. Sci. Semicond. Process. **78**, 132 (2018)
5. S. Fujita, Jpn. J. Appl. Phys. **54**, 030101 (2015)
6. M. Higashiwaki, H. Murakami, Y. Kumagai, A. Kuramata, Jpn. J. Appl. Phys. **55**, 1202A1 (2016)
7. T. Hibiya, K. Hoshikawa, in *Bulk Crystal Growth of Electronic, Optical and Optoelectonic Materials*, ed. P. Capper (Wiley, 2005), pp. 1–42
8. J. Friedrich, W. Ammon, G. Muller, in *Bulk Crystal Growth: Basic Techniques*, ed. by P. Rudolph (Elsevier, 2015), pp. 45–104
9. A. Muizunieks, J. Virbulis, A. Ludge, H. Riemann, N. Werner, in *Bulk Crystal Growth: Basic Techniques*, ed. by P. Rudolph (Elsevier, 2015), pp. 241–279
10. T.S. Sudarshan, D. Cherednichenko, R. Yakimova, in *Bulk Crystal Growth of Electronic, Optical and Optoelectonic Materials*, ed. by P. Capper (Wiley, 2005), pp. 433–449
11. D. Hofmann, M.H. Müller, Mater. Sci. Eng. **B61–62**, 29 (1999)
12. H. Daikoku, M. Kado, H. Sakamoto, H. Suzuki, T. Bessho, K. Kusunoki, N. Yashiro, N. Okada, K. Moriguchi, K. Kamei, Mater. Sci. Forum **717–720**, 61 (2012)
13. K. Kusunoki, N. Okada, K. Kamei, K. Moriguchi, H. Daikoku, M. Kado, H. Sakamoto, T. Bessho, T. Ujihara, J. Cryst. Growth **395**, 68 (2014)
14. K. Hoshikwa, E. Ohba, T. Kobayashi, J. Yanagisawa, C. Miyagawa, Y. Nakamura, J. Cryst. Growth **447**, 36 (2016)
15. E. Ohba, T. Kobayashi, M. Kado, K. Hoshikwa, Jpn. J. Appl. Phys. 55, 1202BF (2016)
16. N. Ueda, H. Hosono, R. Waseda, H. Kawazoe, Appl. Phys. Lett. **70**(26), 3561 (1997)
17. E.G. Víllora, K. Shimamura, Y. Yoshikawa, K. Aoki, N. Ichinose, J. Cryst. Growth **270**, 420 (2004)
18. E.G. Víllora, K. Shimamura, Y. Yoshikawa, T. Ujiie, K. Aoki, Appl. Phys. Lett. **92**, 202120 (2008)
19. Y. Tomm, P. Reiche, D. Klimm, T. Fukuda, J. Cryst. Growth **220**, 510 (2000)
20. Z. Galazka, R. Uecker, K. Irmscher, M. Albrecht, D. Klimm, M. Pietsch, M. Brützam, R. Bertram, S. Ganschow, R. Fornari, Cryst. Res. Technol. **45**(12), 1229 (2010)
21. Z. Galazka, K. Irmscher, R. Uecker, R. Bertram, M. Pietsch, A. Kwasniewski, M. Naumann, T. Schulz, R. Schewski, D. Klimm, M. Bickermann, J. Cryst. Growth **404**, 184 (2014)
22. H. Aida, K. Nishiguchi, H. Takeda, N. Aota, K. Sunakawa, Y. Yaguchi, Jpn. J. Appl. Phys. **47**, 8506 (2008)
23. E.G. Víllora, S. Arjoca, K. Shimamura, D. Inomata, K. Aoki, Proc. SPIE, 89871U (2014)
24. D. Klimm, S. Ganschow, D. Bertram, R. Uecker, P. Reiche, R. Fornari, J. Cryst. Growth **311**, 534 (2009)
25. Z. Galazka, R. Uecker, D. Klimm, K. Irmscher, M. Naumann, M. Pietsch, A. Kwasniewski, R. Bertram, S. Ganschow, M. Bickermann, ECS J. Solid State Sci. Technol. **6**(2), Q3007 (2017)
26. A. Kuramata, K. Koshi, S. Watanabe, Y. Yamaoka, T. Masui, S. Yamakoshi, Jpn. J. Appl. Phys. 55, 1202A2 (2016)
27. K. Nakai, T. Nagai, K. Noami, T. Futagi, Jpn. J. Appl. Phys. **54**, 051103 (2015)
28. K. Hanada, T. Moribayashi, T. Uematsu, S. Masuya, K. Koshi, K. Sasaki, A. Kuramata, O. Ueda, M. Kasu, Jpn. J. Appl. Phys. **55**, 030303 (2016)
29. A.V. Zhdanov, G.A. Satunkin, V.A. Tatarchenko, N.N. Talyanskaya, J. Cryst. Growth **49**, 659 (1980)
30. V.A. Tatarchenko, J. Cryst. Growth **143**, 294 (1994)
31. K. Wada, K. Hoshikawa, Jpn. J. Appl. Phys. **17**(2), 449 (1978)
32. V.A. Tatarchenko, T.N. Yalovets, G.A. Satunkin, L.M. Zatulovsky, L.P. Egorov, D.Y. Kravetsky, J. Cryst. Growth **50**, 335 (1980)
33. K. Sasaki, A. Kuramata, T. Masui, E.G. Víllora, K. Shimamura, S. Yamakoshi, Appl. Phys. Express **5**, 035502 (2012)
34. S. Geller, J. Chem. Phys. **33**(3), 676 (1960)

35. G. Katz, R. Roy, J. Am. Ceram. Soc. **49**(3), 168 (1966)
36. J. Åhman, G. Svensson, J. Albertsson, Acta Crystallogy. Sect. C **C52**, 1336 (1996)
37. T. Onuma, S. Fujioka, Y. Yamaguchi, M. Higashiwaki, K. Sasaki, T. Masui, T. Honda, Appl. Phys. Lett. **103**, 041910 (2013)

Chapter 4
Floating Zone Method, Edge-Defined Film-Fed Growth Method, and Wafer Manufacturing

Akito Kuramata, Kimiyoshi Koshi, Shinya Watanabe and Yu Yamaoka

Abstract This chapter describes the floating zone growth method of β-Ga_2O_3, the edge-defined film-fed growth of β-Ga_2O_3, and the manufacturing of β-Ga_2O_3 wafers. The floating zone method section briefly mentions the method's history and typical growth conditions. The section on edge-defined film-fed growth method discusses the history, growth sequence, and conditions. It also covers the material properties of edge-defined film-fed grown β-Ga_2O_3 such as twin boundaries, dislocations, nanovoids, residual impurities, intentional doping, and dopant distribution. The wafer manufacturing section describes the basic wafer process and the effect of annealing on carrier concentration.

4.1 Floating Zone Method

The floating zone (FZ) method, which is one of the melt growth methods, is suitable for growing highly pure crystals because a crucible is not used in this method. Figure 4.1 shows a schematic drawing of the FZ method using lump heating. Resistive or inductive heating can also be used as heating sources in the place of lump heating. A sintered compact is usually used as a source. Before starting the crystal growth, one edge of the source is heated by lump and is melted. The initiation of the crystal growth is done by touching a seed crystal to the surface of the melted source. To proceed with the crystal growth, the source is pushed down and the crystal is pulled down. The melt is held between the source and the crystal

A. Kuramata (✉) · K. Koshi · S. Watanabe · Y. Yamaoka
Novel Crystal Technology, Inc., Sayama, Saitama 350-1305, Japan
e-mail: kuramata@novelcrystal.co.jp

K. Koshi
e-mail: k.koshi@novelcrystal.co.jp

S. Watanabe
e-mail: shinya.watanabe@novelcrystal.co.jp

Y. Yamaoka
e-mail: yu.yamaoka@novelcrystal.co.jp

M. Higashiwaki and S. Fujita (eds.), *Gallium Oxide*, Springer Series in Materials Science 293, https://doi.org/10.1007/978-3-030-37153-1_4

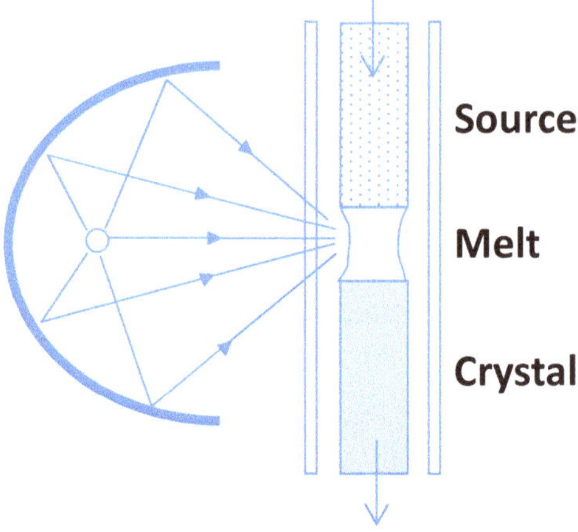

Fig. 4.1 Schematic drawing of FZ method using lump heating

by surface tension. Since no crucible is used, it is very easy to change the composition of the source material from growth to growth. Therefore, this method is suitable for research purpose. Meanwhile, the FZ method is not suitable for making a large-diameter crystal of β-Ga_2O_3 because it is difficult to hold a large amount of melt only by surface tension.

β-Ga_2O_3 growth by FZ method has been reported thoroughly in the literature [1–8]. Vivien et al. grew Cr-doped β-Ga_2O_3 to investigate if it has the potential as a material for a near-infrared tunable laser [1]. Their samples were easily cleaved to slabs with the dimensions of 15 mm × 5 mm × 3 mm. Ueda et al. have reported the use of FZ growth to control the resistivity of Sn-doped β-Ga_2O_3 material [2]. They found that resistivity of an unintentionally doped sample can be controlled by changing the oxygen content in the growth atmosphere and that a Sn-doped sample grown in pure oxygen is conductive. Their samples were also cleaved easily to the sample dimension of 3–5 mm × 5–10 mm × ~0.3 mm. Tomm et al. have reported FZ growth of undoped, Ge-doped, and Ti-doped β-Ga_2O_3 [3]. They reported that monoclinic β-Ga_2O_3 structure was maintained in the doped samples although the incorporated dopants significantly influenced the lattice parameter. They have also reported that the transmission of the samples was strongly influenced by the doping. Ti-doped samples showed deep-level absorption. There was a correlation between infrared transmission and free carrier concentration. Villora et al. have reported the first large-size β-Ga_2O_3 crystal: 1 inch in diameter and 5 cm long [4]. They were able to make substrate wafers with the (100), (010), and (001) surface orientations because stable growth proceeds only along the three crystallographic directions <100>, <010>, and <001>. Since the wafers were highly transparent as well as electrically conductive, they proposed to use β-Ga_2O_3 as a substrate for optoelectric devices with vertical current flow.

An example of growth conditions is shown in Table 4.1. The source and the crystal are rotated in opposite directions. The crystal pull-down speed and the source push-down speed are adjusted so that the amount of melt keeps constant. For

Table 4.1 Typical growth conditions for FZ growth of β-Ga_2O_3

Rotation speed of crystal	10 rpm
Rotation speed of source	5 rpm
Pull-down speed of crystal	1–6 mm/h
Push-down speed of source	1–10 mm/h
Atmosphere	20% N_2 and 80% O_2
Growth orientation	(010)

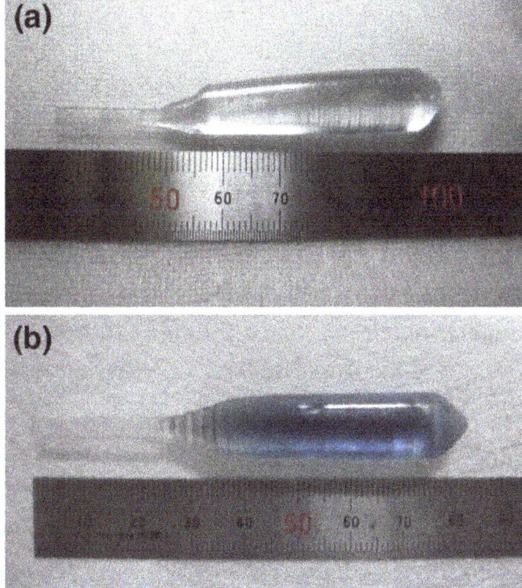

Fig. 4.2 Photographs of **a** unintentionally doped and **b** Sn-doped β-Ga_2O_3 crystals

the atmosphere, a mixture of N_2 and O_2 with appropriate composition is used because too much of O_2 leads to the formation of a lot of voids in the melt. Figure 4.2 shows photographs of unintentionally doped β Ga_2O_3 and Sn doped β-Ga_2O_3.

4.2 Edge-Defined Film-Fed Growth Method

Edge-defined film-fed growth (EFG) is a melt growth method in which a component called a *die* is used. The role of the die is to determine the cross-sectional shape of the crystal that is grown. This method was first described by LaBelle and Mlabsky [9–11] and has been used in the bulk growth of many kinds of materials, such as Al_2O_3 [10], TiO_2 [12], and Si [13, 14].

Shimamura et al. were the first to demonstrate EFG of β-Ga_2O_3 [15]. The bulk plate they grew was of 2-inch size, contained many cracks, and looked as if it had some polycrystalline regions. Aida et al. have improved the quality of β-Ga_2O_3 made using the EFG method [16]. They clarified that the seeding temperature and neck width were the most important factors for growing a plate of single-crystal bulk material. The etch pit density of their crystal was about 10^5 cm^{-2}. Kuramata et al. have further improved the crystal quality of β-Ga_2O_3 made by EFG [17]. They have grown bulk crystal without twin boundaries and with an etch pit density as low as about 10^3 cm^{-2}. They have also demonstrated 4-inch-size β-Ga_2O_3 wafers, and recently, the first 6-inch-size β-Ga_2O_3 wafers have been demonstrated [18]. Although several melt-growth methods for β-Ga_2O_3 have been reported, EFG is used for the mass production of n-type β-Ga_2O_3 wafers because it is presently the only method by which large n-type β-Ga_2O_3 bulk crystal can be grown.

4.2.1 Growth Sequence and Conditions of EFG

Figure 4.3 shows a schematic drawing of EFG of β-Ga_2O_3. Radio frequency (RF) inductive heating is used to heat a crucible containing source powder. When the temperature exceeds the melting point of Ga_2O_3 (about 1800 °C), the powder starts melting. The capillary effect causes the melt to go up through the slit formed in the die and reach the top of the die. Growth is initiated by melt touching a seed crystal at the top of the die. The bulk crystal becomes bigger as the RF power is decreased, and the seed crystal is pulled up. The cross-sectional shape of bulk crystal finally becomes same as the shape of die top and is kept constant during the entire growth. The high growth rate and the stability of the crystal shape are features of EFG. Figure 4.4 shows a photograph of EFG apparatus. The apparatus consists of quartz tube, thermal insulator, RF coil, and pulling-up system. The crucible is set in the

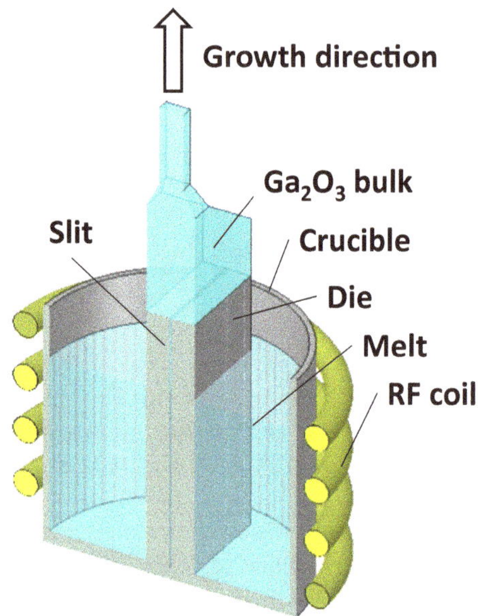

Fig. 4.3 Schematic drawing of EFG

Fig. 4.4 Photograph of EFG apparatus

Table 4.2 Typical growth conditions for EFG of β-Ga_2O_3

Source material	5 N Ga_2O_3 powder
Dopant source (n-type)	SiO_2, SnO_2
Dopant source (insulating)	FeO
Kind of metal for crucible and die	iridium
Dimension of die top surface	3 mm × 60 mm
	18 mm × 60 mm
	12 mm × 110 mm
	6 mm × 160 mm
	30 mm × 30 mm
Pressure	1×10^5 Pa
Atmosphere	98% N_2 and 2% O_2
Pulling rate	15 mm/h
Growth direction	(010)

thermal insulator at the height of the RF coil. The typical growth conditions are summarized in Table 4.2.

4.2.2 *Twin Boundaries*

In the bulk growth of β-Ga_2O_3 using EFG, twin boundaries are easily generated. Figure 4.5a is a photograph of the as-grown crystal with a high density of twin boundaries. A stripe pattern consisting of smooth surfaces and rough surfaces is

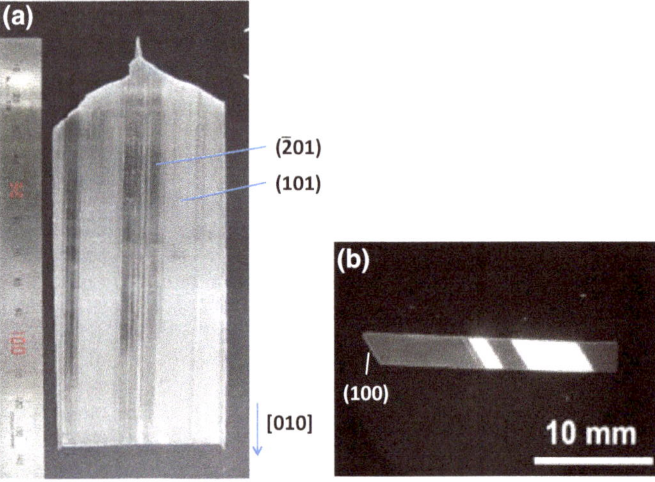

Fig. 4.5 Photographs of crystals with a high density of twin boundaries: **a** photograph of the as-grown crystal. The surface orientation of the smooth surface areas is (101) and that of the rough surface areas is ($\bar{2}$01). A twin boundary exists at the boundary between a smooth area and a rough area. **b** Polarization microscope picture of the (010) cross section. Dark areas and bright area are seen in turn. The difference in brightness is due to the difference in crystallographic orientation which is caused by the presence of twin boundaries between rough and smooth areas. The inclined facet at the left edge is the (100) facet

seen. The surface orientations of the smooth areas and the rough areas were confirmed by X-ray diffraction measurements to be (101) and ($\bar{2}$01), respectively. Figure 4.5b is a polarization microscope picture of the (010) cross section of a similar sample. Dark area and bright area are seen in turn. The difference in brightness is due to the difference in crystallographic orientation. The oblique facet at the left edge is the (100) facet, so we see that twin boundaries are parallel to the (100) plane.

The atomic arrangement near a twin boundary could be inferred from the experimental observations shown in Fig. 4.5. A model of that arrangement is shown in Fig. 4.6, where the green dashed line represents the twin boundary. The model well reproduces the results of the experimental observations. The ($\bar{2}$01) plane of the left-side crystal is parallel to the (101) plane of the right-side crystal. This corresponds to the feature shown in Fig. 4.5a. The boundary in the model consists of the common (100) plane of the crystals on both sides. This is consistent with the results shown in Fig. 4.5b.

Eliminating the twin boundary was the most critical issue for commercialization of β-Ga_2O_3 wafers. As shown in Fig. 4.5a, the twin boundaries are generated during the shouldering process, which is the initial step in the growth process. Once the twin boundary was generated, it remained throughout the growth and never disappeared. Therefore, it was thought that optimizing the shouldering process would be effective in suppressing the formation of twin boundaries. Indeed, after

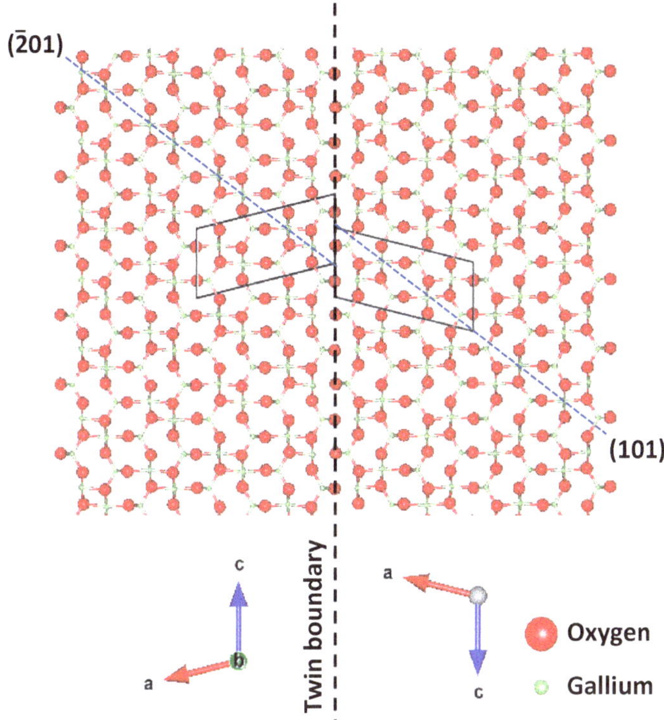

Fig. 4.6 Atomic arrangement model near a twin boundary. The dashed line represents the twin boundary. The boundary consists of the common (100) plane of the crystals on both sides. The ($\bar{2}$01) plane of the left-side crystal is parallel to the (101) plane of the right-side crystal

intensive efforts devoted to optimizing the shouldering process, it has recently become possible to grow β-Ga_2O_3 bulk crystals without forming twin boundaries.

4.2.3 Dislocations and Nanovoids

Most of the defects in the recent commercial β-Ga_2O_3 substrate wafers made by EFG are dislocations. Ueda et al. have reported that a dislocation exists under an etch pit of β-Ga_2O_3 with the ($\bar{2}$01) surface and that its Burgers vector is [010] [19]. The density of dislocations was examined by etch pits which were formed by dipping the samples in H_3PO_4 solution heated at 130 °C [17]. The samples investigated were the wafers with surface orientation of the ($\bar{2}$01), the (010), and the (001). The dislocation density was found to be of the order of 10^3 cm^{-2}. This small dislocation density was consistent with the fact that the full width of half maximum of the ($\bar{2}$01) X-ray rocking curve was as narrow as 17″. An example of the etch pits

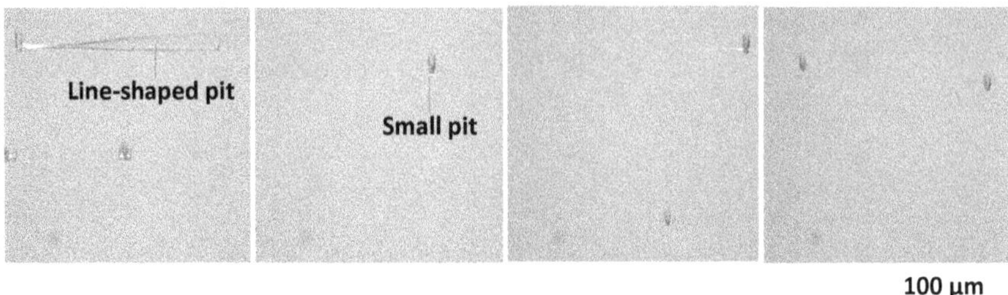

Fig. 4.7 Etch pits formed on the ($\bar{2}01$) surface of β-Ga_2O_3. The samples were dipped in H_3PO_4 heated at 130 °C for 2 h to form etch pits. The origin of a small pit is a dislocation and that of a line-shaped pit is a nanovoid

is shown in Fig. 4.7. In this case, small pits and a line-shaped pit were seen on the sample surface. The pits corresponding to dislocations were the small ones.

In addition to dislocations, there is another type of hollow defect often seen in β-Ga_2O_3 made by EFG. In this chapter, we call it a *nanovoid*. Nakai et al. have reported that the etch pit density related to the nanovoids (in their paper, they call them rodlike defects or nanopipes) is of the order of 10^2 cm^{-2}, and the nanovoids have no surrounding strain [20]. Micropipes seen in SiC are known to be accompanied by very large strain around them. The situation in terms of strain is completely different between nanovoids in β-Ga_2O_3 and micropipes in SiC. Hanada et al. have reported that there are groove-shaped pits (these correspond to the nanovoids) on the as-grown surface, and their dimensions are 40 nm in the [100] direction and hundreds of nanometers in the [001] direction [21]. Meanwhile, the nanovoids are elongated along the [010] direction, and their lengths are hundreds of micrometers. This means that a nanovoid has an extremely slender shape. The origin of the line-shaped pit seen in Fig. 4.7 is thought to be a nanovoid.

4.2.4 Residual Impurities

Residual impurity elements contained in the Ga_2O_3 source powder and the unintentionally doped (UID) and Sn-doped crystals were investigated by glow discharge mass spectrometry (GDMS) measurements [17]. Table 4.3 lists the results. The major residual impurity elements in the crystal were Si and Ir. Since Si was the element with the second largest concentration in the source powder, the impurity Si is thought to have come from the source powder. The Ir probably came from the crucible because the source powder did not contain Ir, and the crucible was made of Ir. Sub-weight ppm levels of Al, Cr, and Fe were also detected in the crystals.

The amounts of Si and Ir were about 2 weight ppm and 3 weight ppm. These values correspond to 2.5×10^{17} cm^{-3} and 0.8×10^{17} cm^{-3} in atomic concentration per unit volume, respectively, considering the difference in atomic weight of

Table 4.3 Concentration (wt. ppm) of impurity elements in Ga_2O_3 source material and β-Ga_2O_3 bulk crystal

	Source material	Bulk crystal (near center)	Bulk crystal (near surface)	Bulk crystal (near center)
		UID	UID	Sn-doped
B	0.02	0.13	<0.01	<0.01
Na	4.8	0.18	0.02	0.03
Mg	0.17	0.04	0.05	0.11
Al	0.1	0.89	0.81	0.9
Si	1.7	1.8	2.3	1.6
S	0.3	0.17	0.04	0.23
Cl	<0.05	1.5	0.1	0.13
Ca	<0.1	<0.1	<0.1	<0.1
Ti	0.006	0.03	0.04	0.02
Cr	0.07	0.31	0.27	0.34
Fe	0.74	0.24	0.81	0.51
Ni	<0.05	0.07	<0.05	<0.05
Cu	0.48	<0.05	<0.05	<0.05
Zr	<0.05	0.07	0.21	<0.05
Ir	<0.05	0.47	4.9	2.9
Sn	<0.5	<0.5	<0.5	36
Pb	0.59	<0.05	<0.05	<0.05
Bi	0.07	<0.05	<0.05	<0.05

each atom. Capacitance–voltage measurements showed that the UID crystal had a donor concentration of around 2×10^{17} cm^{-3}. Si is known to act as a shallow donor in β-Ga_2O_3. Although Ir is supposed to act as an acceptor, it is not known how Ir acts electrically in β-Ga_2O_3. It seems reasonable to think that UID crystals become *n*-type because of the presence of residual Si. SIMS measurements at three different points near the center gave Si concentrations of 1.1×10^{17}, 2.2×10^{17}, and 2.6×10^{17} cm^{-3}.

4.2.5 Intentional Doping

Either n-type or insulating doping is possible in β-Ga_2O_3. Highly conductive p-type β-Ga_2O_3 made by intentional doping has not been achieved yet.

Figure 4.8 shows control of net donor concentration, N_d–N_a, by Sn or Si doping [17]. As shown in the figure, N_d–N_a can be controlled by changing the SnO_2 or SiO_2 content in the source powder which is a mixture of Ga_2O_3 power and dopant powder. The upper limit of N_d–N_a is about 2×10^{19} cm^{-3} for both Sn and Si although the reasons for the limitation are different for these two dopants. In the

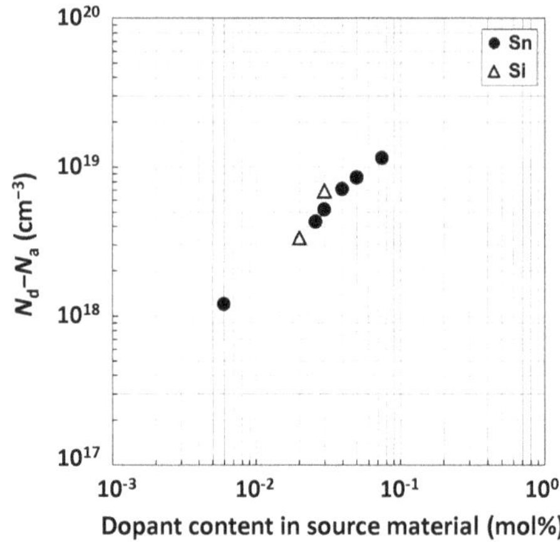

Fig. 4.8 Relationship between SnO_2 content in the source material and N_d–N_a for the grown crystals. It is shown that N_d–N_a can be controlled by varying the SnO_2 content

case of Sn, high vapor pressure of SnO_2 determines the limit. N_d–N_a did not exceed the upper limit even when the dopant content increased further. In the case of Si, generation of large voids or cracks was observed when Si concentration was more than the upper limit. The lower limit seems to be about 2×10^{17} cm^{-3}, but as mentioned in Sect. 5.2.4, its actual value depends on the residual Si concentration.

It is known that doping Fe or Mg makes β-Ga_2O_3 insulating. Figure 4.9 shows the relationship between resistivity and Fe concentration. The arrow in the figure indicates the background *n*-type carrier concentration of UID crystals. When Fe concentration was smaller than the background carrier concentration, the crystal was semiconducting with resistivity of 10^{-1} Ω cm order. When the Fe concentration exceeded the background concentration, the crystal became insulating (resistivity > 10^{11} Ω cm). Since Fe is assumed to act as a deep acceptor in β-Ga_2O_3, the possible mechanism is that Fe compensates the residual Si and makes the crystal highly insulating.

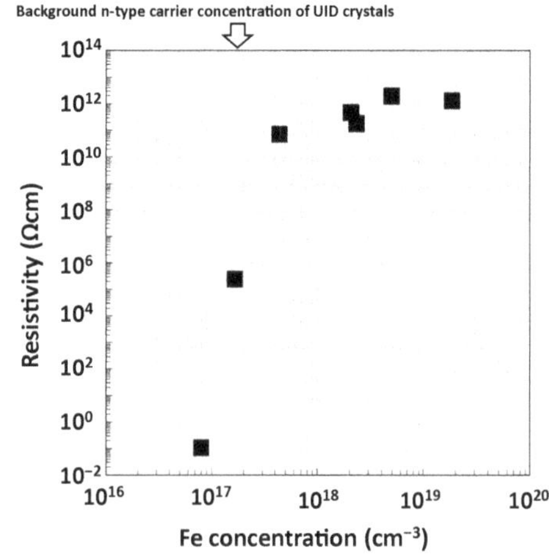

Fig. 4.9 Relationship between resistivity and Fe concentration in Fe-doped β-Ga_2O_3 crystals. Crystals become highly resistive when Fe concentration is more than background *n*-type carrier concentration

4.2.6 Distribution of Dopants

In general, inhomogeneous dopant distribution, which is called *segregation*, exists in crystals grown from melt when the segregation coefficient of the dopant in the melt is less than 1. The dopant distribution is given by the equation of normal freezing which is

$$C_s^* = kC_0(1 - f_s)^{(k-1)},$$

where C_s^* is the dopant concentration in the crystal, k is the segregation coefficient, C_0 is the initial dopant concentration in the melt, and f_s is the solidification ratio. The segregation occurs because the dopant not incorporated during solidification remains in the melt. The degree of condensation increases as the solidification proceeds.

In the EFG of crystals containing dopants, segregation occurs only in the thickness direction, not in the longer direction. The reason can be explained using schematic model of EFG shown in Fig. 4.10. In the EFG method, the melt flows uniformly in one direction at a time: first from the bottom to the top of the die, and then from the center to the edge of the die. The dopant concentration of the melt in the crucible, therefore, does not change during the growth, so the dopant distribution is basically uniform in the length direction. Meanwhile, dopant concentration of the melt on the top of the die depends on the lateral position, according to the equation of normal freezing. The melt supplied on top of the slit was solidified at the concentration of $C_s^* = kC_0$. The dopant concentration in the melt increases from center to the edge, so dopant distribution emerges in the thickness direction. The value of the normalized position from the center to the edge which is defined as distance X over L is equal to the value of the solidification ratio f_s.

Figure 4.11 shows the distribution of Si in EFG of Si-doped β-Ga_2O_3 in the thickness direction. The filled marks show the data obtained using a die with a single slit at the position of 0 mm, and the open marks show data obtained using a die with three slits at the positions of −1.5, 0, and 1.5 mm. By looking at the single-slit data,

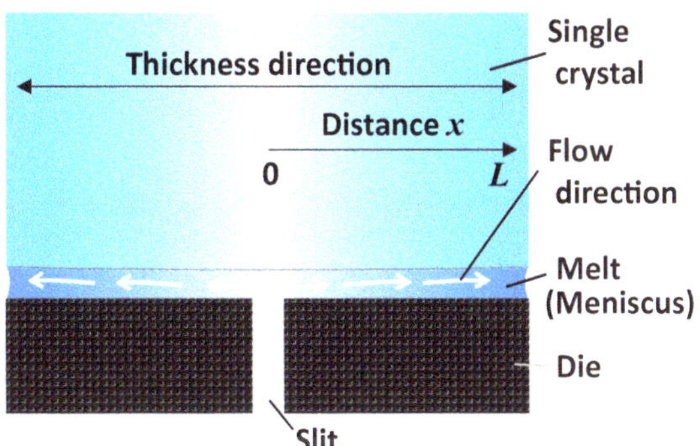

Fig. 4.10 Schematic model of EFG

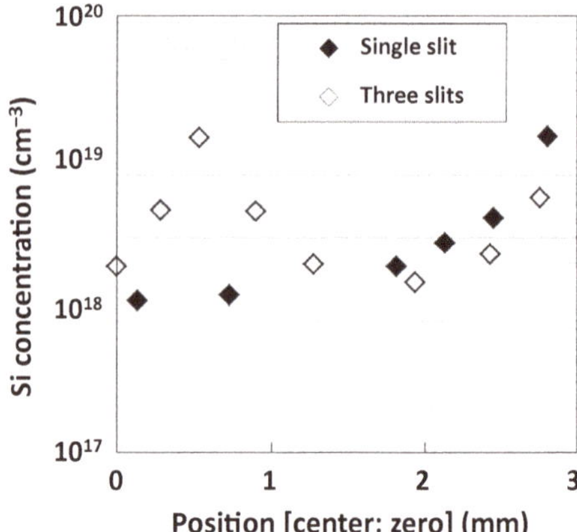

Fig. 4.11 Si concentration as a function of the position

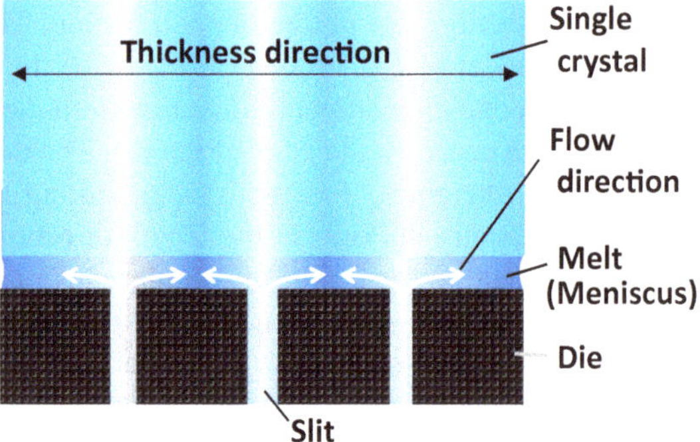

Fig. 4.12 Schematic model of EFG using a die with three slits

one sees that the Si concentration increases from the center to the edge. This corresponds to the basic image of the normal freezing model. In the three-slit data, however, the Si concentration is low at the positions of the slits (0 and ±1.5 mm) and is high between the slits and between the outer slits and the edge. The distribution can be well explained by the model shown in Fig. 4.12.

The segregation coefficient k can be estimated by fitting the data using the normal freezing equation. Figure 4.13 is an example. In this case, Si distribution data in Si-doped $\beta\text{-}Ga_2O_3$ obtained using a 6-mm-thick die and an 18-mm-thick die were used. The data were fitted well when k was between 0.25 and 0.35. Therefore, the segregation coefficient of Si in $\beta\text{-}Ga_2O_3$ was estimated to be around 0.3. In the same manner, the segregation coefficients of Fe and Sn were estimated to be 0.25 and 0.2, respectively. Note that the fitting was not done well in the case of Sn, probably because a lot of SnO_2 evaporates from the edge of the melt because of its high vapor pressure.

Figure 4.14 shows the distribution of dopant in longer direction in EFG of $\beta\text{-}Ga_2O_3$. As expected by the normal freezing model applied on EFG, the dopant concentration does not vary in longer direction in the case of Si and Fe. However, in

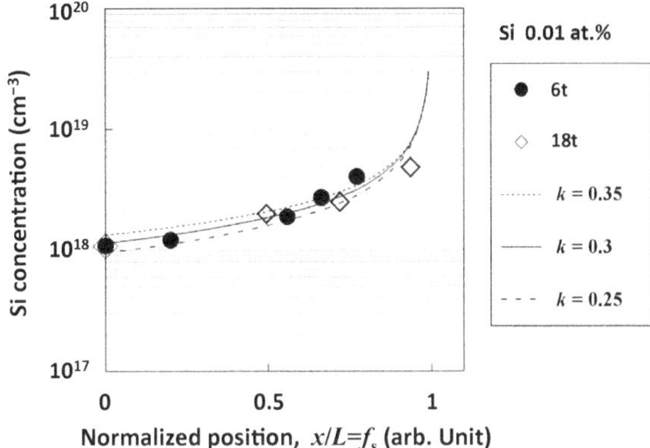

Fig. 4.13 Estimation of segregation coefficient for Si in β-Ga_2O_3

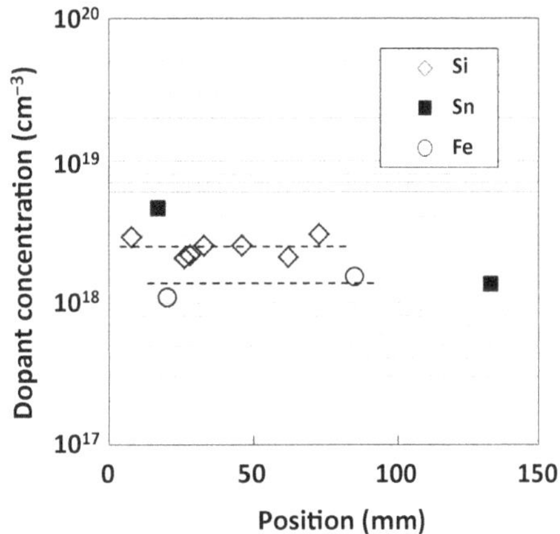

Fig. 4.14 Dopant distribution in length direction in β-Ga_2O_3

the case of Sn, the concentration decreased as the thickness increased. This is also probably because of a large amount of SnO_2 evaporation from crucible.

4.3 Wafer Manufacturing

In this section, wafer manufacturing process and the effect of annealing on carrier concentration are described.

4.3.1 Wafer Manufacturing Process

Figure 4.15 shows the wafer manufacturing process used for the production of 2-inch β-Ga_2O_3 wafers. After the bulk growth by EFG, the ingot grown is cut into blocks with square principal surfaces by using a rotary cutting machine with peripheral teeth. Figure 4.16 shows photographs of blocks with (a) a ($\bar{2}$01) principal surface and (b) a (001) principal surface. The size of the principal surface is 55 mm × 55 mm, and the blocks are 18 mm thick. The growth direction is [010]. On both kinds of blocks, (100) facets are spontaneously formed at the edge because the growth rate perpendicular to that plane is very slow. To change the surface orientation of the principal surface, the seed crystal is just rotated appropriately while the growth direction is kept the same, in the [010] direction. The blocks are annealed in nitrogen ambient at 1450 °C to achieve the full activation of donors and to reduce residual stress. This donor activation is based on the knowledge described in Sect. 4.3.2. The blocks are then sliced into plates by using a multiple-wire slicing machine. The plates are cut into circular wafers by using an electric discharge wire saw. The circular wafers are chamfered and polished using a grinder, lapping machine, and chemical mechanical polishing machine.

Figure 4.17 shows photographs of commercially available 2-inch β-Ga_2O_3 wafers and a 6-inch β-Ga_2O_3 wafers made for demonstration.

Figure 4.18 shows another manufacturing process used to make wafers with a surface orientation different from that of the principal plane of the bulk crystal. The

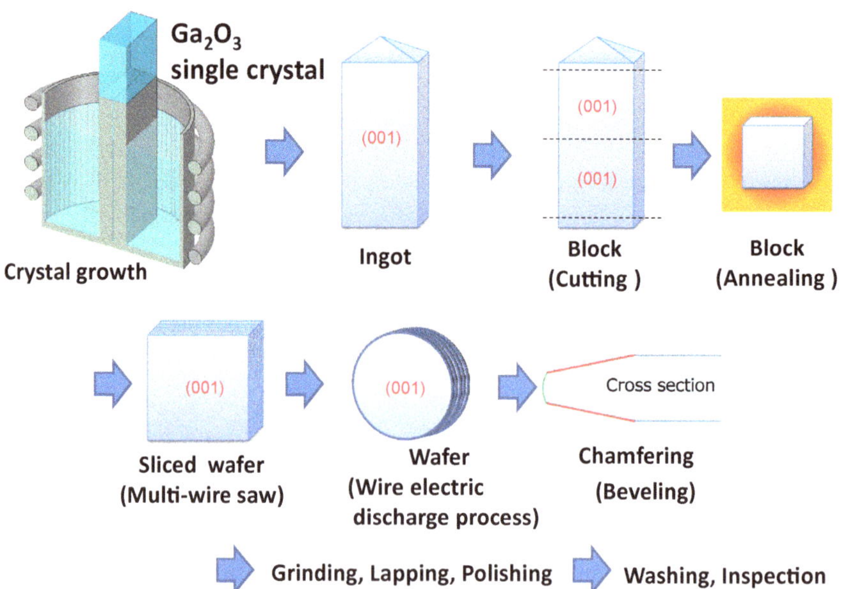

Fig. 4.15 Schematic drawing of process for manufacturing 2-inch wafers

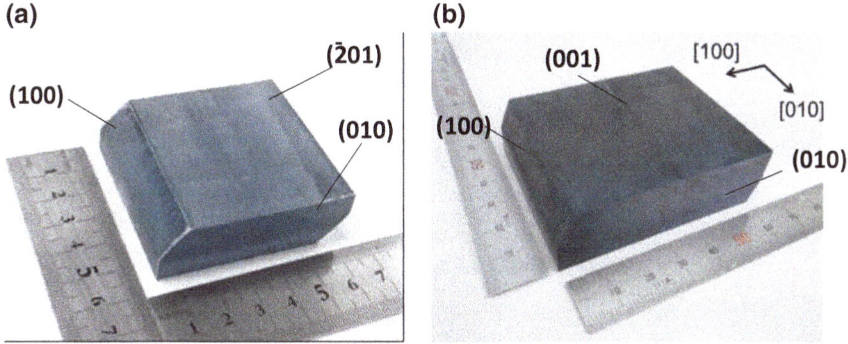

Fig. 4.16 Blocks of β-Ga_2O_3 with **a** ($\bar{2}$01) and **b** (001) principal surfaces

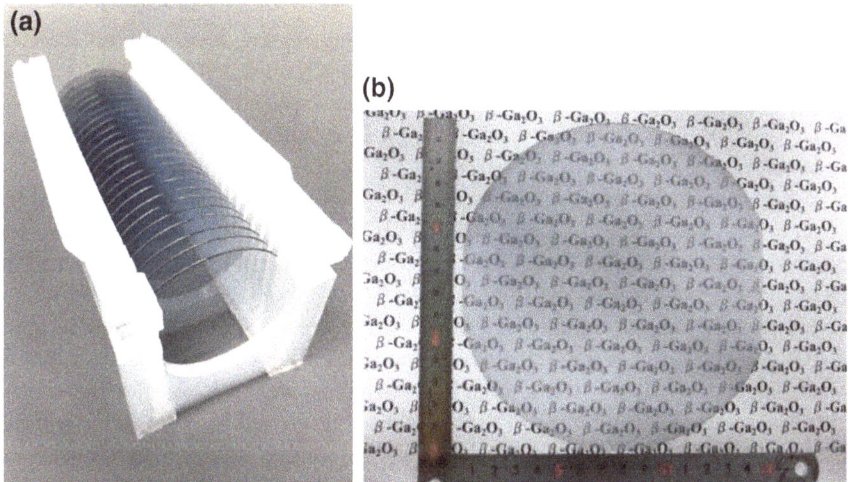

Fig. 4.17 Photographs of β-Ga_2O_3 wafers: **a** commercially available 2-inch wafers, **b** 6-inch wafer made for demonstration

blocks are first cut into smaller blocks and then are sliced into thin plates with the desired surface orientation.

4.3.2 *Effect of Annealing on Carrier Concentration*

The carrier concentration of the β-Ga_2O_3 varies depending on the atmosphere in thermal annealing. Understanding the phenomena is important if we are to control the carrier concentration of the substrate wafers precisely.

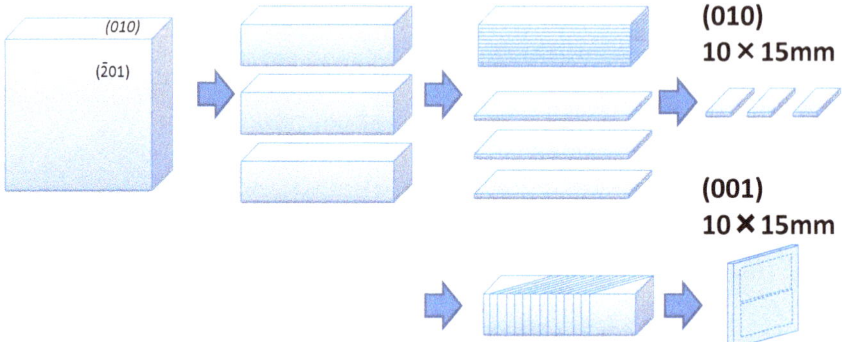

Fig. 4.18 Schematic drawing of process for manufacturing wafers with a surface orientation different from that of the principal plane of the bulk crystal

Oshima et al. first reported the formation of an insulating layer with thickness on the order of 100 nm on the surface of *n*-type β-Ga_2O_3 annealed in oxygen at around 1000 °C [22]. Galazuka et al. later reported the same phenomenon and also showed a reduction of carrier concentration in the bulk crystal after the formation of semi/insulating layer [23].

To visualize the phenomenon caused by oxygen annealing, the following observation was done. A cubic bulk β-Ga_2O_3 crystal with 10-mm sides was annealed in oxygen at 1150 °C for 30 min. Then, it was cut into two pieces, and the cross section was observed using an optical microscope. Figure 4.19 shows the result. It was seen that there is an area of whitish color with the thickness of about 1 mm, which corresponds to the area of reduced carrier concentration reported by Galazuka et al. [23]. It was also seen that there is an area of blue color in the inner part of the sample. Combining this observational result and the knowledge reported previously [22, 23], it was concluded that the sample consisted of three areas. Area I is the thin surface insulating layer with the thickness of a few hundred nanometers. Area II is the layer with reduced carrier concentration with the thickness of 1 mm. Area III is the inner part that still has the initial carrier concentration. Oshima et al. thought that the surface insulating layer (Area I) was formed by the diffusion of oxygen from the surface. Figure 4.19 indicates that the layer with reduced carrier concentration (Area II) is also formed by the diffusion of something.

Figure 4.20 shows the relationship between N_d–N_a and Si concentration in Area II for samples annealed in nitrogen at 800 or 1000 °C for 30–180 min as well as for samples annealed in oxygen at 1150 °C for 30–180 min. For samples annealed in nitrogen, N_d–N_a had almost the same value as the Si concentration, whereas for the samples annealed in oxygen, N_d–N_a was around 1×10^{17} cm^{-3} and independent of the Si concentration. These effects of the annealing ambient on the relationship between N_d–N_a and dopant concentration in Area II were the same for Sn-doped samples.

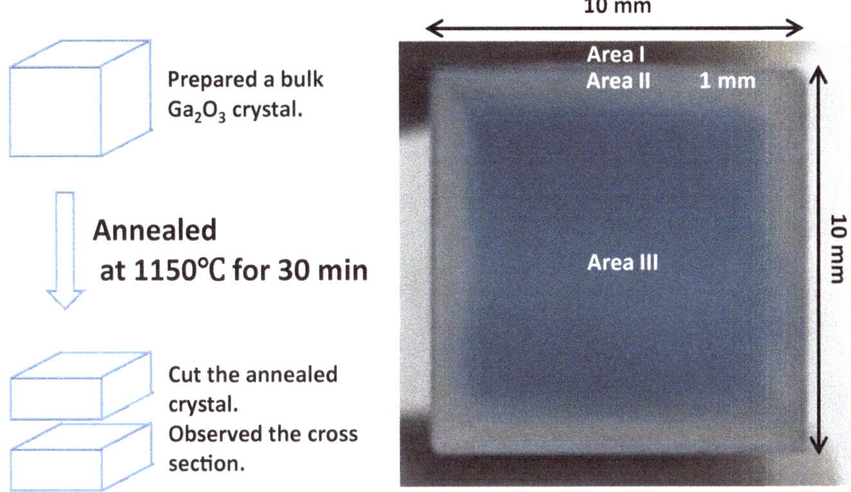

Fig. 4.19 Cross-sectional optical microscopic photograph of β-Ga_2O_3 annealed in oxygen atmosphere

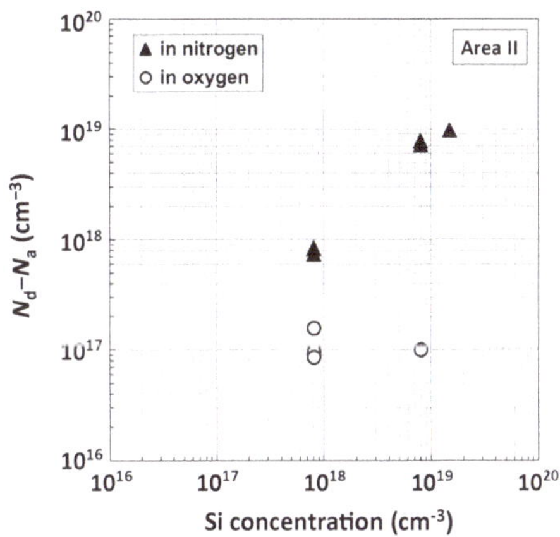

Fig. 4.20 Relationship between donor concentration and Si concentration for samples annealed in nitrogen and in oxygen

Oxygen annealing reduced the donor concentration, and nitrogen annealing increased the donor concentration. Figure 4.21 shows the changes in N_d–N_a observed when a sample with a Si concentration of 8 × 10^{17} cm^{-3} was annealed at 1100 °C in nitrogen for 1 h, then in oxygen for 6 h, and finally in nitrogen again for 3 h. After the first annealing in nitrogen, N_d–N_a had a high value of about 8 × 10^{17} cm^{-3}. Then, after 1-h annealing in oxygen, N_d–N_a decreased to 1 × 10^{17} cm^{-3} and remained at this low value during the rest of the annealing in oxygen. As a result of the third annealing in nitrogen, N_d–N_a increased and again reached its initial value of about 8 × 10^{17} cm^{-3}. Galazka et al. reported similar results, although their

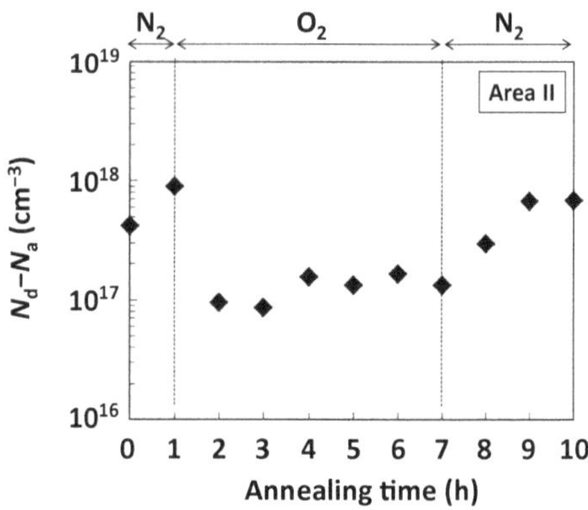

Fig. 4.21 Changes in N_d–N_a accompanying annealing at 1100 °C. Annealing in oxygen decreased N_d–N_a, and annealing in nitrogen increased N_d–N_a

annealing ambients were oxygen and hydrogen [23]. They observed a reduction in carrier concentration after annealing in oxygen and an increase after annealing in hydrogen.

It seems that oxygen vacancies and gallium vacancies are related to these phenomena. Further investigation is needed to clarify their origins.

References

1. D. Vivien, B. Viana, A. Revcolevschi, J.D. Barrie, B. Dunn, P. Nelson, O.M. Stafsudd, J. Lumin. **39**, 29 (1987)
2. N. Ueda, H. Hosono, R. Waseda, H. Kawazoe, Appl. Phys. Lett. **70**, 3561 (1997)
3. Y. Tomm, J.M. Ko, A. Yoshikawa, T. Fukuda, Sol. Energy Mater. Sol. Cells **66**, 369 (2001)
4. E.G. Víllora, K. Shimamura, Y. Yoshikawa, K. Aoki, N. Ichinose, J. Cryst. Growth **270**, 420 (2004)
5. J. Zhang, B. Li, C. Xia, G. Pei, Q. Deng, Z. Yang, W. Xu, H. Shi, F. Wu, Y. Wu, J. Xu, J. Phys. Chem. Solids **67**, 2448 (2006)
6. N. Suzuki, S. Ohira, M. Tanaka, T. Sugawara, K. Nakajima, T. Shishido, Phys. Status Solidi C **4**, 2310 (2007)
7. S. Ohira, N. Suzuki, N. Arai, M. Tanaka, T. Sugawara, K. Nakajima, T. Shishido, Thin Solid Films **516**, 5763 (2008)
8. E.G. Víllora, K. Shimamura, Y. Yoshikawa, T. Ujiie, K. Aoki, Appl. Phys. Lett. **92**, 202120 (2008)
9. H.E. LaBelle Jr., A.I. Mlavsky, Mat. Res. Bull. **6**, 571 (1971)
10. H.E. LaBelle Jr., Mat. Res. Bull. **6**, 581 (1971)
11. B. Chalmers, H.E. LaBelle Jr., A.I. Mlavsky, Mat. Res. Bull. **6**, 681 (1971)
12. H. Machida, K. Hoshikawa, T. Fukuda, Jpn. J. Appl. Phys. **31**, L974 (1992)
13. K.V. Ravi, J. Cryst. Growth **39**, 1 (1977)
14. T.F. Ciszek, G.H. Schwuttke, K.H. Yang, J. Cryst. Growth **50**, 160 (1980)
15. K. Shimamura, E.G. Víllora, K. Muramatsu, K. Aoki, M. Nakamura, S. Takekawa, N. Ichinose, K. Kitamura, Nihon Kessho Seicho Gakkaishi **33**, 147 (2006). [in Japanese]
16. H. Aida, K. Nishiguchi, H. Takeda, N. Aota, K. Sunakawa, Y. Yaguchi, Jpn. J. Appl. Phys. **47**, 8506 (2008)

17. A. Kuramata, K. Koshi, S. Watanabe, Y. Yamaoka, T. Masu, and S. Yamakoshi, Jpn. J. Appl. Phys. **55**, 1202A2 (2016)
18. A. Kuramata to be submitted
19. O. Ueda, N. Ikenaga, K. Koshi, K. Iizuka, A. Kuramata, K. Hanada, T. Moribayashi, S. Yamakoshi, and M. Kasu, Jpn. J. Appl. Phys. **55**, 1202BD (2016)
20. K. Nakai, T. Nagai, K. Noami, T. Futagi, Jpn. J. Appl. Phys. **54**, 051103 (2015)
21. K. Hanada, T. Moribayashi, K. Koshi, K. Sasaki, A. Kuramata, O. Ueda, and M. Kasu, Jpn. J. Appl. Phys. **55**, 1202BG (2016)
22. T. Oshima, K. Kaminaga, A. Mukai, K. Sasaki, T. Masui, A. Kuramata, S. Yamakoshi, S. Fujit, A. Ohtomo, Jpn. J. Appl. Phys. **52**, 051101 (2013)
23. Z. Galazka, K. Irmscher, R. Uecher, R. Bertram, M. Pietsch, A. Kwasniewski, M. Naumann, T. Schulz, R. Schewski, D. Klimm, M. Bickermann, J. Cryst. Growth **404**, 184 (2014)

Part II
Epitaxial Growth

Chapter 5
Plasma-Assisted Molecular Beam Epitaxy 1

Growth, Doping, and Heterostructures

Akhil Mauze and James Speck

Abstract Plasma-assisted molecular beam epitaxy has been used to grow the highest quality β-Ga_2O_3 thin films and has shown potential to realize various efficient device structures. Growth of β-Ga_2O_3 is defined by the suboxide desorption that limits growth rates at high temperatures and Ga fluxes. Growth in various orientations has been demonstrated with the (010) b-plane in particular showing promise for homoepitaxy due to high realized growth rates and materials quality. N-type doping with Sn, Ge, and Si has allowed for device structures that utilize electron conduction in this materials system. Heterostructures with β-$(Al_xGa_{1-x})_2O_3$ have been used for modulation doped field effect transistors; however, thermodynamic limitations of maximum achievable Al content have limited device performance. Expanding the growth regime through metal-oxide catalyzed epitaxy using In could help improve heterostructure growth for future devices.

5.1 Introduction

Among the many thin film deposition techniques for β-Ga_2O_3, molecular beam epitaxy (MBE) has been the most investigated due to its ability to produce high quality, controllable thin films. While study of the materials system is still in its infancy, plasma-assisted molecular beam epitaxy (PAMBE) has been able to produce β-Ga_2O_3 with various doping and heterostructure schemes.

MBE is an ultra-high vacuum deposition technique that relies on the slow deposition of gas, evaporated, or sublimed sources onto a substrate. MBE growth temperatures allow for surface diffusion of constituent adatoms to attach at their crystal sites but limit bulk diffusion to allow abrupt films. This along with the high

A. Mauze (✉) · J. Speck
Materials Department, University of California Santa Barbara,
Santa Barbara, CA 93106, USA
e-mail: akhilmauze@ucsb.edu

J. Speck
e-mail: speck@ucsb.edu

M. Higashiwaki and S. Fujita (eds.), *Gallium Oxide*, Springer Series
in Materials Science 293, https://doi.org/10.1007/978-3-030-37153-1_5

vacuum and high purity source materials allows for minimal impurity incorporation and sharp interfaces. PAMBE uses a gas source to produce an active flux of an element, in this case presumably atomic oxygen. Modern radio frequency (rf) plasma sources confine nearly all ions within the plasma and produce a flux of active atomic oxygen (O) and more inert molecular oxygen (O_2). The atomic oxygen reacts with the other evaporated sources like Ga at the surface of the substrate to produce β-Ga_2O_3 and its alloys. In PAMBE, growth rates are often limited by the ability to provide a large active flux from the plasma source and growth rates for β-Ga_2O_3 tend to be on the order of 1–5 nm/min [1–3]. Other materials systems like nitrides have used PAMBE to grow the highest quality films and devices in the past [4]. Similar fundamentals can be applied to oxide MBE with a high purity O_2 source that can produce atomic O via the rf plasma source. The contents of these rf plasmas have been studied in detail to determine atomic O to be a primary active oxygen species [5, 6]. For typical rf-PAMBE sources used in studies discussed later in this chapter, only about 1–2% of the oxygen is active atomic O. Due to the use of O_2 instead of ozone, PAMBE is typically a safer technique than ozone MBE. With ozone techniques, extensive pumping and safety measures to minimize the buildup of ozone in the MBE chamber are necessary. Furthermore, the higher boiling point of ozone means it is liquid at liquid nitrogen temperature of 77 K. Thus, safety concerns prohibit the use of liquid nitrogen [1].

The other necessary source for β-Ga_2O_3, Ga, is melted and heated to a point where it can be slowly evaporated. Typical fluxes yielded, as measured in beam equivalent pressure (BEP) by a nude ion gage in at the substrate position are around 10^{-7} Torr, whereas typical chamber pressures are $\sim 10^{-5}$ Torr due to the high flux of molecular oxygen from the rf plasma source. For the growth of β-Ga_2O_3, the active oxygen is necessary as a pure O_2 flux will not yield any growth. The high background O_2 pressure in the chamber can be problematic for oxidation of reactive sources such as Al or dopants. Typically, in PAMBE, the growth is performed close to stoichiometry when considering the Ga and active O fluxes. Additional solid source elements like In and Al are used for heterostructures and Si, Sn, and Ge are used for n-type doping. Generally, Ga and In are relatively robust sources in the oxygen environment, whereas Al can oxidize after extensive exposure to oxygen. A schematic of a vertical PAMBE setup is shown in Fig. 5.1 with the plasma source, substrate and heater, and effusion cells [7]. A vertical MBE system is ideal to maximize both amount of source material loaded and number of available ports for sources that can be liquid at operating cell temperatures.

5.2 Characterization

One benefit of using a high vacuum process like MBE is the accessibility of in situ characterization techniques. Reflection high-energy electron diffraction (RHEED) is one such technique that allows for monitoring of the surface morphology and growth rates. RHEED uses an electron beam incident at a small angle to interact with the

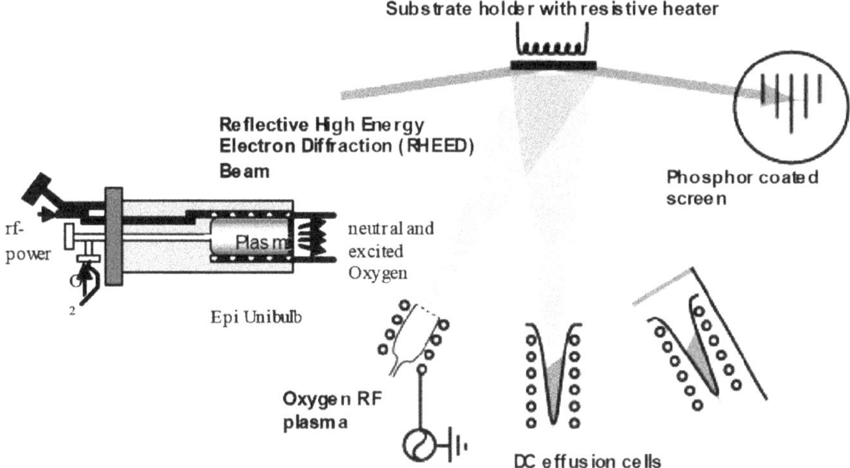

Fig. 5.1 Typical plasma-assisted MBE setup [7]

sample surface and produce diffraction patterns that are highly sensitive to surface morphology. Rough surface morphology results in a spotty RHEED pattern while relatively smooth surfaces characteristic of primarily two-dimensional growth produce streaky RHEED patterns. This is extremely important for understanding growth modes of thin films as optimization of these can provide the smoothest surfaces necessary for thin film devices. Additionally, oscillations in spot intensity of a specific reflection in this RHEED pattern can be temporally monitored to determine growth rates. A typical RHEED setup in situ is also shown in Fig. 5.1 [7].

Quadrupole mass spectroscopy (QMS) has also been used in PAMBE to measure desorbing fluxes from the growth surface during growth and etching (also referred to as desorption mass spectroscopy). QMS measures charge to mass ratios of ionized elements and compounds coming from the substrate to determine particle fluxes. QMS has been applied to the study of decomposition of MBE grown films in situ for materials systems like the nitrides in the past [8]. For β-Ga_2O_3 growth, suboxide desorption can be monitored for analysis of growth chemistry and reactions [9, 10]. More QMS studies will be discussed in Chap. 6.

5.3 β-Ga_2O_3 PAMBE Growth

Interest in β-Ga_2O_3 as a transparent conductive oxide led to early investigation of the material system and its growth by PAMBE. Initial attempts involved heteroepitaxy of $(\bar{2}01)$ β-Ga_2O_3 on c-plane sapphire substrates and homoepitaxy of (100) β-Ga_2O_3 [1, 11]. $(\bar{2}01)$ oriented growth on sapphire is common because of the similar oxygen arrangement in sapphire and $(\bar{2}01)$ β-Ga_2O_3. Still for this heteroepitaxy there is lattice mismatch and three rotational domains for possible epitaxy making homoepitaxy more desirable for high film quality [1].

Villora et al. compared $(\bar{2}01)$ and (100) heteroepitaxy with (100) homoepitaxy, revealing smoother surface morphology and relatively high growth rate for (100) homoepitaxial growth [11]. Additionally, $(\bar{2}01)$ oriented growth demonstrated by Tsai et al. had a crystal mosaic of $\sim 1°$ (full width at half-maximum, FWHM) as determined by X-ray ω-rocking curves, whereas (100) homoepitaxial growth demonstrated mosaic of only 0.02°, which is similar to that of the substrate, indicating a higher quality of material for this homoepitaxial growth. The availability of high quality bulk substrates along with the higher epitaxial materials quality for homoepitaxy makes it the primary focus of current MBE research of β-Ga_2O_3 [1].

Tsai et al. demonstrated phase pure β-Ga_2O_3 grown in these orientations in a Veeco 620 oxide PAMBE system. This study determined an optimal growth temperature of 700 °C as measured by pyrometry to produce quality films in both orientations. (100) homoepitaxial growth demonstrated elongated island formation along the [010] direction at lower temperatures and rougher surfaces at higher temperatures than 700 °C, the optimal temperature at which root mean square (RMS) roughness was 0.67 nm. No Ga droplets or adlayers were observed for this O-rich growth. Homoepitaxial growth mode and rate were monitored using RHEED, with each oscillation corresponding to a monolayer of β-Ga_2O_3, or half a unit cell. A typical growth rate of 46 nm/h was found at optimal conditions. Streaky RHEED throughout growth indicated two-dimensional growth with monolayer nucleation and coalescence characteristic of layer-by-layer growth [1].

In addition, growth rate dependence on Ga fluxes and suboxide desorption were investigated. Results found that for a fixed oxygen flux and growth temperature, the growth rate would increase with increasing Ga flux in the O-rich regime and would then decrease for very high Ga fluxes as the growth transitioned to Ga-rich. This is shown in Fig. 5.2. Similar growth rate dependences had been seen in SnO_2 growth which demonstrated volatile suboxide desorption at higher Sn fluxes [12]. QMS studies revealed that Ga_2O was the primary suboxide desorbing at high Ga fluxes [1, 9, 10]. The suggested reaction is demonstrated in 5.1.

$$Ga_2O_3 + 4Ga \rightarrow 3Ga_2O \tag{5.1}$$

This same volatile oxide has been seen in the desorption of GaAs oxides in MBE. The limited maximum growth rate is the result of the competition between Ga_2O_3 formation (growth) and Ga_2O suboxide formation (etching reaction for metal rich growth) [1]. A more detailed explanation of these growth mechanisms was studied by Vogt et al. and will be given in the Chap. 6 [10].

The dependence of surface morphology on Ga flux was also investigated by Tsai et al. [1]. These studies demonstrated that lower Ga fluxes yielded 3D columnar growth while a stepped terraced structure was observed at higher Ga fluxes. This

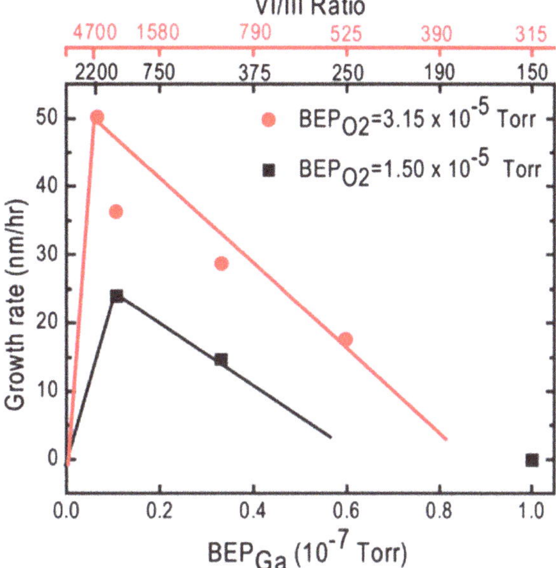

Fig. 5.2 Growth rate dependence on Ga flux for (100) β-Ga_2O_3. Reproduced from [1] with the permission of the American Vacuum Society

stepped structure was also observed by Oshima et al. [13]. For very high Ga fluxes with no β-Ga_2O_3 growth, the surface morphologies showed a stepped structure similar to the bare substrate. Additional study of (100) β-Ga_2O_3 growth by PAMBE demonstrated the occurrence of twin domains [14]. This phenomenon was also observed in (100) films grown by metal organic chemical vapor deposition (MOCVD) recently [15].

After demonstration of β-Ga_2O_3 as a potential wide bandgap semiconductor for power electronics [16], efforts into MBE growth of the material and associated heterostructures increased. Our understanding of β-Ga_2O_3 growth regimes and orientations, and the potential to produce high quality electronics is still in the early stages, but initial demonstrations suggest there is much promise for epitaxially grown β-Ga_2O_3 devices.

Sasaki et al. demonstrated device quality β-Ga_2O_3 grown by ozone MBE [17]. Room temperature mobilities of more than 100 $cm^2\ V^{-1}\ s^{-1}$ were achieved, which is on par with other thin film deposition techniques. Different substrate orientations were investigated for homoepitaxial growth revealing a strong growth rate dependence on orientation. Specifically, high suboxide desorption of cleavage planes like (100) led to significantly lower growth rates in these orientations, whereas (010) oriented growth yielded the highest growth rate [16]. This effect of substrate orientation inspired investigation of (010) growth in PAMBE. Now, this is the leading orientation for PAMBE research because it has demonstrated the best quality material and devices with the highest growth rates.

Okumura et al. demonstrated β-Ga_2O_3 (010) growth at rates of 2.2 nm/min, more than twice the (100) growth rate [9]. Growth rates were determined by high resolution X-ray diffraction (HRXRD) thickness fringes of homoepitaxially grown material. Despite it being homoepitaxy, the slight rigid body displacement of the film from the substrate often allows thickness fringes to be observed for homoepitaxially grown films. This was first demonstrated in other oxide materials systems by LeBeau et al. [18]. The enhanced growth rate for the same oxygen source utilizes higher Ga fluxes for growth before suboxide desorption limits growth rate [9]. This further proved the strong dependence of suboxide desorption and growth rate on substrate orientation. In general, the cleavage planes like $(\bar{2}01)$, (100), and (001) have demonstrated lower growth rates than (010) due to higher suboxide desorption. This study demonstrated growth at a wide variety of temperatures and investigated Ga flux and O-flux conditions for (010) growth. Without striking the plasma, no growth of β-Ga_2O_3 is observed, suggesting that the active oxygen species is atomic O and not O_2 [9].

Ahmadi et al. demonstrated the dependence of growth rate on Ga flux for (010) growth and demonstrated similar trends as with other orientations [2]. An O-rich regime existed in which growth rate was proportional to Ga flux and a plateau regime at higher Ga fluxes existed for which the growth rate remained constant with change in Ga flux. Very high Ga fluxes were not investigated so reduced growth rate in the Ga-rich growth conditions was not observed in this study. In this plateau regime, for a constant Ga and O flux, the growth rate decreases at higher growth temperatures, likely due to increased suboxide desorption. Smooth surfaces measured by atomic force microscopy (AFM) were achieved at growth temperatures of 500–700 °C as shown in Fig. 5.3 [9]. At very high temperatures, there was little growth. Optimal growth conditions were found to be in the 600–750 °C range for good surface morphology and high growth rates. In addition, due to large miscuts of these (010) substrates, smooth growth was achieved with some elongated features along the [100] direction. These features are common in (010) growth even without miscut [9].

Later studies demonstrated that surface treatment with the O-plasma [19] and Ga flux [20] further improved surface morphologies to achieve RMS roughness less than 0.3 nm for slightly Ga-rich growth, whereas pitting and rougher surfaces were seen in very O-rich conditions [2].

Other orientations have also been pursued for growth by PAMBE. (001) β-Ga_2O_3 in particular shows some promise due to its scalability with the edge-defined film fed growth (EFG) method and its initial PAMBE growth demonstration was performed by Oshima et al. [3]. Growth rates were limited to about 1 nm/min for the same oxygen source that yielded 3.3 nm/min for (010) growth. This was attributed to the higher suboxide decomposition seen in cleavage planes, which is consistent with ozone MBE studies of Sasaki et al. [17] and low growth rates achieved in other orientations by Tsai et al. [1]. This study demonstrated the three typical growth regimes seen in PAMBE growth of β-Ga_2O_3 with growth rates initially increasing with Ga flux in the O-rich regime, then leveling off in the plateau regime, and finally decreasing for very Ga-rich conditions. In addition, etching of (001) β-Ga_2O_3 was shown without an oxygen plasma,

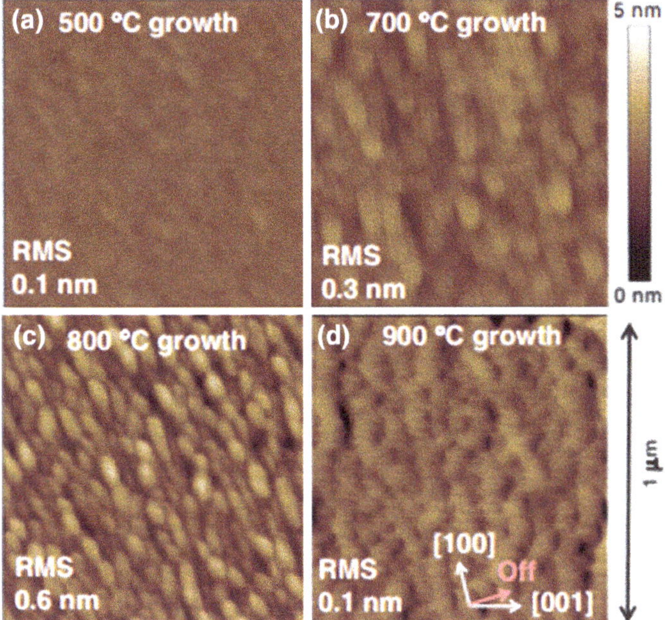

Fig. 5.3 Surface morphology measured by AFM of (010) β-Ga_2O_3 grown at different temperatures [9]

further suggesting Ga_2O desorption. This etching did not significantly increase the surface roughness of (001) β-Ga_2O_3. A comparison of these growth and etching rates with growth rates in the (010) orientation is shown in Fig. 5.4. (001) β-Ga_2O_3 growth rates also demonstrated some temperature dependence with temperatures above 750 °C yielding progressively reduced growth rates with increasing growth temperature [3].

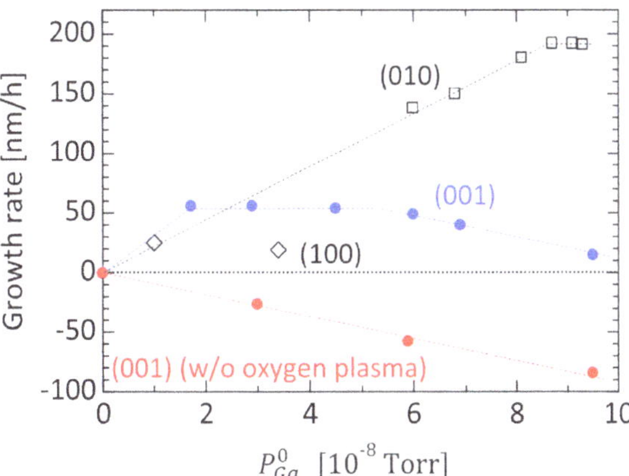

Fig. 5.4 Growth and etch rates of β-Ga_2O_3 for different Ga fluxes and orientations [3]

Further study of (001) β-Ga_2O_3 PAMBE was demonstrated by Han et al. Here, it was shown that higher growth temperatures near 750 °C yielded smoother surfaces than 675 °C, with typical RMS roughness of 2.0 nm under best conditions. While higher growth temperatures for (001) β-Ga_2O_3 are ideal for surface morphology, they also bring up limitations in doping to be discussed later [21].

5.4 Doping

Controllable doping of semiconductors is necessary for nearly all device technologies. For transistors, low-doped films can be used as active channel regions in lateral devices or drift layers in vertical devices. Highly doped films can be used for ohmic contacts with low resistance. An ideal dopant would be controllable over a wide range of concentrations with high activation while maintaining good surface morphology and minimal defects in the film. P-type doping for efficient hole conduction is unrealizable in β-Ga_2O_3 due to high acceptor ionization energies, very high effective hole masses (due to negligible valence band dispersion), and the likelihood of strong polaron formation in the event a hole is created [22]. In contrast, n-type doping for unipolar devices is a viable option for β-Ga_2O_3 transistors with theoretical bulk electron mobilities predicted to be about 200 $cm^2\ V^{-1}\ s^{-1}$ as limited by optical phonon scattering [23]. Higher mobilities have been predicted for two-dimensional electron gases due to high screening at higher sheet charge densities in these structures [24]. For n-type doping in PAMBE of β-Ga_2O_3, the dopant should ideally be robust to various growth conditions, easily controllable, and be easy to use in the high vacuum chamber without oxidizing in the growth environment.

Sn, Si, and Ge have all been shown to be shallow donors in β-Ga_2O_3 and potential options for doping of PAMBE grown films [25]. As a source, Si shows the most difficulty due to oxidation of the solid Si in the high background oxygen environments in PAMBE growth (i.e., the chamber pressure during growth is ~10^{-5} torr—nearly all due to molecular oxygen). However, successful pulsed delta doping has been used to realize high n-type doping concentrations and employed in various devices [26]. This scheme involves only opening the Si shutter for about one second over the span of a minute and continuing this throughout the growth of a doped layer. This limits the Si source's exposure to oxygen while obtaining large fluxes of Si over a short time to achieve a target average doping concentration. Pulsing the shutter can achieve uniform doping concentrations over hundreds of nanometers. Alternatively, Si can be used to achieve high concentrations for δ-doping and has been applied to modulation doped field effect transistors (MODFETs) as well [27]. Sn and Ge have demonstrated a wider range of doping achievable with Ge being the most investigated in (010) β-Ga_2O_3 PAMBE [2].

Ahmadi et al. investigated doping with Ge over various growth regimes. Ge incorporation into the films had a high dependence on growth temperature with Ge concentration dropping from 10^{19} to 10^{17} cm^{-3} for a growth temperature increase from 600 to 700 °C. Furthermore, the Ge incorporation depends strongly on growth regime with oxygen rich growth yielding much higher Ge doping levels of greater than 10^{20} cm^{-3} for conditions that would achieve 10^{18} cm^{-3} in plateau regime growth. This suggests a site competition between Ge and Ga. Additionally, at very high Ge cell temperatures, and therefore fluxes, Ge incorporation into β-Ga_2O_3 decreases. This along with the growth temperature dependence of Ge incorporation has been tentatively attributed to Ge oxide desorption. At higher Ge cell temperatures, pitting and rougher surface morphologies were also measured by AFM. Ultimately, a range of 10^{17} to 10^{20} cm^{-3} can be achieved controllably at typical cell temperatures ranging from 550 to 650 °C and a growth temperature of 600 °C. While a variety of Ge doping levels can be obtained with (010) β-Ga_2O_3, its growth regime is limited for achieving this range [2].

Sn doping was also shown to achieve a wide range of doping concentrations from 5×10^{17} to 10^{20} cm^{-3} achieving a mobility of 120 cm^2 V^{-1} s^{-1} for the lowest doping conditions. Sn doping has also shown robustness to growth temperature and group III flux in the typical growth regimes of (010) β-Ga_2O_3. This makes it appealing for its application to growth at varying growth conditions. Typical growth rates were 3.3 nm/min at 600 °C as limited by the active oxygen flux of the plasma source [2].

In the (001) oriented growth, Ge dopant incorporation demonstrated a strong growth temperature dependence with higher growth temperatures limiting maximum Ge concentrations to $\sim 3 \times 10^{18}$ cm^{-3} while a growth temperature of 675 °C could yield Ge concentrations of $\sim 10^{20}$ cm^{-3} for otherwise similar conditions. This trend was also observed in (010) growth and seems to be a feature of Ge doping of β-Ga_2O_3 by PAMBE. The limited incorporation for the higher temperatures that are needed for smooth surface morphologies is a challenge for the use of Ge in (001) β-Ga_2O_3 PAMBE growth. Ge doping dependence on cell temperature was also investigated. For lower Ge cell temperatures, and therefore fluxes, a clear Arrhenius relationship between cell temperature and doping concentration can be seen, yielding an activation energy in agreement with solid Ge vapor pressure curves. This suggests flux limited doping for these lower cell temperatures. At higher Ge cell temperatures, this incorporation saturates at a level dependent on the limitations of growth temperatures showing surface reaction limited doping. This further suggests Ge could be leaving as a suboxide as seen in (010); however, proper mass spectroscopy studies for confirmation are yet to be performed. This cell temperature dependence on doping concentration is shown in Fig. 5.5a. One final limitation of Ge doping in (001) growth was observed in a doping delay as initial incorporation of Ge into a film occurs slowly and is dependent on surface roughness and possibly Ge surface coverage [21].

Sn doping seems more robust as it is not affected by growth temperature in the typical range of (001) growth. Furthermore, Sn doping is flux limited for a range of concentrations from 10^{17} to 10^{21} cm^{-3} and does not demonstrate a doping delay. For this reason, Sn seems to be the ideal dopant for (001) growth. The dependence of Sn concentration on Sn cell temperature is shown in Fig. 5.5b. Mobilities for Sn and Ge doped films were limited to around 26 cm^2 V^{-1} s^{-1}. Because of its lower

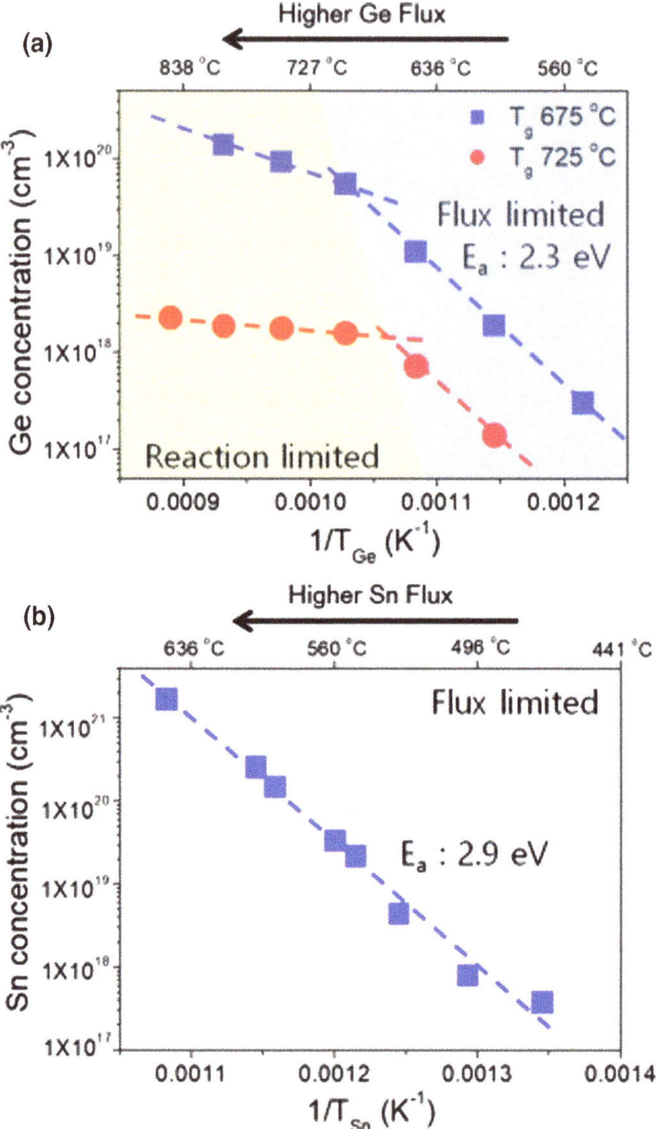

Fig. 5.5 Ge (**a**) and Sn (**b**) doping versus cell temperature for (001) β-Ga_2O_3 [21]

mobility and higher surface roughness, (001) seems to be a less ideal orientation that (010) for growth by PAMBE. However, if film quality can improve, its scalability would make it a promising orientation [21].

5.5 Heterostructures

Heterostructures have been utilized for bandgap and lattice constant engineering in thin films. Coherent epitaxially grown semiconductor heterostructures have been demonstrated for the AlGaAs/GaAs and AlGaN/GaN system to produce devices that utilize the high conduction band offset between these two materials with high Al incorporation into these structures. This allows for maximum charge density in a channel just below the heterostructure interface in the binary material that is spatially separated from the donors in the ternary material. In the past, MBE has produced the highest quality and highest mobility heterostructures in these materials systems. The analogous heterostructure in gallium oxide is β-$(Al_xGa_{1-x})_2O_3/Ga_2O_3$ and PAMBE growth has been focused on producing high quality, coherent β-$(Al_xGa_{1-x})_2O_3/Ga_2O_3$ heterostructures with high Al content. More about MODFETs that utilize this heterostructure will be discussed in Chap. 33.

The initial demonstration and analysis of coherent epitaxially grown (010) β-$(Al_xGa_{1-x})_2O_3$ was done by Kaun et al. [28]. Further X-ray and compositional analysis was performed by Oshima et al. [29]. Films with controllable Al contents up to 17% were grown with thicknesses of 130 nm. X-ray analysis of (010) β-$(Al_xGa_{1-x})_2O_3$ determined the relationship between out of plane lattice parameter (b_c) in Å and Al fraction on the group III site (x) to be:

$$x \cong 15.923 - 5.238 \times b_c \tag{5.2}$$

Conversion of this to on-axis peak separation in degrees ($\Delta\theta_{020}$) between the (020) β-$(Al_xGa_{1-x})_2O_3$ and the (020) β-Ga_2O_3 peak in an ω-2θ HRXRD scan with Cu Kα X-rays results in the following relationship:

$$x \cong 0.4727 \times \Delta\theta_{020} \tag{5.3}$$

The coherent strain was consistent with both HRXRD data and theoretical elastic constants of the alloy that determine this strain. Atom probe tomography (APT) also determined composition of β-$(Al_xGa_{1-x})_2O_3$ films in this study to further confirm this relation. Typical growth temperatures of 650 °C were used to achieve these Al contents. Al/(Ga + Al) BEP ratios measured by an ion gage in the MBE chamber prior to growth were around 0.104 for Al contents of 17%. At this growth temperature, increasing Al flux can achieve around 20% Al, with coherent crystallinity compromised at higher Al fluxes. This maximum Al content has been shown to depend on growth temperature with a maximum of ~16% at 600 °C and concentrations of 25% achieved at 725 °C for fully coherent, single crystalline β-$(Al_xGa_{1-x})_2O_3$.

Doping of heterostructures has been demonstrated with Si for δ-doped β-$(Al_xGa_{1-x})_2O_3$ MODFETs [27]. This can achieve a high Si concentration with an ideal δ-doped structure for these devices. Ge doping has limitations in heterostructures. In addition to heavy competition with Ga for the group III site, Ge also competes with Al in β-$(Al_xGa_{1-x})_2O_3$ decreasing its concentration with increasing Al content. At concentrations of 20% Al in the group III site, Ge incorporation decreases by an order of magnitude from that of similar growth conditions of β-Ga_2O_3. In addition, for higher growth temperatures needed for maximum Al content, the maximum Ge incorporation further decreases. Sn, on the other hand, is not as heavily outcompeted by Al, achieving high doping concentrations in 20% Al content β-$(Al_xGa_{1-x})_2O_3$.

The dependence of maximum Al incorporation on growth temperature is consistent with the Al_2O_3–Ga_2O_3 phase diagram produced by ceramic analysis by Hill et al. in 1952, shown in Fig. 5.6 [30]. This strong agreement with the expected thermodynamically stable phases is an interesting feature of β-$(Al_xGa_{1-x})_2O_3$ growth, but also makes it difficult to achieve high Al contents at typical (010) β-$(Al_xGa_{1-x})_2O_3$ growth temperatures of 600–750 °C. The phase diagram predicts a congruent point around 800 °C above which Al contents of more than 60% can be stable in the β phase [30]. Achieving a high Al content is essential to high conduction band offset and therefore better heterostructure devices [31], but the conventional β-$(Al_xGa_{1-x})_2O_3$ PAMBE growth regime is limited to temperatures of 600–750 °C because of the reduced growth rate at high temperatures due to suboxide formation.

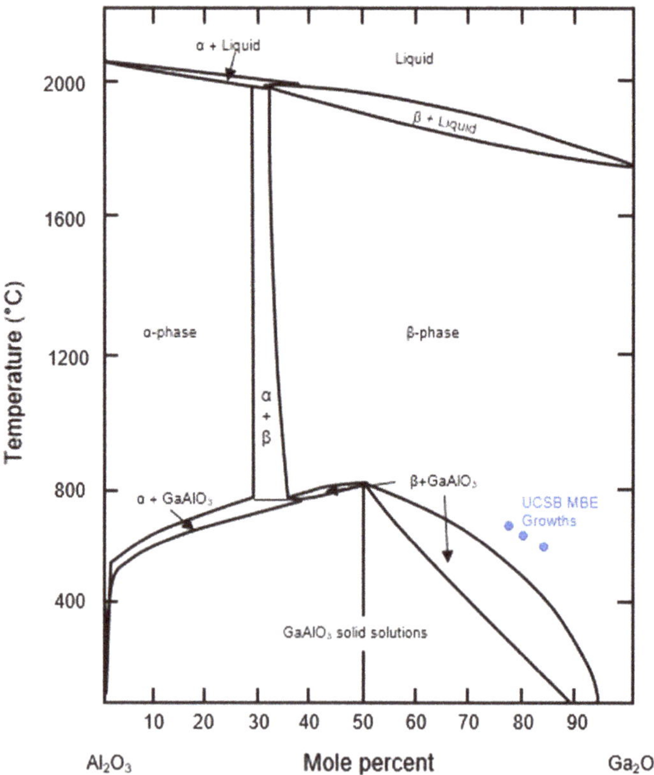

Fig. 5.6 Al_2O_3–Ga_2O_3 phase diagram based on data from Hill et al. [30] along with UCSB PAMBE data

Recently the use of In-catalyzed growth has demonstrated a remarkable expansion of the PAMBE growth regime available for β-Ga_2O_3 [32]. This involves an exchange mechanism between Ga and In in In_2O_3 formed during growth with a supplied In flux. Ga favorably incorporates over In and at high Ga fluxes almost no In is incorporated into the single crystalline (010) β-Ga_2O_3. Through this mechanism, growth can be performed at higher temperatures and higher Ga fluxes while achieving higher growth rates that are less limited by Ga_2O suboxide desorption. Specifics of the thermodynamics and kinetics of this growth will be discussed later in Chap. 6. This metal-oxide catalyzed epitaxy (MOCATAXY) can be expanded to β-$(Al_xGa_{1-x})_2O_3$ growth as shown in Fig. 5.7 [33].

With the Al–O bond being even stronger than Ga–O and In–O, Al can favorably incorporate over the other group III metals allowing for the growth of β-$(Al_xGa_{1-x})_2O_3$ with very limited In incorporation as demonstrated by Vogt et al. [33]. Despite supplying an In flux that was higher than the Al flux in III-rich conditions, β-$(Al_xGa_{1-x})_2O_3$ was still formed. APT confirmed random distribution of Al with only about 1% In incorporation despite similar Ga and In fluxes during growth. The compositions of constituents on the group III sites as a function of depth are shown in Fig. 5.8. This expands the range of growth temperatures at which coherent β-$(Al_xGa_{1-x})_2O_3$ can be grown to above 900 °C and even improves the growth rate. The mechanism by which the growth rate is improved is not fully understood, but the formation of In_2O_3 seems to access more oxygen, which is the limitation of growth rates as III fluxes are easily increased with increasing cell temperatures. Furthermore, suboxide desorption is limited at higher growth temperatures with In. In_2O_3 could grow from molecular oxygen, whereas β-Ga_2O_3 requires atomic oxygen to form; however, further studies of In catalyzed growth are needed to verify this. β-Ga_2O_3 has been shown to not form without striking the plasma, suggesting that it requires atomic oxygen to grow [1, 32].

Coherent β-$(Al_xGa_{1-x})_2O_3$ has been demonstrated with Al contents of 20%; however, much work into exploring the highly complex quaternary growth space involving In, Al, Ga, and O to improve maximum Al content still needs to be performed. The heterostructure was coherent and of high quality, demonstrated by

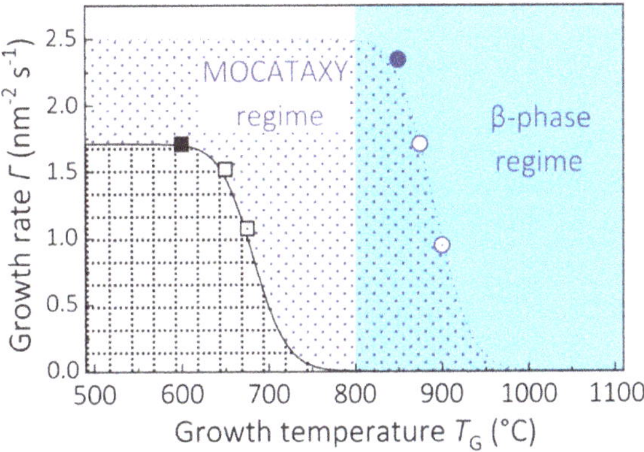

Fig. 5.7 Expansion of (010) β-$(Al_xGa_{1-x})_2O_3$ growth regime with MOCATAXY versus regular growth without In [33]

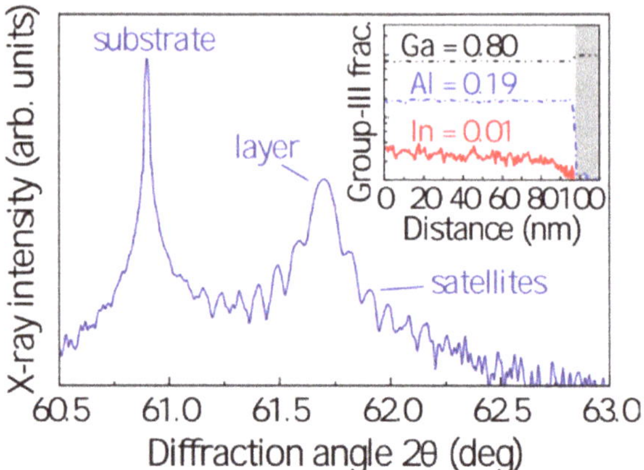

Fig. 5.8 HRXRD of MOCATAXY (020) β-$(Al_xGa_{1-x})_2O_3$/β-Ga_2O_3 heterostructure and Ga, Al, and In content on the group III sites determined by APT [33]

HRXRD rocking curve FWHMs similar to those of the bare substrate and thickness fringes in ω-2θ scans as shown in Fig. 5.8. In addition, cross-sectional transmission electron microscopy (TEM) analysis revealed no evidence of dislocations and twinning. Surface morphologies were good with RMS roughness of 0.3 nm and no step-bunching observed [33]. Initial attempts to dope β-$(Al_{0.2}Ga_{0.8})_2O_3$ with Sn have been able to produce doping concentrations on the order of 3×10^{18} cm^{-3}, which is desirable for future MODFET structures and further suggests Sn's capability as a dopant.

There is great potential for improvement of β-Ga_2O_3 and β-$(Al_xGa_{1-x})_2O_3$ growth by MBE. Mobilities achieved are still far from theoretical room temperature mobilities limited by optical phonon scattering. The best dopant, or ideal applications for each dopant are still not determined and β-$(Al_xGa_{1-x})_2O_3$ must achieve higher Al contents to achieve viable heterostructure based devices. Furthermore, most PAMBE growth has focused on the (010) orientation, but other orientations like (001) or (100) could still see improvement in materials quality or maximum Al content and prove to be the best orientation for future β-Ga_2O_3 devices.

References

1. M.Y. Tsai, O. Bierwagen, M.E. White, J.S. Speck, J. Vac. Sci. Technol. A **28**, 354 (2010)
2. E. Ahmadi, O.S. Koksaldi, S.K. Kaun, Y. Oshima, D.B. Short, U.K. Mishra, J.S. Speck, Appl. Phys. Express **10**, 041102 (2017)
3. Y. Oshima, E. Ahmadi, S. Kaun, F. Wu, J.S. Speck, Semicond. Sci. Technol. **33**, 015013 (2018)
4. G. Koblmuller, F. Wu, T. Mates, J.S. Speck, Appl. Phys. Lett. **91**, 221905 (2007)
5. M.A. Liebermann, A.J. Lichtenberg, *Principles of Plasma Discharges and Materials Processing* (Wiley, New York, 2005)
6. E. Stoffels, W.W. Stoffels, D. Vender, M. Kando, G.M.W. Kroesen, F.J. de Hoog, Phys. Rev. E **51**, 2425 (1995)
7. M.Y. Tsai, *Dissertation* (University of Califronia, Santa Barbara, 2010)

8. S. Fernandez-Garrido, G. Koblmuller, E. Calleja, J.S. Speck, J. Appl. Phys. **104**, 033541 (2008)
9. H. Okumura, M. Kita, K. Sasaki, A. Kuramata, M. Higashiwaki, J.S. Speck, Appl. Phys. Express **7**, 9 (2014)
10. P. Vogt, O. Bierwagen, Appl. Phys. Lett. **106**, 081910 (2015)
11. E.G. Villora, K. Shimamura, K. Kitamura, K. Aoki, Appl. Phys. Lett. **88**, 031105 (2006)
12. M.Y. Tsai, M.E. White, J.S. Speck, J. Cryst. Growth **310**, 4256 (2008)
13. T. Oshima, N. Arai, N. Suzuki, S. Ohira, S. Fujita, Thin Solid Films **516**, 5768 (2008)
14. Z. Cheng, M. Hanke, Z. Galazka, A. Trampert, Nanotechnology **29**, 39570 (2018)
15. R. Schewski, M. Baldini, K. Irmscher, A. Fiedler, T. Markurt, B. Neushulz, T. Remmele, T. Schulz, G. Wagner, Z. Galazka, M. Albrecht, J. Appl. Phys. **120**, 225308 (2016)
16. M. Higashiwaki, K. Sasaki, A. Kuramata, T. Masui, S. Yamakoshi, Appl. Phys. Lett. **100**, 013504 (2012)
17. K. Sasaki, A. Kuramata, T. Masui, E.G. Villora, K. Shimamura, S. Yamakoshi, Appl. Phys. Express **5**, 035502 (2012)
18. J.M. LeBeau, R. Engel-Herbert, B. Jalan, J. Cagnon, P. Moetakef, S. Stemmer, Appl. Phys. Let. **95**, 142905 (2009)
19. E. Ahmadi, Y. Oshima, F. Wu, J.S. Speck, Semicond. Sci. Technol. **32**, 035004 (2017)
20. E. Ahmadi, O.S. Koksaldi, X. Zheng, T. Mates, Y. Oshima, U.K. Mishra, J.S. Speck, Appl. Phys. Express **10**, 07110 (2017)
21. S.H. Han, A. Mauze, E. Ahmadi, T. Mates, Y. Oshima, J.S. Speck, Semicond. Sci. Technol. **33**, 0450001 (2018)
22. H. Paelaars, C.G. Van de Walle, Phys. Status Solidi B **252**, 4 (2015)
23. Y. Kang, K. Krishnaswamy, H. Peelaers, C.G. Van de Walle, J. Phys.: Condens. Matter **29**, 234001 (2017)
24. K. Ghosh, U. Singisetti, Mater. Res. **32**, 4142 (2017)
25. J.B. Varley, J.R. Weber, A. Janotti, C.G. Van de Walle, Appl. Phys. Lett. **97**, 142106 (2010)
26. S. Krishamoorthy, Z. Xia, S. Bajaj, M. Brenner, S. Rajan, Appl. Phys. Express **10**, 051102 (2017)
27. S. Krishamoorthy, Z. Xia, C. Joishi, Y. Zhang, J. McGlone, J. Johnson, M. Brenner, A. Arehart, J. Hwang, S. Lodha, S. Rajan, Appl. Phys. Lett. **111**, 023502 (2017)
28. S.W. Kaun, F. Wu, J.S. Speck, J. Vac. Sci. Technol. A **33**, 041508 (2015)
29. Y. Oshima, E. Ahmadi, S.C. Bedescu, F. Wu, J.S. Speck, Appl. Phys. Express **9**, 061102 (2016)
30. V.G. Hill, R. Rustom, E.F. Osborn, J. Am. Ceram. Soc. **35**, 135 (1952)
31. H. Peelaers, J.B. Varley, J.S. Speck, C.G. Van de Walle, Appl. Phys. Lett. **112**, 242101 (2018)
32. P. Vogt, O. Brandt, H. Riechert, J. Lahnemann, O. Bierwagen, Phys. Rev. Lett. **119**, 196001 (2017)
33. P. Vogt, A. Mauze, F. Wu, B. Bonef, J.S. Speck, Appl. Phys. Express **11**, 11 (2018)

Chapter 6
Plasma-Assisted Molecular Beam Epitaxy 2

Fundamentals of Suboxide-Related Growth Kinetics, Thermodynamics, Catalysis, Polymorphs, and Faceting

Oliver Bierwagen, Patrick Vogt and Piero Mazzolini

Abstract This chapter sheds light on various fundamental aspects of the O plasma-assisted molecular beam epitaxy of Ga_2O_3. It discusses the volatile suboxide-related growth kinetics of Ga_2O_3 explaining the observed growth rate behavior as function of all growth parameters. The binary growth kinetics of Ga_2O_3 is then compared to that of its related oxides In_2O_3 and SnO_2. During the ternary growth of $(In_xGa_{1-x})_2O_3$, thermodynamic aspects based on different metal-oxygen bond strengths become important, which will be shown to govern the Ga- versus In-incorporation into the $(In_xGa_{1-x})_2O_3$ thin film. More importantly, we describe how the collaborative effect of the different growth kinetics and thermodynamics of the binary oxides can lead to a strong growth rate enhancement of Ga_2O_3 by metal-exchange catalysis (MEXCAT) using In_2O_3 or SnO_2 as catalyst. A brief overview of the polymorphs of Ga_2O_3 stabilized by different substrates is given. The surface morphology obtained by homoepitaxy will be discussed in relation to thermodynamically induced faceting. Finally, open questions will be identified that require further research.

6.1 Introduction

The MBE growth rate of Ga_2O_3 is rather limited and adjusting the composition of the ternary oxides $(In,Ga)_2O_3$ and $(Al,Ga)_2O_3$ by selecting appropriate growth parameters (metal and oxygen fluxes, and substrate temperature) is not straightforward. Depending on growth conditions, the growth rate of Ga_2O_3 can be negative—as used

---

O. Bierwagen (✉) · P. Vogt · P. Mazzolini
Paul-Drude-Institut für Festkörperelektronik,
Leibniz Institut im Forschungsverbund Berlin e.V,
Hausvogteiplatz 5-7, 10117 Berlin, Germany
e-mail: bierwagen@pdi-berlin.de

M. Higashiwaki and S. Fujita (eds.), *Gallium Oxide*, Springer Series in Materials Science 293, https://doi.org/10.1007/978-3-030-37153-1_6

in etching/decomposing the layer by a Ga flux in the absence of O at sufficiently high temperature—but its native growth can also be strongly enhanced by an additional metal flux in a process called metal-exchange catalysis. In homoepitaxy, the film orientation equals that of the substrate. This crystal orientation, however, does not necessarily reflect that of the surface which can be facetted. Finally, different polymorphs (α, β, γ, ε/κ) of Ga_2O_3 can be synthesized by heteroepitaxy.

In this chapter, we explore and explain the underlying physical and chemical mechanisms during the plasma-assisted MBE of III-O compounds. We identify the limiting role of the suboxide-related growth kinetics for the growth of the binary oxides (Ga_2O_3 and In_2O_3) and demonstrate how the thermodynamics of the metal-oxygen bonds becomes dominant during growth of the ternary compound $(In, Ga)_2O_3$. We further discuss the metal-exchange catalysis in the ternary In–Ga–O system that circumvents the kinetic limits observed during binary Ga_2O_3 formation. A final part of this chapter will be dedicated to the formation of polymorphs by the proper choice of substrate and growth conditions during heteroepitaxy of Ga_2O_3 and to aspects of surface faceting during the homoepitaxy. For simplicity, the thermodynamically stable β-phase of Ga_2O_3 will be named "Ga_2O_3" in the following while other phases will be named by their phase name, e.g., rhombohedral "α-Ga_2O_3" or hexagonal "ε/κ-Ga_2O_3".

6.2 MBE Growth Chamber and in Situ Analytics to Investigate Growth Kinetics

A schematics of the plasma-assisted MBE (PAMBE) growth chamber used for most results [1–7] discussed in this chapter is shown in Fig. 6.1.

The metals (Me = Ga, In) are evaporated from effusion cells and activated oxygen is provided by passing a controlled flow of molecular oxygen (in units of standard cubic centimeter per minute, SCCM) through an RF plasma source run at a certain power (typically 200–300 W).

Reflection high-energy electron diffraction (RHEED) provides information about the surface crystallinity, morphology, and faceting by the diffraction of an electron beam that is reflected from the sample surface. The reflected electrons impinge on a phosphorous screen where they produce characteristic patterns [8]. When growth proceeds in a layer-by-layer mode, RHEED intensity oscillations with a periodicity of one monolayer can in principle be observed and be utilized to measure the growth rate in situ [8]. To date RHEED oscillations have only been observed during the homoepitaxy of $Ga_2O_3(100)$ [9, 10]. Therefore, alternative techniques to determine the growth rate need to be used for other surface orientations.

Central quantities to understand the growth kinetics and thermodynamics are the growth rate (= metal incorporation rate), the provided metal and oxygen fluxes, and the flux of desorbing species. The law of conservation dictates that the rate of incorporated metal atoms per area equals the difference of provided metal flux and

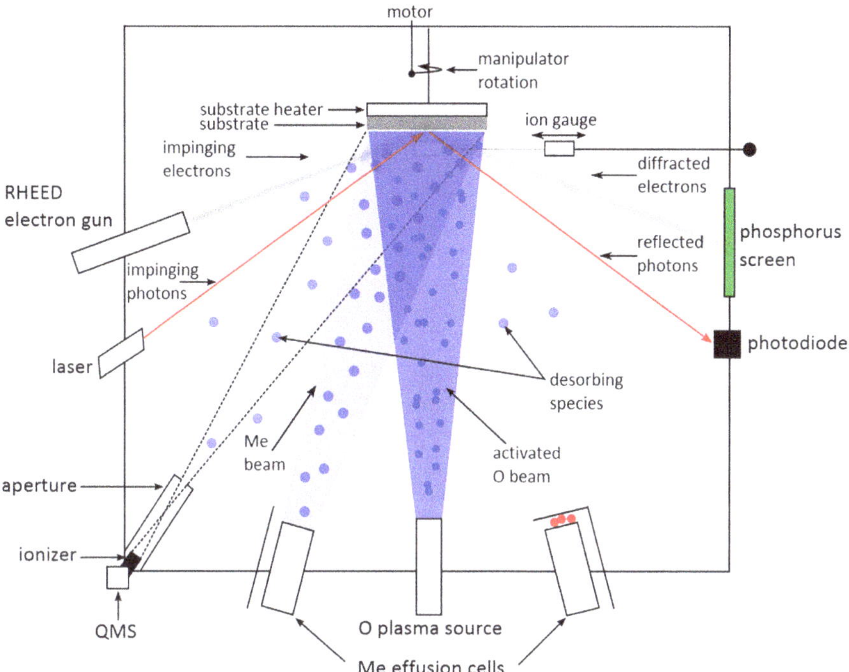

Fig. 6.1 Schematic of a PAMBE growth chamber. The species fluxes from the metal (Me) effusion cells and O plasma source, the substrate and substrate heater, as well as all installed in situ analysis methods are schematically represented. Chamber, devices, and number of species are not to scale. Reprinted from [7]

the flux of desorbing metal-containing species. Metal and oxygen fluxes from the sources are measured as "beam equivalent pressure" (BEP) by a retractable, nude ion gauge that can be placed in the growth position. The measured BEP is directly proportional to the flux from the sources. It can be calibrated as discussed later (section "Quantifying the metal-, oxygen-, and desorbing fluxes"). Two additional in situ tools discussed next provide a means to efficiently determine the growth rate and flux of desorbing species.

6.2.1 *Line-of-Sight Quadrupole Mass Spectrometry (QMS)*

Line-of-sight quadrupole mass spectrometry is used to detect and quantify the flux of desorbing species from the substrate in the growth chamber [11]. It is based on a quadrupole mass spectrometer (QMS) that is mounted on a source flange. An aperture between the substrate and the QMS (shown in Fig. 6.1) ensures that only the substrate is visible in the line-of-sight of the QMS, thus excluding detection of desorbing species from the substrate holder outside the substrate. This method has been used by us with 2 inch substrates to ensure a sufficient amount of species to reach the QMS. QMS measurements on smaller substrates, e.g., typically available Ga_2O_3 substrates with an area of ≈ 1 cm^2, have not been attempted by us yet as it would require a precise adjustment of the aperture and result in a significantly decreased QMS signal. The calibration of measured QMS signal and desorbing flux will be described later.

6.2.2 Laser Reflectometry (LR)

Heteroepitaxy of Ga_2O_3 typically results in rough films that do not allow observation of RHEED oscillations to determine the growth rate. As an alternative, laser reflectometry can be used to measure the growth rate in situ during heteroepitaxy. It is based on the interference of a laser beam reflected at the film surface and substrate-film interface as shown in Fig. 6.2a. This method cannot be used with homoepitaxy as reflection at the substrate-film interface requires a contrast of the refractive index *n*. The continuously increasing film thickness during growth leads to alternating constructive and destructive interference over time, resulting in a sinusoidal intensity of the reflected beam as shown in Fig. 6.2b. Using Braggs law and Snells law, the thickness change *d* that results in one period of this intensity oscillation can be calculated as [7]:

$$d = \lambda/[2n\cos(\arcsin(\sin(\alpha)/n))].$$

With the used laser wavelength of λ = 650 nm, angle of incidence α = 60°, and refractive index of Ga_2O_3 of $n \approx 2.0$, one oscillation period corresponds to a

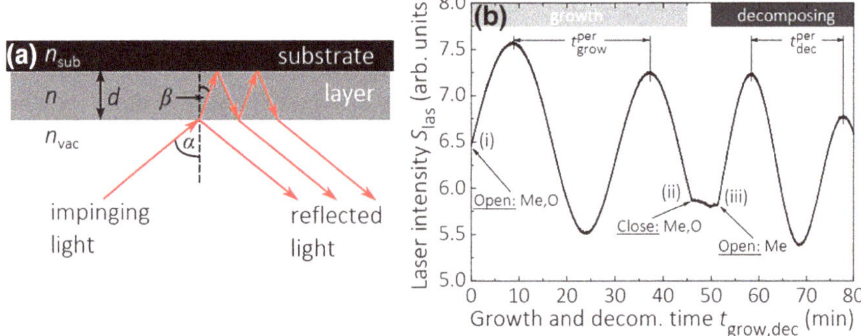

Fig. 6.2 Laser reflectometry: **a** Depicts the principle of the measurement using laser reflectometry. **b** Detected laser signal while growing and decomposing a Ga_2O_3(-201) layer on a c-plane sapphire substrate a layer as a function of time *t*. Position (i) indicates the start of growth, (ii) its end, and (iii) refers to the start of layer decomposition by Ga-etching. **a**, **b** Reprinted from [7]

change of film thickness by $d \approx 180$ nm. The data in Fig. 6.2b additionally show that a change from layer growth to etching can be identified by a changing sign of the slope of the LR signal.

6.2.3 *Quantifying the Metal-, Oxygen-, and Desorbing Fluxes*

A quantitative analysis of the growth kinetics requires a precise knowledge of the involved source and desorbing fluxes which can be determined as follows. To compare the provided metal flux to the growth rate, we are often expressing the growth rate as a metal incorporation rate, i.e., metal atoms $nm^{-2}\ s^{-1}$.

The BEP of a metal (cation) measured by an ion gauge (e.g., expressed in Torr) is directly proportional to the measured growth rate (e.g., expressed in nm/s) of the corresponding oxide under conditions where full metal incorporation is guaranteed. As described in subsection "3.1 Flux stoichiometry and growth vs. etching" an excess of O and sufficiently low growth temperatures provide such conditions. Multiplying the obtained growth rate by the cation number density in the layer (e.g., expressed in cations nm^{-3}), we obtain the incorporated cation flux expressed as cations $nm^{-2}\ s^{-1}$. By this approach, the BEP of all cations can now be connected to the incorporated cation flux. In the experiments by Vogt et al. [1–7], a measured BEP of 1 x 10^{-7} Torr for Ga, In, and Sn corresponded to growth rates and metal fluxes of 0.30 Å/s Ga_2O_3 and 1.15 Ga $nm^{-2}\ s^{-1}$, 0.18 Å/s In_2O_3 and 0.55 In $nm^{-2}\ s^{-1}$, and 0.36 Å/s SnO_2 and 1.00 Sn $nm^{-2}\ s^{-1}$, respectively. Note that different cation ionization cross sections (or sensitivity factors of the ion gauge to the different cations) and potentially different cell-to-BEP geometries for the different metal sources lead to different ratios of BEP to particle flux. This is particularly important for the growth of ternary oxides if the cation stoichiometry should be adjusted by BEP measurements of the two different cations.

The active O flux at a given O_2 flow (SCCM) and plasma RF power (W) can be determined by multiplying the oxygen atom number density of the oxide film by the growth rate under conditions of highest possible O incorporation into the oxide film, i.e., at the peak shown in Fig. 6.3 as described in the following subsection "3.1 Flux stoichiometry and growth vs. etching". This active O flux incorporation also depends on the specific oxidation efficiencies of the metal as discussed in subsection "3.1 Flux stoichiometry and growth vs. etching" [1].

In this chapter, we are consistently using the flux units of $nm^{-2}\ s^{-1}$ with an oxygen-to-metal flux ratio of 3/2 corresponding to a stoichiometric flux (that allows full incorporation of both species into the Me_2O_3 film). Note that in the literature quantitative oxygen and metal fluxes are sometimes given in units of growth rate (e.g., Å/s) with corresponding stoichiometric metal-to-oxygen flux ratio of unity.

Desorbing metal fluxes were calibrated by relating the measured QMS signal (in pressure units or units of counts per second) of the metal atoms to a known source

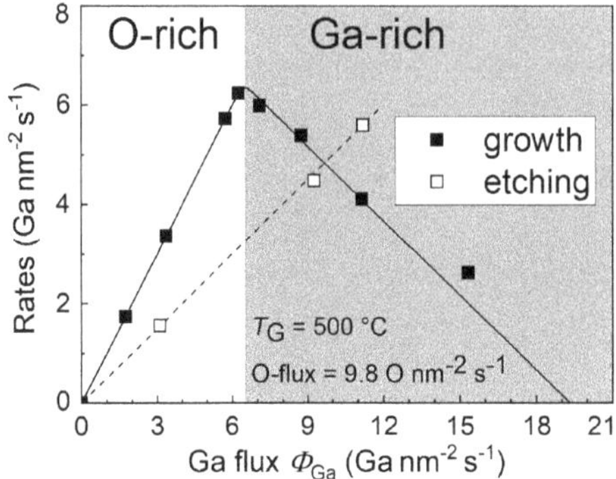

Fig. 6.3 Growth rate of Ga_2O_3(-201) as a function of the Ga flux (solid squares). The O-rich and Ga-rich regimes are indicated as white and gray areas, respectively, while the stoichiometric point is corresponding to the growth rate peak at their intersection. Solid line shows the model prediction as given in [6]. Open squares depict the etch rate as obtained by supplying a Ga flux to a Ga_2O_3 surface in the absence of an O flux. The dashed line is a linear fit to this data Adapted from [7]

metal flux onto a substrate that provides full metal reflection. This is achieved by a sufficiently high substrate temperature to grant full metal desorption and no metal accumulation as verified by RHEED and LR.

Desorbing suboxide fluxes were calibrated by relating the measured QMS signal of the suboxide molecules to the known suboxide flux derived from the measured etch rate by metal induced oxide etching as described in subsection "3.1 Flux stoichiometry and growth vs. etching".

6.3 Suboxide-Related Kinetics During Binary Growth

The growth rate behavior of III-O compounds differs fundamentally from the well-known one of II-O or III-V compounds. A typically observed behavior for the latter ones has been demonstrated, for example, with the growth of GaN [12] and ZnO [13]: The growth rate proportionally increases with the metal flux (Ga or Zn) in the anion-rich (N or O) regime (at otherwise constant growth parameters), i.e.,it is limited by the metal flux itself in presence of excess anions. In the metal-rich regime, the growth rate saturates as it is limited by the supplied anion flux and excess metal that cannot contribute to growth either adsorbs on or desorbs off the growth surface, depending on growth conditions. These compounds form via a single-step reaction mechanism where one cation (Ga or Zn) reacts directly with one anion (N or O) to the desired compound.

Different to III-V or II-VI compounds, which possess two reactants in their solid form, III-VI possess five reactants in their solid form. Consequently, the growth kinetics during III-VI MBE is much more complex than for II-O or III-V MBE.

During the MBE of Ga_2O_3, for instance, the suboxide Ga_2O is formed—having a lower oxidation state of Ga. It is shown in the following, how and why the formation and desorption of this suboxide is the rate-limiting factor for the growth of Ga_2O_3.

6.3.1 Flux Stoichiometry and Growth Versus Etching

The heteroepitaxy of Ga_2O_3(-201) on Al_2O_3 (0001), studied by Vogt et al., yielded the growth rate diagram shown in Fig. 6.3 [1]. At a relatively low growth temperature T_G of 500 °C, a linearly increasing growth rate with full incorporation of the impinging Ga flux into the film can be observed. This situation, including the desorption of excess O corresponds to O-rich growth conditions as schematically described in Fig. 6.4a.

The maximum growth rate at a stoichiometric Ga-to-activated O flux ratio of 2/3 results in full Ga and O incorporation into Ga_2O_3 [schematic Fig. 6.4a], as given by reaction

$$2\mathrm{Ga(a)} + 3\mathrm{O(a)} \rightarrow \mathrm{Ga_2O_3(s)}, \tag{6.1}$$

with (a) and (s) denoting the adsorbed and the solid phase, respectively.

In the Ga-rich regime, and increasing Ga flux, a linear decrease of the growth rate [schematic Fig. 6.4b] with a slope of −0.5 is observed. The modulus of this slope equals the half of the increasing growth rate in the O-rich regime. A microscopic explanation of this decreasing growth rate in the Ga-rich regime is

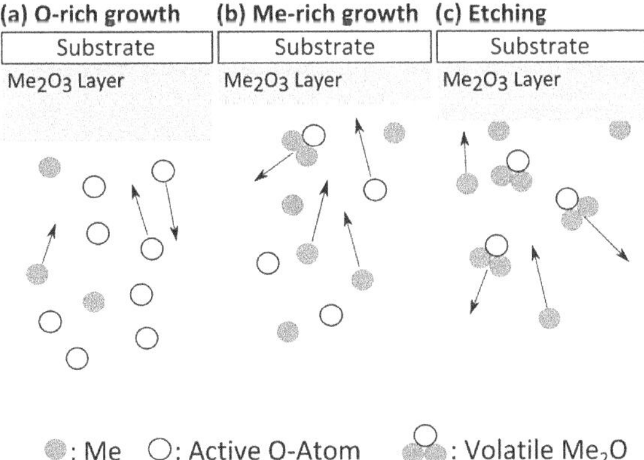

Fig. 6.4 Schematic of the formation or etching of a Ga_2O_3 layer: **a** In the oxygen-rich regime, all Ga is incorporated and excess O desorbs. **b** In the Ga-rich regime, Ga is incorporated and excess Ga forms the volatile suboxide Ga_2O, which desorbs. **c** Etching of the film by only applying Ga vapor that reduces the Ga_2O_3 surface to Ga_2O, which desorbs subsequently. Reprinted from [1], with the permission of AIP Publishing

given in subsection "3.3 Suboxide-mediated two-step growth mechanism: A general explanation".

Supplying a Ga flux, in the absence of an O flux, leads to a decomposition/etching of the Ga_2O_3 thin film [schematic Fig. 6.4c]. During both, Ga-rich growth and etching, but not during O-rich growth, a desorbing flux of Ga_2O was measured by QMS as schematically shown in Fig. 6.4b, c. The chemical reaction leading to Ga_2O formation by Ga-etching of Ga_2O_3 is:

$$4\mathrm{Ga(a)} + \mathrm{Ga_2O_3(s)} \rightarrow 3\mathrm{Ga_2O(g)}, \tag{6.2}$$

with (g) denoting the gaseous phase. The slope of the increasing etch rate equals half of the slope of the increasing growth rate during growth in the O-rich regime. This behavior can be quantitatively understood with the above-mentioned reactions in which reaction (6.2) requires four Ga atoms to etch one formula unit of Ga_2O_3 compared to two Ga atoms required for its growth by reaction (6.1).

The findings presented here are not limited to the (-201) surface of β-Ga_2O_3, as the triangular growth rate diagram has also been demonstrated by Kracht et al. during heteroepitaxy of α-Ga_2O_3 [14]. Moreover, Ga-induced layer etching (also termed Ga-polishing) has also been demonstrated by Ahmadi et al. [15] and Mazzolini et al. [16] on Ga_2O_3(010), as well as on Ga_2O_3(001) by Oshima et al. [17].

In both cases, the etching and Ga-rich growth at low growth temperature, the Ga_2O formation and desorption is caused by the metal-rich flux ratio that contains too little O to oxidize all provided Ga to Ga_2O_3. Nonetheless, as it will be thoroughly further explained (see section "3.3 Suboxide-mediated two-step growth mechanism: A general explanation"), it is important to stress that the Ga-etching (6.2) is not the responsible mechanism of the decreasing growth rate found in the Ga-rich growth regime (gray shaded area in Fig. 6.3). A qualitatively similar evolution of growth and etch rate has been observed during the binary MBE of In_2O_3 and SnO_2 and related to the suboxides In_2O and SnO [1].

6.3.2 Growth Temperature Dependence and Growth Window

At a higher growth temperature of 610 °C, a quantitatively different growth rate evolution of Ga_2O_3(-201) has been observed, as depicted as squares in Fig. 6.5. At these growth conditions, a growth rate plateau appears between Ga fluxes of 4 and 9 nm^{-2} s^{-1}. In comparison with the low-temperature growth rate shown in Fig. 6.3, a reduced growth rate is now observed even under an O-rich flux ratio, coinciding with a desorbing Ga_2O flux as shown by the blue discs in Fig. 6.5. This behavior indicates a thermally induced Ga_2O desorption, since in the O-rich regime enough active O is available to fully oxidize all Ga to Ga_2O_3. The quantitative analysis by

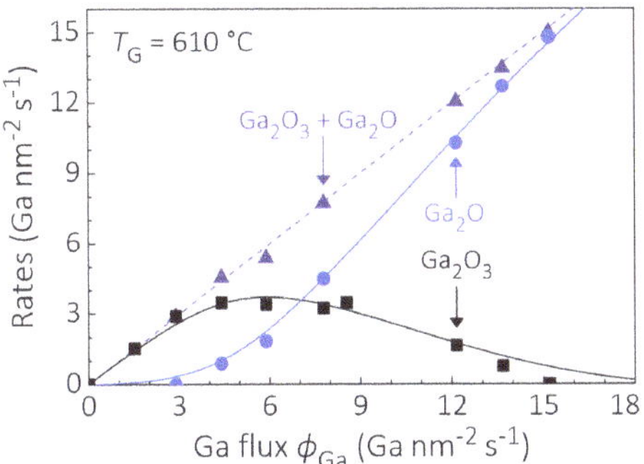

Fig. 6.5 Growth rate of Ga_2O_3(-201) (black squares), desorbing Ga_2O flux (blue discs), and their sum (dark blue triangles), as a function of the Ga flux. The active O flux was set to 9.8 nm^{-2} s^{-1}. A growth rate plateau is visible between Ga fluxes of 4 and 9 nm^{-2} s^{-1}. The lines are model predictions according to [6]. Adapted from [2, 7]

QMS (blue discs in Fig. 6.5) confirms that the desorbing Ga_2O flux accounts for any Ga that is not incorporated into the Ga_2O_3 film, as growth rate and desorbing flux add up (black triangles in Fig. 6.5) to the total impinging Ga flux, irrespective of the growth regime [2]. Since the Ga_2O removes two Ga ad-atoms but just one O ad-atom from the growth surface, the plateau spans into the Ga-rich flux regime, as well. Due to this desorption, the growth surface remains O-rich up to a critical Ga-to-O flux ratio. Once this critical ratio is reached (at the end of the plateau), the growth rate decreases again with increasing Ga flux as the growth surface becomes Ga-rich and O-deficiency induced suboxide desorption takes place like in the low-temperature regime shown in Fig. 6.3 [7].

The same growth rate evolution, including the decreasing growth rate after the plateau, has been reported by Y. Oshima et al. for the homoepitaxy of Ga_2O_3(001) (growth temperature 750 °C) [17]. A growth rate plateau has already been observed (and attributed to Ga-rich growth conditions) by T. Oshima et al. (heteroepitaxy of Ga_2O_3(-201) at 800 °C) [18], Okumura et al. (homoepitaxy of Ga_2O_3(010) at 700 ° C) [19], and Ahmadi et al. (homoepitaxy of Ga_2O_3(010) at 600 °C) [20]. These observations indicate that the formation and desorption of Ga_2O is not limited to a specific surface orientation of Ga_2O_3.

By merging all measured growth data (e.g., as the ones depicted in Figs. 6.3 and 6.5) as a function of all growth parameters: the Ga-to-O flux ratio and the growth temperature, the growth window of Ga_2O_3(-201) was determined [2, 7]. Figure 6.6 shows this growth domain in the two-dimensional parameter space spanned by Ga-to-O flux ratio and growth temperature. All major and minor growth regimes are indicated in the figure caption.

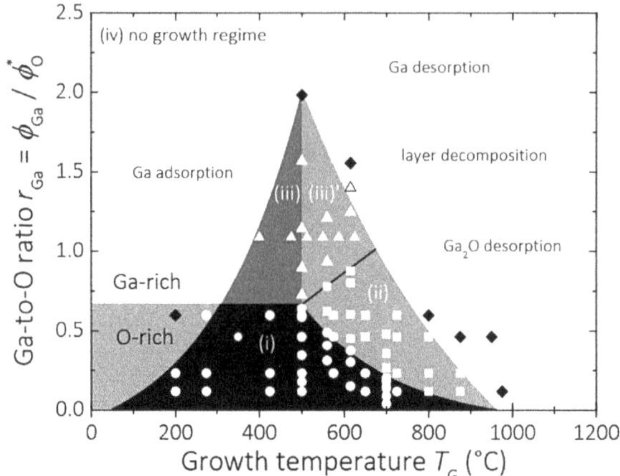

Fig. 6.6 MBE growth domain of Ga_2O_3(-201) grown on Al_2O_3(0001). It depicts the measured growth rate behavior projected onto the two-dimensional parameter space spanned by the Ga-to-O flux ratio and growth temperature. The growth domain is divided in two major growth regimes, the O-rich and Ga-rich regimes. The different symbols relate to four minor growth regimes: (i) discs—complete Ga-incorporation, (ii) squares—plateau of growth rate, (iii) triangles—decreasing growth rate in Me-rich regime, (iv) rhombs—no growth regime. Reprinted from [7]

6.3.3 Suboxide-Mediated Two-Step Growth Mechanism a General Explanation

A general model for the growth that accounts for both, (i) the flux-stoichiometrically induced Ga_2O desorption in the Ga-rich regime, as well as the (ii) additionally thermally induced Ga_2O desorption at elevated growth temperature in all flux regimes, has been recently developed by Vogt et al. [6, 7]. This growth model is based on a two-step reaction mechanism:

In the first reaction step, all Ga ad-atoms rapidly react with the available O species on the growth surface to the suboxide Ga_2O, through the reaction

$$2\text{Ga(a)} + \text{O(a)} \rightarrow \text{Ga}_2\text{O(a)}. \tag{6.3}$$

Note that layer etching during growth through reaction (6.2) can be kinetically and thermodynamically excluded [6, 7] and significant suboxide formation by decomposition of the Ga_2O_3 layer at growth temperature in the absence of Ga- and O fluxes has been excluded in a control experiment [6].

Depending on growth temperature and Ga-to-O flux ratio, this Ga_2O can either desorb off the growth surface—causing the decreasing growth rate—or be further oxidized to the full metal oxide in a second step through the reaction

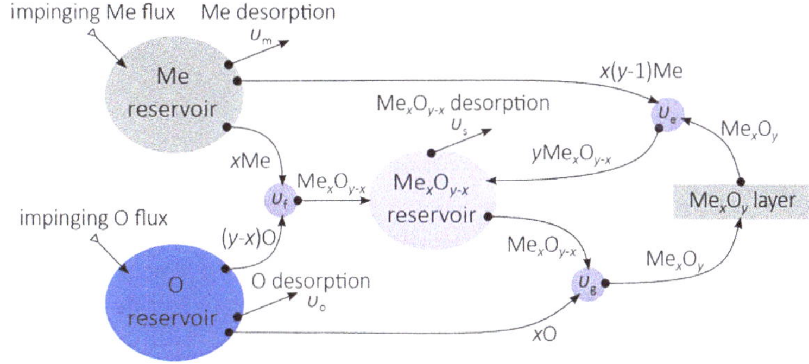

Fig. 6.7 Growth scheme for III-O and IV-O PAMBE showing impinging fluxes, resulting reservoirs, and a Me_xO_y layer. Possible chemical reactions and desorption rates occurring are indicated by respective rate constants υ_r ($r = m, o, f, s, g, e$). The stoichiometric coefficients of the anions (x) and cations (y) in the respective III-VI and IV-VI compound are $x = 2$ and $y = 3$, as well as $x = 1$ and $y = 2$, respectively. Reprinted with permission from [6]. Copyright (2018) by the American Physical Society

$$Ga_2O(a) + 2O(a) \rightarrow Ga_2O_3(s). \tag{6.4}$$

The competition between the suboxide desorption and its oxidation by reaction (6.4) is shifted toward Ga_2O_3 formation under a sufficiently O-rich flux ratio even at a high growth temperature (cf. Fig. 6.5 at Ga flux of 3 $nm^{-2}\ s^{-1}$). Conversely, the absence of sufficient oxygen in the source flux can lead to a complete stop of the growth even at low growth temperature [cf. Fig. 6.3 at a Ga flux of 18 $nm^{-2}\ s^{-1}$ where all O is already consumed in the Ga_2O formation by reaction (6.3)].

A general schematics of this reaction process is shown in Fig. 6.7.

The rate equations corresponding to this model based on reactions (6.3), (6.4), and suboxide desorption and the resulting growth rate as their solution are given in [6] for various III-VI and IV-VI compounds. The derived model is mathematically described by three material dependent parameters: a pre-exponential factor and an activation energy with linear dependence on flux stoichiometry. Fixing the former one, the latter two are obtained by fitting experimental growth rate data to the model.

The resulting model predictions based on the fixed set of three material dependent parameters describe the experimentally obtained Ga_2O_3(-201) growth rates as a function of all growth parameters well, as shown in Fig. 6.8 [6].

The trend of a decreasing growth rate with increasing growth temperature has also been reported for the homoepitaxy on Ga_2O_3(001) [17] and (010) [21]. An observed initial increase of the growth rate before the expected decrease with increasing growth temperature [16, 19] of the (010) surface under slightly Ga-rich growth conditions (and for In_2O_3 growth at high In fluxes and low growth temperature in [3]) is likely due to an increase of the reactivity of reaction (6.3)—i.e., due to thermal activation of the suboxide formation that has not been taken into account in the rate equation model.

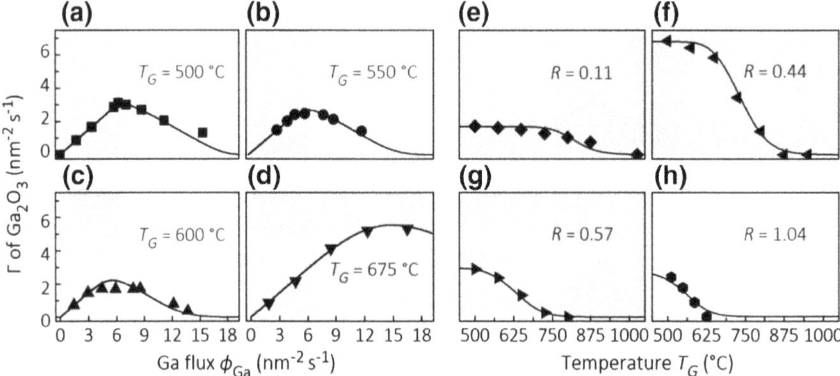

Fig. 6.8 **a–d** and **e–h** Dependence of the growth rate of Ga_2O_3 on the Ga flux at different growth temperatures T_G, and on T_G at different Ga-to-O flux ratios R, respectively. Corresponding parameters are indicated in the figures. Symbols represent experimental data and solid lines are model predictions [6]. In (**a**), (**b**), (**c**), (**g**), (**h**) and in (**d**), (**e**), (**f**) the active O flux was 10.2 and 30.6 $nm^{-2}\,s^{-1}$, respectively. Reprinted with permission from [6]. Copyright (2018) by the American Physical Society

The activation energy of the model is very likely surface dependent as a strong growth rate dependence on surface orientation has been reported for the homoepitaxy by ozone-MBE [22] and plasma-assisted MBE [17] of Ga_2O_3. This dependence has been related to different adhesion energies [22] of the surfaces or different catalytic activities for the Ga_2O formation [17]. Both studies indicate the highest growth rates (>100 nm/h) for the (010) orientation, intermediate growth rates for (001) and extremely low growth rates (well below 100 nm/h) for (100), which is the easiest cleavage plane. (As an exception, an early work by Villora et al. reports a high growth rate of 750 nm/h [23], likely related to the exceptionally close proximity of the O plasma source to the substrate [24]).

Note that the two-step reaction mechanism described in [6] is fundamentally different from the relatively simple single-step reaction mechanism during II-VI [13, 25] or V-III [12, 26] MBE. The formation of an intermediate subcompound is an inherent property for the formation of III-VI and IV-VI materials and desorption of this subcompound is found to be the growth rate limiting step during their MBE growth.

A potential avenue to a more simple, single-step growth of Ga_2O_3 is the use of Ga_2O sources instead of Ga-sources, which would circumvent reaction (6.3). This approach has been recently demonstrated by Ghose et al. [27, 28] using a high-temperature effusion cell to decompose Ga_2O_3 powder into a mixture of Ga_2O vapor and O_2 at 1750 °C. As the Ga_2O already contains oxygen, less additional oxygen is required for growth, which should be beneficial for the lifetimes of filaments in the growth chamber. Indeed, Ghose et al. have demonstrated with this approach the growth of thin Ga_2O_3 even without additionally supplying O plasma [28].

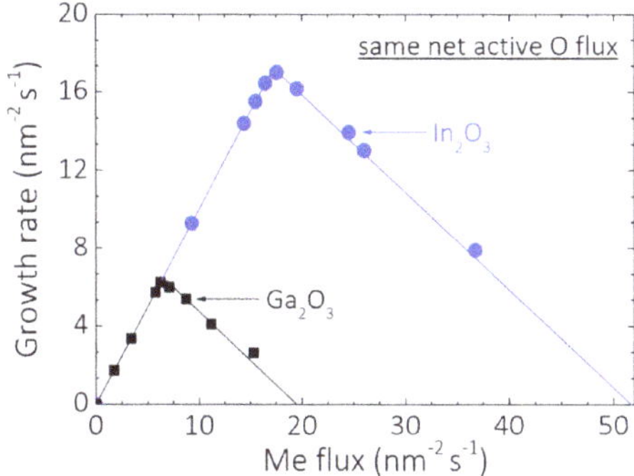

Fig. 6.9 Comparison of flux-stoichiometrically induced suboxide desorption (oxidation efficiencies): growth rate of In_2O_3(111) and Ga_2O_3(-201) as a function of the In flux and Ga flux, respectively. For both sets of experiments, the O flux was 0.75 SCCM with an O plasma power of 300 W. Adapted from supplement to [5]

6.3.4 *Comparing Ga_2O_3, In_2O_3, and SnO_2 Growth Oxidation Efficiency and Suboxide Vapor Pressure*

In_2O_3 possesses the same stoichiometry as Ga_2O_3 and is an interesting material to engineer the band gap of Ga_2O_3 by growing $(In_xGa_{1-x})_2O_3$ layers. SnO_2 is a relevant reference material to understand the Sn incorporation for Sn-doping of Ga_2O_3. As shown in Figs. 6.9 and 6.10 the growth kinetics of In_2O_3 is qualitatively the same as that of Ga_2O_3, which is related to the existence of the volatile suboxide In_2O. Likewise, SnO_2 possesses the volatile suboxide SnO that leads to qualitatively similar growth rate diagrams [1, 29].

Quantitatively, however, higher growth rates for In_2O_3 than for Ga_2O_3 under identical growth conditions suggest kinetic advantages for In_2O_3 for two reasons:

(i) Figure 6.9 shows that at low growth temperature under identical O flux, the maximum growth rate for In_2O_3 is roughly three times that of Ga_2O_3, indicating reduced flux-stoichiometrically induced suboxide desorption of In_2O compared to Ga_2O [7]. The growth rate (= metal incorporation) peaks at 4.5 Ga nm^{-2} s^{-1} and 12.8 In $nm^{-2}s^{-1}$ correspond to the stoichiometric flux ratio at which all activated O is incorporated. The applied O flux at the peak hence corresponds to an activated O flux (O-incorporation rate) of 6.75 O $nm^{-2}s^{-1}$ and 19.2 $nm^{-2}s^{-1}$ for the oxidation of Ga and In, respectively. To quantify the effect, Vogt et al. defined the ratio of the O-incorporation rate into the film to the total flux of oxygen (derived from the measured oxygen BEP using kinetic gas theory) at the stoichiometric point (peak) of the growth rate diagram as oxidation efficiency. The oxidation efficiency was found to be $\approx$ 10% for Ga, $\approx$ 26% for In, and $\approx$ 20% for Sn [1, 7].

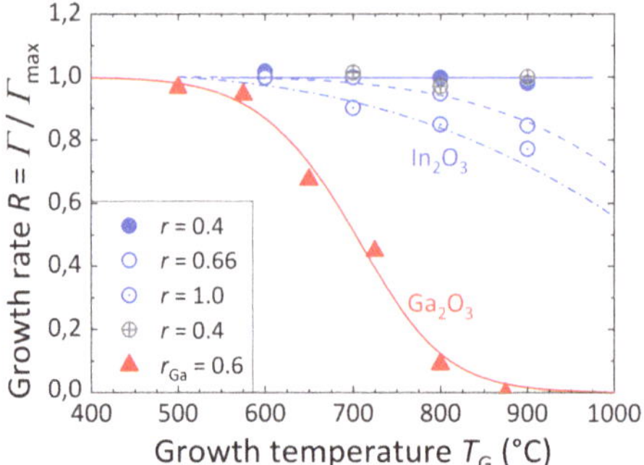

Fig. 6.10 Comparison of thermally induced suboxide desorption (suboxide vapor pressure): Growth rate normalized by its maximum value as a function of growth temperature for different metal-to-oxygen flux ratios r. $r < 2/3$ corresponds to O-rich conditions, $r = 2/3$ corresponds to a stoichiometric flux ratio (peak in Fig. 6.9), and $r > 2/3$ corresponds to metal-rich conditions. Adapted from [3]

(ii) The evolution of the growth rate with increasing growth temperature shown in Fig. 6.10 also exhibits a stronger decrease for Ga_2O_3 than for In_2O_3 even under O-rich flux stoichiometries ($r < 2/3$) [3]. This behavior is related to a stronger thermally induced Ga_2O desorption than In_2O desorption due to the higher vapor pressure of Ga_2O than In_2O [3, 7]. For example, at a typical growth temperature of 700 °C the vapor pressure of Ga_2O [30] is ≈ 10 times higher than that of In_2O [31] and ≈ 1000 times higher than that of SnO [32].

6.4 Thermodynamics Aspects

The MBE growth of binary oxides is predominantly governed by the kinetics described above. As soon as two metals come into play, however, thermodynamics plays a significant role that can even override the kinetics growth preferences of the individual binary oxides—an effect already known from the growth of II-VI compounds [33]. This effect will be demonstrated with the example of $(In_xGa_{1-x})_2O_3$ growth focusing on the stoichiometry control. We will show how both, thermodynamics and kinetics, collaborate in a catalytic effect that strongly enhances the Ga_2O_3 growth window. Thermodynamics further plays a role for the formation of the crystal phase and surface faceting as discussed at the end of this section.

6.4.1 $(In_xGa_{1-x})_2O_3$ Growth: Ga–O Bonds Versus In–O Bonds

As discussed above, the growth of In_2O_3 is kinetically preferred over that of Ga_2O_3, i.e., In_2O_3 can grow without noticeable In_2O desorption under growth conditions that lead to significant Ga_2O desorption during growth of Ga_2O_3 [1, 3]. This situation is drastically changed when a flux of both, In and Ga, is present to grow $(In_xGa_{1-x})_2O_3$ as illustrated in Fig. 6.11 for the situation of approximately equal In- and Ga fluxes. The Ga- and In-incorporation expressed as Ga_2O_3 and In_2O_3 pseudo-binary growth rates is shown as a function of growth temperature for different active O fluxes.

Under metal-rich conditions, where both metals compete for the available oxygen only Ga is incorporated, resulting in a Ga_2O_3 layer. Even at conditions where twice the oxygen required to oxidize both metals is provided, the In-incorporation drops more rapidly with growth temperature than the Ga-incorporation [4].

Both these results are in stark contrast to the strong kinetic preference for the growth of In_2O_3. Vogt et al. explained these findings by the thermodynamics of stronger Ga–O bonds compared to In–O bonds supported by thermochemical calculations and reference experiments trying to etch an In_2O_3 layer by Ga vapor

$$6\mathrm{Ga(a)} + \mathrm{In_2O_3(s)} \rightarrow 3\mathrm{Ga_2O(g)} + 2\mathrm{In(a\,or\,g)} \tag{6.5}$$

or

$$4\mathrm{Ga(a)} + \mathrm{In_2O_3(s)} \rightarrow 2\mathrm{Ga_2O(g)} + 2\mathrm{In_2O(a\,or\,g)} \tag{6.6}$$

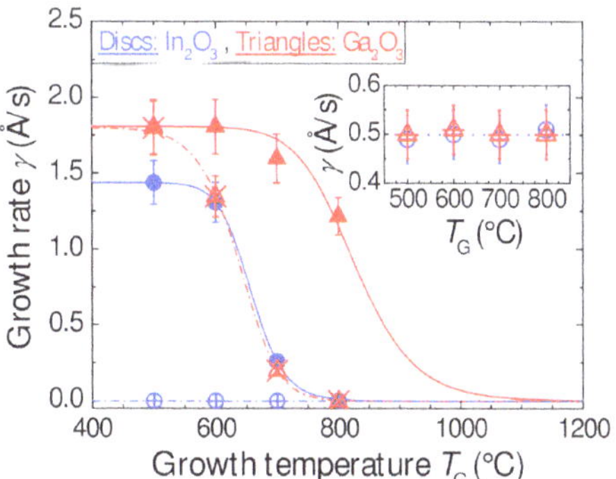

Fig. 6.11 Comparison of In- and Ga-incorporation during $(In_xGa_{1-x})_2O_3$ growth: The pseudo-binary growth rates for In_2O_3 (blue discs) and Ga_2O_3 (red triangles) as a function of growth temperature are depicted for approximately equal In- and Ga fluxes and different metal-to-oxygen flux ratios r. Filled symbols: $r = 0.31$ (O-rich); crossed symbols: $r = 4/3$ (Me-rich, sufficient O only for one of the metals); inset: $r = 0.066$ (strongly O-rich). Reprinted from [4]

and to etch a Ga_2O_3 layer by In-vapor

$$6\mathrm{In}(\mathrm{a}) + \mathrm{Ga_2O_3}(\mathrm{s}) \rightarrow 3\mathrm{In_2O}(\mathrm{g}) + 2\mathrm{Ga}(\mathrm{a\,or\,g}) \tag{6.7}$$

or

$$4\mathrm{In}(\mathrm{a}) + \mathrm{Ga_2O_3}(\mathrm{s}) \rightarrow 2\mathrm{In_2O}(\mathrm{g}) + 2\mathrm{Ga_2O}(\mathrm{a\,or\,g}) \tag{6.8}$$

Experimentally they found Ga to etch In_2O_3 whereas a Ga_2O_3 layer was stable against In-vapor, in agreement with a negative Gibbs free enthalpy calculated for reactions (6.5, 6.6) and positive one for reactions (6.7, 6.8) [4]. Strongly O-rich growth conditions, however, were able to overcome the thermodynamics limitations and enforce the full incorporation of all In and Ga as shown in the inset of Fig. 6.11.

The generality of the thermodynamic impact of bond strength on cation composition becomes apparent also with other growth methods or ternary systems: For the pulsed laser deposition of $(In_xGa_{1-x})_2O_3$ layers, the same trend of reduced In- versus Ga-incorporation, that can be mitigated by a sufficiently high oxygen partial pressure, can be seen in Fig. 1 of the work of Kranert et al. [34] Y. Oshima et al. have reported a higher Al-content than expected from the Al-to-Ga beam flux ratio in MBE-grown $(Al_xGa_{1-x})_2O_3$ films, [35] and attributed it to Ga_2O formation. Similar to the case of $(In_xGa_{1-x})_2O_3$ the stronger Al–O bonds than Ga–O bonds [36] suggest a thermodynamic explanation but further experiments are needed to compare the Al_2O_3 growth kinetics to that of Ga_2O_3.

6.4.2 *Metal-Exchange Catalysis (MEXCAT) Enhances the Growth Window of Ga_2O_3: Kinetics Collaborating with Thermodynamics*

A strong enhancement of the growth rate of Ga_2O_3 has been observed, using either an additional In flux [5] or Sn flux [37] during growth. In either case, no significant In or Sn incorporation into the obtained Ga_2O_3 layers was detected. Figures 6.12 and 6.13 show examples of the growth rate of Ga_2O_3 as a function of the In flux and Sn flux, respectively. In both cases, an approximately sigmoidal growth rate evolution with In or Sn flux occurs. More importantly, a widening of the growth window is observed in Figs. 6.12c, d and 6.13 in which growth is enabled by adding the corresponding element (In or Sn) to the Ga–O system—indicating a similar underlying reaction mechanism.

Vogt et al. explained this reaction mechanism by a simplified two-step (net) reaction that circumvents the Ga_2O containing two-step process (prone to Ga_2O desorption) [5]: In the first step In_2O_3 is predominantly formed through reaction

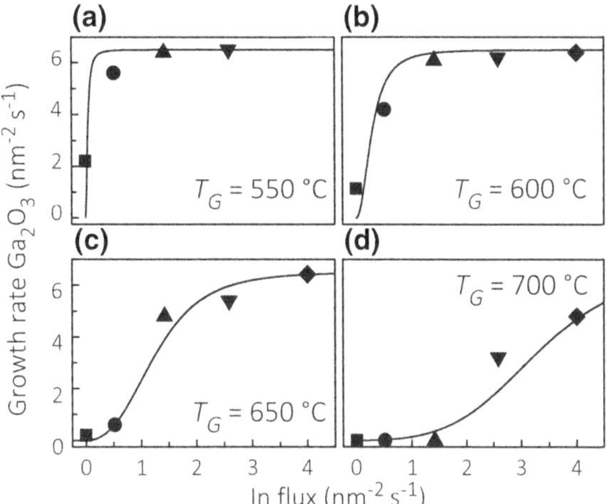

Fig. 6.12 Growth rate dependence of Ga_2O_3 as a function of the In flux. The used Ga flux was 6.5 nm^{-2} s^{-1}. The applied active O flux was 6.7 O nm^{-2} s^{-1} and 19.2 O nm^{-2} s^{-1} for the oxidation of Ga and In, respectively. Reprinted with permission from [5]. Copyright (2017) by the American Physical Society

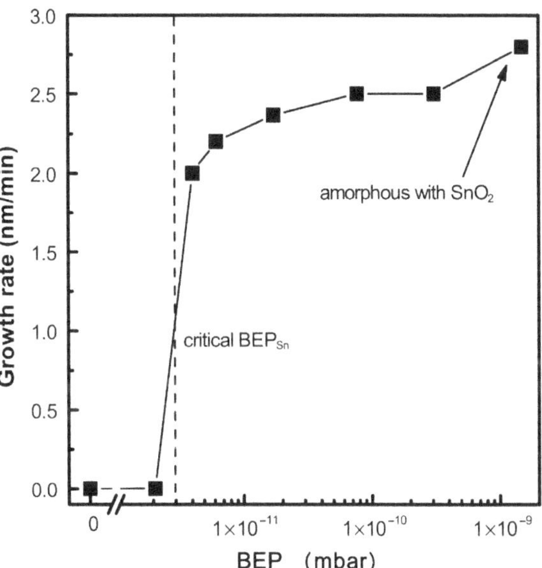

Fig. 6.13 Growth rate dependence of Ga_2O_3 as a function of the Sn flux at a set Ga BEP of 1.77×10^{-7} mbar. The used growth temperature, O flux, and O plasma power were 700 °C, 0.5 SCCM, and 150 W. Reprinted with permission from [37]. Copyright (2017) by the American Physical Society

$$2\text{In(a)} + 3\text{O(a)} \rightarrow \text{In}_2\text{O}_3\text{(s)}, \tag{6.9}$$

due to the kinetic advantages (higher oxidation efficiency, lower suboxide vapor pressure) over Ga_2O_3 [1, 3, 7]. In a second reaction step, Ga replaces the In atoms in In_2O_3 via an interatomic In–Ga exchange, through reaction

$$2\text{Ga(a)} + \text{In}_2\text{O}_3\text{(s)} \rightarrow \text{Ga}_2\text{O}_3\text{(s)} + 2\text{In(a)}, \tag{6.10}$$

due to the thermodynamically stronger Ga–O than In–O bonds [4]. Since the In released through reaction (6.10) can be re-oxidized by other O ad-atoms, the net Ga_2O_3 growth rate can exceed the supplied In flux. This re-oxidation step makes In_2O_3 a catalyst for the oxidation of Ga_2O_3. The resulting metal-exchange catalysis is hence based on the collaborative interplay of the kinetics and thermodynamics of In_2O_3 and Ga_2O_3.

Similarly, Kracht et al. explained their observed growth rate enhancement upon adding Sn by the thermodynamically preferred oxidation of Ga_2O over SnO_2 formulating a similar exchange reaction to (6.10):

$$\text{Ga}_2\text{O(a)} + \text{SnO}_2\text{(s)} \rightarrow \text{Ga}_2\text{O}_3\text{(s)} + \text{Sn(a)}, \tag{6.11}$$

in agreement with thermochemical calculations [37].

The kinetic advantage of Sn over Ga to be oxidized [1, 7]

$$\text{Sn(a)} + 2\text{O(a)} \rightarrow \text{SnO}_2\text{(s)} \tag{6.12}$$

(due to higher oxidation efficiency of Sn and lower vapor pressure of SnO) can explain the predominant SnO_2 formation over Ga_2O_3 in a first reaction step.

The significantly larger In flux than Sn flux required for the catalysis (cf. Figs. 6.12 and 6.13) suggests a significantly stronger In- than Sn- or SnO desorption at the chosen growth temperatures, in agreement with a significantly lower vapor pressure of Sn and SnO than In and In_2O [7].

Both Vogt et al. [5] and Kracht et al. [37] reported the formation of the ε-phase of Ga_2O_3 by the catalytic growth on Ga_2O_3(-201) on Al_2O_3(0001), as further discussed below.

The metal-exchange catalysis by adding In to the Ga–O system has been also recently applied to drastically increase the growth rate during homoepitaxy of Ga_2O_3(010) [16]. In this work, Mazzolini et al. demonstrate that in the case of Ga_2O_3(010) homoepitaxial growth, the monoclinic phase of the substrate is maintained, i.e., no ε/k-phase is formed. Figure 6.14 compares the symmetric out-of-plane XRD 2Θ-ω scans around the (020) reflection of two Ga_2O_3(010) homoepitaxial layers grown under identical conditions for the same growth time, except for an additional In flux for the catalyzed one. The absence of a diffraction peak besides the Ga_2O_3(020) one confirms the absence of In-incorporation in the catalyzed layer. The thickness fringes of the Ga_2O_3(020)-peak confirm the formation of pure β-Ga_2O_3(010) in both cases. The periodicity of the fringes can be used to determine the film thickness and the higher periodicity for the In-catalyzed layer clearly indicates its larger thickness (corresponding to a threefold growth rate increment) [16]. The possibility to employ metal-exchange catalysis in Ga_2O_3 homoepitaxy while preserving the monoclinic structure of the substrate could allow to investigate the MBE growth on crystalline orientations previously neglected due to the extremely low growth rates [e.g., (100)].

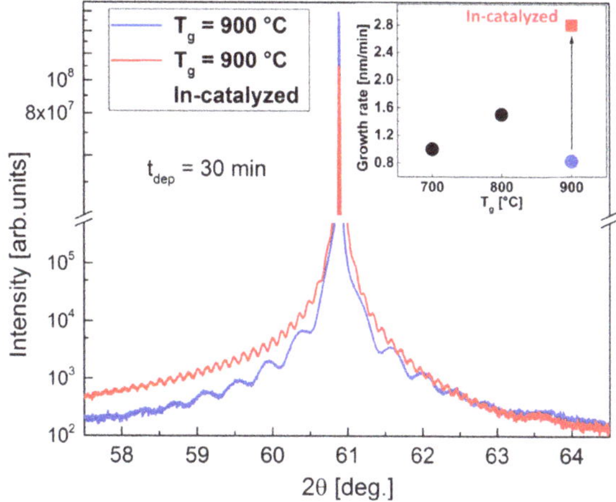

Fig. 6.14 XRD 2Θ-ω scan of the (020) reflection of homoepitaxially grown layers with (red curve) and without (blue) an additional In flux (1/3 x Ga flux) during the deposition process at a growth temperature T_g of 900° C. A Ga flux of 2.2 nm^{-2} s^{-1} (corresponding to a growth rate of 3.5 nm/min at full incorporation), and slightly Ga-rich growth conditions (O flux = 0.33 sccm, 300 W plasma power) were employed. The inset shows the corresponding growth rates extracted from the thickness fringes as well as those of lower growth temperatures without additional In flux. Reprinted from [16]

An extension from the binary to ternary compound formation using metal-exchange catalysis is shown by Vogt et al. [38]. Here, by adding In to Al–Ga–O system during growth a drastic enhancement of the $(Al_xGa_{1-x})_2O_3$ growth rate during PAMBE is observed as described in the previous chapter.

6.4.3 *Stabilizing Different Phases*

Besides the thermodynamically stable β-phase, the α-, ε/κ-, and γ-phases (polymorphs, polytypes) have been stabilized by MBE using heteroepitaxy.

The frequently used substrate sapphire (Al_2O_3) and the polymorph α-Ga_2O_3 share the same corundum crystal structure, albeit with a lattice mismatch of 4.7 and 3.4% along the *c*- and *a*-axis, respectively. Due to the large lattice mismatch, the growth of coherently strained (pseudomorphic) α-Ga_2O_3 on Al_2O_3 is limited to layer thicknesses of ~1 nm, as discussed below.

The MBE growth on (c-plane) Al_2O_3(0001) substrates frequently results in the formation of β-Ga_2O_3(-201) layers with rotational domains [9, 23, 39]. A close inspection of the interface to the substrate by Schewski et al., however, revealed the presence of a ~1 nm thick, pseudomorphic α-Ga_2O_3(0001) interfacial layer that is likely stabilized by the lattice-mismatch induced strain [40]. Al_2O_3 substrates with other orientations enabled the formation of thicker α-Ga_2O_3 layers before growth

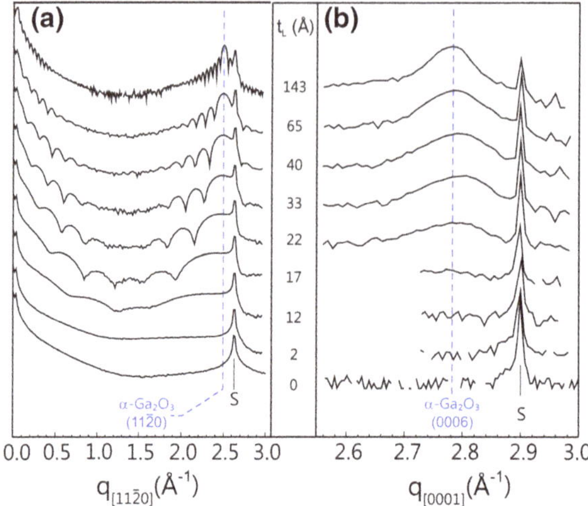

Fig. 6.15 Intensity evolution along out-of-plane and in-plane crystal truncation rods, during the growth of α-Ga_2O_3(11–20) layer on Al_2O_3(11–20): t_L denotes the deposited layer thickness. **a** The (11–20) diffraction peak of α-Ga_2O_3 appears on its bulk position left of the sapphire (11–20) contribution. **b** Grazing incidence diffraction along the substrate's [0001] direction monitors the change of the in-plane lattice parameter. The azimuth scan of the in-plane Bragg peak reveals the epitaxial relation Ga_2O_3[0001]||Al_2O_3[0001]. Reprinted from [41], with the permission of AIP Publishing

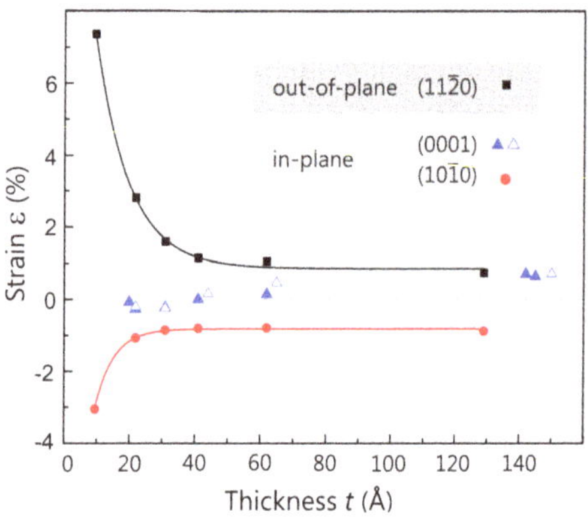

Fig. 6.16 Extracted strain relaxation of α-Ga_2O_3(11–20) layer on Al_2O_3(11–20) during growth. Reprinted from [41], with the permission of AIP Publishing

proceeded by formation of β-Ga_2O_3 on top: a combined growth and in situ X-ray diffraction study by Cheng et al. demonstrated the growth of a 14 nm-thick α-Ga_2O_3(11–20) layer on (a-plane) Al_2O_3(11–20) that rapidly relaxed the in-plane strain within the first 2 nm of growth [41]. Figure 6.15 shows the evolution of the related out-of-plane and in-plane XRD-diffractograms with increasing film thickness. The extracted anisotropic strain evolution of the layer during growth is shown in Fig. 6.16.

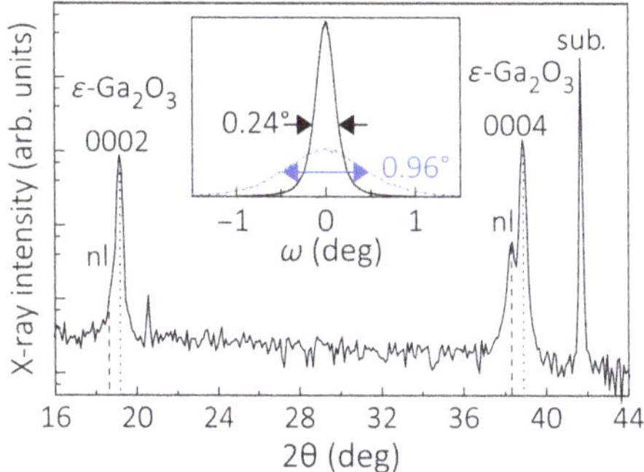

Fig. 6.17 XRD of ε-Ga_2O_3(0001) layer on β-Ga_2O_3(-201) nucleation layer (nl): Out-of-plane, symmetric 2Θ-ω scan indicating the presence of the ε-Ga_2O_3(0001), the nucleation layer with (-201) and (-402) reflections as well as the c-plane sapphire substrate (sub.) with (0006) reflection. Inset: Omega rocking curve of the ε-Ga_2O_3(0004) (solid black line) and ε-Ga_2O_3(10–14) (dashed-dotted blue line) reflections. The full width at half maxima of the reflections is indicated in the figure. Reprinted with permission from [5]. Copyright (2017) by the American Physical Society

Kracht et al. grew over 200 nm-thick α-Ga_2O_3(10–12) layers on (r-plane) Al_2O_3(10–12) [14] before the growth of β-Ga_2O_3(-201) nucleated on the (0001)-facets that emerged by roughening of the α-Ga_2O_3(10–12) layer. Oshima et al. grew α-Ga_2O_3(10–12)/α-Al_2O_3(10–12) superlattices on (r-plane) Al_2O_3(10–12) achieving coherent interfaces without misfit dislocations for up to ~1 nm-thick α-Ga_2O_3(10–12) layers [42].

Another example of substrate-stabilized phase formation is that of the cubic defective-spinel γ-Ga_2O_3(001) on spinel $MgAl_2O_4$(001) substrates [43].

The growth of ε/κ-Ga_2O_3(0001) by metal-exchange catalysis mediated through an additional In flux [5] or Sn flux [37] has been reported by Vogt et al. and Kracht et al., respectively. As an example, Fig. 6.17 shows an out-of-plane X-ray diffraction 2Θ-ω scan indicating the presence of the (0001)-oriented ε-Ga_2O_3 layer as well as the underlying β-Ga_2O_3(-201) nucleation layer (nl) on top of the Al_2O_3(0001) substrate. The underlying mechanism that stabilizes the ε-phase is still to be understood. It is conceivable that the conditions for Ga_2O_3 growth catalyzed by In or Sn at high growth temperature (that prevents growth in the absence of In or Sn) thermodynamically promote the formation of the ε/κ phase. This idea is supported by the relatively low difference in formation energy between the β and ε/κ phase and its decrease with increasing growth temperature [44].

6.4.4 *Surface Faceting*

The β-Ga_2O_3 nominal substrate/layer orientation [e.g., (010), (100), (001), (-201)] is not itself unambiguously defining the crystal planes that are constituting its

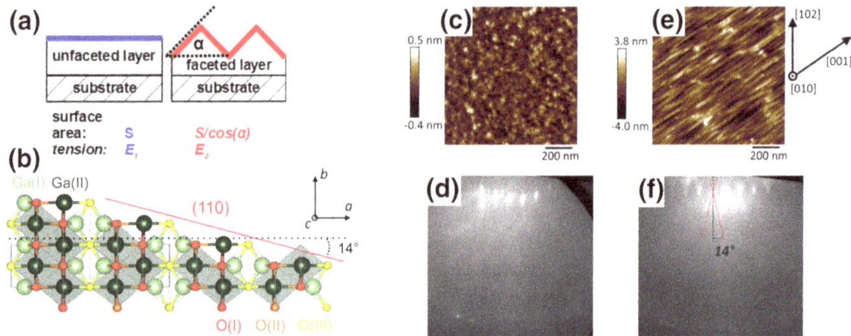

Fig. 6.18 **a** Schematics of a general unfaceted and faceted surface. **b** β-Ga_2O_3 atomic model. **c**–**f** Atomic force microscopy (AFM) micrographs and RHEED patterns along the [001] azimuth of an O plasma treated (**c**) and (**d**), respectively, as well as a Ga-etched (**e**) and (**f**), respectively, β-Ga_2O_3(010) substrate. **a** Adapted from [45]. **b**–**f** Adapted from [16]

surface termination. In fact, the combined effect of temperature and chemical potential (e.g., reducing/oxidizing atmosphere) can result in the formation of facets, i.e., surfaces not parallel to the substrate surface which result to be more stable despite forming a higher surface area with respect to the unfaceted case (see schematic representation of Fig. 6.18a) [45]. The thermodynamic driving force is the lower surface tension E_2 of the facet orientation compared to that of the surface orientation E_1, which overcompensates the area increase by a factor $1/\cos(\alpha)$ as schematically shown in Fig. 6.18a. As an example for the class of oxides, a recent work by Bierwagen et al. on heteroepitaxial In_2O_3 thin films [45] experimentally demonstrated a change of surface faceting of the deposited layers while passing from oxygen to metal-rich MBE deposition conditions. These findings were in agreement with the theoretically predicted dependence of surface tension for the different surface orientations on oxygen chemical potential [46].

In the case of β-Ga_2O_3 homoepitaxy, the substrate orientation has been shown to play a major role in both the determination of the growth rate [22] and the resulting electrical properties [47] of the deposited layer. Therefore, the achievement of a deep control over the synthesis of Ga_2O_3 should also rely on clear experimental and theoretical assessments on its surface stability over different crystal orientations and in different synthesis conditions. In this section, we briefly summarize the main evidence so far reported in literature in this regard. Being (100) and (001) the most well known β-Ga_2O_3 cleavage planes, they are both supposed to be rather stable surfaces, with the (100) having the lowest calculated surface energy between the two [48].

Nonetheless, recent experimental works focusing on (001)-oriented β-Ga_2O_3 homoepitaxy (by PAMBE) reported on the presence on the thin film surface of stripe-like elongated features oriented along the [010] direction [17, 49] Y. Oshima et al. [17] pointed out that the presence of these elongated features is not related to the miscut direction of the substrate; therefore, such morphology should be attributed to the faceting of the nominally (001)-oriented surface. Nevertheless, experimental evidence for a clear identification of the probable facets formation after (001)-homoepitaxy is missing to date.

Even in the case of the (100)-oriented Ga_2O_3 homoepitaxial layers, different experimental works are reporting on the formation of elongated features oriented along the [010] direction for films grown by PAMBE [9, 10, 50] and metal organic vapor phase epitaxy (MOVPE) [47, 51]. Nevertheless, experimental or theoretical evidence is missing in order to clarify the possible presence of facets on the layer surfaces.

The (010)-orientation is so far the most investigated growth direction of β-Ga_2O_3. The (010) is not a cleavage plane, and its surface energy has been calculated to be higher than the (001) and the (100) ones [48, 52]. In a recent work, Mazzolini et al. [16] reported on the formation of (110) and ($\bar{1}$10)-facets (equivalent surfaces, Fig. 6.19b) on the nominally oriented (010) surface independently induced by both Ga-etching and Ga-rich MBE homoepitaxial deposition conditions. Similarly to the previously mentioned case of In_2O_3, [45] it is noteworthy that also for (010)-oriented Ga_2O_3 different surfaces have been found to be thermodynamically stable while passing from oxygen-rich to metal-rich synthesis conditions [16]. At a substrate temperature high enough to allow surface rearrangements (i.e., $T \approx 800$ °C), the exposure of an (010)-Ga_2O_3 substrate to an oxygen plasma (O_2-flow = 1 SCCM, P = 300 W, t = 30 min) resulted in an atomically flat featureless surface (Fig. 6.18c) [16]; this experimental evidence highlights the (010) surface stability under oxygen-rich conditions.

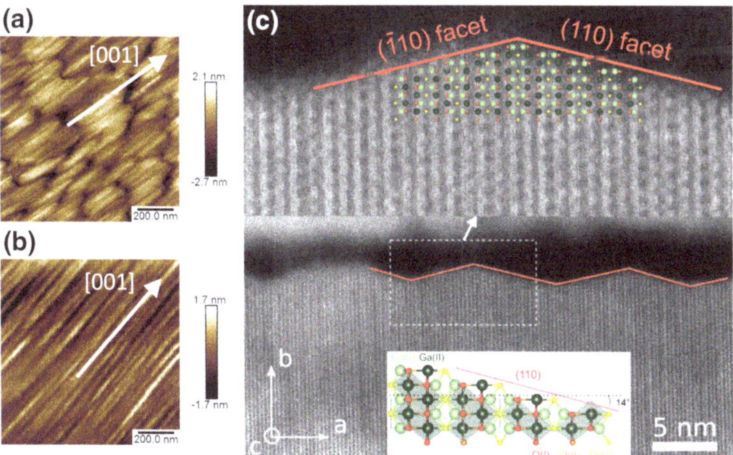

Fig. 6.19 AFM acquisitions of homoepitaxial (010)-oriented β-Ga_2O_3 films deposited under the same Ga-rich conditions (t = 30 min, growth T = 900 °C) without (**a**) and with (**b**) an additional In flux. **c** HAADF-STEM images in the c-projection of the In-catalyzed layer reported in (**b**) with an overlay of the Ga_2O_3 atomic model. **a–d** Adapted from [16]

Differently, if the same surface was exposed at the same T to a Ga flux (i.e., Ga-etching, $BEP_{Ga} = 1.9 \times 10^{-7}$mbar, t = 30 min), the resulting surface exhibited the appearance of elongated features oriented along the [001] direction leading to an increased roughness (see Fig. 6.18e) [16]. The RHEED patterns acquired along the [001] azimuthal direction (i.e., along the direction of the elongated features) after the O plasma (Fig. 6.18d) and the Ga-etching treatments allowed to evidence the formation of wedges after the exposure of the surface to the Ga flux (highlighted by the red segmented lines in Fig. 6.18f) [16]. These weak oblique streaks point to the presence of facets oriented parallel to the investigated RHEED azimuth and normal to the oblique streaks [8]; therefore, the angle of 14° between the oblique streaks and the substrate surface normal, allowed to identify these facets to be the (110) and ($\bar{1}$10) (see Fig. 6.18b) [16]. A similarly oriented morphology has already been observed in literature on (010)-Ga_2O_3 by Sasaki et al. [53] after a substrate annealing in N_2 atmosphere at 1000 °C. Moreover, the formation of facets on the (010)-oriented surface has been also observed for homoepitaxial layers deposited by both MOVPE [54] and plasma-assisted MBE [16]. Coherently with the reported results on the substrate treatments, also for (010) β-Ga_2O_3 homoepitaxy the (110)—($\bar{1}$10) facets are formed on the sample surface when the layer is deposited under metal-rich deposition conditions (see Fig. 6.19a) and the substrate temperature is sufficiently high ($T \geq$ 700 °C) [16]. While maintaining the same metal-rich deposition conditions, the faceted morphology was preserved (Fig. 6.19b) even with the employment of an additional In flux during homoepitaxial growth, i.e., with a tripled growth rate induced by the metal-exchange catalysis [16]. High-angle annular dark-field scanning TEM (HAADF-STEM) images acquired in cross section view perpendicular to the [001] direction (Fig. 6.19c) confirmed the presence of the (110)—($\bar{1}$10) facets in homoepitaxial layers deposited under metal-rich conditions; moreover, it also allowed to identify the presence of exclusively tetrahedrally coordinated Ga atoms on the surface (light green atoms in the overlay of the β-Ga_2O_3 atomic structure in Fig. 6.19c) differently from what would be expected from a pure (110) surface (see atomic model reported in Fig. 6.18b) [16].

Despite the presence of several works focusing on β-Ga_2O_3 homoepitaxy, definitive experimental evidence on surface faceting has been so far just provided for the (010) substrate orientation.

6.5 Open Questions

Further research is needed to improve our understanding of the PAMBE of Ga_2O_3 and other oxides. The following (not necessarily complete) list presents open questions related to the findings presented in this chapter.

- The underlying physics/chemistry of the determined oxidation efficiencies needs to be better understood. In particular, what are the oxidation efficiencies of the individual reaction steps: metal toward the suboxide and suboxide toward the full oxide?
- What is the influence of surface roughness/faceting on growth rates (as different facets exhibit different desorption behavior)?
- Are suboxides involved in the individual reactions during metal-exchange catalysis?
- What are the physical reasons for the stabilization of different phases by choice of substrate and growth conditions?
- Regarding faceting, the experimental findings should be combined with theoretical studies aiming at the calculation of the surface free energies as a function of different chemical potentials and at the evaluation of possible surface reconstructions.

Acknowledgements The research work by the chapter authors was performed in the framework of GraFOx, a Leibniz-Science Campus partially funded by the Leibniz Association.

References

1. P. Vogt, O. Bierwagen, Appl. Phys. Lett. **106**, 081910 (2015)
2. P. Vogt, O. Bierwagen, Appl. Phys. Lett. **108**, 072101 (2016)
3. P. Vogt, O. Bierwagen, Appl. Phys. Lett. **109**, 062103 (2016)
4. P. Vogt, O. Bierwagen, APL Mater. **4**, 086112 (2016)
5. P. Vogt, O. Brandt, H. Riechert, J. Lähnemann, O. Bierwagen, Phys. Rev. Lett. **119**, 196001 (2017)
6. P. Vogt, O. Bierwagen, Phys. Rev. Mater. **2**, 120401(R) (2018)
7. P. Vogt, Dissertation, Humboldt-Universität zu Berlin, 2017
8. A. Ichimiya, P.I. Cohen, *Reflection High-Energy Electron Diffraction* (Cambridge University Press, Cambridge, 2004)
9. M.-Y. Tsai, O. Bierwagen, M.E. White, J.S. Speck, J. Vac. Sci. Technol. A **28**, 354 (2010)
10. Z. Cheng, M. Hanke, Z. Galazka, A. Trampert, Nanotechnology **29**, 395705 (2018)
11. G. Koblmüller, R. Averbeck, H. Riechert, P. Pongratz, Phys. Rev. B **69**, 035325 (2004)
12. S. Fernández-Garrido, G. Koblmüller, E. Calleja, J.S. Speck, J. Appl. Phys. **104**, 033541 (2008)
13. H. Kato, M. Sano, K. Miyamoto, T. Yao, Jpn. J. Appl. Phys. **42**, 2241 (2003)
14. M. Kracht, A. Karg, M. Feneberg, J. Bläsing, J. Schörmann, R. Goldhahn, M. Eickhoff, Phys. Rev. Appl. **10**, 024047 (2018)
15. E. Ahmadi, O.S. Koksaldi, X. Zheng, T. Mates, Y. Oshima, U.K. Mishra, J.S. Speck, Appl. Phys. Express **10**, 071101 (2017)
16. P. Mazzolini, P. Vogt, R. Schewski, C. Wouters, M. Albrecht, O. Bierwagen, APL Mater. **7**, 022511 (2019)
17. Y. Oshima, E. Ahmadi, S. Kaun, F. Wu, J.S. Speck, Semicond. Sci. Technol. **33**, 015013 (2017)
18. T. Oshima, T. Okuno, S. Fujita, Jpn. J. Appl. Phys. **46**, 7217 (2007)
19. H. Okumura, M. Kita, K. Sasaki, A. Kuramata, M. Higashiwaki, J.S. Speck, Appl. Phys. Express **7**, 095501 (2014)

20. E. Ahmadi, O.S. Koksaldi, S.W. Kaun, Y. Oshima, D.B. Short, U.K. Mishra, J.S. Speck, Appl. Phys. Express **10**, 041102 (2017)
21. K. Sasaki, M. Higashiwaki, A. Kuramata, T. Masui, S. Yamakoshi, J. Cryst. Growth **392**, 30 (2014)
22. K. Sasaki, A. Kuramata, T. Masui, E.G. Villora, K. Shimamura, S. Yamakoshi, Appl. Phys. Express **5**, 035502 (2012)
23. E.G. Villora, K. Shimamura, K. Kitamura, K. Aoki, Appl. Phys. Lett. **88**, 031105 (2006)
24. E.G. Villora, Private Communication (2018)
25. Ü. Özgür, Y.I. Alivov, C. Liu, A. Teke, M.A. Reshchikov, S. Doagan, V. Avrutin, S.-J. Cho, H. Morkoç, J. Appl. Phys. **98**, 041301 (2005)
26. K. Ploog, Annu. Rev. Mater. Sci. **12**, 123 (1982)
27. S. Ghose, M.S. Rahman, J.S. Rojas-Ramirez, M. Caro, R. Droopad, A. Arias, N. Nedev, J. Vac. Sci. Technol. B **34**, 02L109 (2016)
28. S. Ghose, S. Rahman, L. Hong, J.S. Rojas-Ramirez, H. Jin, K. Park, R. Klie, R. Droopad, J. Appl. Phys. **122**, 095302 (2017)
29. M.Y. Tsai, M.E. White, J.S. Speck, J. Appl. Phys. **106**, 024911 (2009)
30. C.J. Frosch, C.D. Thurmond, J. Phys. Chem. **66**, 877 (1962)
31. J. Valderrama-N, K.T. Jacob, Thermochim. Acta **21**, 215 (1977)
32. R. Colin, J. Drowart, G. Verhaegen, Trans. Faraday Soc. **61**, 1364 (1965)
33. S.V. Ivanov, S.V. Sorokin, and I.V. Sedova, in *Molecular Beam Epitaxy*, ed. by M. Henini (Elsevier, 2013), p. 614
34. C. Kranert, J. Lenzner, M. Jenderka, M. Lorenz, H. von Wenckstern, R. Schmidt-Grund, M. Grundmann, J. Appl. Phys. **116**, 013505 (2014)
35. Y. Oshima, E. Ahmadi, S.C. Badescu, F. Wu, J.S. Speck, Appl. Phys. Express **9**, 061102 (2016)
36. I.D. Brown, R.D. Shannon, Acta Cryst. A **29**, 266 (1973)
37. M. Kracht, A. Karg, J. Schörmann, M. Weinhold, D. Zink, F. Michel, M. Rohnke, M. Schowalter, B. Gerken, A. Rosenauer, P.J. Klar, J. Janek, M. Eickhoff, Phys. Rev. Appl. **8**, 054002 (2017)
38. P. Vogt, A. Mauze, F. Wu, B. Bonef, J.S. Speck, Appl. Phys. Express **11**, 115503 (2018)
39. S. Nakagomi andY. Kokubun, J. Cryst. Growth **349**, 12 (2012)
40. R. Schewski, G. Wagner, M. Baldini, D. Gogova, Z. Galazka, T. Schulz, T. Remmele, T. Markurt, H. von Wenckstern, M. Grundmann, O. Bierwagen, P. Vogt, M. Albrecht, Appl. Phys. Express **8**, 011101 (2015)
41. Z. Cheng, M. Hanke, P. Vogt, O. Bierwagen, A. Trampert, Appl. Phys. Lett. **111**, 162104 (2017)
42. T. Oshima, Y. Kato, M. Imura, Y. Nakayama, M. Takeguchi, Appl. Phys. Express **11**, 065501 (2018)
43. T. Oshima, Y. Kato, M. Oda, T. Hitora, M. Kasu, Appl. Phys. Express **10**, 051104 (2017)
44. S. Yoshioka, H. Hayashi, A. Kuwabara, F. Oba, K. Matsunaga, I. Tanaka, J. Phys.: Condens. Matter **19**, 346211 (2007)
45. O. Bierwagen, J. Rombach, J.S. Speck, J. Phys.: Condens. Matter **28**, 224006 (2016)
46. P. Agoston, K. Albe, Phys. Rev. B **84**, 045311 (2011)
47. R. Schewski, M. Baldini, K. Irmscher, A. Fiedler, T. Markurt, B. Neuschulz, T. Remmele, T. Schulz, G. Wagner, Z. Galazka, M. Albrecht, J. Appl. Phys. **120**, 225308 (2016)
48. V.M. Bermudez, Chem. Phys. **323**, 193 (2006)
49. S.-H. Han, A. Mauze, E. Ahmadi, T. Mates, Y. Oshima, J.S. Speck, Semicond. Sci. Technol. **33**, 045001 (2018)
50. T. Oshima, N. Arai, N. Suzuki, S. Ohira, S. Fujita, Thin Solid Films **516**, 5768 (2008)
51. G. Wagner, M. Baldini, D. Gogova, M. Schmidbauer, R. Schewski, M. Albrecht, Z. Galazka, D. Klimm, R. Fornari, Phys. Status Solidi A **211**, 27 (2014)

52. R. Schewski, K. Lion, A. Fiedler, C. Wouters, A. Popp, S.V. Levchenko, T. Schulz, M. Schmidbauer, S. Bin Anooz, R. Grüneberg, Z. Galazka, G. Wagner, K. Irmscher, M. Scheffler, C. Draxl, and M. Albrecht, APL Mater. **7**, 022515 (2019)
53. K. Sasaki, M. Higashiwaki, A. Kuramata, T. Masui, S. Yamakoshi, Appl. Phys. Express **6**, 086502 (2013)
54. M. Baldini, M. Albrecht, A. Fiedler, K. Irmscher, R. Schewski, G. Wagner, ECS J. Solid State Sci. Technol. **6**, Q3040 (2017)

Chapter 7
Ozone-Enhanced Molecular Beam Epitaxy

Kohei Sasaki, Shigenobu Yamakoshi and Akito Kuramata

Abstract This chapter explains homoepitaxial growth of β-Ga_2O_3 by using ozone-enhanced molecular beam epitaxy (MBE). First, in order to reveal the suitable surface orientation for β-Ga_2O_3 homoepitaxial MBE growth, we investigate the surface orientation dependence of the growth rate and surface morphology. The results show that the (010) plane is the most suitable one for growth. Next, we optimize the growth conditions of films grown on the (010) plane. In the case of unintentionally doped films, we find that smooth surface films can be obtained by optimization of the surface migration of Ga and O adatoms. On the other hand, in the case of Sn-doped films, segregation of Sn leads to roughening of the surface. A highly O-rich condition is good for obtaining a smooth surface, because it reduces segregation of Sn. By optimizing the growth conditions in this manner, one can fabricate device-quality β-Ga_2O_3 homoepitaxial films with precisely controllable donor concentrations over a wide range (10^{16}–10^{18} cm^{-3}) and atomically flat surfaces.

7.1 Introduction

In order to fabricate high-performance β-Ga_2O_3 devices, it is important to ensure the high crystal quality of not only the substrate but also the epitaxial film grown on the substrate. In the case of power-device applications, films are mainly used for the drift layer. Their typical thicknesses range from several to tens of microns. This means a growth rate of at least 1 μm/h is required. Moreover, the donor concentration of the films must be precisely controlled in the range of 10^{15}–10^{17} cm^{-3}. To meet these demands, conditions such as growth temperature (T_g), growth rate, or material supply ratio must be optimized. The surface orientation is also an

K. Sasaki (✉) · S. Yamakoshi · A. Kuramata
Novel Crystal Technology, Inc., Sayama, Saitama 350-1305, Japan
e-mail: sasaki@novelcrystal.co.jp

S. Yamakoshi
Tamura Corporation, Sayama, Saitama 350-1305, Japan

M. Higashiwaki and S. Fujita (eds.), *Gallium Oxide*, Springer Series in Materials Science 293, https://doi.org/10.1007/978-3-030-37153-1_7

important condition. However, since the crystal structure of β-Ga_2O_3 is a special one called beta-gallia, we cannot use the extensive knowledge that has been gathered on the relationship between the surface orientation and crystal quality of Si (which has a diamond cubic lattice), GaAs (zinc blende), and GaN (wurtzite). Moreover, it is necessary to gather data on the orientation dependence in relation to the growth [1]. Here, to reveal the suitable surface orientation for β-Ga_2O_3 molecular beam epitaxy (MBE) growth, the relationship between the surface orientation and film quality has been examined using ozone-enhanced MBE (ozone MBE). In particular, it has been found that the (010) plane is a good one for MBE growth of β-Ga_2O_3. In this chapter, we show how to optimize the growth conditions on the (010) plane and describe the technique of growing high-quality films with smooth surfaces and precisely controlled donor dopants (Sn).

7.2 Surface Orientation Dependence of β-Ga_2O_3 Homoepitaxial Growth by Ozone MBE

The surface orientation is one of the most important aspects for obtaining high-quality films. In this section, we shall discuss it from the viewpoints of growth rate, surface morphology, and crystal quality. In particular, we will discuss how the orientation affects the quality of β-Ga_2O_3 homoepitaxial films fabricated using an ozone MBE system.

7.2.1 Experimental Methodology

7.2.1.1 Ozone MBE System

A schematic illustration of the ozone MBE system is shown in Fig. 7.1. A gallium flux is generated from a Knudsen cell (K-cell) containing gallium metal with a purity of 99.99999%. The donor dopant, tin dioxide (SnO_2) with a purity of 99.99%, is supplied from another K-cell. The crucible is made from pyrolytic boron nitride. The oxygen source is a gas mixture consisting of ozone (5%) and oxygen (95%). Note that oxygen is not effective for growth; hence, we refer to it as an "ozone and oxygen gas mixture" or simply "ozone gas." The concentration of ozone in the mixture is limited to 5% by the generation rate of the ozone generator. The β-Ga_2O_3 substrate is bonded to a silicon dummy wafer with indium metal before being set in the growth chamber. T_g is measured with a pyrometer. The use of a liquid nitrogen shroud helps to decrease the background pressure and provides heat insulation for the K-cells. During growth, the growth chamber is kept evacuated with a turbo molecular pump.

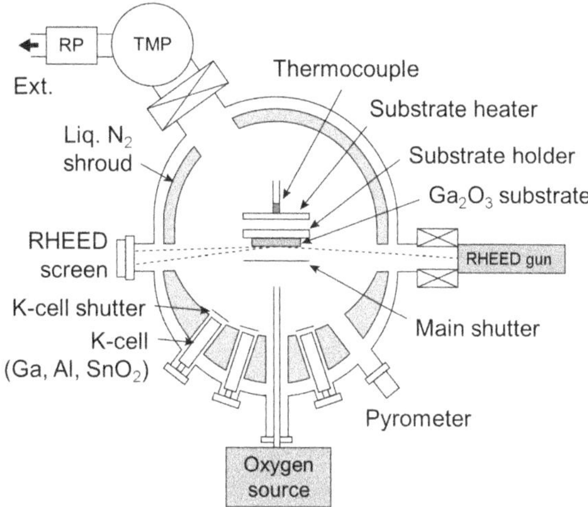

Fig. 7.1 Schematic illustration of β-Ga_2O_3 ozone MBE system

7.2.1.2 Sample Preparation and Evaluation

Sn-doped single-crystal bulk is grown by edge-defined film-fed growth (EFG). The raw material is Ga_2O_3 powder with a purity of 99.999% and 0.05 mol% of SnO_2 as a donor dopant. The growth direction is <010>. The grown β-Ga_2O_3 bulk crystals are cut in various directions, and the cut surface is treated by polishing and acid cleaning.

To investigate the dependence of the *b*-axis and *c*-axis rotation planes, the orientation of the substrate's surface is represented by the angle formed by the (100) plane and the substrate surface, i.e., the rotation angle from the (100) plane. Figure 7.2 shows the relationship between the major orientation, that is, of the major low-index mirror plane and the rotation angle from the (100) plane. The black dotted line indicates the substrate surface. The positive rotation direction of the *b*-axis rotation is defined as shown in Fig. 7.2a. T_g is set at 750 °C. The beam equivalent pressure (BEP) of the Ga flux is 2.1×10^{-4} Pa. The flow rate of the ozone gas is 5.0 sccm, and growth duration is 30 min. Note that the experiments reported here did not use substrate with the principal surface of the *b*-axis rotation 160–170° planes, because it was difficult to get smooth polished surface due to cleavage caused by the (100) plane. The thickness of the grown films is determined using electrochemical capacitance-voltage (ECV) or secondary ion mass spectrometry (SIMS) [2]. The crystal quality is estimated by X-ray diffraction (XRD), and atomic force microscopy (AFM) is used to evaluate the surface morphology.

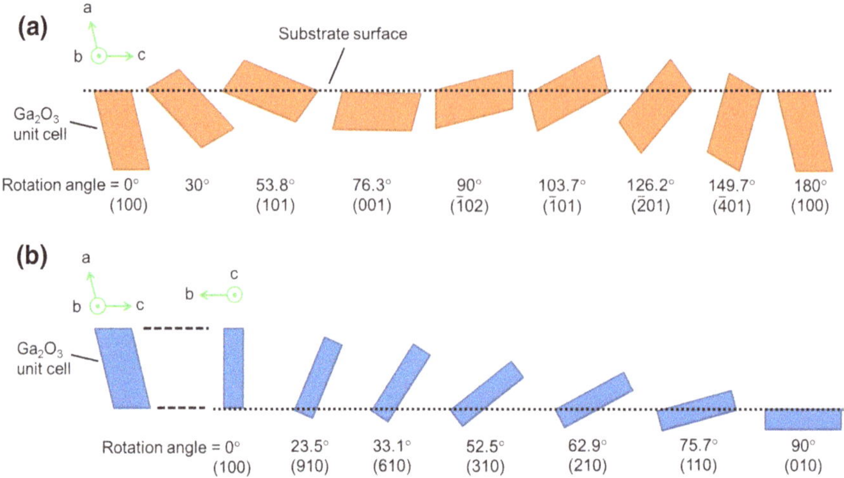

Fig. 7.2 Relationship between surface orientation and rotation angle from (100) plane on *b*-axis rotation planes (**a**) and *c*-axis rotation planes (**b**)

7.2.2 Orientation Dependence of Growth Rate

Figures 7.3a, b show the growth rate as a function of the rotation angle from the (100) plane of (a) the *b*-axis and (b) *c*-axis rotation plane. The (100) plane shows a very small growth rate, less than 0.01 μm/h. This value is one order or more in magnitude smaller than that of the other planes. The growth rate increases with rotation angle, and the rate mostly saturates at 10° or more from the (100) plane. The (001) plane has a slightly small growth rate. The (100) and (001) planes are strong cleavage planes, and the surface energy is smaller than that of the other planes [3]. Therefore, re-evaporation of the materials supplied to these two planes is likely to be large because the dangling-bond density and/or the binding energy of these surfaces is small. The average growth rate of the *b*-axis rotation planes is about 0.4–0.5 μm/h. On the other hand, the average growth rate of the *c*-axis rotation planes is about 40% larger, i.e., 0.7 μm/h. Its origin is likely related to the dangling-bond density and/or binding energy of the surface.

It has been shown that a plane rotated 10° or more from the (100) plane is suitable for growth. Moreover, *c*-axis rotation planes are better than *b*-axis rotation planes because they have about a 40% larger growth rate. Studying a low-index plane is considered convenient for initial research. Accordingly, we think the (010) plane is the most suitable plane for β-Ga_2O_3 MBE growth.

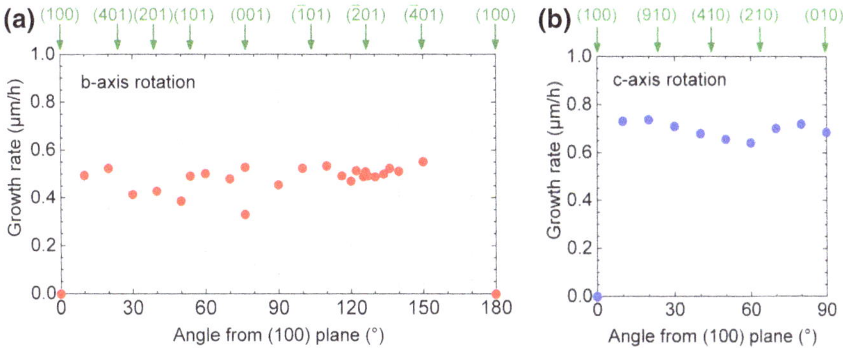

Fig. 7.3 Growth rate of MBE-grown β-Ga_2O_3 homoepitaxial films as a function of surface orientation on *b*-axis rotation planes (**a**) and *c*-axis rotation planes (**b**)

7.2.3 *Orientation Dependence of Surface Morphology*

Next, let us investigate the relationship between the surface morphology and surface orientation. The RMS surface roughness, as estimated from AFM images, is shown in Fig. 7.4. As shown in Fig. 7.4a, in the case of the *b*-axis rotation plane, a film with a smooth surface can be grown by using 30°, 76.3–110°, or 150° rotation planes from the (100) plane. In contrast, all the films grown on the *c*-axis rotation planes have smooth surfaces (Fig. 7.4b). As in the case of the surface orientation dependence of the growth rate, the *c*-axis rotation planes also seem suitable for β-Ga_2O_3 MBE growth from the viewpoint of low surface roughness.

Figure 7.5 shows surface AFM images of the films grown on *b*-axis rotation planes. The planes have small RMS values, and as shown in Fig. 7.4a, they show a clear step-terrace structure. On the other hand, many large hillocks occur on the

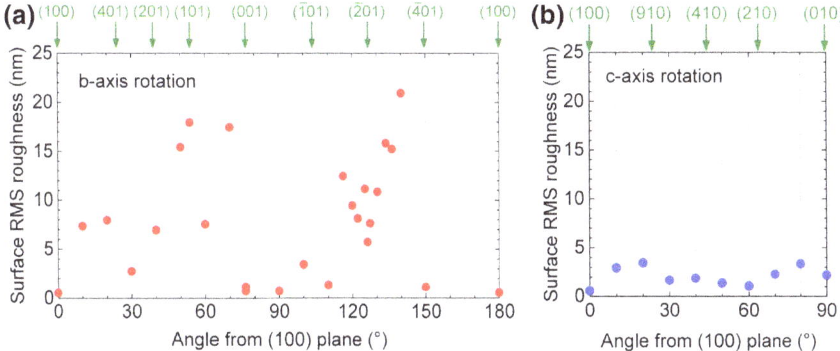

Fig. 7.4 Surface RMS roughness of MBE-grown β-Ga_2O_3 homoepitaxial films as a function of surface orientation on *b*-axis rotation planes (**a**) and *c*-axis rotation planes (**b**)

films grown on the 60, 70, 120, or 126.2° rotation planes. Since the shapes of the hillocks are irregular, there is a chance that crystal defects are generated instead of step bunching. The next section considers their origin from the viewpoint of crystal quality.

7.2.4 Orientation Dependence of Crystal Quality

To reveal the origin of surface roughening, we can measure 2θ-ω XRD for various surface orientations of films grown on the *b*-axis rotation planes. Figure 7.6 shows one such measurement and the XRD spectra of a (001) plane asymmetric 2θ-ω scan. The incident direction of the X-rays is parallel to the *b*-axis. The right axis of this figure shows the surface orientation (rotation angle from the (100) plane of the samples). The spectra of the 50–70° rotation planes (red lines) include diffractions from the crystal defect at 2θ, about 30.5°, and the $(\bar{4}01)$ plane of the crystal defect is parallel to the (001) plane of the substrate. The spectra of the 120–140° rotation plane (blue lines) include the diffraction from crystal defect at 2θ of about 30.0°, and the (400) plane of the crystal defect is parallel to the (001) plane of the substrate. These rotation angles correspond to the angles that give a rough surface. Therefore, the origin of surface roughening is crystal defects.

The low-index plane near the 120–140° rotation planes is $(\bar{2}01)$, and the low-index plane near 50–70° rotation planes is (101). These planes are potentially

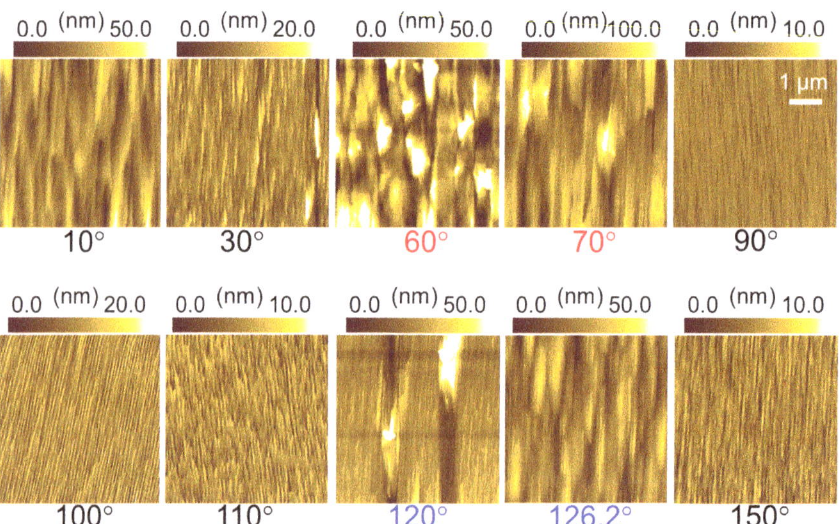

Fig. 7.5 Surface AFM images of β-Ga_2O_3 homoepitaxial films grown on *b*-axis rotation planes

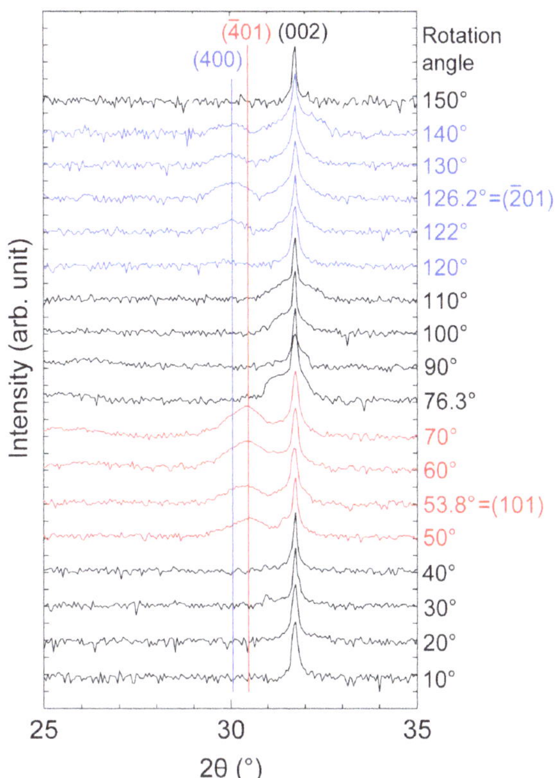

Fig. 7.6 XRD spectra of (001) plane asymmetric 2θ-ω scan for β-Ga_2O_3 homoepitaxial films grown on *b*-axis rotation planes

responsible for the crystal defects. The next section discusses the mechanism by which crystal defects occur in relation to the ($\bar{2}01$) and (101) plane.

7.2.4.1 Crystal Defect Generation Mechanism on the ($\bar{2}01$) Plane

Figure 7.7 shows the orientation relationship between the crystal defect and substrate, as determined from XRD measurements. The left side of the figure shows the normal crystal structure of β-Ga_2O_3 substrate in which the ($\bar{2}01$) plane is horizontal. The right side shows the crystal structure of a crystal defect in the normal crystal. The green dotted line indicates the ($\bar{2}01$) plane of the substrate. The blue and red circles correspond to Ga and O atoms. The black lines show the unit cell of β-Ga_2O_3. It can be seen that the ($\bar{2}01$) planes of the substrate and crystal defect are parallel, and each crystal is rotated 180° on the ($\bar{2}01$) plane. In other words, the crystal defects on the ($\bar{2}01$) plane are ($\bar{2}01$)-plane twin defects.

Transmission electron microscopy (TEM) can be used to observed atomic arrangement and thereby determine the orientation relationship. Figure 7.8 shows (a) low- and (b) high-magnification cross-sectional TEM images of the epitaxial

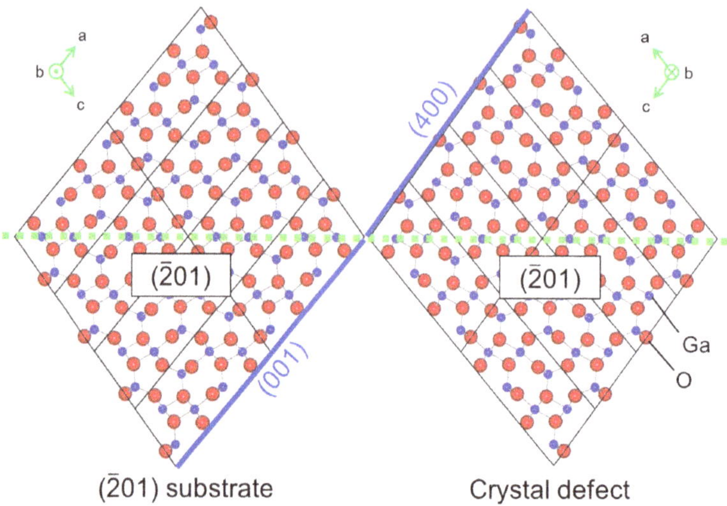

Fig. 7.7 Schematic illustration of crystal structure of β-Ga_2O_3 ($\bar{2}01$) plane and crystal defect on *b*-axis 120–140° rotation planes

film grown on the ($\bar{2}01$) plane, while Fig. 7.8c shows the diffraction pattern of the sample. Note that the electron beam spot diameter is about 80 nm, so the diffraction pattern includes contributions from both the normal crystal and the defect. The mirror indices are estimated from the distance between the diffraction patterns and are indicated in Fig. 7.8c.

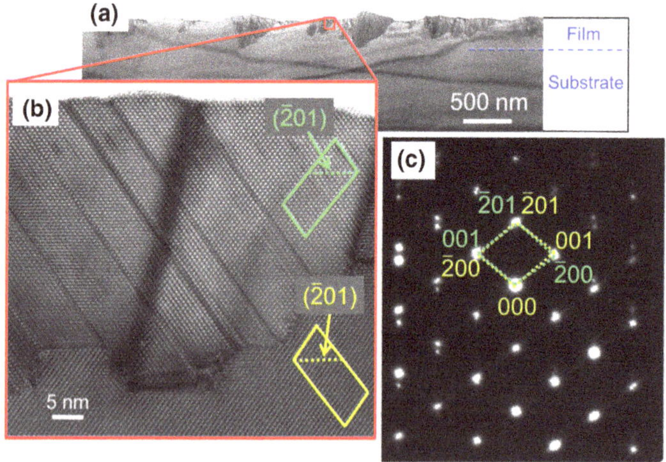

Fig. 7.8 **a** Low- and **b** high-magnification cross-sectional TEM images of epitaxial film grown on (($\bar{2}01$)) plane. **c** Diffraction pattern of that sample

As can be seen in Fig. 7.8a, the epitaxial film includes regions with different contrasts. The high-resolution image of this region in Fig. 7.8b indicates that the crystal orientation is different from the orientation of the substrate, and the crystal defect is inverted on the $(\bar{2}01)$ plane. The diffraction patterns from normal and defect regions mostly overlap (Fig. 7.8c). This means the $(\bar{2}01)$ planes of the normal and defect crystals are almost parallel. These TEM results also suggest that the crystal defects are $(\bar{2}01)$-plane twin defects.

O atoms are aligned in an almost hexagonal closest packed arrangement on the $(\bar{2}01)$ plane. Consequently, the O layer has high in-plane symmetry. In contrast, the arrangement of Ga atoms on the O layer is not closest packed. Therefore, it is reasonable to suppose that misalignments of Ga atoms easily occur on the highly symmetric O layer, and thereby $(\bar{2}01)$-plane twin defects are generated.

7.2.4.2 Crystal Defect Generation on the (101) Plane

A similar method as used for examining the $(\bar{2}01)$-plane twin defects can be used to examine how crystal defects are generated at a rotation angle of 50–70° around the (101) plane.

Figure 7.9 shows the orientation relationship between the crystal defect and substrate, as estimated using XRD. The left side of the figure shows the crystal structure of β-Ga_2O_3 substrate in which the (101) planes are horizontal. The right side shows the crystal structure of the crystal defect inside the normal crystal. The green dotted line, black lines, and blue and red circles have the same meanings as in

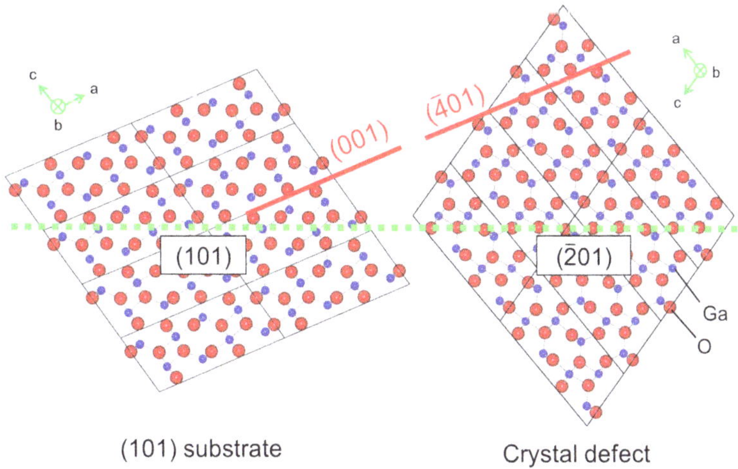

Fig. 7.9 Schematic illustration of crystal structure of β-Ga_2O_3 (101) plane and crystal defect on *b*-axis 50–70° rotation planes

Fig. 7.7. This figure shows that the (101) plane of the substrate and the $(\bar{2}01)$ planes of the crystal defect are parallel.

Figure 7.10 shows (a) low- and (b) high-magnification cross-sectional TEM images of the film grown on the 60° rotation plane. The contrast of these films is different from that of the substrate. This means that almost all of the film is composed of crystal defects. The high-resolution image of the interface between the defect and normal crystal can be used to estimate the mirror indices from the plane distance; in this case, the (101) plane of the substrate and $(\bar{2}01)$ plane of the defect are parallel.

In Fig. 7.10a, the interface between the defect and normal crystal is almost parallel to the (101) plane. Therefore, the defects probably grew on the (101) plane. The layer of O atoms of the (101) plane is hexagonal closest packed, so it has high in-plane symmetry. On the other hand, the Ga atoms on the (101) plane are not hexagonal closest packed. These features are similar to those of the $(\bar{2}01)$ plane. Moreover, the positions of half of the Ga atoms on the (101) plane are almost the same as on the $(\bar{2}01)$ plane. These features indicate that Ga atoms can easily become misaligned on the symmetric O layer, and the (101) plane changes to the $(\bar{2}01)$ plane.

These defects are stacking faults. Stacking faults usually have a bad influence on device characteristics. Moreover, as discussed above, $(\bar{2}01)$ plane twin defects and the $(\bar{2}01)$ growth on the (101) plane defects are easily generated by misalignment of a few Ga atoms. Therefore, it is difficult to prevent these defects, and areas around $(\bar{2}01)$ and (101) planes are not suitable for growth.

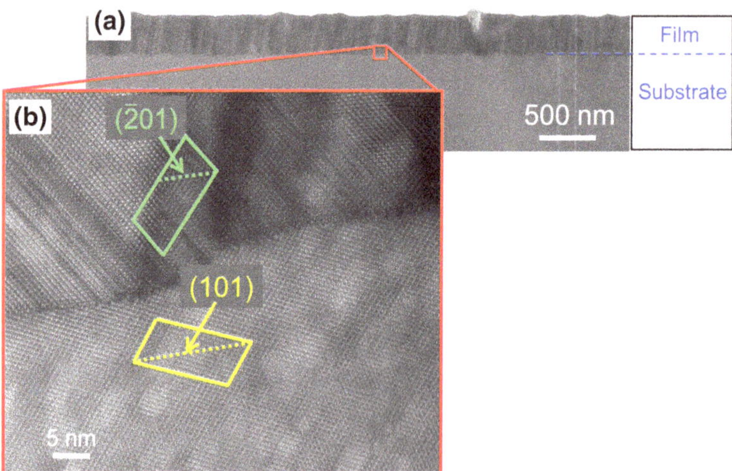

Fig. 7.10 **a** Low- and **b** high-magnification cross-sectional TEM images of epitaxial film grown on *b*-axis 60° rotation plane

7.3 β-Ga_2O_3 Ozone MBE Growth on (010) Plane

The above investigation of the dependence of crystal quality on the surface orientation revealed that the (010) plane is a suitable orientation for β-Ga_2O_3 MBE growth. This section describes how the growth conditions can be optimized to realize device-quality β-Ga_2O_3 single-crystal films on the (010) plane.

7.3.1 Experimental Methodology

Sections 7.2.1.1 and 7.2.1.2 describe the EFG of Sn-doped n-type β-Ga_2O_3 (010) substrates and the method of loading sample in the MBE chamber. Here, T_g is set at 560, 650, and 750 °C, and the growth duration at 30 min. The Ga flux is fixed at 2.1×10^{-4} Pa. The VI/III ratio is varied by changing the ozone gas flow rate in the range of 0.5–5.0 sccm. To vary the VI/III ratio over a wider range than would be possible by changing on the ozone gas flow rate, the distance between the ozone gas pipe head and the substrate (hereinafter called the "pipe distance") is also varied (15 or 30 mm).

The effective ozone concentration at the substrate surface (C_{ozone}) depends on the ozone gas flow rate and pipe distance. Ozone gas is injected from the pipe head with a certain spread angle. Here, suppose that the sprayed area (S_{std}) of the ozone gas at a pipe distance of 30 mm is 1 arb. unit. Moreover, suppose that the ozone concentration per unit area (C_{std}) at the center of the sprayed region is 1 arb. unit. If we change the pipe distance from 30 mm to 15 mm, S_{std} changes from 1 to 1/4. Thus, C_{std} changes from 1 to 4. Accordingly, C_{ozone} can be calculated using the following equation.

$$\begin{aligned}&\text{Ozone gas flow rate } (0.5-5.0\text{ sccm}) \times C_{std}(1\text{ or }4\text{ arb. unit})\\&\quad = C_{ozone}(0.5-20\text{ arb. unit})\end{aligned}$$

Therefore, C_{ozone} can be varied over a range of 40 times in magnitude by changing the ozone gas flow rate and pipe distance.

7.3.2 Growth Rate as a Function of VI/III Ratio

Here, we will define the VI/III ratio as the ratio of adatoms instead of the material supply ratio. As it is almost impossible to measure the adatom ratio directly, we estimate it from the relationship between the material supply ratio and growth rate.

Figure 7.11 plots the growth rate as a function of C_{ozone} at a constant Ga flux. Blue circles, red squares, and green triangles indicate the growth rates at T_g of 560, 650, and 750 °C, respectively. The dotted lines are guides for the eye.

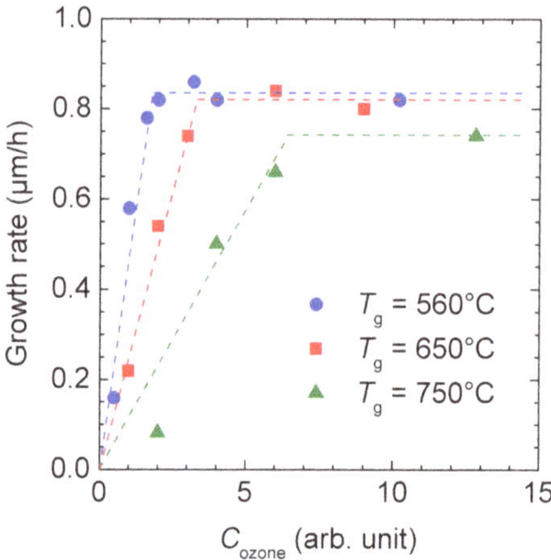

Fig. 7.11 Growth rate of β-Ga_2O_3 homoepitaxial films grown on the (010) plane as a function of C_{ozone} for various T_g

The growth rate increases with C_{ozone} for all three values of T_g and saturates at a different value of C_{ozone} for each T_g. The adatom ratio on the growth surface is stoichiometric at the saturation points. The left side of each saturation point is the Ga-rich range, while the right side is the O-rich range. Note that, in the case of Ga_2O_3, the VI/III ratio is 2:3 at those saturated points. However, we say that the VI/III ratio is 1:1 at those points for simplicity.

As shown in Fig. 7.11, the growth rate dramatically decreases with increasing T_g under Ga-rich conditions. There are several possible reasons for this. One is that the re-evaporation ratio of supplied ozone increases with T_g. On the other hand, the growth rate also changes with T_g under O-rich conditions. The growth rates at T_g of 560 and 650 °C are similar, while the rate at T_g of 750 °C is lower. We think the origin of this phenomenon is re-evaporation of supplied Ga atoms.

7.3.3 Growth Temperature and VI/III Ratio Dependence of Surface Morphology of Unintentionally Doped β-Ga_2O_3 Homoepitaxial Films

Here, we examine the growth temperature and VI/III ratio dependences of the surface morphology of unintentionally doped β-Ga_2O_3 homoepitaxial films. Figure 7.12 shows 1-μm-square AFM images of film surfaces grown under various growth conditions. The vertical and horizontal axes are T_g and the VI/III ratio, respectively.

The images outlined in blue in Fig. 7.12 are of the films fabricated with a VI/III ratio of around 1 and T_g of 560 °C; many pits are visible on their surfaces. In contrast, the images outlined in green show smooth surfaces that were obtained using conditions slightly different from those that caused pitting. The images

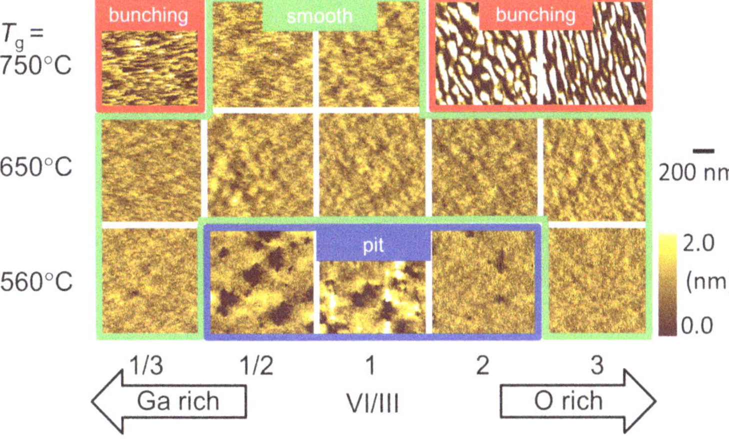

Fig. 7.12 Growth temperature or VI/III ratio dependence of surface morphology of unintentionally doped β-Ga_2O_3 homoepitaxial films grown on (010) plane

outlined in red indicate that step bunching occurred when the growth conditions were varied even more.

The above dependence on growth conditions can be explained by using the surface diffusion model of adatoms. In the case of ozone-MBE growth of β-Ga_2O_3, surface diffusion of adatoms increases with T_g and under Ga- or O-rich conditions. The schematic illustration of the model in Fig. 7.13a indicates that at low T_g, surface diffusion of adatoms decreases because the thermal energy of the adatoms is low. Therefore, the surface becomes rough. In contrast, the smooth surface obtained at high T_g is due to the high thermal energy. At low T_g such as 560 °C and with a VI/III ratio of around 1, adatoms are captured near the first contact area of the supplied materials because the dangling bonds of the growth surface are those of Ga or O atoms in a similar ratio (Fig. 7.13b). Under such conditions, few adatoms diffuse over the surface. On the other hand, under Ga-rich conditions, the growth surface is terminated by Ga atoms, and surface diffusion becomes large because the supplied Ga atoms migrate on the surface to reach O dangling bonds. The same is true for O-rich conditions. Moreover, step bunching is likely induced by too much surface diffusion under Ga-rich or O-rich conditions at high T_g. Moreover, this investigation shows that unintentionally doped films with atomically smooth surfaces are produced by optimizing the surface diffusion of adatoms.

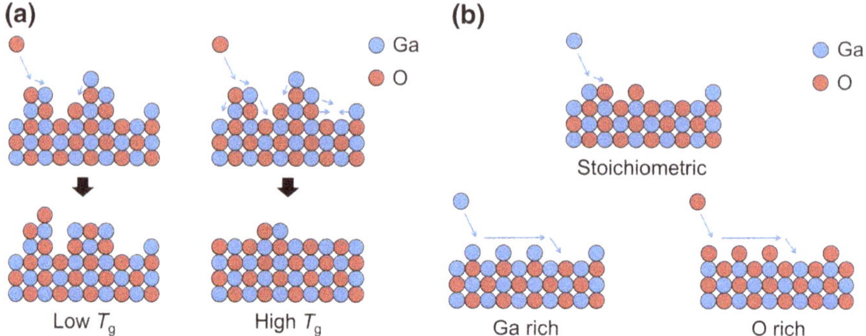

Fig. 7.13 Schematic illustrations of **a** growth temperature and **b** VI/III ratio dependence of surface diffusion of adatoms

7.3.4 Growth Temperature and VI/III Ratio Dependence of Surface Morphology of Sn-Doped β-Ga_2O_3 Homoepitaxial Films

The experiments described in the preceding sections show a clear relationship between growth conditions and the surface morphology of unintentionally doped β-Ga_2O_3 homoepitaxial films. This section describes the growth of Sn-doped films with smooth surfaces under those conditions (e.g., T_g of 560 °C and VI/III of 3). As the resulting surfaces are very rough, we investigate the VI/III ratio dependence of the surface morphology of Sn-doped films.

Figure 7.14 shows surface AFM images of unintentionally doped and Sn-doped films grown on Sn-doped n-type β-Ga_2O_3 (010) substrate at a T_g of 560 °C. The horizontal axis is the VI/III ratio. The Ga-(O-) rich condition is to the left(right) of the VI/III ratio of 1. The value under each image indicates the surface RMS roughness. The unintentionally doped films grown with VI/III ratios over 2 or under 1/3 have smooth surfaces. In contrast, the Sn-doped films grown with a small VI/III ratio have very rough surfaces. However, the roughness decreases as the ratio increases, and a smooth surface similar to the unintentionally doped films is obtained at a VI/III ratio of about 10.

To determine why the suitable conditions vary between the Sn-doped and unintentionally doped cases, Fig. 7.15 shows Sn concentration depth profiles as measured by SIMS for the films grown with a VI/III ratio of 1, 2, and 10. The horizontal and vertical axes indicate the depth and Sn concentration, respectively. The interface between the substrate and epitaxial films is at a depth of about 350 nm. The profile for the films grown with a VI/III ratio of 10 is almost flat. In contrast, the films grown with a VI/III ratio of 1 or 2 are affected by a large doping delay in the initial stage of growth. The AFM and SIMS measurements suggest the following mechanism of surface roughening when the VI/III ratio is small (see also Fig. 7.16).

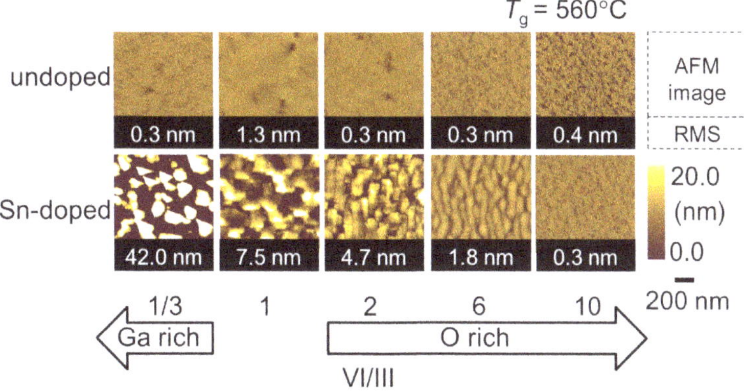

Fig. 7.14 VI/III ratio dependence of surface morphology of unintentionally doped and Sn-doped β-Ga_2O_3 homoepitaxial films grown on (010) plane

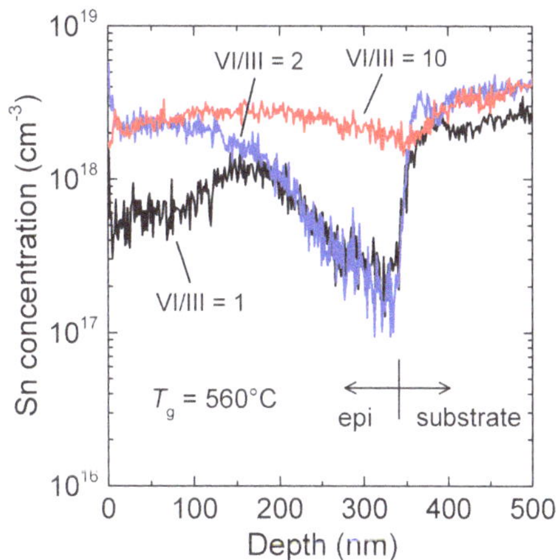

Fig. 7.15 Sn concentration depth profiles of Sn-doped β-Ga_2O_3 homoepitaxial films grown with different VI/III ratios

(1) SnO_2, Ga, and O_3 are supplied to the growth surface.
(2) SnO_2 decomposes into Sn and O_2.
(3) Sn segregates on the growth surface.
(4) The segregated Sn retards the growth of Ga_2O_3.
(5) A rough surface forms.

Thus, O-rich conditions suppress decomposition of SnO_2, i.e., reduce segregation of Sn and enable a smooth film surface with a uniform depth profile to be obtained. Sn segregation in MBE growth has also been reported for GaAs growth [5]. Therefore, this seems to be an intrinsic characteristic of Sn. This analysis

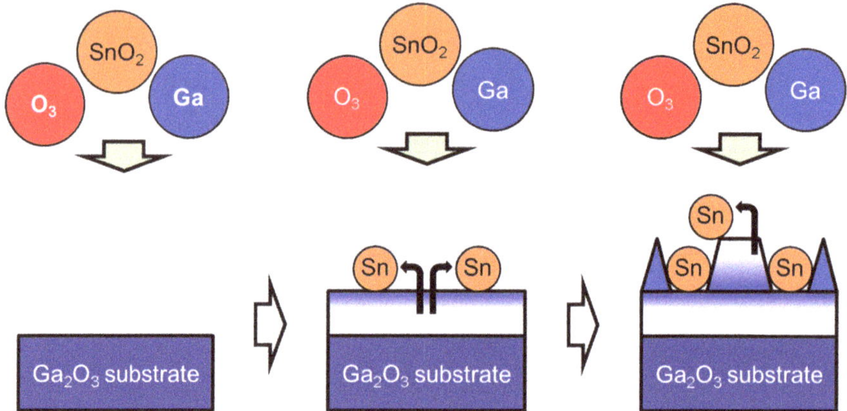

Fig. 7.16 Schematic illustration of surface roughening mechanism in Sn segregation model

presented in this section indicates that Sn-doped β-Ga_2O_3 films with smooth surfaces and uniform depth profiles can be obtained by using a large VI/III ratio (over 10).

7.3.5 *Sn Doping of β-Ga_2O_3 Homoepitaxial Films*

7.3.5.1 Sn-Dopant Control

Finally, we investigate the controllability of doping by examining the results of an experiment in which β-Ga_2O_3 films were doped with various concentrations of Sn. In the experiment, Sn-doped β-Ga_2O_3 (010) substrates were fabricated with the floating-zone method. T_g was 540 or 570 °C. The BEP of Ga was 2.1 × 10^{-4} Pa. The ozone gas flow rate was 5.0 sccm, and the pipe distance was 15 mm. The VI/III ratio was about 10. The growth duration was 30 min, yielding a film thickness of about 300–350 nm. The SnO_2 cell temperature was varied in the range of 650–735 °C. The effective donor concentration ($N_d - N_a$) was measured by ECV.

Figure 7.17 shows the $N_d - N_a$ of the grown films and the vapor pressure curve of SnO_2 [6] as a function of the temperature of the SnO_2 K-cell. The red squares and blue circles indicate the $N_d - N_a$ of the grown films at T_g of 540 and 570 °C, respectively. The black dotted line shows the vapor pressure curve of SnO_2. $N_d - N_a$ and the vapor pressure curve show good agreement. This demonstrates that the Sn-dopant concentration can be precisely controlled in a wide range (10^{16}–10^{18} cm^{-3}) by changing the temperature of the SnO_2 K-cell.

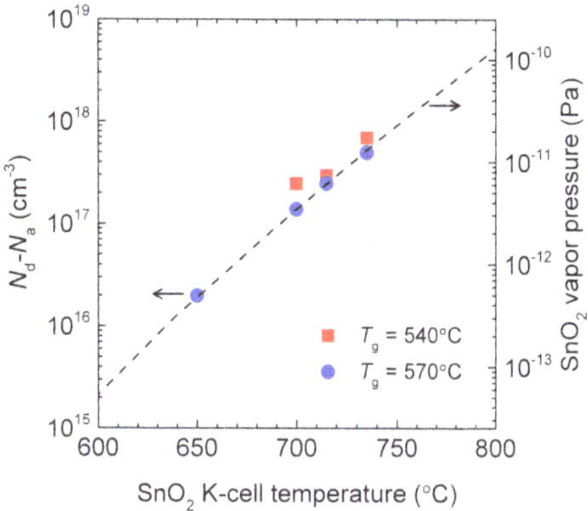

Fig. 7.17 $N_d - N_a$ of Sn-doped β-Ga_2O_3 homoepitaxial films or SnO_2 vapor pressure as a function of SnO_2 K-cell temperature

7.3.5.2 Carrier Concentration and Electron Mobility of Sn-Doped β-Ga_2O_3 Films

Hall measurements were made to investigate the relationship between the electron mobility and carrier concentration of Sn-doped β-Ga_2O_3 films. Mg-doped semi-insulating β-Ga_2O_3 substrates fabricated by the floating-zone method were used. The substrates were about 5–10 mm in diameter. The Sn-doped β-Ga_2O_3 homoepitaxial film was grown by ozone MBE. The film thickness was about 700 nm. Si-doped β-Ga_2O_3 bulk substrates were also measured for reference. The van der Pauw method was used to make the Hall measurements. Ohmic electrodes were prepared by using indium metal balls.

Figure 7.18 shows the electron mobility as a function of the carrier concentration of the Sn-doped β-Ga_2O_3 films (blue circles) and Si-doped β-Ga_2O_3 bulk substrates (red squares). In both cases, the electron mobility apparently monotonically increases with decreasing carrier concentration. If we were to use these films as the

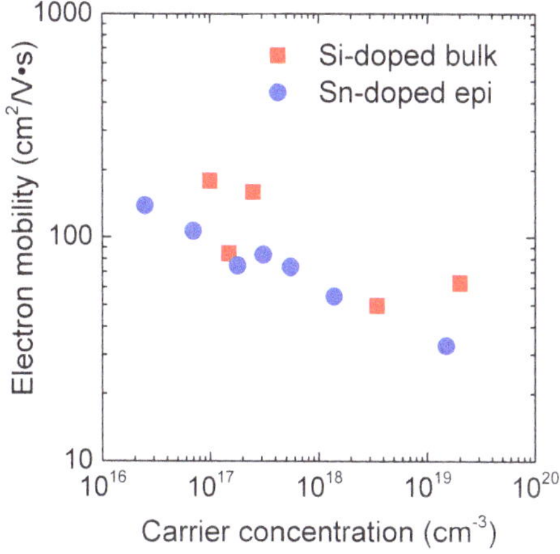

Fig. 7.18 Electron mobility of Sn-doped β-Ga_2O_3 homoepitaxial films or Si-doped bulk substrates as a function of carrier concentration at room temperature

blocking layer of a power device, for instance, the carrier concentration (to be accurate, donor concentration) would about 10^{16} cm^{-3}. Hence, the electron mobility of such films would be about 200–300 cm^2/Vs.

7.4 Summary

This chapter described homoepitaxial growth of β-Ga_2O_3 by ozone MBE. First, the dependence of the crystal characteristics on the surface orientation was investigated. The reported experiments reveal that the (010) plane is especially suitable for β-Ga_2O_3 MBE growth. Moreover, it was found that stacking faults are easily generated on the (101) or $(\bar{2}01)$ plane. Next, we described how to optimize the growth conditions for the (010) plane to achieve the device-quality β-Ga_2O_3 films. Unintentionally doped films with atomically smooth surfaces were obtained by optimizing the surface diffusion of adatoms. In the case of Sn-doped films, serious surface roughening occurs at small VI/III ratios because of Sn segregation. On the other hand, Sn-doped films with smooth surfaces and uniform doping depth profiles can be achieved under very O-rich conditions, such as a VI/III ratio of 10 or more. Finally, it was shown that the Sn-dopant concentration can be precisely controlled over a wide range (10^{16}–10^{18} cm^{-3}) by changing the temperature of the SnO_2 K-cell.

References

1. K. Sasaki, A. Kuramta, T. Masui, E.G. Vìllora, K. Shimamura, S. Yamakoshi, Appl. Phys. Express **5**, 035502 (2012)
2. D.K. Schroder, *Semiconductor Material and Device Characterization* (Wiley, New Jersey, 2006) 3rd ed., p. 77
3. V.M. Bermudez Chem. Phys. **323**, 193 (2012)
4. G. Wagner, M. Baldini, D. Gogova, M. Schmidbauer, R. Schewski, M. Albrecht, Z. Galazka, D. Klimm, R. Fornari, Phys. Status Solidi A **211**, 27 (2014)
5. A.Y. Cho, J. Appl. Phys. **46**, 1733 (1975)
6. R.H. Lamoreaux, D.L. Hildenbrand, L. Brewer, J. Phys. Chem. Ref. Data **16**, 419 (1987)

Chapter 8
Metalorganic Chemical Vapor Deposition 1

Homoepitaxial and Heteroepitaxial Growth of Ga_2O_3 and Related Alloys

Fikadu Alema and Andrei Osinsky

Abstract This chapter is devoted to the growth of Ga_2O_3 and its alloys by metalorganic chemical vapor deposition (MOCVD) or, equivalently, metalorganic vapor phase epitaxy (MOVPE). MOCVD is a standard epitaxial growth technique used for nitride, III–V, and oxide-based power devices as well as LEDs and laser diodes. It would, therefore, seem that MOCVD would be the most appropriate growth method to accelerate the development and commercialization of Ga_2O_3. However, molecular beam epitaxy (MBE) and halide vapor phase epitaxy (HVPE) were commonly used in the earlier growth studies of Ga_2O_3 epitaxial films. Fortunately, the broad range of knowledge available on the hardware and control systems of the MOCVD tool for the growth of device quality epitaxial films makes it easily adaptable to the growth of epitaxial Ga_2O_3. Device quality films grown at ~ 10 µm/h were demonstrated [1, 2], evidencing the ability of MOCVD to achieve the high throughput Ga_2O_3 epitaxial layer growth needed for high voltage power device and deep ultraviolet solar-blind photodetector commercial applications. This chapter discusses the following topics: the selection of suitable metalorganic precursors and oxygen sources used for the growth of Ga_2O_3 and $(Al, Ga)_2O_3$ alloys; the need for the careful design of the MOCVD reactors, homoepitaxial and heteroepitaxial growth on c-plane sapphire and native Ga_2O_3 substrates, donor and acceptor doping; and the origin and methods for mitigating or reducing unintentional impurities.

F. Alema (✉) · A. Osinsky
Agnitron Technology, Inc., Chanhassen, MN 55317, USA
e-mail: fikadu.alema@agnitron.com

A. Osinsky
e-mail: andrei.osinsky@agnitron.com

M. Higashiwaki and S. Fujita (eds.), *Gallium Oxide*, Springer Series in Materials Science 293, https://doi.org/10.1007/978-3-030-37153-1_8

8.1 Metalorganic Precursors and Oxygen Sources

The success of the MOCVD method for the growth of Ga_2O_3 thin films depends on the availability of suitable volatile metalorganic precursors. For Ga, traditional gallium precursors, including trimethylgallium (TMGa) and triethylgallium (TEGa), have been used [1, 3–5]. At room temperature, the physical state of these precursors is liquid, and they are typically delivered using stainless steel bubblers. Sufficient and reproducible concentration of these precursors can be introduced into an MOCVD reactor by controlling the carrier gas (e.g., Argon) flow rate, bubbler pressure, and temperature [6]. Another Ga precursor that has been used for the growth of Ga_2O_3 thin films is Ga $(DPM)_3$ (DPM = dipivaloylmethanate) [1]. However, unlike the TEGa and TMGa precursors, $Ga(DPM)_3$ is a solid precursor which requires it to be sublimated ($\sim$155 °C) to generate a Ga vapor. Because of the high source temperature, the inlet and outlet valves, gas delivery lines, and showerheads are heated $\sim$40–120 °C above the sublimation temperature to prevent metalorganic condensation and clogging. This makes it very challenging to use Ga $(DPM)_3$ as a precursor for the growth of Ga_2O_3.

One of the long-standing obstacles limiting the growth of Ga_2O_3 by MOCVD is the extremely slow growth rate (<0.5 μm/h) [5]. This is mainly due to problems with growth kinetics such as undesirable gas phase nucleation that depletes the precursors before they reach the substrate surface. The depletion of the precursors can be reduced by separately injecting the metalorganic and oxygen precursors in close proximity to the substrates. Recently, high growth rates were achieved by using a close injection showerhead (CIS) MOCVD reactor that separately injects the metalorganic precursors and oxygen source, preventing the premature oxidation of the metalorganic precursors. In particular, high growth rates for β-Ga_2O_3 thin films using TMGa, TEGa, and Ga $(DPM)_3$ metalorganic precursors and pure oxygen have been realized [1, 7]. Figure 8.1 shows the growth rate of $(\bar{2}01)$-oriented β-Ga_2O_3 thin films grown on c-plane sapphire substrates using a CIS-MOCVD (substrate-to-showerhead separation of $\sim$1.0 cm) [1, 8, 9]. In Fig. 8.1a, the growth rate dependence on substrate temperature (T_S) for films grown using Ga $(DPM)_3$, TEGa and TMGa metalorganic precursors is shown. The growth rate of the films grown using Ga $(DPM)_3$ increases with the increase in T_S, where $\sim$5 μm/h was obtained at 900 °C. The precise cause of the increase in the growth rate with T_S is unclear and requires additional investigation as there is only a limited amount of available research on the use of the Ga $(DPM)_3$ precursor for the MOCVD growth of Ga_2O_3. However, deducing from the observed trend, the increase in the growth rate with T_S may be attributed to an increase in the efficiency of Ga $(DPM)_3$ pyrolysis, making it readily able to react with the oxygen on the substrate. A similar increase in growth rate with T_S had been reported for an MOCVD-grown Bi–Sr–Ca–Cu–O thin film using triphenylbismuth, $Sr(DPM)_2$, Ca $(DPM)_2$, and $Cu(DPM)_2$ MO sources. The increase was attributed to the slow thermal decomposition of the individual sources on the substrate rather than in the gas phase [10], which could also likely be the case here.

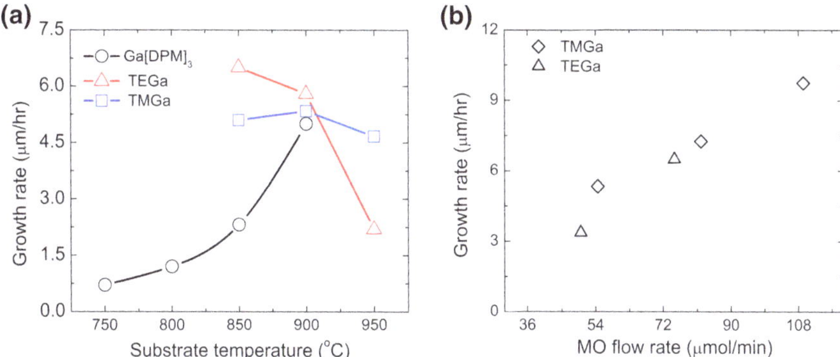

Fig. 8.1 Growth rate versus **a** T_S and **b** metalorganic (MO) flow rate for β-Ga_2O_3 films grown using CIS-MOCVD on c-plane sapphire substrates at 40 Torr. The Ga $(DPM)_3$ was sublimated at 155 °C; the TEGa and TMGa flow rates were 75.0 and 55 μmol/min, respectively, for the corresponding bubbler temperatures of 22 and 5 °C. In (**b**), the films were grown using TEGa and TMGa at 850 °C and 900 °C, respectively [1, 8]

For TEGa, the trend of the film growth rate as a function of T_S is in complete contrast to that for films grown using Ga $(DPM)_3$. With an increase of T_S from 850 to 950 °C, the growth rate of the film quickly drops from 6.5 to 2.2 μm/h. Such a decrease in the growth rate is likely due to depletion of the TEGa source by parasitic reactions given its low decomposition temperature of ~300 °C [11]. Unlike the Ga $(DPM)_3$ and TEGa precursors, the effect of T_S on the growth rate of films grown using TMGa is negligible. The reason for this is not entirely clear, but the need for an elevated temperature (>500 °C) to effectively crack the TMGa may lead to a supersaturation of Ga species near the substrate for each T_S considered above, which is likely the case for CIS-MOCVD reactors. Figure 8.1b presents the growth rate as a function of the concentration of the TEGa and TMGa metalorganic precursors. With an increase in the molar flow rate, the growth rate increases linearly, reaching ~10.0 μm/h at a TMGa molar flow of ~110 μmol/min [1, 2, 8].

In MOCVD, the only reaction that plays a significant role in the growth of the film is the one that occurs on the surface of the substrates. Since the pure oxygen precursor is very reactive with the metalorganic precursors, unintended gas phase reactions are unavoidable. This problem can be reduced to some degree by adjusting the growth conditions. For example, growing the films at a reduced chamber pressure and low oxygen flow rate can decrease the parasitic reactions [12]. However, to significantly reduce the depletion of the precursors resulting from pre-reactions, using oxygen sources that are less reactive with the metalorganic precursors in the gas phase is crucial. This, combined with a careful design of the MOCVD system that includes the separate injection of the metalorganic and oxygen precursors at close proximity to the substrate, mitigates the challenges created by gas phase reactions and, thus, increases the growth rate of the films.

Alternative oxygen sources that can be potentially used in the growth of Ga_2O_3 include water vapor, tert-butanol, isopropanol, and nitrous oxide (laughing gas). Some of these oxidants have been successfully employed in the growth of similar oxide materials, including CdO, ZnO, and MgZnO [13–15]. The effects of pure oxygen, water vapor, and tert-butanol on the growth of Ga_2O_3 thin films were investigated by virtual reactor MOCVD (STR Inc.) simulation. TMGa was used as the gallium precursor in these simulations. A comparison of the simulation results is presented in Fig. 8.2. In Fig. 8.2a, the simulated growth rate of Ga_2O_3 as a function of growth temperature at 40 Torr chamber pressure is shown. With increasing temperature, the growth rate decreases for the pure oxygen source, an indication of increasing premature reactions which depletes the metalorganic source. For the H_2O source, the most efficient reaction occurs at ~740 °C, and once the growth temperature exceeds this value, the growth rate declines in the same manner as for pure oxygen. On the other hand, the tert-butanol source shows a strikingly different result in which the growth rate increases with increasing substrate temperature. The gas phase reactions that appear to dominate at the high temperature ranges for O_2 and H_2O play a lesser role for the tert-butanol source, resulting in a faster growth rate. Figure 8.2b presents the growth rate of Ga_2O_3 as a function of growth pressure while keeping the substrate temperature at 800 °C. A narrow operating pressure range is observed for O_2 and H_2O. However, for tert-butanol the growth rate increases with chamber pressure. The simulation results indicate that tert-butanol is a suitable oxygen source leading to a faster growth rate for Ga_2O_3 thin films. This could be due to the slow gas phase reactions which normally dominate at high temperature and chamber pressure for O_2 and H_2O sources [13, 15]. The other oxygen source that can be used for the growth of Ga_2O_3 is nitrous oxide (N_2O) which has the twofold benefit of being able to grow both pure and nitrogen-doped layers (see details in Sect. 8.3.2) depending on the growth conditions used.

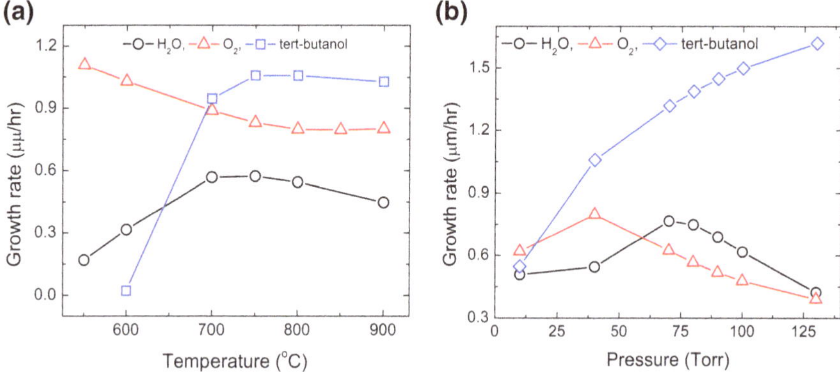

Fig. 8.2 Virtual reactor MOCVD simulation results showing the growth rate dependence of Ga_2O_3 thin films on the substrate temperature (**a**) and chamber pressure (**b**) using TMGa as the gallium precursor, and H_2O (black), O_2 (red) and tert-butanol (C_4H_9OH, blue) as oxidants. Simulation results at 40 Torr and 800 °C are shown in (**a**) and (**b**), respectively

8.2 Homoepitaxial Ga_2O_3 Growth by MOCVD

The availability of native β-Ga_2O_3 substrates with various cleavage planes, including (100), ($\bar{2}$01), (001), and (010) grown by traditional melt growth methods enables the growth of homoepitaxial layers. These substrates are either unintentionally doped (UID), conducting, or semi-insulating. In the UID substrates, Si is the dominant impurity with a typical concentration of $\sim 2 \times 10^{17}$ cm^{-3} [16], making it an n-type material. This background n-type conductivity is compensated by deep acceptors such as Mg and Fe producing semi-insulating substrates [17]. The conducting substrates are n-type doped either with Si or Sn dopants [18, 19]. β-Ga_2O_3 was found to cleave easily on the (100), ($\bar{2}$01),and (001) planes, making large diameter wafer production possible [20, 21]. However, in early growth experiments by MBE, growth on these planes suffered from slow growth rates due to the decomposition of β-Ga_2O_3 at the growth temperature [22]. On the other hand, the non-cleave planes, particularly the (010) plane did not suffer from the same decomposition problem [23], but producing large area substrates from these planes is tricky due to challenges with wafering. For MOCVD growth, β-Ga_2O_3 substrates with (100) and (010) orientations have been widely used.

Ga_2O_3 films can be grown using one of the Ga metalorganic precursors discussed in Sect. (8.1). So far, though, the homoepitaxial layers have been generally grown using TEGa and pure oxygen [3, 4]. The main reason for this is that TEGa produces smooth and continuous Ga_2O_3 layers with low background impurity levels. In MOCVD-grown Ga_2O_3 layers, carbon and hydrogen are the major impurities which incorporate into the growing layers, thereby reducing the carrier mobility. The main source of these impurities is the metalorganic precursor itself. In the TEGa pyrolysis, the by-products are a relatively stable ethylene (C_2H_4) group which flows to the exhaust with limited involvement in the substrate surface reactions, and, therefore, lower incorporation of C and H impurities into the growing layer [24]. This is in contrast to the case for the TMGa precursor which produces highly reactive methyl (CH_3) radicals that lead to the incorporation of C and H impurities into the growing layers [3, 24]. It is also well established that the C contamination in MOCVD-grown GaN [25] and GaAs [26] layers using TEGa is lower than that for TMGa.

Another limitation of TMGa growth of epitaxial β-Ga_2O_3 as compared to TEGa is its incompatibility with the pure oxygen precursor. This was investigated in detail by Wagner et al. [5, 27]. Growing Ga_2O_3 with TMGa and pure oxygen resulted in wires or agglomerates in the epilayers irrespective of the growth parameters. But, by replacing pure oxygen with water vapor, the growth of smooth epitaxial β-Ga_2O_3 was achieved. This was attributed to the formation of a high concentration of hydrogen molecules as a by-product which promotes the layer-by-layer growth of the film [5]. The resulting films are dominated by a high density of stacking faults and twins in which, irrespective of the n-type dopants (Sn or Si) used, no reproducible conductivity was achieved.

8.2.1 MOCVD Growth of β-Ga_2O_3/(100) β-Ga_2O_3

The MOCVD growth of β-Ga_2O_3 on (100)-oriented β-Ga_2O_3 substrates has been almost exclusively investigated by the IKZ group [3, 5, 28, 29]. Prior to the growth of the epilayers on the (100) surface, a thorough substrate surface preparation was required to create a surface suitable for the growth. This surface treatment effectively removes damage from subsurface layers which would otherwise lead to various defects in the grown layer. The surface preparation involves cleaning the substrates using acetone and isopropanol organic solvents followed by annealing at 900–1000 °C in an oxygen atmosphere. Figure 8.3 shows a typical AFM image of a (100) β-Ga_2O_3 substrate that underwent the substrate preparation process. The cleaned substrate shows a terraced surface with step heights of ~0.6 nm, corresponding to half the unit cell along the (100) direction, and terrace widths of ~70 −100 nm [5, 27].

Epitaxial layers grown on the (100) β-Ga_2O_3 substrates suffer from a high density of twin and stacking fault defects [31]. These defects are parallel to the (100) plane and are characterized by a *c*/2 glide reflection twin relation. The intersection of the randomly oriented crystal domains along the c-plane leads to the formation of incoherent twin boundaries that result in dangling bonds. This results in ineffective doping or leads to charge carrier compensation, thereby reducing doping efficiency and the electron mobility of the grown layers. The maximum electron mobility obtained for layers grown on this substrate was ~41 cm^2/Vs [3]. For Sn and Si dopants, a doping concentration >10^{18} cm^{-3} is required to obtain a measurable free carrier concentration [3, 27]. This is shown in Fig. 8.4a by plotting the room temperature Hall measured free carrier concentration as a function of the SIMS measured doping concentration [27]. Because of extended defects in the layers grown on the (100) substrates, an increase in the doping concentration by more than two orders of magnitude led only to a narrow range of change in the free carrier concentration.

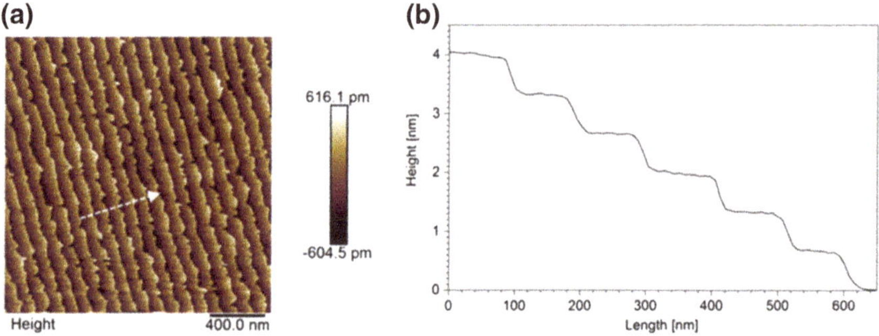

Fig. 8.3 **a** AFM image and **b** line scan of a (100) β-Ga_2O_3 surface after cleaning in organic solvents and followed by a 60 min annealing at 1000 °C in an oxygen environment [27]

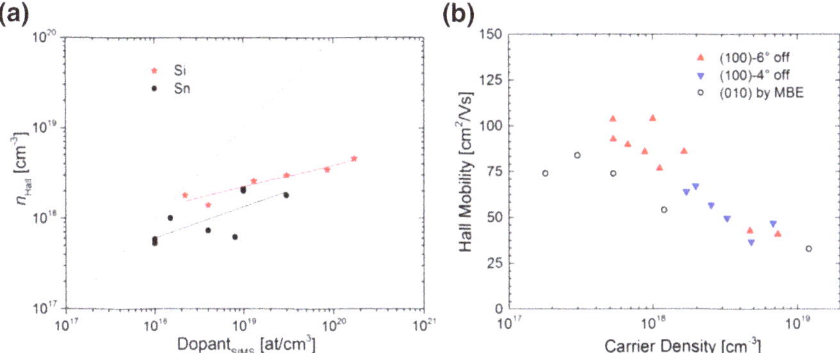

Fig. 8.4 **a** RT Hall measured free carrier concentration as a function of SIMS measured dopant concentration for β-Ga_2O_3/(100) β-Ga_2O_3 layers doped with various concentration of Sn and Si dopants. The solid lines are the linear regression of experimental data, and the dotted line is the limit of 100% electrically active donors without compensation. **b** Hall mobility versus free carrier density for layers grown on (100) β-Ga_2O_3 substrates with 4° and 6° miscut angles (triangles) [27]. The data shown by open circles are values for layers grown on (010) β-Ga_2O_3 substrates by MBE [30]

The extended defects in the layers grown on (100) β-Ga_2O_3 substrates result from the width of the substrate terraces and the diffusion length of adatoms during the growth process. The two parameters determine whether the growth process of the layer is 2D island or step-flow dominated. In a situation where the 2D island growth dominates, stacking faults and twins form by following the double positioning growth mechanisms [27, 31]. Alternatively, if the adatom diffusion length is either comparable to or greater than the terrace width, the grown layer follows the lattice periodicity of the substrates and prevents the formation of stacking faults. The latter is dominated by the step-flow growth mechanisms. Baldini et al. [31] studied the density of the 2D islands and associated twins by growing β-Ga_2O_3 epitaxial layers on misoriented (100) β-Ga_2O_3 substrates with various miscut angles between the surface normal with respect to the [001] direction. Figure 8.5 compares the AFM images of substrate surfaces with miscut angles of 0.1°, 2°, 4°, and 6° along with 0.2-μm-thick homoepitaxial layers grown on the respective substrates. The substrates' surface is characterized by well-defined steps aligned along [010] where the terrace width decreases with the increase in the miscut angle. For substrates with miscut angles <4°, the 2D island growth mechanism dominates, whereas for those substrates with higher miscut angles the step-flow growth mechanism dominates. The density of the stacking faults in the grown epilayers decreases with the increase in the miscut angle of the substrates. A TEM investigation confirmed the decrease in the stacking fault density from $\sim 10^{17}$ cm^{-3} for the layer grown on the 0.1° miscut angle substrate to 0 for the layer grown on the 6° miscut angle substrate [31]. The decrease in the stacking fault density resulted in an improvement of the electron Hall mobility (Fig. 8.4b), with a maximum value of ~ 100 cm^2/Vs at n $\sim 10^{18}$ cm^{-3} for the epitaxial film grown on the 6° miscut angle

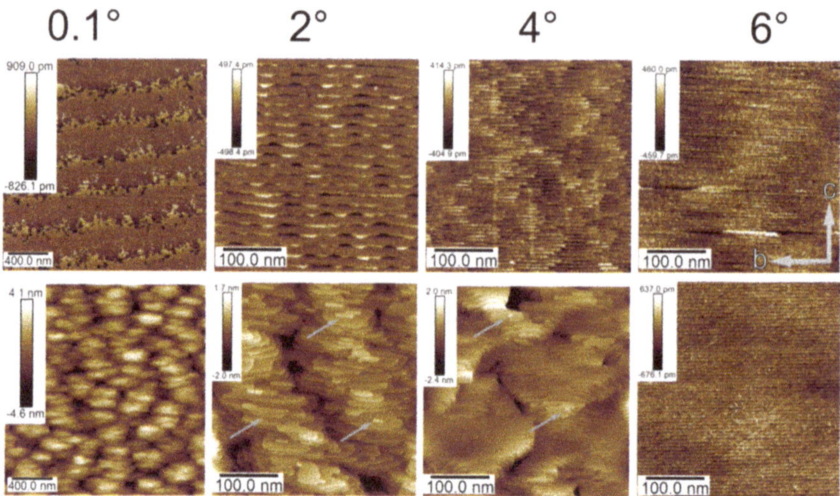

Fig. 8.5 AFM images of substrates with miscut angles of 0.1°, 2°, 4°, and 6° toward *c*-axis (upper row) and epitaxial grown layers on them (lower row). The surface morphology undergoes a transition from 2D island growth to step-flow growth with increasing miscut angle. The arrows indicate the presence of two-dimensional islands on the terraces [31]

substrate [27]. A high-quality β-Ga_2O_3 layer was also grown on (100) β-Ga_2O_3 using indium as a surfactant [32]. By using indium in this way, the concentration of structural defects, including stacking faults and twins, was significantly reduced and the step-flow growth mode was achieved.

8.2.2 MOCVD Growth of β-Ga_2O_3/(010) β-Ga_2O_3

The MOCVD growth of β-Ga_2O_3 on the (010) β-Ga_2O_3 substrate has been widely studied. Unlike the (100) plane, the (010) is a non-cleavage plane so that it is challenging to obtain large area substrates. Typical size for a (010) β-Ga_2O_3 substrate is 10 × 15 mm^2. The thermal conductivity of β-Ga_2O_3 is anisotropic with the largest value being ~30 W/mK in the [010] direction and the lowest being ~11 W/mK in the [100] direction [33]. Though not very significant, the higher thermal conductivity in the [010] direction still offers the advantage of better heat dissipation in power devices fabricated on the (010) β-Ga_2O_3 substrates. Moreover, as opposed to layers grown on (100) β-Ga_2O_3 substrates, there are a reduced number of film/substrate crystalline defects. The growth mechanism on the (010) plane is dominated by step-flow, and the grown layers have a very high degree of crystallinity [27]. The substrate surface preparation involves cleaning the substrates in methanol and acetone organic solvents, followed by piranha etching using the stoichiometry of $1H_2O:1H_2O_2:4H_2SO_4$.

MOCVD growth of high-quality β-Ga_2O_3/(010) β-Ga_2O_3 layers grown at a substantially faster growth rate without compromising the material quality critical for device applications was reported by researchers at Agnitron Technology [8, 9]. In this section, the properties of β-Ga_2O_3 layers grown at 0.5, 1.0, 2.0, and 4 μm/h on (010) β-Ga_2O_3 semi-insulating and conducting substrates are presented. The growth of the films was conducted using TEGa and pure oxygen precursors, and Si doping was achieved by using silane diluted in helium (SiH_4/He). Figure 8.6 presents the 2D AFM images of the grown films. The morphology of the layers is dominated by distinct multi-atomic steps with the root mean square (RMS) roughness estimated from the 2D AFM images in Fig. 8.6a–d of 0.28, 0.30, 0.38, and 0.42 nm, respectively, over a 5 × 5 μm^2 scan area. The UID layer (Fig. 8.6a) has the smoothest surface which is comparable to that of bulk substrates. Doping the layers with Si showed no impact on the surface morphology of the layers (Fig. 8.6b). The effect of growth rate on the surface roughness of the layers is subtle with a RMS roughness value of only 0.42 nm for the layer grown at 4.0 μm/h (Fig. 8.6d). This value shows that the grown layer is smoother than or comparable to the roughness of the β-Ga_2O_3 layers grown by MBE [22, 34] and MOVPE [29], despite a more than 30 times higher growth rate.

The high-resolution XRD (HRXRD) for the (020) peaks of the layers are shown in Fig. 8.7a. The XRD full width at half maximum (FWHM) of the (020) peak for

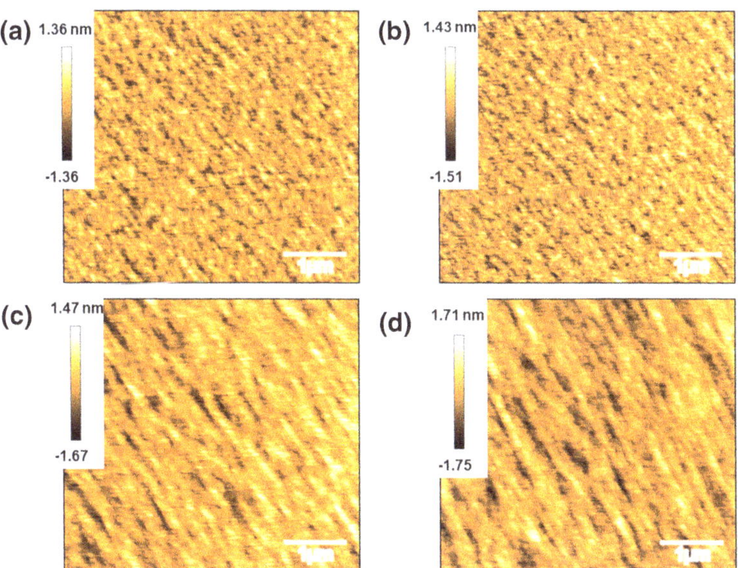

Fig. 8.6 2D AFM images of β-Ga_2O_3 layers grown on semi-insulating (**a** and **b**) and conducting (**c** and **d**) (010) β-Ga_2O_3 substrates using a CIS-MOCVD system. The layers in (**a**), (**b**), (**c**), and (**d**) were grown at 0.5 μm/h, 1.0 μm/h, 2.0 μm/h, and 4.0 μm/h, respectively, by introducing [SiH4/He]/[TEGa] molar ratios of 0, 6.3 × 10^{-3}, 3.1 × 10^{-4}, and 1.7 × 10^{-4}, respectively [8]

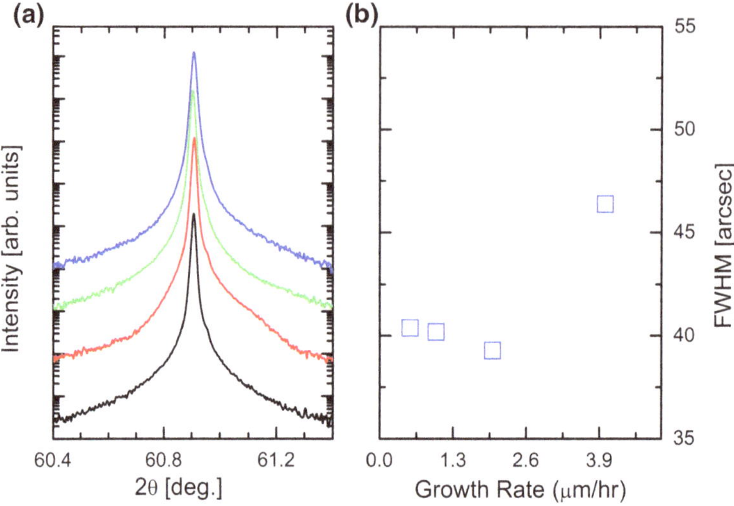

Fig. 8.7 **a** HRXRD scan in the ω-2θ mode of the (020) reflection of UID (black) and Si-doped (red, green and blue) homoepitaxial β-Ga_2O_3 layers grown on (010) β-Ga_2O_3 substrates by CIS-MOCVD. The growth rate and [SiH_4/He]/[TEGa] flux ratio of the layers labeled by black, red, green and blue traces are (0.5 μm/h, 0.0), (1.0 μm/h, 6.3×10^{-4}), (2.0 μm/h, 3.1×10^{-4}), and (4.0 μm/h, 1.7×10^{-4}), respectively. **b** The FWHM versus growth rate of layers [8]

all the films is ~40 arcsec, except for the layer grown at 4 μm/h for which it increases to 46 arcsec. Figure 8.7b shows the FWHM dependence on the growth rate of the layers. Increasing the film's growth rate from 0.5 to 4.0 μm/h had little effect on the crystal quality of the layers. FWHM values for β-Ga_2O_3 homoepitaxial layers grown on (001), (100), and (010) β-Ga_2O_3 substrates by the HVPE, MBE, and LPCVD methods range between 40 and 72 arcsec [35–38], indicating that the FWHM of the homoepitaxial layers grown at 4 μm/h is either comparable to or better than the values reported in the literature using other growth methods. The narrow XRD line width demonstrates that the as-grown films are high-quality epitaxial films with low dislocations densities. The growth of layers at such a high growth rate with XRD line width <50 arcsec and RMS roughness of <0.5 nm, strongly indicates the suitability of the MOCVD method for realizing epitaxial layers with excellent structural and surface quality grown at a high growth rate [1].

8.3 Doping and Background Impurities in an MOCVD-Grown Ga_2O_3 Thin Films

8.3.1 Unintentional and N-Type Doping

Due to its wide bandgap, pure β-Ga_2O_3 is an insulating material. Controlling the conductivity via extrinsic doping and managing defects, such as oxygen vacancies [39] and gallium vacancies [40] which act as deep donors and deep acceptors in Ga_2O_3, respectively, is extremely important in the development of device applications for this material. Korhonen et al. [41] have shown by a positron annihilation study that gallium vacancies (V_{Ga}) are efficiently formed in the MOCVD growth method with an estimated concentration of 5×10^{18} cm^{-3}. This concentration is high and accounts for the electrical compensation observed in both UID and intentionally doped layers, which effectively limits the efficiency of n-type doping [39]. The two defects noted above have also been frequently used to explain the electrical and optical properties of bulk and thin film β-Ga_2O_3 crystals [17, 39]. Process conditions optimization is crucial for reducing the concentration of both vacancies in the MOCVD-grown films.

For extrinsic doping, n-type conductivity in Ga_2O_3 can be achieved controllably over a wide range of free carrier concentration, i.e., $\sim 10^{15}$–10^{20} cm^{-3}. β-Ga_2O_3 is particularly receptive to Si [8, 9, 29], Sn [3, 29], and Ge [42] shallow donors to achieve an excellent n-type conductivity. Si and Sn are the most common doping impurities for β-Ga_2O_3 grown by MOCVD. The Si precursor is available in the form of silane diluted in inert gasses (e.g., 40 ppm SiH_4 diluted in He or Ar) or tetraethyl orthosilicate (TEOS). Similarly, for Sn doping, a tetraethyltin (TESn) bubbler is used. So far, there is no report on the use of Ge as a dopant for β-Ga_2O_3 grown by MOCVD, but the availability of suitable Ge precursors, including germane (GeH_4) diluted in inert gasses and isobutylgermane, makes it a potential n-type dopant in the same manner that it has been for n-type doping in similar semi-conductor materials, including GaN [43, 44] and GaAs [45] epilayers.

Baldini et al. used TEOS and TESn precursors to grow $\sim$0.15–0.20 μm homoepitaxial n-type β-Ga_2O_3 layers [29] in which the Si and Sn showed a clear difference in doping β-Ga_2O_3. With Si, it was possible to achieve doping concentrations ranging between 5×10^{16} and 5×10^{19} cm^{-3} by systematically changing the molar flow rate of the TEOS precursor. The free electron carrier concentration in the Si-doped layers was comparable with the atomic doping concentration estimated by SIMS. The use of Sn as a dopant for β-Ga_2O_3 grown by MOCVD, however, is complicated for two main reasons. The first is a pronounced memory effect. The second issue is the incoherency of Sn incorporation in β-Ga_2O_3 at high Sn doping levels ($>10^{19}$ cm^{-3}) where it leads to a deterioration of the quality of the crystal and its surface. Moreover, the ionization efficiency of the Sn in the film is poor. As a result of these limitations, the use of Si appears to be the logical choice for n-type doping of MOCVD-grown β-Ga_2O_3.

The doping of Si into the Ga_2O_3 layer using SiH_4/He is very convenient, and a wide range of doping concentrations can be achieved. Figure 8.8 shows the electron mobility (300 K) as a function of free carrier concentration for UID and Si-doped β-Ga_2O_3 layers grown on semi-insulating (010) β-Ga_2O_3 substrates. With the increase in Si doping concentration, the electron mobility decreases from ~176 cm^2/Vs at n = 7.4 × 10^{15} cm^{-3} to ~13 cm^2/Vs at n = 8.0 × 10^{19} cm^{-3} [8, 46–48]. The decrease in electron mobility with the Si doping concentration is due to the increase in the effect of ionized impurity scattering. The mobility values of 176 cm^2/Vs and 150 cm^2/Vs were measured for UID layers grown using different growth conditions. As seen from the figure, the RT electron mobility of ~150 cm^2/Vs can be regularly obtained by using the MOCVD growth technique with a wide range of free carrier concentration varying between 3.4 × 10^{15} cm^{-3} and 2.9 × 10^{16} cm^{-3}. Furthermore, by using nitrous oxide (N_2O) as an oxygen source, the growth of UID β-Ga_2O_3 epitaxial film with RT electron mobility of 153 cm^2/Vs and free carrier concentration of 2.4 × 10^{14} cm^{-3} is demonstrated. Such a low background concentration with high electron mobility is the first to be demonstrated and is attributed to the use of N_2O as an oxygen source, which results in high-quality films (see Sect. 8.3.2). The result obtained using N_2O as an oxidant is promising in the realization high-performance β-Ga_2O_3 power devices, where low background carrier concentration and high electron mobilities are the key parameters.

The background impurity concentration and the compensation levels in the UID β-Ga_2O_3 layers were estimated using temperature-dependent Hall measurements [46, 47]. This analysis was performed on the samples with RT electron mobilities of 176 cm^2/Vs and 150 cm^2/Vs and the corresponding free carrier densities of 7.4 × 10^{15} cm^{-3} and ~2.9 × 10^{16} cm^{-3}, respectively (see Fig. 8.8). Figure 8.9

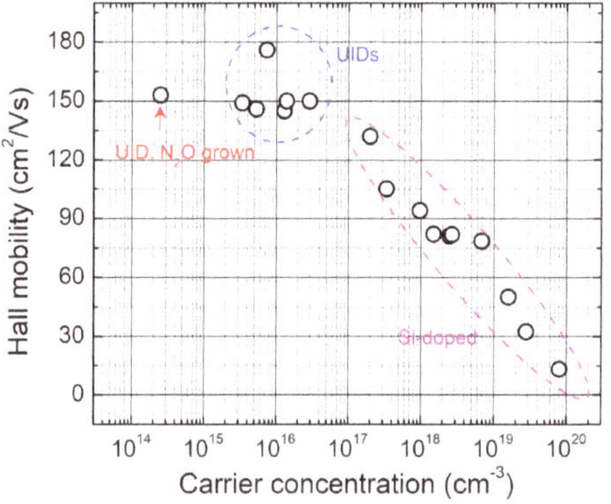

Fig. 8.8 RT Hall electron mobility versus free carrier concentration for β-Ga_2O_3/(010)β-Ga_2O_3 epitaxial films grown by MOCVD. The data points shown for carrier concentrations above 10^{17} cm^{-3} are for intentionally Si-doped films, while those data points shown for carrier concentration between 3 × 10^{15} cm^{-3} and 3 × 10^{16} cm^{-3} are for UID films grown using pure oxygen source as oxidant. The data point shown at carrier concentration of ~2 × 10^{14} cm^{-3} is for a UID film grown using N_2O as oxidant

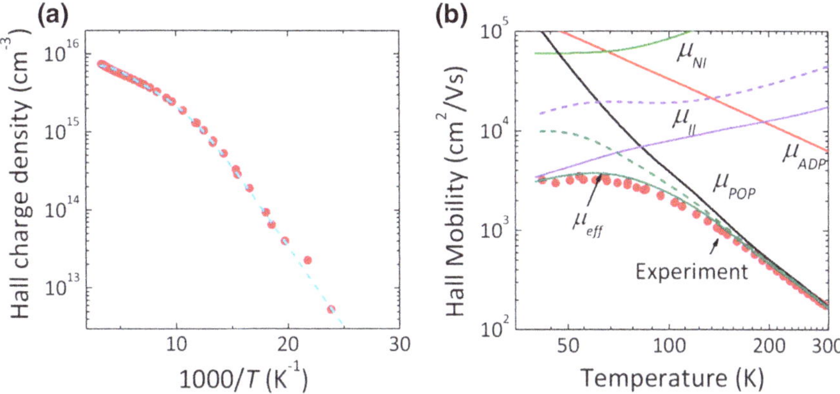

Fig. 8.9 **a** Hall carrier density (log scale) versus 1000/T of the UID β-Ga_2O_3 layer with a RT mobility of 176 cm^2/Vs. The symbols are experimental data and the dashed line represents the best fit to the charge density calculated from the charge neutrality equation. **b** Experimental and calculated dependence of Hall mobility on the temperature. The dashed lines in (b) are the predicted mobility values (μ_{II} and μ_{eff}) when $N_{d1} = 5 \times 10^{15}$ cm^{-3}, $N_{d2} = 0$, and $N_a = 1 \times 10^{15}$ cm^{-3}. Polar optical phonon scattering (μ_{POP}), ionized impurity scattering (μ_{II}), neutral impurity scattering (μ_{NI}), and acoustic deformation potential scattering (μ_{ADP}) scattering mechanisms are included in the mobility calculations [47]

shows the Hall carrier charge density and mobility as a function of temperature for the UID β-Ga_2O_3 layer that showed a RT electron mobility of 176 cm^2/Vs. The temperature-dependent Hall charge density can be fitted to the charge neutrality (1) [49], out of which the donors' energy levels and concentrations, as well as compensation accepter values in the layers can be estimated.

$$n + N_a = \frac{N_{d1}}{1 + 2\exp\left(-\frac{E_{d1} - E_F}{k_B T}\right)} + \frac{N_{d2}}{1 + 2\exp\left(-\frac{E_{d2} - E_F}{k_B T}\right)}, \quad (1)$$

The n, N_a, N_{d1}, and N_{d2} in (1) represent the electron density, compensating acceptor concentration, and donor concentrations with activation energies of E_{d1} and E_{d2}, respectively. E_F is the Fermi level at the measurement temperature (T), and k_B is the Boltzmann constant. The Fermi level can be estimated by $n = N_c e^{-(E_c - E_F)/k_B T}$, where N_c is the conduction band density of states estimated using an electron effective mass of $m^* = 0.313\ m_0$ [50]. The extracted fitting parameters are summarized in Table 8.1.

For the layer with the highest electron mobility (176 cm^2/Vs at RT), the best fit to the experimental data is shown as the dashed line in Fig. 8.9a. Two donors with energy levels that lie at 35.2 and 80 meV below the conduction band edge with corresponding donor concentrations of 9.5×10^{15} cm^{-3} and 3×10^{15} cm^{-3}, respectively, were determined [47]. A similar fitting of the layer with 150 cm^2/Vs RT mobility resulted in two donor levels at 26 and 60 meV below the conduction

Table 8.1 Extracted parameters from the fitting of the temperature-dependent Hall charge density for UID β-Ga_2O_3 layers with RT mobility of 176 cm^2/Vs (A) and 150 cm^2/Vs (B). The peak electron mobility for the layers is also presented

Samples	N_{d1} (cm^{-3})	E_c-E_{d1} (meV)	N_{d2} (cm^{-3})	$Ec-E_{d2}$ (meV)	N_a (cm^{-3})	Peak μ_e (cm^2/Vs)
A	9.5×10^{15}	35.2	3.0×10^{15}	80.0	5.2×10^{15}	3481@54 K
B	1.8×10^{16}	26.0	5.0×10^{16}	60.0	1.0×10^{16}	1350@75 K

band edge with corresponding donor concentrations of 1.8×10^{16} cm^{-3} and 5×10^{16} cm^{-3}, respectively [46]. The origin of the donor states with energy level of 60–80 meV below the conduction band edge in the UID β-Ga_2O_3 films is unclear and needs to be explored further. However, the shallow donors could have contributions from unintentional Si incorporation during growth [51]. The compensation acceptor densities for the layers with RT mobilities of 176 cm^2/Vs and 150 cm^2/Vs were extracted to be 5.2×10^{15} cm^{-3} and 1×10^{16} cm^{-3}, giving respectively, ~42 and 15% compensation of the donors. The high compensation levels could originate from native defects such as Ga vacancies [41]. Further optimization to reduce the compensation charge density is crucial for improving the electron mobility in lightly doped drift layers [52].

In Fig. 8.9b, the temperature-dependent Hall mobility for the sample with RT mobility of 176 cm^2/Vs is presented. The result showed a peak value of 3481 cm^2/Vs at 54 K. For the sample with the RT mobility of 150 cm^2/Vs, the peak mobility was measured to be 1350 cm^2/Vs at 75 K. In comparison, previously reported RT mobility values in bulk β-Ga_2O_3 were lower than 150 cm^2/Vs [30], while the highest LT mobility reported was less than 890 cm^2/Vs [53]. The significant improvement in the electron mobility is attributed to the low background impurity concentration obtained through MOCVD growth [47]. In Fig. 8.9b, the contributions of various scattering mechanisms, including polar optical phonon (μ_{POP}) scattering, acoustic deformation potential (μ_{ADP}) scattering, and ionized (μ_{II}) and neutral impurity (μ_{NI}) scattering have been investigated [54, 55].

Ultra-high purity Ga_2O_3 epitaxial films with RT electron mobility of ~150 cm^2/Vs have been grown routinely by MOCVD in dry oxygen using optimal growth conditions (see Fig. 8.8). The films have n-type conductivity with background concentration in the range of 3×10^{15} to 3×10^{16} cm^{-3}. An ~2.5-μm-thick Ga_2O_3 film, lightly doped with silicon, showed a record high mobility of 11,200 cm^2/Vs at 50 K. The temperature-dependent Hall mobility and charge density of the film are presented in Fig. 8.10a, b, respectively. The temperature-dependent Hall charge density (Fig. 8.10b) has been fitted using the charge neutrality (1) to extract the donor's energy levels and concentrations as well as compensation acceptor values. For the best fit to the experimental data, three donor models were used. Accordingly, three donors with energy levels of 35.2, 60, and 140 meV below the conduction band edge with corresponding donor concentrations of 2.6×10^{15} cm^{-3}, 2×10^{15} cm^{-3}, and 3×10^{15} cm^{-3}, respectively, were

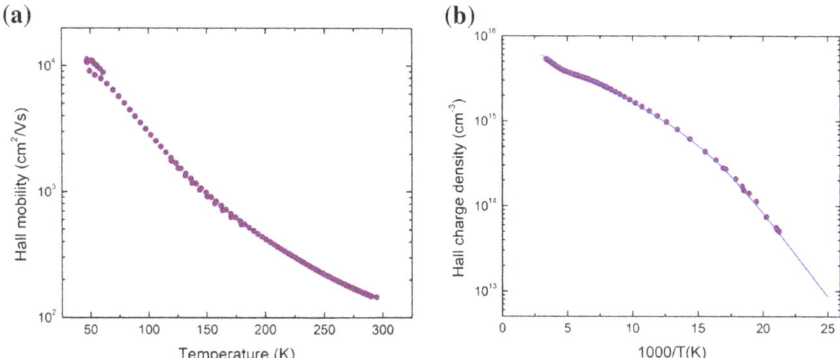

Fig. 8.10 **a** Hall mobility versus temperature (log scale) for the lightly Si-doped β-Ga_2O_3 layer grown using MOCVD. The RT Hall mobility is ~146 cm^2/Vs with the free carrier concentration of ~5×10^{15} cm^{-3}. The peak LT mobility of >11,000 cm^2/Vs is measured at 50 K. **b** Hall charge density versus 1000/T. The symbols are experimental data, and the solid line represents the best fit to the charge density calculated from the charge neutrality equation using three donor models. This temperature-dependent Hall mobility and charge density data are measured at UCSB, Materials Department, by Dr. Yuewei Zhang and Prof. James S. Speck

extracted as the fitting parameters. Similarly, the acceptor density was extracted to be 8×10^{14} cm^{-3}. The obtained high LT electron mobility shows the purity of the MOCVD-grown material and is attributed to the low acceptor compensation density in the film, which is only ~10% compensation.

Further lowering of the background impurity concentration has been achieved by adjusting the MOCVD process conditions and using a thin insulating buffer layer grown at the UID film/substrate interfaces [56–58]. One of the successful insulating buffer layers used in the reduction of the background impurity concentration of the UID layers was obtained by growing the buffer layer using N_2O as an oxygen precursor. The use of N_2O as an oxygen source incorporates nitrogen into the epilayers [57], resulting in the insulation of thc epilayer as demonstrated by doping of nitrogen into the Ga_2O_3 layer by ion implantation [59]. This will be further discussed in Sect. 8.3.2. Figure 8.11c compares the background impurity concentration of ~3.2-μm-thick UID layers grown by MOCVD on semi-insulating (010) Ga_2O_3 substrates. The UID layers are grown either directly on the substrate (A and B, Fig. 8.11a) or using an insulating buffer layer (C and D, Fig. 8.11b) grown with N_2O as an oxygen precursor. The charge distribution ($N_d - N_a$) depth profiles for all the UID layers are below 10^{16} cm^{-3}, but for those layers grown using the insulating buffer layer, the background impurity concentration dropped significantly reaching as low as ~2.0×10^{15} cm^{-3} [57].

The use of water vapor as an oxygen source was revealed to be beneficial as it promotes a layer-by-layer growth of the Ga_2O_3 thin film. This was attributed to the generation of high H concentration from water dissociation which has a positive effect on the kinetics at the surface of the substrate that supports the layer-by-layer growth [5]. DFT calculations also predicted that H impurities are strongly attracted

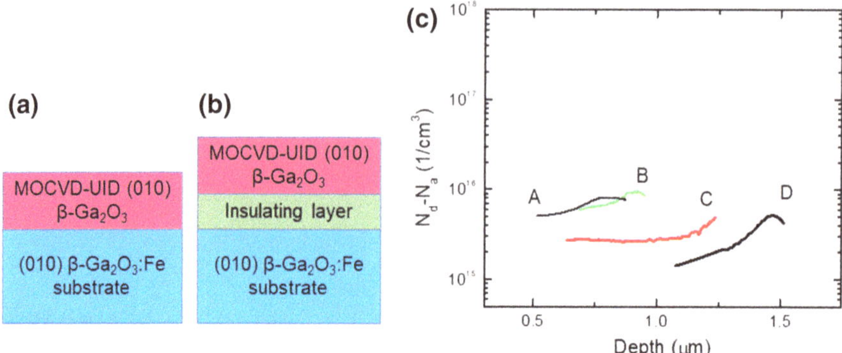

Fig. 8.11 Schematics of UID β-Ga_2O_3 films grown by MOCVD **a** directly on semi-insulating (010) β-Ga_2O_3 substrates and **b** using an N_2O grown insulating buffer layer at the film/substrate interface. **c** The charge distribution ($N_d - N_a$) versus depth profile for samples A through D

to deep acceptor, V_{Ga}, resulting in the formation of stable (up to ~800 °C) V_{Ga}–H complexes, thereby limiting its electrical activities. Besides, the H incorporation reduces the charge of negatively charged scattering centers and enhances carrier mobility the film [60]. Nevertheless, the concentration of H introduced into the layer should be controlled as too much of it increases the conductivity of the material since it acts as a shallow donor.

The effect of water vapor on the surface quality and background impurity concentration was investigated by growing Ga_2O_3 epitaxial layers by MOCVD. The growth was conducted by introducing small amount of H_2O vapor ~250 ppm into the growth reactor along with oxygen during the film growth. Figure 8.12 shows the surface morphologies of the films grown without (Fig. 8.12a) and with (Fig. 8.12b) H_2O vapor. The H_2O-assisted grown film showed a rough surface with an RMS roughness over a 5 × 5 μm^2 scan area of 13.4 nm, which is significantly high compared to the film grown without H_2O (RMS = 0.96 nm). The electron mobility and carrier concentration of the Ga_2O_3 layer grown without water were ~150 cm^2/Vs at 2.9 × 10^{16} cm^{-3} [46]. While no change in electron mobility was observed, the free carrier concentration for the H_2O assisted grown film dropped to 1.4 × 10^{16} cm^{-3}, showing an improvement in background impurity concentration resulting from H_2O vapor. The increase of H_2O concentration ~25 × has shown no effect on electron mobility and background concentration but the RMS roughness increased to ~17 nm (see Fig. 8.12c).

The effect of the O_2/TEGa ratio on the background concentration of Si, C, and H, as well as its effect on the incorporation of intentionally doped Si was investigated. Figure 8.13a shows the layers of Ga_2O_3 doped with various Si concentrations separated by UID layers. The growth rate for the layers is ~1 μm/h, and various O_2/TEGa ratios and SiH_4 flow rates were used. The layers (A and B), (C and D), (E and F), and (G, H and I) are grown at O_2/TEGa ratios of 430, 1070, 2150, and 3220, respectively. The layers A, C, E, G, and I are UID layers while the

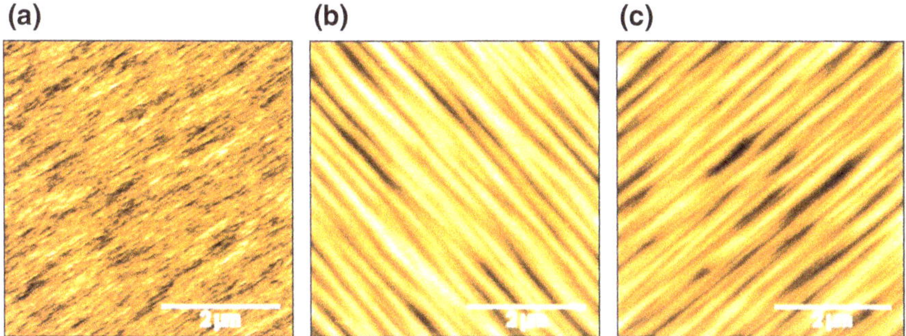

Fig. 8.12 AFM images of MOCVD-grown Ga_2O_3 using the mixture of oxygen and H_2O. The concentration of H_2O vapor is 0 (**a**), 250 ppm (**b**) and 6000 ppm (**c**). The RMS roughness for the layer in (**a**), (**b**, and (**c**) is 0.96 nm, 13.4 nm, and 16.4 nm, respectively

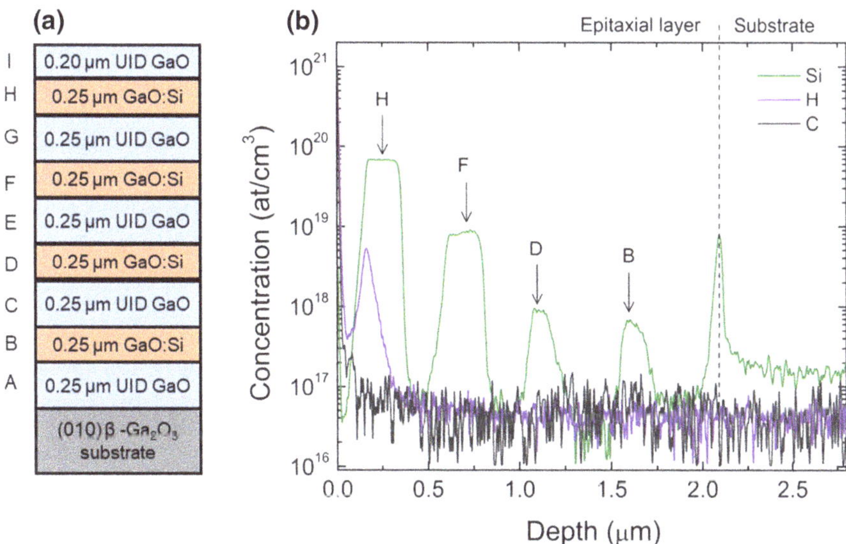

Fig. 8.13 **a** Schematics of MOCVD-grown multilayers of β-Ga_2O_3 doped with various concentrations of Si and O_2/TEGa ratios separated by UID layers. (A and B), (C and D), (E and F), and (G, H and I) were grown at O_2/TEGa ratios of 430, 1070, 2150, and 3220, respectively. Layers A, C, E, G, and I are UID layers while the layers B, D, F, and H are intentionally Si-doped layers with SiH_4/He molar flow rates of 2.3×10^{-10} mol/min, 2.3×10^{-10} mol/min, 2.3×10^{-9} mol/min, and 2.3×10^{-8} mol/min, respectively. **b** SIMS depth profile of Si, H, and C for the multilayered film shown in (**a**). The layers that show the incorporation of Si are marked with an arrow and the letter used to label the doped layers in (**a**)

layers B, D, F, and H are intentionally Si-doped layers with SiH_4/He molar flow rates of 2.3×10^{-10} mol/min, 2.3×10^{-10} mol/min, 2.3×10^{-9} mol/min, and 2.3×10^{-8} mol/min, respectively. The SIMS depth profiles for Si, C, and H in the layers are shown in Fig. 8.13b. At the film/substrate interface, the accumulation of

Si is observed [8, 61]. This accumulation of Si at the interface is similar to the phenomenon commonly encountered in MBE-grown β-Ga_2O_3, β-$(Al, Ga)_2O_3$/Ga_2O_3 heterostructures and GaN which is frequently attributed to the adsorption of ambient contaminants onto the exposed substrate surface [61–63]. A similar contamination of the substrate surface might lead to the accumulation of Si at the film/substrate interface for the MOCVD-grown β-Ga_2O_3 layers. The accumulation of Si at the interface has a deleterious effect on device performance such as the lack of a sharp current pinch-off in field-effect transistors (FETs). One possible method to mitigate this problem is to dope the first few layers of the film/substrate interface with compensating dopants such as Fe or Mg as is often done in the growth of semi-insulating GaN on sapphire [64]. Additionally, the use of a N_2O grown insulating buffer layer could be an effective way to compensate for the interface Si accumulation. The SiH_4/He flow rate in layers B and D is the same, but the concentration of Si atoms incorporated into the two layers is different. An increase in the O_2/TEGa ratio from 430 to 1073, i.e., an increase in oxygen flow led to the increase of Si atomic concentration from $\sim 6 \times 10^{17}$–9×10^{17} cm^{-3}. This clearly indicates greater Si incorporation resulting from the increased oxygen flow. The change in the O_2/TEGa ratio, on the other hand, yielded a similar background impurity concentration for the UID layers separating the intentionally doped layers. With the increase of the SiH_4/He and O_2/TEGa ratio, a Si concentration up to 7×10^{19} cm^{-3} was achieved. As indicated in the SIMS profile, a well-controlled Si doping of the Ga_2O_3 epilayer can be achieved by using SiH_4/He.

The incorporation of carbon and hydrogen into the MOCVD-grown Ga_2O_3 layers is anticipated due to the metalorganic precursors used for the growth of the films. DFT calculations based on a hybrid functional study have shown that carbon [65–67] and hydrogen [39, 68] act as shallow donors in Ga_2O_3. However, the C and H incorporation throughout the layers grown by MOCVD in Fig. 8.13a appears to be unaffected (concentration ~ 3–4×10^{16} cm^{-3}) by the change in the oxygen flow. The measured SIMS values for the C and H are close to the detection limit (DL) of the SIMS instrument, effectively showing that no incorporation of the two impurities takes place even though it is hardly possible to avoid them. On the other hand, near the top surface of the layers an accumulation of H is observed. This could be attributed to two possible factors, the diffusion of H to the upper surface of the film during the growth process and the higher flow of SiH_4/He used for the heavy doping of the top layer (in addition to the contribution coming from the TEGa precursor). The concentration of hydrogen incorporated into the grown Ga_2O_3 can be reduced or completely eliminated by annealing the films at reduced pressure in an O_2 environment [48]. This annealing potentially results in an out-diffusion of H from the film.

8.3.2 *Deep Acceptor Doping*

In contrast to n-type conductivity, the realization of p-type conductivity in β-Ga_2O_3 remains a serious roadblock to achieving PN junction device architecture. The primary challenge is the lack of available impurities that can be used as shallow acceptors for achieving significant free hole concentrations. Despite the lack of consensuses on their behavior, numerous acceptor dopants in Ga_2O_3 have been suggested by theoretical studies. Mg, Fe, and N acceptors are some of such candidates [59, 69]. The other challenge is the presence of self-trapped holes, small dispersion, and a large effective mass in the valence band state which leads to very low hole mobilities making the development of an efficient p-type doping technique challenging. In bulk Ga_2O_3 substrates, Si is the major background impurity with a typical free carrier concentration of $\sim 10^{17}$ cm^{-3} [16] making it an n-type material. This background n-type conductivity is compensated to produce semi-insulating substrates by introducing Mg or Fe dopants which act as deep acceptors. These dopants can also be used to compensate unintentional n-type conductivity in the homoepitaxial Ga_2O_3 films. As noted above, one of the major growth problems in MOCVD is the accumulation of Si at the film/substrate interface which acts as a parasitic conducting channel hampering the realization or performance of power devices such as FETs. In this section, the doping of Ga_2O_3 with Fe and nitrogen will be discussed.

The doping of Ga_2O_3 thin films using Fe was demonstrated by using ferrocene (Cp_2Fe) precursor as a source for Fe [57]. Shown in Fig. 8.14a is the schematics of multiple, $\sim$0.3-μm-thick, layers grown with various concentrations of Fe-doped layers separated by UID layers. The doped layers B, D, F, and H were grown with Cp_2Fe concentrations of 2.4 nmol/min, 4.4 nmol/min, 8.4 nmol/min, and 12.8 nmol/min, respectively. Figure 8.14b shows the SIMS depth profile of the film showing the incorporation of Fe in the MOCVD-grown layers. With the increase in the concentration of ferrocene flow into the growth reactor, an increase in the incorporation of Fe from $\sim 2 \times 10^{16}$ cm^{-3} (layer B) to $\sim 8 \times 10^{17}$ cm^{-3} (layer H) is observed. However, the interfaces between the Fe-doped layers presented a V-shaped characteristic, showing the diffusion of Fe into the UID layers deposited to separate them. This diffusion of Fe into the UID layers is a common problem in GaN, and Fe appears to behave in a similar fashion in Ga_2O_3 as well. In Ga_2O_3 thin films, the Fe diffusion tail can extend to a depth of $\sim$1.5 μm [57], requiring the deposition of thick UID layers to block its effect on the active channels.

The other deep accepter doping impurity used for Ga_2O_3 is nitrogen [59, 70]. Peelaers et al. [71] reported that N incorporates in the sites of both Ga and O, but the substitution of N on an O site acts as a compensating center with the deep energy level at >1.3 eV above the valance band maximum. Experimentally, Wong et al. demonstrated the doping of N into the Ga_2O_3 bulk substrates by ion implantation [59]. The ion beam carries substantial energy and causes damage to the crystal structure of the target material (Ga_2O_3 substrate in this case). To repair the damage, post-implantation thermal annealing is required. This not only enables

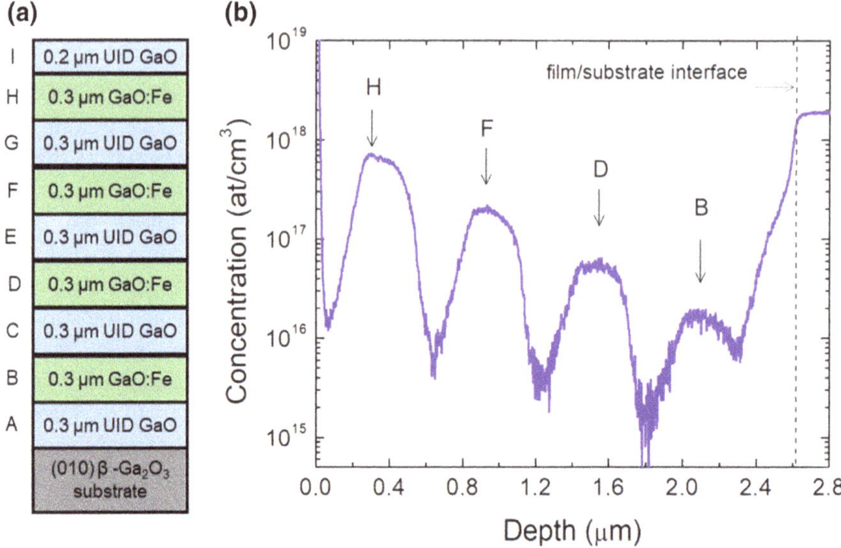

Fig. 8.14 **a** Schematic of MOCVD-grown Ga_2O_3:Fe multilayers with various Fe concentrations separated by UID layers. The layers B, D, F, and H were doped by introducing Cp_2Fe of 2.4 nmol/min, 4.4 nmol/min, 8.4 nmol/min, and 12.8 nmol/min, respectively. **b** SIMS depth profiles of Fe for the layers grown in (**a**)

removal of the implantation damage, but it also drives the impurities to their intended site. This method, however, is unsuitable for doping of the homoepitaxial layers. Efficient doping of the film with no post-growth thermal treatment can be achieved by *in-situ* doping of the film with an impurity precursor during the growth process of the homoepitaxial film.

At Agnitron Technology, the doping of N into the MOCVD-grown homoepitaxial Ga_2O_3 was achieved by using nitrous oxide (N_2O) gas [57]. The N_2O is essentially used as an alternative source of oxygen for the growth of Ga_2O_3. To efficiently crack the N_2O for the growth of the film, an elevated growth pressure (>100 Torr) is required. The incorporation of N into the growing layer is significantly influenced by the process conditions, including the chamber pressure and flow rate of N_2O. Figure 8.15a shows the schematics of a multilayered Ga_2O_3 thin film grown with N_2O and pure oxygen. The N_2O-grown layers (labeled as *i*-GaO) are grown by changing the growth pressures and N_2O/TEGa ratio and separated by UID layers grown using pure oxygen. The *i*-GaO-I and *i*-GaO-II layers were grown at a chamber pressure of P_1 and N_2O/TEGa ratio of 4288 and 2144, respectively. Similarly, the *i*-GaO-III and *i*-GaO-IV layers were grown at a chamber pressure of P_2 ($P_1 > P_2 >$ 100 Torr) and N_2O/TEGa ratio of 4288 and 2144, respectively.

The SIMS depth profiles of N, H, and C impurities in the layers grown with N_2O and oxygen are shown in Fig. 8.15b. The N_2O grown layers show an incorporation of N into the Ga_2O_3 films which drop to the detection limit (DL) of the SIMS

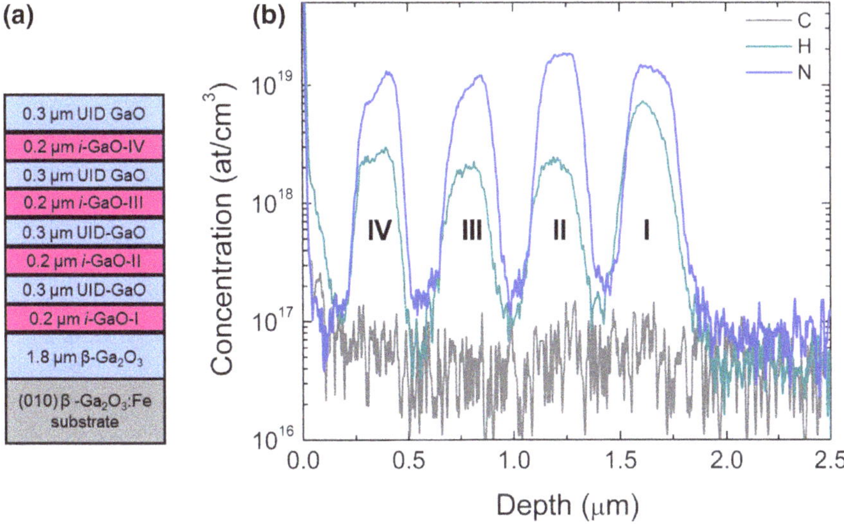

Fig. 8.15 **a** Schematics of MOCVD-grown multilayers of β-Ga_2O_3 doped with N using N_2O, separated by pure oxygen grown UID layers. The *i*-GaO-I and *i*-GaO-II layers were grown at a growth pressure of P_1 using N_2O/TEGa ratios of 4288 and 2144, respectively. The *i*-GaO-III and *i*-GaO-IV layers were grown at a growth pressure of P_2 using N_2O/TEGa ratios of 4288 and 2144, respectively. $P_1 > P_2 > 100$ Torr. **b** SIMS depth profile of N, H, and C for the multilayered film shown in (**a**)

instrument in the UID Ga_2O_3 layers grown using pure oxygen. The highest N incorporation ($\sim 1.9 \times 10^{19}$ cm^{-3}) is achieved for the *i*-GaO-II layer which is grown at a higher growth pressure (P_1) and lower N_2O/TEGa ratio of 2144, which seems counterintuitive (lower N_2O/TEGa leads to higher incorporation of N). With the decrease in growth pressure, the incorporation of N into the N_2O grown layer is reduced. The UID layers between the N_2O grown layers do not contain any N, indicating that there is either very little or no diffusion of N out of the N_2O grown layers, which is consistent with the lower diffusion of N observed in an implantation doped Ga_2O_3 substrate [64]. However, the N doping is accompanied by the incorporation of H which virtually follows the same profile as N (see Fig. 8.15b). This contrasts with the film grown using a pure oxygen precursor where there is no H incorporation observed. The use of N_2O may create a chemical by-product that keeps the H in the growth chamber, but the exact cause of the H incorporation has yet to be understood. The incorporation of H along with N is undesirable as it effectively cancels the effect of N in the film given the shallow donor behavior of H. This requires the elimination, or at least the reduction, of H in the film while maintaining the N incorporation. This can be done either by annealing the film in vacuum or adjusting the growth conditions. Interestingly, the incorporation of the two impurities shows opposing trends as seen from the SIMS profile of the *i*-GaO-II layer. By changing the N_2O/TEGa ratio from 4288 (*i*-GaO-I) to 2144 (*i*-GaO-II),

Table 8.2 Peak SIMS values for H, N, and C of the Ga_2O_3 layers grown by using N_2O gas. The C impurity is at the detection limit of the instrument for the entire layer

Impurities	*i*-GaO-I	*i*-GaO-II	*i*-GaO-III	*i*-GaO-IV
N	1.41×10^{19}	1.85×10^{19}	1.23×10^{19}	1.27×10^{19}
H	7.44×10^{18}	2.36×10^{18}	1.92×10^{18}	2.57×10^{18}
C	1×10^{17} (DL)	1×10^{17} (DL)	1×10^{17} (DL)	1×10^{17} (DL)

the H incorporation was reduced by 3.2 times while at the same time the N incorporation increased by 1.3 times. Therefore, reducing the flow of N_2O leads to a high N incorporation with the removal of H. Peak SIMS values for N, H, and C in the N_2O grown layers are given in Table 8.2. In a separate (unpublished) study, the proper tuning of the growth temperature, N_2O flow, and chamber pressure led to the complete removal of both N and H from the grown films. An ~3.2-μm-thick β-Ga_2O_3 grown using N_2O with no H and N in the film as measured by SIMS (effectively UID β-Ga_2O_3 grown using N_2O) showed a RT electron mobility of 153 cm^2/Vs with a free carrier concentration of 2.4×10^{14} cm^{-3}. But the films grown using N_2O generally remained resistive when compared to UID layers grown by using pure oxygen.

8.4 β-$(Al_xGa_{1-x})_2O_3$/β-Ga_2O_3 Heterostructures and Superlattices

In order to design and optimize modern electronic and optoelectronic devices with great flexibility, the realization of bandgap engineering to create barrier layers and quantum wells in device structures is crucial. The bandgap modulation in Ga_2O_3 has been demonstrated by the development of $(Al_xGa_{1-x})_2O_3$ alloys [72, 73] for larger bandgap material and $(Ga_{1-y}In_y)_2O_3$ alloys [28, 74, 75] for smaller bandgap materials. The alloying of In_2O_3 and Al_2O_3 with Ga_2O_3 allows for the development of a material system with bandgap tuning from 2.9 eV to 8.8 eV. β-$(Al_xGa_{1-x})_2O_3$/β-Ga_2O_3 heterostructures are uniquely suited for such applications as high power high electron mobility transistors (HEMTs) and modulation-doped field-effect transistors (MODFETs) [76–78], because alloying Ga_2O_3 with Al_2O_3 (E_g ~ 8.8 eV) enables the widening of the bandgap of Ga_2O_3. To realize the formation of two-dimensional electron gas (2DEG) in β-$(Al_xGa_{1-x})_2O_3$/β-Ga_2O_3 heterojunctions, a high Al content in the β-$(Al_xGa_{1-x})_2O_3$ barrier layer is critical. According to the phase diagram of the Al_2O_3–Ga_2O_3 binary system, the solubility of Al_2O_3 in Ga_2O_3 is drastically reduced below 800 °C due to the formation of $GaAlO_3$ intermediate compounds [79]. The growth of β-$(Al_xGa_{1-x})_2O_3$ (AlGaO) has typically been done by molecular beam epitaxy (MBE) with the state-of-the-art Al composition of ~20% [72, 78, 80]. However, the MBE method operates under ultra-high vacuum environment and requires 600–700 °C substrate temperatures to reduce the decomposition of Ga_2O_3 into volatile suboxides. This limits the maximum Al_2O_3 incorporation into β-Ga_2O_3 at ~20%. By contrast, the low to medium operating pressure in the MOCVD system enables the growth of AlGaO at substrate temperatures above 800 °C, improving the solubility of Al_2O_3 in β-Ga_2O_3, and to potentially allow up to 60% Al_2O_3 content incorporation into β-Ga_2O_3.

8.4.1 β-$(Al_xGa_{1-x})_2O_3$/β-Ga_2O_3 Heterostructures

In the growth of AlGaO by MOCVD, either trimethylaluminum (TMAl) or triethylaluminum (TEAl) can be used as a metalorganic source for Al while TEGa or TMGa can be used as Ga precursors. Figure 8.16 shows the UV-Vis transmission spectra of AlGaO layers grown on sapphire using TEAl and TEGa as the precursor for Al and Ga, respectively. The blue shift in the optical absorption edge confirms the incorporation of Al into the AlGaO alloys. The Al content in the films was estimated using the relationship between the bandgap (E_g) and the Al composition (x) as: $E_g(x) = 1.54x + 4.9\,\text{eV}$ [81], where E_g is estimated from the Tauc plots shown in Fig. 8.16 (inset) to be 14 and 21% for the blue and magenta traces, respectively. In this analysis, the β-$(Al_xGa_{1-x})_2O_3$ is assumed to be a direct bandgap material since practically there is no differences between the direct and indirect energy values [39]. On the other hand, Miller et al. demonstrated an MOCVD-grown β-$(Al_xGa_{1-x})_2O_3$ alloy on sapphire using TMAl and TEGa precursors with an Al content of ~43% as confirmed by Rutherford backscattering spectroscopy (RBS) [8].

MOCVD was also used to grow β-$(Al_xGa_{1-x})_2O_3$ alloys on native β-Ga_2O_3 substrates. Figure 8.17 shows the high-resolution x-ray diffraction (HRXRD) ω-2θ

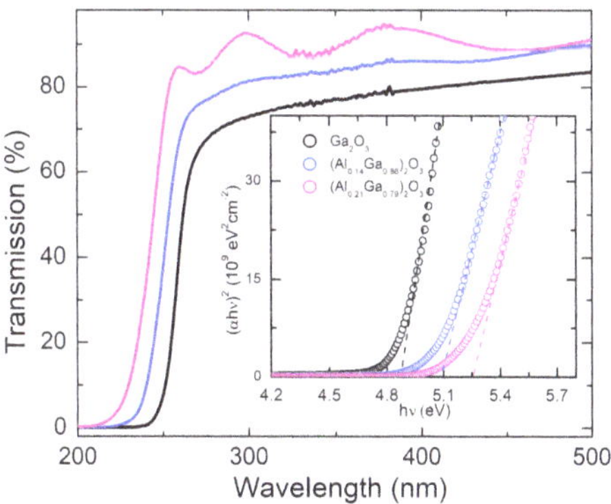

Fig. 8.16 UV-visible transmission spectra for AlGaO layers grown on c-plane sapphire using TEAl and TEGa metalorganic precursors as sources for Al and Ga, respectively. The inset shows the plot of $(\alpha h\nu)^2$ versus $h\nu$ used to estimate the bandgap of AlGaO films, which in turn is used for determining the Al content in the alloy

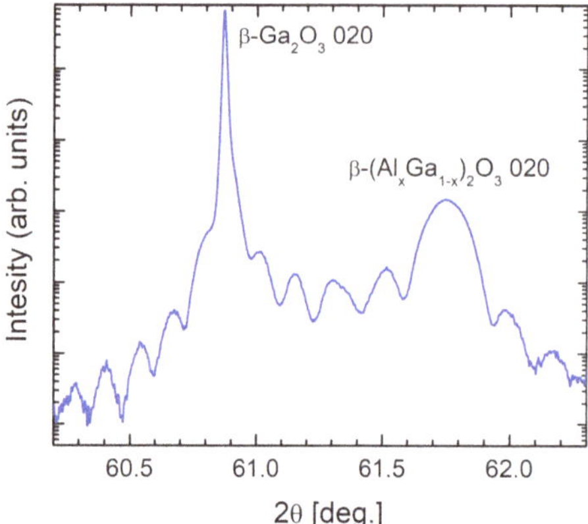

Fig. 8.17 XRD ω-2θ scan profiles of β-$(Al_xGa_{1-x})O_3$ films grown on semi-insulating (010) β-Ga_2O_3 substrate [2]. The Al content is estimated to be 21%

scan around the β-Ga_2O_3 (020) reflection for the β-$(Al_xGa_{1-x})_2O_3$ layer. The XRD pattern shows the as-grown film is a high-quality coherently strained β-$(Al_xGa_{1-x})_2O_3$ layer grown on a (010) β-Ga_2O_3 substrate [2, 8]. In this growth, the TEAl and TEGa precursors were introduced with an Al mole fraction (i.e., Al/(Al+Ga)) of ~0.16. The profile exhibits a distinct peak corresponding to the β-$(Al_xGa_{1-x})_2O_3$ layer with Pendellösung fringes, which shows an estimated layer thickness of ~70 nm. The growth of high-quality coherently stained β-$(Al_xGa_{1-x})_2O_3$ is generally process condition sensitive, but the effect of the chamber pressure is significant [8]. High chamber pressure and high oxygen flow rate leads to a poor-quality film. The Al content in the films was estimated to be 21% using the ω-2θ peak separation between the β-Ga_2O_3 and β-$(Al_xGa_{1-x})_2O_3$ peak positions [61].

8.4.2 β-$(Al_xGa_{1-x})_2O_3$/β-Ga_2O_3 Superlattices

β-$(Al_xGa_{1-x})_2O_3$/β-Ga_2O_3 superlattice structures (SL) were grown by MOCVD method. The growth conditions used to grow coherently strained β-$(Al_xGa_{1-x})_2O_3$ alloys (Fig. 8.17) were also used to grow β-$(Al_{0.21}Ga_{0.79})_2O_3$/β-$Ga_2O_3$ SL structures as schematically shown in Fig. 8.18 (left). The β-$(Al_{0.21}Ga_{0.79})_2O_3$ barrier and β-Ga_2O_3 well thicknesses were 5 nm and 10 nm, respectively, and the SL structure was grown on a ~130-nm-thick UID β-Ga_2O_3 layer. Figure 8.18 (right) shows the HRXRD ω-2θ scans of three SL structures around the β-Ga_2O_3 (020) plane. The different SL structures were grown by varying the SL periods (**N**) and substrate temperature but keeping the other growth parameters the same. The XRD pattern shows superlattices with related satellite peaks of −1 and +1 orders. The XRD profiles shown in Fig. 8.18 (right, a) and 8.18 (right, c) are for SLs grown under the same

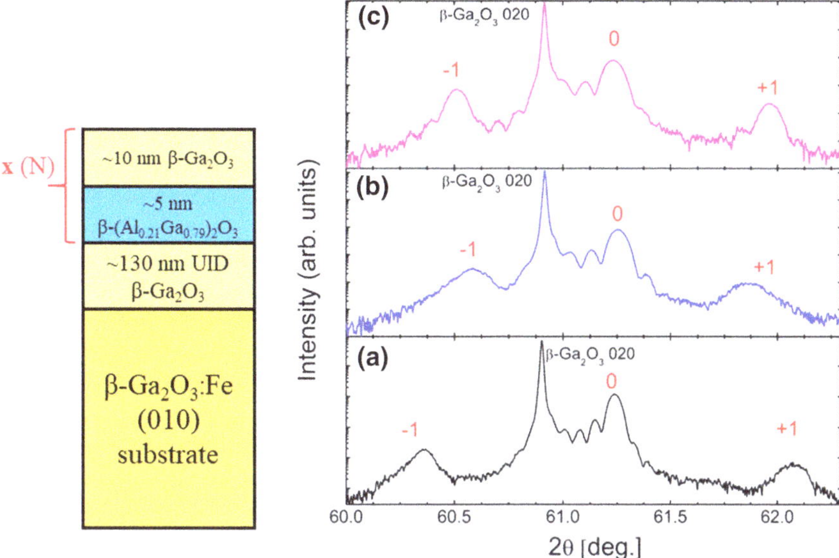

Fig. 8.18 (Left) schematic of β-$(Al_{0.21}Ga_{0.79})_2O_3$/β-$Ga_2O_3$ (010) superlattices grown by MOCVD with periods (*N*) of 8 or 14. (Right) (020) XRD *ω-2θ* scan for β-$(Al_{0.21}Ga_{0.79})_2O_3$/β-$Ga_2O_3$ superlattices: **a** grown at 800 °C for N = 14, **b** grown at 825 °C for *N* = 8, **c** grown at 800 °C with *N* = 8 [8]

conditions except for the SL period. For a fixed substrate temperature of 800 °C, the decrease in the SL period from 14 to 8 leads to the growth of SLs with improved quality as can be seen from the intense first-order peaks (Fig. 8.18 (right, c)). The profiles shown in Figs. 8.18 (right, b) and 8.18 (right, c) are for SLs grown with *N* = 8, but different substrate temperatures. The increase of the substrate temperature by 25 °C (Fig. 8.18 (right, b)) leads to a relaxed SL structure which is likely due to an increase in growth rate with temperature. The increase in growth rate leads to the increase of the total thickness of the SL as well as the thicknesses of the β-$(Al_{0.21}Ga_{0.79})_2O_3$ barrier and β-Ga_2O_3 well. Thus, good quality SL growth is achieved using the optimal growth conditions of 800 °C substrate temperature and a period of 8.

8.5 Conclusions

We discussed the MOCVD growth of Ga_2O_3 thin films and related alloys. The use of various Ga precursors, including solid Ga $[DPM]_3$ and liquid TEGa and TMGa metalorganic sources for the growth of Ga_2O_3 thin films, was discussed. The solid metalorganic precursor is sublimated at an elevated temperature which in turn requires the gas delivery lines to be heated, making it difficult to handle.

The advantages and drawbacks of using TEGa and TMGa for the growth of Ga_2O_3 thin film from the point of view of carbon incorporation were discussed. Unlike TMGa which forms a highly reactive methyl radical, TEGa forms a stable ethylene group and, subsequently, incorporates less C into the growing film. Oxygen precursors used for the growth of Ga_2O_3 were identified by using a virtual reactor (VR) modeling. The VR simulation was carried out with water vapor, pure oxygen, and tert-butanol as oxygen sources and TMGa as the Ga source for the growth Ga_2O_3. Tert-butanol was found to be an effective oxygen source for the realization of high growth rates of Ga_2O_3. High temperature and pressure growth conditions were found to produce the best results for the tert-butanol source as it efficiently cracks the source molecules that then become readily available for the reactions on the substrate surface. In addition to assisting with the selection of suitable oxygen sources, the VR simulation also highlighted the need for the careful design of the MOCVD reactor to prevent premature oxidation of the metalorganic precursors. The key aspect of the MOCVD system design is that it should be able to inject the reacting species (metalorganic precursors and oxygen sources) separately near the substrate surface. By using TMGa as the Ga precursor, β-Ga_2O_3 layers with growth rates up to 10 μm/h were obtained using a close injection showerhead (CIS) MOCVD reactor that meets the requirement for separate injection of TMGa and oxygen in close proximity of the substrate surface.

The growth of device quality homoepitaxial β-Ga_2O_3 films on (100) and (010) β-Ga_2O_3 substrates by MOCVD was discussed. The films grown on the (100) plane β-Ga_2O_3 substrates suffer from high density twin and stacking fault defects which result from the width of the substrate terraces and diffusion length of adatoms during the growth process. These extended defects were seen to significantly reduce the electron mobilities of the films. To address these issues, (100) plane substrates with miscut angle $>2°$ can be used. On the other hand, the growth of Ga_2O_3 on (010) plane substrates does not face the same challenges, and epitaxial layers with reduced defects can be achieved without requiring special treatments or substrates with miscut angles. The effect of growth rate and doping concentration on the surface and crystal qualities of the films were studied. The crystal quality and surface morphologies of the layers were nearly independent of the growth rate and doping concentration. For β-Ga_2O_3 layers with growth rates up to 4.0 μm/h, the XRD FWHM and RMS roughness were measured to be <50.0 arcsec and <0.5 nm, respectively. Epitaxial UID β-Ga_2O_3 films with RT electron mobility ranging from ~ 150 to 176 cm^2/Vs have been grown routinely using MOCVD with a wide range of free carrier concentration (3×10^{15} to 3×10^{16} cm^{-3}) depending on growth conditions. An ~ 2.5-μm-thick Ga_2O_3 film, lightly doped with silicon, showed a record high mobility of 11,200 cm^2/Vs at 50 K.

The doping of Ga_2O_3 with Si, Fe, and N was discussed. Si doping can be achieved by using SiH_4/He or TEOS precursors. By using SiH_4/He, controllable Si doping between low-10^{15} to high-10^{19} was demonstrated. The Fe and N dopants acted as deep acceptors that can be used for compensation. Fe doping was conducted using ferrocene (Cp_2Fe) precursor. The dependence of Fe incorporation (into the growing Ga_2O_3 film) on the molar flow rate of the ferrocene precursor was investigated. With the increase in ferrocene flow, an increase in the Fe incorporation was observed and followed a linear relationship. However, the diffusion of Fe into the adjacent layers during growth requires the growth of thick layers (up to 1.5 μm) to block its effect on active layers. The doping of N into Ga_2O_3 was conducted using nitrous oxide (N_2O). The use of N_2O for doping Ga_2O_3 with N is a novel technique as it contains both oxygen for growth of the film and nitrogen for doping —no additional oxygen source is needed. However, the incorporation of N into Ga_2O_3 is highly dependent on process conditions. By tuning the process conditions, Ga_2O_3 epitaxial films can be grown with no doping or with N doping concentrations $>3 \times 10^{19}$ cm^{-3}. In contrast to the behavior of Fe in the grown film, the diffusion of N into adjacent layers is limited, making it a suitable dopant. However, the incorporation of N is coupled with the introduction of H into the film which requires annealing and process conditions adjustment to remove it.

The use of MOCVD for the growth of β-$(Al_xGa_{1-x})_2O_3$ alloys and superlattices was also discussed. In the growth of AlGaO, TMAl or TEAl was used as the precursor for Al while TEGa was used as the precursor for Ga. β-$(Al_xGa_{1-x})_2O_3$ alloys with Al content x up to 43% were grown on sapphire substrates using TMAl and TEGa precursors. With TEAl and TEGa, AlGaO films with Al composition ~21% were demonstrated both on sapphire and (010) β-Ga_2O_3 substrates. The β-$(Al_xGa_{1-x})_2O_3$/β-Ga_2O_3 heterostructures and superlattices were of high quality with coherently strained layers.

Acknowledgements This work was partially performed under the sponsorships AFOSR and ONR programs through grant numbers FA9550-17-P-0029 and N00014-16-P-2058, respectively. Collaborations and discussions with James S. Speck, Y. Zhang, A. Mauze, B. Hertog, P. Mukhopadhyay, W. V. Schoenfeld, T. Vogt, Adam T. Neal, and Shin Mou are acknowledged.

References

1. F. Alema, B. Hertog, A. Osinsky, P. Mukhopadhyay, M. Toporkov, W.V. Schoenfeld, J. Cryst. Growth **475**, 77 (2017)
2. R. Miller, F. Alema, A. Osinsky, Compd. Semicond. Mag. **24**(4), 18 (2018)
3. M. Baldini, M. Albrecht, A. Fiedler, K. Irmscher, D. Klimm, R. Schewski, G. Wagner, J. Mater. Sci. **51**(7), 3650 (2015)
4. F. Alema, B. Hertog, O. Ledyaev, D. Volovik, G. Thoma, R. Miller, A. Osinsky, P. Mukhopadhyay, S. Bakhshi, H. Ali, W.V. Schoenfeld, Phys. Status Solidi A **214**, 1600688 (2017)
5. G. Wagner, M. Baldini, D. Gogova, M. Schmidbauer, R. Schewski, M. Albrecht, Z. Galazka, D. Klimm, R. Fornari, Phys. Status Solidi A **211**(1), 27 (2014)
6. A.C. Jones, M.L. Hitchman (ed.), *Chemical Vapour Deposition* (Cambridge, UK, 2009), p. 297
7. S.J. Pearton, J. Yang, P.H. Cary, F. Ren, J. Kim, M.J. Tadjer, M.A. Mastro, Appl. Phys. Rev. **5**(1), 011301 (2018)
8. R. Miller, F. Alema, A. Osinsky, IEEE Trans. Semicond. Manuf. **31**(4), 467 (2018)

9. R. Miller, F. Alema, A. Osinsky, Advances toward industrial compatible epitaxial growth of β-Ga_2O_3 and alloys for power electronics. Paper presented at the CS-MANTECH, Austin, TX, 7–10 May 2018
10. T. Sugimoto, S. Yuhya, Y. Yamada, H. Hayashi, K. Kikuchi, M. Yoshida, K. Sugawara, Y. Shiohara, in *Advances in Superconductivity II*, ed. by T. Ishiguro, K. Kajimura (Springer Japan, Tokyo, 1990), p. 907
11. C.A. Larsen, N.J. Buchan, S.H. Li, G.B. Stringfellow, J. Cryst. Growth **102**, 103 (1990)
12. F. Alema, O. Ledyaev, R. Miller, B. Hertog, A. Osinsky, W.V. Schoenfeld, Epitaxial growth of Ga_2O_3 by metal organic chemical vapor deposition from various precursors. Paper presented at the 18th International Conference on Metalorganic Vapor Phase Epitaxy, San diego, CA, 10–15 July 2016
13. C. Kirchner, T. Gruber, F. Reuß, K. Thonke, A. Waag, Ch. Gießen, M. Heuken, J. Cryst. Growth **248**, 20 (2003)
14. K. Ogata, K. Maejima, S. Fujita, S. Fujita, J. Electron. Mater. **30**(6), 659 (2001)
15. A. Waag, T. Gruber, K. Thonke, R. Sauer, R. Kling, C. Kirchner, H. Ress, J. Alloys Compd. **371**(1), 77 (2003)
16. A. Kuramata, K. Koshi, S. Watanabe, Y. Yamaoka, T. Masui, S. Yamakoshi, Jpn. J. Appl. Phys. **55**(12), 1202A2 (2016)
17. T. Onuma, S. Fujioka, T. Yamaguchi, M. Higashiwaki, K. Sasaki, T. Masui, T. Honda, Appl. Phys. Lett. **103**, 041910 (2013)
18. J. Zhang, C. Xia, Q. Deng, W. Xu, H. Shi, F. Wu, J. Xu, J. Phys. Chem. Solids **67**(8), 1656 (2006)
19. E.G. Víllora, K. Shimamura, Y. Yoshikawa, T. Ujiie, K. Aoki, Appl. Phys. Lett. **92**(20), 202120 (2008)
20. Y. Tomm, P. Reiche, D. Klimm, T. Fukuda, J. Cryst. Growth **220**(4), 510 (2000)
21. E.G. Víllora, Y. Murakami, T. Sugawara, T. Atou, M. Kikuchi, D. Shindo, T. Fukuda, Mater. Res. Bull. **37**(4), 769 (2002)
22. K. Sasaki, M. Higashiwaki, A. Kuramata, T. Masui, S. Yamakoshi, J. Cryst. Growth **378**, 591 (2013)
23. H. Okumura, M. Kita, K. Sasaki, A. Kuramata, M. Higashiwaki, J.S. Speck, Appl. Phys. Express **7**(9), 095501 (2014)
24. K.-H. Lee, P.-C. Chang, S.-J. Chang, Y.-K. Su, S.-L. Wu, M. Pilkuhn, Mater. Chem. Phys. **134**(2–3), 899 (2012)
25. K.-M. Song, D.-J. Kim, Y.-T. Moon, S.-J. Park, J. Cryst. Growth **233**, 439 (2001)
26. H. Ito, K. Kurishima, J. Cryst. Growth **165**(3), 215 (1996)
27. M. Baldini, Z. Galazka, G. Wagner, Mater. Sci. Semicond. Process. **78**, 132 (2018)
28. M. Baldini, D. Gogova, K. Irmscher, M. Schmidbauer, G. Wagner, R. Fornari, Cryst. Res. Technol. **49**, 552 (2014)
29. M. Baldini, M. Albrecht, A. Fiedler, K. Irmscher, R. Schewski, G. Wagner, ECS J. Solid State Sci. Technol. **6**(2), Q3040 (2016)
30. K. Sasaki, A. Kuramata, T. Masui, E.G. Víllora, K. Shimamura, S. Yamakoshi, Appl. Phys. Express **5**(3), 035502 (2012)
31. R. Schewski, M. Baldini, K. Irmscher, A. Fiedler, T. Markurt, B. Neuschulz, T. Remmele, T. Schulz, G. Wagner, Z. Galazka, M. Albrecht, J. Appl. Phys. **120**(22), 225308 (2016)
32. M. Baldini, M. Albrecht, D. Gogova, R. Schewski, G. Wagner, Semicond. Sci. Tech. **30**(2), 024013 (2015)
33. Z. Guo, A. Verma, X. Wu, F. Sun, A. Hickman, T. Masui, A. Kuramata, M. Higashiwaki, D. Jena, T. Luo, Appl. Phys. Lett. **106**(11), 111909 (2015)
34. K. Sasaki, M. Higashiwaki, A. Kuramata, T. Masui, S. Yamakoshi, J. Cryst. Growth **392**, 30 (2014)
35. M.-Y. Tsai, O. Bierwagen, M. E. White, J. S. Speck, J. Vac. Sci. Technol., A **28**, 354 (2010)
36. H. Murakami, K. Nomura, K. Goto, K. Sasaki, K. Kawara, Q.T. Thieu, R. Togashi, Y. Kumagai, M. Higashiwaki, A. Kuramata, S. Yamakoshi, B. Monemar, A. Koukitu, Appl. Phys. Express **8**(1), 015503 (2015)

37. S. Rafique, L. Han, M.J. Tadjer, J.A. Freitas, N.A. Mahadik, H. Zhao, Appl. Phys. Lett. **108** (18), 182105 (2016)
38. S. Rafique, M.R. Karim, J.M. Johnson, J. Hwang, H. Zhao, Appl. Phys. Lett. **112**, 052104 (2018)
39. J.B. Varley, J.R. Weber, A. Janotti, C.G. Van de Walle, Appl. Phys. Lett. **97**, 142106 (2010)
40. B.E. Kananen, L.E. Halliburton, K.T. Stevens, G.K. Foundos, N.C. Giles, Appl. Phys. Lett. **110**, 202104 (2017)
41. E. Korhonen, F. Tuomisto, D. Gogova, G. Wagner, M. Baldini, Z. Galazka, R. Schewski, M. Albrecht, Appl. Phys. Lett. **106**, 242103 (2015)
42. E. Ahmadi, O.S. Koksaldi, S.W. Kaun, Y. Oshima, D.B. Short, U.K. Mishra, J.S. Speck, Appl. Phys. Express **10**(4), 041102 (2017)
43. D. Armin, B. Jürgen, D. Annette, K. Alois, Appl. Phys. Express **4**(1), 011001 (2011)
44. S. Fritze, A. Dadgar, H. Witte, M. Bügler, A. Rohrbeck, J. Bläsing, A. Hoffmann, A. Krost, Appl. Phys. Lett. **100**(12), 122104 (2012)
45. T. Moriizumi, K. Takahashi, Jpn. J. Appl. Phys. **8**(3), 348 (1969)
46. A.T. Neal, S. Mou, F. Alema, A. Osinsky, J.D. Blevins, K.D. Chabak, G.H. Jessen, Transport studies of unintentionally doped MOCVD grown β-Ga_2O_3. Paper presented at *GOX:* 3rd US workshop on Gallium oxide, Columbus, Ohio, 15–16 August 2018
47. Y. Zhang, F. Alema, A. Mauze, O.S. Koksaldi, R. Miller, A. Osinsky, J.S. Speck, APL Mater. **7**, 022506 (2019)
48. F. Alema, R. Miller, A. Osinsky, A. Mauze, J.S. Speck, A.T. Neal, S. Mou, T. Vogt, P. Mukhopadhyay, W.V. Schoenfeld, Device quality β-$(Al_xGa_{1-x})_2O_3$/β-Ga_2O_3 heterostructures grown by MOCVD. Paper presented at *GOX:* 3rd US workshop on Gallium oxide, Columbus, Ohio, 15–16 August, 2018
49. K. Goto, K. Konishi, H. Murakami, Y. Kumagai, B. Monemar, M. Higashiwaki, A. Kuramata, S. Yamakoshi, Thin Solid Films **666**, 182 (2018)
50. Y. Zhang, A. Neal, Z. Xia, C. Joishi, J.M. Johnson, Y. Zheng, S. Bajaj, M. Brenner, D. Dorsey, K. Chabak, G. Jessen, J. Hwang, S. Mou, J.P. Heremans, S. Rajan, Appl. Phys. Lett. **112**, 173502 (2018)
51. A.T. Neal, S. Mou, R. Lopez, J.V. Li, D.B. Thomson, K.D. Chabak, G.H. Jessen, Sci. Rep. **7** (1), 13218 (2017)
52. Y. Kang, K. Krishnaswamy, H. Peelaers, C.G. Van de Walle, J. Phys. Cond. Mater. **29**(23), 234001 (2017)
53. T. Oishi, Y. Koga, K. Harada, M. Kasu, Appl. Phys. Express **8**(3), 031101 (2015)
54. M. Lundstrom, *Fundamentals of Carrier Transport*, 2nd ed. (Cambridge university press, 2009)
55. N. Ma, N. Tanen, A. Verma, Z. Guo, T. Luo, H. Xing, D. Jena, Appl. Phys. Lett. **109**(21), 212101 (2016)
56. F. Alema, B. Hertog, P. Mukhopadhyay, Y. Zhang, A. Mauze, A. Osinsky, W.V. Schoenfeld, J.S. Speck, T. Vogt, APL Mater. **7**(2), 022527 (2019)
57. F. Alema, A. Osinsky, Y. Zhang, A. Mauze, J.S. Speck, Device quality β-Ga_2O_3 and β-$(Al_xGa_{1-x})_2O_3$ heterostructures—control of doping and impurity incorporation in MOCVD process. Paper presented at Material Research Society, Boston, MA, 25–28 November 2018
58. F. Alema, B. Hertog, A. Osinsky, Y. Zhang, A. Mauze, J.S. Speck, P. Mukhopadhyay, W.V. Schoenfeld, High performance β-Ga_2O_3 based vertical solar blind schottky photodiode. Paper presented at Material Research Society, Boston, MA, 25–28 November 2018
59. M.H. Wong, C.-H. Lin, A. Kuramata, S. Yamakoshi, H. Murakami, Y. Kumagai, M. Higashiwaki, Appl. Phys. Lett. **113**(10), 102103 (2018)
60. J.B. Varley, H. Peelaers, A. Janotti, C.G. Van de Walle, J. Phys.: Cond. Mater. **23**(33), 334212 (2011)
61. T. Oshima, Y. Kato, N. Kawano, A. Kuramata, S. Yamakoshi, S. Fujita, T. Oishi, M. Kasu, Appl. Phys. Express **10**(3), 035701 (2017)
62. M.H. Wong, K. Sasaki, A. Kuramata, S. Yamakoshi, M. Higashiwaki, Jpn. J. Appl. Phys. **55** (12), 1202B9 (2016)

63. M.J. Manfra, N.G. Weimann, J.W.P. Hsu, L.N. Pfeiffer, K.W. West, S. Syed, H.L. Stormer, W. Pan, D.V. Lang, S.N.G. Chu, G. Kowach, A.M. Sergent, J. Caissie, K.M. Molvar, L. J. Mahoney, R.J. Molnar, J. Appl. Phys. **92**(1), 338 (2002)
64. S. Heikman, S. Keller, S.P. DenBaars, U.K. Mishra, Appl. Phys. Lett. **81**, 439 (2002)
65. J.L. Lyons, D. Steiauf, A. Janotti, C.G. Van de Walle, Phys. Rev. Appl. **2**(6), 064005 (2014)
66. S. Lany, APL Mater. **6**(4), 046103 (2018)
67. S.-S. Huang, R. Lopez, S. Paul, A.T. Neal, S. Mou, M.-P. Houng, J.V. Li, Jpn. J. Appl. Phys. **57**(9), 091101 (2018)
68. Y. Wei, X. Li, J. Yang, C. Liu, J. Zhao, Y. Liu, S. Dong, Sci. Rep. **8**(1), 10142 (2018)
69. J.L. Lyons, Semicond. Sci. Tech. **33**(5), 05LT02 (2018)
70. L. Dong, R. Jia, C. Li, B. Xin, Y. Zhang, J. Alloys. Compd. **712**, 379 (2017)
71. H. Peelaers, J.L. Lyons, J.B. Varley, C.G. Van de Walle, APL Mater. **7**(2), 022519 (2019)
72. T. Oshima, T. Okuno, N. Arai, Y. Kobayashi, S. Fujita, Jpn. J. Appl. Phys. **48**(7), 070202 (2009)
73. F. Zhang, K. Saito, T. Tanaka, M. Nishio, M. Arita, Q. Guo, Appl. Phys. Lett. **105**(16), 162107 (2014)
74. M.B. Maccioni, F. Ricci, V. Fiorentini, J. Phys: Conf. Ser. **566**(1), 012016 (2014)
75. F. Zhang, K. Saito, T. Tanaka, M. Nishio, Q. Guo, Solid State Commun. **186**, 28 (2014)
76. S. Krishnamoorthy, Z. Xia, C. Joishi, Y. Zhang, J. McGlone, J. Johnson, M. Brenner, A.R. Arehart, J. Hwang, S. Lodha, Appl. Phys. Lett. **111**(2), 023502 (2017)
77. Y. Zhang, A. Neal, Z. Xia, C. Joishi, J. Johnson, Y. Zheng, S. Bajaj, M. Brenner, D. Dorsey, K. Chabak, G. Jessen, J. Hwang, S. Mou, J. Heremans, S. Rajan, Appl. Phys. Lett. **112**(17), 173502 (2018)
78. E. Ahmadi, O.S. Koksaldi, X. Zheng, T. Mates, Y. Oshima, U.K. Mishra, J.S. Speck, Appl. Phys. Express **10**(7), 071101 (2017)
79. V.G. Hill, R. Roy, E.F. Osborn, J. Am. Cer. Soc. **74**, 719 (1952)
80. S.W. Kaun, F. Wu, J.S. Speck, J. Vac. Sci. Technol., A **33**(4), 041508 (2015)
81. F. Zhang, K. Saito, T. Tanaka, M. Nishio, M. Arita, Q. Guo, Appl. Phys. Lett. **105**, 162107 (2014)

Chapter 9
Metal Organic Chemical Vapor Deposition 2

Heteroepitaxial MOCVD Growth of α-, β-, and ε-Ga_2O_3 Thin Films on Sapphire Substrates

Yao Yao, Robert F. Davis and Lisa M. Porter

Abstract This chapter reviews the heteroepitaxial growth of α-, β-, and ε-gallium oxide (Ga_2O_3) films, with a focus on those grown using the metalorganic chemical vapor deposition (MOCVD) technique. Variations in growth conditions and substrates result in the growth of different polymorphs of Ga_2O_3 or combinations of them. β-Ga_2O_3 is consistently reported as the dominant phase to grow at high substrate temperatures >700 °C. At lower substrate temperatures, α- and ε- metastable phases have been observed. Other growth conditions and substrates that have yielded α- and ε-Ga_2O_3 epitaxial films are also discussed. Doping of MOCVD-grown β-Ga_2O_3 is also briefly reviewed, where Si and Sn are the most commonly used dopants. Doping concentrations between 1×10^{17} and 8×10^{19} cm^{-3} have been achieved, with corresponding electron mobility values between ~130 and 50 cm^2/Vs.

9.1 Introduction

As discussed in detail in other chapters of this book, gallium oxide (Ga_2O_3) has exceptional properties for high-power and radio frequency (RF) electronic devices [1] and also for ultraviolet optoelectronic devices [2]. These properties, in combination with the recent availability of β-Ga_2O_3 single-crystal substrates, have fueled the rapidly growing interest in this wide bandgap semiconductor material. In order to make devices based on Ga_2O_3, the production of device-quality Ga_2O_3 epitaxial films (i.e., single-crystal layers grown on β-Ga_2O_3 substrates or on other single-crystal substrates) at commercially viable growth rates is essential.

Many different techniques have been employed to grow Ga_2O_3 epitaxial layers. Methods include chemical solution deposition [3], spray pyrolysis [4, 5], sol-gel [6], MIST CVD [7], RF sputtering [8], pulsed laser deposition (PLD) [9], molecular beam

Y. Yao · R. F. Davis · L. M. Porter (✉)
Department of Materials Science & Engineering, Carnegie Mellon University, Pittsburgh, PA 15213, USA
e-mail: lporter@andrew.cmu.edu

M. Higashiwaki and S. Fujita (eds.), *Gallium Oxide*, Springer Series in Materials Science 293, https://doi.org/10.1007/978-3-030-37153-1_9

epitaxy (MBE) [10], halide vapor phase epitaxy (HVPE) [11], and metalorganic chemical vapor deposition (MOCVD) [12, 13]. Most Ga_2O_3 epitaxial films are the monoclinic β-phase, which is the thermodynamically stable (equilibrium) phase at all temperatures up to its melting point [14]. However, epitaxial films comprising metastable Ga_2O_3 phases have also been reported, primarily via growth on other substrates, such as silicon or sapphire. This chapter focuses on the heteroepitaxial growth of α-, β-, and ε-phases of Ga_2O_3 using MOCVD. A comprehensive survey of published literature on the subject will be presented, and some perspectives on future development will be offered.

9.2 Metalorganic Chemical Vapor Deposition (MOCVD)

MOCVD is one of a number of common techniques used to deposit Ga_2O_3 epitaxial layers (Table 9.1). Among its advantages, which is central to this chapter, is the capability of growing epitaxial films comprised of different Ga_2O_3 phases. It can also be used to grow films with a wide range of doping levels. Table 9.1 shows a comparison chart among common Ga_2O_3 epitaxial growth techniques. Common precursors that have been demonstrated for Ga_2O_3 epitaxial film growth include trimethyl gallium (TMGa) and triethyl gallium (TEGa), with oxygen, ozone, O plasma, and H_2O as the common oxidizers.

Table 9.1 Comparison between different Ga_2O_3 growth methods. Revised from [15]

Parameter	MIST CVD	MBE	ALD	MOCVD	HVPE
P (Torr)	10–600	10^6 to 10^{10}	~ 1	10–600	100–600
T (°C)	400–500	600–900	100–400	450–900	450–900
Ga Source	$GaBr_3$, GaI_3, $GaCl_3$, Ga	Ga	TMGa, TEGa, $Ga(aca)_3$, $((CH_3)_2GaNH_2)_3$, $Ga_2(N(CH_3)_2)_6$	TMGa, TEGa	GaCl, $GaCl_3$
O Source	Alcohol, H_2O, O_2	RF plasma, O_3	H_2O, H_2O_2, IPA, O_2, O_3	H_2O, O_2, O_3	H_2O, O_2, O_3
Growth Rate (μm/h)	~0.1	~0.1	~0.05	~1	~10
Cons	Slow dep. rate	Slow dep. rate, ultra-high vacuum	Slow dep. rate	Toxic sources, memory effect, high temp.	Poorer film quality and lower doping modulation
Pros	Alloyed films possible	Precise thickness control	Low temp., precise thickness control	Faster dep. rate, good crystal quality, low pressure, alloyed films possible	Fastest dep. rate

TMGa trimethylgallium, *TEGa* triethylgallium

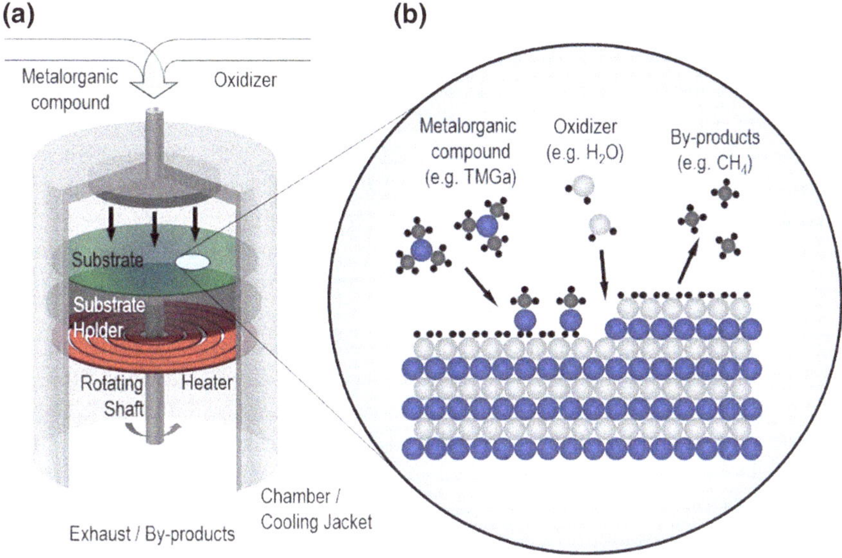

Fig. 9.1 **a** Schematic of a vertical MOCVD chamber. Precursors enter the chamber from a showerhead and react on the substrate. Substrate is heated by a heating element and rotated to maintain uniform growth. **b** Schematic of a MOCVD process to grow Ga_2O_3. Precursors react on substrate and film grows epitaxially

MOCVD reactors can be horizontal or vertical. The metalorganic compound—typically flowed in by running a carrier gas through a metalorganic bubbler—and other precursors, are introduced into the reaction chamber over a substrate. The substrate is heated to enable the desired chemical reaction to overcome the activation energy, resulting in film growth on the surface of the substrate. The reaction by-products are then removed through the exhaust. To prevent undesirable reactions from occurring in the gas phase, vertical reactors where the precursors are flowed into the chamber separately and allowed to mix very close to the substrate can be used. Rotating the substrate helps ensure uniform distribution of precursors over the substrate surface and results in more uniform and controlled growth rate, composition, and doping concentration. Vertical, high speed, rotating disk reactors (RDR) have successfully been used to grow Ga_2O_3 epitaxial films from highly reactive chemistries. Figure 9.1 shows a schematic of a vertical RDR, and a typical reaction during the growth of Ga_2O_3 epitaxial layers.

Current research is directed toward increasing the growth rates of Ga_2O_3 epitaxial layers grown by MOCVD and other techniques to produce thick layers (e.g., ten's of μm) that can withstand high breakdown voltages for power device applications. While MOCVD growth rates are faster than those of ALD and MBE, they are typically $\sim$1 μm/h. In comparison, HVPE has been shown to produce higher growth rates (up to 250 μm/h) [12, 16], but questions about sacrificing control over film quality and doping remain.

9.3 Heteroepitaxy of Ga_2O_3 Using MOCVD

Ga_2O_3 is commonly described to exist in five distinct polymorphs—the α-, β-, δ-, ε-, and γ-phases [17]. However, there has recently been some disagreement over the distinction between ε-, δ- and a new κ-Ga_2O_3 phase, with some evidence that δ-Ga_2O_3 is a nanocrystalline form of ε-Ga_2O_3; and that the hexagonal ε-Ga_2O_3 phase consists of the six twins of the rhomohedral κ-Ga_2O_3 [18].

In this chapter, we assume the traditional five polymorphs, but focus on the α, β, and ε polymorphs. Their free energies of formation have been calculated to be $\beta < \varepsilon < \alpha < \delta < \gamma$ at low temperatures [19].

β-Ga_2O_3 has a base-centered monoclinic structure with space group C2/m. It is the thermodynamically stable phase up to the melting point [14], and the subject of most early studies. As such, most of the epitaxial films are the β-phase. Because of its wide bandgap, undoped β-Ga_2O_3 behaves as an electrical insulator; doping with Si or Sn yields films with *n*-type conductivity. Doping will be discussed in more detail in Sect. 9.4.

ε-Ga_2O_3 is the second most stable polymorph of Ga_2O_3. It is commonly reported to have a hexagonal crystal structure, making it better matched with other hexagonal wide bandgap semiconductors such as GaN and SiC. ε-Ga_2O_3 is also reported to be ferroelectric with a large spontaneous polarization, shown in experimental results using dynamic hysteresis measurements [20]. Heterostructures with ε-Ga_2O_3 can be grown to produce a two-dimensional electron gas (2DEG), a phenomenon that can be exploited in high electron mobility transistors (HEMTs).

α-Ga_2O_3 is the third most stable polymorph of Ga_2O_3. However, it is subject to numerous studies because it shares the same corundum structure as sapphire. MIST CVD® is currently the most advanced method for growing α-Ga_2O_3 heteroepitaxial films on sapphire [7]. α-Ga_2O_3 films grown by MOCVD have also been reported [13, 16, 21] but are limited. Further investigations of the MOCVD process conditions to grow pure-phase α-Ga_2O_3 are needed.

As far as the authors are aware, there are no reports of γ-Ga_2O_3 growth using MOCVD. It has, however, been reported to be grown on spinel substrates using mist CVD [22].

Highlights from the literature on MOCVD growth of β-, α-, and ε-Ga_2O_3 epitaxial layers are shown in Table 9.2. It is observed that the growth of α and ε phases tend to occur at growth temperatures <800 °C.

9.3.1 *Growth of* β-Ga_2O_3

The first reported growth of Ga_2O_3 using the MOCVD method was on Si(100) in 2004 [23]. The resulting film was predominantly amorphous but contained small crystallites of Ga_2O_3.

Table 9.2 Summary of heteroepitaxy of Ga_2O_3 using MOCVD

Substrate	P (Torr)	T (°C)	Ga Source	O Source	Growth rate (μm/h)	Film grown
p-type Si (100)	N.R.	500–600	TMGa (Ar as carrier gas)	O_2	N.R.	Amorphous with small crystallites [23]
a-, *c*- *m*-, and *r*-plane sapphire	N.R.	600–850	TEGa (N_2 as carrier gas)	N_2O	0.7	$(\bar{2}01)$ β-Ga_2O_3 ‖ (0001) Al_2O_3. $(\bar{2}01)$ β-Ga_2O_3 ‖ $(11\bar{2}0)$ Al_2O_3 and $(11\bar{2}0)$ α-Ga_2O_3 ‖ $(11\bar{2}0)$ Al_2O_3. $(10\bar{1}0)$ α-Ga_2O_3 ‖ $(10\bar{1}0)$ Al_2O_3. $(01\bar{1}2)$ α-Ga_2O_3 ‖ $(01\bar{1}2)$ Al_2O_3 [13]
GaAs $(\bar{1}11)_{As}$, $(111)_{Ga}$ and (100)	Atmospheric pressure	600–850	TEGa (N_2 as carrier gas)	N_2O	0.7	β-Ga_2O_3 ‖ $(\bar{1}11)$ GaAs [13]
c-plane sapphire	Low pressure	800–850	TMGa (Ar as carrier gas)	H_2O	N.R.	Pure-phase $(\bar{2}01)$ β-Ga_2O_3 ‖ (0001) Al_2O_3 [24]
c-plane sapphire		500–600	$Ga(DPM)_3$, TEGa, TMGa (Ar as carrier gas)	O_2	Up to 10	Pure-phase $(\bar{2}01)$ β-Ga_2O_3 ‖ (0001) Al_2O_3 [12]
c-plane sapphire	3.75–37.5	800	TMGa	H_2O	N.R.	3 monolayer pseudomorphic (0001) α-Ga_2O_3 ‖ (0001) Al_2O_3 followed by $(\bar{2}01)$ β-Ga_2O_3 ‖ (0001) α-Ga_2O_3 [21]

(continued)

Table 9.2 (continued)

Substrate	P (Torr)	T (°C)	Ga Source	O Source	Growth Rate (μm/h)	Film grown
c-plane sapphire		650	TMGa	H_2O	Average 1.2	ε-Ga_2O_3 with some γ-Ga_2O_3 at substrate interface [18, 20, 25]
c-plane sapphire	45	600	TEGa (Ar as carrier gas)	O_2	0.1–1	Combinations of α-, β- and ε-Ga_2O_3 [16]
6H-SiC	26	500	TEGa	O_2	0.8	Pure-phase ε-Ga_2O_3 [26]
c-plane sapphire	9.1	450–570	TEGa (Ar as carrier gas)	O_2	N.R.	$(\bar{2}01)$ β-Ga_2O_3 ∥ (0001) Al_2O_3 and (0001) ε-Ga_2O_3 ∥ (0001) Al_2O_3 [27]

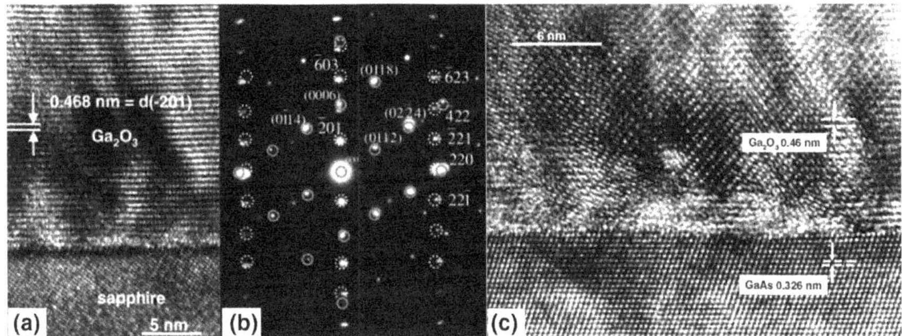

Fig. 9.2 **a** Cross-sectional HRTEM cross-section image and **b** SAD pattern of β-Ga_2O_3 on *c*-plane sapphire. The electron beam direction is parallel to [$2\bar{1}\bar{1}0$] sapphire. The SAD pattern from the substrate is marked by closed circles and indexed with brackets, whereas the pattern from the epilayer is marked by dotted circles and indexed without brackets. **c** HRTEM cross-section image of β-Ga_2O_3 grown on $(\bar{1}10)_{As}$ GaAs. The electron beam direction is parallel to [110] GaAs and [10] β-Ga_2O_3. Reprinted with permission from publisher [13]

The first successful epitaxial growth of β-Ga_2O_3 was reported in 2009 [13]. Using a non-commercial horizontal reactor operating at atmospheric pressure, they studied the growth of Ga_2O_3 on *a*-, *c*-, *m*- and *r*-plane sapphire. Figure 9.2 shows high-resolution transmission electron microscopy (HRTEM) images of the epitaxial films on sapphire (0001) and GaAs $(\bar{1}10)_{As}$.

β-Ga_2O_3 epitaxial layers on (0001) sapphire by MOCVD often display columnar morphology as we have observed (see Fig. 9.3), which would suggest that the growth mechanism follows the Volmer-Weber growth mode, or island growth mode. The out-of-plane orientation relationship is normally ($\bar{2}01$) β-Ga_2O_3 ∥ (001) sapphire. The corresponding θ-2θ x-ray diffraction scans display the set of $\bar{2}01$ β-Ga_2O_3 peaks; however, the ($\bar{4}02$) peaks are typically larger than the ($\bar{6}03$) and ($\bar{8}04$) peaks. Gogova et al. suggest that this could be due to a high density of stacking faults, which leads to large vertical disorder in diffracting lattice planes [24].

Increased growth rates of epitaxial β-Ga_2O_3 thin films on c-plane sapphire substrates using a close coupled showerhead vertical MOCVD reactor were recently reported [12]. This system has a small separation between the showerhead gas entry and substrate of only approximately 1 cm. They examined the use of various Ga precursors, including $Ga(DPM)_3$ (DPM = dipivaloylmethanate), triethylgallium (TEGa), and trimethylgallium (TMGa), with separate injection of molecular oxygen as the oxidizer. The $Ga(DPM)_3$ was sublimated at 155 °C, whereas the TEGa and TMGa sources were kept at or slightly below room temperature and Ar was used as the carrier gas. The sapphire substrates were rotated at 170 rpm to enhance deposition uniformity. Growth rates up to 10 μm/h were achieved using TMGa at a substrate temperature of 900 °C.

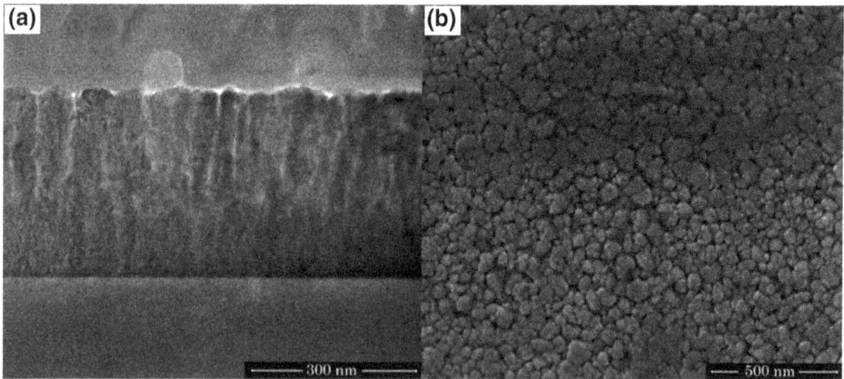

Fig. 9.3 **a** Cross-section and **b** plan view SEM of β-Ga_2O_3 on sapphire showing columnar morphology of the film

9.3.2 *Growth of α- and ε-Ga_2O_3*

As stated previously, MIST CVD® is the established method for producing pure-phase α-Ga_2O_3 heteroepitaxial layers, with sapphire being the preferred substrate, since both materials share the corundum crystal structure. α-Ga_2O_3 heteroepitaxial layers grown by MOCVD tend to be in mixed-phase form. In fact, α-Ga_2O_3 is sometimes observed as a thin interlayer between the substrate and an overlayer of a different phase. For example, Schewski et al. observed three monolayers of pseudomorphic α-Ga_2O_3 between β-Ga_2O_3 and a sapphire (0001) substrate across three different growth techniques, namely MOCVD, PLD, and MBE. A high-angle annular dark field (HAADF) cross-sectional STEM image of this interlayer is shown in Fig. 9.4. Yao et al. also reported the growth of a thin (∼10 nm) interlayer of α-Ga_2O_3 in films grown by HVPE [28]; in this case, the α-Ga_2O_3 layer was located underneath an epitaxial layer of ε-Ga_2O_3. As shown in Fig. 9.5, the epitaxial layer relationship was determined to be $[\bar{1}100]$ ε-Ga_2O_3 || $[11\bar{2}0]$ α-Ga_2O_3 || $[11\bar{2}0]$ α-Al_2O_3.

ε-Ga_2O_3 heteroepitaxial layers have been grown by HVPE, MOCVD, ALD, and MIST CVD® [16, 18, 20, 25, 27–32]. ε-phase growth has been demonstrated on a variety of substrates including GaN, AlN, β-Ga_2O_3, SiC, MgO, yttria-stabilized zirconia, and sapphire. The ε-phase was reported to nucleate and grow on sapphire as hexagonal islands up to ∼400 nm before coalescing into a continuous film [25]. This growth morphology resembles that shown in Fig. 9.6. A recent report indicates that a NiO buffer layer on sapphire can be used to promote the growth of pure-phase ε-Ga_2O_3 using mist CVD [31].

Sun et al. observed growth of three different phases of Ga_2O_3 (α, β, and ε) films on *c*-plane sapphire by introducing different flow rate of HCl to the MOCVD growth process [16]. They observed a three-fold increase in the growth rate of pure-phase β-Ga_2O_3 at HCl flow rate of 5 sccm. When the HCl flow rate is increased to 10 sccm, the film transitioned from pure-phase β-Ga_2O_3 to a mixture of β- and ε-Ga_2O_3. At

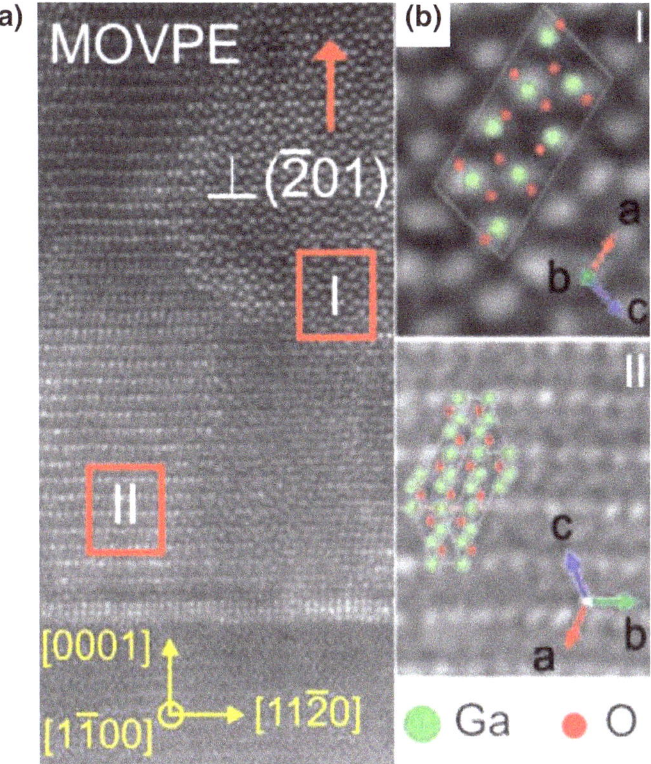

Fig. 9.4 **a** STEM HAADF image of β-Ga_2O_3 grown on Al_2O_3 (0001) substrate by MOCVD. The bottom part of the image shows the sapphire substrate with a [$1\bar{1}00$] direction parallel to the electron beam; in the upper part, the textured β-Ga_2O_3 is visible. The orientation of two rotational domains are highlighted with (I) <010> and (II) <nearly> <132> directions parallel to the [1100] direction of the sapphire. Three atomic layers of α-Ga_2O_3 are observed between β-Ga_2O_3 and the substrate. **b** Stick-and-ball models of the two β-Ga_2O_3 domains (region I and II) are overlaid to their respective projection. Reprinted with permission from publisher [21]

a HCl flow rate of 30 sccm, which yielded the highest growth rate of ~1 μm/hr, only ε-Ga_2O_3 was observed. Further, increase in HCl flow rate produced a mixture of α- and ε-Ga_2O_3 and a rougher surface morphology. The authors proposed that this could be due to the higher growth rate of ε-Ga_2O_3 compared to α-Ga_2O_3.

Certain MOCVD growth conditions have also been demonstrated to yield β-Ga_2O_3 and ε-Ga_2O_3 on *c*-plane sapphire. The phase composition was found to be a function of both growth temperature and VI/III ratio, with some conditions leading to a mixture of β- and ε-Ga_2O_3. The structural quality of ε-Ga_2O_3 was superior to that of β-Ga_2O_3 under similar growth conditions. With decreasing growth temperature, the MOCVD layers went from pure β-Ga_2O_3 to a mixture of the β- and ε-Ga_2O_3 polymorphs and finally to microcrystalline structures at the lowest temperatures investigated [27].

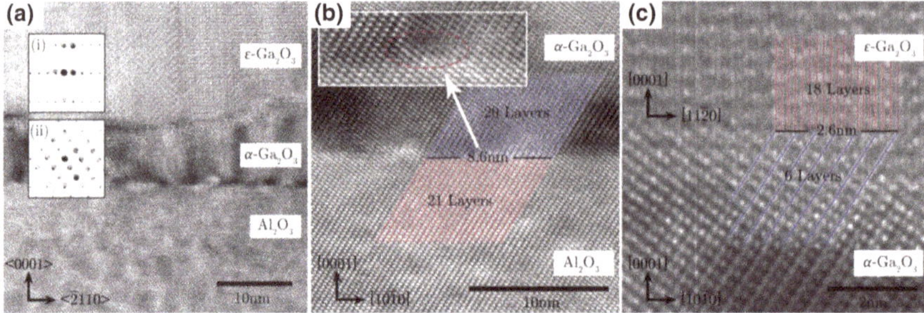

Fig. 9.5 **a** Cross-section HRTEM of Ga_2O_3 *c*-plane sapphire along the Al_2O_3 [11$\bar{2}$0] zone axis showing an ~10 nm thick interfacial layer of α-Ga_2O_3 grew directly on the substrate, followed by a thicker layer of ε-Ga_2O_3. Insets show selected area diffraction corresponding to (i) ε-Ga_2O_3 [$\bar{1}$100] and (ii) α-Ga_2O_3 [11$\bar{2}$0]. The epitaxial relationship was found to be [$\bar{1}$100] ε-Ga_2O_3 ‖ [11$\bar{2}$0] α-Ga_2O_3 ‖ [11$\bar{2}$0] α-Al_2O_3. **b** Cross-section HRTEM of α-Ga_2O_3 on *c*-plane sapphire along the α-Al_2O_3 [11$\bar{2}$0] zone axis. Inset is magnified image of misfit dislocation. **c** Cross-section HRTEM of ε-Ga_2O_3 on α-Ga_2O_3 along the α-Ga_2O_3 [11$\bar{2}$0] zone axis. Reprinted with permission from publisher [28]

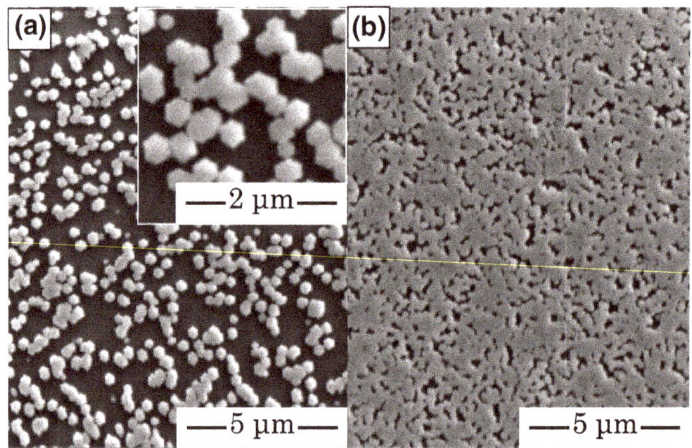

Fig. 9.6 **a** SEM images showing the growth mode of ε-Ga_2O_3 on *c*-plane sapphire with two growth conditions—**a** formation of hexagonal islands and **b** almost complete coalescence hexagonal islands to form continuous film, where hexagon-shape edges are still visible. This resembles the morphology reported by Boschi et al. showing two different stages of film growth [25]. Reprinted with permission from publisher [28]

9.3.3 *Effect of Growth Condition on Film Growth*

Growth temperatures >800 °C favor the thermodynamically stable β-Ga_2O_3 phase. MOCVD growth rates in the temperature range that are typically used to grow β-Ga_2O_3 epitaxial films have been known to decrease with increasing temperature. An example is shown in Fig. 9.7 [33], which represents β-Ga_2O_3 film growth using TMGa and H_2O as precursors. The decreasing growth rate was attributed to desorption of gallium suboxide (Ga_2O) at higher growth temperatures. The choice of growth temperature then becomes a compromise between increased growth rates at lower temperatures and improved crystal quality at higher temperatures.

Another effect of lower growth temperatures is that metastable polymorphs such as α- or ε-Ga_2O_3 may nucleate and grow, either as pure- or mixed-phase films. In addition to lower temperatures, faster growth rates [28], compressive stress [28, 34], and lattice-matched substrates [31] have been proposed as conditions that favor the growth of metastable Ga_2O_3 phases.

On *c*-plane sapphire, which is the most common substrate of choice, the lattice mismatch between β-Ga_2O_3 is larger than that of α- and ε-Ga_2O_3. Heteroepitaxy of β-Ga_2O_3 may therefore be kinetically impeded at lower temperatures and higher growth rates, thereby "locking in" the growth of a better lattice-matched metastable phase to reduce strain.

Other growth conditions also appear to have an effect on the formation of metastable phases. Zhuo et al. found that the growth of ε-Ga_2O_3 is favored at low VI/III ratios. In fact, they hypothesized that ε-Ga_2O_3 is more stable in weakly oxidizing or even reducing environments. They also found that reducing the growth rate by reducing the precursor flow rate (while keeping the VI/III ratio constant) favors the growth of ε-Ga_2O_3 [27]. In HVPE grown films, the HCl flow rate was also reported to determine the phases that grew [16]. The authors concluded that Cl acts as a catalyst for nucleation of α- and ε-Ga_2O_3. Further investigations are necessary to gain a conclusive understanding of the effects of growth conditions on the formation of various phases of Ga_2O_3.

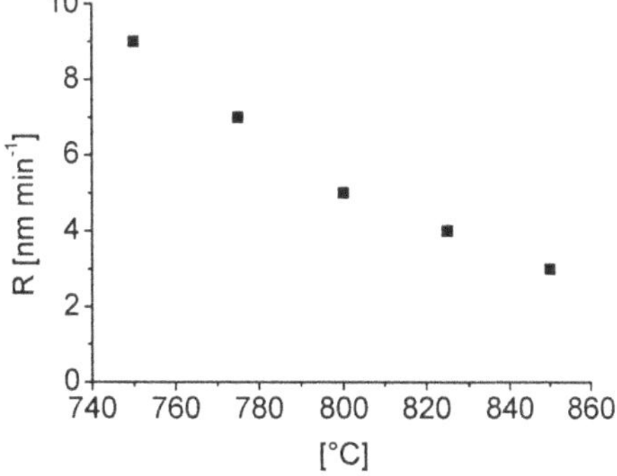

Fig. 9.7 Dependence of the growth rate on the substrate temperature during homoepitaxial growth of β-Ga_2O_3. Chamber pressure is kept at 20 mbar. Reprinted with permission from publisher [33]

9.4 Doping in MOCVD-Grown Heteroepitaxial Ga_2O_3 Films

Unintentionally doped β-Ga_2O_3 is often found to have *n*-type conductivity. Although this unintentional doping has sometimes been attributed to oxygen vacancies, Varley et al. showed through hybrid density functional theory (DFT) calculations that oxygen vacancies are in fact deep donors in Ga_2O_3 and therefore do not contribute to its conductivity. The unintentional doping has instead been connected with other impurities, such as Si and H [35]. Aside from Si, other potential shallow donors include Ge, Sn, F, Cl, and Nb [35, 36]. Si and Ge can substitute at tetrahedrally coordinated Ga(I) sites; Sn can substitute at octahedrally coordinated Ga(II) sites; F and Cl can substitute at three-fold coordinated O(I) sites; and Nb can substitute at either tetrahedrally coordinated Ga(I) sites or octahedrally coordinated Ga(II) sites, with the octahedral site being the preferred one.

Although *n*-type β-Ga_2O_3 is readily obtained, *p*-type doping has proved to be elusive. Varley et al. calculated that holes in Ga_2O_3 preferentially form localized self-trapped holes (STH) with low mobility. These results indicate that achieving conductive, *p*-type Ga_2O_3 is not expected.

The two most common *n*-type dopants in β-Ga_2O_3 grown by MOCVD are Si and Sn. Electron mobilities of 130 and 100 cm^2/Vs for Si and Sn doping, respectively, at carrier densities of $\sim 10^{17}$ cm^{-3} (Fig. 9.8b), have been reported [37]. The data in Fig. 9.8a show that there is good correlation between doping and carrier concentration up to 8×10^{19} cm^{-3} and 10^{19} cm^{-3} for Si and Sn, respectively. Another study of Si-doped β-Ga_2O_3 grown on *c*-plane sapphire indicated that the structural properties did not degrade up to carrier concentration of 10^{18} cm^{-3} [24]. In fact, the densities of twins and stacking faults were found to be an order of magnitude lower in doped films than in undoped films.

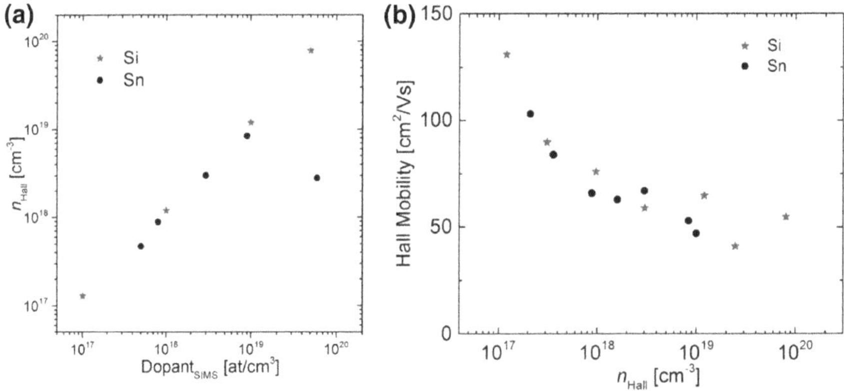

Fig. 9.8 **a** Hall free carrier concentration versus the dopant (Si and Sn) concentration obtained by SIMS. **b** Hall mobility versus Hall free carrier concentration. Reprinted with permission from publisher [37]

9.5 Summary and Conclusions

This chapter reviewed the growth of α-, β-, and ε-Ga_2O_3 heteroepitaxial layers using MOCVD, and included discussion of other deposition methods where relevant. It presented an overview of the range of growth conditions that have been reported in the literature to yield epitaxial layers of these different polymorphs. Although β-Ga_2O_3 is the stable phase up to the melting point, the energy differences among the three polymorphs are sufficiently small to enable growth of the metastable phases by tuning growth conditions, such as substrate temperature, growth rate, and precursor chemistry and/or flow rates. To date, *c*-plane sapphire has been the most common substrate for these heteroepitaxial layers, although many different substrates have been employed, with certain ones reported to lead to preferred growth of the α- or ε-phases.

N-type doping of MOCVD-grown β-Ga_2O_3 films using Si and Sn is well established. In short, the doping concentration ranges between 1×10^{17} and 8×10^{19} cm^{-3}, with corresponding electron mobility values between ~130 and 50 cm^2/Vs. Theory and experiments indicate that achieving effective *p*-type conductivity in Ga_2O_3 is not feasible. To realize commercial devices based on Ga_2O_3 in the future, advancements in a number of areas will be needed: e.g., increasing growth rates while improving crystallinity, understanding the role of defects on electrical performance, and controlling the growth conditions that lead to particular phases of Ga_2O_3.

References

1. M. Higashiwaki, K. Sasaki, A. Kuramata, T. Masui, S. Yamakoshi, Phys. Status Solidi A **211**(1), 21 (2014)
2. T. Oshima, T. Okuno, N. Arai, N. Suzuki, S. Ohira, S. Fujita, Appl. Phys. Express **1**(1), 011202 (2008)
3. M. Bartic, M. Ogita, M. Isai, C.L. Baban, H. Suzuki, J. Appl. Phys. **102**(2), 023709 (2007)
4. H.G. Kim, W.T. Kim, J. Appl. Phys. **62**(5), 2000 (1987)
5. A. Ortiz, J. Alonso, E. Andrade, C. Urbiola, J. Electrochem. Soc. **148**(2), F26 (2001)
6. Y. Kokubun, K. Miura, F. Endo, S. Nakagomi, Appl. Phys. Lett. **90**(3), 031912 (2007)
7. D. Shinohara, S. Fujita, Jpn. J. Appl. Phys. **47**(9R), 7311 (2008)
8. C.I. Baban, Y. Toyoda, M. Ogita, Jpn. J. Appl. Phys. **43**(10R), 7213 (2004)
9. M. Orita, H. Ohta, M. Hirano, H. Hosono, Appl. Phys. Lett. **77**(25), 4166 (2000)
10. M. Higashiwaki, K. Sasaki, A. Kuramata, T. Masui, S. Yamakoshi, Appl. Phys. Lett. **100**(1), 013504 (2012)
11. H. Murakami, K. Nomura, K. Goto, K. Sasaki, K. Kawara, Q.T. Thieu, R. Togashi, Y. Kumagai, M. Higashiwaki, A. Kuramata, S. Yamakoshi, B. Monemar, A. Koukitu, Appl. Phys. Express **8**(1), 015503 (2014)
12. F. Alema, B. Hertog, A. Osinsky, P. Mukhopadhyay, M. Toporkov, W.V. Schoenfeld, J. Cryst. Growth **475**, 77 (2017)
13. V. Gottschalch, K. Mergenthaler, G. Wagner, J. Bauer, H. Paetzelt, C. Sturm, U. Teschner, Phys. Status Solidi A **206**(2), 243 (2009)
14. M. Mohamed, C. Janowitz, I. Unger, R. Manzke, Z. Galazka, R. Uecker, R. Fornari, J. Weber, J. Varley, C. Van de Walle, Appl. Phys. Lett. **97**(21), 211903 (2010)

15. S. Okur, G.S. Tompa, N. Sbrockey, T. Salagaj, V. Blank, B. Henniger, M. Baldini, G. Wagner, Z. Galazka, Y. Yao, J. Rokholt, R.F. Davis, L.M. Porter, Vac. Tech. & Coating, pp. 31–39 (May 2017)
16. H. Sun, K.H. Li, C.T. Castanedo, S. Okur, G.S. Tompa, T. Salagaj, S. Lopatin, A. Genovese, X. Li, Cryst. Growth Des. **18**(4), 2370 (2018)
17. R. Roy, V. Hill, E. Osborn, J. Am. Chem. Soc. **74**(3), 719 (1952)
18. I. Cora, F. Mezzadri, F. Boschi, M. Bosi, M. Čaplovičová, G. Calestani, I. Dódony, B. Pécz, R. Fornari, Cryst Eng Comm **19**(11), 1509 (2017)
19. S. Yoshioka, H. Hayashi, A. Kuwabara, F. Oba, K. Matsunaga, I. Tanaka, J. Phys. C: Solid State Phys. **19**(34), 346211 (2007)
20. F. Mezzadri, G. Calestani, F. Boschi, D. Delmonte, M. Bosi, R. Fornari, Inorg. Chem. **55**(22), 12079 (2016)
21. R. Schewski, G. Wagner, M. Baldini, D. Gogova, Z. Galazka, T. Schulz, T. Remmele, T. Markurt, H. von Wenckstern, M. Grundmann, O. Bierwagen, P. Vogt, M. Albrecht, Appl. Phys. Express **8**(1), 011101 (2014)
22. T. Oshima, T. Nakazono, A. Mukai, A. Ohtomo, J. Cryst. Growth **359**, 60 (2012)
23. H.W. Kim, N.H. Kim, Mater. Sci. Eng. B **110**(1), 34 (2004)
24. D. Gogova, G. Wagner, M. Baldini, M. Schmidbauer, K. Irmscher, R. Schewski, Z. Galazka, M. Albrecht, R. Fornari, J. Cryst. Growth **401**, 665 (2014)
25. F. Boschi, M. Bosi, T. Berzina, E. Buffagni, C. Ferrari, R. Fornari, J. Cryst. Growth **443**, 25 (2016)
26. X. Xia, Y. Chen, Q. Feng, H. Liang, P. Tao, M. Xu, G. Du, Appl. Phys. Lett. **108**(20), 202103 (2016)
27. Y. Zhuo, Z. Chen, W. Tu, X. Ma, Y. Pei, G. Wang, Appl. Surf. Sci. **420**, 802 (2017)
28. Y. Yao, S. Okur, L.A. Lyle, G.S. Tompa, T. Salagaj, N. Sbrockey, R.F. Davis, L.M. Porter, Mater. Res. Lett. **6**(5), 268 (2018)
29. Y. Oshima, E.G. Víllora, Y. Matsushita, S. Yamamoto, K. Shimamura, J. Appl. Phys **118**(8), 085301 (2015)
30. D. Tahara, H. Nishinaka, S. Morimoto, M. Yoshimoto, Jpn. J. Appl. Phys. **56**(7), 078004 (2017)
31. Y. Arata, H. Nishinaka, D. Tahara, M. Yoshimoto, Cryst Eng Comm **20**(40), 6236 (2018)
32. H. Nishinaka, H. Komai, D. Tahara, Y. Arata, M. Yoshimoto, Jpn. J. Appl. Phys. **57**(11), 115601 (2018)
33. G. Wagner, M. Baldini, D. Gogova, M. Schmidbauer, R. Schewski, M. Albrecht, Z. Galazka, D. Klimm, R. Fornari, Phys. Status Solidi A **211**(1), 27 (2014)
34. Y. Yao, L.A. Lyle, J.A. Rokholt, S. Okur, G.S. Tompa, T. Salagaj, N. Sbrockey, R.F. Davis, L.M. Porter, ECS Trans. **80**(7), 191 (2017)
35. J. Varley, J. Weber, A. Janotti, C. Van de Walle, Appl. Phys. Lett. **97**(14), 142106 (2010)
36. H. Peelaers, C. Van de Walle, Phys. Rev. B **94**(19), 195203 (2016)
37. M. Baldini, M. Albrecht, A. Fiedler, K. Irmscher, R. Schewski, G. Wagner, ECS J. Solid State Sci. Technol. **6**(2), Q3040 (2017)

Chapter 10
Halide Vapor Phase Epitaxy 1

Homoepitaxial Growth of β-Ga_2O_3 on β-Ga_2O_3 Substrates

Yoshinao Kumagai, Keita Konishi, Ken Goto, Hisashi Murakami and Bo Monemar

Abstract Homoepitaxial growth of β-Ga_2O_3 on β-Ga_2O_3 substrates by halide vapor-phase epitaxy (HVPE) using GaCl and O_2 was investigated by both thermodynamic analysis and growth experiments. The thermodynamic analysis clarified that growth of Ga_2O_3 is expected at high temperatures around 1000 °C using an inert carrier gas. The experimental results revealed that homoepitaxial growth of unintentionally doped (UID) layers with a low effective donor concentration ($N_d - N_a$) of less than 10^{13} cm^{-3} is possible at 1000 °C on β-Ga_2O_3 (001) substrates with a high growth rate of up to 28 μm/h. Furthermore, HVPE growth of intentionally Si-doped β-Ga_2O_3 layers was investigated by supplying $SiCl_4$, which revealed that *n*-type carrier density almost equal to the Si-doping concentration can be controlled in the range of 10^{15}–10^{18} cm^{-3}. The carrier mobility decreased with increasing Si impurity concentration and was about 150 cm^2/V·s at room temperature for a layer with a

Y. Kumagai (✉) · H. Murakami · B. Monemar
Institute of Global Innovation Research, Tokyo University of Agriculture and Technology, Tokyo, Japan
e-mail: 4470kuma@cc.tuat.ac.jp

H. Murakami
e-mail: murak@cc.tuat.ac.jp

B. Monemar
e-mail: bo.monemar@liu.se

Y. Kumagai · K. Konishi · K. Goto · H. Murakami
Department of Applied Chemistry, Tokyo University of Agriculture and Technology, Tokyo, Japan
e-mail: keitakonishi@go.tuat.ac.jp

K. Goto
e-mail: gotoken@go.tuat.ac.jp

K. Goto
Novel Crystal Technology, Inc., Saitama, Japan

B. Monemar
Department of Physics, Chemistry and Biology (IFM), Linköping University, Linköping, Sweden

M. Higashiwaki and S. Fujita (eds.), *Gallium Oxide*, Springer Series in Materials Science 293, https://doi.org/10.1007/978-3-030-37153-1_10

carrier density of 3.2 × 10^{15} cm^{-3}. Thus, the intentionally Si-doped homoepitaxial layers grown on β-Ga_2O_3 substrates can be applicable for the production of β-Ga_2O_3-based power devices.

10.1 Introduction

Crystal growth of bulk β-Ga_2O_3 has been investigated by several melt growth methods such as Czochralski (CZ) [1, 2], floating zone (FZ) [3], Bridgman [4], and edge-defined film-fed growth (EFG) [5, 6]. At present, 2-in.-diameter single-crystal β-Ga_2O_3 substrates are mass-produced by the EFG method [6]. However, in order to develop high-performance power devices such as vertical Schottky barrier diodes (SBDs) and field-effect transistors (FETs), a high-speed growth method of conductivity controlled homoepitaxial layers (drift layers) is essential. Homoepitaxial growth of β-Ga_2O_3 has been reported using molecular beam epitaxy (MBE) [7–9], mist chemical vapor deposition (mist CVD) [10], metalorganic CVD (MOCVD) [11–13], and halide vapor-phase epitaxy (HVPE) [14–18]. Among these growth methods, the HVPE method is most suitable for high-speed growth of high-purity layers and also enables conductivity control by intentional doping.

Recently, the authors of this chapter have performed thermodynamic analysis of the HVPE growth system of β-Ga_2O_3 using GaCl and O_2 as precursors and clarified conditions suitable for growth [14]. Based on the analysis results, the authors tried HVPE growth of β-Ga_2O_3 and achieved high-speed growth of high-purity unintentionally doped (UID) homoepitaxial layers with an effective donor concentration ($N_d - N_a$) of less than 10^{13} cm^{-3} [15, 17]. Furthermore, $SiCl_4$ was supplied as a doping gas during the HVPE growth to grow intentionally Si-doped β-Ga_2O_3 layers, and control of *n*-type carrier density in the range of 10^{15}–10^{18} cm^{-3} was achieved [18]. At present, 2-in.-diameter homoepitaxial substrates are mass-produced by homoepitaxial growth of Si-doped layers on β-Ga_2O_3 substrates prepared by the EFG method [16]. With the availability of homoepitaxial substrates fabricated by the HVPE method, the development of vertical SBDs [19–21] and FETs [22–24] has been accelerated.

In this chapter, thermodynamic analysis of the HVPE growth system for β-Ga_2O_3 is first introduced. Then, high-speed homoepitaxial growth of high-purity UID layers and intentionally Si-doped layers reported by the authors are introduced.

10.2 Thermodynamic Analysis of β-Ga_2O_3 Growth by HVPE

10.2.1 Calculation Procedure

HVPE uses halides of metal elements. Here, gallium monochloride (GaCl) or gallium trichloride ($GaCl_3$) is considered as a Ga precursor. On the other hand, oxygen (O_2) or water (H_2O) can be used as an oxygen precursor. Therefore, the following four chemical reactions are considered for the growth of β-Ga_2O_3:

$$2GaCl(g) + (3/2)O_2(g) = \beta\text{-}Ga_2O_3(s) + Cl_2(g), \tag{10.1}$$

$$2GaCl(g) + 3H_2O(g) = \beta\text{-}Ga_2O_3(s) + 2HCl(g) + 2H_2(g), \tag{10.2}$$

$$2GaCl_3(g) + (3/2)O_2(g) = \beta\text{-}Ga_2O_3(s) + 3Cl_2(g), \tag{10.3}$$

$$2GaCl_3(g) + 3H_2O(g) = \beta\text{-}Ga_2O_3(s) + 6HCl(g). \tag{10.4}$$

Figure 10.1 shows the equilibrium constants (*K*) of the β-Ga_2O_3 growth reactions (10.1)–(10.4) with respect to the reciprocal of growth temperature. The values were calculated using thermochemical data [25, 26]. It is seen that the growth reaction (10.1) has the largest equilibrium constant. This indicates that the HVPE using GaCl and O_2 as precursors is favorable for β-Ga_2O_3 growth.

Then, equilibrium partial pressures of gaseous species over the β-Ga_2O_3 substrate after supplying GaCl and O_2 were calculated. As a carrier gas, a mixed gas of an inert gas (IG: N_2, He or Ar) and H_2 was considered. In this case, the following 17 gaseous species coexist: GaCl, $GaCl_2$, $GaCl_3$, $(GaCl_3)_2$, GaO, Ga_2O, Ga, GaH, GaH_2, GaH_3, GaOH, Cl_2, O_2, H_2, HCl, H_2O, and IG, and these species are related to the following 12 simultaneous chemical reactions (10.5)–(10.16) in addition to the reaction (10.1):

$$GaCl(g) + (1/2)Cl_2(g) = GaCl_2(g), \tag{10.5}$$

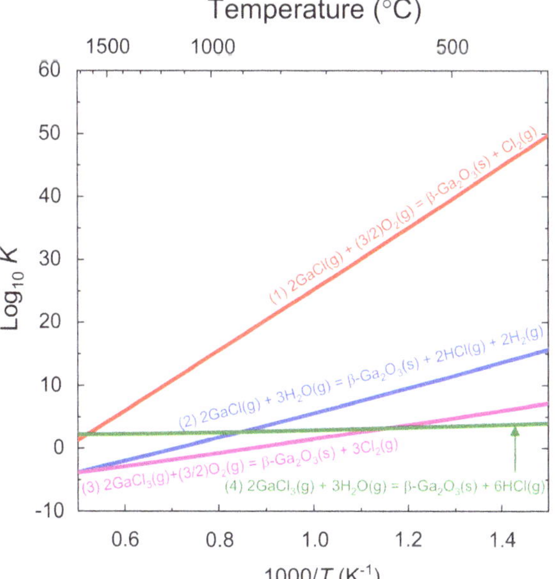

Fig. 10.1 Values of equilibrium constants (*K*) as a function of the reciprocal of growth temperature calculated for the β-Ga_2O_3 growth reactions

$$GaCl(g) + Cl_2(g) = GaCl_3(g), \tag{10.6}$$

$$2GaCl_3(g) = (GaCl_3)_2(g), \tag{10.7}$$

$$GaCl(g) + (1/2)O_2(g) = GaO(g) + (1/2)Cl_2(g), \tag{10.8}$$

$$2GaCl(g) + (1/2)O_2(g) = Ga_2O(g) + Cl_2(g), \tag{10.9}$$

$$GaCl(g) = Ga(g) + (1/2)Cl_2(g), \tag{10.10}$$

$$H_2(g) + Cl_2(g) = 2HCl(g), \tag{10.11}$$

$$H_2(g) + (1/2)O_2(g) = H_2O(g), \tag{10.12}$$

$$Ga(g) + (1/2)H_2(g) = GaH(g), \tag{10.13}$$

$$Ga(g) + H_2(g) = GaH_2(g), \tag{10.14}$$

$$Ga(g) + (3/2)H_2(g) = GaH_3(g), \tag{10.15}$$

$$Ga(g) + H_2O(g) = GaOH(g) + (1/2)H_2(g). \tag{10.16}$$

The equilibrium equations for the chemical reaction (10.1) and (10.5)–(10.16) are expressed as follows:

$$K_1(T) = \frac{P_{Cl_2}}{P^2_{GaCl} P^{3/2}_{O_2}}, \tag{10.17}$$

$$K_5(T) = \frac{P_{GaCl_2}}{P_{GaCl} P^{1/2}_{Cl_2}}, \tag{10.18}$$

$$K_6(T) = \frac{P_{GaCl_3}}{P_{GaCl} P_{Cl_2}}, \tag{10.19}$$

$$K_7(T) = \frac{P_{(GaCl_3)_2}}{P^2_{GaCl_3}}, \tag{10.20}$$

$$K_8(T) = \frac{P_{GaO} P^{1/2}_{Cl_2}}{P_{GaCl} P^{1/2}_{O_2}}, \tag{10.21}$$

$$K_9(T) = \frac{P_{Ga_2O} P_{Cl_2}}{P^2_{GaCl} P^{1/2}_{O_2}}, \tag{10.22}$$

$$K_{10}(T) = \frac{P_{\mathrm{Ga}} P_{\mathrm{Cl_2}}^{1/2}}{P_{\mathrm{GaCl}}}, \tag{10.23}$$

$$K_{11}(T) = \frac{P_{\mathrm{HCl}}^2}{P_{\mathrm{H_2}} P_{\mathrm{Cl_2}}}, \tag{10.24}$$

$$K_{12}(T) = \frac{P_{\mathrm{H_2O}}}{P_{\mathrm{H_2}} P_{\mathrm{O_2}}^{1/2}}, \tag{10.25}$$

$$K_{13}(T) = \frac{P_{\mathrm{GaH}}}{P_{\mathrm{Ga}} P_{\mathrm{H_2}}^{1/2}}, \tag{10.26}$$

$$K_{14}(T) = \frac{P_{\mathrm{GaH_2}}}{P_{\mathrm{Ga}} P_{\mathrm{H_2}}}, \tag{10.27}$$

$$K_{15}(T) = \frac{P_{\mathrm{GaH_3}}}{P_{\mathrm{Ga}} P_{\mathrm{H_2}}^{3/2}}, \tag{10.28}$$

$$K_{16}(T) = \frac{P_{\mathrm{GaOH}} P_{\mathrm{H_2}}^{1/2}}{P_{\mathrm{Ga}} P_{\mathrm{H_2O}}}, \tag{10.29}$$

where P_{i} are the equilibrium partial pressures of the gaseous species, and $K_i(T)$ are the temperature-dependent equilibrium constants. The $K_i(T)$ can be calculated from thermochemical data [25–27] and fitted by the equation:

$$\log_{10} K_i(T) = a + b/T + c \log_{10} T, \tag{10.30}$$

where T is the absolute temperature. The fitting parameters a, b, and c for each chemical reaction are listed in Table 10.1. Since the total pressure in the growth reactor (P_{tot}) is constant, the following equation is obtained:

$$\begin{aligned} P_{\mathrm{tot}} = {} & P_{\mathrm{GaCl}} + P_{\mathrm{GaCl_2}} + P_{\mathrm{GaCl_3}} + P_{\mathrm{(GaCl_3)_2}} + P_{\mathrm{GaO}} + P_{\mathrm{Ga_2O}} + P_{\mathrm{Ga}} + P_{\mathrm{GaH}} + P_{\mathrm{GaH_2}} \\ & + P_{\mathrm{GaH_3}} + P_{\mathrm{GaOH}} + P_{\mathrm{Cl_2}} + P_{\mathrm{O_2}} + P_{\mathrm{H_2}} + P_{\mathrm{HCl}} + P_{\mathrm{H_2O}} + P_{\mathrm{IG}}. \end{aligned} \tag{10.31}$$

Denoting the input partial pressures as $P_{\mathrm{i}}^{\mathrm{o}}$, the following equation is obtained from the stoichiometric relationship for Ga_2O_3 formation:

$$\begin{aligned} & \frac{1}{2}\left[P_{\mathrm{GaCl}}^{\mathrm{o}} - (P_{\mathrm{GaCl}} + P_{\mathrm{GaCl_2}} + P_{\mathrm{GaCl_3}} + 2P_{\mathrm{(GaCl_3)_2}} + P_{\mathrm{GaO}} + 2P_{\mathrm{Ga_2O}} + P_{\mathrm{Ga}} + P_{\mathrm{GaH}} \right. \\ & \left. + P_{\mathrm{GaH_2}} + P_{\mathrm{GaH_3}} + P_{\mathrm{GaOH}})\right] = \frac{1}{3}\left[2P_{\mathrm{O_2}}^{\mathrm{o}} - (P_{\mathrm{GaO}} + P_{\mathrm{Ga_2O}} + P_{\mathrm{GaOH}} + 2P_{\mathrm{O_2}} + P_{\mathrm{H_2O}})\right]. \end{aligned} \tag{10.32}$$

Table 10.1 Fitting parameters for temperature-dependent equilibrium constants for chemical reactions

Reactions	a	b	c
$2GaCl(g) + \frac{3}{2}O_2(g) = \beta\text{-}Ga_2O_3(s) + Cl_2(g)$	-3.72×10^1	5.03×10^4	4.05×10^0
$GaCl(g) + \frac{1}{2}Cl_2(g) = GaCl_2(g)$	-3.00×10^0	7.91×10^3	1.71×10^{-1}
$GaCl(g) + Cl_2(g) = GaCl_3(g)$	-9.67×10^0	1.91×10^4	8.50×10^{-1}
$2GaCl_3(g) = (GaCl_3)_2(g)$	-1.40×10^1	5.34×10^3	2.00×10^0
$GaCl(g) + \frac{1}{2}O_2(g) = GaO(g) + \frac{1}{2}Cl_2(g)$	-1.30×10^0	-1.12×10^4	4.05×10^{-1}
$2GaCl(g) + \frac{1}{2}O_2(g) = Ga_2O(g) + Cl_2(g)$	-4.33×10^0	-2.05×10^3	1.08×10^{-1}
$GaCl(g) = Ga(g) + \frac{1}{2}Cl_2(g)$	1.55×10^{-1}	-1.78×10^4	6.93×10^{-1}
$H_2(g) + Cl_2(g) = 2HCl(g)$	2.70×10^0	9.58×10^3	-5.82×10^{-1}
$H_2(g) + \frac{1}{2}O_2(g) = H_2O(g)$	2.28×10^{-2}	1.26×10^4	-8.37×10^{-1}
$Ga(g) + \frac{1}{2}H_2(g) = GaH(g)$	8.60×10^{-2}	2.97×10^3	-7.01×10^{-1}
$Ga(g) + H_2(g) = GaH_2(g)$	-1.55×10^0	5.60×10^3	-9.08×10^{-1}
$Ga(g) + \frac{3}{2}H_2(g) = GaH_3(g)$	9.45×10^0	7.72×10^3	-5.78×10^0
$Ga(g) + H_2O(g) = GaOH(g) + \frac{1}{2}H_2(g)$	-4.36×10^0	9.15×10^3	4.50×10^{-1}

Furthermore, since it can be considered that chlorine and hydrogen are not incorporated in the solid phase, the following two constraints are obtained:

$$\frac{P^{o}_{GaCl}}{2P^{o}_{H_2} + 2P^{o}_{IG}} = \frac{P_{GaCl} + 2P_{GaCl_2} + 3P_{GaCl_3} + 6P_{(GaCl_3)_2} + 2P_{Cl_2} + P_{HCl}}{P_{GaH} + 2P_{GaH_2} + 3P_{GaH_3} + P_{GaOH} + 2P_{H_2} + P_{HCl} + 2P_{H_2O} + 2P_{IG}}, \tag{10.33}$$

$$\frac{P^{o}_{H_2}}{P^{o}_{H_2} + P^{o}_{IG}} = \frac{P_{GaH} + 2P_{GaH_2} + 3P_{GaH_3} + P_{GaOH} + 2P_{H_2} + P_{HCl} + 2P_{H_2O}}{P_{GaH} + 2P_{GaH_2} + 3P_{GaH_3} + P_{GaOH} + 2P_{H_2} + P_{HCl} + 2P_{H_2O} + 2P_{IG}}. \tag{10.34}$$

Equations (10.33) and (10.34) represent the ratios of the number of chlorine and hydrogen molecules to the number of hydrogen and IG molecules, respectively.

The equilibrium partial pressures of gaseous species were obtained by solving the simultaneous equations (10.17)–(10.29) and (10.31)–(10.34) under the growth condition specified by the growth temperature, P_{tot}, P^{o}_{GaCl}, input VI/III ratio $(2P^{o}_{O_2}/P^{o}_{GaCl})$, and mole fraction of H_2 in the carrier gas $(F^{o} = P^{o}_{H_2}/(P^{o}_{H_2} + P^{o}_{IG}))$. After obtaining the equilibrium partial pressures of gaseous species, the resultant driving force of β-Ga_2O_3 growth $(\Delta P_{Ga_2O_3})$ was estimated from the difference between P^{o}_{GaCl} and the sum of the P_i of Ga-containing gaseous species:

$$\Delta P_{Ga_2O_3} = \frac{1}{2}[P^o_{GaCl} - (P_{GaCl} + P_{GaCl_2} + P_{GaCl_3} + 2P_{(GaCl_3)_2} + P_{GaO} + 2P_{Ga_2O} + P_{Ga} + P_{GaH} + P_{GaH_2} + P_{GaH_3} + P_{GaOH})]. \tag{10.35}$$

When the growth is limited by mass transportation, the growth rate (GR) is given by the equation:

$$GR = K_g \cdot \Delta P_{Ga_2O_3}, \tag{10.36}$$

where K_g is the mass transfer coefficient.

10.2.2 Calculation Results

Figure 10.2 shows the equilibrium partial pressures of the gaseous species (P_i) over β-Ga_2O_3 and the resultant driving force for β-Ga_2O_3 growth ($\Delta P_{Ga_2O_3}$) as a function of the input VI/III ratio at 1000 °C under an inert carrier gas ($F^o = 0$). Values of P_{tot} and P^o_{GaCl} were 1.0 atm and 1 × 10^{-3} atm, respectively. It can be seen that the equilibrium partial pressures of the gaseous species except for $GaCl_3$, $(GaCl_3)_2$, and IG are significantly changed at an input VI/III ratio of about 1. If the growth system is governed only by the reaction (10.1), the significant change of the gaseous species is thought to occur at an input VI/III ratio of 1.5. Therefore, the above calculation result indicates that a part of GaCl is consumed by reactions other than reaction (10.1). When the input VI/III ratio is less than 1, $\Delta P_{Ga_2O_3}$ increases as the input VI/III ratio increases, but $\Delta P_{Ga_2O_3}$ is almost saturated when the input VI/III ratio 1 or above is employed. Therefore, it is preferable to employ an input VI/III ratio above 1 to maintain uniformity of the growth.

Equilibrium partial pressures of gaseous species over β-Ga_2O_3 and the resultant $\Delta P_{Ga_2O_3}$ as a function of the growth temperature are shown in Fig. 10.3. Values of P_{tot}, P^o_{GaCl}, and input VI/III ratio were 1.0 atm, 1 × 10^{-3} atm, and 10, respectively. An IG was used as the carrier gas ($F^o = 0$). It is seen that a sufficiently large $\Delta P_{Ga_2O_3}$ can be obtained at round 1000 °C. However, $\Delta P_{Ga_2O_3}$ decreases as the growth temperature increases and becomes negative (etching) above 1600 °C. Therefore, growth of β-Ga_2O_3 is expected at 1000 °C. At the growth temperatures around 1000 °C, the equilibrium partial pressure of $GaCl_3$ is the highest, except for IG and O_2. This indicates that the supplied GaCl reacts not only with O_2 by the reaction (10.1) but also with Cl_2 [by-product of the reaction (10.1)] to form $GaCl_3$ by the reaction (10.6).

Finally, the influence of the mole fraction of H_2 in the carrier gas (F^o) on the growth temperature dependence of $\Delta P_{Ga_2O_3}$ is shown in Fig. 10.4. If all of the supplied GaCl is consumed for the growth of β-Ga_2O_3, the value of $\Delta P_{Ga_2O_3}$ is expected to be half of P^o_{GaCl}. In Fig. 10.4, it is found that $\Delta P_{Ga_2O_3}$ almost equal to

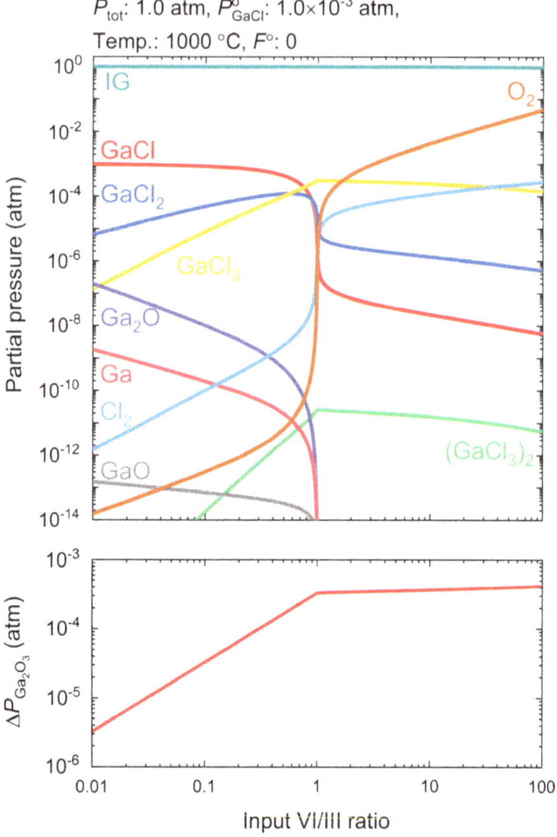

Fig. 10.2 Equilibrium partial pressures of the gaseous species and the resultant driving force for β-Ga_2O_3 growth ($\Delta P_{Ga_2O_3}$) as a function of the input VI/III ratio at 1000 °C with an inert carrier gas ($F^o = 0$)

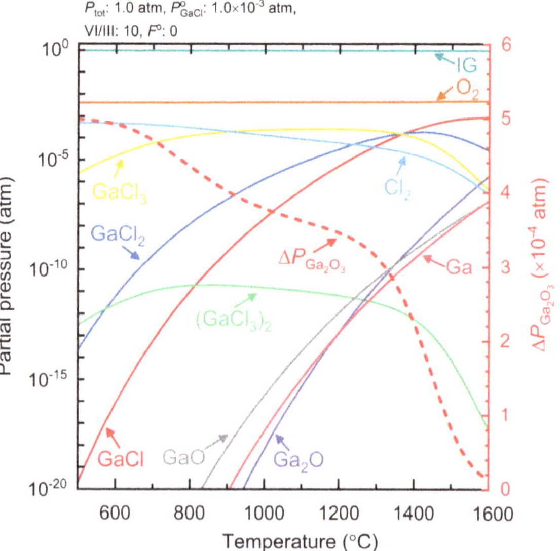

Fig. 10.3 Equilibrium partial pressures of the gaseous species and the resultant driving force for β-Ga_2O_3 growth ($\Delta P_{Ga_2O_3}$) as a function of growth temperature with an input VI/III ratio of 10 and an inert carrier gas ($F^o = 0$)

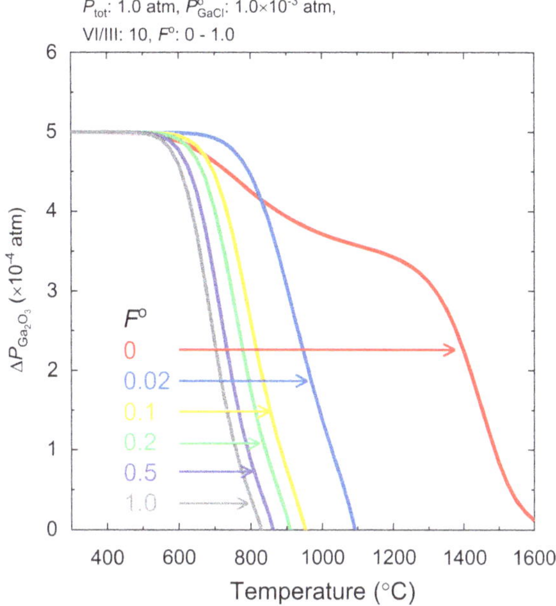

Fig. 10.4 Growth temperature dependence of driving force for β-Ga_2O_3 growth ($\Delta P_{Ga_2O_3}$) calculated for various F^o values

half of the P^o_{GaCl} is obtained at temperatures below 500 °C for all F^o. The value of $\Delta P_{Ga_2O_3}$ decreases with the growth temperature, but decreases faster as F^o becomes larger. For high-speed growth of β-Ga_2O_3 at around 1000 °C, it is essential to keep F^o less than 2% ($F^o < 0.02$).

10.3 Homoepitaxial Growth of UID β-Ga_2O_3 Layers by HVPE

Based on the results of the thermodynamic analysis, an atmospheric-pressure HVPE system as schematically shown in Fig. 10.5 was constructed. The carrier gas used was N_2 ($F^o = 0$). GaCl was generated in the upstream region maintained at 850 °C by the reaction of high-purity Ga metal (6N grade) with Cl_2. Since Cl_2 introduced over the Ga metal reacts almost completely with Ga to preferentially generate GaCl, it can be regarded that input partial pressures of GaCl twice the input partial pressure of Cl_2 is obtained. As the oxygen precursor, O_2 was used. GaCl and O_2 were separately introduced into the downstream region (growth zone), where homoepitaxial growth of UID β-Ga_2O_3 layer on Sn-doped n-type (N_d − N_a = 2.5 × 10^{18} cm^{-3}) or Fe-doped semi-insulating (SI) β-Ga_2O_3 (001) substrate prepared by EFG was carried out. For intentional Si doping of the grown layer, $SiCl_4$ can be introduced separately into the growth zone (the results are described in Sect. 10.4).

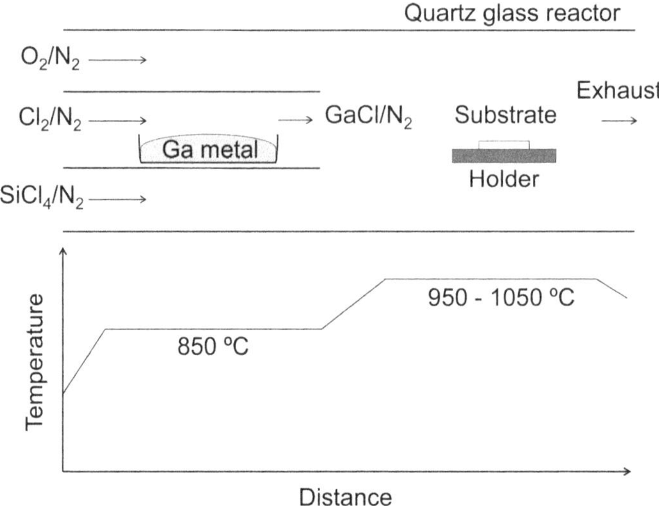

Fig. 10.5 Schematic diagram and the temperature profile of atmospheric-pressure HVPE system for β-Ga_2O_3 growth

After setting the substrate in the growth reactor, the substrate was raised to the growth temperature ranging from 950 to 1050 °C under supply of O_2 at the same input partial pressure as that for growth to suppress the decomposition of the substrate surface. Then, homoepitaxial growth was started by supplying GaCl. The input VI/III ratio during growth was fixed at 10.

Figure 10.6 shows the experimental growth rate of β-Ga_2O_3 at 1000 °C plotted as a function of P^{o}_{GaCl} together with the calculated growth rate using K_g of 2.3×10^4 μm/h·atm in the thermodynamic analysis (solid line). The experimental growth rate can be well fitted by the thermodynamic analysis results and proportionally increases with increase of P^{o}_{GaCl}. As for the crystalline quality of the homoepitaxial layer grown at 28 μm/h, the full-width at half-maximum (FWHM) values of high-resolution X-ray diffraction rocking curves for the symmetric (002) and skew-symmetric (400) reflections were almost the same as those of the substrate used. Therefore, it is found that the homoepitaxial growth of β-Ga_2O_3 at 1000 °C by HVPE using GaCl and O_2 is thermodynamically controlled, and it is possible to achieve the high-speed growth of several 10 μm/h without deterioration of crystalline quality.

The temperature dependence of the homoepitaxial growth rate was investigated in the range of 950 to 1050 °C and shown in Fig. 10.7. In the experiments, P^{o}_{GaCl} and input VI/III ratio were 1.0×10^{-3} atm and 10, respectively. The experimentally obtained growth rate was nearly constant in the growth temperature range studied and well fitted by multiplying the value of $\Delta P_{Ga_2O_3}$ obtained by the thermodynamic analysis by the same K_g value as in Fig. 10.6.

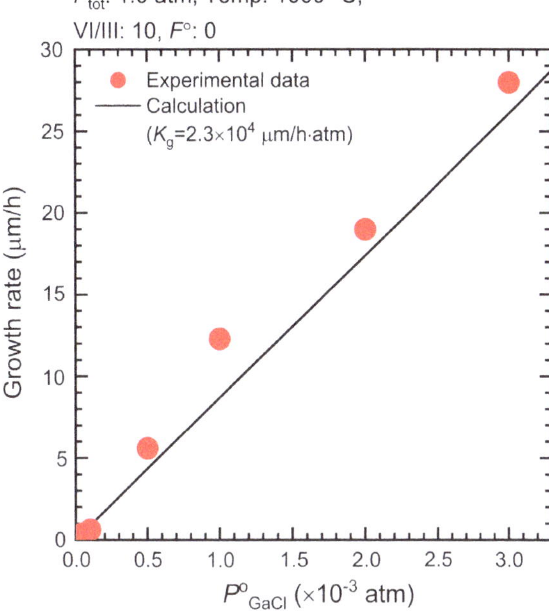

Fig. 10.6 Experimental and calculated (solid line) growth rate of β-Ga_2O_3 at 1000 °C as a function of P^o_{GaCl}

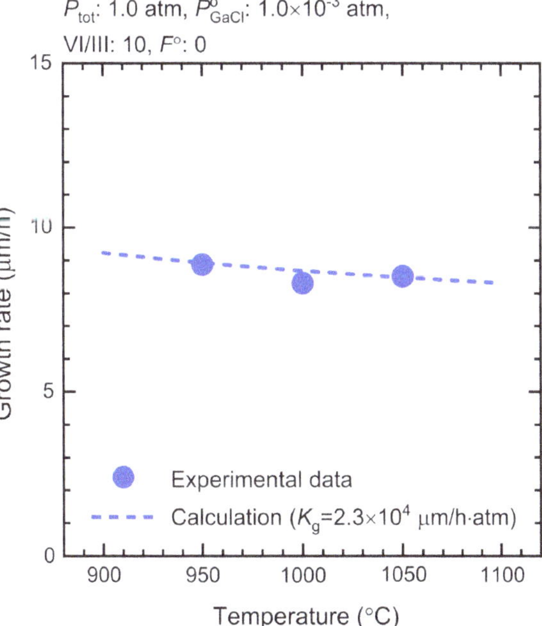

Fig. 10.7 Growth temperature dependence of experimental and calculated (dotted line) growth rate of β-Ga_2O_3

Figure 10.8 shows surface micrographs of the homoepitaxial layers grown at various temperatures for 1 h observed with confocal laser 3D profile microscopy. The growth conditions are the same as those in Fig. 10.7, and the numerical value in each image is the peak-to-valley height of the surface. The grooves running parallel to [010] direction were observed on all sample surfaces, which is thought to be due to the fact that the (100) plane is the cleavage plane and the growth rate in the [100] direction is slow [8]. Although the density of the grooves decreased as the growth temperature increased, it was difficult to completely suppress the formation of grooves on the growing surface. Since the depth of the grooves is shallower than the thickness of the homoepitaxial layer, epitaxial substrates for device fabrication are being produced at present by chemical mechanical polishing (CMP) of the surface of the homoepitaxial layers [20].

Impurity and effective donor concentrations in the homoepitaxial layer grown at 1000 °C with a growth rate of 8 μm/h were evaluated. After growing a 16-μm-thick homoepitaxial layer on the Sn-doped *n*-type β-Ga_2O_3 (001) substrate, grooves on the surface were removed by CMP, which yielded a 12.5-μm-thick flat homoepitaxial layer. Figure 10.9 shows depth profiles of impurities measured by secondary ion mass spectrometry (SIMS). It was confirmed that concentrations of all impurities were below the background level in the HVPE-grown UID homoepitaxial layer, which indicates that a high-purity UID layer can be grown by HVPE. Furthermore, SBDs were fabricated using the same UID homoepitaxial layer, and the effective donor concentration ($N_d - N_a$) in the homoepitaxial layer was evaluated. The capacitance–voltage (*C*–*V*) characteristics and cross-sectional schematic view of the fabricated SBD are shown in Fig. 10.10 and its inset, respectively. After ohmic cathode formation by evaporation of Ti(20 nm)/Au(230 nm) on the back of the substrate, the Schottky anode (diameter: 200 μm) of Pt(15 nm)/Ti(5 nm)/Au (250 nm) was fabricated on the homoepitaxial layer surface using photolithography. The SBD showed a constant capacitance in the applied voltage range from −200 to +200 V. This means that the 12.5-μm-thick homoepitaxial layer is completely depleted, which stops at the interface between the substrate and the homoepitaxial layer. By using the relative permittivity of 10 [28] and the zero-bias built-in potential of 1.0 eV [29] for β-Ga_2O_3, the $N_d - N_a$ was estimated to be less than 10^{13} cm^{-3}. Thus, the UID homoepitaxial layer grown by HVPE is high purity with a low $N_d - N_a$ of less than 10^{13} cm^{-3}.

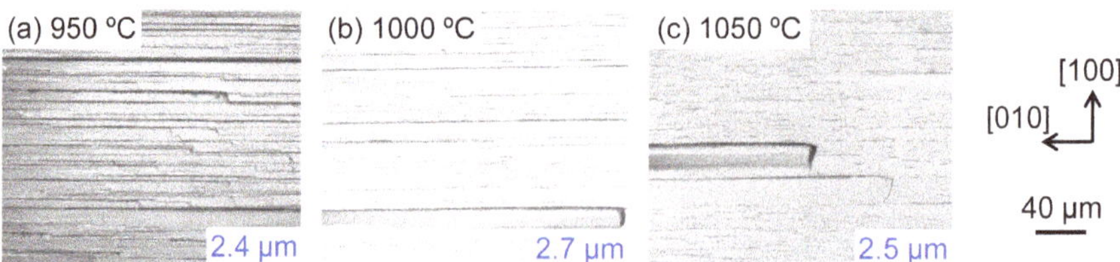

Fig. 10.8 Confocal laser 3D profile microscopy micrographs of the homoepitaxial layer surfaces grown at various temperatures for 1 h with growth conditions same as those in Fig. 10.7. The numerical value in each image is the peak-to-valley variation of height of the surface

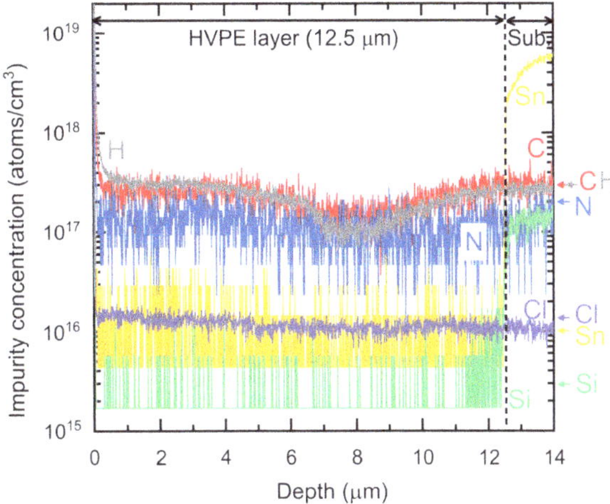

Fig. 10.9 SIMS depth profiles of impurities measured after CMP for a β-Ga_2O_3 homoepitaxial layer grown at 1000 °C with a growth rate of 8 μm/h. Arrows represent background levels of each element in the SIMS system

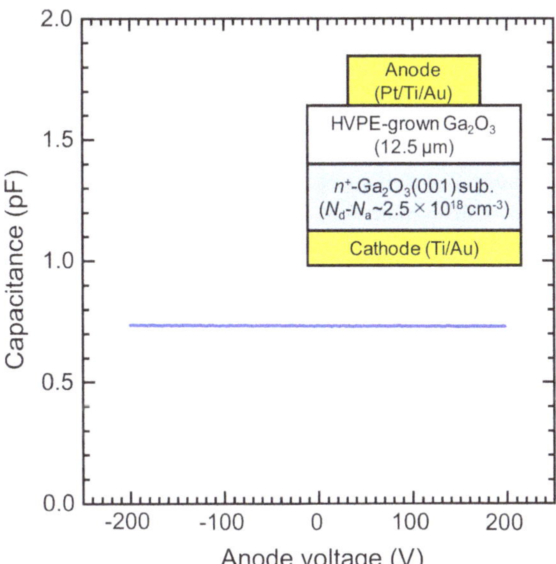

Fig. 10.10 C–V characteristics of the SBD using an UID homoepitaxial layer grown at 1000 °C with a growth rate of 8 μm/h by HVPE. The inset shows the cross-sectional schematic view of the fabricated SBD

10.4 Growth of Intentionally Si-Doped β-Ga_2O_3 Layers by HVPE

As a next step, growth and characterization of intentionally Si-doped β-Ga_2O_3 homoepitaxial layers by HVPE were investigated. An about 1-μm-thick UID homoepitaxial layer was first grown at 1000 °C on the Fe-doped SI β-Ga_2O_3 (001) substrate to prevent Fe diffusion from the substrate to the Si-doped homoepitaxial layer [30]. Then, 6–8-μm-thick Si-doped homoepitaxial layers were grown continuously at 1000 °C. The values of P^{o}_{GaCl} and input VI/III ratio were fixed at 1.0×10^{-3} atm and 10, respectively. The doping concentration of Si was varied by changing the input partial pressure ratio of $SiCl_4$ to GaCl using R_{Si} $\left[R_{Si} = P^{o}_{SiCl_4}/(P^{o}_{GaCl} + P^{o}_{SiCl_4})\right]$ as a control parameter. The growth rate of 10 μm/h was obtained. After the growth, an activation anneal of the Si dopant was performed at 1150 °C for 60 min in flowing N_2. Finally, about 2 μm of the surface was polished by CMP. The doping concentration of Si in the Si-doped homoepitaxial layers was measured by SIMS with a background level of 5×10^{15} cm^{-3}. The electrical properties of the Si-doped homoepitaxial layers were investigated by van der Pauw–Hall measurements, which were performed in the temperature range 80–350 K using a magnetic field of 0.57 T and a DC sample current of 0.1 mA. For the Hall measurements, ohmic contacts were formed on each corner of the sample using indium metal.

Table 10.2 summarizes Si impurity concentration, n-type carrier density (n), mobility (μ), and resistivity at room temperature of Si-doped homoepitaxial layers grown with various R_{Si}. The concentrations of impurities other than Si were below the background levels of the SIMS system. Figure 10.11 shows the SIMS Si concentration of the Si-doped β-Ga_2O_3 homoepitaxial layers grown with various R_{Si}, together with the carrier density shown in Table 10.2. It was found that the Si concentration can be changed in proportion to R_{Si} and its value is approximately equal to the value obtained by multiplying the Ga-site density in β-Ga_2O_3 crystal (3.8×10^{22} atoms/cm^3) with R_{Si}. This indicates that the incorporation ratio of Si into the Ga site is almost 1. The carrier density also changed in proportion to R_{Si} and is almost equal to the Si concentration. These results mean that the activation rate of Si dopant is nearly 1. In our study, the n-type carrier density could be linearly controlled in the range of 10^{15}–10^{18} cm^{-3} by changing R_{Si}.

For the samples (1) to (4) shown in Table 10.2 and Fig. 10.11, the temperature dependence of the carrier density was examined. The results are shown in Fig. 10.12 with fitting the results using the charge neutrality equation:

Table 10.2 SIMS Si concentration and electrical properties of intentionally Si-doped homoepitaxial β-Ga_2O_3 layers grown with various R_{Si}

Sample	R_{Si}	SIMS [Si] (atoms/cm^3)	n (cm^{-3})	μ (cm^2/V·s)	ρ (Ω·cm)	E_a (meV)
1	2.5×10^{-8}	B. G.	3.18×10^{15}	149	13.2	44.7
2	1.9×10^{-7}	9.37×10^{15}	9.62×10^{15}	145	4.48	45.6
3	3.1×10^{-6}	1.65×10^{17}	1.85×10^{17}	111	0.30	31.2
4	3.1×10^{-5}	1.15×10^{18}	1.18×10^{18}	88	0.06	16.7

Electrical properties were measured at room temperature (298 K). B. G. in the table indicates below the background level

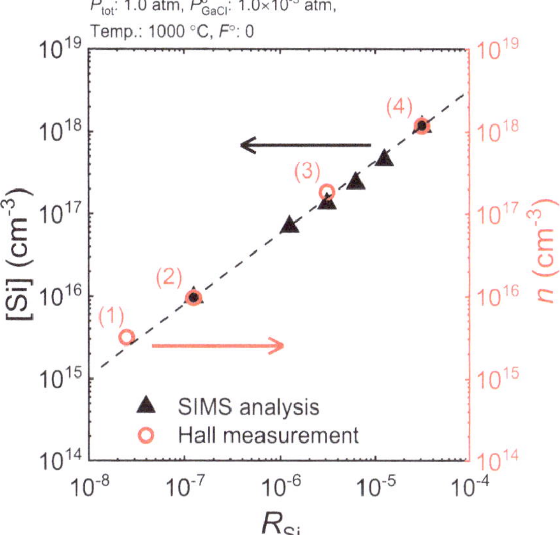

Fig. 10.11 Si concentration measured by SIMS and carrier density in intentionally Si-doped β-Ga_2O_3 homoepitaxial layers grown at various R_{Si}. The dashed line is the linear fit

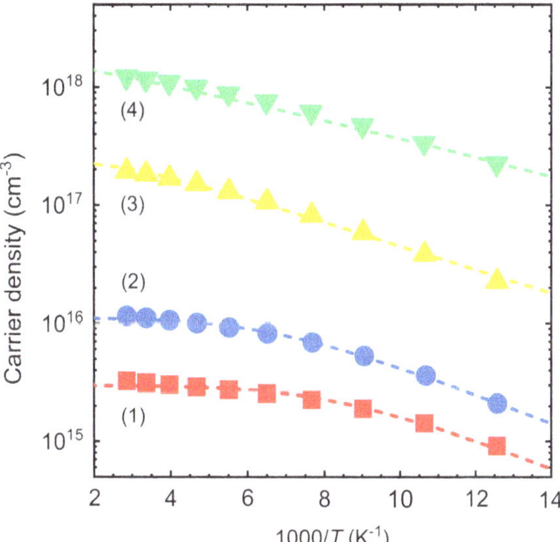

Fig. 10.12 Temperature dependence of the carrier density for samples (1) to (4) in Table 10.2. The dashed lines are the fitting results using the charge neutrally equation

$$\frac{n(n+N_a)}{N_d - N_a - n} = \frac{N_C}{2} \cdot \exp\left(-\frac{E_a}{k_B T}\right), \tag{10.37}$$

where k_B is the Boltzmann constant and N_C is the effective density of states in the conduction band as

$$N_C = 2\left(\frac{m_e k_B T}{2\pi\hbar^2}\right)^{\frac{3}{2}}. \quad (10.38)$$

Here, m_e is the effective mass of the electron, and the value of $0.28m_0$ was used for β-Ga_2O_3 [31]. The donor density (N_d), the acceptor density (N_a), and the activation energy of donors (E_a) were estimated as fitting parameters. As shown in Table 10.2, the obtained E_a decreased as the Si concentration increased, and it is roughly equal to the value reported so far [32, 33]. The obtained values of N_a were all less than 10^{14} cm^{-3}.

Focusing on the carrier mobility shown in Table 10.2, the carrier mobility increased as the carrier density decreased and the mobility reached about 150 cm^2/V·s for the lowest carrier density [sample (1)]. This value agrees well with the value reported for high-quality bulk β-Ga_2O_3 crystals [2, 32], suggesting that the crystallinity of the HVPE-grown Si-doped β-Ga_2O_3 layers is also high enough. Finally, the temperature dependence of the carrier mobility of the samples (1) to (4) in Table 10.2 is investigated as shown in Fig. 10.13 to clarify the scattering mechanism of carriers. In the layer with the highest carrier density [sample (4)], saturation is seen of the increase in mobility with decreasing temperature. This is thought to be due to the influence of impurity scattering proportional to $T^{3/2}$. It is expected that a decrease in mobility will be observed at lower temperatures than the temperature range in Fig. 10.13. On the other hand, for the low carrier density layers [samples (1) and (2)], the mobility is proportional to $T^{-5/2}$, which indicates that the optical phonon scattering is dominant. The mobility of sample (1) exceeded 5000 cm^2/V·s at 80 K, which was very close to the predicted value of the theoretical calculation [34]. These temperature-dependent characteristics are typical semiconductor

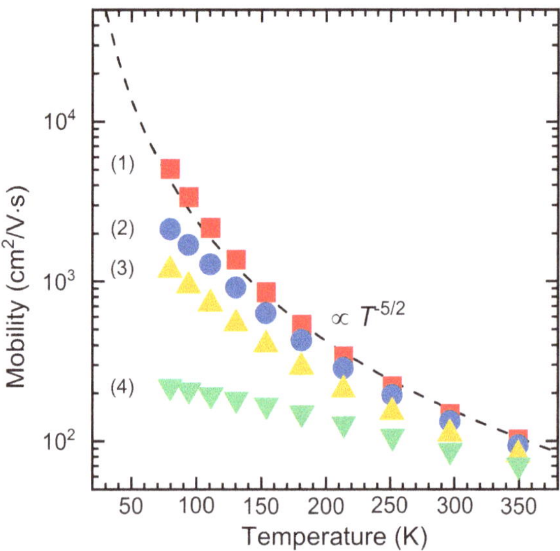

Fig. 10.13 Temperature dependence of the carrier mobility for samples (1) to (4) in Table 10.2. The dashed line is a curve of $\mu \propto T^{-5/2}$ showing the behavior of optical phonon scattering

behavior, indicating that the intentionally Si-doped homoepitaxial $\beta\text{-}Ga_2O_3$ layers are applicable in the fabrication of power devices.

10.5 Summary

In this chapter, the results of our study on homoepitaxial growth of $\beta\text{-}Ga_2O_3$ by the HVPE method were introduced. Thermodynamic analysis revealed the possibility of growth at high temperatures around 1000 °C by using GaCl and O_2 as precursors and inert carrier gas. Growth experiments on $\beta\text{-}Ga_2O_3$ (001) substrates revealed that homoepitaxial growth of UID layers is possible at 1000 °C with a high growth rate of up to 28 μm/h. It was also found that a high-purity layer with low effective donor concentration ($N_d - N_a$) of less than 10^{13} cm^{-3} is obtained by HVPE although post-growth CMP is required to remove surface grooves before fabrication of devices. Growth of intentionally Si-doped homoepitaxial layers is possible using $SiCl_4$ as a doping gas, and an n-type carrier density almost equal to the Si concentration was obtained. Control of carrier density in the range of 10^{15}–10^{18} cm^{-3} has been demonstrated. Currently, these results are the foundation of mass-production technology of $\beta\text{-}Ga_2O_3$ homoepitaxial substrates indispensable for fabricating vertical structure power devices.

Acknowledgements This work was partially supported by the Council for Science, Technology and Innovation (CSTI), Cross-ministerial Strategic Innovation Promotion Program (SIP), "Next-generation power electronics" (funding agency: NEDO). The authors would like to express their sincere thanks to Dr. M. Higashiwaki of NICT, Japan, for his help in the electrical measurements and for fruitful discussions.

References

1. Y. Tomm, P. Reiche, D. Klimm, T. Fukuda, J. Cryst. Growth **220**, 510 (2000)
2. Z. Galazka, R. Uecker, K. Irmscher, M. Albrecht, D. Klimm, M. Pietsch, M. Brützam, R. Bertram, S. Ganschow, R. Fornari, Cryst. Res. Technol. **45**, 1229 (2010)
3. E.G. Víllora, K. Shimamura, Y. Yoshikawa, K. Aoki, N. Ichinose, J. Cryst. Growth **270**, 420 (2004)
4. K. Hoshikawa, E. Ohba, T. Kobayashi, J. Yanagisawa, C. Miyagawa, Y. Nakamura, J. Cryst. Growth **447**, 36 (2016)
5. H. Aida, K. Nishiguchi, H. Takeda, N. Aota, K. Sunakawa, Y. Yaguchi, Jpn. J. Appl. Phys. **47**, 8506 (2008)
6. A. Kuramata, K. Koshi, S. Watanabe, Y. Yamaoka, T. Masui, S. Yamakoshi, Jpn. J. Appl. Phys. **55**, 1202A2 (2016)
7. T. Oshima, N. Arai, N. Suzuki, S. Ohira, S. Fujita, Thin Solid Films **516**, 5768 (2008)
8. K. Sasaki, A. Kuramata, T. Masui, E.G. Víllora, K. Shimamura, S. Yamakoshi, Appl. Phys. Express **5**, 035502 (2012)
9. H. Okumura, M. Kita, K. Sasaki, A. Kuramata, M. Higashiwaki, J.S. Speck, Appl. Phys. Express **7**, 095501 (2014)

10. S. Lee, K. Kaneko, S. Fujita, Jpn. J. Appl. Phys. **55**, 1202B8 (2016)
11. G. Wagner, M. Baldini, D. Gogova, M. Schmidbauer, R. Schewski, M. Albrecht, Z. Galazka, D. Klimm, R. Fornari, Phys. Status Solidi A **211**, 27 (2014)
12. X. Du, W. Mi, C. Luan, Z. Li, C. Xia, J. Ma, J. Cryst. Growth **404**, 75 (2014)
13. M. Baldini, M. Albrecht, A. Fiedler, K. Irmscher, R. Schewski, G. Wagner, ECS J. Solid State Sci. Technol. **6**, Q3040 (2017)
14. K. Nomura, K. Goto, R. Togashi, H. Murakami, Y. Kumagai, A. Kuramata, S. Yamakoshi, A. Koukitu, J. Cryst. Growth **405**, 19 (2014)
15. H. Murakami, K. Nomura, K. Goto, K. Sasaki, K. Kawara, Q.T. Thieu, R. Togashi, Y. Kumagai, M. Higashiwaki, A. Kuramata, S. Yamakoshi, B. Monemar, A. Koukitu, Appl. Phys. Express **8**, 015503 (2015)
16. Q.T. Thieu, D. Wakimoto, Y. Koishikawa, K. Sasaki, K. Goto, K. Konishi, H. Murakami, A. Kuramata, Y. Kumagai, S. Yamakoshi, Jpn. J. Appl. Phys. **56**, 110310 (2017)
17. K. Konishi, K. Goto, R. Togashi, H. Murakami, M. Higashiwaki, A. Kuramata, S. Yamakoshi, B. Monemar, Y. Kumagai, J. Cryst. Growth **492**, 39 (2018)
18. K. Goto, K. Konishi, H. Murakami, Y. Kumagai, B. Monemar, M. Higashiwaki, A. Kuramata, S. Yamakoshi, Thin Solid Films **666**, 182 (2018)
19. K. Konishi, K. Goto, H. Murakami, Y. Kumagai, A. Kuramata, S. Yamakoshi, M. Higashiwaki, Appl. Phys. Lett. **110**, 103506 (2017)
20. J. Yang, S. Ahn, F. Ren, S.J. Pearton, S. Jang, J. Kim, A. Kuramata, Appl. Phys. Lett. **110**, 192101 (2017)
21. K. Sasaki, D. Wakimoto, Q.T. Thieu, Y. Koishikawa, A. Kuramata, M. Higashiwaki, S. Yamakoshi, IEEE Electron Device Lett. **38**, 783 (2017)
22. K. Sasaki, Q.T. Thieu, D. Wakimoto, Y. Koishikawa, A. Kuramata, S. Yamakoshi, Appl. Phys. Express **10**, 124201 (2017)
23. Z. Hu, K. Nomoto, W. Li, N. Tanen, K. Sasaki, A. Kuramata, T. Nakamura, D. Jena, H.G. Xing, IEEE Electron Device Lett. **39**, 869 (2018)
24. M.H. Wong, K. Goto, H. Murakami, Y. Kumagai, M. Higashiwaki, IEEE Electron Device Lett. **40**, 431 (2019)
25. M.W. Chase Jr. (ed.), *NIST-JANAF Thermochemical Tables* (The American Chemical Society and the American Institute of Physics for the National Institute of Standards and Technology, Gaithersburg, 1998)
26. L.V. Gurvich, I.V. Veyts, C.B. Alcock (eds.), *Thermodynamic Properties of Individual Substances* (USSR Academy of Sciences, Institute for High Temperatures and State Institute of Applied Chemistry in cooperation with the National Standard Reference Data Service of the U.S.S.R., Moscow, 1994)
27. M. Tirtowidjojo, R. Pollard, J. Cryst. Growth **77**, 200 (1986)
28. M. Passlack, N.E.J. Hunt, E.F. Schubert, G.J. Zydzik, M. Hong, J.P. Mannaerts, R.L. Opila, R.J. Fischer, Appl. Phys. Lett. **64**, 2715 (1994)
29. M. Higashiwaki, K. Konishi, K. Sasaki, K. Goto, K. Nomura, Q.T. Thieu, R. Togashi, H. Murakami, Y. Kumagai, B. Monemar, A. Koukitu, A. Kuramata, S. Yamakoshi, Appl. Phys. Lett. **108**, 133503 (2016)
30. M.H. Wong, K. Sasaki, A. Kuramata, S. Yamakoshi, M. Higashiwaki, Appl. Phys. Lett. **106**, 032105 (2015)
31. J.B. Varley, J.R. Weber, A. Janotti, C.G. Van de Walle, Appl. Phys. Lett. **97**, 142106 (2010)
32. T. Oishi, Y. Koga, K. Harada, M. Kasu, Appl. Phys. Express **8**, 031101 (2015)
33. N.T. Son, K. Goto, K. Nomura, Q.T. Thieu, R. Togashi, H. Murakami, Y. Kumagai, A. Kuramata, M. Higashiwaki, A. Koukitu, S. Yamakoshi, B. Monemar, E. Janzén, J. Appl. Phys. **120**, 235703 (2016)
34. N. Ma, N. Tanen, A. Verma, Z. Guo, T. Luo, H. Xing, D. Jena, Appl. Phys. Lett. **109**, 212101 (2016)

Chapter 11
Halide Vapor Phase Epitaxy 2

Heteroepitaxial Growth of α- and ε-Ga_2O_3

Yuichi Oshima

Abstract Halide vapor phase epitaxy of metastable α- and ε-Ga_2O_3 is reviewed. The both polymorphs were grown using GaCl and O_2 as precursors. Phase-pure corundum α-Ga_2O_3 was heteroepitaxially grown on (0001) sapphire at temperatures of approximately 550 °C or lower. The *n*-type electrical conductivity was controlled by Ge doping using $GeCl_4$ as the dopant source, and very low resistivity of 8.6 mΩ cm was achieved. Epitaxial lateral overgrowth was shown to be effective in improving the crystal quality, and the dislocation density was reduced from 10^{10} cm^{-2} to less than 5×10^6 cm^{-2} in the laterally grown wing region. Morphology of α-Ga_2O_3 islands was controlled such that inclined facets well develop, and dislocation density above mask openings remarkably decreased due to dislocation bending caused by the inclined facets. Orthorhombic ε-Ga_2O_3 was grown on (0001) GaN and (0001) AlN using virtually the same growth recipe as that used for α-Ga_2O_3. Fundamental material properties of ε-Ga_2O_3, such as the optical bandgap energy, thermal stability, and thermal expansion coefficient, were investigated using the epitaxial film.

11.1 Introduction

Ga_2O_3 has been reported to crystalize into various polymorphs such as α-, β-, δ-, ε-, and γ-phases [1]. Among these polymorphs, the β-phase is thermodynamically the most stable at atmospheric pressure, and melt-grown high-quality single crystal substrates are available [2–4]. Accordingly, high-quality homoepitaxial growth is possible, and promising β-Ga_2O_3-based power devices have been demonstrated [5–7].

On the other hand, other polymorphs are metastable and transform into the β-phase at high temperatures beyond specific thresholds. Among the metastable

Y. Oshima (✉)
Optical Single Crystals Group, National Institute for Materials Science, 1-1 Namiki, Tsukuba, Ibaraki 305-0044, Japan
e-mail: OSHIMA.Yuichi@nims.go.jp

M. Higashiwaki and S. Fujita (eds.), *Gallium Oxide*, Springer Series in Materials Science 293, https://doi.org/10.1007/978-3-030-37153-1_11

Table 11.1 Polymorphs of Ga_2O_3 and HVPE growth conditions

Polymorph	β-Ga_2O_3	α-Ga_2O_3	ε-Ga_2O_3
Substrate	(0001) Sapphire		(0001) GaN, etc.
Typical growth temperature	1050 °C	550 °C	

Ga_2O_3 phases, the corundum α-phase and orthorhombic ε-phase, which are the target materials of this review, have attracted great attention as wide-bandgap semiconductors. These materials are promising for power device applications because of their superior properties, including their larger bandgaps compared with that of β-Ga_2O_3, the ease of growing $(Al_xIn_yGa_{1-x-y})_2O_3$ solid solutions, and spontaneous polarization.

To fabricate α-/ε-Ga_2O_3 devices, it is essential to establish a thin-film growth technique. Unfortunately, the melt-grown native substrates are not available for α- or ε-Ga_2O_3. Accordingly, these materials must be grown heteroepitaxially. However, the development of such heteroepitaxial growth techniques remains in its infancy. To obtain high-quality phase-pure films with controlled electrical properties, we must overcome many technical challenges such as the selection of appropriate growth methods and substrates, nucleation control, defect control, and doping control. Moreover, it is desirable to establish a high-speed epitaxial technique for thick film growth, which will be useful for the growth of drift layers in power devices or the fabrication of freestanding α-/ε-Ga_2O_3 substrates. We have employed halide vapor phase epitaxy (HVPE) as the growth method and reported the first HVPE growth of α- and ε-Ga_2O_3 [8, 9].

Table 11.1 summarizes the HVPE growth conditions and resulting polymorphs [8–10]. The key points for the phase selection are the growth temperature and substrate. For example, α-Ga_2O_3 grows on sapphire at low temperatures, whereas β-Ga_2O_3 grows on the same substrate at high temperatures under the same gas supply conditions. If GaN is used instead of sapphire as the substrate under the same growth conditions for α-Ga_2O_3, then ε-Ga_2O_3 grows. Note that this tendency may not be universal and may be dependent on the growth method. Further study is required to clarify the essential factors for phase selection. In this chapter, we describe our HVPE technologies for α-/ε-Ga_2O_3 and the properties of the grown epilayers.

11.2 HVPE of α-Ga_2O_3

11.2.1 Features of α-Ga_2O_3

α-Ga_2O_3 is a corundum-structured wide-bandgap semiconductor with a reported bandgap energy of 5.2–5.3 eV [8, 11], which is even larger than that of β-Ga_2O_3 (4.5–4.9 eV). The crystal structure is the same as that of sapphire (α-Al_2O_3);

therefore, α-Ga_2O_3 can be grown on sapphire epitaxially. It is possible to grow α-$(Al_xGa_{1-x})_2O_3$ solid solutions without compositional limitation, which is useful for band engineering and the fabrication of high-performance hetero-structured devices [12]. In addition, α-Ir_2O_3 and α-Rh_2O_3 have been shown to exhibit clear *p*-type conduction; therefore, hetero-pn-junction bi-polar devices can be expected [13]. Hence, α-Ga_2O_3 is a promising wide-bandgap semiconductor for power device applications. Indeed, Schottky barrier diodes with very low on-resistance and normally-off metal-oxide semiconductor field-effect transistors (MOSFETs) using a hetero-pn-junction have already been demonstrated [14].

The first epitaxial growth of α-Ga_2O_3 was demonstrated by Shinohara and Fujita using the mist chemical vapor deposition (CVD) technique [11]. The mist CVD technique also enables *n*-type conductivity control [15–17] and growth of α-$(Al_xIn_yGa_{1-x-y})_2O_3$ solid solutions [12]. These achievements of the mist CVD technique enabled the fabrication of the α-Ga_2O_3–based devices described above. See Chap. 12 for details. The first HVPE of α-Ga_2O_3 was demonstrated by Oshima et al. [8]. In the following sections, we provide an outline of the HVPE of α-Ga_2O_3, the properties of the epilayers, conductivity control, and technologies for crystal quality improvement.

11.2.2 Growth Apparatus and Growth Conditions

In this section, a brief description of the growth apparatus and basic growth conditions is presented. See Chap. 10 for a detailed explanation of the principle of HVPE and its thermodynamical basis.

We used a lab-made horizontal quartz reactor for the HVPE of α-Ga_2O_3. The growth was performed on (0001) sapphire under atmospheric pressure at 430–650 °C using GaCl and O_2 as precursors. Figure 11.1 presents a schematic illustration of the HVPE reactor. GaCl was synthesized upstream in the reactor via the chemical reaction between metal Ga (>99.99999% pure) and HCl gas (>99.999% pure). The reaction temperature was fixed at 570 °C unless otherwise stated. The GaCl and O_2 were transferred together with N_2 carrier gas, and injected on the substrate located

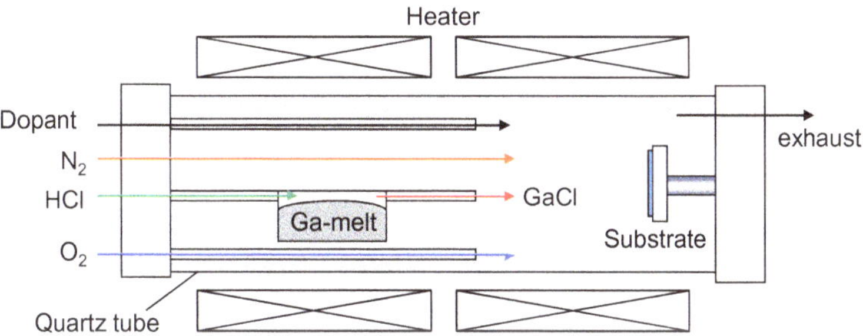

Fig. 11.1 Schematic illustration of an HVPE reactor for Ga_2O_3

downstream in the reactor. The partial pressures of the GaCl and O_2 supply were 4×10^{-2} to 7.5×10^{-1} kPa and 0.5–6.0 kPa, respectively. The growth was performed directly on sapphire without any buffer layer or slow growth.

11.2.3 HVPE Growth Characteristics of α-Ga_2O_3

Figure 11.2 shows the relative growth rate of α-Ga_2O_3 as a function of growth temperature T_g under a constant supply of the precursors. The growth rate increased with increasing temperature, which indicates that HVPE of α-Ga_2O_3 occurs in the chemical-reaction-limited regime. Nonetheless, it is possible to achieve a high growth rate by increasing the supply of precursors. Figure 11.3 shows the growth rate at 550 °C as a function of the partial pressures of the supply of precursors. The growth rate increased monotonically with increasing precursor supply, reaching over 100 μm/h.

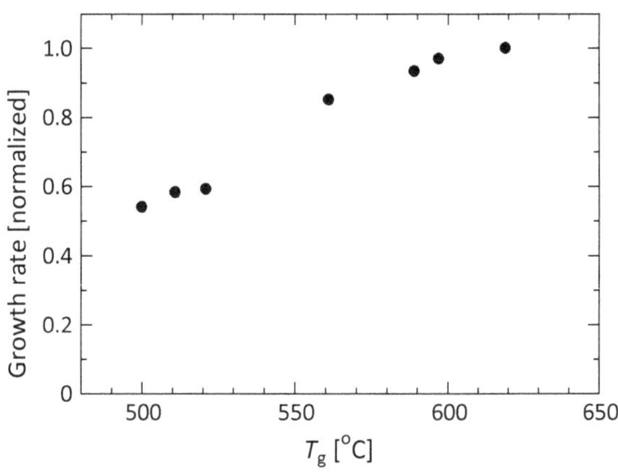

Fig. 11.2 Growth rate of α-Ga_2O_3 as a function of growth temperature

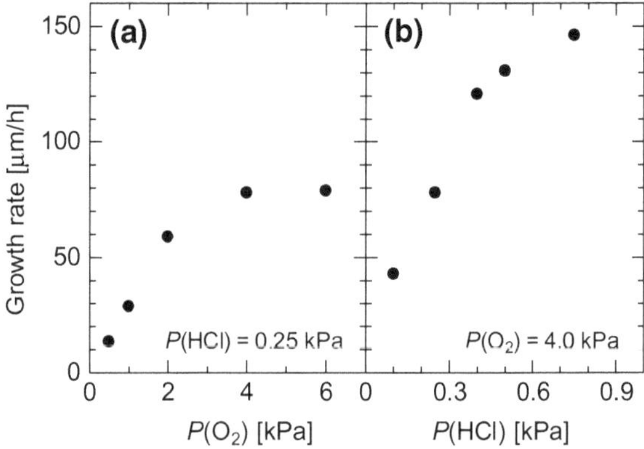

Fig. 11.3 Growth rate of α-Ga_2O_3 as functions of **a** O_2 partial pressure and **b** HCl partial pressure [8]

11.2.4 Properties of HVPE-Grown α-Ga_2O_3

In this section, we describe the properties of a typical HVPE-grown α-Ga_2O_3 film on (0001) sapphire.

Phase Purity and Growth Temperature T_g

Figure 11.4a–d present photographs of Ga_2O_3 films grown at various temperatures under the same precursor supply conditions. For T_g = 650 °C, the film was translucent (Fig. 11.4a). The fraction of the specular part increased with decreasing T_g, and became mirror-like throughout the wafer when T_g was less than 575 °C (Fig. 11.4d).

Figure 11.5a, b present X-ray diffraction (XRD) 2θ–ω scan profiles of an α-Ga_2O_3 film grown at 550 °C. Only diffraction peaks from the *c*-plane of α-Ga_2O_3 are observed except for those from the substrate. In contrast, diffraction peaks from β-Ga_2O_3 were dominant for T_g = 650 °C (not shown). Thus, phase-pure α-Ga_2O_3 can be grown at appropriately low temperatures.

Figure 11.6a, b present surface and cross-sectional scanning electron microscopy (SEM) images of an α-Ga_2O_3 film grown at 550 °C, respectively. The surface was smooth. The film thickness was 3.6 μm for 7-min growth; therefore, the growth rate was approximately 30 μm/h.

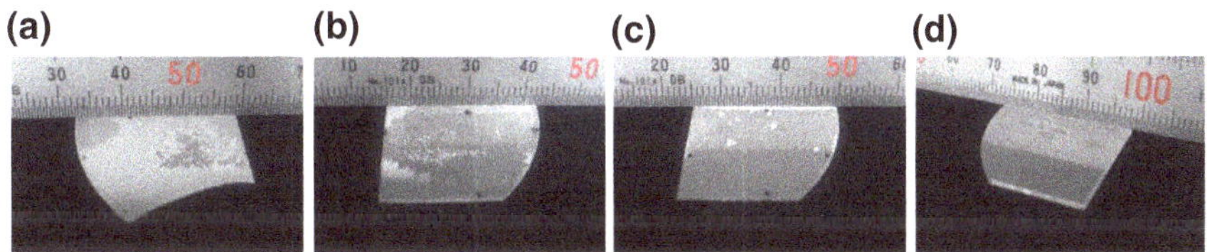

Fig. 11.4 Photographs of Ga_2O_3 films grown at **a** 650 °C, **b** 600 °C, **c** 575 °C, and **d** 550 °C

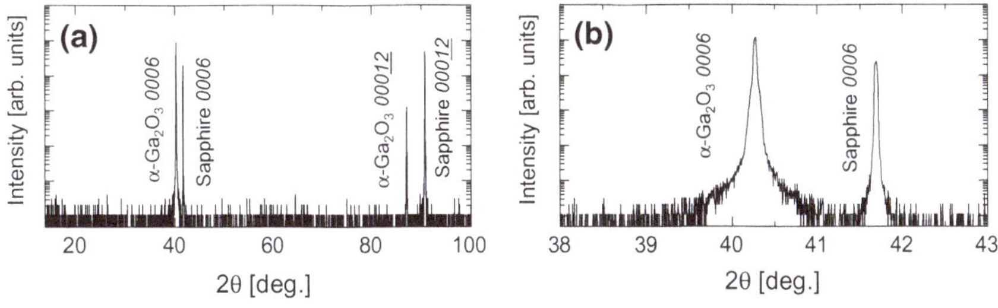

Fig. 11.5 XRD 2θ–ω profiles of an α-Ga_2O_3 layer: **a** wide-scan profile and **b** narrow-scan profile near *0006* diffraction [8]

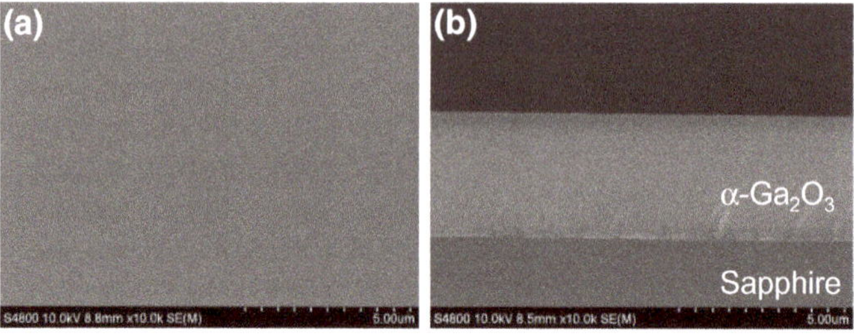

Fig. 11.6 SEM images of an α-Ga_2O_3 layer: **a** surface image and **b** cross-sectional image [8]

Crystal Orientation

Figure 11.7a, b present XRD pole figures of an α-Ga_2O_3 film and the sapphire substrate, respectively. Only three diffraction spots are observed in the pole figure, which were expected for the single-crystalline corundum structure. The epitaxial relationships were $[10\bar{1}0]_{\alpha-Ga_2O_3} \| [10\bar{1}0]_{\alpha-Al_2O_3}$ and $(0001)_{\alpha-Ga_2O_3} \| (0001)_{\alpha-Al_2O_3}$.

Crystal Quality

α-Ga_2O_3 epilayers grown on sapphire exhibit large mosaicity (represented by tilting of the *c*-plane and twisting around the *c*-axis) because of the large lattice mismatch ($\Delta a/a$ = 4.5%, $\Delta c/c$ = 3.3%). The mosaicity was estimated using the full-width half-maximum (FWHM) values of X-ray rocking curves (XRCs) of the *0006* and $10\bar{1}2$ diffractions measured in symmetric and skew-symmetric geometry, respectively (Fig. 11.8). In most cases, both the tilt angle and twist angles were broad. The tilt angle was sometimes very narrow (less than 100 arcsec). In this case, the twist angle tended to be very broad. Thus, the tilt and twist angles appear to have a

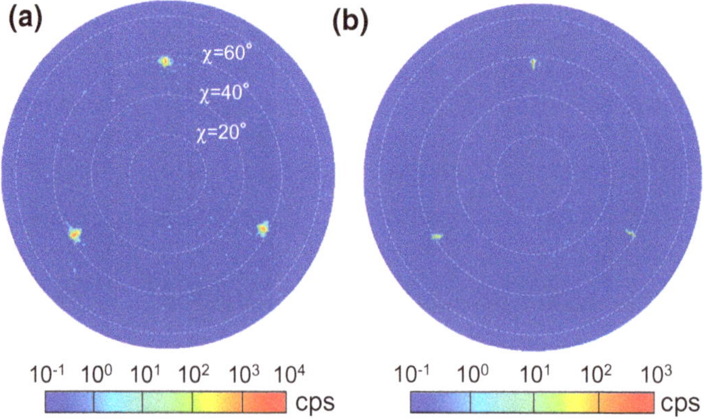

Fig. 11.7 X-ray $10\bar{1}2$ pole figures (log scale) of **a** α-Ga_2O_3 layer and **b** sapphire substrate [8]

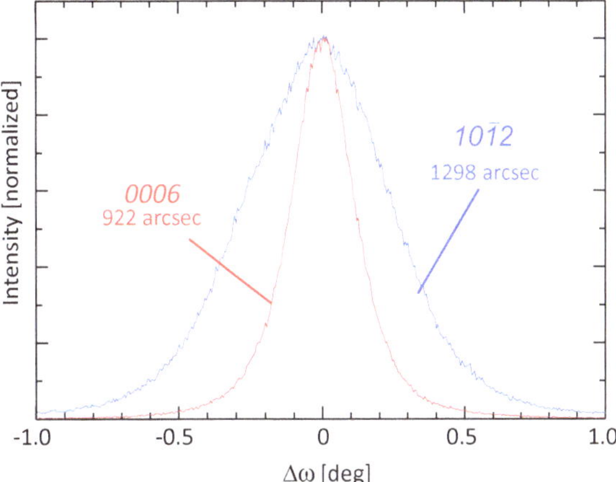

Fig. 11.8 XRCs of a conventional α-Ga_2O_3 film grown by HVPE [18]

trade-off relationship. Therefore, both the tilt and twist angles must be measured to estimate the crystal quality of α-Ga_2O_3 layers.

Figure 11.9a, b present cross-sectional and plan-view transmission electron microscopy (TEM) images of an α-Ga_2O_3 film, respectively. A high density of dislocations along [0001] was observed, and the density was on the order of 10^{10} cm^{-2}.

Impurity Analysis

Table 11.2 summarizes the results of impurity analysis of an undoped α-Ga_2O_3 film performed using secondary mass spectrometry (SIMS). The concentrations were below the detection limits except for [Cl]. The chlorine impurity source would be GaCl. We speculate that the low growth temperature led to the incomplete dissociation of the Ga–Cl bond.

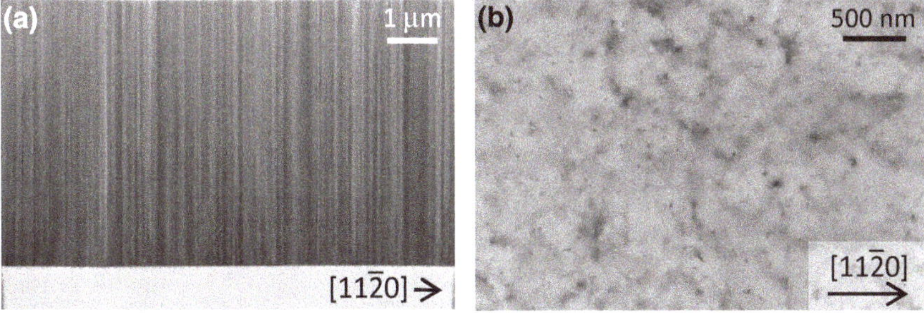

Fig. 11.9 TEM images of a conventional α-Ga_2O_3 film grown by HVPE: **a** cross-sectional image and **b** plan-view image [18]

Table 11.2 Impurity concentrations in α-Ga_2O_3 measured by SIMS

Element	Concentration (cm^{-3})	Element	Concentration (cm^{-3})
H	$<6 \times 10^{16}$	Al	$<6 \times 10^{16}$
C	$<6 \times 10^{16}$	Cr	$<1 \times 10^{14}$
Si	$<3 \times 10^{15}$	Fe	$<4 \times 10^{14}$
Cl	1×10^{16}	Ni	$<3 \times 10^{14}$
S	$<3 \times 10^{15}$	Mo	$<1 \times 10^{15}$

Optical Bandgap

Figure 11.10 presents an optical transmittance spectrum of an α-Ga_2O_3 film. Although the transition type of α-Ga_2O_3 remains under discussion, the fitting result was better when direct transition was assumed (inset of Fig. 11.10). The optical bandgap energy was determined to be 5.15 eV, which is close to the value reported for mist-CVD-grown material [11].

Thermal Stability and Thermal Expansion Coefficients

High-temperature XRD measurements were performed under air to estimate the thermal stability of an HVPE-grown α-Ga_2O_3 film. Each 2θ–ω scan was performed at constant temperature after 30 min to allow for temperature stabilization, and the temperature was elevated stepwise from room temperature (RT). Figure 11.11 presents the results. Apart from the diffraction peaks of the Pt sample holder and sapphire substrate, only the *0006* peak of α-Ga_2O_3 was observed from RT to

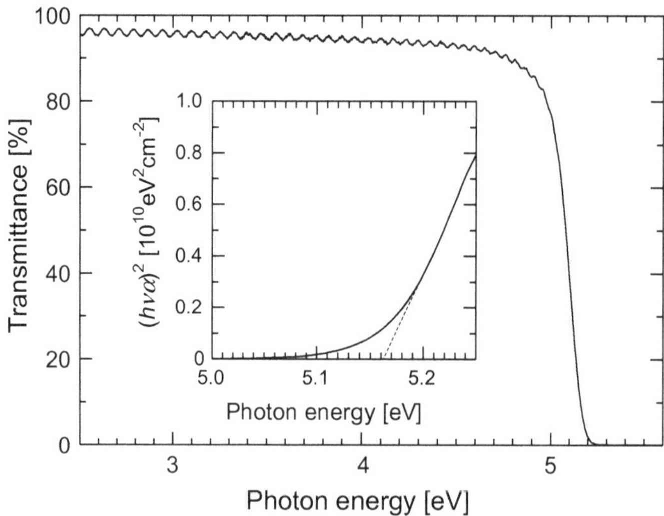

Fig. 11.10 Transmittance spectra of α-Ga_2O_3. The inset shows the absorption coefficient in $(h\nu\alpha)^2$ versus $h\nu$ [8]

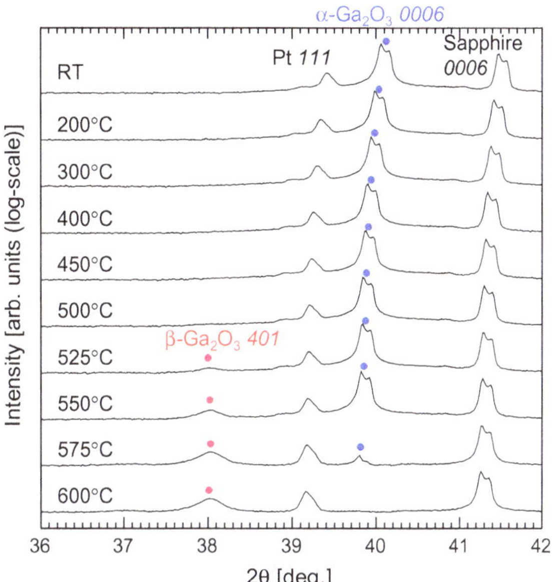

Fig. 11.11 XRD 2θ–ω scan profiles for an HVPE-grown α-Ga_2O_3 film measured at RT and elevated temperatures

500 °C. When the temperature reached 525 °C, the *401* peak of β-Ga_2O_3 appeared, and the peak intensity of the *401* peak increased with increasing temperature while that of *0006* of α-Ga_2O_3 decreased. This result indicates that α-Ga_2O_3 is thermally stable up to approximately 500 °C. Note that this threshold would be dependent on the crystal quality and strain, and the intrinsic threshold could be much higher. Indeed, it has been reported that α-Ga_2O_3 grown under special conditions using mist CVD was stable up to 800 °C. See Chap. 12 for further details.

It is possible to calculate the thermal expansion coefficients (TECs) from the peak shift values in Fig. 11.11. The TECs for α-Ga_2O_3 and sapphire along [0001] were 1.1×10^{-5} K^{-1} and 8.6×10^{-6} K^{-1}, respectively. The value for α-Ga_2O_3 was in excellent agreement with that reported for powder material (1.1×10^{-5} K^{-1}) [19]. The values reported for sapphire range from 7.7×10^{-6} K^{-1} to 9.06×10^{-6} K^{-1} [20], and our value is in this range. Note that these values may not be exactly the same as those for freestanding materials because of the thermal stress resulting from the difference in TECs between α-Ga_2O_3 and sapphire.

11.2.5 n-Type Doping Control

11.2.5.1 Motivation

It is essential to establish electrical conductivity control of α-Ga_2O_3 to utilize this material for semiconductor device applications. *n*-type doping control of α-Ga_2O_3

has been reported for mist-CVD-grown materials using Sn or Si as the dopant [15–17]. *n*-type conductivity control by HVPE has been demonstrated recently by Oshima et al. using Ge as the dopant [21]. In this section, we describe the current state of *n*-type doping control of α-Ga_2O_3 by HVPE.

11.2.5.2 Experimental Methods

Selection of Dopant Element

We selected Ge as the dopant. *n*-type doping of α-Ga_2O_3 was performed by substituting Ga sites with group IV materials, such as Si, Ge, or Sn. Of these materials, the ionic radius of Ge should be the closest to that of Ga. Accordingly, the use of Ge could minimize the negative impacts of the doping, such as increasing strain and decreasing thermal conductivity, especially upon heavy doping.

Experimental Methods

$GeCl_4$ (>99.999% pure) was used as the dopant source. $GeCl_4$ is a liquid phase near RT (melting point: −50 °C), and the bubbling technique using N_2 was used to transport the vapor into the HVPE reactor. The HVPE growth conditions of α-Ga_2O_3 were similar to those described in Sect. 2.2. The temperatures of the Ga source and substrate were both 560 °C. The partial pressures of the GaCl and O_2 supplies were 0.25 and 1.0 kPa, respectively.

The impurity concentrations in Ge-doped α-Ga_2O_3 were estimated using SIMS. The influence of Ge doping on the crystal quality was investigated using XRC measurements. The electrical properties at both RT and elevated temperatures were investigated using Hall measurements and the van der Pauw method.

11.2.5.3 Properties of Ge-Doped α-Ga_2O_3

Impurity Analysis

Figure 11.12 presents a SIMS depth profile of a Ge-doped α-Ga_2O_3 film. An undoped thin layer was grown on the sapphire substrate first, and then Ge doping was performed on the undoped layer. [Ge] rose steeply, and no significant doping delay or diffusion was observed. [Cl] also increased at the same time. The source of the chlorine impurity is most likely $GeCl_4$. [Ge] was below the detection limit in a sample grown directly after the growth with the highest $GeCl_4$ supply, indicating that the memory effect of $GeCl_4$ was not significant. [Ge] could be linearly controlled by adjusting the $GeCl_4$ bubbling rate, as shown in Fig. 11.13.

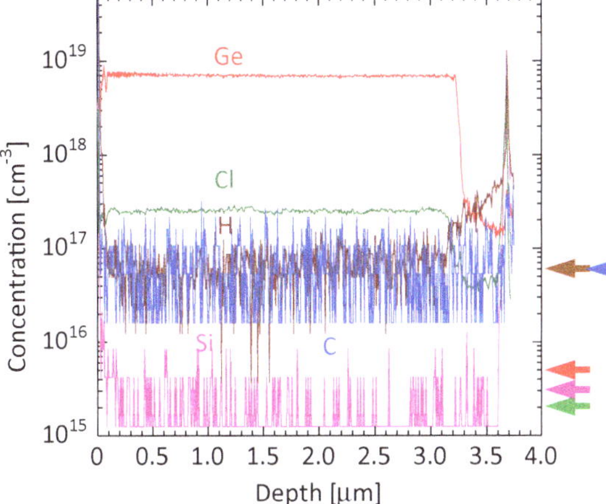

Fig. 11.12 SIMS depth profile of a Ge-doped α-Ga_2O_3 film. The arrows indicate the detection limits

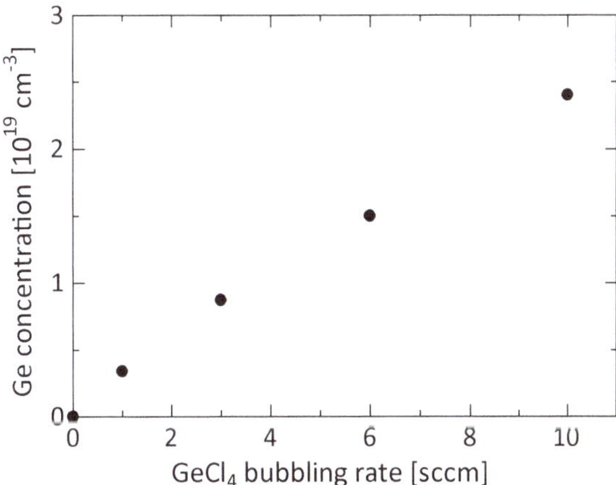

Fig. 11.13 Ge concentration in α-Ga_2O_3 as a function of $GeCl_4$ bubbling rate

Influence of Ge Doping on Crystal Quality

XRC measurements of the Ge-doped samples were performed, and the results were compared with those for an undoped reference sample to assess the influence of the Ge doping on the structural quality. The Ge doping had no significant effect on the tilt or twist angle.

Electrical Properties

Figure 11.14a, b present the results of Hall measurements at RT. The carrier concentration first increased and then decreased at a $GeCl_4$ bubbling rate of 6 sccm, most likely because of the increase in the point defect density. The carrier mobility

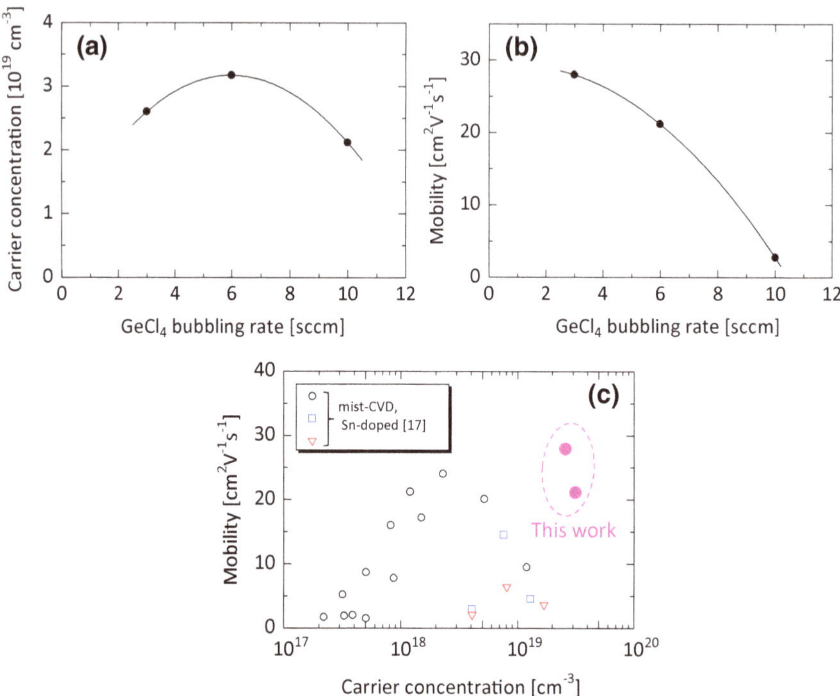

Fig. 11.14 **a** Carrier concentration and **b** electron mobility at RT as functions of $GeCl_4$ bubbling rate. **c** Relationship between electron mobility and carrier concentration at RT for Ge-doped α-Ga_2O_3. Data from the literature [17] are also shown for comparison

decreased monotonically with increasing $GeCl_4$ supply. Figure 11.14c shows the relationship between the carrier mobility and carrier concentration. Data for Sn-doped α-Ga_2O_3 grown using mist CVD are also shown for comparison [17]. The electron mobility in this work was much greater than previously reported values. The lowest resistivity was 8.6 mΩ cm, which is less than half of that of typical commercially available conductive SiC wafers.

Figure 11.15a, b present the results of the Hall measurements at elevated temperatures. The ionization energy of Ge in α-Ga_2O_3 was determined to be 18 meV from the slope of the temperature dependence of the carrier concentration (Fig. 11.15a); this value is similar to that for Sn [17]. The carrier mobility decreased monotonically with increasing temperature, most likely because of the scattering by both optical phonons and ionized donors. To clarify the detailed scattering mechanism, the low-temperature behavior must also be investigated.

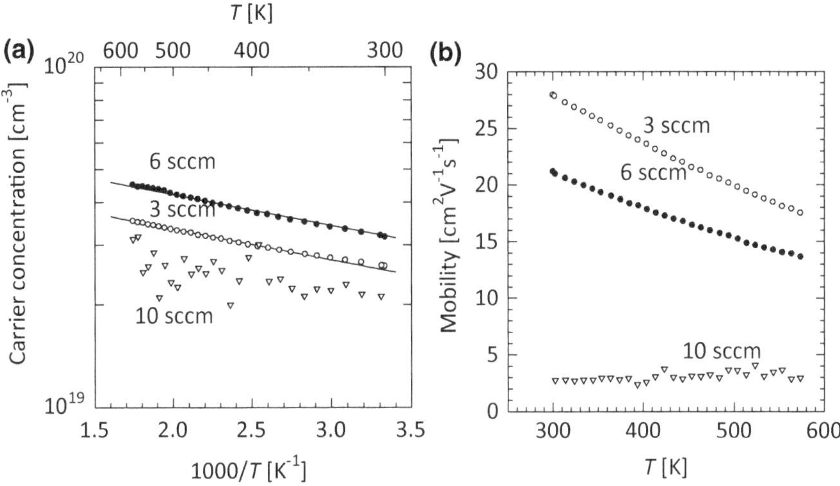

Fig. 11.15 **a** Carrier concentration and **b** electron mobility as functions of temperature

11.2.6 Improvement of Crystal Quality by Epitaxial Lateral Overgrowth

11.2.6.1 Motivation

Need for Improvement of Crystal Quality

As described in Sect. 2.4, heteroepitaxial α-Ga_2O_3 films contain high density of crystal defects. Accordingly, the crystal quality should be improved because such defects could lead to deterioration of the device performance.

Principle of Epitaxial Lateral Overgrowth Technique

To improve the crystal quality of heteroepitaxial films, a technique called epitaxial lateral overgrowth (ELO) has been established mainly for III–V semiconductors such as GaN. Figure 11.16a–f show the steps in the ELO process. First, a seed layer of the target material is grown on a substrate, and photolithography is used to fabricate periodic masks on the template (Fig. 11.16a). SiO_x and SiN_x are used as the mask material in many cases. The mask dimensions are typically on the micro-meter scale, and a stripe- or dot-patterned mask is usually employed. Next, the regrowth process is performed on the template. The regrowth process begins selectively from the windows of the mask, and isolated stripes or islands of the target crystal are formed (Fig. 11.16b). These crystals grow vertically and laterally and coalesce (Fig. 11.16c, d), finally forming a flat surface (Fig. 11.16e). During the regrowth process, dislocations in the seed layer propagate into the regrown crystal only through the mask windows; therefore, the dislocation density in the laterally grown wing region is very low. In addition, to minimize the elastic strain energy, dislocations in the window area bend toward the lateral direction when the stripes or islands are grown such that they have inclined facets on the top by controlling the regrowth condition. As a result, the dislocation density above the windows can also be reduced (Fig. 11.16f). This type of ELO is called facet-initiated ELO (FIELO) [22].

The ELO technique was first proposed as a method to produce mesa structures of electronic devices of Si or GaAs [23, 24], and later, the technique was used by Nishinaga et al. to improve the crystal quality of GaAs [25]. Usui et al. applied this technique to GaN on sapphire and established the FIELO technique [22]. Today, the FIELO technique is recognized as an essential technique for the production of high-quality freestanding GaN wafers [22, 26, 27].

The effectiveness of the ELO technique for α-Ga_2O_3 has already been demonstrated by mist CVD. In the demonstration, α-Ga_2O_3 was grown on a (0001) sapphire substrate using mist CVD with stripe-patterned SiO_2 masks (mask/window = 2 μm/2 μm) on the surface. Although coalescence of α-Ga_2O_3 was not achieved, cross-sectional TEM analysis revealed that no dislocations presented in the wing region.

Purpose of the Study

To reduce the dislocation density effectively, it is desirable to use a mask pattern with a small mask fill factor, i.e., a small window size and wide window spacing. In that case, however, island coalescence requires thick film growth. Accordingly, it is preferable to employ a high-speed epitaxial technique from the viewpoint of cost effectiveness. We thus used HVPE as the growth method for ELO of α-Ga_2O_3.

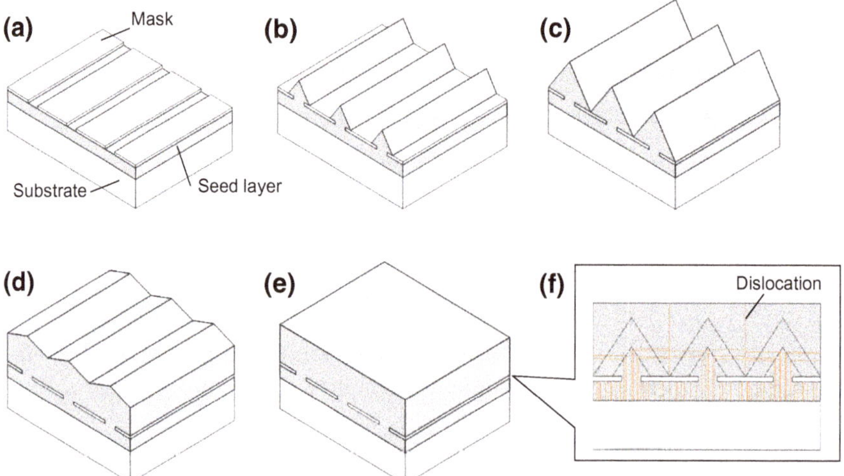

Fig. 11.16 Procedure of FIELO technique

In the following parts, we describe the current state of the development of the ELO technique using HVPE.

11.2.6.2 Experimental Methods

A (0001) α-Ga_2O_3 template grown on a sapphire substrate was used as the seed substrate for the investigation. Dot- or stripe-patterned SiO_x masks were prepared on the template using RF sputtering and conventional photolithography. The dot-patterned mask was designed such that 5-μm-diameter circular windows formed a triangular lattice pattern. The window spacing (the distance between the mask edges of the nearest windows) was 5 μm unless otherwise specified. The widths of the mask/window of the stripe-patterned mask were 5 μm/5 μm.

The HVPE growth conditions used for this investigation are described in Sect. 2.2. The partial pressures of the GaCl and O_2 supplies were 1.25×10^{-1} and 1.25 kPa, respectively, unless otherwise mentioned.

The morphology of the grown samples was examined using SEM, the mosaicity was estimated using XRC, and the behavior of the dislocations was investigated using TEM.

11.2.6.3 Morphological Characterization of ELO-Grown α-Ga_2O_3

Effect of Window Spacing and Growth Rate

Figure 11.17a–c present SEM images of the samples grown under the same conditions on a dot-patterned mask with a window spacing of 5–20 μm. For the spacing of 5 μm, α-Ga_2O_3 islands were selectively grown at the mask openings, forming a regular array (Fig. 11.17a). When the spacing was wider, however, additional crystal grains appeared around each α-Ga_2O_3 island (Fig. 11.17b, c). In this type of selective area growth, the precursors are consumed virtually only in the window regions. Therefore, the effective precursor supply increases with increasing window spacing and decreasing window density. Indeed, the island size increases with increasing window spacing. It is likely that this increase in the driving force for the growth resulted in the undesired nucleation. To confirm this speculation, slower growth was performed on the 20-μm-wide mask by decreasing only the GaCl supply such that the nominal growth rates (growth rate for flat films) decreased from 12 to 7 and 5 μm/h. As a result, nucleation of additional crystal grains was markedly suppressed, as observed in Fig. 11.18a, b.

Effect of Growth Temperature T_g

Figure 11.19a–c present SEM images of the samples grown at 540, 500, and 460 °C. For T_g = 540 °C, the island morphology is primarily a hexagonal pillar with a well-developed c-plane and inclined $(10\bar{1}1)$ planes cutting off the top edge

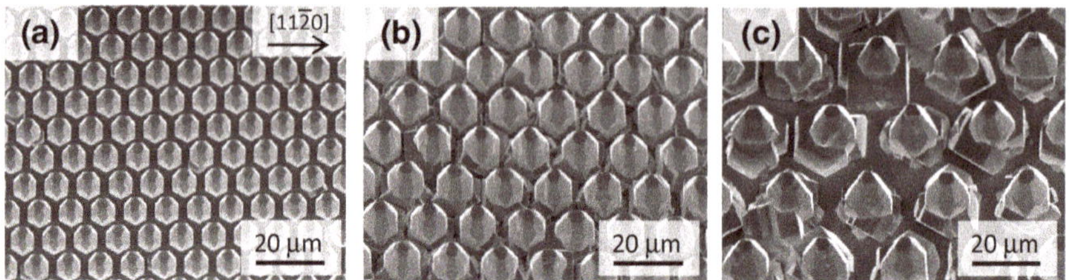

Fig. 11.17 SEM images of α-Ga_2O_3 islands grown on dot-patterned mask with window spacings of **a** 5 µm, **b** 10 µm, and **c** 20 µm (bird's eye view) [18]

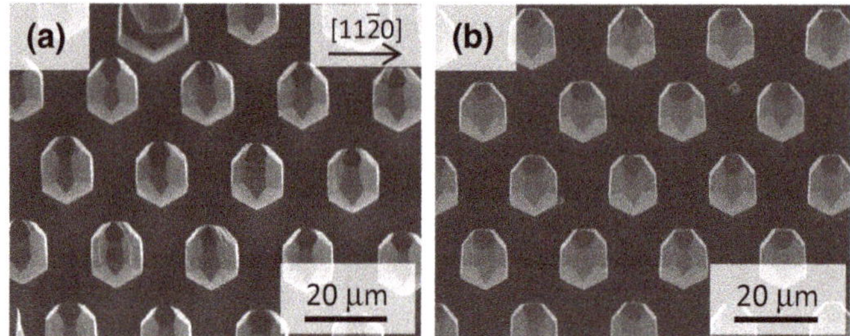

Fig. 11.18 SEM images of α-Ga_2O_3 islands grown at nominal growth rates of **a** 7 µm/h and **b** 5 µm/h (bird's eye view) [18]

(Fig. 11.19a). When T_g was decreased to 500 °C, $(10\bar{1}1)$ planes developed well, whereas the (0001) plane become unstable (Fig. 11.19b). When T_g was further decreased to 460 °C, the (0001) plane disappeared and $(10\bar{1}4)$ planes appeared instead (Fig. 11.19c). Thus, the island morphology is sensitive to the growth temperature and can be controlled. This nature is essential to perform the FIELO process, as described in Sect. 2.6.1.

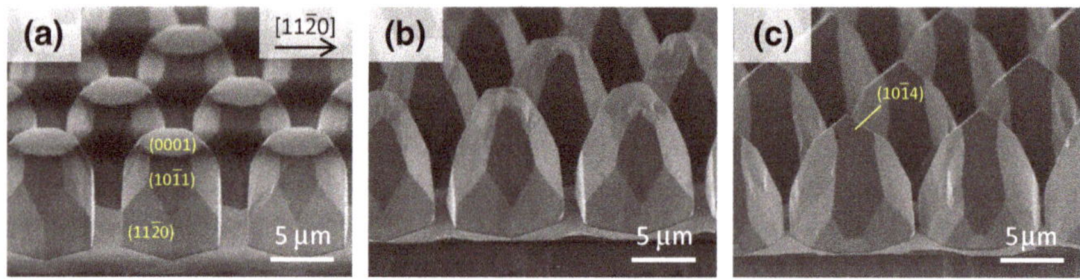

Fig. 11.19 SEM images of α-Ga_2O_3 islands grown at **a** 540 °C, **b** 500 °C, and **c** 460 °C (bird's eye view) [18]

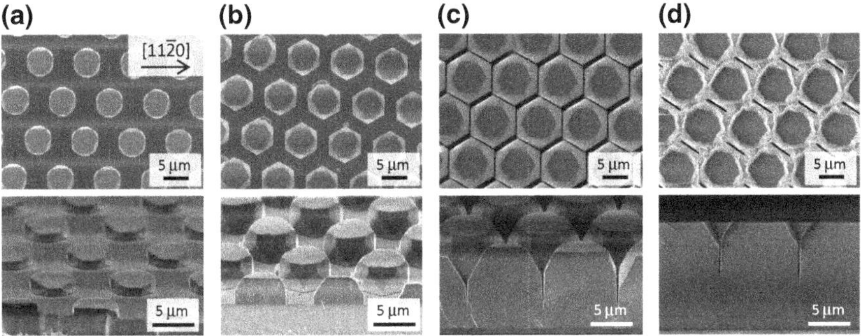

Fig. 11.20 SEM images of α-Ga_2O_3 islands with nominal thickness of **a** 0.5 μm, **b** 1.6 μm, **c** 8 μm, and **d** 12 μm (plan-view and bird's eye view) [18]

Time Evolution of the Growth

Figure 11.20a–d show the time evolution of the ELO process of α-Ga_2O_3. The degree of growth is indicated by the nominal thickness (thickness for a flat film). At the beginning of the regrowth, the shape of each island was similar to that of a circular window (Fig. 11.20a). Then, the crystal habit became clear, and hexagonal pillars were formed (Fig. 11.20b). Under these growth conditions, vertical growth was faster than lateral growth. Island coalescence started at the bottom of the islands (Fig. 11.20c, d), and finally, a flat film was obtained.

11.2.6.4 Structural Characterization of ELO-Grown α-Ga_2O_3

XRC-FWHM

XRC measurements of the ELO-grown samples were performed to verify the effectiveness on the improvement of the crystal quality. Figure 11.21 plots the tilt and twist angles as functions of the nominal thickness. Both the tilt and twist angles decreased with increasing nominal thickness, which indicates the increase of high-quality area during the 3D growth. Figure 11.22 presents XRC profiles of the sample with a nominal thickness of 12 μm. Multiple peaks were not observed for the *0006* diffraction in contrast to the case for the ELO-grown GaN. It has been reported that out-of-plane XRC of ELO-grown GaN contains multiple peaks because of the formation of small-angle grain boundaries [28, 29]. This inclination of the laterally grown wing region was attributed to edge dislocation array formed above the mask edge, which was clearly visible in cross-sectional TEM image [28]. Accordingly, the absence of peak splitting for ELO-grown α-Ga_2O_3 indicates that character of the dislocations and their response to crystal strain in ELO-grown α-Ga_2O_3 can be different from those in GaN.

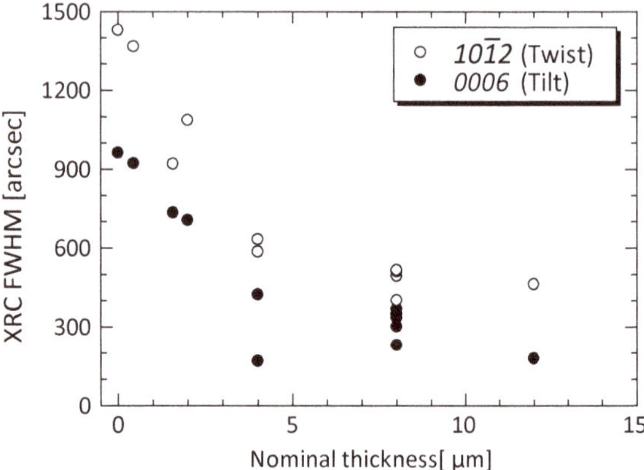

Fig. 11.21 XRC FWHMs of ELO-grown α-Ga_2O_3 as a function of nominal thickness [18]

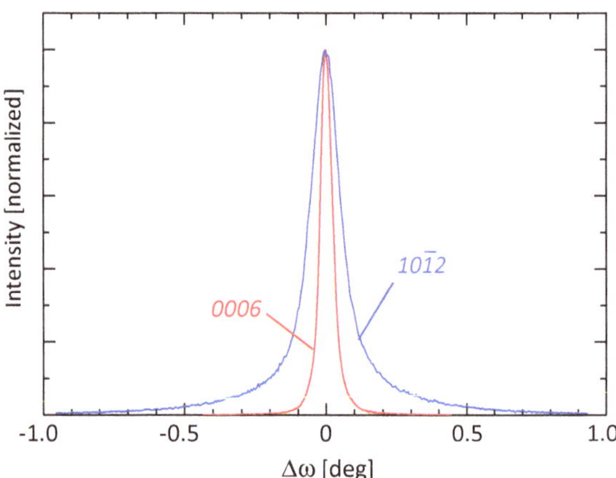

Fig. 11.22 XRC profiles of ELO-grown α-Ga_2O_3 with nominal thickness of 12 μm [18]

Effect of Island Morphology on the Behavior of Dislocations

Figure 11.23a, b present cross-sectional TEM images of α-Ga_2O_3 islands grown at 540 and 460 °C; the corresponding SEM images are presented in Fig. 11.19a, c, respectively. The 540 °C-grown island had a well-developed (0001) plane on the top, whereas the 460 °C-grown island had well-developed inclined facets. In both cases, dislocations in the seed layer propagated into the island through the window. For T_g = 540 °C, dislocation density in the wing area was clearly lower than that in the seed layer; however, dislocations in the window area extended to the island top (Fig. 11.23a). However, dislocations in the 460 °C-grown island bent such that the dislocation line was nearly normal to the free surface (Fig. 11.23b). Thus, the FIELO process can be used not only for GaN but also for α-Ga_2O_3. To estimate the dislocation density in the wing region, plan-view TEM analysis was performed for a sample grown on a stripe-patterned mask under growth conditions similar to those used for the 540 °C-grown sample. Figure 11.24 presents the results. The α-Ga_2O_3

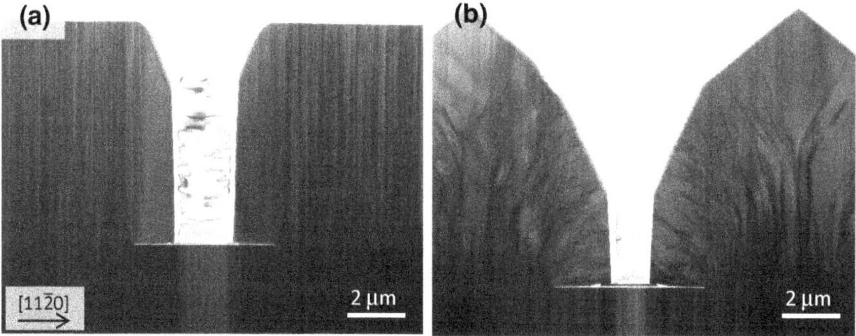

Fig. 11.23 Cross-sectional TEM images of α-Ga_2O_3 islands grown at **a** 540 °C and **b** 460 °C

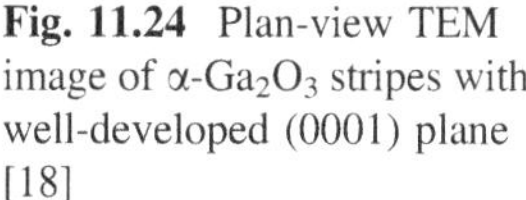

Fig. 11.24 Plan-view TEM image of α-Ga_2O_3 stripes with well-developed (0001) plane [18]

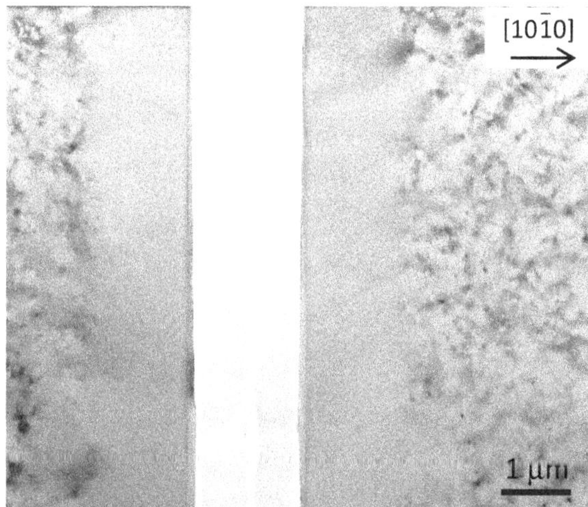

stripes did not coalesce with each other yet, and low-dislocation-density regions were observed on both sides of the gap. No dislocations were observed in the low-dislocation-density regions of approximately 22 μm^2; therefore, the dislocation density should be less than 5×10^6 cm^{-2}.

Behavior of Dislocations in a Coalesced Film

Cross-sectional TEM analysis was performed on a coalesced film to clarify the behavior of dislocations in an ELO-grown α-Ga_2O_3. The sample was grown for 2 h at 520 °C; therefore, the film should have been formed through inclined facet growth. Figure 11.25a presents a plan-view SEM image of the sample. The surface was still bumpy, with protruding parts corresponding to the window areas. Figure 11.25b presents a schematic illustration of the sample cross-section. TEM analysis was performed in the dashed-line rectangle area. Figure 11.25c presents

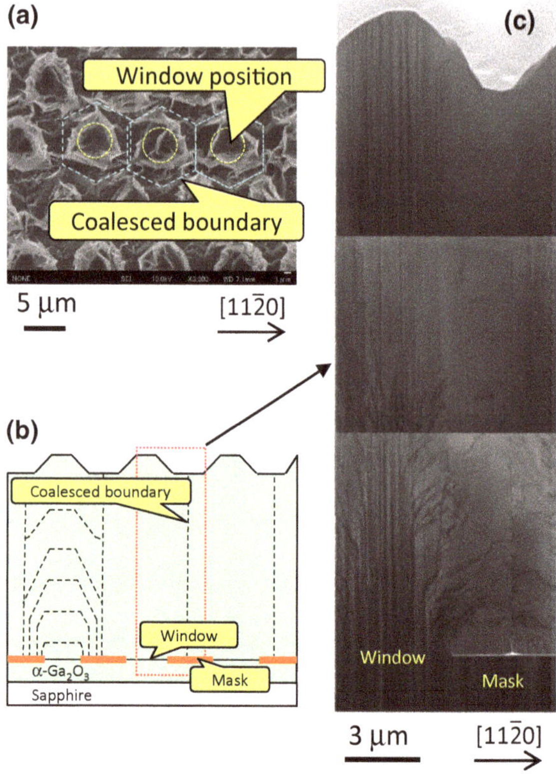

Fig. 11.25 **a** Plan-view SEM image of a coalesced α-Ga_2O_3 film. **b** Schematic illustration of the cross section. **c** Cross-sectional TEM image of the film [18]

the TEM image. The dislocations above the window bent because of the faceted growth, and the dislocation density of the top part was much lower than that of the seed layer. No dislocation array was observed above the mask edge, which was observed for ELO-GaN. This result is consistent with the XRC results described in 2.6.4. At the coalesced boundary, we can see the dislocation contrast in the vicinity of the mask. The density decreased in the upper part, and no dislocation contrast was observed in the top part. Thus, ELO of α-Ga_2O_3 by HVPE is promising for improvement of the crystal quality. It would be possible to further reduce the dislocation density by increasing the duration of the 3D growth. Double-ELO would be also effective if the second mask is aligned to cover the window positions of the first mask. Note that in general, acceptable defect density is strongly dependent on the material, device structure, and driving condition. Accordingly, it is desirable to simultaneously perform device investigations to clarify the effect of crystal defects and to specify the target quality.

11.3 HVPE of ε-Ga_2O_3

11.3.1 Features and Potential Applications of ε-Ga_2O_3

ε-Ga_2O_3 is a metastable phase of Ga_2O_3 as well as α-Ga_2O_3. The first synthesis of ε-Ga_2O_3 was reported by Roy et al. [1]. They obtained a mixture powder of ε-Ga_2O_3 and β-Ga_2O_3 through annealing of $Ga(NO_3)_3$. The crystal structure of ε-Ga_2O_3 should to be addressed carefully. Playford et al. performed structural analysis of their powder material by neutron diffraction, and concluded that their ε-Ga_2O_3 had hexagonal crystal structure with space group $P6_3mc$ (PDF# 01–082-3196) [30]. On the other hand, Cora et al. examined the microstructure of their MOCVD-grown pseudo-hexagonal "ε-Ga_2O_3" film on (0001) sapphire by TEM, and found that the film was consisted of three-fold in-plane rotational nano-scale domains of orthogonal Ga_2O_3 with space group $Pna2_1$, which is referred to as κ-Ga_2O_3 [31]. Nishinaka et al. also concluded their "ε-Ga_2O_3" films grown on (0001) GaN or (111) $SrTiO_3$ by mist CVD were orthorhombic [32]. We investigated the microstructure of our HVPE-grown "ε-Ga_2O_3" on (0001) GaN, (0001) AlN, and $(\bar{2}01)$ β-Ga_2O_3 by TEM, and confirmed that the crystal structure was orthorhombic. This polymorph is also called "orthorhombic ε-Ga_2O_3". The term should be unified, but careful discussion will be required. In this article, we temporarily employ "orthorhombic ε-Ga_2O_3" or simply "ε-Ga_2O_3".

The first synthesis of phase-pure ε-Ga_2O_3 and its epitaxial growth was demonstrated by HVPE, and the band gap energy was reported to be 4.9 eV [9]. The crystal structure of ε-Ga_2O_3 does not have inversion symmetry along the *c*-axis. Spontaneous polarization is therefore expected [33], and ferroelectric behavior was observed [34]. The existence of polarization would lead to the formation of a high concentration of two-dimensional electron gas [33]. Thus, ε-Ga_2O_3 is also a promising material for power device applications.

To fabricate ε-Ga_2O_3 devices, it is essential to establish epitaxial growth techniques. As described above, the epitaxial growth technique of ε-Ga_2O_3 was first demonstrated by HVPE, and then other techniques, such as MOCVD and mist CVD, were also shown to be effective [35, 36]. In this section, HVPE of ε-Ga_2O_3 and the characteristics of the grown layers are presented.

11.3.2 Growth Methods and Conditions of ε-Ga_2O_3

The HVPE apparatus used in this study is described in Sect. 2.2, and the growth conditions were also similar. The growth temperature was 550 °C, and the partial pressures of GaCl and O_2 were 0.25 and 1.0 kPa, respectively. (0001) GaN and (0001) AlN were used as the substrates. The in-plane lattice mismatches between ε-Ga_2O_3 and these substrates are summarized in Table 11.3. Note that the mismatches are calculated approximating the crystal structure of ε-Ga_2O_3 as pseudo-hexagonal.

Table 11.3 Lattice constants of ε-Ga_2O_3 and its substrates and lattice mismatches between them

Material	Lattice constant (nm)	Mismatch (%)
(001) ε-Ga_2O_3	$a = 0.2904$	–
(0001) GaN	$a = 0.3189$	8.8
(0001) AlN	$a = 0.3112$	6.6

11.3.3 Properties of HVPE-Grown ε-Ga_2O_3

XRD Analysis of HVPE-grown Ga_2O_3 Films

Figure 11.26a, b present XRD 2θ–ω scan profiles of the films grown by HVPE on the two types of substrates. In the both cases, only diffraction peaks from the *c*-plane of ε-Ga_2O_3 were observed in addition to those from the substrates. Thus, ε-Ga_2O_3 with no contamination by β-Ga_2O_3 was obtained.

SEM Observation of Typical Films

Figure 11.27a–d present plan-view SEM images of the ε-Ga_2O_3 films. The sample surfaces were specular to human eyes; however, 3D grains are visible in the SEM images. The grain size was uniform on each sample. The density of the grains did not change by increasing the growth time, although the grain size increased. These results indicate that the grains nucleate at a certain moment during the growth. Figure 11.28a, b present cross-sectional SEM images of the ε-Ga_2O_3 films grown for 2 and 7 min, respectively. From these images, the growth rate was estimated to be approximately 20 μm/h. Figure 11.28b shows that the grain nucleated at the

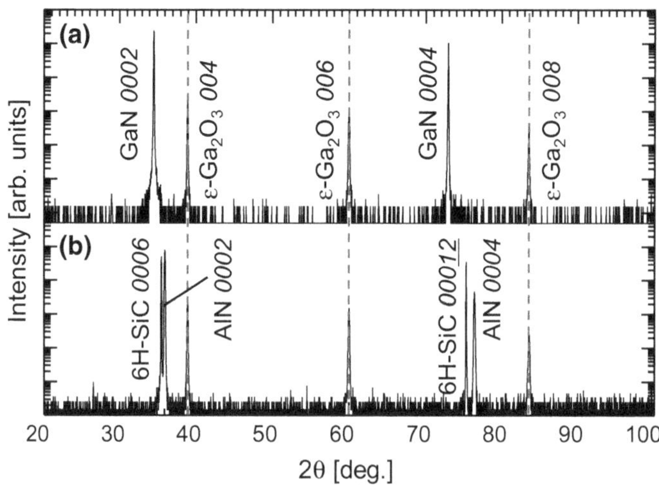

Fig. 11.26 XRD 2θ–ω scan profiles of ε-Ga_2O_3 layers grown on **a** (0001) GaN and **b** (0001) AlN [9]

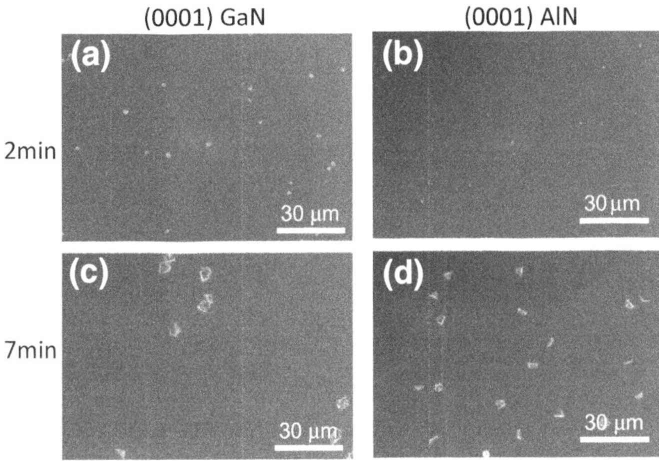

Fig. 11.27 Plan-view SEM images of ε-Ga_2O_3 layers grown on (0001) GaN and (0001) AlN with growth times of 2 and 7 min [9]

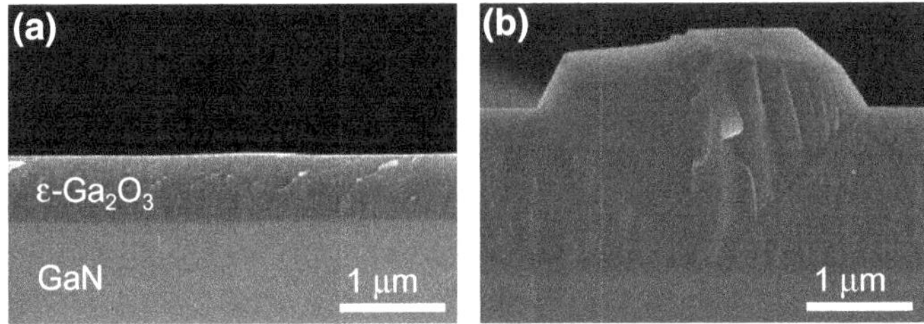

Fig. 11.28 Cross-sectional SEM images of ε-Ga_2O_3 layers grown on (0001) GaN with growth times of **a** 2 min and **b** 7 min [9]

film/substrate boundary. Accordingly, growth optimization is necessary at the beginning of the growth to suppress the grain nucleation.

Crystal Orientation

Figure 11.29a, b show the XRD pole figures of ε-Ga_2O_3 and the GaN substrate, respectively. The pole figures for ε-Ga_2O_3 (Fig. 11.29a) and (0001) GaN (Fig. 11.29b) exhibited six-fold symmetry, indicating that (001) ε-Ga_2O_3 was grown epitaxially. The result for ε-Ga_2O_3 grown on (0001) AlN was similar (not shown).

Crystal Quality

The mosaicity of the HVPE-grown ε-Ga_2O_3 films were estimated from XRC measurements of *004* and *131/201* diffractions in symmetric and skew-symmetric

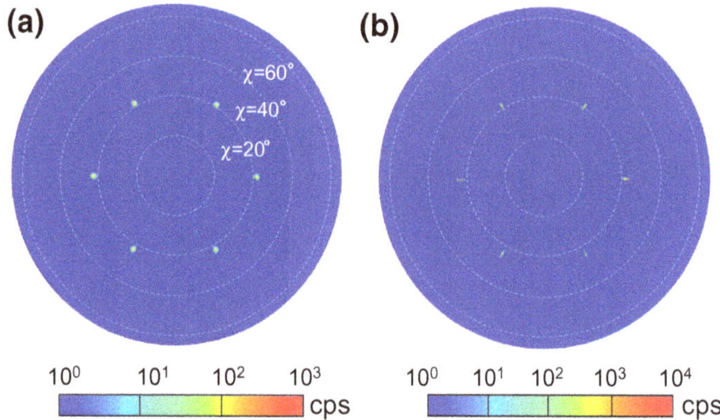

Fig. 11.29 X-ray pole figures (log scale) of **a** ε-Ga_2O_3 *134/204* and **b** GaN *1 0 $\bar{1}$ 2* [9]

geometry. Figure 11.30a, b present XRC profiles of the ε-Ga_2O_3 films grown on (0001) GaN and (0001) AlN, respectively. The FWHMs tended to be narrow when the in-plane lattice mismatch, which is summarized in Table 11.3, was small. However, for the both cases, the mosaicity was very large. Accordingly, the crystal quality should be improved by optimizing the growth conditions, introducing buffer layers, and/or utilizing the ELO technique.

Impurity Analysis

Table 11.4 summarizes the impurity analysis results obtained using SIMS of an HVPE-grown ε-Ga_2O_3 film. Although the growth conditions for ε-Ga_2O_3 were similar to those for α-Ga_2O_3, the concentrations of H and Cl were much higher than those in α-Ga_2O_3. Further study is necessary to clarify the reason for this difference,

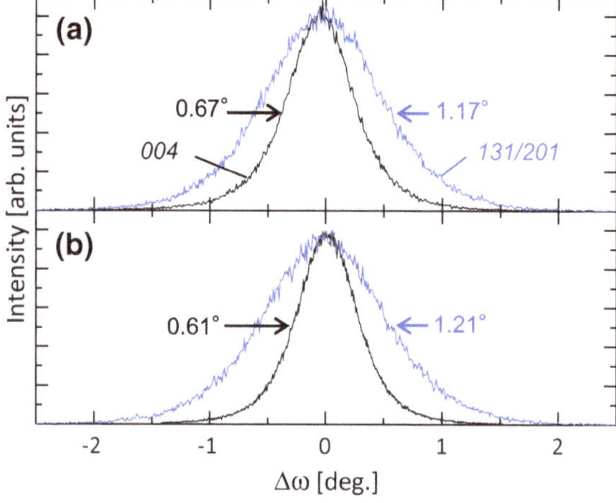

Fig. 11.30 XRCs of ε-Ga_2O_3 layers grown on **a** (0001) GaN and **b** (0001) AlN [9]

Table 11.4 Impurity concentrations in ε-Ga_2O_3 measured by SIMS

Element	Concentration (cm^{-3})
H	1×10^{18}
C	$<6 \times 10^{16}$
N	$<5 \times 10^{16}$
Si	$<1 \times 10^{16}$
Cl	2×10^{18}
Al	$<3 \times 10^{15}$
Cr	$<4 \times 10^{14}$
Fe	$<8 \times 10^{14}$
Ni	$<3 \times 10^{15}$

which likely originates from the difference in the surface structures of these polymorphs.

Optical Bandgap of ε-Ga_2O_3

Until recently, ε-Ga_2O_3 was obtained only as the mixture powder with β-Ga_2O_3; therefore, the optical bandgap was unknown. Now phase-pure ε-Ga_2O_3 has been successfully grown epitaxially, and the optical bandgap energy was estimated for the first time using transmittance measurements. Figure 11.31 presents the optical transmittance of ε-Ga_2O_3 grown by HVPE on (0001) AlN. Although the optical transition type of ε-Ga_2O_3 remains under discussion, better fitting was obtained when direct transition was assumed (inset of Fig. 11.31). The optical band gap energy was estimated to be 4.9 eV, which is close to the value for β-Ga_2O_3.

Thermal Stability and TEC

ε-Ga_2O_3 is a metastable phase and should transform into β-Ga_2O_3 above a certain threshold temperature. The thermal stability of ε-Ga_2O_3 was investigated using the

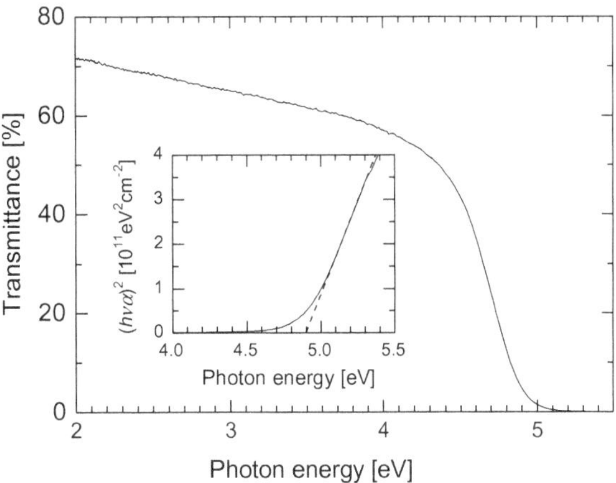

Fig. 11.31 Transmittance spectrum of ε-Ga_2O_3. The inset shows the absorption coefficient in $(h\nu\alpha)^2$ versus $h\nu$ [9]

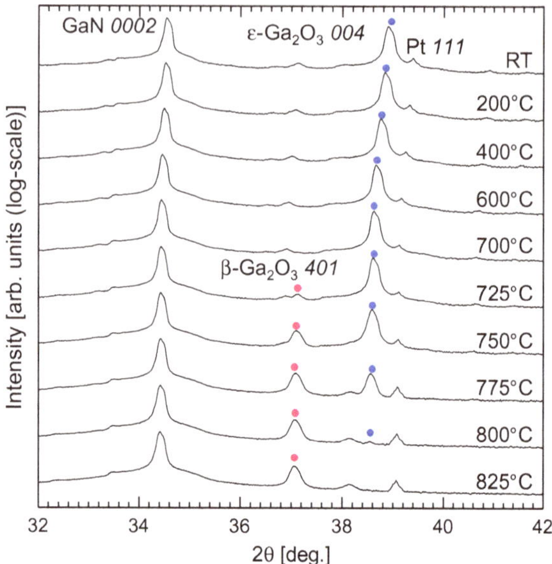

Fig. 11.32 XRD 2θ–ω scan profiles for an HVPE-grown ε-Ga_2O_3 film measured at RT and elevated temperatures [9]

same high-temperature XRD system described in Sect. 2.6. Figure 11.32 presents the result for ε-Ga_2O_3 grown on (0001) GaN. Apart from the diffraction peaks of the Pt sample holder and GaN substrate, only the *004* peak of ε-Ga_2O_3 was observed from RT to 700 °C. When the temperature reached 725 °C, the *401* peak of β-Ga_2O_3 appeared, and the peak intensity increased with increasing temperature, whereas that of *004* of ε-Ga_2O_3 decreased. This result indicates that ε-Ga_2O_3 is thermally stable up to approximately 700 °C.

It is possible to calculate the TECs from the peak shift values in Fig. 11.32. The TECs along [001] for ε-Ga_2O_3 grown on (0001) GaN was determined to be 1.1×10^{-5} K^{-1}. Again, note that the value may not be the same as that for freestanding materials because of the thermal stress resulting from the difference in TECs between ε-Ga_2O_3 and the substrates.

11.4 Summary and Future Prospects

The current state of HVPE technologies for α-Ga_2O_3 and ε-Ga_2O_3 were reviewed.

Regarding α-Ga_2O_3, the device technology is developing steadily. Engineering samples of SBDs with very low R_{on} are commercially available, and normally-off MOSFETs have been demonstrated. To further improve the device performance, the establishment of growth technologies to obtain high-quality α-Ga_2O_3 and the realization of conductive freestanding α-Ga_2O_3 wafers are indispensable. The techniques described in this chapter, such as high-speed growth, conductivity control, and the ELO technique, will be essential in achieving these objectives.

Further development of these techniques is required, and study of scientific fundamentals of the related phenomena is also necessary.

Currently, ε-Ga_2O_3 is attracting considerable attention because of its unique properties, including spontaneous polarization, and is potentially useful for power device applications. However, research on ε-Ga_2O_3 still remains in a very primitive stage, and further accumulation of fundamental knowledge is required.

Acknowledgements Part of this work is based on results obtained from a project commissioned by the New Energy and Industrial Technology Development Organization (NEDO). This work was also partly supported by a Grant-in-Aid for Scientific Research (C) Nos. 25420307 and 17K05047 from the Japan Society for the Promotion of Science (JSPS). The author would like to acknowledge K. Kawara, T. Shinohe, T. Hitora, Prof. M. Kasu, and Prof. S. Fujita for their contributions to the NEDO project. The author would also like to thank Drs. E.G. Villora, Y. Matsushita, S. Yamamoto, and K. Shimamura for their support in the JSPS projects.

References

1. R. Roy, V.G. Hill, E.F. Osborn, J. Am. Chem. Soc. **74**, 719 (1952)
2. H. Aida, K. Nishiguchi, H. Takeda, N. Aota, K. Sunakawa, Y. Yaguchi, Jpn. J. Appl. Phys. **47**, 8506 (2008)
3. E.G. Víllora, K. Shimamura, Y. Yoshikawa, K. Aoki, N. Ichinose, J. Cryst. Growth **270**, 420 (2004)
4. Z. Galazka, K. Irmscher, R. Uecker, R. Bertram, M. Pietsch, A. Kwasniewski, M. Naumann, T. Schulz, R. Schewski, D. Klimm, M. Bickermann, J. Cryst. Growth **404**, 184 (2014)
5. K. Sasaki, A. Kuramata, T. Masui, E.G. Víllora, K. Shimamura, S. Yamakoshi, Appl. Phys. Express **5**, 035502 (2012)
6. M. Higashiwaki, K. Sasaki, A. Kuramata, T. Masui, S. Yamakoshi, Appl. Phys. Lett. **100**, 013504 (2012)
7. M. Higashiwaki, K. Sasaki, T. Kamimura, M.H. Wong, D. Krishnamurthy, A. Kuramata, T. Masui, S. Yamakoshi, Appl. Phys. Lett. **103**, 123511 (2013)
8. Y. Oshima, E.G. Villora, K. Shimamura, Appl. Phys. Express **8**, 055501 (2015)
9. Y. Oshima, E.G. Villora, Y. Matsushita, S. Yamamoto, K. Shimamura, J. Appl. Phys. **118**, 085301 (2015)
10. Y. Oshima, E.G. Villora, K. Shimamura, J. Cryst. Growth **410**, 53 (2015)
11. D. Shinohara, S. Fujita, Jpn. J. Appl. Phys. **47**, 7311 (2008)
12. S. Fujita, K. Kaneko, J. Cryst. Growth **401**, 588 (2014)
13. K. Kaneko, S. Fujita, T. Hitora, Jpn. J. Appl. Phys. **57**, 02CB18 (2018)
14. M. Oda, R. Tokuda, H. Kambara, T. Tanikawa, T. Sasaki, T. Hitora, Appl. Phys. Express **9**, 021101 (2016)
15. T. Kawaharamura, G.T. Dang, M. Furuta, Jpn. J. Appl. Phys. **51**, 040207 (2012)
16. K. Akaiwa, S. Fujita, Jpn. J. Appl. Phys. **51**, 070203 (2012)
17. K. Akaiwa, K. Kaneko, K. Ichino, S. Fujita, Jpn. J. Appl. Phys. **55**, 1202BA (2016)
18. Y. Oshima, K. Kawara, T. Shinohe, T. Hitora, M. Kasu, S. Fujita, APL mater. **7**, 022503 (2019)
19. L.J. Eckert, R.C. Bradt, J. Amer. Cer. Soc. **56**, 229 (1973)
20. W.M. Yim, R.J. Paff, J. Appl. Phys. **45**, 1456 (1974)
21. Y. Oshima, K. Kawara, M. Kasu, T. Shinohe, T. Hitora, Ge Doping of α-Ga_2O_3 by Halide Vapor Phase Epitaxy, in *Paper Presented at 60th Electronic Materials Conference* (University of California, Santa Barbara, June 2018), pp. 27–29

22. A. Usui, H. Sunakawa, A. Sakai, A.A. Yamaguchi, Jpn. J. Appl. Phys. **36**, L899 (1997)
23. B.D. Joyce, J.A. Baldrey, Nature **195**, 486 (1962)
24. F.W. Tausch Jr., A.G. Lapierre III, J. Electrochem. Soc. **112**, 706 (1965)
25. T. Nishinaga, T. Nakano, S. Zhang, Jpn. J. Appl. Phys. **27**, L964 (1988)
26. Y. Oshima, T. Eri, M. Shibata, H. Sunakawa, K. Kobayashi, T. Ichihashi, A. Usui, Jpn. J. Appl. Phys. **42**, L1 (2003)
27. K. Motoki, T. Okahisa, N. Matsumoto, M. Matsushima, H. Kimura, H. Kasai, K. Takemoto, K. Uematsu, T. Hirano, M. Nakayama, S. Nakahata, M. Ueno, D. Hara, Y. Kumagai, A. Koukitu, H. Seki, Jpn. J. Appl. Phys. **40**, L140 (2001)
28. A. Sakai, H. Sunakawa, A. Usui, Appl. Phys. Lett. **73**, 481 (1998)
29. P. Fini, C. Thompson, G.B. Stephenson, J.A. Eastman, M.V. Ramana Murty, O. Auciello, L. Zhao, S.P. Denbaars, J.S. Speck, Appl. Phys. Lett. **76**, 3893 (2000)
30. H.Y. Playford, A.C. Hannon, E.R. Barney, R.I. Walton, Chem. Eur. J. **19**, 2803 (2013)
31. I. Cora, F. Mezzadri, F. Boschi, M. Bosi, M. Čaplovičová, G. Calestani, I. Dódony, B. Pécz, R. Fornari, Cryst. Eng. Comm **19**, 1509 (2017)
32. H. Nishinaka, H. Komai, D. Tahara, Y. Arata, M. Yoshimoto, Jpn. J. Appl. Phys. **57**, 115601 (2018)
33. M.B. Maccioni, V. Fiorentini, Appl. Phys. Express **9**, 041102 (2016)
34. F. Mezzadri, G. Calestani, F. Boschi, D. Delmonte, M. Bosi, R. Fornari, Inorg. Chem. **55**, 12079 (2016)
35. F. Boschi, M. Bosi, T. Berzina, E. Buffagni, C. Ferrari, R. Fornari, J. Cryst. Growth **443**, 25 (2016)
36. H. Nishinaka, D. Tahara, M. Yoshimoto, Jpn. J. Appl. Phys. **55**, 1202BC (2016)

Chapter 12
Mist Chemical Vapor Deposition 1

Heteroepitaxial Growth of α-Ga_2O_3 Thin Films on Sapphire Substrates

Shizuo Fujita

Abstract This chapter summarizes fundamental issues of corundum-structured gallium oxide (α-Ga_2O_3), which is obtained by heteroepitaxy on sapphire substrates and is featured by its large bandgap (~5.3 eV), bandgap engineering (3.7 to ~9 eV), and existing corundum-structured *p*-type oxide such as α-Ir_2O_3, as well as low epitaxy cost on inexpensive sapphire substrates.

12.1 Dawn of Corundum-Structured α-Ga_2O_3

12.1.1 Semistable Phases of Ga_2O_3

It is well known that Ga_2O_3 takes at least five different phases (α, β, γ, ε, and δ), in which orthorhombic β-gallia-structure (β-Ga_2O_3) is thermodynamically the most stable phase [1, 2]. This has allowed the growth of single-crystalline β-Ga_2O_3 bulks and enhanced the world-wide research on β-Ga_2O_3 materials and devices. In the case of heterospitaxy on sapphire (α-Al_2O_3) by molecular-beam epitaxy (MBE), we experienced that β-Ga_2O_3 was grown without following the corundum crystal structure of sapphire [3].

Later we achieved the growth of single-phase corundum-structured α-Ga_2O_3, in spite of that it is thermodynamically a semistable phase, on sapphire substrates by the use of mist chemical vapor deposition (CVD) technology [4]. This opened the history of α-Ga_2O_3 materials and devices.

S. Fujita (✉)
Kyoto University, Katsura, Nishikyo-Ku, Kyoto 615-8520, Japan
e-mail: fujitasz@kuee.kyoto-u.ac.jp

M. Higashiwaki and S. Fujita (eds.), *Gallium Oxide*, Springer Series in Materials Science 293, https://doi.org/10.1007/978-3-030-37153-1_12

12.1.2 Mist Chemical Vapor Deposition for Oxide Growth

In the earlier stage of the evolution of oxide semiconductors, they had been grown by MBE, pulsed-laser deposition (PLD), and metalorganic chemical vapor deposition (MOCVD). These vacuum-based growth technologies have been essential for the growth of traditional non-oxide semiconductors such as GaAs, InP, and GaN in order to completely eliminate leakage of flammable and/or toxic sources (such as organometallics, AsH_3, PH_3, and NH_3) and contamination of oxygen impurities which seriously degrade electrical and optical properties. On the other hand, for oxide semiconductors, oxygen is not an impurity but a constituent element. In this viewpoint, the author paid attention to mist CVD as a growth technology for oxide semiconductors capable of being applied to the active layer of devices.

Mist CVD has originally been developed as a technology to grow oxide films of ferroelectric materials and transparent conductors [5–8]. The potential for growing device-quality single-crystalline oxide semiconductors was evidenced for ZnO by achieving the electron concentration of the order of 10^{17} cm^{-3} and the mobility of 90 cm^2/Vs for unintentionally doped samples [9]. The mist CVD is considered to be effective to reduce oxygen vacancies, because the growth is carried out under sufficient overpressure of oxygen, by the use of water or alcohol solution, with respect to the metal sources.

12.1.3 Growth of α-Ga_2O_3 on Sapphire Substrates

Figure 12.1 shows a schematic illustration of a mist CVD system designed for the growth of single-crystalline oxide semiconductor films. In order to grow Ga_2O_3, a water solution of gallium acetylacetonate [$Ga(C_5H_8O_2)_3$], for example, was prepared as a reaction source. It was ultrasonically atomized with transducers, and

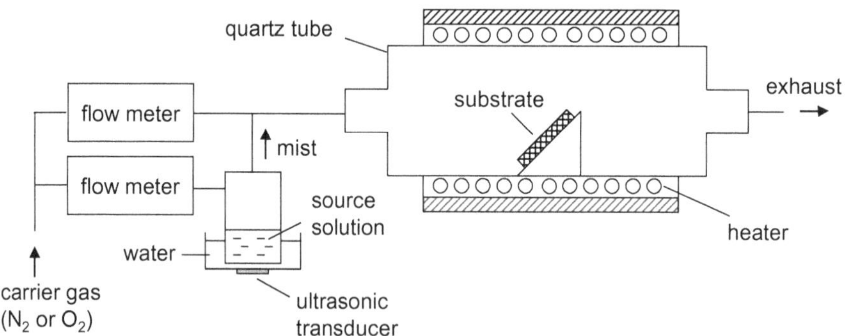

Fig. 12.1 Schematic illustration of a mist CVD system designed for the growth of single-crystalline oxide semiconductor films at high (>400 °C) temperatures

Table 12.1 SIMS analysis results for concentrations of plausible impurities in an α-Ga_2O_3 film together with the detection limit (DL)

Element	Detection limit (cm^{-3})	Concentration (cm^{-3})
H	8×10^{16}	2×10^{17}
C	7×10^{16}	<DL
F	2×10^{15}	<DL
Cl	7×10^{14}	<DL
I	2×10^{14}	<DL
Na	1×10^{14}	<DL
Mg	3×10^{14}	<DL
K	7×10^{13}	<DL
Ca	8×10^{14}	<DL

generated mist particles ($\sim$3 μm^{ϕ}), containing Ga elements, were transferred by carrier gas (O_2 or N_2) to the reaction area, where Ga_2O_3 was formed by chemical reaction. It is possible to use carbon-free Ga sources such as gallium chloride ($GaCl_3$) in the mist CVD. With the use of carbon-free sources, we can eliminate carbon contamination from the source, which is inevitable in MOCVD.

We investigated the growth of Ga_2O_3 by mist CVD on sapphire substrates, and found that well-aligned corundum-structured α-Ga_2O_3, which is thermodynamically a semistable phase, was grown at 470 °C instead of stable β-Ga_2O_3 [4]. The growth of α-Ga_2O_3 was attributed to the similarity of lattice structure between the epilayer and the substrate, in spite of the lattice mismatch (3.5% and 4.8% along *c*- and *a*-axes, respectively). α-Ga_2O_3 on sapphire exhibits very narrow spectrum in (0006) symmetric X-ray ω-scan diffraction. Typical full-widths at half maximum (FWHMs) of the spectra were as small as 30–60 arcsec. However, for antisymmetric ($10\bar{1}4$) X-ray ω-scan diffraction, the FWHMs were as large as $\sim$2000 arcsec for the α-Ga_2O_3 thicknesses of 300–2500 nm.

Since a mist CVD system is much simpler compared to an MOCVD system, one may speculate that a lot of impurities are unintentionally incorporated in grown films. However, efforts have been made to reduce the impurities by careful modification of sources and growth equipment. The use of carbon-free sources was effective to reduce carbon contamination. By the secondary ion mass spectroscopy (SIMS), as shown in Table 12.1, it was revealed that impurities in α-Ga_2O_3 grown by mist CVD were successfully below the detection limit except for hydrogen (H) [10].

12.2 Fundamental Properties of α-Ga_2O_3

12.2.1 Growth Characteristics

Figure 12.2 shows the cross-sectional transmission electron microscope (TEM) images near the α-Ga_2O_3/sapphire interface observed along the [$11\bar{2}0$] and [$10\bar{1}0$] zone axes [11, 12]. It is seen that defects are confined at the α-Ga_2O_3/sapphire

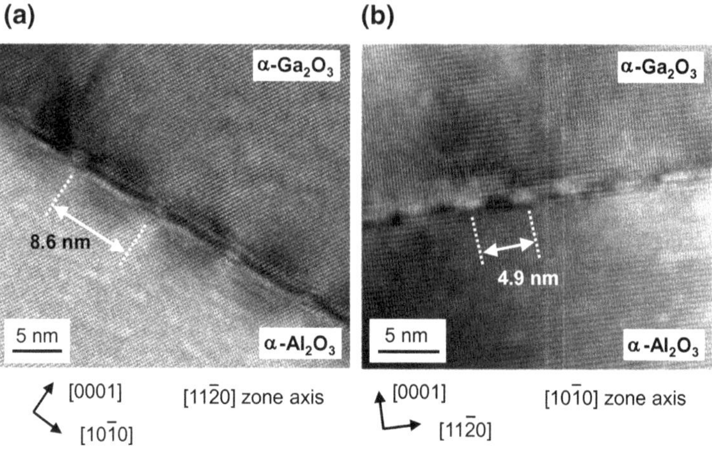

Fig. 12.2 Cross-sectional TEM images near the α-Ga_2O_3/sapphire interface observed along [11$\bar{2}$0] and [10$\bar{1}$0] zone axes [11, 12]

interface region and periodic structures were observed at the interface region with the period of 8.6 and 4.9 nm along 10$\bar{1}$0 and 11$\bar{2}$0, respectively. It should be noted that the lattice constants of α-Ga_2O_3 are 0.43 and 0.249 nm while those of sapphire are 0.41 and 0.238 nm along 10$\bar{1}$0 and 11$\bar{2}$0, respectively. This means that the period of the observed periodic structure coincides with 20 lattices of α-Ga_2O_3 and 21 lattices of α-Al_2O_3 both for 10$\bar{1}$0 and 11$\bar{2}$0, suggesting domain matching epitaxy [13].

Figure 12.3 is the reciprocal space map for X-ray 10$\bar{1}$$\underline{10}$ diffraction of a thin (estimated to be 6 nm from the growth rate) α-Ga_2O_3/sapphire sample [10]. The peaks A and B correspond to diffractions from sapphire and α-Ga_2O_3, respectively. The reciprocal space coordinate (Q_x, Q_z) for the peak B (α-Ga_2O_3) was found to be almost the same as that of thick α-Ga_2O_3. This indicates that the α-Ga_2O_3 is already free standing at the very initial stage of the growth, and this supports the discussions for TEM observation.

From the TEM observation shown in Fig. 12.2, edge dislocation density was estimated as 7×10^{10} cm^{-2} while screw dislocation density was below the detectable limit ($<10^7$ cm^{-2}) [11]. The lattice constant of α-Ga_2O_3 is larger than that of sapphire, and the α-Ga_2O_3 film is subjected to be in-plane compressive strain being strongly rocked by the sapphire crystal at the beginning of the growth. Therefore, it is necessary to introduce extra-half-planes along {$\bar{2}$110} to escape from the strain. The large FWHM values of XRD ω-scanning for antisymmetric (10$\bar{1}$4) and high-density of edge dislocation may be associated with the introduction of the extra-half-planes. The extra-half-planes appear as in-plane disorder and resulted in large FWHM values in the in-plane XRD.

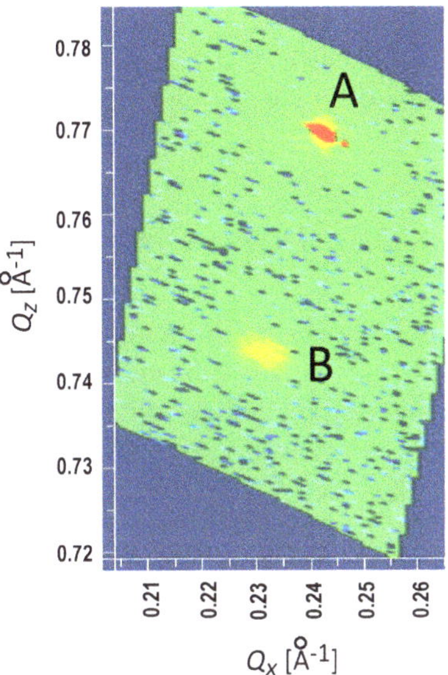

Fig. 12.3 Reciprocal space map for X-ray $10\bar{1}10$ diffraction of a thin (estimated to be 6 nm from the growth rate) α-Ga_2O_3/sapphire sample [10]

12.2.2 Electrical Properties

Unintentionally doped (UID) α-Ga_2O_3 showed very high resistivity, but doping of tin (Sn) [14, 15] or silicon (Si) [16] achieved *n*-type conductivity. In the mist CVD, tin(II) chloride ($SnCl_2$) and chloro-(3-cyanopropyl)-dimethylsilane [ClSi $(CH_3)_2((CH_2)_2CN)$], for example, were used for Sn and Si sources, which were solved in the solvent with the Ga source. Figure 12.4 shows the carrier (electron) concentration in Sn-doped α-Ga_2O_3 against the Sn concentration in the source solution [15]. The carrier concentration was successfully controlled in the range of

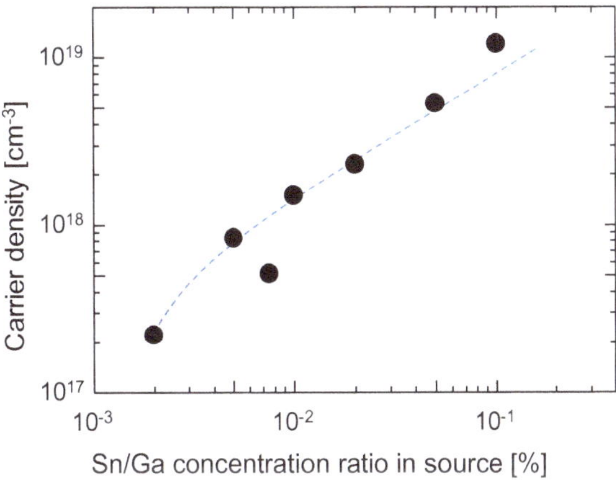

Fig. 12.4 Carrier (electron) concentration in Sn-doped α-Ga_2O_3 against the Sn concentration in the source solution [15]

10^{17} to 10^{19} cm^{-3}. However, the mobility was lower than 24 cm^2/Vs and crystal defects such as dislocations severely restricted the mobility. Improvement of the crystallinity is an important issue for the better electrical properties.

12.2.3 Defect Control

It is necessary to reduce the dislocation density in α-Ga_2O_3 layers for future high-performance device applications. For this purpose, the introduction of strained buffer layers and the epitaxial layer overgrowth (ELO) have been investigated.

An example of the buffer layers is quasi-graded buffer layers, constituted with α-$(Al_{0.9}Ga_{0.1})_2O_3/(Al_{0.2}Ga_{0.8})_2O_3$ multilayers as shown in Fig. 12.5a [17]. Here, by changing the thicknesses of $(Al_{0.9}Ga_{0.1})_2O_3$ and $(Al_{0.2}Ga_{0.8})_2O_3$ layers, the averaged composition was gradually changed from $(Al_{0.9}Ga_{0.1})_2O_3$ to $(Al_{0.2}Ga_{0.8})_2O_3$. The TEM image shown in Fig. 12.5b demonstrates that many dislocation defects are confined in the buffer layers [17]. The edge dislocation density was found to be 6×10^8 cm^{-2}, which was less by two orders in magnitude compared to that without buffer layers (7×10^{10} cm^{-2}, as shown above). However, screw dislocations, which were hardly seen in the α-Ga_2O_3 layer grown without the buffer layers, existed with the density of 3×10^8 cm^{-2}. This may be due to residual strain in the α-Ga_2O_3 layer as an effect of buffer layers, and further optimization of the buffer layer structure will much improve the quality of the films.

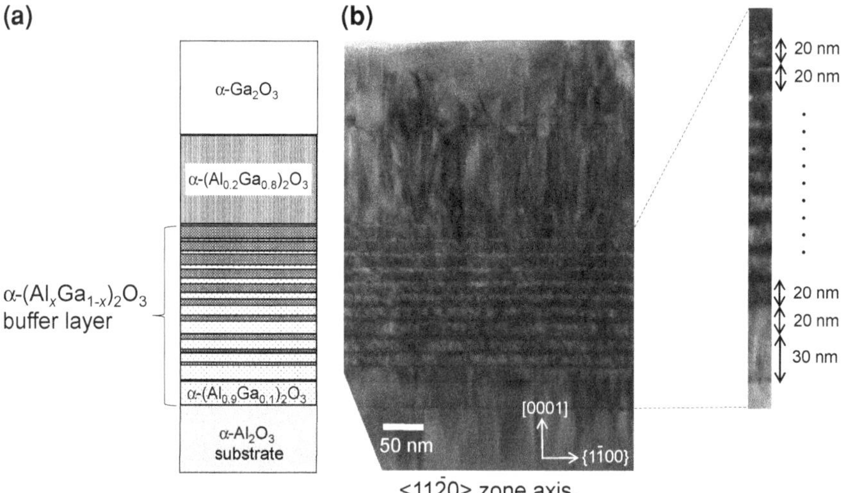

Fig. 12.5 **a** Sample structure with quasi-graded buffer layers. **b** Cross-sectional TEM image of the sample [17]

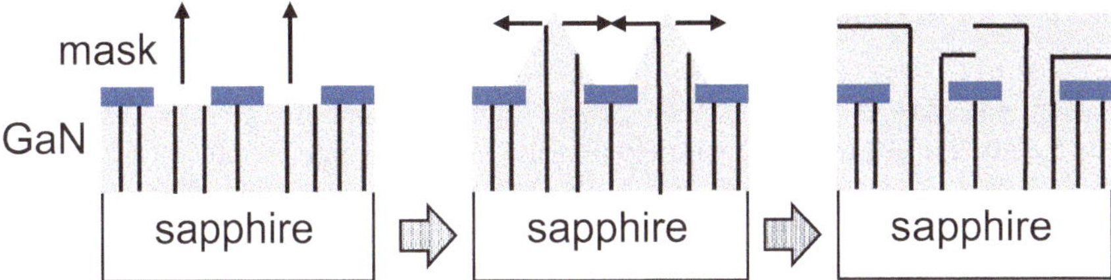

Fig. 12.6 Growth model of ELO

The ELO, as shown in Fig. 12.6, is known to be effective to reduce dislocation defects in gallium nitride (GaN) on sapphire. This technology was also found to be effective for α-Ga_2O_3 grown on sapphire. Figure 12.7 shows the cross-sectional TEM image of an α-Ga_2O_3/sapphire sample, where SiO_2 masks were formed on the sapphire substrate before the growth of α-Ga_2O_3. Threading dislocations are extended from the α-Ga_2O_3/sapphire interface, but they are hardly seen in α-Ga_2O_3 on the SiO_2 masks.

By the combination of appropriate buffer layers and ELO, it is expected that the defects in α-Ga_2O_3 on sapphire, in spite of heteroepitaxial growth, are markedly reduced. Further, high-quality α-Ga_2O_3 grown in this manner can be a seed layer, being lifted off from sapphire, to grow thick α-Ga_2O_3 on it; this allows the formation of α-Ga_2O_3 bulk substrates like GaN in the future.

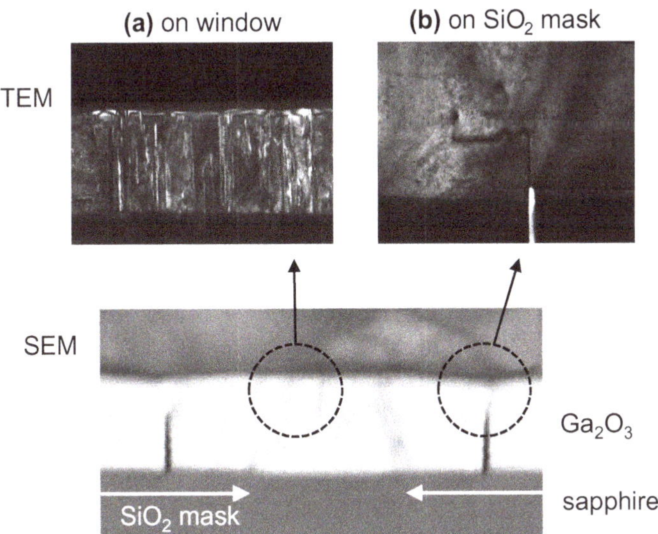

Fig. 12.7 Cross-sectional TEM and SEM images of ELO Ga_2O_3 on sapphire with SiO_2 masks

12.2.4 Thermal Stability

Since α-Ga_2O_3 is a semistable phase, high temperature growth (for example, >550 °C) and high temperature annealing (for example, >600 °C) of α-Ga_2O_3 lead inclusion of β-phase. This seriously limits the growth and process windows. However, slight ($\sim$1%) doping of Al markedly improved the thermal stability. For example, single-phase α-Ga_2O_3:Al was grown at 550 °C and stable at 650 °C [18]. More ($\sim$2.5%) inclusion of Al resulted in more stable films, that is, the α-phase was kept up to 750 °C, without marked widening of the bandgap. These phenomena may be similar to the solution hardening known for GaAs doped with indium (In), and can be a useful technology to keep the α-phase at high temperature processes after the growth.

More recently, it was reported by investigating the crystalline properties of ELO α-Ga_2O_3 that the α-Ga_2O_3 layer maintained corundum structure at higher than 750 °C [19]. This suggests the enhanced thermal stability by reducing defects and compressive stress in α-Ga_2O_3. In spite of the semistable phase of α-Ga_2O_3, the appropriate growth conditions will enhance the thermal stability and allow high temperature processes.

12.3 Growth and Properties of Corundum-Structured III-Oxide Alloys

12.3.1 Bandgap Engineering

Corundum-structured α-In_2O_3 is also a semistable phase, but it was successfully grown on sapphire substrates by the mist CVD. α-Al_2O_3 is a stable phase. Therefore, together with α-Ga_2O_3, it is possible to form corundum-structured alloy semiconductors with α-Al_2O_3, α-Ga_2O_3, and α-In_2O_3 [α-$(Al,Ga,In)_2O_3$]. This is a quaternary alloy system with which the bandgap is tuned from 3.7 to $\sim$9 eV. Figure 12.8 shows the bandgap engineering by α-$(Al,Ga)_2O_3$ and α-$(In,Ga)_2O_3$ alloys [20]. The bandgap can be tuned almost the entire range, but phase separation of α-$(In,Ga)_2O_3$ was recognized at middle of the composition range. This phenomenon is similar to that in InGaN.

The growth of alloys over wide compositional range allows fabrication of heterostructures. It should be noted that uniform single-phase β-$(Al_xGa_{1-x})_2O_3$ is not grown for high Al composition x, because of the inclusion of α-phase which is the stable structure for Al_2O_3 [21]. On the other hand, a variety of heterostructures are realized by α-$(Al_xGa_{1-x})_2O_3$ with the entire compositional range of x. The band lineup of α-$(Al_xGa_{1-x})_2O_3/Ga_2O_3$ heterostructure was evidenced by X-ray photoemission spectroscopy (XPS) as type-I [22], which is favorable for applications to heterostructure transistors and multiple quantum wells.

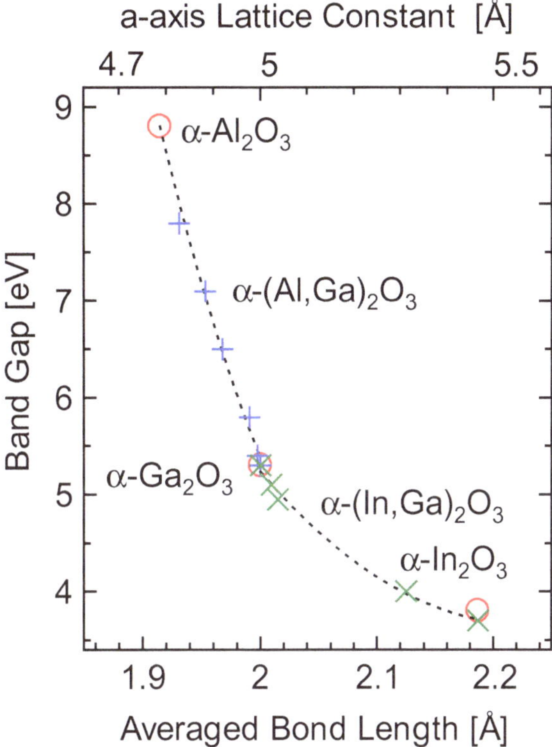

Fig. 12.8 Bandgaps of α-$(Al,Ga)_2O_3$ and α-$(In,Ga)_2O_3$ alloys [20]

12.3.2 Corundum-Structured p-Type Layers

The lack of *p*-type oxide semiconductors has obstructed evolution of oxide semiconductors toward novel and high-performance devices. It is speculated that the fabrication of *p*-type Ga_2O_3 is a difficult task. However, there are corundum-structured *p*-type oxides which may make pn heterojunctions with α-Ga_2O_3. Among them, α-Rh_2O_3 and α-Ir_2O_3, which have been reported to show *p*-type conductivity by Seebeck effect and thermoelectric power measurement, respectively [23–26], are the promising materials.

Using the mist CVD, it became possible to grow single-crystalline α-Rh_2O_3 and α-Ir_2O_3, showing *p*-type conductivity by Hall effect measurements [27]. The *p*-type conductivity is attributed to that the Rh 4d and Ir 5d orbitals, which supply holes, are located above the O 2p orbital. Formation of alloys with α-Ga_2O_3, that is, α-$(Rh,Ga)_2O_3$, and α-$(Ir,Ga)_2O_3$, are effective to widen the bandgaps and to control the hole concentration. Rectification characteristics were demonstrated for the pn junction of α-Ir_2O_3/α-Ga_2O_3, where the lattice mismatches are as small as 0.3 and 0.6% along the *a*- and *c*-axes, respectively [28].

12.4 Future Prospects of α-Ga_2O_3

Corundum-structured α-Ga_2O_3 has been grown by the mist CVD, but recently the growth by MOCVD and halide vapor-phase epitaxy (HVPE) are being reported [29, 30]. HVPE, offering high growth rate, is useful to grow thin films for devices with high-voltage endurance. Defect control by ELO is also successfully demonstrated by HVPE [31].

Device-oriented research and development of α-Ga_2O_3 are making rapid progress. Metal-semiconductor field-effect transistors (FETs) [32], Schottky-barrier diodes with low on-resistance[10], crack-free thin films [33], and normally off metal-oxide-semiconductor FETs [34] are the most up-to-date results, and the details are given in another chapter. The use of sapphire substrates allows low-cost devices and the growth of α-$(Al_xGa_{1-x})_2O_3$ over the entire range of x from 0 to 1 is preferable for designing a variety of heterostructure devices. Corundum-structured *p*-type materials will contribute to demonstrate devices compatible with those of SiC and GaN.

References

1. R. Roy, V.G. Hill, E.F. Osborn, J. Am. Chem. Soc. **74**, 719 (1952)
2. H.H. Tippins, Phys. Rev. **140**, A316 (1965)
3. T. Oshima, T. Okuno, S. Fujita, Jpn. J. Appl. Phys. **46**, 7217 (2007)
4. D. Shinohara, S. Fujita, Jpn. J. Appl. Phys. **47**, 7311 (2008)
5. S. Kawasaki, S. Motoyama, T. Tatsuta, O. Tsuji, S. Okamura, T. Shiosaki, Jpn. J. Appl. Phys. **43**, 6562 (2004)
6. T. Kawaharamura, H. Nishinaka, K. Kametani, Y. Masuda, M. Tanigaki, S. Fujita, J. Soc. Mater. Sci. Jpn. **55**, 153 (2006). (in Japanese)
7. J.G. Lu, S. Fujita, T. Kawaharamura, H. Nishinaka, Y. Kamada, T. Ohshima, Z.Z. Ye, Y. J. Zeng, Y.Z. Zhang, L.P. Zhu, H.P. He, B.H. Zhao, J. Appl. Phys. **101**, 083705 (2008)
8. H. Nishinaka, T. Kawaharamura, S. Fujita, Jpn. J. Appl. Phys. **46**, 6811 (2007)
9. H. Nishinaka, Y. Kamada, N. Kameyama, S. Fujita, Jpn. J. Appl. Phys. **48**, 121103 (2009)
10. M. Oda, R. Tokuda, H. Kambara, T. Tanikawa, T. Sasaki, T. Hitora, Appl. Phys. Express **9**, 021101 (2016)
11. K. Kaneko, H. Kawanowa, H. Ito, S. Fujita, Jpn. J. Appl. Phys. **51**, 020201 (2012)
12. S. Fujita, M. Oda, K. Kaneko, T. Hitora, Jpn. J. Appl. Phys. **55**, 1202A3 (2016)
13. J. Narayan, J. Appl. Phys. **93**, 278 (2003)
14. T. Kawaharamura, G.T. Dang, M. Furuta, Jpn. J. Appl. Phys. **51**, 040207 (2012)
15. K. Akaiwa, K. Kaneko, K. Ichino, S. Fujita, Jpn. J. Appl. Phys. **55**, 1202BA (2016)
16. T. Uchida, K. Kaneko, S. Fujita, MRS Adv. **3**, 171 (2018)
17. R. Jinno, T. Uchida, K. Kaneko, S. Fujita, Appl. Phys. Express **9**, 071101 (2016)
18. S.-D. Lee, Y. Ito, K. Kaneko, S. Fujita, Jpn. J. Appl. Phys. **54**, 030301 (2015)
19. R. Jinno, K. Kaneko, S. Fujita, Presented at Compound Semiconductor Week 2018; *45th International Symposium on Compound Semiconductors and 30th International Conference on Indium Phosphide and Related Materials*, Boston, USA (2018)
20. S. Fujita, K. Kaneko, J. Cryst. Growth **401**, 588 (2014)
21. T. Oshima, T. Okuno, N. Arai, Y. Kobayashi, S. Fujita, Jpn. J. Appl. Phys. **48**, 070202 (2009)

22. T. Uchida, R. Jinno, S. Takemoto, K. Kaneko, S. Fujita, Jpn. J. Appl. Phys. **57**, 040314 (2018)
23. F.P. Koffyberg, J. Phys. Chem. Solids **53**, 1285 (1992)
24. R.K. Kawar, P.S. Chigare, P.S. Patil, Appl. Surf. Sci. **206**, 90 (2003)
25. Y.B. He, A. Stierle, W.X. Li, A. Farkas, N. Kasper, H. Over, J. Phys. Chem. C **112**, 11946 (2008)
26. W.H. Chung, D.S. Tsai, L.J. Fan, Y.W. Yang, Y.S. Huang, Surf. Sci. **606**, 1965 (2012)
27. K. Kaneko, S. Takemoto, S. Kan, T. Shinohe, S. Fujita, Presented at Compound Semiconductor Week 2018; *45th International Symposium on Compound Semiconductors and 30th International Conference on Indium Phosphide and Related Materials*, Boston, USA (2018)
28. S. Kan, S. Takemoto, K. Kaneko, I. Takahashi, M. Sugimoto, T. Shinohe, S. Fujita, Appl. Phys. Lett. **113**, 212104 (2018)
29. Y. Oshima, E.G. Víllora, K. Shimamura, Appl. Phys. Express **8**, 055501 (2015)
30. Y. Yao, S. Okur, L.A.M. Lyle, G.S. Tompa, T. Salagaj, N. Sbrockey, R.F. Davis, L.M. Porter, Mat. Res. Lett. **6**, 268 (2017)
31. Y. Oshima, K. Kawara, T. Shinohe, T. Hitora, M. Kasu, S. Fujita, APL Mater. **7**, 022503 (2019)
32. G.T. Dang, T. Kawaharamura, M. Furuta, M.W. Allen, I.E.E.E. Trans, Electron Devices **62**, 3640 (2015)
33. M. Oda, K. Kaneko, S. Fujita, T. Hitora, Jpn. J. Appl. Phys. **55**, 1202B4 (2016)
34. Press release from FLOSFIA Inc. and Kyoto Univ. July 13 (2018)

Chapter 13
Mist Chemical Vapor Deposition 2

Heteroepitaxial Growth of ε-Ga_2O_3 on Various Substrates

Hiroyuki Nishinaka

Abstract ε-Ga_2O_3 is one of the five polymorphs of Ga_2O_3, which has attracted considerable attention because it exhibits unique polarization and ferroelectric properties. In this chapter, we describe the growth of ε-Ga_2O_3 thin films via mist chemical vapor deposition (CVD). In the epitaxial growth of ε-Ga_2O_3 thin films, we demonstrated that various substrates allow ε-Ga_2O_3 growth via mist CVD because of the atomic arrangement between the surface of the substrate and the ε-Ga_2O_3 growth surface. Further, a method for distinguishing the hexagonal and orthorhombic structures of ε-Ga_2O_3 using X-ray diffraction (XRD) φ-scans was explained in detail. In our experiments, all ε-Ga_2O_3 thin films grown via mist CVD exhibited orthorhombic crystal structure, which permits rotational domains. Furthermore, the growth mechanism of these rotational domains was explained using the atomic arrangement of ε-Ga_2O_3 and the crystal structures of the substrates. Finally, bandgap engineering from 4.5 to 5.9 eV was demonstrated via mist CVD with the incorporation of In and Al.

13.1 Introduction

In this chapter, ε-Ga_2O_3 growth via mist chemical vapor deposition (CVD) is described. Ga_2O_3 has five polymorphs, namely the α-, β-, γ-, δ-, and ε-phases [1]. Among them, ε-Ga_2O_3 has attracted considerable attention owing to its unique polarization and ferroelectric properties. In particular, based on theoretical calculations, ε-Ga_2O_3 polarization has been estimated at approximately eight times that of GaN [2]. In addition, the ferroelectric properties of ε-Ga_2O_3 thin films were characterized using dynamic hysteresis measurements [3]. These properties paved the way for the high electron mobility transistor (HEMT) using high-concentration two-dimensional electron gas (2DEG) without doping [4]. Epitaxial growth methods for ε-Ga_2O_3 that have already been demonstrated include hydride chloride

H. Nishinaka (✉)
Kyoto Institute of Technology, Matsugasaki Sakyo-ku, Kyoto 606-8585, Japan
e-mail: nisinaka@kit.ac.jp

M. Higashiwaki and S. Fujita (eds.), *Gallium Oxide*, Springer Series in Materials Science 293, https://doi.org/10.1007/978-3-030-37153-1_13

epitaxy (HVPE) [5], metalorganic CVD (MOCVD) [3, 6–13], atomic layer deposition (ALD) [6], mist CVD [14–22], molecular beam epitaxy (MBE) [23, 24], and pulsed laser deposition (PLD) [25]. These growth methods are classified into two groups: CVD or physical vapor deposition (PVD) methods. Most reports of ε-Ga_2O_3 thin film growth are categorized into CVD through chemical reactions. There exist only few reports on ε-Ga_2O_3 thin film growth via PVD, and the methods therein often have additional requirements, such as a Tin (Sn) surfactant. A surfactant is unnecessary for ε-Ga_2O_3 growth through a chemical reaction, and it is often possible to grow ε-Ga_2O_3 by simply controlling the growth temperature.

Mist CVD has several advantages for materials' growth in early-stage research. The mist CVD process differs from conventional CVD processes that require handling atmospherically active precursors; in that the mist, including precursors, is stable in the atmosphere. Therefore, a large number of growth experiments can be easily implemented because vacuum is not required. Further, alloying and impurities doping can be achieved by simply mixing these precursors in a solution. Utilizing these advantages, our group has reported on the growth of ε-Ga_2O_3 thin films on various substrates and bandgap engineering by alloying. In this chapter, the ε-Ga_2O_3 growth on various substrates and bandgap engineering of ε-Ga_2O_3 with the incorporation of Al and In via the mist CVD method are described in detail.

13.2 Mist CVD

Mist CVD has been developed as a growth method for oxide semiconductors by the Kyoto University group. This method is a spray pyrolysis technique using the mist of the solution containing precursors. Spray pyrolysis methods are primarily used for growing polycrystalline thin films. Applying the mist for epitaxial growth was not considered because it is incomparably larger than the atoms and requires precise atomic alignment. Our research group (including this chapter's author) at the Kyoto University successfully grew ZnO homoepitaxial thin films with an atomically flat surface via step-flow growth using mist CVD [26] and demonstrated high-quality epitaxial growth of oxide semiconductors. Furthermore, high-quality α-Ga_2O_3 epitaxial growth was reported [27], leading to the wide use of mist CVD methods for the epitaxial growth of Ga_2O_3.

The application of mist CVD for ε-Ga_2O_3 epitaxial growth is reported herein for the first time. Figure 13.1 shows a schematic illustration of the hot-wall-type mist CVD apparatus [17]. For ε-Ga_2O_3 growth, gallium chloride or gallium acetylacetonato dissolved in water was used as Ga_2O_3 precursor solutions. In the mist CVD process, mist atomized from the solution using ultrasonic transducers (2.4 MHz) is transferred into a heated tube, where the substrate is set. Their epitaxial growth can be achieved through this simple process.

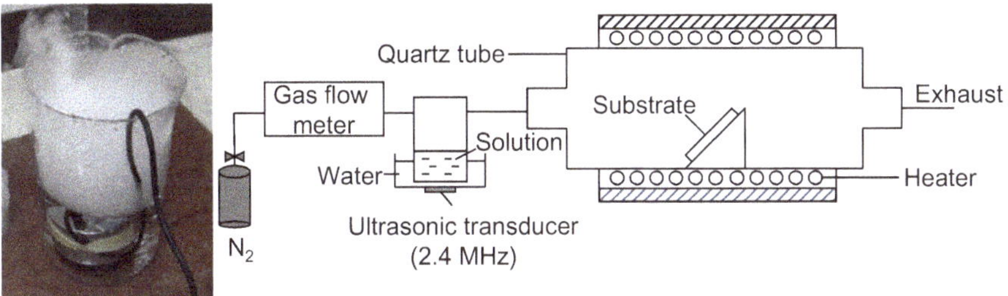

Fig. 13.1 Schematic of hot-wall-type mist CVD apparatus [17] (Reproduced from [17] with permission from The Royal Society of Chemistry)

13.3 Crystal Structure

In this section, we examine the growth of ε-Ga_2O_3 thin films with a hexagonal or orthorhombic crystal structure through the mist CVD method (Fig. 13.2). To determine the crystal structure of ε-Ga_2O_3, understanding the crystal growth mechanism is important. The first study aiming to distinguish between hexagonal and orthorhombic structures via an XRD φ-scan for {122} reflex originating from orthorhombic structured crystals was implemented by Kracht et al. [24]. They provided a simple description of how the crystal structure can be determined via XRD φ-scans. The hexagonal structure possesses no reflexes at the 2θ and χ of the {122} reflex of the orthorhombic structure. Here, we explain the method for distinguishing the two structures and why they are confused. Figure 13.3 shows the relationships between the 2θ and χ angles in the hexagonal and orthorhombic

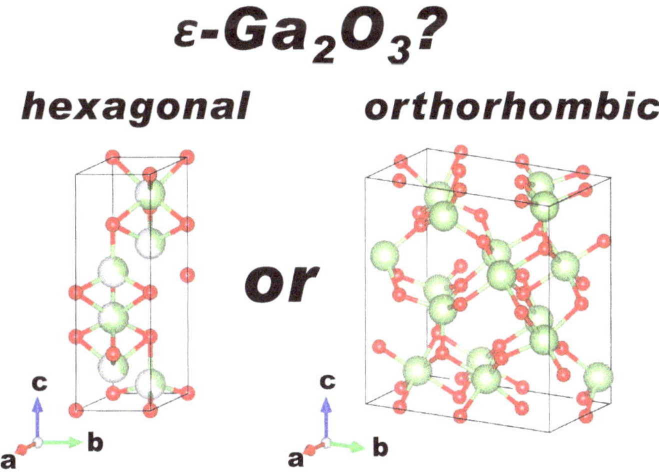

Fig. 13.2 Crystal structure of ε-Ga_2O_3

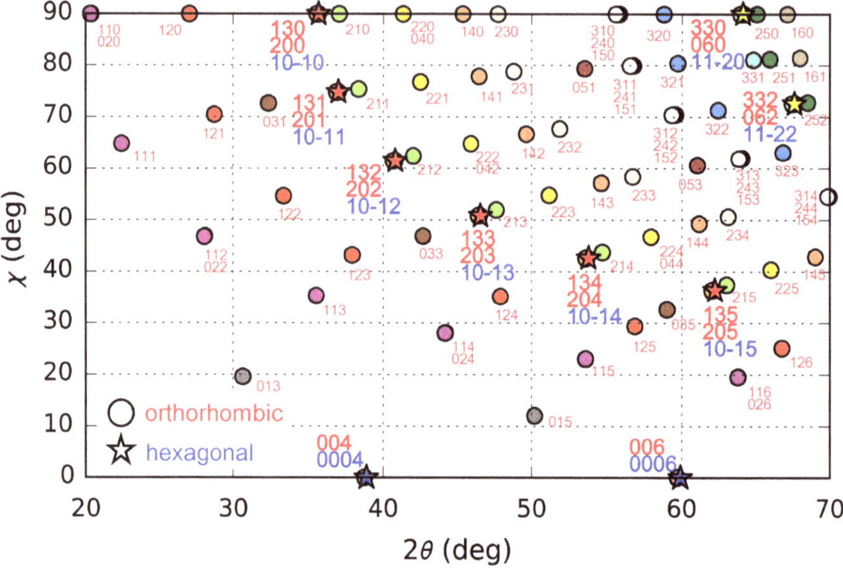

Fig. 13.3 XRD angle relationships (2θ and χ) in the orthorhombic and hexagonal structures of ε-Ga_2O_3 [21] (Copyright (2018) The Japan Society of Applied Physics)

structures of ε-Ga_2O_3 [21]. All diffraction positions of hexagonal structures correspond to specific positions on the orthorhombic diffraction spectrum. In short, the crystal structure of ε-Ga_2O_3 cannot be distinguished using the angles of the hexagonal structures. For example, analyzing 2θ-ω scans and indicating χ are meant to be 0°, (0004) for the hexagonal structure is consistent with (004) for the orthorhombic structure; thus, the crystal structure cannot be distinguished. Similarly, {10-1*l*} reflexes for the hexagonal structure, which are often used to investigate the epitaxial in-plane relationship, cannot be utilized for the same reason. Instead, to determine the crystal structure, only angles of the orthorhombic structure should be analyzed, such as the {122} reflex, as shown in Fig. 13.3. Figure 13.4 shows the typical φ-scan result of the {122} reflex for the orthorhombic ε-Ga_2O_3 thin film. Twelve peaks are observed at the angles of 2θ = 33.32° and χ = 54.63°, indicating that the thin film exhibits an orthorhombic structure. Measuring the φ-scan of {122} or other reflexes except for {20*l*}, {13*l*}, {06*l*}, and {33*l*}, the diffraction peaks are not observed; therefore, we consider that the thin film has a hexagonal structure. Of course, noise could be hiding any weak diffraction peaks present. Moreover, although single-crystal ε-Ga_2O_3 analysis should only yield four peaks for the {122} reflex for orthorhombic structures, and twelve peaks appeared for the {122} reflex (Fig. 13.4), indicating that there are three rotational domains in the ε-Ga_2O_3 thin film. The occurrence of rotational domains in ε-Ga_2O_3 will be discussed in section *occurrence of rotational domains in orthorhombic ε-Ga_2O_3*.

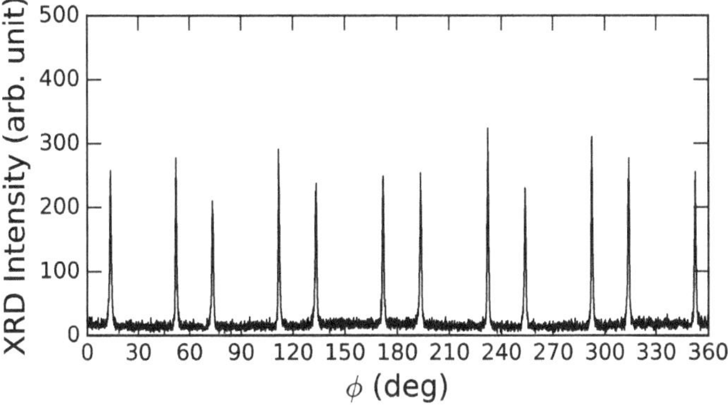

Fig. 13.4 XRD φ-scans of {122} in orthorhombic ε-Ga_2O_3 thin film

13.4 Substrates

For the growth of ε-Ga_2O_3, different crystal structure substrates have been proposed by several research groups. First, the atomic arrangement relationships, which are considerably significant for epitaxial growth, are described. The oxygen arrangement of orthorhombic ε-Ga_2O_3 (001) exhibits a pseudo-hexagonal structure, as shown in Fig. 13.5a. Similarly, most substrates utilized for the epitaxial growth of ε-Ga_2O_3 possess hexagonal and pseudo-hexagonal atomic arrangements, such as (0001) α-Al_2O_3 [6–13, 17, 23, 24], α-$(Al_xGa_{1-x})O_3$ [22] (rhombohedral), (0001) GaN [5, 21], AlN [5, 15, 18], 6H-SiC [8] (hexagonal), (111) MgO, YSZ [14], STO [21], NiO [20], GGG (cubic) [16], and (-201) β-Ga_2O_3 [5] (monoclinic). Furthermore, orthorhombic ε-Ga_2O_3 (001) exhibits a rectangular shape; wafers possessing similar shapes can also be utilized as substrates [19]. The atomic arrangements of typical substrates are depicted in Fig. 13.5b–e. All oxygen arrangements of the substrates are almost consistent with that of ε-Ga_2O_3.

Our group demonstrated ε-Ga_2O_3 epitaxial growth on hexagonal and pseudo-hexagonal substrates, such as (0001) GaN, AlN, α-Al_2O_3, (111) YSZ, MgO, STO, NiO, GGG, and a rectangular structure, such as (100) SnO_2. Figure 13.6 shows the XRD 2θ-ω scans of ε-Ga_2O_3 on all these substrates via the mist CVD method. As shown in Fig. 13.6, single-phase ε-Ga_2O_3 epitaxial growth can be achieved via the mist CVD method by selecting appropriate substrates based on the atomic arrangement. It is assumed that achieving ε-Ga_2O_3 epitaxial thin films through growth methods via chemical reaction, such as MOCVD, HVPE, ALD, or mist CVD, is easier than using PVD, such as MBE or PLD. In the MBE and PLD process for ε-Ga_2O_3 growth, adding a surfactant material is often necessary, such as Sn. Conversely, most CVD processes for ε-Ga_2O_3 do not need a surfactant. The cause of this difference is unclear.

Next, the difficulty of growing ε-Ga_2O_3 thin films on c-plane sapphire substrates is discussed. C-plane sapphire substrates are the most widely used substrates for

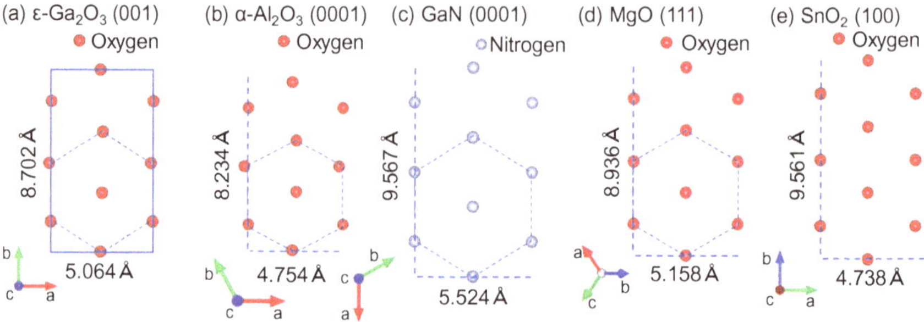

Fig. 13.5 Schematic representations of the atomic arrangement in (**a**) orthorhombic ε-Ga_2O_3 (001) and (**b**)–(**e**) typical substrates

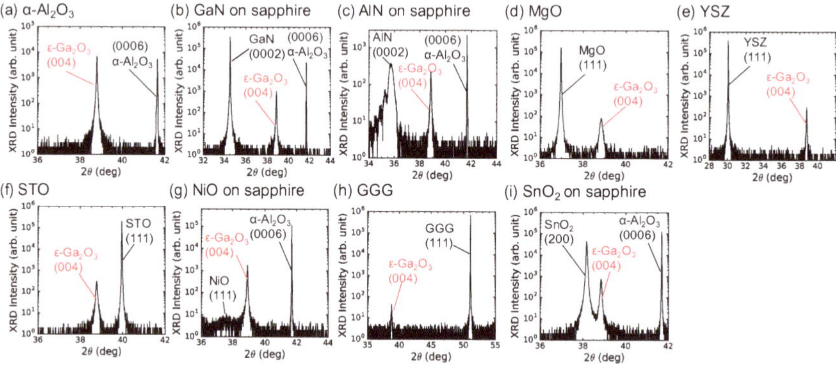

Fig. 13.6 XRD 2θ-ω scans of ε-Ga_2O_3 epitaxial thin films on various substrates via mist CVD

semiconductor growth, and they have been commercialized as they are cost friendly. However, the crystal structure of α-Al_2O_3 complicates Ga_2O_3 growth. C-plane sapphire substrates allow Ga_2O_3 to form in three crystal phases, namely α-, β-, and ε-Ga_2O_3. Therefore, depending on the growth conditions, these three crystal phases grow in single or mixed phases. For example, α-Ga_2O_3 is a metastable phase possessing the same corundum structure as α-Al_2O_3, and α-Ga_2O_3 epitaxial growth on sapphire substrates via mist CVD [27, 28], HVPE [29], and MBE [30] has been reported. Furthermore, controlling the flow rate of a reaction species, such as HCl via MOCVD [10] and HVPE [31], enables the synthesis of the three phases. Thus, if growth parameters, such as the growth temperature and conditions of the substrates, i.e., the off-axis, growth rates, and precursors compositions, exceed specific ranges, ε-Ga_2O_3 mixed with α- or β-Ga_2O_3 may be grown on c-plane sapphire substrates. Hence, we proposed inserting buffer layers with a different crystal structure to control the growth of single-phase ε-Ga_2O_3 thin films [20]. The buffer layer utilized is (111) NiO thin films, and it can be used to achieve growth via the mist CVD method (Fig. 13.7a). Figure 13.7b, c shows the XRD 2θ-ω of Ga_2O_3 thin

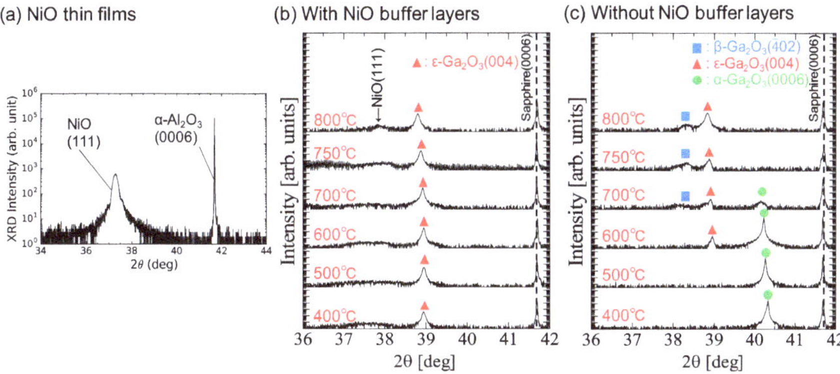

Fig. 13.7 XRD 2θ-ω scans of (**a**) NiO thin films via mist CVD, Ga_2O_3 thin films (**b**) with, and (**c**) without NiO buffer layer via mist CVD [20] (Reproduced from [20] with permission from The Royal Society of Chemistry)

films with and without such NiO buffer layers, respectively. As shown in Fig. 13.7b, by inserting NiO buffer layers, only diffraction peaks originating from (004) ε-Ga_2O_3 were observed, grown at 400–800 °C, and no peaks related to α- and β-Ga_2O_3 were observed. In contrast, without NiO buffer layers, (0006) α-Ga_2O_3 peaks at 400–700 °C, (004) ε-Ga_2O_3 peaks at 600–800 °C, and (-402) β-Ga_2O_3 peaks at 700–800 °C were observed. Based on the thermodynamics of Ga_2O_3, the α-phase is stable at low temperatures, ε-Ga_2O_3 is stable at medium temperatures, and β-Ga_2O_3 is stable at high temperatures; the results herein are consistent with the thermodynamics of Ga_2O_3. Furthermore, ε-Ga_2O_3 positioning at the medium temperature region is inevitably mixed with other polymorphs. Thus, (111) NiO enables the growth of single-phase ε-Ga_2O_3 epitaxial thin films on c-plane sapphire substrates and prevents α-Ga_2O_3 growth at low temperatures.

13.5 Occurrence of Rotational Domains in Orthorhombic ε-Ga_2O_3

To achieve ε-Ga_2O_3-based HEMTs with 2DEG, investigating the mechanism of the occurrence of rotational domains is important, which are common in thin films. Rotational domains may generally cause poor electron transport because they are aggregated in thin films. Cora et al. reported a ε-Ga_2O_3 thin film containing an aggregation of 5–10 nm small domains via a plane-view transmission electron microscopy (TEM) observation [12]. The existence of the rotational domains can be evaluated via XRD φ-scan or a pole figure.

The insertion of the rotational domains must be prevented to achieve ε-Ga_2O_3 HEMT. In this section, the mechanism of occurrence of rotational domains is described in detail. We hypothesize that there are two possible mechanisms underlying the occurrence of rotational domains: One involves the orthorhombic structure of ε-Ga_2O_3 allowing a rotation of 120°, and the other originates from the crystal structure of the substrates.

First, the methods of evaluating rotational domains using XRD φ-scans are described. Figure 13.8 shows the typical φ-scan of ε-Ga_2O_3 {122} and {204} (using GaN template). Although a single ε-Ga_2O_3 crystal should exhibit two peaks in {204} and four peaks in {122}, ε-Ga_2O_3 thin films exhibit six peaks in {204} and twelve peaks in {122}. The result for {122} indicates the occurrence of three rotational domains in orthorhombic ε-Ga_2O_3. In contrast, those rotational domains are not suggested by the result for {204}. When the φ-scan of {204} diffractions is evaluating, {134} diffractions are also observed at the evaluating angle. In Fig. 13.2, the positions of {204} and {134} are in agreement, and {134} of ε-Ga_2O_3 exhibits four peaks. Thus, when two peaks from {204} and four peaks from {134} are combined, the result of the φ-scan shows six peaks without the three rotational domains. Therefore, to investigate the rotational domains, a single diffraction plane in Fig. 13.2, such as {122}, should be evaluated.

Figure 13.9a shows a schematic illustration of the orthorhombic ε-Ga_2O_3 oxygen arrangement viewed along the c-axis [21]. Regarding the single unit cells, the [110] direction corresponding to the (130) plane approximately coincides with the [100] direction corresponding to the (100) plane, when the [100] direction is rotated by 60°. To explain the occurrence of the rotational domains based on ε-Ga_2O_3 itself, the three unit cells shown in blue are depicted with rotations (1) 0°, (2) 120°, and (3) 240°. The positions of the blue oxygen atoms with rotations 120° and 240° almost coincide with those of the red oxygen atoms. Furthermore, the lattice spacing of (200), i.e., d_{200} = 0.2523 nm, which is half the lattice spacing of (100), is similar to that of (130) (d_{130} = 0.2515 nm). Therefore, the structure of the orthorhombic ε-Ga_2O_3 itself allows the occurrence of three rotational domains.

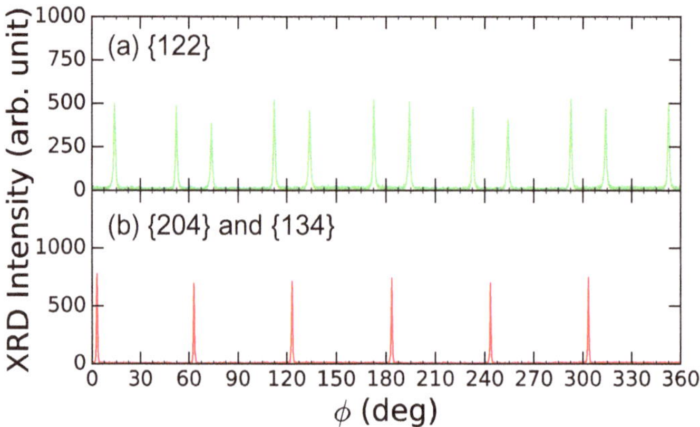

Fig. 13.8 XRD φ-scans of (**a**) {122} and (**b**) {204} and {134} of orthorhombic ε-Ga_2O_3 [21] (Copyright (2018) The Japan Society of Applied Phyics)

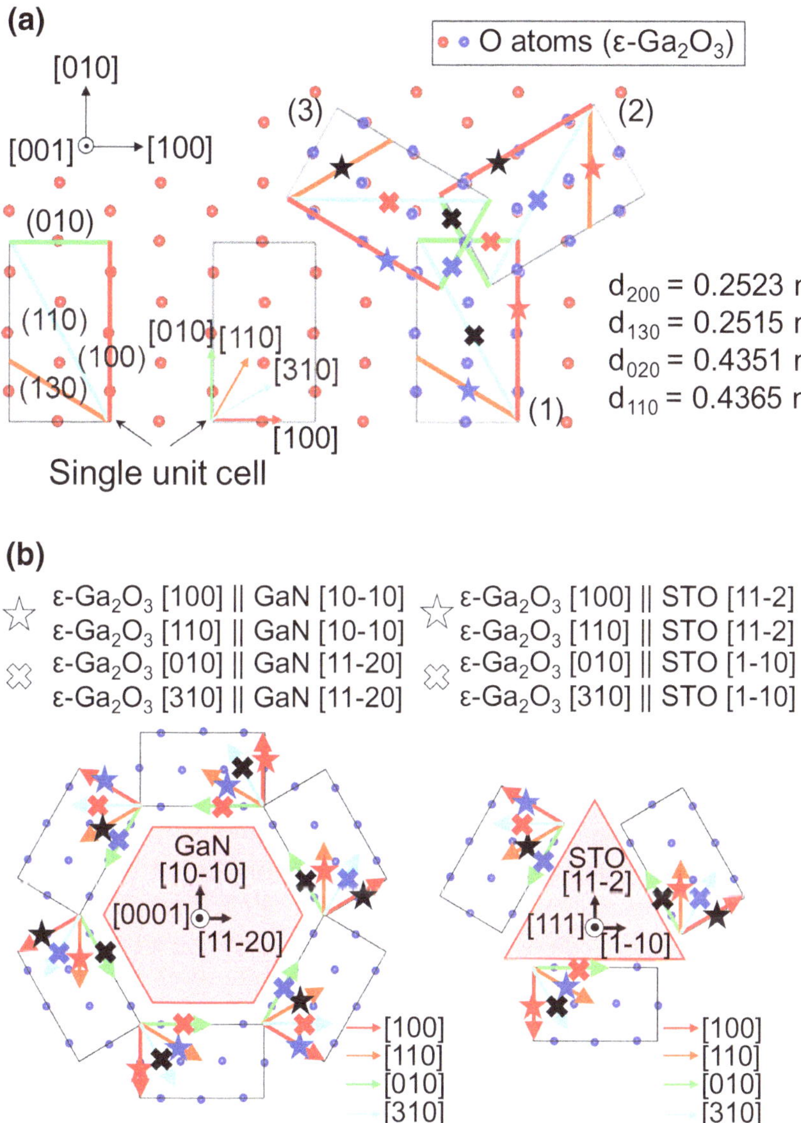

Fig. 13.9 **a** Schematic of the orthorhombic ε-Ga_2O_3 atomic arrangement viewed along the *c*-axis. Single unit cells are placed (1) without rotation, (2) with a 120° rotation (2), and (3) with a 240° rotation on the atomic arrangement of ε-Ga_2O_3. **b** Schematics of the occurrence of the three rotational domains for ε-Ga_2O_3 due to the crystal structure of the (**a**) GaN template and (**b**) STO substrate [21] (Copyright (2018) The Japan Society of Applied Physics)

Next, the occurrence of the rotational domains caused by the crystal structure of the substrates is explained by considering (0001) hexagonal (GaN) and (111) cubic (STO) substrates as an example. The schematic illustrations show the rotational domains of orthorhombic ε-Ga_2O_3 on (0001) GaN and (111) STO. As shown in Fig. 13.9b [21], for GaN and STO, three rotational domains are possible. Furthermore, the in-plane relationship indicated from the result of the φ-scan is consistent with the single-cell arrangements in Fig. 13.9b.

The structure of the small domains caused by the rotations was observed using cross-sectional TEM. Figure 13.10 shows the cross-sectional TEM observations of ε-Ga_2O_3 thin films on GaN, STO [21], and NiO [20]. The vertical lines in the growth direction are observed at the positions indicated by the arrows in all samples. The small column divided by the vertical lines is assumed as one of the domains. To estimate the sizes of these domains, a plane-view TEM observation should be used. The rotational domains occurred on all substrates grown by our group and a growth technique for ε-Ga_2O_3 thin films comprising a single domain are required.

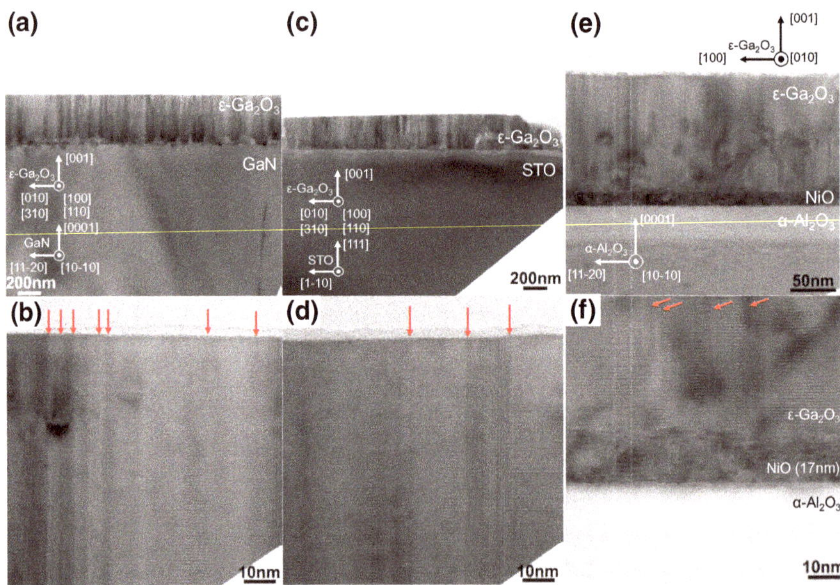

Fig. 13.10 TEM images of a cross section of the ε-Ga_2O_3 thin films on (**a**), (**b**) GaN, (**c**), (**d**) STO, and (**e**), (**f**) NiO. Several vertical lines dividing the domains are indicated by red arrows [20] (Reproduced from [20] with permission from The Royal Society of Chemistry), [21] (Copyright (2018) The Japan Society of Applied Phyics)

13.6 Bandgap Engineering

For heterojunction electronic device application, bandgap engineering by alloying with other elements is one of the most important techniques. Bandgap control in Ga_2O_3 has been achieved by incorporating group-III elements, such as In and Al. For the bandgap control of β-Ga_2O_3, the incorporation of these group-III elements into β-Ga_2O_3 enables the bandgap to be tuned from 4.0 eV [32] to 6.0 eV [33]. For the bandgap control of α-Ga_2O_3, the bandgap energies have been tuned from 3.8 to 8.8 eV by incorporating In and Al [34]. The bandgap energy of γ-Ga_2O_3 has been tuned from 5.0 to 7.0 eV by incorporating Al [35]. For the bandgap control of ε-Ga_2O_3, incorporating In and Al into ε-Ga_2O_3 enables the bandgap energies to be tuned from 4.5 eV [17] to 5.9 eV [18] using mist CVD. Mist CVD possesses a unique feature, wherein any source materials that can be dissolved in a solvent can be utilized as CVD precursors. Therefore, mist CVD enables alloying through simple dissolution of multiple source materials in the solvent. Figure 13.11 shows the XRD 2θ-ω results of the incorporation of In and Al into ε-Ga_2O_3 thin films. The (004) peak shifts toward lower and higher angles with increasing In and Al compositions, respectively. Conversely, the phase separation of (400) bcc-$(In_xGa_{1-x})_2O_3$ occurs at $x > 0.2$. Thus, the solubility limit of In incorporation into ε-Ga_2O_3 via mist

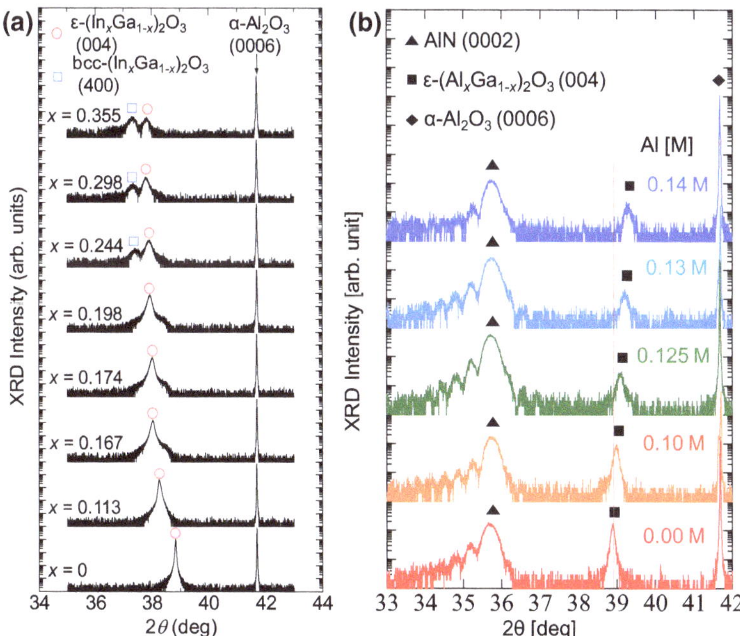

Fig. 13.11 XRD 2θ-ω scans of **a** ε-$(In_xGa_{1-x})_2O_3$ thin films and **b** ε-$(Al_xGa_{1-x})_2O_3$ thin films via mist CVD [17] (Reproduced from [17] with permission from The Royal Society of Chemistry), [18] (Reprinted from [18], with the permission of AIP Publishing)

CVD on sapphire substrates is $x = 0.2$. By contrast, the phase separation is not observed during Al incorporation. We cannot find the solubility limit of Al incorporation into ε-Ga_2O_3 thin films owing to the solubility limit of the precursor solutions in our experimental conditions. The optical bandgaps are evaluated using the direct transition relationship $(\alpha h\nu)^2 \propto (h\nu - E_g)$ plots shown in Fig. 13.12a, b. Based on these results, we obtain the bandgap energy dependence on the In and Al compositions of the films, as shown in Fig. 13.12c, d. The bandgap energies vary from 4.5 (In = 0.2) to 5.9 (Al = 0.4) eV. These bandgap engineering results are expected to pave the way for the application of ε-Ga_2O_3-based heterojunction devices.

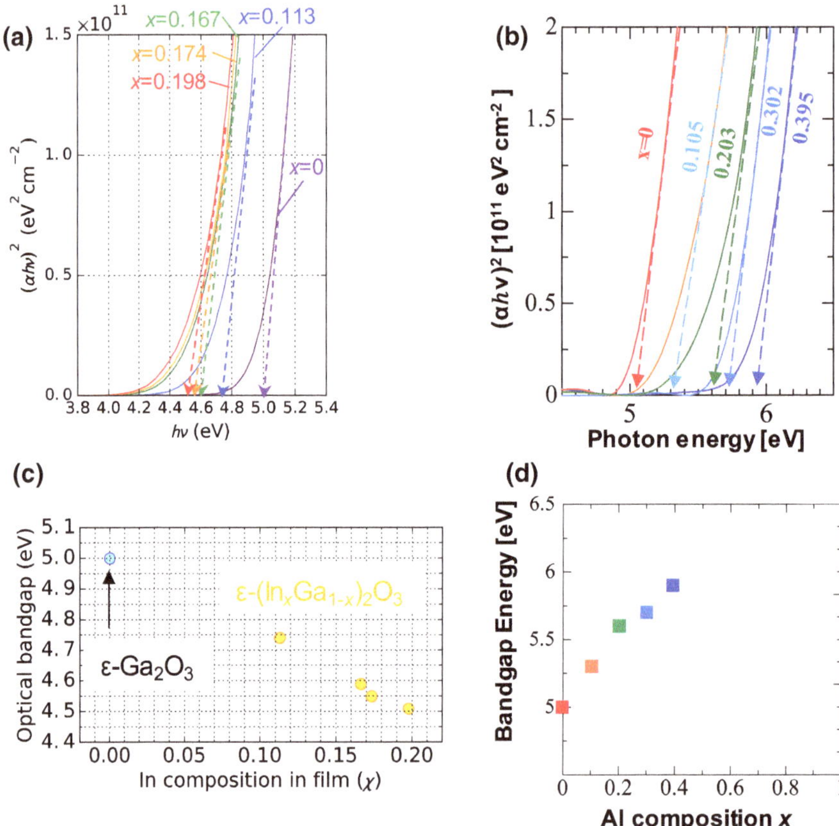

Fig. 13.12 $(\alpha h\nu)^2$ as a function of $h\nu$ (solid lines) and linear extrapolation (dashed lines) for estimating the direct bandgaps of **a** ε-$(In_xGa_{1-x})_2O_3$ thin films and **b** ε-$(Al_xGa_{1-x})_2O_3$ films. Optical bandgaps of **c** ε-$(In_xGa_{1-x})_2O_3$ thin films and **d** ε-$(Al_xGa_{1-x})_2O_3$ films as a function In and Al composition of the films [17] (Reproduced from [17] with permission from The Royal Society of Chemistry) [18] (Reprinted from [18], with the permission of AIP Publishing)

References

1. R. Roy, V.G. Hill, E.F. Osborn, J. Am. Chem. Soc. **74**, 719 (1952)
2. M.B. Maccioni, V. Firorentini, Appl. Phys. Express **9**, 041102 (2016)
3. F. Mezzadori, G. Calestani, F. Boschi, D. Delmonte, M. Bosi, R. Fornari, Inorg. Chem. **55**, 12079 (2016)
4. S.B. Cho, R. Mishra, Appl. Phys. Lett. **112**, 162101 (2018)
5. Y. Oshima, E.G. Víllora, Y. Matsushita, S. Yamamoto, K. Shimamura, J. Appl. Phys. **118**, 085301 (2015)
6. F. Boschi, M. Bosi, T. Berzina, E. Buffagni, C. Ferrari, R. Fornari, J. Cryst. Growth **443**, 25 (2016)
7. Y. Chen, X. Xia, H. Liang, Q. Abbas, Y. Liu, G. Du, Cryst. Growth Des. **18**, 1147 (2018)
8. X. Xia, Y. Chen, W. Tu, H. Liang, P. Tao, M. Xu, G. Du, Appl. Phys. Lett. **108**, 202103 (2016)
9. Y. Zhuo, Z. Chen, W. Tu, X. Ma, Y. Pei, G. Wang, Appl. Surf. Sci. **420**, 802 (2017)
10. H. Sun, K.H. Li, C.T. Castanedo, S. Okur, G.S. Tompa, T. Salagaj, S. Lopatin, A. Genovese, X. Li, Cryst. Growth Des. **18**, 2370 (2018)
11. M. Pavesi, F. Fabbri, F. Boschi, G. Piacentini, A. Baraldi, M. Bosi, E. Gombia, A. Parisini, R. Fornari, Mater. Chem. Phys. **205**, 502 (2018)
12. I. Cora, F. Mezzadri, F. Boschi, M. Bosi, M. Čaplovičová, G. Calestani, I. Dódony, B. Pécza, R. Fornari, CrystEngComm **19**, 1509 (2017)
13. Z. Chen, Z. Li, Y. Zhuo, W. Chen, X. Ma, Y. Pei, G. Wang, Appl. Phys. Express **11**, 101101 (2018)
14. H. Nishinaka, D. Tahara, M. Yoshimoto, Jpn. J. Appl. Phys. **55**, 1202BC (2016)
15. D. Tahara, H. Nishinaka, S. Morimoto, M. Yoshimoto, Jpn. J. Appl. Phys. **56**, 078004 (2017)
16. D. Tahara, H. Nishinaka, S. Morimoto, M. Yoshimoto, in *2017 IEEE International Meeting for Future of Electron Devises, Kansai*, Ryukoku University Avanti Kyoto Hall, Kyoto, 29–30 June 2017
17. H. Nishinaka, N. Miyauchi, D. Tahara, S. Morimoto, M. Yoshimoto, CrystEngComm **20**, 1882 (2018)
18. D. Tahara, H. Nishinaka, S. Morimoto, M. Yoshimoto, Appl. Phys. Lett. **112**, 152102 (2018)
19. D. Tahara, H. Nishinaka, M. Noda, M. Yoshimoto, Mater. Lett. **232**, 47 (2018)
20. Y. Arata, H. Nishinaka, D. Tahara, M. Yoshimoto, CrystEngComm **20**, 6236 (2018)
21. H. Nishinaka, H. Komai, D. Tahara, Y. Arata, M. Yoshimoto, Jpn. J. Appl. Phys. **57**, 115601 (2018)
22. R. Jinno, T. Uchida, K. Kaneko, S. Fujita, Phys. Status Solidi B **255**, 1700326 (2018)
23. X. Bi, Z. Wu, Y. Huang, W. Tang, AIP Adv. **8**, 025008 (2018)
24. M. Kracht, A. Karg. J. Schörmann, M. Weinhold, D. Zink, F. Michel F, M. Rohnke, M. Schowalter, B. Gerken, A. Rosenauer, P. J. Klar, J. Janek, M. Eickhoff, Phys. Rev. Appl. **8**, 054002 (2017)
25. M. Orita, H. Hiramatsu, H. Ohta, M. Hirano, H. Hosono, Thin Solid Films **411**, 134 (2002)
26. H. Nishinaka, S. Fujita, J. Cryst. Growth **310**, 5007 (2008)
27. D. Shinohara, S. Fujita, Jpn. J. Appl. Phys. **47**, 7311 (2008)
28. S. Fujita, M. Oda, K. Kaneko, T. Hitora, Jpn. J. Appl. Phys. **55**, 1202A3 (2016)
29. Y. Oshima, E.G. Víllora, K. Shimamura, Appl. Phys. Express **8**, 055501 (2015)
30. T. Oshima, Y. Kato, M. Imura, Appl. Phys. Express **11**, 065501 (2018)
31. Y. Yao, S. Okur, L. Lyle, G.S. Tompa, T. Salagaj, N. Sbrockey, R.F. Davis, L.M. Porter, Mater. Res. Lett. **6**, 268 (2018)
32. B. Krueger, C. Dandeneau, E. Nelson, S.T. Dunham, F.S. Ohuchi, M.A. Olmstead, J. Am. Ceram. Soc. **99**, 2467 (2016)
33. T. Oshima, S. Fujita, Phys. Status Solidi C **5**, 3113 (2008)
34. S. Fujita, K. Kaneko, J. Cryst. Growth **401**, 588 (2014)
35. T. Oshima, Y. Kato, M. Oda, T. Hitora, M. Kasu, Appl. Phys. Express **10**, 051104 (2017)

Chapter 14
Pulsed Laser Deposition 1

Homoepitaxial Growth of β-Ga_2O_3 on β-Ga_2O_3 Substrates

Kevin D. Leedy

Abstract This chapter examines homoepitaxial β-Ga_2O_3 thin films fabricated by pulsed laser deposition. The recent availability of high-quality native substrates affords investigations using a wide range of growth conditions to achieve single crystalline films by pulsed laser deposition. Deposition parameter optimization and structural, electrical and chemical film characterization are presented for undoped and impurity-doped films. Films grown on commercially available (010) Fe compensation-doped substrates fabricated by the edge-defined film-fed growth technique and a developing Czochralski method are examined. Hall effect results from PLD Si-doped β-Ga_2O_3 films are compared with data produced from other vapor phase epitaxial growth techniques. The influence of substrate orientation on film properties is also studied with (010) and (001) crystals. Implementation of a PLD n+ regrowth layer in a transistor device is demonstrated to achieve low ohmic contact resistance. Results from β-$(Al_xGa_{1-x})_2O_3$ film studies show the potential utility of heterostructures for field-effect transistor 2DEG formation.

14.1 Introduction

The benefits of using native substrates for epitaxial β-Ga_2O_3 film growth with multiple vapor phase deposition techniques have been discussed in previous chapters. A range from intrinsic, unintentionally doped to fully degenerative-doped β-Ga_2O_3 films have been demonstrated by molecular beam epitaxy (MBE), metal organic vapor phase epitaxy (MOVPE), mist chemical vapor deposition (Mist CVD) and low-pressure chemical vapor deposition (LPCVD). Another technique for β-Ga_2O_3 homoepitaxial growth is pulsed laser deposition (PLD). PLD offers a unique deposition capability through a wide deposition pressure and process gas range, film composition control through ablation

K. D. Leedy (✉)
Air Force Research Laboratory, Sensors Directorate, 2241 Avionics Circle, Wright-Patterson Air Force Base, OH 45433, USA
e-mail: kevin.leedy@us.af.mil

M. Higashiwaki and S. Fujita (eds.), *Gallium Oxide*, Springer Series in Materials Science 293, https://doi.org/10.1007/978-3-030-37153-1_14

target chemistry and more modest growth temperatures compared to other vacuum deposition techniques owing to reactive oxygen species generated from a laser source operating externally from the growth chamber. Investigation of multiple PLD deposition variables offers substantial insight into the electrical and structural film properties that can be exploited for device applications. Electrical characterization of films by Hall effect provides a direct indication of film quality and is augmented by structural characterization including X-ray diffraction (XRD), atomic force microscopy (AFM), secondary ion mass spectrometry (SIMS) and transmission electron microscopy (TEM).

14.2 Unintentionally Doped β-Ga_2O_3 Homoepitaxial Films

There are few reports of unintentionally doped homoepitaxial β-Ga_2O_3 films by PLD, with analyses focused on structural characterization. Wakabayashi [1] performed homoepitaxial PLD growth of β-Ga_2O_3 assisted by oxygen radicals on (010) β-Ga_2O_3 substrates. Compared to conventional PLD in an oxygen ambient, the addition of oxygen radicals was found to increase deposition rate as shown in Fig. 14.1, attributed to more complete oxidation of the Ga suboxides and associated significant reduction in Ga sublimation at high deposition temperatures. Film surface roughness also improved with oxygen radical use. A 1.1 nm RMS surface roughness was measured from a film deposited at 800 °C and 9.3×10^{-3} Pa of O_2 with oxygen radicals vs. a 7.6 nm RMS roughness for a similar film without oxygen radicals at 13.3 Pa O_2. Li [2] demonstrated homoepitaxial growth of β-Ga_2O_3 on (100) substrates as a function of oxygen deposition pressure. The (400) rocking curve FWHM values increased from 198 arcsec to 360 arcsec as oxygen deposition pressure increased, although explicit oxygen pressures were not reported. Interpretation of XRD reciprocal space maps also indicated higher defect densities with higher oxygen deposition pressures. In this instance, the film AFM surface roughness increased from 0.44 to 2.12 nm as oxygen deposition pressure decreased.

Fabrication of undoped β-Ga_2O_3 films by AFRL using a 99.99% pure Ga_2O_3 ablation target yielded homoepitaxial β-Ga_2O_3 films on (010) substrates.

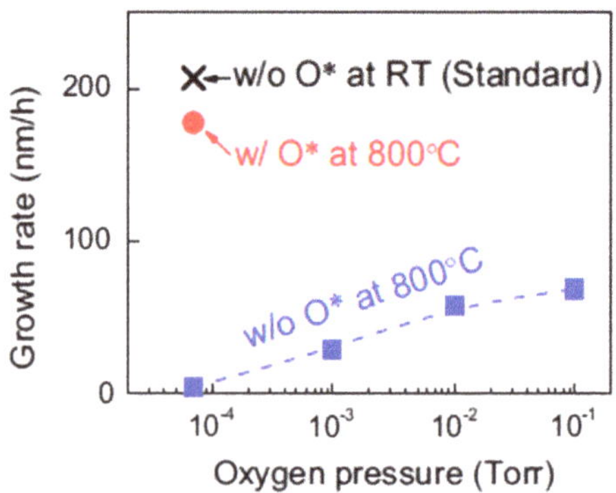

Fig. 14.1 Growth rates of homoepitaxial films grown at T_g = 800 °C with and without O*. Standard growth rate (*X*) is also plotted for comparison. From [1]

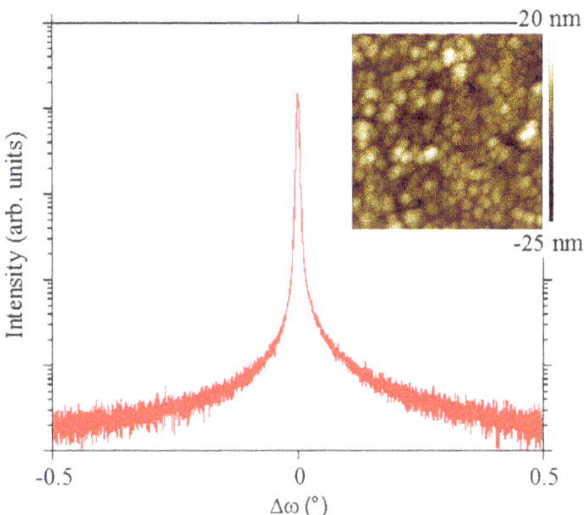

Fig. 14.2 XRD rocking curve of (020) reflection from Ga_2O_3 film deposited at 550 °C on (010) β-Ga_2O_3 substrate. Inset shows 1 μm × 1 μm AFM image of film surface

A 440-nm-thick film was deposited at 550 °C in 27 Pa of O_2 with a deposition rate of 5.3 nm/min. XRD rocking curve analysis of the (020) orientation revealed a single peak with no delineation between the film and underlying crystal reflections, as shown in Fig. 14.2. A FWHM of 19 arcsec was measured, equivalent to the substrate FWHM of 19 arcsec. In the Fig. 14.2 inset, an AFM image of a 1 μm × 1 μm surface area is included showing a granular structure with an RMS surface roughness of 4.9 nm. Lower-pressure O_2 depositions resulted in a film XRD (020) RC peak adjacent to the substrate reflection indicating a strained lattice, but with a lower surface RMS roughness of 0.14 nm. The enhanced surface roughening observed in higher-pressure PLD growth is consistent with other studies [1]. Sheet resistance of the undoped β-Ga_2O_3 films was measured in the GΩ range by Hall effect, indicative of minimal unintentional impurity doping.

14.3 Impurity-Doped β-Ga_2O_3 Homoepitaxial Films

Group IV elements of Sn, Si and Ge are the primary investigated shallow dopants in vapor phase homoepitaxial growth of β-Ga_2O_3 as discussed in previous chapters. Typically, depositions are performed on Fe compensation-doped (010) β-Ga_2O_3-oriented crystals. Initial PLD studies of Sn, Si and Ge dopants in β-Ga_2O_3 thin films on (010) native substrates revealed high-quality films with the Si dopant only. An analysis of the influence of substrate material on Si-doped β-Ga_2O_3 thin films showed that native substrates resulted in films with superior structural and electrical properties compared to films grown on (0001) sapphire. Furthermore, a range of deposition parameters were identified that optimized the films. The influence of different crystal orientations and substrate fabrication methods on film properties is also under investigation.

Deposition rate is an important factor when assessing epitaxial vacuum growth techniques. Although fast deposition rates are desired for several μm-thick vertical

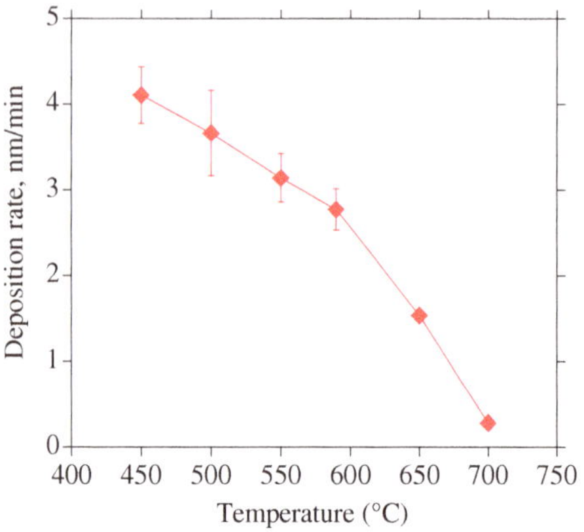

Fig. 14.3 Homoepitaxial deposition rate of Ga_2O_3:Si films on (010) β-Ga_2O_3 substrates by PLD as a function of deposition temperature in a fixed 1.33 Pa 5% O_2/95% Ar atmosphere

device structures, more moderate rates are suitable for thinner epitaxial films in lateral device applications with appropriate consideration of the processing maximum thermal budget. As the (010) β-Ga_2O_3 plane was demonstrated to yield the highest MBE deposition rates [3], subsequent reports with other deposition techniques followed with the same (010) substrate orientation. The PLD β-Ga_2O_3 deposition rate on (010) Ga_2O_3 substrates was found to decrease with increasing deposition temperature as shown in Fig. 14.3. The measured rate of 3–4 nm/min in the 500–600 °C range is consistent with rates reported for MBE at 600 °C [3] and MOVPE at 700 °C [4]. Si-doped films fabricated by LPCVD produced a deposition rate of 31 nm/min at a 900 °C substrate temperature [5]. A sharply decreased PLD rate at temperatures >600 °C is attributed to the enhanced desorption of depositing species and resulting in insufficient oxidation at higher temperatures [6]. However, as previously shown, Wakabayashi [1] achieved a 3 nm/min PLD deposition rate of undoped β-Ga_2O_3 with supplemental oxygen radicals at an 800 °C deposition temperature.

A first analysis comparing homoepitaxial vs. heteroepitaxial Si-doped β-Ga_2O_3 thin films resulted in superior electrical and structural properties in the homoepitaxial films [7]. Films were fabricated concurrently on Fe compensation-doped (010) β-Ga_2O_3 and (0001) Al_2O_3 substrates with depositions performed between 450 °C and 590 °C and with a Ga_2O_3- 1 wt% SiO_2 ablation target. For heteroepitaxial films on Al_2O_3 substrates, an average Hall effect mobility of 0.37 $cm^2\ V^{-1}\ s^{-1}$ and carrier concentration of 1.81×10^{19} cm^{-3} yielding a conductivity of 1 S cm^{-1} was obtained, in line with other PLD studies of Sn-doped Ga_2O_3 [8] and Si-doped Ga_2O_3 films [9–11]. The corresponding homoepitaxial films produced a significantly higher average Hall effect mobility of 26.5 $cm^2\ V^{-1}\ s^{-1}$ and carrier concentration of 1.74×10^{20} cm^{-3} yielding a conductivity of 732 S cm^{-1}. The high active Si carrier incorporation in the β-Ga_2O_3 lattice was further supported by a uniform Si chemical concentration of 4.15×10^{20} atoms cm^{-3} as measured by SIMS depth profiling. The approximately two orders of magnitude improvement in mobility, and conductivity, directly demonstrated the benefits of using native substrates.

Indeed, the improved film electrical properties with the native substrates were partially attributed to the single crystal film morphology achieved. Structural analyses by XRD of the Ga_2O_3:Si films on (010) β-Ga_2O_3 substrates confirmed epitaxial growth. Figure 14.4 shows 2Θ XRD scans of films on (010) β-Ga_2O_3 and (0001) Al_2O_3 substrates. While multiple (-201) β-Ga_2O_3 reflections indicative of a single phase were observed on the (0001) Al_2O_3, only a single (020) β-Ga_2O_3 reflection was measured from a film on the (010) β-Ga_2O_3 crystal. Clearly, the native substrate was vital in this case to achieve a single crystal film. The associated (020) rocking curve of the homoepitaxial film shown in Fig. 14.5 had a FWHM of 27 arcsec from a deposition conducted at 550 °C, slightly larger than the substrate FWHM of 19 arcsec. Further indications of a higher-quality homoepitaxial film structure were observed by surface roughness analyses. AFM of homoepitaxial film surfaces revealed a smooth 0.2 nm RMS roughness, close to the 0.1 nm RMS roughness measured on a bare substrate. In contrast, heteroepitaxial films had a 4.7 nm RMS roughness with a distinct granular morphology.

Additional assessments of film microstructure were obtained with cross-sectional TEM imaging of a Ga_2O_3:Si film deposited at 590 °C shown in Fig. 14.6. In Fig. 14.6a, the β-Ga_2O_3 substrate in the lower left corner is shown with a 218-nm-thick Ga_2O_3:Si film containing high contrast striations perpendicular to the substrate. These features could be attributed to Si segregation in the film, the presence of micro-twins or in-plane domains. In Fig. 14.6b, an atomic resolution image confirms that homoepitaxial film growth was achieved; a continuity of the (020) substrate lattice planes extends through the film/substrate interface and into the film.

After the initial work on Si-doped β-Ga_2O_3 thin films by PLD, which focused on identifying the benefits of native substrates, a subsequent investigation of homoepitaxial Si-doped β-Ga_2O_3 thin films examined in more detail the influence of film Si content and deposition gas species and pressure on the electronic and

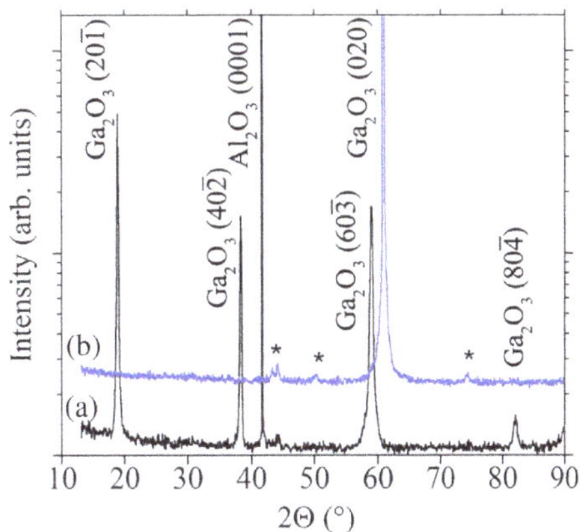

Fig. 14.4 2Θ XRD patterns of Ga_2O_3:Si films deposited at 590 °C on (**a**) (0001) Al_2O_3 and (**b**) (010) β-Ga_2O_3 substrates. *Instrument stage reflections. From [7]

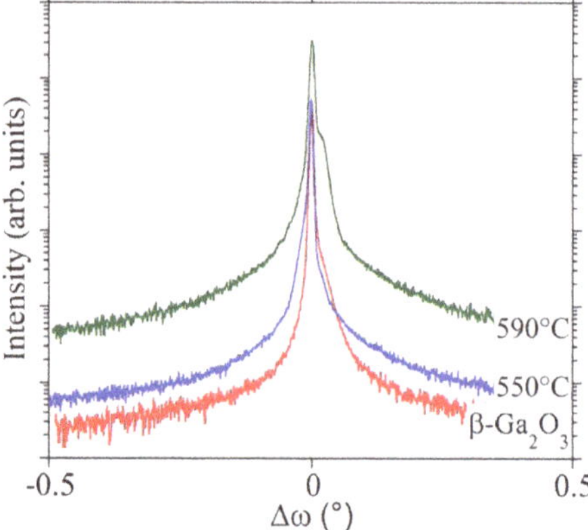

Fig. 14.5 XRD rocking curves of a (010) β-Ga_2O_3 substrate and (020) reflection from Ga_2O_3:Si films deposited at 550 and 590 °C on (010) β-Ga_2O_3 substrates. From [7]

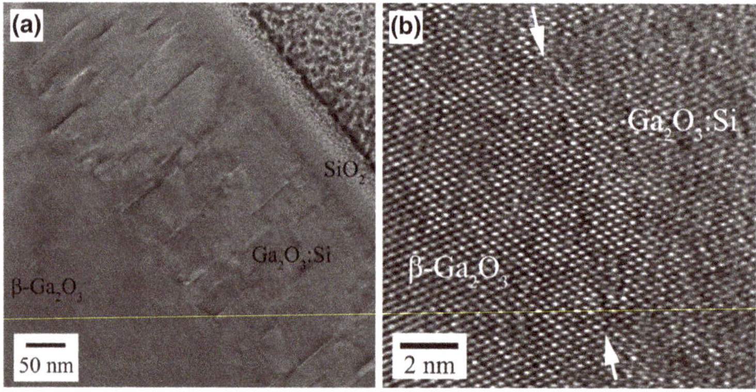

Fig. 14.6 Cross-sectional TEM images of Ga_2O_3:Si film deposited at 590 °C on a (010) β-Ga_2O_3 substrate. Overall film thickness is shown in (**a**) and higher magnification of film–substrate area with substrate on left in (**b**). Interface is indicated by arrows. From [7]

structural film properties [12]. First, a series of depositions were performed in a range of O_2 pressures using a fixed 1 wt% SiO_2 target and a fixed 590 °C deposition temperature. Second, a set of depositions was conducted with a fixed deposition pressure with varied O_2/Ar ratios with the same fixed 1 wt% SiO_2 target and 590 °C deposition temperature. Finally, with an optimized deposition pressure and O_2/Ar ratio, a series of depositions were performed with varied Si contents and deposition temperatures.

Because the optimization of deposition parameters was focused primarily on achieving films with mobilities suitable for device applications, electron transport properties provided the key performance evaluation criteria of the deposited films. Figure 14.7 shows the effect of O_2 deposition pressure on Hall effect measurements. A maximum mobility of ~30 $cm^2\ V^{-1}\ s^{-1}$ was found at 0.13 Pa O_2 pressure with

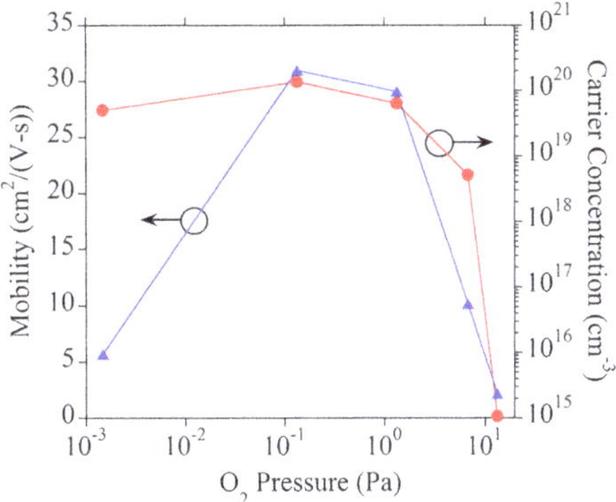

Fig. 14.7 Mobility and carrier concentration of Ga_2O_3:Si films from a Ga_2O_3- 1 wt% SiO_2 target deposited at 590 °C as a function of O_2 deposition pressure. From [12]

a carrier concentration of 1.37×10^{20} cm^{-3}. A low mobility observed at the lowest pressure (1.5×10^{-3} Pa with no O_2 added) was attributed to oxygen vacancies, as insufficient oxidation during growth likely occurred. For the higher O_2 pressures, current flow within the matrix becomes more hindered by the inherent Si dopant inhomogeneity, resulting in lower mobilities. As a 0.13 Pa O_2 deposition pressure allowed unintentional film deposition on the chamber laser window and hence, a lower net fluence on the ablation target, a higher-pressure regime of 1.3 Pa was preferred to mitigate dynamic fluence loss. The introduction of an inert gas such as Ar added to the O_2 process gas allowed a series of depositions with O_2/Ar ratios to be conducted at a total 1.3 Pa. A balanced high mobility and high carrier concentration using a 5% O_2/95% Ar mixture was obtained.

Using the 1.3 Pa, 5% O_2/95% Ar deposition atmosphere, a series of homoepitaxial films were fabricated with varied deposition temperatures and Si contents to assess the transport properties. The deposition temperatures spanned 450–590 °C while the Si contents were adjusted via ablation targets of Ga_2O_3 with 0.01, 0.025, 0.05, 0.1 and 1 wt% SiO_2. Film conductivity shown in Fig. 14.8 as a function of deposition temperature revealed an overall range between 6 and 798 S cm^{-1}. Films from targets with nominally 0.05, 0.1 and 1 wt% SiO_2 had similar conductivities that mostly increased with increasing deposition temperature up to 590 °C, to a

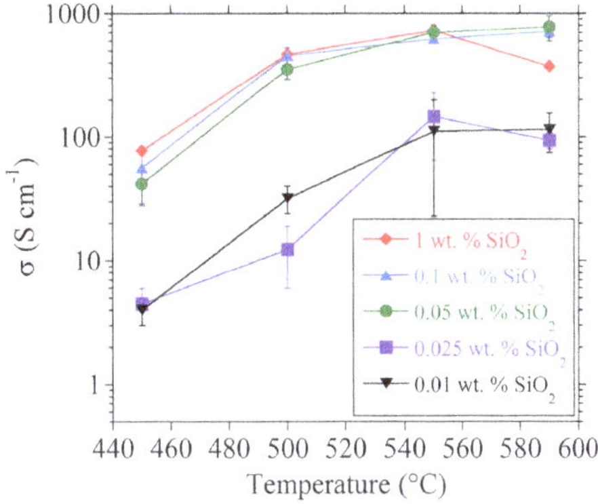

Fig. 14.8 Conductivity of Ga_2O_3:Si thin films from Ga_2O_3 targets with different SiO_2 contents as a function of deposition temperature. From [12]

maximum of 798 S cm^{-1}. Films from 0.01 and 0.025 wt% SiO_2 targets had consistently lower conductivities over the same temperature range with a maximum 108 S cm^{-1}. The highest conductivities were observed from depositions at 550 °C with some decline evident at 590 °C attributed to film desorption and more substantial decline at lower temperatures associated with diffusion limited growth. A carrier concentration and mobility range of 3.25×10^{19} cm^{-3} – 1.75×10^{20} cm^{-3} and 20 cm^2 V^{-1} s^{-1} – 27 cm^2 V^{-1} s^{-1} were obtained for films from 0.025 wt% and 1 wt% SiO_2 targets, respectively.

The two groupings of conductivities observed in Fig. 14.8 with different SiO_2 target compositions were attributed to imprecise SiO_2 dopant incorporation in the ablation targets. As the target vendor did not guarantee uniform doping in compositions with <0.5 wt% SiO_2, the four lower-doped targets reflected this lack of dopant control. SIMS depth profiles shown in Fig. 14.9 from representative thin film samples deposited at 550 °C confirmed the similar Si doping levels. The 1 wt% SiO_2 target yielded a film with 4.15×10^{20} atoms cm^{-3}. The Si concentration in films from the 0.1 and 0.05 wt% SiO_2 targets both had 3.08×10^{20} atoms cm^{-3} while the lower SiO_2-doped targets of 0.025 and 0.01 wt% SiO_2 produced films with 9.25×10^{19} atoms cm^{-3} and 9.78×10^{19} atoms cm^{-3}, respectively. Although the target composition amounts of SiO_2 are designated as nominal in this work due to compositional control limitations, the transport properties were, in general, consistent with the observed Si chemical compositions.

Additional characterization of the homoepitaxial films was conducted with XRD, TEM and AFM to examine the film structural properties. For depositions at 550 °C, XRD rocking curves on the (020) β-Ga_2O_3 peak showed slight tensile strain in films fabricated from the 1 wt% SiO_2 target while films from all other lower Si-doped targets had compressive in-plane film strain. Films from lower Si content targets deposited between 500 and 590 °C showed slightly increasing in-plane compressive strain. Cross-sectional TEM images collected from representative films at 450 and 550 °C and Ga_2O_3- 0.025 wt% SiO_2 and Ga_2O_3- 1 wt% SiO_2 targets are shown in Fig. 14.10. Both film compositions deposited at 450 °C reveal coarse morphologies with a transition from smooth to rough growth occurring with the lower Si-doped film (Fig. 14.10a, b). In comparison, the films

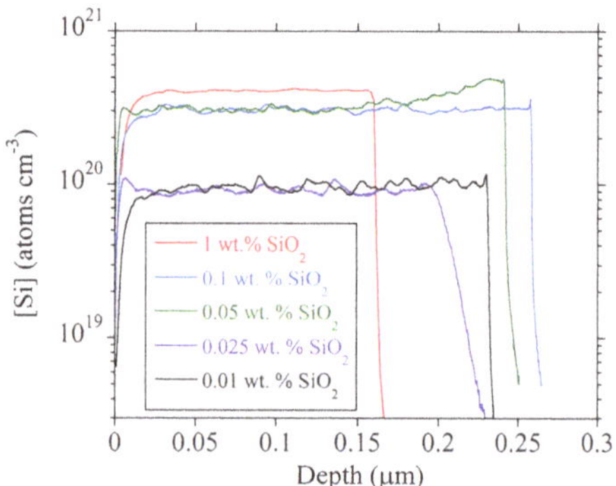

Fig. 14.9 SIMS depth profiles of Si in Ga_2O_3:Si films deposited at 550 °C from Ga_2O_3 targets with different SiO_2 contents. From [12]

deposited at 550 °C displayed smoother morphologies and strain localized defect regions. For the low Si-doped sample in Fig. 14.10c, strain contrast was observed at the film/substrate interface (identified with the white circle in Fig. 14.10c) with some dislocations extending into the film (indicated by white arrows). However, in the film from the 1 wt% SiO_2 target in Fig. 14.10d, the dark spots are more uniformly distributed throughout the thickness. Figure 14.10e shows an additional view where the sample was imaged with a different orientation, and the dark spots in Fig. 14.10d now appear as split dark lobes. This lobe contrast has been reported in other materials with localized strain fields [13–15]. X-ray energy-dispersive spectrometry indicated no significant enhancement or depletion of Si in these areas, indicating that this contrast is primarily due to strain-induced defect clusters. RMS surface roughness values from AFM were <0.2 nm for all films deposited between 500 and 590 °C reflecting high-quality homoepitaxial growth. Films deposited at 450 °C were considerably coarser between 5 and 8 nm RMS roughness as also observed in cross-sectional TEM images and attributed to diffusion limitations at the lowest temperature.

A summary of Hall effect mobility vs. electron concentration data plotted in Fig. 14.11 includes reports from multiple vapor deposited, homoepitaxial (010) β-Ga_2O_3 films impurity doped with Sn, Ge or Si. All of the studies employed either Mg or Fe compensation-doped (010) β-Ga_2O_3 substrates from Tamura Corporation. The deposition technique is indicated for each data set as well as the

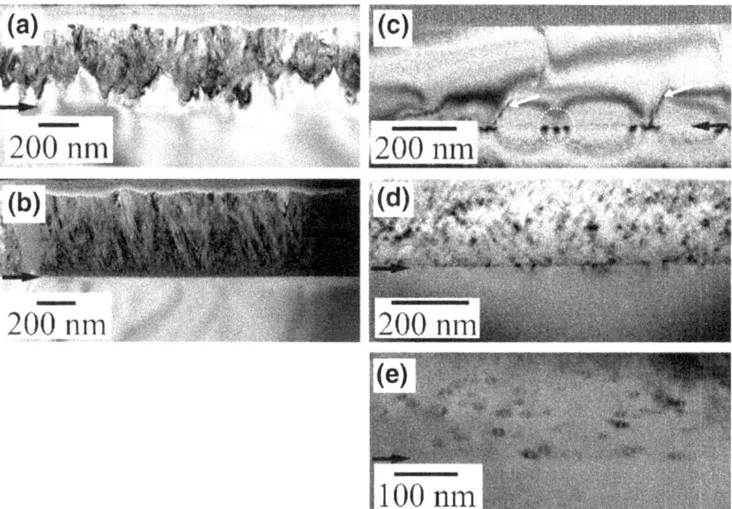

Fig. 14.10 Cross-sectional TEM images of Ga_2O_3:Si films deposited on (010) β-Ga_2O_3 substrates: **a** Ga_2O_3- 0.025 wt% SiO_2 target at 450 °C, **b** Ga_2O_3- 1 wt% SiO_2 target at 450 °C, **c** Ga_2O_3- 0.025 wt% SiO_2 target at 550 °C, **d** and **e** Ga_2O_3- 1 wt% SiO_2 target at 550 °C. Note that images **a–d** were imaged with the diffraction vector g along [020], viz. normal to the film/substrate interface; whereas, image **e** was imaged with g along [201], viz. parallel to the film/substrate interface. Black arrows indicate the film/substrate interface. From [12]

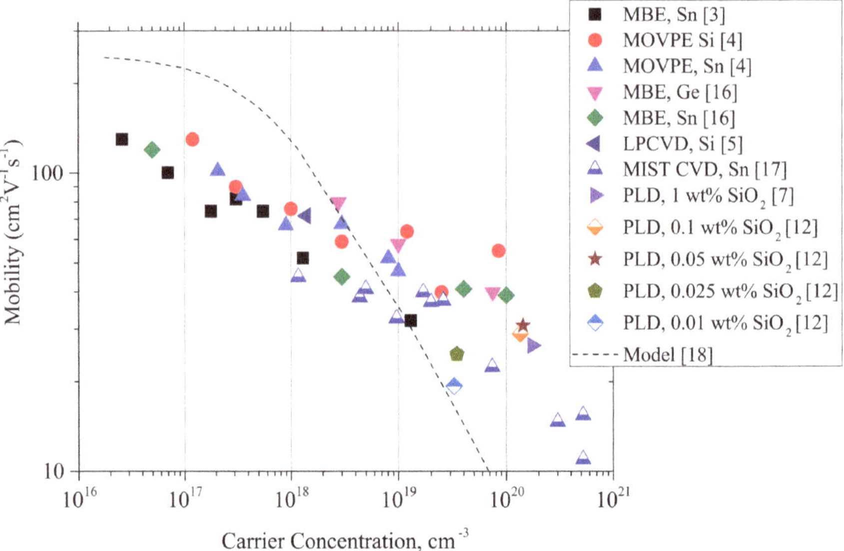

Fig. 14.11 Mobility versus carrier concentration of homoepitaxial β-Ga_2O_3 films deposited on (010) substrates using different deposition techniques and dopants. After [12]

dopant species in the legend. The first published report by Sasaki [3] used Sn dopant by MBE, and a 700 °C deposition temperature with a subsequent MBE study by Ahmadi [16] conducted with Ge dopant at 600 °C and Sn dopant at 650 °C. Baldini [4] utilized MOVPE with Si and Sn dopants at 850 °C while Rafique [5] deposited by LPCVD at 900 °C with Si dopant. Lee fabricated Sn-doped films by Mist CVD at 700 °C [17]. AFRL PLD results at 550 °C using a 1 wt% SiO_2 target are shown [7] along with lower-doped Si films by PLD at 550 °C [12]. The included model results from Ma [18] show polar optical phonon scattering limiting mobility to ~250 $cm^2\ V^{-1}\ s^{-1}$ at low doping levels. For the degenerate case with $n >$ ~10^{19} cm^{-3}, the Coulomb interaction between the electron and ionized dopant atoms is the dominant scattering mechanism leading to a mobility $\mu_{ii}(n) = \mu_{ii0}(n)(1 - K)/(1 + K)$, where the compensation ratio $K = N_A/N_D$. Thus, μ cannot be predicted based on n alone unless $K = 0$, i.e., $N_A = 0$. At that limit, $\mu_{ii0}(10^{20}) = 182$ $cm^2\ V^{-1}\ s^{-1}$, and including phonon scattering, $\mu_{ii}(10^{20}) = 55$ $cm^2\ V^{-1}\ s^{-1}$ at 300 K [12]. Therefore, no experimental point in Fig. 14.11 is above the theoretical limit if degeneracy is properly considered. PLD produced some of the highest carrier concentrations of any deposition technique, and also at the lowest deposition temperature. However, compared to other deposition techniques, PLD results were limited to carrier concentrations >3 × 10^{19} cm^{-3}, attributed to unintentionally high Si incorporation from the ablation targets.

While the previous Hall effect results discussed were all from as-deposited films, the influence of post deposition anneals on film properties was also studied. An anneal of PLD-grown, Si-doped β-Ga_2O_3 for 10 min at 525 °C in forming gas (3%

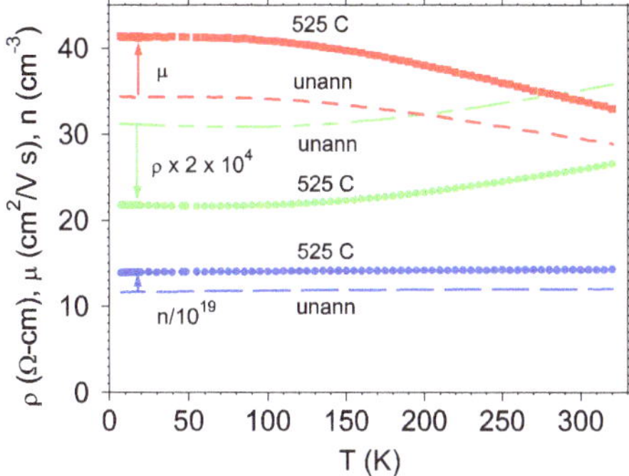

Fig. 14.12 Temperature-dependent Hall effect results of a Si-doped β-Ga_2O_3 film as deposited (labeled unann) and after a 525 °C forming gas anneal

H_2 in Ar) was found to improve the already low resistivity by 43%, to 1.2×10^{-3} Ω cm as shown in Fig. 14.12. While this is not unusual for oxide semiconductors, β-Ga_2O_3 is unique in that both mobility and concentration increased, not just one or the other. This means the principal mechanism must be destruction, or at least passivation, of acceptors. Higher temperatures are under investigation to assess if more improvement is possible. However, these results bode well for β-Ga_2O_3 as a transparent conducting oxide or ohmic contact material.

To demonstrate the ohmic contact utility, highly conductive, PLD Si-doped β-Ga_2O_3 films were implemented in a FET to improve device operation through lowered ohmic contact resistance. A 25 nm PLD regrowth layer using a 1 wt% SiO_2 target deposited at 550 °C was incorporated into ohmic trenches of a β-$(Al_{0.14}Ga_{0.86})_2O_3/Ga_2O_3$ modulation-doped FET. An extracted contact resistance, including layer resistance, of 1 Ω mm was measured, comparable to the lowest achieved by regrown ohmics (1.1 Ω mm) on a heterostructure [19]. No anneal of the ohmic contacts was employed, differentiating the PLD process from typical evaporated ohmic metal and ion implantation processes that require alloying or activation anneals, respectively, to obtain minimal contact resistance [19–21].

Although Si has shown the most promise for impurity doping in PLD β-Ga_2O_3, other Group IV elements are under consideration. Preliminary studies of Ge and Sn-doped PLD β-Ga_2O_3 films on native substrates showed lower-quality electrical properties compared to Si-doped films. With deposition parameters identical to Si-doped films (550 °C in 1.3 Pa of 5% O_2/95% Ar) and (010) β-Ga_2O_3 substrates, films deposited from a Ga_2O_3- 1 mol% SnO_2 target had a mobility of 4 $cm^2\ V^{-1}\ s^{-1}$ and electron concentration of 3.22×10^{17} cm^{-3}. Films fabricated from a Ga_2O_3- 0.015 wt% Ge target were found to be highly resistive. One study of Sn-doped films deposited at 450 °C in 6.67×10^{-4} Pa from a Ga_2O_3- 2 wt% Sn-doped target on (-201) Ga_2O_3 yielded XRD pole figures indicating epitaxial growth. Transport properties were not reported for the film nor the substrate [22]. Further PLD investigations are necessary to establish deposition conditions conducive to more effective electrical activation of the Sn and Ge dopants.

Another critical variable in fabricating epitaxial films is the quality of the substrate crystal. Most published β-Ga_2O_3 epitaxial film results to date utilized Fe compensation-doped β-Ga_2O_3 (010) substrates from Tamura Corporation primarily due to limited commercial substrate availability. In addition to the EFG crystal fabrication technique used by Tamura, the Czochralski (CZ) method has also been employed to produce substrates for epitaxial film growth. Homoepitaxial Si-doped PLD films on Fe compensation-doped β-Ga_2O_3 (010) CZ substrates fabricated by Northrop Grumman Synoptics, along with witness (010) EFG Tamura substrates, were fabricated and characterized. Hall effect results were used to quantify film electrical properties and by extension, substrate quality which can be manifested in multiple variables including subsurface damage, surface roughness, twinning and defect density. Deposition rates were found to be similar, independent of the substrate fabrication technique. The Hall effect mobility and electron concentration are reported in Table 14.1 for four PLD β-Ga_2O_3 films fabricated from a Ga_2O_3- 1 wt% SiO_2 ablation target at 550 °C in a 1.33 Pa 5% O_2/95% Ar atmosphere. The XRD (020) rocking curve FWHM was 428 arcsec from a representative film on the CZ substrate, appreciably higher than that measured from films on EFG substrates. RMS surface roughness of films deposited on the CZ substrates was 0.14 nm, consistent with the <0.2 nm RMS roughness measured on films deposited on EFG substrates. Both mobility and concentration values were also comparable to EFG substrate results for the four depositions reported in Table 14.1, indicating that the Synoptics substrates yield epitaxial films of similar quality for device applications.

Alternative substrate crystal orientations to β-Ga_2O_3 (010) are also being explored despite the early assessment of lower MBE deposition rates on orientations such as (100) due to potentially lower energy growth planes. As discussed in a previous chapter, (100) substrates have shown significant promise for epitaxial growth with crystal miscuts of up to 6° [23, 24]. Another orientation with minimal reports of epitaxial film growth is (001). PLD Si-doped Ga_2O_3 was fabricated on Fe compensation-doped β-Ga_2O_3 (001) crystals from Tamura. A 550 °C deposition temperature and 1.33 Pa 5% O_2/95% Ar atmosphere with a Ga_2O_3- 1 wt% SiO_2 ablation target yielded films with mobility of 5 $cm^2\ V^{-1}\ s^{-1}$ and electron concentration of 7.97×10^{19} cm^{-3}. 2Θ XRD scans showed a mixed phase: a strong β-Ga_2O_3 [002] peak at 31.722° with a FWHM of 137 arcsec and a secondary peak at 30.238° attributed to the β-Ga_2O_3 [110] reflection with a FWHM of 623

Table 14.1 Hall effect mobility and carrier concentration of PLD Si-doped β-Ga_2O_3 films grown at 550 °C on (010) β-Ga_2O_3 Tamura EFG and Synoptics CZ substrates

Mobility ($cm^2\ V^{-1}\ s^{-1}$)		Concentration (cm^{-3})	
Tamura EFG	Synoptics CZ	Tamura EFG	Synoptics CZ
32.9	31.2	1.57×10^{20}	1.34×10^{20}
26.9	29.1	7.51×10^{19}	1.09×10^{20}
32.7	22.2	7.95×10^{19}	8.39×10^{19}
30	24.8	1.29×10^{20}	7.42×10^{19}

arcsec. AFM showed an RMS roughness of 5.3 nm. The degraded structure identified by XRD and AFM results indicates that crystal miscuts found useful for the (100) orientation may also be beneficial for the (001) orientation.

14.4 β-Ga_2O_3 Heterostructures

Extending beyond impurity doping with the Group IV elements, alteration of the β-Ga_2O_3 bandgap through alloying with elements such as Al has the potential to create useful heterostructures for device applications. For β-$(Al_xGa_{1-x})_2O_3$, a bulk solubility limit of $x = 0.8$ with a tunable bandgap range of 1.8 eV was demonstrated by solution combustion synthesis [25]. Wakabayashi [1] utilized oxygen radical-assisted PLD with an $(Al_{0.05}Ga_{0.95})_2O_3$ target to fabricate β-$(Al_xGa_{1-x})_2O_3$ films at 800 °C on (010) β-Ga_2O_3 substrates using the same O_2 plasma methodology discussed previously with homoepitaxial β-Ga_2O_3 film growth. Al incorporation in an O_2 plasma-assisted growth was found to be $x = 0.06$, very close to the ablation target stoichiometry. In Fig. 14.13, XRD of films deposited with and without O_2 plasma assist is shown. The (020) XRD rocking curve FWHM value for the O_2 plasma sample was 108 arcsec. Samples deposited without oxygen radicals exhibited a significantly higher Al content ($x = 0.86$) along with the presence of a cubic Al_2O_3 phase. Feng [26] deposited $(Al_{0.08}Ga_{0.92})_2O_3$ films on Sn-doped (010) β-Ga_2O_3 substrates from a $(Al_{0.05}Ga_{0.95})_2O_3$ target at 650 °C in 1 Pa of O_2. Schottky diodes were fabricated with Ni/Au contacts. The ideality factor and barrier height data indicated barrier inhomogeneity at the metal–semiconductor interface of the PLD films. Wakabayashi [27] grew $(Al_xGa_{1-x})_2O_3$ films with oxygen radical-assisted PLD by alternating ablation on β-Ga_2O_3 and α-Al_2O_3 single crystal targets. Films with Al content, x, between 0 and 0.37 were fabricated on

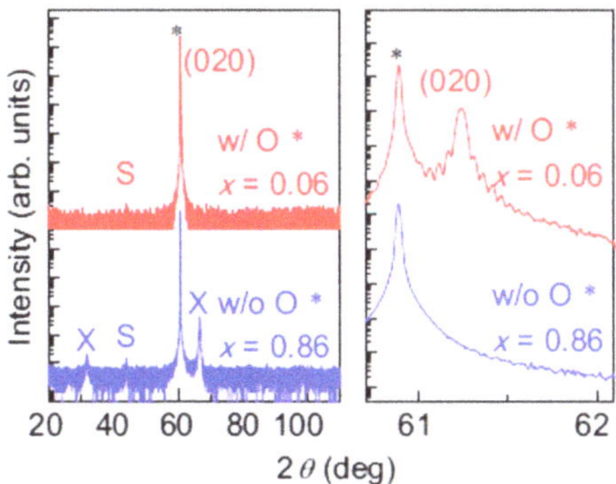

Fig. 14.13 Out-of-plane XRD patterns of $(Al_xGa_{1-x})_2O_3$ films grown with O* ($x = 0.06$) and without O* ($x = 0.86$) using the $(Al_{0.05}Ga_{0.95})_2O_3$ ceramic target. The patterns shown in (**a**) were taken at wide scan range and those in (**b**) were taken near (020) reflection of the substrates (*). The two reflections labeled *X* were assignable to (220) and (440) reflections from a cubic Al_2O_3 (γ- or η-Al_2O_3). The reflection labeled S was from the sample stage. From [1]

(100) β-Ga_2O_3 substrates. Conduction and valence band offsets of 0.52 and 0.13 eV, respectively, for an $x = 0.37$ composition were measured. The conduction band offset was found to increase with increasing Al content while the valence band offset did not change.

References

1. R. Wakabayashi, T. Oshima, M. Hattori, K. Sasaki, T. Masui, A. Kuramata, S. Yamakoshi, K. Yoshimatsu, A. Ohtomo, J. Cryst. Growth **424**, 77 (2015)
2. F. Li, Q. Feng, X. Huang, Y. Hao, Homoepitaxial growth of beta-Ga_2O_3 films by pulsed laser deposition, in *Paper Presented at the International Conference on the Physics and Semiconductors*, Beijing, China, 31 July-5 August 2016
3. K. Sasaki, A. Kuramata, T. Masui, E.G. Villora, K. Shimamura, S. Yamakoshi, Appl. Phys. Express **5**(3), 035502 (2012)
4. M. Baldini, M. Albrecht, A. Fiedler, K. Irmscher, R. Schewski, G. Wagner, ECS, J. Sol. State Sci. Technol. **6**(2), Q3040 (2017)
5. S. Rafique, M.R. Karim, J.M. Johnson, J. Hwang, H. Zhao, Appl. Phys. Lett. **112**(5), 052104 (2018)
6. K. Matsuzaki, H. Hiramatsu, K. Nomura, H. Yanagi, T. Kamiya, M. Hirano, H. Hosono, Thin Solid Films **496**(1), 37 (2006)
7. K.D. Leedy, K.D. Chabak, V. Vasilyev, D.C. Look, J.J. Boeckl, J.L. Brown, S.E. Tetlak, A. J. Green, N.A. Moser, A. Crespo, D.B. Thomson, R.C. Fitch, J.P. McCandless, G.H. Jessen, Appl. Phys. Lett. **111**(1), 012103 (2017)
8. M. Orita, H. Ohta, M. Hirano, H. Hosono, Appl. Phys. Lett. **77**(25), 4166 (2000)
9. S. Müller, H. von Wenckstern, D. Splith, F. Schmidt, M. Grundmann, Phys. Status Solidi A **211**(1), 34 (2014)
10. F.B. Zhang, K. Saito, T. Tanaka, M. Nishio, Q.X. Guo, J. Mater. Sci.: Mater. Electron. **26** (12), 9624 (2015)
11. F. Zhang, M. Arita, X. Wang, Z. Chen, K. Saito, T. Tanaka, M. Nishio, T. Motooka, Q. Guo, Appl. Phys. Lett. **109**(10), 102105 (2016)
12. K.D. Leedy, K.D. Chabak, V. Vasilyev, D.C. Look, K. Mahalingam, J.L. Brown, A.J. Green, C.T. Bowers, A. Crespo, D.B. Thomson, G.H. Jessen, APL Mater. **6**(10), 101102 (2018)
13. T. Benabbas, P. François, Y. Androussi, A. Lefebvre, J. Appl. Phys. **80**(5), 2763 (1996)
14. M.F. Ashby, L.M. Brown, Philos. Mag. A **8**(91), 1083 (1963)
15. M.L. Jenkins, M.A. Kirk, *Characterisation of Radiation Damage by Transmission Electron Microscopy* (Institute of Physics Publishing, Bristol, 2000), p. 29
16. E. Ahmadi, O.S. Koksaldi, S.W. Kaun, Y. Oshima, D.B. Short, U.K. Mishra, J.S. Speck, Appl. Phys. Express **10**(4), 041102 (2017)
17. S. Lee, K. Kaneko, S. Fujita, Jap. J. Appl. Phys. **55**, 1202B8 (2106)
18. N. Ma, N. Tanen, A. Verma, Z. Guo, T. Luo, H.G. Xing, D. Jena, Appl. Phys. Lett. **109**(21), 212101 (2016)
19. Y. Zhang, C. Joishi, Z. Xia, M. Brenner, S. Lodha, S. Rajan, Appl. Phys. Lett. **112**(23), 233503 (2018)
20. M. Higashiwaki, K. Sasaki, T. Kamimura, M.H. Wong, D. Krishnamurthy, A. Kuramata, T. Masui, S. Yamakoshi, Appl. Phys. Lett. **103**(12), 123511 (2013)
21. A.J. Green, K.D. Chabak, M. Baldini, N. Moser, R. Gilbert, R.C. Fitch, G. Wagner, Z. Galazka, J. McCandless, A. Crespo, K. Leedy, G.H. Jessen, IEEE Electron Device Lett. **38**(6), 790 (2017)
22. L.M. Garten, A. Zakutayev, J.D. Perkins, B.P. Gorman, P.F. Ndione, D.S. Ginley, MRS Commun. **6**(4), 348 (2016)

23. R. Schewski, M. Baldini, K. Irmscher, A. Fiedler, T. Markurt, B. Neuschulz, T. Remmele, T. Schulz, G. Wagner, Z. Galazka, M. Albrecht, J. Appl. Phys. **120**(22), 225308 (2016)
24. M. Baldini, Z. Galazka, G. Wagner, Mater. Sci. Semicond. Process. **78**, 132 (2018)
25. B.W. Krueger, C.S. Dandeneau, E.M. Nelson, S.T. Dunham, F.S. Ohuchi, M.A. Olmstead, J. Am. Ceram. Soc. **99**(7), 2467 (2016)
26. Q. Feng, Z. Feng, Z. Hu, X. Xing, G. Yan, J. Zhang, Y. Xu, X. Lian, Y. Hao, Appl. Phys. Lett. **112**(7), 072103 (2018)
27. R. Wakabayashi, M. Hattori, K. Yoshimatsu, K. Horiba, H. Kumigashira, A. Ohtomo, Appl. Phys. Lett. **112**(23), 232103 (2018)

Chapter 15
Pulsed Laser Deposition 2

Heteroepitaxial Growth of Ga_2O_3 and Related Alloys

Holger von Wenckstern, Daniel Splith and Marius Grundmann

Abstract We review heteroepitaxial growth of Ga_2O_3 and related alloys by pulsed laser deposition (PLD). First, we briefly summarize the history of PLD and discuss its evolution and development since its breakthrough in the 1980s with the focus on combinatorial material synthesis. Then, the impact of strain on the lattice constant of rhombohedral, pseudomorphic $(Al, Ga)_2O_3$ thin films is introduced and the determination of thin film composition from X-ray diffraction measurements is outlined. For monoclinic Ga_2O_3 layers the influence of key growth parameters on growth rate and surface morphology is discussed. Electrical transport properties of monoclinic thin films doped by silicon or tin are presented and compared to that of homoepitaxial layers. For ternary thin films growth parameters strongly influence the chemical composition in addition to growth rate and morphology. High oxygen pressures and/or low growth temperatures are necessary for a stoichiometric transfer of the target composition to the epilayer which is explained by the desorption of gallium suboxides occurring otherwise. Further, we resume solubility limits and the dependence of structural, optical and vibrational properties on the alloy composition of monoclinic $(In, Ga)_2O_3$ and $(Al, Ga)_2O_3$ thin films.

H. von Wenckstern (✉) · D. Splith · M. Grundmann
Felix Bloch Institute for Solid State Physics, Linnéstrasse 5, 04103 Leipzig, Germany
e-mail: wenckst@uni-leipzig.de

D. Splith
e-mail: daniel.splith@uni-leipzig.de

M. Grundmann
e-mail: grundmann@physik.uni-leipzig.de

M. Higashiwaki and S. Fujita (eds.), *Gallium Oxide*, Springer Series
in Materials Science 293, https://doi.org/10.1007/978-3-030-37153-1_15

15.1 Short Introduction and History of Pulsed Laser Deposition

The history of pulsed laser deposition is strongly connected to the invention and developments of the laser starting with the theory of stimulated emission established by Einstein in 1917 [1]. Some thirty years later, the concept of stimulated emission was explored by Alexander Prokhorov and Nikolai Basov at the Lebedev Laboratories and by Charles Townes at Columbia University. In 1964, the three of them received the Nobel Price in Physics "for fundamental work in the field of quantum electronics, which has led to the construction of oscillators and amplifiers based on the maser-laser principle." The first operating laser was designed by Theodore Maiman at Hughes Research Laboratories in 1960 and used ruby as a gain medium. It was working in pulsed operation. The first continuous-wave laser was demonstrated in 1962 by Willard Boyle at the Bell Laboratories. It was also in 1962 that Breech and Cross used a ruby laser to vaporize atoms from solids and characterized the formed plasma by optical spectroscopy [2]. In 1965, Smith and Turner deposited first thin films by using a ruby laser for ablation of powder or single crystalline materials [3]. The deposited inorganic and organic layers included ZnTe, Te, MoO_3, fuchsine and Nidimethylglyoxime. However, the layer quality was inferior compared to that of other deposition methods. First, $Hg_{0.7}Cd_{0.3}Te$ layers with quality similar to films grown by molecular beam epitaxy were reported by Cheung in 1983 [4]. It was also within the 1980s as first pulsed laser deposition (PLD) systems were installed in research labs for thin film epitaxy. Prerequisite was the development of reliable electronic Q-switches and high-efficiency second-harmonic generators necessary for generating very short and low wavelength pulses, respectively.

Selected Milestones in the development of PLD for growth of inorganic thin films	
1916	Einstein's theory of stimulated emission [1]
1954	Townes fabricated first MASER [5]
1958	Bell labs filed patent application for optical maser Schawlow and Townes published calculations on optical maser [6]
1959	Gould published paper: *The LASER, Light Amplification by Stimulated Emission of Radiation* at The Ann Arbor Conference on Optical Pumping
1960/03	Bell Labs (Schwalow, Townes) was granted a patent for the optical maser
1960/05	First operating laser by Maiman at Hughes Research Laboratories Nature publication in August 1960 [7]
1962	Breech and Cross used ruby laser to vaporize atoms from a solid [2]
1964	Nobel Price in Physics awarded to Townes, Basov and Prokhorov
1965	Smith and Turner used ruby laser to deposit thin films [3]
1970s	Development of (i) Reliable electronic Q-switches for generating very short pulses; (ii) High-efficiency second-harmonic generators for shorter wavelength
1983	PLD of $Hg_{0.7}Cd_{0.3}Te$ by Cheung [4]
1987	PLD in Bellcore group used successfully to grow HTS YBCO [8]
1990	Development of large-area deposition by Greer et al. [9–11] with substrate diameter up to 8 in.
1993	Cross-beam PLD (CBPLD) introduced by Strikovsky et al. [12]
1994	Eclipse-PLD introduced by Kinoshita et al. [13]
1998	Introduction of combinatorial PLD by Xiang et al. (discrete libraries) [14]
1999	Ultrafast pulsed laser deposition established by Rode et al. [15]
1999	Introduction of CCS-PLD (continuous library) by Fukumura et al. [16]
2000	Scanning multi-component PLD introduced by Jacquot et al. [17]
2003	Improved CCS-PLD method by Christen et al. [18]
2004	PLD of nano- and micropillars by Lorenz et al. [19, 20]
2007	Quantum Hall Effect in Polar Oxide Heterostructure by Tsukasaki et al. [21]
2013	CCS-PLD with segmented targets introduced by von Wenckstern et al. [22]
2015	Multiple-plume PLD introduced by Mao et al. [23]
2018	Vertical CCS-PLD (VCCS-PLD) introduced by Kneiß et al.

The actual breakthrough of PLD came as the discovery of the high-temperature superconductor LaBaCuO in 1986 (Nobel Prize for Georg Bednorz and Alexander Müller in 1987 "for their important breakthrough in the discovery of superconductivity in ceramic materials") initiated a boost in the deposition and investigation of perovskite cuprate oxides with transition temperature above the boiling point of nitrogen such as $YBa_2Cu_3O_{7-\delta}$ thin films that could be grown with high quality by PLD [8, 24]. The main advantage in growing complex oxides by PLD compared to other growth methods is the stoichiometric transfer of the target constituents to the substrate (or at least a stoichiometric flux generation) being a non-trivial task with, e.g., evaporation techniques requiring an individual source for each cation to be incorporated. Hence, it is demanding to achieve the desired cation composition in an epitaxial layer and in some cases not realizable. Another advantage of PLD is the very large pressure window for film deposition. It ranges from ultra-high vacuum conditions to 100 mbar and above. In principle, PLD is capable to grow very pure thin films (strongly dependent on the purity of the target material) since the target holder (being the source of material to be deposited) is not heated such as source containers during evaporation. Further, the method does not require pre-cursors that might lead to contamination. PLD chambers are typically equipped with multi-target holders allowing the growth of heterostructures and superlattices with sharp interfaces. Disadvantages include the likelihood of droplet ablation and incorporation and the fact that homogeneous thin films are only obtained on comparatively small substrate sizes (typically restricted to substrate diameters up to 4 in. on lab scale and about 8 in. in commercial systems [25, 26]).

15.2 The PLD Process

The basic processes occurring during PLD are schematically depicted in Fig. 15.1 and can be shortly described by (i) absorption of the laser irradiation by the target material, (ii) formation of a laser-induced plasma having the stoichiometry of the target and interaction of laser radiation with the plasma, (iii) expansion of the plasma in the vacuum/low-pressure environment and (iv) condensation of the plasma constituents on the substrate. Additionally, incorporation of atoms/molecules from the background gas or from another plasma source can assist film growth. Typically, a sub-monolayer is deposited for a single laser pulse. The process is repeated until the desired film thickness is obtained. Multiple target systems allow the growth of heterostructures and superlattices.

A more detailed description of processes (i)–(iv) shall be provided for the case of solid-state target material. The absorption of the typically focused laser irradiation (commonly the target is not within the focal plane such that a certain area of the target surface is irradiated instead of a point) initially leads to a melting of the target material, the melting front is moving into the target. A few nanoseconds later vaporization occurs, and the interaction between laser radiation and vapor begins, leading to plasma formation. The vapor has the stoichiometry of the target. The plasma expands into the vacuum/low-pressure environment in a direction perpendicular to the target surface. The expanding plasma is highly forward-directed, and the degree of directionality depends on the background pressure (higher pressures lead to a broadening of the plasma). The plasma constituents (typically molecules, atoms, ions and electrons) may interact with the constituents of the ambient gas. Finally, the particle flux reaches the substrate where it condensates. Depending on growth conditions, re-evaporation may play a role influencing the stoichiometry of the resulting thin film.

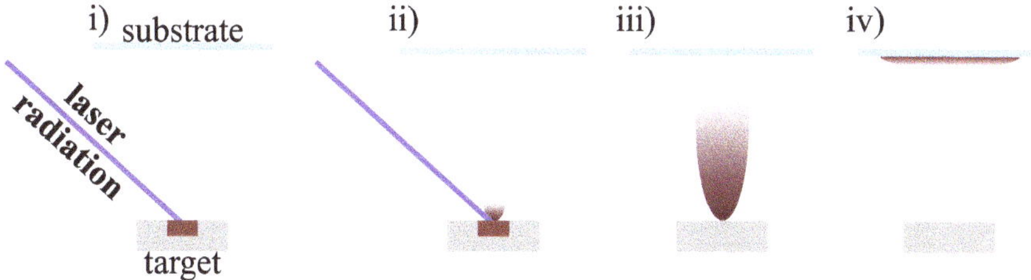

Fig. 15.1 Physical processes involved in pulsed laser deposition: (i) interaction of laser radiation with target, (ii) interaction of formed plasma with laser radiation, (iii) plasma expansion, (iv) thin film condensation

15.3 Creation of Continuous Cation Composition Spreads by PLD

PLD is highly suited for the creation of material libraries since the target is typically stoichiometrically ablated. The usage of multiple targets and shadowing masks with motion synchronized to the target exchange enables a defined lateral variation of the cation composition [16, 27]. Christen et al. used a 180° rotation of the substrate during target exchange within an off-axis PLD approach such that shadowing was not required [28]. Later, Christen et al. introduced an advanced approach relying on a lateral movement of the substrate behind a shadow masks [18]. From film thickness simulations 450 lateral positions of the substrate were defined each triggering the firing of one laser pulse. For all of these approaches, the growth rate is tremendously lower compared to conventional PLD due to repeated target exchange. This exchange can be avoided by ablation of a segmented PLD target. In an off-axis configuration, a defined continuous composition spread is obtained if the target and the substrate rotation are synchronized [22]. Further it is straightforward to obtain various cation gradients by usage of a multifold segmented target [22]. So far, the method was used to realize material gradients in (Mg, Zn)O:Al/Ga [22, 29–31], zinc-tin-oxide [32], $(In, Ga)_2O_3$ [33–36] and $(Al, Ga)_2O_3$ [37, 38].

15.4 Growth of $(In, Ga, Al)_2O_3$ by PLD

15.4.1 Pseudomorphic Growth of α-$(Al, Ga)_2O_3$ Thin Films on R-Plane Sapphire

The calculation of the strain state of pseudomorphic epilayers in the framework of continuous medium elasticity theory assuming coherent growth in the interface plane was previously successfully applied to cubic and hexagonal materials such as zincblende and wurtzite heterostructures. For the sesquioxide systems, pseudomor-

Table 15.1 Hexagonal lattice constants a and c (in nm) and elastic constants (in 10^{11} Pa) as compiled from the literature in [39] for Ga_2O_3 and Al_2O_3 in corundum phase. C_{14} is given with positive sign for both materials

Parameter	Ga_2O_3	Al_2O_3
a	0.49825	0.4759
c	1.3433	1.2991
C_{11}	3.815	4.97
C_{12}	1.736	1.63
C_{13}	1.260	1.16
C_{33}	3.458	5.01
C_{14}	0.173	0.22
C_{44}	0.797	1.47

phic strain in heteroepitaxial structures needs to be considered also for monoclinic and rhombohedral heterostructures based on β-Ga_2O_3 and α-Al_2O_3 (corundum), respectively. As usual, the stress in a perpendicular direction and the related shear stresses are assumed zero. From these conditions, using the (different) tensors of elastic constants for monoclinic and rhombohedral crystals, the strain tensor of the epilayer and subsequently lattice constants and tilts (relative to the same lattice plane of the substrate) can be calculated. We note that the mechanics of strained rhombohedral heterostructure differs from hexagonal ones, well known from GaN- and ZnO-based heterostructures [40], by the fact that the elastic constant C_{14} is nonzero. For further details of the theory, see [41]. An important ingredient into the theory is the material constants that enter the calculation. We use the parameters listed in Table 15.1 for the binary materials Ga_2O_3 and Al_2O_3 in corundum phase and linear interpolations of these parameters for the $(Ga_xAl_{1-x})_2O_3$ alloy system. Nd-doped pseudomorphic α-$(Ga_xAl_{1-x})_2O_3$ thin films with $0.39 \leq x \leq 1$ were grown by molecular beam epitaxy on r-, a- and m-plane Al_2O_3 for envisaged application as graded-index layers for solid-state waveguide lasers [42]. In [39], a detailed study is reported on PLD growth of α-phase $(Ga_xAl_{1-x})_2O_3$ thin films on r-plane Al_2O_3. The lattice constants of substrate and epilayer and relative tilts were evaluated using high-resolution X-ray diffraction for various symmetric, skew and asymmetric reflexes [(02.4), (04.8), (00.6), (00.12), (40.10), (10.16), and the six reflections of the {41.6}-family]. The pseudomorphic distortion of the epilayers agrees very well between the elastic theory and experiment [39]. From the data, the elemental Ga concentration can be determined and is plotted in Fig. 15.2 as a function of the growth temperature for ablation from two targets with 10 at.% and 20 at.% gallium, respectively. It is found that the gallium incorporation decreases with increasing growth temperature, and for both targets at 800 °C, the stoichiometric transfer of gallium is only about one-third.

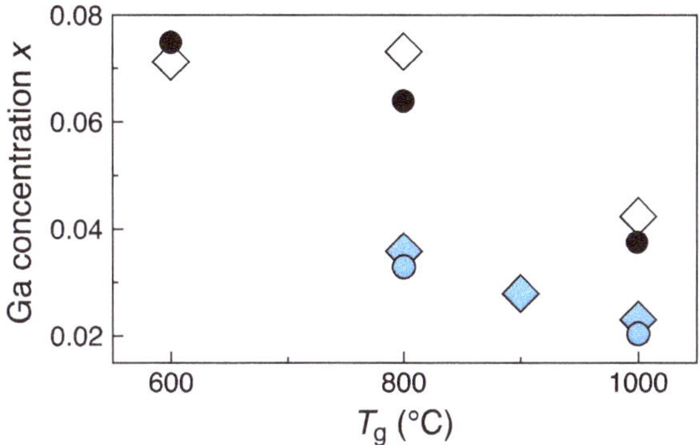

Fig. 15.2 Ga-concentration x from fits to experimental peak splitting (circles) and EDX analysis (diamonds) in dependence of PLD growth temperature, for films grown from PLD targets with 10 at.% Ga (blue) and 20 at.% Ga (black). Size of EDX symbols corresponds to their experimental uncertainty

15.4.2 Pulsed Laser Deposition of Heteroepitaxial Ga_2O_3 Thin Films

Pulsed laser deposition was employed in 1994 by Phillips et al. and Cava et al. to deposit $GaInO_3$ thin films in β-gallia structure for potential application as transparent conductive oxide (TCO) material [43, 44]. Highest conductivity of about 365 $(\Omega\,cm)^{-1}$ was obtained for tin-doped thin films with an optical band gap of about 3.3 eV and a refractive index of 1.65, matching well that of commonly used glass substrates, which is favorable compared to tin-doped indium oxide. However, the conductivity was more than a factor of ten lower and hence application as TCO material was not progressed. The search for a new deep UV-transparent conducting oxide material motivated PLD growth of binary Ga_2O_3:Sn by Orita et al. in 2000 [45, 46], marking the starting point of systematic investigations on heteroepitaxy of β-Ga_2O_3 by PLD. The most commonly used substrate for heteroepitaxy of Ga_2O_3 thin films by PLD is (00.1)Al_2O_3. Growth on (00.1)GaN, (100)MgO, (100)$MgAl_6O_{10}$ and homoepitaxial growth were investigated as well; epitaxial relations and the occurrence of rotational domains [47, 48] are summarized in Table 15.2.

15.4.2.1 Growth Temperature and Oxygen Pressure Dependence of Growth Rate

Heteroepitaxial growth of crystalline Ga_2O_3 thin films by pulsed laser deposition requires growth temperatures of 380 °C or higher [46, 51–53, 56]. Orita et al. reported that the repetition rate of the laser has an influence on epitaxial growth. They obtained crystalline thin films on c-plane sapphire for a growth pressure of 5×10^{-8} mbar, a growth temperature of 380 °C and a repetition rate of 1 Hz, whereas amorphous layers

Table 15.2 Epitaxial relationship of β-Ga_2O_3 thin films on substrates commonly used for heteroepitaxy. nRD: number of rotational domains. The epitaxial relationship is summarized/determined in the provided references

Substrate	Growth direction	Epitaxial relationship	Rotation domains	References
$Al_2O_3(00.1)$	$[\bar{2}01]$	$\langle 010\rangle_{Ga_2O_3} \| \langle 10.0\rangle_{Al_2O_3}$	6RD	[49]
GaN(00.1)	$[\bar{2}01]$	$\langle 010\rangle_{Ga_2O_3} \| \langle 10.0\rangle_{GaN}$	6RD	[50]
MgO(100)	[100]	$[001]_{Ga_2O_3} \| \langle 011\rangle_{MgO}$	4RD	[49]
$MgAl_6O_{10}(100)$	[100]	$[001]_{Ga_2O_3} \| \langle 011\rangle_{MgAl_6O_{10}}$	4RD	[49]

Table 15.3 Parameters of PLD chambers used for growth of Ga_2O_3 thin film on c-plane sapphire. d_{T-S}, T_g, $p(O_2)$, f_r and e_t correspond to taget-to-substrate distance, growth temperature, oxygen pressure during growth and energy density of the laser radiation incident to the target, respectively

Author	d_{T-S} (mm)	T_g (°C)	$p(O_2)$ (mbar)	f_r (Hz)	e_t (J/cm^{-2})	Reference
Orita et al.	25	325–490	5×10^{-8}	1	3.5	[46]
Ou et al.	50	400–1000	6.66	10	2.4	[51]
Müller et al.	100	400–650	3×10^{-4}–0.024	15	2	[52]
Zhang et al.	30	200–600	1×10^{-3}	1		[53]
Yu et al.	50	400–1000	2.66	10	2.4	[54]
Zhang et al.	40	RT + anneal	1×10^{-3}	1		[55]
Zhang et al.	40	290–500				[56]

are obtained if the repetition rate was increased to 10 Hz [46]. The oxygen pressure influences the minimal temperature necessary for epitaxial growth $T_{g,min}^{epi}$ as well. For an oxygen pressure of 1×10^{-3} mbar Zhang et al. obtained crystalline thin films only if $T_g \geq 500\,°C$ [53] for a target-to-substrate distance $d_{T-S} = 30$ mm being similar to the 25 mm used by Orita et al. (more information on chamber geometry and typical growth parameters used for Ga_2O_3 heteroepitaxy are summarized in Table 15.3). The influence of the oxygen pressure on crystalline properties was also investigated by Müller et al. for various growth temperatures [52]. Similar to the elevation $T_{g,min}^{epi}$ with $p(O_2)$, the crystalline quality of epitaxial layers is lowered for increasing $p(O_2)$ at a given growth temperature. For instance, for $T_g = 650\,°C$, polycrystalline layers are obtained for $p(O_2) \geq 2 \times 10^{-3}$ mbar, while growth in $[\bar{2}01]$-direction only is observed elsewise. Generally speaking, crystallinity of heteroepitaxial β-Ga_2O_3 thin films is best for growth at low oxygen pressure and high temperature, which are conflicting growth conditions as outlined below.

The growth rate and stoichiometry of Ga_2O_3 depend strongly on the deposition parameters. Matsuzaki et al. reported in 2006 that for oxygen pressures below 1×10^{-6} mbar, oxygen-deficient $Ga_2O_{3-\delta}$ thin films are obtained independent of the growth temperature examined [57]. Such samples are electrically conducting

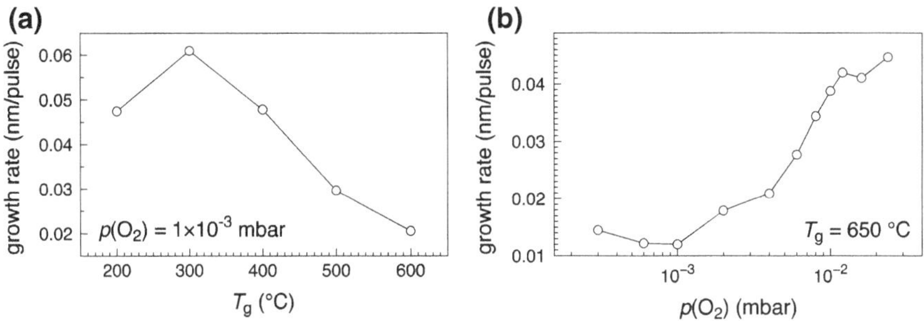

Fig. 15.3 Dependence of growth rate of PLD β-Ga_2O_3 thin films on **a** temperature and **b** oxygen growth pressure. In **a** crystalline thin films in monoclinic phase are obtained for $T_g \geq 500\,°C$; otherwise samples are X-ray amorphous. The data was adopted from **a** Zhang et al. [53] and **b** Müller et al. [52]

but appear blackish. For $T_g > 450\,°C$ and $p(O_2) \geq 1 \times 10^{-6}$ mbar, crystalline and transparent layers are obtained. For $p(O_2) \leq 1 \times 10^{-4}$ mbar and growth temperatures above 600 °C, thin film growth did not occur [57]. In Fig. 15.3a, the growth rate of Ga_2O_3 on (00.1)Al_2O_3 is depicted for a 100 times higher oxygen growth pressure $p(O_2) = 1 \times 10^{-3}$ mbar in dependence on the growth temperature T_g using data from Zhang et al. [53]. As stated above, for higher $p(O_2)$ $T_{g,min}^{epi}$ is higher, such that crystalline layers were only obtained for $T_g \geq 500\,°C$. The growth rates were calculated from thin film thickness and total number of laser pulses [53]. The highest growth rate was achieved for $T_g = 300\,°C$. An increase in growth temperature leads to an inversely proportional decrease of growth rate [53]. The dependence of the growth rate on the oxygen growth pressure is visualized in Fig. 15.3b for a growth temperature of 650 °C [52]. Lowest growth rates are obtained for low oxygen pressures. For $p(O_2) > 1 \times 10^{-3}$ mbar, a systematic increase in the growth rate is observed. Both effects can be explained by insufficient oxidation of Ga leading to the desorption of volatile gallium sub-oxides. This effect was studied and modeled in detail for thin film growth by molecular beam epitaxy [58]. For PLD growth, there exists for each oxygen pressure a corresponding maximal growth temperature for which the incident flux of Ga atoms is fully incorporated in the growing thin film. Higher growth temperatures lead to a decrease in growth rate due to sub-oxide desorption. Vice versa, for each growth temperature exists a minimal oxygen pressure for which a full incorporation of the incident Ga flux occurs; lower oxygen pressures lead to a decrease in growth rate. The growth window for a stoichiometric transfer of the target constituents to the thin film can be altered by using a reactive oxygen ambient instead of molecular oxygen as illustrated for $(Al, Ga)_2O_3$ below.

Growth of ternary alloys impacts the growth rate compared to binary β-Ga_2O_3 thin films. Von Wenckstern et al. reported the growth rate of $(In_xGa_{1-x})_2O_3$ thin film as a function of the alloy composition for $T_g = 650\,°C$ and $p(O_2) = 3 \times 10^{-4}$ mbar [34]. Within the monoclinic modification, the growth rate roughly doubles if the indium content increases from $x = 0$ to 0.2. Wang et al. reported a decrease

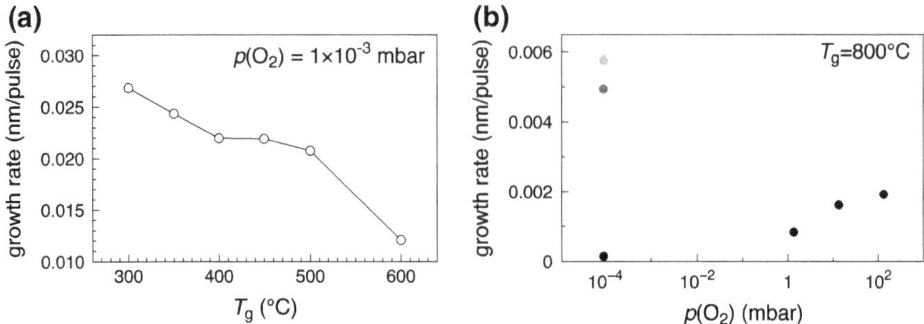

Fig. 15.4 Dependence of growth rate of PLD β-(Al, Ga)$_2$O$_3$ thin films on **a** temperature and **b** oxygen growth pressure. The data was adopted from **a** Wang et al. [60] and **b** Wakabayashi et al. [59]. The thin films were deposited on **a** (00.1)Al$_2$O$_3$ and **b** (010)Ga$_2$O$_3$. The Al content in the target was **a** 0.17 and **b** 0.05. In **b** black and blue markers correspond to growth in oxygen ambient, sample depicted by blue marker was grown at room temperature (RT). Plasma-assisted growth (providing oxygen radicals O*) at 800 °C was used for the sample depicted by red marker

in growth rate with increasing growth temperature for (Al, Ga)$_2$O$_3$ (Al content x in target was about 0.17) similar to β-Ga$_2$O$_3$ layers (cf. Fig. 15.3a) deposited using identical growth parameters as Zhang et al. [53]. For $T_g = 600\,°C$, the growth rate is about a factor of two lower for (Al, Ga)$_2$O$_3$ compared to binary thin films. The dependence of the growth rate on temperature is depicted in Fig. 15.4a. Wakabayashi et al. investigated the influence of radical oxygen species (O*) on the growth of (Al, Ga)$_2$O$_3$ thin films on (010)Ga$_2$O$_3$ substrates [59]. The growth rate at 800 °C is depicted as a function of the oxygen pressure for the case without O*. Similarly to binary gallium oxide, the growth rate increases with increasing oxygen pressure. The standard growth rate was determined for a thin film grown at room temperature and $p(O_2) = 9.3 \times 10^{-5}$ mbar to be about a factor of 40 higher than for the sample grown at 800°C and the same oxygen pressure. Growth at 800 °C and $p(O_2) = 9.3 \times 10^{-5}$ mbar with (without) O* leads to Al contents of 0.06, corresponding to the target stoichiometry (0.86, a factor of 17 higher than in the target). Further, for plasma-assisted growth the growth rate is about 85% of the standard growth rate. Both facts indicate clearly that the desorption of gallium sub-oxides is effectively suppressed for growth with O* [59]. In addition, plasma-assisted growth yields much lower surface roughness compared to growth without O* [59].

15.4.2.2 Growth Temperature and Oxygen Pressure Dependence of Surface Roughness

The surface roughness and grain size of PLD thin films on c-plane sapphire was investigated in dependence on the growth temperature by Ou et al. using an oxygen pressure of 6.66 mbar [51]. The data, depicted in Fig. 15.5a, shows that the increasing grain size for increasing growth temperature is connected to an increase of the root

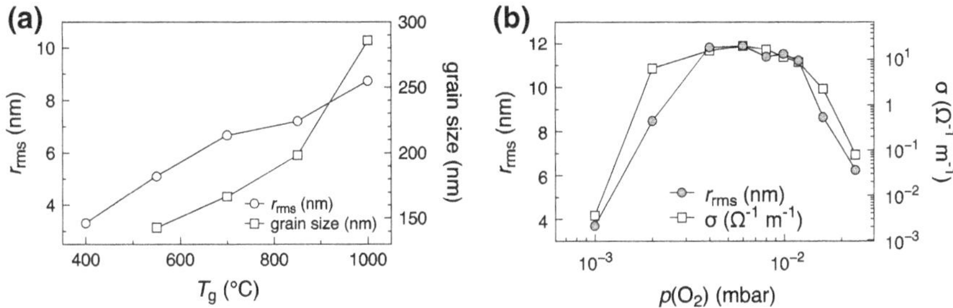

Fig. 15.5 **a** Dependence of root mean square surface roughness r_{rms} and grain size on growth temperature for an oxygen pressure of 6.66 mbar. The data was adapted from Ou et al. [51]. **b** Dependence of r_{rms} and the electrical conductivity on oxygen pressure for a growth temperature of 650 °C. The data was adopted from Müller et al. [52]

mean square surface roughness r_{rms}. For $T_g < T_{g,min}^{epi}$, a step-like decrease in the r_{rms} from about 6 nm and 3.5 nm to values below 1 nm was reported for binary Ga_2O_3 [53] and for ternary $(Al, Ga)_2O_3$ thin films [60], respectively. The surface roughness is also dependent on the oxygen pressure during growth as shown exemplarily for a growth temperature of 650 °C in Fig. 15.5b. For low oxygen pressure, the thin films have $(\bar{2}01)$ orientation and lowest r_{rms}. With increasing $p(O_2)$, the grain size increases and additional crystallographic orientations appear, which results in an increase of r_{rms} to values of about 12 nm for $p(O_2) = 6 \times 10^{-3}$ mbar [52]. A further increase in $p(O_2)$ induces a change from circularly shaped crystallites to more elongated shapes which is connected to a decrease of r_{rms}.

15.4.3 *Doping and Alloying of $(In, Ga, Al)_2O_3$ PLD Thin Films*

15.4.3.1 Room Temperature Electric Transport Properties of Binary Layers

For n-type doping of β-Ga_2O_3, the group IV elements Si and Ge, being preferentially incorporated at tetrahedrally coordinated cation sites, and transition metals such as Sn or Zr, being preferentially incorporated at octahedrally coordinated cation sites, are suited. High doping levels of tin will, however, facilitate growth in the orthorhombic κ-Ga_2O_3 phase [46]. Hence, for obtaining highly conductive thin films doping by the group IV element Si is most commonly reported. In Fig. 15.5b, the logarithm of the electrical conductivity of β-Ga_2O_3:Si thin films follows the trend discussed for the surface roughness; highest conductivity is observed for samples with largest grain size and with that highest r_{rms} values. Further, the Si amount incorporated in the thin films depends on $p(O_2)$ [52]. The highest Si cation share of 3.5% was determined

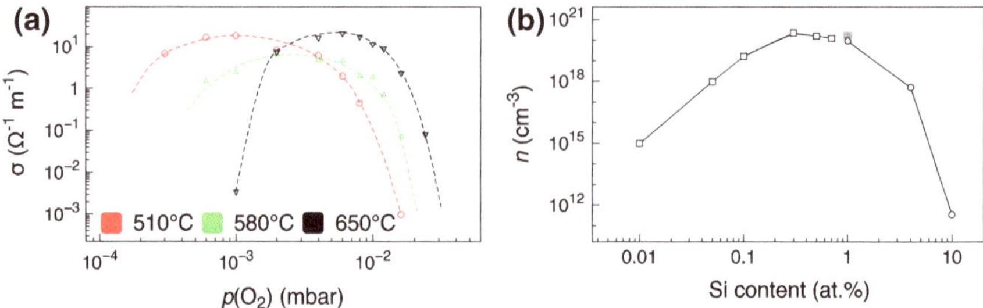

Fig. 15.6 **a** Dependence of electrical conductivity on oxygen pressures for β-Ga_2O_3:Si thin films on c-plane sapphire for growth temperatures as labeled. The data was adopted from Müller et al. [52]. **b** Dependence of the free electron concentration at room temperature on the Si doping level for β-Ga_2O_3:Si thin film on c-plane sapphire. The gray marker indicates the maximal free electron concentration reported for homoepitaxial (010)Ga_2O_3 thin films. The Data was adopted from Zhang et al. [62] (squares), Zhang et al. [63] (circles) and Leedy et al. [64] (gray marker)

for $p(O_2) = 1 \times 10^{-3}$ mbar, for higher $p(O_2)$ it decreases systematically to about 1.25%.

The growth pressure dependence of the electrical conductivity σ was investigated by Müller et al. for various growth temperatures as depicted in Fig. 15.6. In general, similar dependence namely increase of σ with increasing $p(O_2)$ to a maximal value σ_{max} and then a decrease with further increase in $p(O_2)$ was observed for all T_g investigated. The pressure $p(\sigma_{max})$, for which the highest conductivity occurred, decreases with increasing growth temperature. For all samples deposited at the $p(\sigma_{max})$ for the three growth temperatures considered (510, 580 and 650 °C), the structural properties as well as the surface morphology were alike [61]. There exist a combination of structural and morphological properties (large grain size resulting in comparable high surface roughness) which are beneficial for obtaining highly conducting β-Ga_2O_3:Si thin films on c-plane sapphire. These structural and morphological properties are realized at lower oxygen pressure if the growth temperature is reduced.

The doping efficiency for Ga_2O_3:Si thin films on (00.1)Al_2O_3 is depicted in Fig. 15.6b for a growth temperature of 500 °C and an oxygen pressure of 1×10^{-3} mbar. For these growth conditions, the Si content in the PLD target is transferred stoichiometrically to the epilayer [62]. The free electron concentration increases from about 1×10^{15} cm^{-3} for a doping level of 0.01 at.% to a maximum value of about 1×10^{20} cm^{-3} for a doping level of 0.3 at.%. Higher Si doping levels will not result in higher free electron density. For doping levels above 1 at.% the free carrier density decreases strongly which was attributed to lower crystalline quality and compensation by Ga vacancies [63]. Due to these compensating mechanisms, the doping efficiency is about 22% for a doping level of 1.1 at.%. The gray marker in Fig. 15.6b indicates the highest free electron density obtained for homoepitaxial (010)Ga_2O_3:Si PLD thin films. The layers were deposited at 550 °C and working pressure of 0.13 mbar (95%Ar/5%O_2) [64]. The doping efficiency of Si in these homoepitaxial layers is about 50%. The free electron density of control samples,

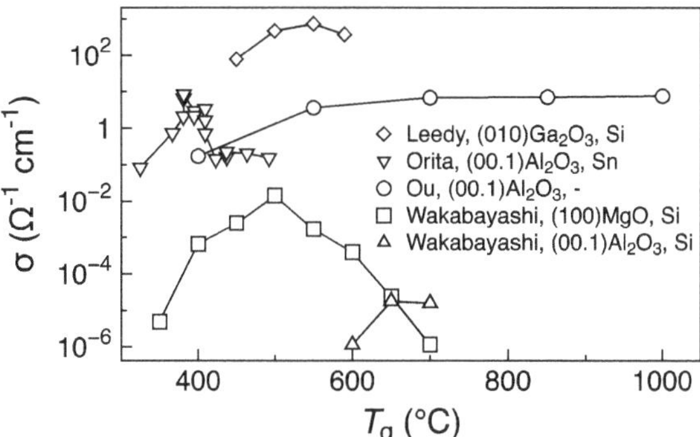

Fig. 15.7 Dependence of room temperature conductivity on growth temperature for Ga_2O_3 thin films grown on substrates with n-type dopant as labeled. The data was compiled from Leedy et al. [64], Orita et al. [46], Ou et al. [51], and Wakabyashi et al. [65]. We note, that the thin films on (100)MgO contain domains of γ-Ga_2O_3

deposited on (00.1)Al_2O_3, was about a factor of 10 lower than for homoepitaxial films. Besides the higher n-values the free carrier mobility of 26.5 cm^2/Vs within the homoepitaxial samples was about 70 times higher than in the control samples [64]. Therefore, the conductivity of such homoepitaxial layers is considerably higher than that of heteroepitaxial thin films. This is illustrated in Fig. 15.7. Here, room temperature conductivity values of β-Ga_2O_3 thin film deposited on various substrates are plotted versus the growth temperature. Highest conductivity of 732 S/cm was achieved for homoepitaxial (010)Ga_2O_3:Si thin films grown at 550°C [64]. Among heteroepitaxial PLD layers, the maximal conductivity of 8.2 S/cm was reported for Sn-doped thin films on (00.1)Al_2O_3 for T_g = 380 °C [46]. For T_g > 410 °C crystallization in the orthorhombic phase occurred. Highest conductivity of heteroepitaxial Ga_2O_3:Si is 2 S/cm and was obtained for T_g = 500 °C and 0.001 mbar by Zhang et al. [63]. Ou et al. reported very high conductivity for nominally undoped thin films. Here, values above 7 S/cm are measured for $T_g \geq$ 700 °C [51]. These exceptionally high conductivities were achieved for $p(O_2)$ = 6.66 mbar and d_{T-S} = 50 mm. Wakabyashi et al. observed a strong correlation between crystalline and electrical properties of Ga_2O_3:Si thin films on (100)MgO. The crystallinity as well as the conductivity increase with increasing T_g. For T_g = 500 °C, highest crystallinity is obtained, and the electrical conductivity reaches a maximal value of 1.4×10^{-2} S/cm (cf. Fig. 15.7). A further increase in T_g induces the formation of twin domains lowering the conductivity [65]. Moreover, the thin films contained γ-Ga_2O_3 as a side phase reducing the overall conductivity. Control samples deposited on c-plane sapphire were amorphous and electrically insulating for T_g < 500 °C. The conductivity of layers deposited at higher temperature, included in Fig. 15.7, is much lower than that of the layers on (100)MgO, which was attributed to the inferior crystalline quality [65].

15.4.3.2 Properties of Ternary $(In, Ga)_2O_3$ and $(Al, Ga)_2O_3$

In_2O_3, Ga_2O_3 and Al_2O_3 crystallize in different crystal structures under ambient conditions [49], and hence, phase separation is expected and will set limits for band gap engineering. For In_2O_3 (Al_2O_3), the cubic bixbyite (rhombohedral corundum) structure is most stable under standard conditions. The solubility limit of In_2O_3 (Al_2O_3) in β-Ga_2O_3 is about 43% [66] (78% [67]).

Highest indium content of as-grown monoclinic $(In_xGa_{1-x})_2O_3$ on c-plane sapphire and (100)MgO substrate material was obtained for thin films with lateral composition spread (CCS) to be 0.2 and 0.3, respectively [33] for a growth temperature of 650 °C. Higher amounts of In may induce the formation of hexagonal $InGaO_3$ II [33–35] belonging to the non-polar space group $P6_3/mmc$. So far this phase was detected only in thin films with continuous variation of the cation. This is most likely due to the comparatively large variation of alloy composition Δx between neighboring samples of an alloy sample series (which can be considered as "discrete composition library"). Phase pure thin films with cubic bixbyite phase are typically obtained for $x > 0.5$ [33, 68]. Zhang *et al.* observed segregation of In_2O_3 if crystalline thin films ($T_g \geq 300\,°C$) are grown on c-plane sapphire using a target with $x = 0.3$ [69]. These In_2O_3 clusters remain after annealing in air at 800 °C. Amorphous layers ($T_g < 300\,°C, x = 0.3$) have a homogeneous In distribution and result in monoclinic $(In, Ga)_2O_3$ after annealing at 800 °C in oxygen ambient [55]. Annealing in vacuum, argon or nitrogen did not induce crystallization.

The dependence of the absorption edge energy $E_{g,abs}$ on the alloy composition was determined from transmission measurements on a CCS-PLD thin film on c-plane sapphire [34]. Within the monoclinic phase the band gap decreases linearly according to $E_{g,abs} = (4.90 - 2.42 \times x)\,eV$, which agrees well to a result obtained by fitting literature data compiled from thin films obtained by various growth methods for $x \leq 0.3$: $E_{g,abs} = (4.965 - 2.9 \times x)\,eV$ [49].

The change of phonon mode frequencies in the ternary alloys was investigated by Kranert et al. [33, 37] and is depicted for the low frequency $A_g^{(3)}$- and the high frequency $A_g^{(9)}$-mode in Fig. 15.8. For comparison, data obtained from bulk ceramic samples are included. A strictly linear compositional dependence is observed, independent of the sample considered, for the $A_g^{(9)}$-mode and high-frequency modes in general, which enables non-destructive determination of the alloy composition [33]. Low-frequency phonon modes as, e.g., the $A_g^{(3)}$-mode are sensitive to lattice distortions such that deviations from a linear x-dependence and from the data obtained from the bulk ceramic samples are also reflected in a nonlinear behavior of the dependence of the lattice constants on alloy composition [33, 37].

Positron annihilation spectroscopy on CCS-PLD $(In_xGa_{1-x})_2O_3$ thin films revealed an increasing cation vacancy concentration with increasing x for the monoclinic polymorph [36]. For low indium content, the cation vacancy concentration is in the mid $10^{16}\,cm^{-3}$. Independent of the In content in monoclinic thin films, growth on (100)MgO results in lower cation vacancy concentration than growth on $(00.1)Al_2O_3$. For cubic $(In_xGa_{1-x})_2O_3$, the cation vacancy concentration is in the $10^{17}\,cm^{-3}$ range

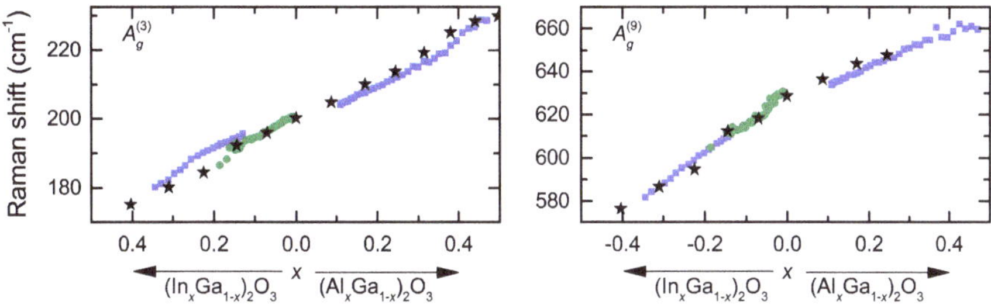

Fig. 15.8 Dependence of $A_g^{(3)}$ and $A_g^{(9)}$ Raman mode on alloy composition for monoclinic $(In_xGa_{1-x})_2O_3$ and $(Al_xGa_{1-x})_2O_3$. Data includes measurements on PLD thin films on c-Al_2O_3 (blue marker), and (100)MgO (green marker) and ceramic bulk samples (black marker) and is adopted from [33, 37]. The figure was created by Christan Kranert

[36], reflecting higher free carrier density and with that higher energy of the Fermi level with respect to the conduction band minimum.

First demonstrator devices on $(In_xGa_{1-x})_2O_3$ thin films include Schottky barrier diodes [34, 35], photoconductors [70] and photodetectors [34, 54, 71, 72]. Schottky barrier diodes (SBD) with high rectification ratio (ratio of forward and absolute reverse current at, e.g., ± 2 V) require extrinsic doping for reasonable low values of the series resistance [34]. This is especially important for low indium contents. Properties of reactively sputtered Pt contacts on a $(In_xGa_{1-x})_2O_3$ CCS-PLD thin film have been reported for $0.01 \leq x \leq 0.85$ [35]. Highest rectification was observed within the monoclinic structure and for low indium content.

Zhang et al. reported properties of metal-semiconductor-metal photodetectors (MSM-PDs) fabricated on a 2 in. in diameter $(In_xGa_{1-x})_2O_3$ thin film grown by CCS-PLD for a wide range of alloy compositions ($0 < x < 0.83$) [71]. The thin film contains regions with phase pure monoclinic structure and phase pure bixbyite structure. In between areas with mixed crystallographic phases are contained and include a region, in which the monoclinic, the bixbyite and the hexagonal $InGaO_3$ II phase coexists and a region that exhibits monoclinic and the hexagonal $InGaO_3$ II phase. Howsoever, MSM-PDs with high photoresponse were obtained on the entire sample independent of the crystallinity. For $x > 0.2$, the detectors exhibit gain which was traced back to a lowering of the Schottky barrier height due to the accumulation of holes within the space charge region. The energy of absorption onset was found to decrease linearly with increasing In content. These MSM-PDs covered the UVA to UVC spectral range from about 3.2–4.85 eV.

The highest aluminum content in as-grown β-$(Al_xGa_{1-x})_2O_3$ is about 0.5 for PLD thin films on c-plane sapphire [73, 74] and a CCS-PLD thin film on (100)MgO [37]. The a-lattice constant of the CCS-PLD thin film on (100)MgO increases with x for $x \leq 0.22$. This is in contradiction to results on bulk ceramic samples, for which a linear decrease of the lattice constants is observed [37]. For $x > 0.22$ the a-lattice constant decreases; the relative change observed was similar to that of bulk $(Al_xGa_{1-x})_2O_3$. Regions with Al content above 50% exhibited the spinel γ-

$(Al_xGa_{1-x})_2O_3$ phase [37]. Similar to $(In,Ga)_2O_3$ higher frequency phonon modes show a linear variation with increasing Al incorporation (shift to higher frequency) and can be used for a non-destructive determination of the Al content; low-frequency modes are strain sensitive [37]. The refractive index decreases linearly with increasing Al content [38].

The growth rate as well as the stoichiometry of heteroepitaxial thin films strongly depend on T_g and $p(O_2)$. The Al–O bond is considerably stronger than the Ga-O bond, therefore high $p(O_2)$ is required for stoichiometric incorporation of Ga. Further, the desorption rate of volatile gallium sub-oxides increases with T_g. Altogether, a stoichiometric transfer of the target composition only occurs for low T_g and high $p(O_2)$. Plasma-assisted growth strongly increases this range toward higher T_g and lower $p(O_2)$ [65]. The temperature dependence of the growth rate was already illustrated in Fig. 15.4a for $p(O_2) = 1 \times 10^{-3}$ mbar and an aluminum content of 17% in the target. The Al content in the thin films increases with increasing growth temperature, and hence, the band gap energy increases with T_g [60].

A problem remaining to be solved is fabrication of electrically conductive (β-$(Al_xGa_{1-x})_2O_3$ thin films by PLD. So far, an effective doping strategy does not exist. For modulation doping within heterostructures, it is necessary to incorporate electrically active shallow donors in the material with higher band gap. This is possible for β-Ga_2O_3/$(In_xGa_{1-x})_2O_3$ but not for β-$(Al_xGa_{1-x})_2O_3$/Ga_2O_3 heterostructures grown by pulsed laser deposition.

15.5 Summary and Outlook

Pulsed laser deposition is a well-suited method for growth of heteroepitaxial as well as homoepitaxial β-Ga_2O_3 thin films. We have elucidated fundamental relations between important growth parameters and growth rate and their impact on thin film properties such as crystallinity, surface morphology or electrical conductivity. Highest growth rates are obtained, if the desorption of gallium sub-oxides is prevented, for which (i) low growth temperatures and (ii) high oxygen pressures are beneficial. Further, growth using (iii) an oxygen plasma source is advantageous for increasing the range of these growth parameters. In general, the crystallization temperature is lowest for (i) low laser repetition rates and (ii) low oxygen background pressure. The surface roughness of heteroepitaxial films increases with growth temperature due to larger grain sizes. Further, a correlation between surface roughness and with that grain size and the electrical conductivity was revealed and traced back to the influence of the oxygen pressure and growth temperature on sample crystallinity. Silicon is a well-suited shallow donor for PLD growth; the highest free electron density is obtained for a Si doping level of about 0.3 at.% resulting in free electron density of 1×10^{20} cm^{-3}. For exploring the properties of the ternary alloys β-$(In_xGa_{1-x})_2O_3$ and β-$(Al_xGa_{1-x})_2O_3$, CCS-PLD has proven to be very efficient. In principle, entire phase diagrams can be realized on single wafers allowing mapping of material properties with unprecedented resolution with respect to chemical composition. Using high-throughput methods, structural, vibrational, optical and electrical properties have been determined and key results were summarized in this chapter.

Acknowledgements This work was financially supported by the European Social Fund within the Young Investigator Group "Oxide Heterostructures" (SAB 100310460) and partly by Deutsche Forschungsgemeinschaft in the Framework of Sonderforschungsbereich 762 "Functionality of Oxide Interfaces."

References

1. A. Einstein, Phys. Z. **18**, 121 (1917)
2. F. Breech, L. Cross, Appl. Spectrosc. **16**, 59 (1962)
3. H.M. Smith, A.F. Turner, Appl. Opt. **4**(1), 147 (1965)
4. J.T. Cheung, Appl. Phys. Lett. **43**(3), 255 (1983)
5. J.P. Gordon, H.J. Zeiger, C.H. Townes, Phys. Rev. **95**(1), 282 (1954)
6. A.L. Schawlow, C.H. Townes, Phys. Rev. **112**(6), 1940 (1958)
7. T.H. Maiman, Nature **187**(4736), 493 (1960)
8. D. Dijkkamp, T. Venkatesan, X.D. Wu, S.A. Shaheen, N. Jisrawi, Y.H.M. Lee, W.L. McLean, M. Croft, Appl. Phys. Lett. **51**(8), 619 (1987)
9. J.A. Greer, H. Jerrold Van Hook, MRS Proc. **169**, 463 (1989)
10. J.A. Greer, J. Vac. Sci. Technol. A **10**(4), 1821 (1992)
11. J.A. Greer, M.D. Tabat, J. Vac. Sci. Technol. A **13**(3), 1175 (1995)
12. M.D. Strikovsky, E.B. Klyuenkov, S.V. Gaponov, J. Schubert, C.A. Copetti, Appl. Phys. Lett. **63**(8), 1146 (1993)
13. K. Kinoshita, H. Ishibashi, T. Kobayashi, Jpn. J. Appl. Phys. **33**(3), L417 (1994)
14. X.D. Xiang, Mater. Sci. Eng. B **56**(2–3), 246 (1998)
15. A.V. Rode, B. Luther-Davies, E.G. Gamaly, J. Appl. Phys. **85**(8), 4222 (1999)
16. T. Fukumura, M. Ohtani, M. Kawasaki, Y. Okimoto, T. Kageyama, T. Koida, T. Hasegawa, Y. Tokura, H. Koinuma, Appl. Phys. Lett. **77**(2), 3426 (2000)
17. A. Jacquot, M.O. Boffoué, B. Lenoir, A. Dauscher, Appl. Surf. Sci. **156**(1–4), 169 (2000)
18. H.M. Christen, C.M. Rouleau, I. Ohkubo, H.Y. Zhai, H.N. Lee, S. Sathyamurthy, D.H. Lowndes, Rev. Sci. Instrum. **74**(9), 4058 (2003)
19. T. Nobis, E.M. Kaidashev, A. Rahm, M. Lorenz, M. Grundmann, Phys. Rev. Lett. **93**, 103903 (2004)
20. T. Nobis, E.M. Kaidashev, A. Rahm, M. Lorenz, J. Lenzner, M. Grundmann, Nano Lett. **4**(5), 797 (2004)
21. A. Tsukazaki, A. Ohtomo, T. Kita, Y. Ohno, H. Ohno, M. Kawasaki, Science **315**(5817), 1388 (2007)
22. H. von Wenckstern, Z. Zhang, F. Schmidt, J. Lenzner, H. Hochmuth, M. Grundmann, Crystengcomm **15**(46), 10020 (2013)
23. S.S. Mao, X. Zhang, Engineering **1**(3), 367 (2015)
24. A. Inam, M.S. Hegde, X.D. Wu, T. Venkatesan, P. England, P.F. Miceli, E.W. Chase, C.C. Chang, J.M. Tarascon, J.B. Wachtman, Appl. Phys. Lett. **53**(10), 908 (1988)
25. J. Greer, in *Pulsed Laser Deposition of Thin Films* (Wiley Inc, Hoboken, NJ, USA, 2006), pp. 191–213
26. J.A. Greer, J. Phys. D: Appl. Phys. **47**(3), 034005 (2013)
27. I. Takeuchi, W. Yang, K.S. Chang, M.A. Aronova, T. Venkatesan, R.D. Vispute, L.A. Bendersky, J. Appl. Phys. **94**(11), 7336 (2003)
28. H.M. Christen, S.D. Silliman, K.S. Harshavardhan, Rev. Sci. Instrum. **72**(6), 2673 (2001)

29. Z. Zhang, H. von Wenckstern, M. Grundmann, IEEE, J. Sel. Top. Quantum Electron. **20**(6), 106 (2014)
30. A. Mavlonov, S. Richter, H. von Wenckstern, R. Schmidt Grund, J. Lenzner, M. Lorenz, M. Grundmann, Phys. Status Solidi A **212**(12), 2850 (2015)
31. Z. Zhang, H. von Wenckstern, J. Lenzner, M. Grundmann, Appl. Phys. Lett. **108**(24), 243503 (2016)
32. S. Bitter, P. Schlupp, M. Bonholzer, H. von Wenckstern, M. Grundmann, A.C.S. Comb, Sci. **18**(4), 188 (2016)
33. C. Kranert, J. Lenzner, M. Jenderka, M. Lorenz, H. von Wenckstern, R. Schmidt-Grund, M. Grundmann, J. Appl. Phys. **116**(1), 013505 (2014)
34. H. von Wenckstern, D. Splith, M. Purfürst, Z. Zhang, C. Kranert, S. Müller, M. Lorenz, M. Grundmann, Semicond. Sci. Technol. **30**(2), 024005 (2015)
35. H. von Wenckstern, D. Splith, A. Werner, S. Müller, M. Lorenz, M. Grundmann, A.C.S. Comb, Sci. **17**(12), 710 (2015)
36. V. Prozheeva, R. Hölldobler, H. von Wenckstern, M. Grundmann, F. Tuomisto, J. Appl. Phys. **123**(12), 125705 (2018)
37. C. Kranert, M. Jenderka, J. Lenzner, M. Lorenz, H. von Wenckstern, R. Schmidt-Grund, M. Grundmann, J. Appl. Phys. **117**(12), 125703 (2015)
38. R. Schmidt-Grund, C. Kranert, H. von Wenckstern, V. Zviagin, M. Lorenz, M. Grundmann, J. Appl. Phys. **117**(16), 165307 (2015)
39. M. Lorenz, S. Hohenberger, E. Rose, M. Grundmann, Atomically stepped, pseudomorphic, corundum-phase $(Al_{1-x}Ga_x)_2O_3$ thin films ($0 \leq x \leq 0.08$) grown on r-plane sapphire (2018)
40. M. Grundmann, J. Zuniga-Pérez, Phys. Status Solidi B **253**, 351 (2016)
41. M. Grundmann, J. Appl. Phys. **124**(18), 185302 (2018)
42. R. Kumaran, T. Tiedje, S.E. Webster, S. Penson, W. Li, Opt. Lett. **35**, 3793 (2010)
43. R.J. Cava, J.M. Phillips, J. Kwo, G.A. Thomas, R.B. van Dover, S.A. Carter, J.J. Krajewski, W.F. Peck, J.H. Marshall, D.H. Rapkine, Appl. Phys. Lett. **64**(16), 2071 (1994)
44. J.M. Phillips, J. Kwo, G.A. Thomas, S.A. Carter, R.J. Cava, S.Y. Hou, J.J. Krajewski, J.H. Marshall, W.F. Peck, D.H. Rapkine, R.B.v. Dover, Appl. Phys. Lett. **65**(1), 115 (1994)
45. M. Orita, H. Ohta, M. Hirano, H. Hosono, Appl. Phys. Lett. **77**(25), 4166 (2000)
46. M. Orita, H. Hiramatsu, H. Ohta, M. Hirano, H. Hosono, Thin Solid Films **411**(1), 134 (2002)
47. M. Grundmann, T. Böntgen, M. Lorenz, Phys. Rev. Lett. **105**, 146102 (2010)
48. M. Grundmann, Phys. Status Solidi B **248**(4), 805 (2011)
49. H. von Wenckstern, Adv. Electron. Mater. **3**(9), 1600350 (2017)
50. S.A. Lee, J.Y. Hwang, J.P. Kim, S.Y. Jeong, C.R. Cho, Appl. Phys. Lett. **89**(18), 182906 (2006)
51. S.L. Ou, D.S. Wuu, Y.C. Fu, S.P. Liu, R.H. Horng, L. Liu, Z.C. Feng, Mater. Chem. Phys. **133**(2–3), 700 (2012)
52. S. Müller, H. von Wenckstern, D. Splith, F. Schmidt, M. Grundmann, Phys. Status Solidi A **211**(1), 34 (2014)
53. F.B. Zhang, K. Saito, T. Tanaka, M. Nishio, Q.X. Guo, J. Cryst. Growth **387**, 96 (2014)
54. F.P. Yu, S.L. Ou, D.S. Wuu, Opt. Mater. Express **5**(5), 1240 (2015)
55. F. Zhang, H. Jan, K. Saito, T. Tanaka, M. Nishio, T. Nagaoka, M. Arita, Q. Guo, Thin Solid Films **578**, 1 (2015)
56. F. Zhang, H. Li, Y.T. Cui, G.L. Li, Q. Guo, AIP Adv. **8**(4), 045112 (2018)
57. K. Matsuzaki, H. Hiramatsu, K. Nomura, H. Yanagi, T. Kamiya, M. Hirano, H. Hosono, Thin Solid Films **496**(1), 37 (2006)
58. P. Vogt, O. Bierwagen, Appl. Phys. Lett. **108**(7), 072101 (2016)
59. R. Wakabayashi, T. Oshima, M. Hattori, K. Sasaki, T. Masui, A. Kuramata, S. Yamakoshi, K. Yoshimatsu, A. Ohtomo, J. Cryst. Growth **424**, 77 (2015)
60. X. Wang, Z. Chen, F. Zhang, K. Saito, T. Tanaka, M. Nishio, Q. Guo, Ceram. Int. **42**(11), 12783 (2016)
61. S. Müller, Schottky-Kontakte auf Zinkoxid- und β-Galliumoxid-Dünnfilmen: Barrierenformation, elektrische Eigenschaften und Temperaturstabilität. Ph.D. thesis, Universität Leipzig, Leipzig (2016). http://nbn-resolving.de/urn:nbn:de:bsz:15-qucosa-206386

62. F. Zhang, M. Arita, X. Wang, Z. Chen, K. Saito, T. Tanaka, M. Nishio, T. Motooka, Q. Guo, Appl. Phys. Lett. **109**(10), 102105 (2016)
63. F. Zhang, K. Saito, T. Tanaka, M. Nishio, Q. Guo, J. Mater. Sci.: Mater. Electron. **26**(12), 9624 (2015)
64. K.D. Leedy, K.D. Chabak, V. Vasilyev, D.C. Look, J.J. Boeckl, J.L. Brown, S.E. Tetlak, A.J. Green, N.A. Moser, A. Crespo, D.B. Thomson, R.C. Fitch, J.P. McCandless, G.H. Jessen, Appl. Phys. Lett. **111**(1), 012103 (2017)
65. R. Wakabayashi, K. Yoshimatsu, M. Hattori, A. Ohtomo, Appl. Phys. Lett. **111**(16), 162101 (2017)
66. D.D. Edwards, T.O. Mason, J. Am. Ceram. Soc. **81**(12), 3285 (1998)
67. A.L. Jaromin, D.D. Edwards, J. Am. Ceram. Soc. **88**(9), 2573 (2005)
68. F. Zhang, K. Saito, T. Tanaka, M. Nishio, Q. Guo, Solid State Commun. **186**, 28 (2014)
69. F. Zhang, K. Saito, T. Tanaka, M. Nishio, Q. Guo, J. Alloys Compd. **614**, 173 (2014)
70. F. Zhang, H. Li, M. Arita, Q. Guo, Opt. Mater. Express **7**(10), 3769 (2017)
71. Z. Zhang, H. von Wenckstern, J. Lenzner, M. Lorenz, M. Grundmann, Appl. Phys. Lett. **108**(12), 123503 (2016)
72. X.H. Chen, S. Han, Y.M. Lu, P.J. Cao, W.J. Liu, Y.X. Zeng, F. Jia, W.Y. Xu, X.K. Liu, D.L. Zhu, J. Alloys Compd. **747**, 869 (2018)
73. F. Zhang, K. Saito, T. Tanaka, M. Nishio, M. Arita, Q. Guo, Appl. Phys. Lett. **105**(16), 162107 (2014)
74. Z. Hu, Q. Feng, J. Zhang, F. Li, X. Li, Z. Feng, C. Zhang, Y. Hao, Superlattices Microstruct. **114**, 82 (2018)

Chapter 16
Low-Pressure Chemical Vapor Deposition

Hongping Zhao

Abstract This chapter reviews the growth and material properties of β-Ga_2O_3 grown via the low-pressure chemical vapor deposition (LPCVD) method. The growth of β-Ga_2O_3 thin films, with Si as an effective and controllable n-type dopant, on off-axis sapphire and native Ga_2O_3 substrates is discussed. LPCVD growth of β-Ga_2O_3 rod structures on 3C-SiC substrates is also discussed. From the crystal structural characterization and electron transport measurements, LPCVD-grown β-Ga_2O_3 materials exhibit high quality with great promises for high-power electronic and short-wavelength optoelectronic device applications.

16.1 Introduction

Ga_2O_3 (E_g ~4.5–4.9 eV) not only has a significantly larger gap than GaN (3.4 eV), ZnO (3.4 eV), and 4H-SiC (3.2 eV) but also n-type dopable and exhibits semiconducting rather than insulating properties. The monoclinic β-phase Ga_2O_3 represents the thermodynamically stable crystal among the known five phases (α, β, γ, δ, and ε). The breakdown field of β-Ga_2O_3 is estimated to be 6–8 MV/cm, which is about three times larger than that of 4H-SiC and GaN. More importantly, Ga_2O_3 native substrates have been commercialized by melting-based method, which represents a low-cost, scalable method used to produce commercial sapphire substrates. Thus, β-Ga_2O_3 is a promising material for high-voltage, high-temperature, and high-frequency electronic device and short-wavelength optoelectronic device applications.

The synthesis and fundamental understanding of β-Ga_2O_3 are still very much at an early stage. The exploration of native defects such as vacancies, interstitials, antisites, and impurities or intentional dopants is complex because three different

H. Zhao (✉)
205 Dreese Laboratory, Department of Electrical and Computer Engineering, Department of Materials Science and Engineering, The Ohio State University, 2015 Neil Ave., Columbus, OH 43210, USA
e-mail: zhao.2592@osu.edu

M. Higashiwaki and S. Fujita (eds.), *Gallium Oxide*, Springer Series in Materials Science 293, https://doi.org/10.1007/978-3-030-37153-1_16

oxygen sites and two different gallium sites are needed to be considered. At the forefront of efforts in the synthesis of β-Ga_2O_3 thin films, the traditional thin-film epitaxy methods such as molecular-beam epitaxy (MBE) [1] and metalorganic vapor-phase epitaxy (MOVPE) [2] are currently under active exploration. Halide vapor-phase epitaxy (HVPE) has also been used to grow β-Ga_2O_3 films with fast growth rates [3].

In this chapter, the growth of β-Ga_2O_3 using the low-pressure chemical vapor deposition (LPCVD) method will be reviewed. The use of low pressure instead of atmospheric pressure is based on the following reasons: (1) Low pressure growth can significantly suppress the precursor reaction at gas phase, which enables fast growth rates of Ga_2O_3; (2) low-pressure growth can reduce the adsorption of impurities on the grown surface at a lower growth temperature than is possible at atmospheric pressure, which has been previously used for silicon growth; (3) low-pressure growth can produce more uniform growth rate over the entire substrate; (4) low-pressure system features with fast source gas speed that is expected to provide more abrupt interfaces at heterojunctions.

For the LPCVD growth of β-Ga_2O_3, high purity gallium pellets (99.99999%) and pure oxygen are selected as precursors, and argon (Ar) is used as the carrier gas [4]. The Ga source is located at the upstream in a multi-temperature zone tube, and the Ga vapor is transferred to downstream at the substrate by the carrier gas with controlled temperature and pressure. Key growth parameters including growth temperature, growth pressure, oxygen flow rate, and substrate surface treatment all play important roles for the LPCVD growth of β-Ga_2O_3. Diluted $SiCl_4$ has been used as the precursor for n-type doping of β-Ga_2O_3 [5]. Recently, LPCVD growths of Si-doped β-Ga_2O_3 thin films with room temperature mobility of ~110 cm^2/Vs (β-Ga_2O_3 on off-axis c-sapphire substrate) and ~120 cm^2/Vs (β-Ga_2O_3 on (010) Ga_2O_3 substrate) were demonstrated. The growth rate can be widely tuned between <1 and >35 μm/h via the effective tuning of the growth condition.

16.2 LPCVD of β-Ga_2O_3 Thin Films on Off-Axis Sapphire Substrates

Small off-angled substrates are often utilized to improve the crystallinity of epitaxial thin films [6], because they can enhance step-flow growth of thin films and control domain structures. The atomic steps on the surface of the off-angled substrates act as the preferential binding sites for the incoming adatoms, promoting the step-flow growth. The same concept was applied to the growth of n-type Si-doped β-Ga_2O_3 thin films on c-plane sapphire substrates with different off-axis angles toward <11–20> direction by LPCVD [7]. It was found that the film crystalline quality, surface morphology, and electrical conductivity were remarkably sensitive to the off-axis angles. For thin film grown on substrate with 0° off-axis angle, there are no preferential sites on the substrate surface for Ga adatoms to incorporate into. On the other hand, when off-axis substrate is used, steps act as the preferred incorporation sites for the Ga adatoms. This suppresses the random nucleation on the surface, and step-flow growth occurs along the off-cut direction. Note that the surface diffusion length of Ga, which depends on the growth condition, needs to be comparable to the step-terrace width in order to sustain stable step morphology. The surface mobility of the adatoms can also be affected by the accommodation of the lattice mismatch between Ga_2O_3 and sapphire with the use of off-axis substrates.

The crystalline quality of the β-Ga_2O_3 films grown on off-axis c-sapphire substrates has been evaluated by different characterization techniques. From the XRD rocking curve measurements for the symmetric (−402) diffraction peak, the FWHM of the rocking curves is measured as 1.633° ($\Delta_a = 0°$), 0.485° ($\Delta_a = 3.5°$), and 0.47° ($\Delta_a = 6°$), respectively. The FWHM value decreased significantly with increasing off-axis angle, indicating that larger off-cut improves the crystal quality of the β-Ga_2O_3 thin films. However, as the off-axis angle was further increased to 8° and 10°, broadening of the XRD FWHM value was observed. Such broadening has occurred due to the step bunching, and it indicates degradation of the material quality of the thin films for off-axis angle larger than 6°.

X-ray phi-scan measurements were used to evaluate the in-plane rotational domains due to the substrate symmetry. For the thin film grown on 0° off-axis substrate, six strong diffraction peaks with similar intensity were measured (Fig. 16.1a). The peaks are separated by 60°. The origin of these six peaks is due to the diploid symmetry of monoclinic β-Ga_2O_3 and three kinds of equivalent in-plane orientations of Ga_2O_3 grains on the substrate. With the introduction of off-axis substrates, the domain structure changed dramatically. A 3.5° off-axis substrate promotes three orientations and suppresses the other three orientations as can be seen from Fig. 16.1b. Increasing off-cut angle from 3.5° to 6° further strengthens the preferential orientation. For the thin film grown on 6° off-axis substrate, only one peak appears in the phi scan (Fig. 16.1c). Such change in domain structure occurs because off-cut substrates introduce steps along a specified direction and break the equivalency on the substrate surface.

From the cross-sectional TEM characterization, the thin film grown on 0° off-axis substrate has a high density of defects, which propagate from the interface to the top of the film along the growth direction. With the increase of the off-axis angle, the defects became fewer and tilted with respect to the interface. This could be due to the reduction of in-plane rotational domains as seen from the XRD phi scan in Fig. 16.1. Such phenomena have been previously observed for GaN thin films grown on vicinal sapphire substrates by MBE [8]. Therefore, the usage of well-controlled off-axis substrates can be an effective method for reducing defect density in heteroepitaxial β-Ga_2O_3 thin films.

The improved crystalline quality of the β-Ga_2O_3 thin films grown on off-axis c-sapphire substrate was further demonstrated through the electron transport studies. By keeping the growth condition (temperature, pressure, Ar, and O_2 flow rate) the same and varying the doping source flow rate from 0.005 to 0.4 sccm, the carrier concentration can be broadly tuned between low-10^{17} and low-10^{20} cm^{-3} range. Figure 16.2 shows the Hall mobility as a function of the *n*-type carrier

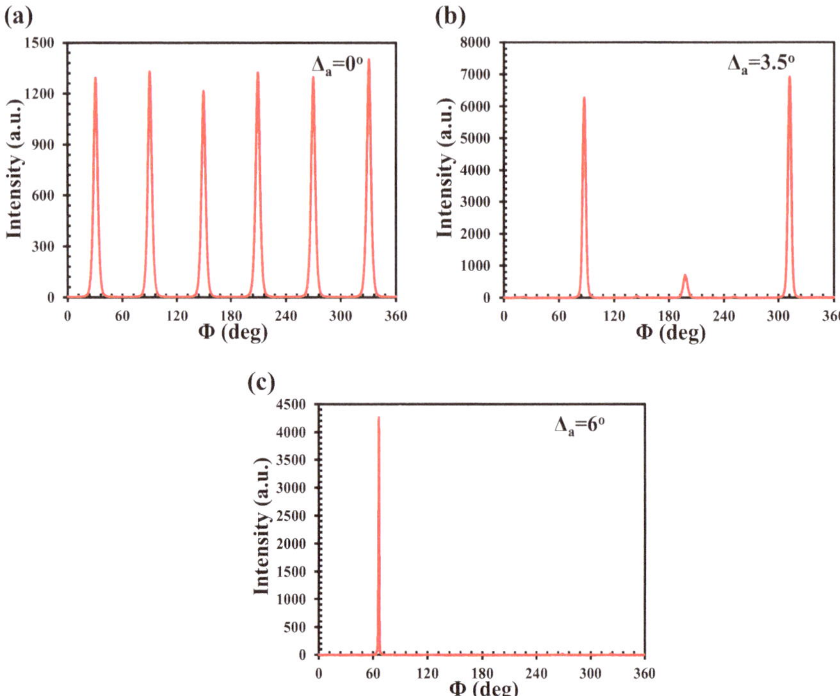

Fig. 16.1 XRD phi scans of (−401) reflections of β-Ga_2O_3 thin films grown on c-plane (0001) sapphire substrates with different off-angles toward <11–20>. **a** Δ_a = 0°. **b** Δ_a = 3.5°. **c** Δ_a = 6°. The thin films have a thickness of ~6 μm. (Reprinted from [7])

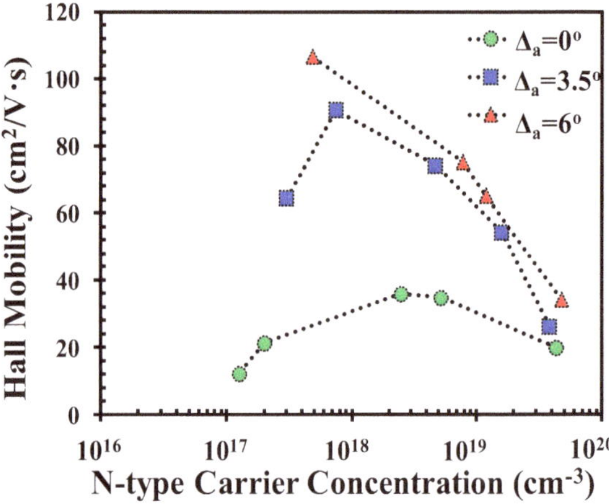

Fig. 16.2 Electron Hall mobility as a function of carrier concentration for Si-doped β-Ga_2O_3 thin films grown on c-plane (0001) sapphire substrates with different off-angles toward <11–20>. The thin films have a thickness of ~6 μm. (Reprinted from [7])

concentration for the Si-doped β-Ga_2O_3 thin films grown on c-sapphire substrates with different off-angle. Improved electrical properties of the thin films were obtained by using the off-axis substrates. The best electrical properties were exhibited by the thin films grown on 6° off-axis sapphire using an oxygen volume percentage of 4.8%. The measured room temperature Hall mobility was 106.6 cm^2/Vs with an n-type carrier concentration of 4.83×10^{17} cm^{-3}. Under the same growth condition, the measured room temperature Hall mobility for the thin films on 3.5° and 0° off-axis sapphire was 64.45 cm^2/Vs and 12.26 cm^2/Vs, respectively. The electrical properties of these LPCVD-grown heteroepitaxial β-Ga_2O_3 thin films on the off-axis sapphire substrates represent the best electrical quality among the reported ones.

In addition to the superior electrical properties of the LPCVD β-Ga_2O_3 films grown on c-sapphire substrates, these films have recently been demonstrated to possess the highest lifetime optical damage performance of any conductive material measured to date [9]. The reported lifetime laser damage threshold for β-Ga_2O_3 is ~10 times higher than that of the benchmarked GaN. This has direct implications for its use as an active component in high-power laser systems and may give insight into its utility for high-power switching applications.

16.3 LPCVD of β-Ga_2O_3 Thin Films on Native Ga_2O_3 Substrates

Development of β-Ga_2O_3 epitaxy on native substrates is technologically important for both lateral and vertical devices. LPCVD technique has been demonstrated as a feasible growth method to produce homoepitaxial β-Ga_2O_3 films with high material quality and controllable n-type doping.

16.3.1 Effects of Growth Temperature on Surface Morphology

LPCVD homoepitaxy of β-Ga_2O_3 on (010) Ga_2O_3 substrate has revealed that the growth temperature has a significant impact on the surface morphology of the as-grown films [10]. It was studied by varying the growth temperature in the range between 780 and 950 °C. The AFM images of 5 × 5 μm^2 scan (Fig. 16.3) for epitaxial layer surfaces grown at different growth temperatures were used to evaluate the surface morphologies. The general trend indicated that the layers became more homogeneous with the increase of growth temperature. All the layers were composed of multi-step arrays resembling terrace-like morphology. The feature size became smaller with the increase of the growth temperature. The lowest surface RMS roughness was 7.02 nm for the epitaxial layer grown at 950 °C, which

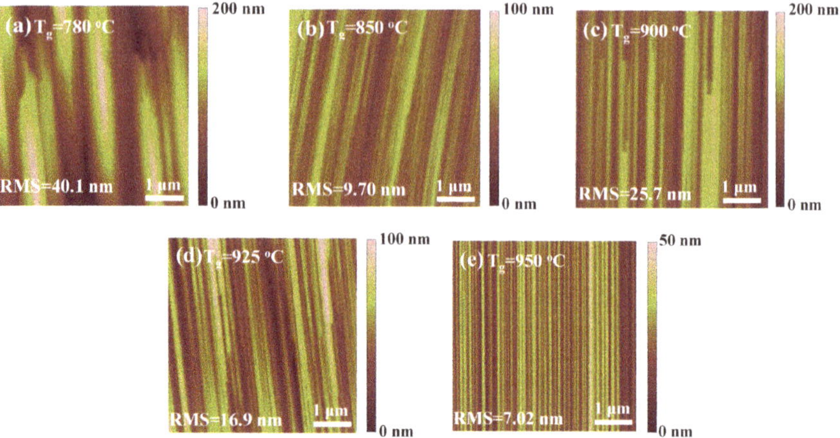

Fig. 16.3 Surface AFM images (5 μm × 5 μm) of β-Ga_2O_3 homoepitaxial layers grown at different growth temperatures. (Reprinted from [10])

represents the highest growth temperature performed in this study. This may have been due to the reduced film growth rate with the increase of growth temperature as a result of the formation and desorption of Ga_2O [11]. Ga_2O is formed at a higher temperature due to the reaction of excess Ga with Ga_2O_3 surface oxides.

16.3.2 Growth of β-Ga_2O_3 on (001) and (010) Ga_2O_3 Substrates

MBE growth of β-Ga_2O_3 thin films prefers the (010) growth direction due to its much faster growth rate as compared to the film grown on other crystal planes such as (100). The lower adhesion energy of the (100) plane results in the re-evaporation of the supplied atoms which in turn lowers the growth rate [12].

LPCVD growth of β-Ga_2O_3 was found to be less sensitive to the crystal orientation. The case study was performed for LPCVD Si-doped n-type β-Ga_2O_3 homoepitaxy on (010) and (001) Fe-doped semi-insulating β-Ga_2O_3 substrates [13]. Prior to the growth, the samples were in situ annealed at 900 °C for 30 min under O_2 atmosphere. The growth rates of the LPCVD β-Ga_2O_3 on both (010) and (001) substrates were demonstrated as >1 μm/h with the growth temperature of 900 °C. The surfaces of the thin films resembled a terrace-like morphology and were composed of multi-step arrays, as shown in the AFM images (Fig. 16.4). The RMS roughnesses for the (001) and (010) homoepitaxial thin films were ~3 nm and ~4 nm, respectively. The surface roughness did not show an obvious dependence on the Si doping level of the thin films.

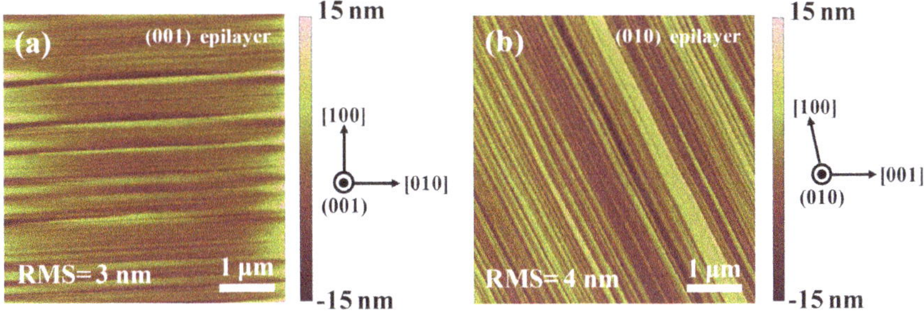

Fig. 16.4 Surface AFM images (5 μm × 5 μm) of β-Ga_2O_3, **a** (001) and **b** (010) homoepitaxial thin films grown at 900 °C. (Reprinted from [13])

The crystal quality of the LPCVD β-Ga_2O_3 homoepitaxial thin films was characterized by the XRD rocking curve measurement and high-magnification HAADF-STEM images. The full width at the half maximum (FWHM) of the asymmetric (400) reflection peak of the β-Ga_2O_3 (001) substrate was 27 arcsec, and the FWHM value of the LPCVD-grown Si-doped thin film was 47 arcsec. The (-42-2) reflection peak of the β-Ga_2O_3 (010) homoepitaxial thin film was taken at a grazing incidence angle of 1.3°. The FWHM values of (-42-2) peaks for the (010) substrate and epi film were 77 and 83 arcsec, respectively. No shift of the XRD peak for the as-grown films relative to the substrates was observed for both cases, indicating that the thin films were free of strain. High-resolution STEM images (Fig. 16.5) of the as-grown (010) and (001) homoepitaxial thin films were acquired using non-rigid registration [14] of 30 fast-scanned images to increase the precision and signal-to-noise ratio. Along the [010] orientation (Fig. 16.5a), Ga atoms have the largest separation between them with an interatomic distance of ~3.3 Å [15]. On the other hand, the [001] orientation has a shorter distance between the columns (Fig. 16.5b). No extended planar defects or dislocations were visible in the view fields of the STEM images for both thin films. However, point defects are expected to be present which resulted in lower carrier mobility than that predicted by theory. Fundamentally, native point defects such as vacancies and interstitials of β-Ga_2O_3 thin films are expected to be complex. Oxygen and gallium vacancies have been predicted to be deep donor and shallow acceptor based on density functional theory (DFT) calculation, respectively [16].

SIMS depth profiles (Fig. 16.6) were used to characterize the impurity concentration for Si-doped β-Ga_2O_3 homoepitaxial thin films grown at 900 °C on (001) and (010) β-Ga_2O_3 substrates. For both films, the Si doping profiles were mostly flat, indicating constant incorporation of the dopants throughout the growth process. A peak in the Si profile is visible at the interface between the epi-layer and the substrate for both films. The chemical Si concentrations in the (001) and (010) thin films measured by SIMS were 9.6×10^{17} cm^{-3} and 2.4×10^{18} cm^{-3}, respectively. The concentrations of other impurities in both thin films grown by LPCVD were comparable to those reported for HVPE-grown films [3].

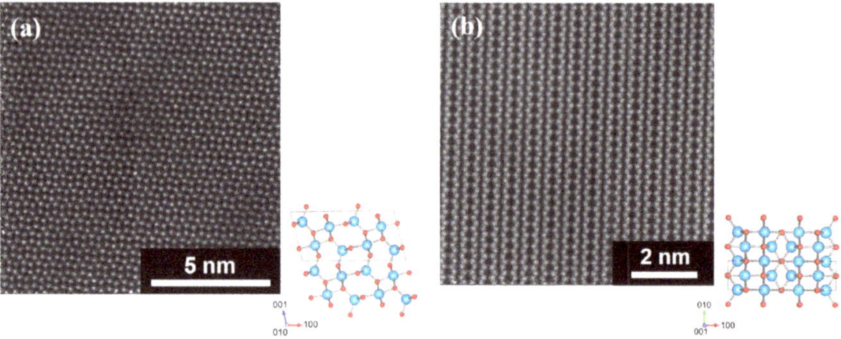

Fig. 16.5 High-magnification HAADF-STEM images of the **a** (010) and **b** (001) orientation view of the homoepitaxial thin films grown by LPCVD. Blue spheres represent Ga atoms, and red spheres represent oxygen atoms. (Reprinted from [13])

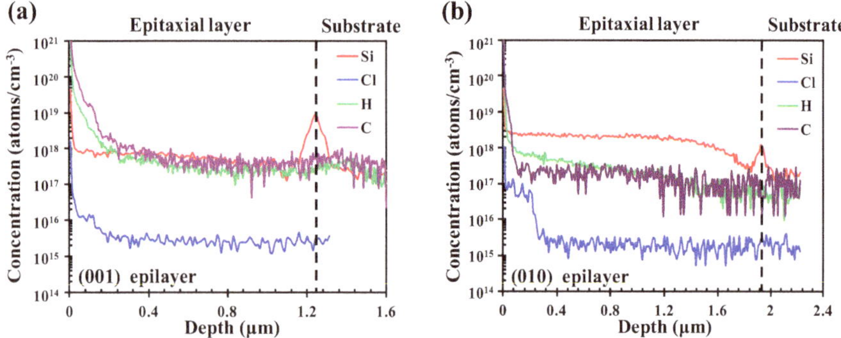

Fig. 16.6 SIMS depth profiles of impurities in the LPCVD-grown Si-doped β-Ga_2O_3 homoepitaxial thin films on **a** (001) and **b** (010) β-Ga_2O_3 substrates. (Reprinted from [13])

Both (001) and (010) β-Ga_2O_3 epitaxial films were demonstrated with effective n-type doping control by varying the doping source flow rate. The carrier concentration was broadly tunable in the range between mid-10^{17} and low-10^{19} cm^{-3} or higher. The free carrier concentrations obtained by Hall measurements for both films were similar to the chemical Si concentrations measured by SIMS, indicating the effective incorporation of the dopants in the electrically active sites and negligible compensation. The thin films were grown under the same growth condition. The difference in the carrier concentration between the two films is due to the different Si incorporation rates along different orientations of β-Ga_2O_3. The activation energies of Si were estimated to be E_a ~ 20.3 and 17.6 meV ($n \sim e^{-Ea/kT}$) from the temperature dependence of the carrier concentration for the (001) and (010) thin films (Fig. 16.7) which were similar to the previously reported values.

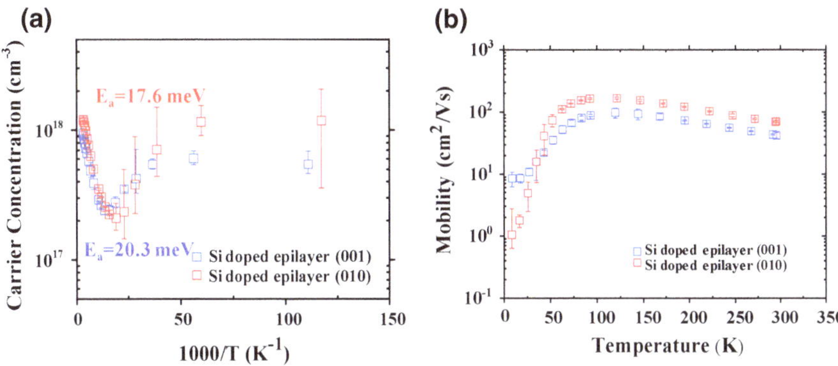

Fig. 16.7 Temperature-dependent Hall measurement of **a** carrier concentration and **b** mobility for the LPCVD-grown Si-doped β-Ga_2O_3 homoepitaxial thin films on (001) and (010) β-Ga_2O_3 substrates. (Reprinted from [13])

Room temperature electron Hall mobilities of ~72 and ~42 cm^2/Vs were measured for Si-doped homoepitaxial (010) and (001) β-Ga_2O_3 thin films with doping concentrations of ~1.2 × 10^{18} cm^{-3} and ~9.5 × 10^{17} cm^{-3}, respectively.

16.3.3 LPCVD β-Ga_2O_3-Based Schottky Barrier Diodes

(010)-oriented β-Ga_2O_3 bevel-field-plated mesa Schottky barrier diodes grown by LPCVD were demonstrated [17]. Schottky diodes with good ideality and low reverse leakage were realized on the epitaxial material. Edge termination using beveled field plates yielded a breakdown voltage of ~190 V, and maximum vertical electric fields of 4.2 MV/cm in the center and 5.9 MV/cm at the edge were estimated, with extrinsic R_{ON} of 3.9 mΩ cm^2 and extracted intrinsic R_{ON} of 0.023 mΩ cm^2.

The two-terminal reverse current breakdown characteristics for identical, fresh devices at various stages of the process were studied (Fig. 16.8). An anode current compliance of 0.2 A/cm^2 (3–4 orders of magnitude lower than I_{ON}) is seen to cause destructive breakdown with a breakdown voltage (V_{BR}) of −138 V for diodes before trench etching and −74 V after trench etching. This indicates that the expected enhancement in breakdown voltage using a bevel edge-termination design was not obtained after BCl_3/Ar etching. One possible reason could be the surface damage caused by BCl_3/Ar etching [18] which leads to surface-related leakage. By fabricating the field plate structure on top of the “after trench” device, V_{BR} was increased from −74 to −190 V. Field-plated Schottky diodes were also found to withstand one order higher current density (>1 A/cm^2) before breaking down catastrophically. The reverse current seen in the curve after −130 V was deduced to be leakage current through SiO_2, which adds to the overall current. Setting a current

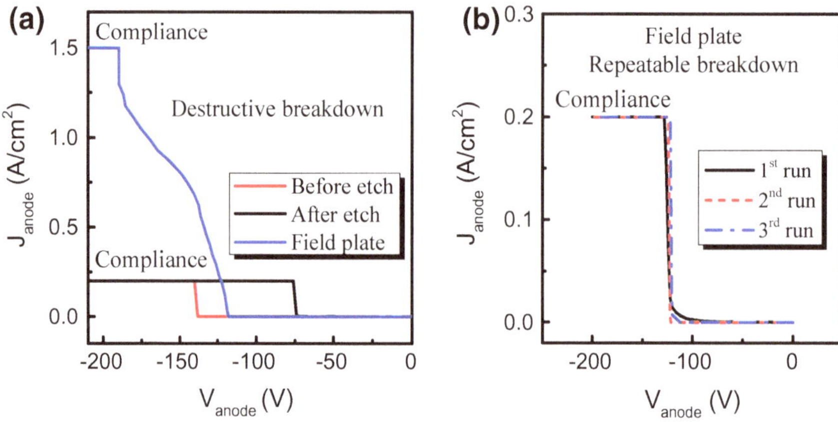

Fig. 16.8 **a** Two-terminal destructive reverse breakdown characteristics. **b** Non-destructive breakdown for the field-plated device. (Reprinted from [17], Copyright (2018) The Japan Society of Applied Physics)

compliance of 0.2 A/cm^2 for the field plates resulted in repeatable breakdown (Fig. 16.8b) with $V_{BR} = -129$ V. The measured V_{BR} values, Schottky contact barrier height (1.5 eV), and doping density (both extracted from the C–V data) were used to calculate the depletion width (W_{dep}) and peak electric field (F_{max}) using one-dimensional Poisson's equation. W_{dep} of 0.8 μm, which corresponds to F_{max} of 3.5 MV/cm, was estimated for the device before trench etching. Since W_{dep} of 0.8 μm is less than the epitaxial layer thickness ($\sim$2 μm), this corresponds to non-punch-through breakdown of the devices.

Two-dimensional device simulations using Silvaco ATLAS were used to estimate the electric field at the center of the diode. 2D Silvaco simulations for destructive breakdown on the field plate were seen to match the calculated 4.2 MV/cm field at the center of the diode with the peak field of 5.9 MV/cm at the corner. The reported results demonstrated the high quality of homoepitaxial LPCVD-grown β-Ga_2O_3 thin films for vertical power electronics applications and showed that this growth method is promising for future β-Ga_2O_3 technology.

16.4 LPCVD of β-Ga_2O_3 Rod Structures

Synthesis of Ga_2O_3-based nanostructures has attracted much attention recently, producing geometric structures which enable investigations relating optical and electrical properties to the quantum confinement effect and surface morphology of the nanostructures. Syntheses of Ga_2O_3-based nanostructures such as nanowires [19], nanoribbons [20], nanosheets [21, 22], nanobelts [23], and nanocolumns have been reported [24]. Most of these nanostructures are synthesized using catalysts, which can potentially introduce impurities into the material. A few catalyst-free synthesis techniques of Ga_2O_3 nanomaterials have been developed using chemical vapor deposition (CVD). These nanostructures can serve as building blocks for novel nanodevices such as deep UV photodetectors, nanophotonics switches, sensors, and field-effect transistors.

LPCVD growth of β-Ga_2O_3 rod structures was developed on 3C-SiC substrates [25]. 3C-SiC (also known as β-SiC) is the only polytype with cubic (or zinc blende) crystal structure. In 3C-SiC, each SiC bilayer can be oriented into only three possible positions with respect to the lattice, while the tetrahedral bonding between the C atom and Si atoms is maintained. The stacking sequence is ABC if these three layers are denoted as A, B, C. Studies indicate that oxygen concentration and growth temperature represent two important parameters in controlling the synthesis of β-Ga_2O_3 rods.

With a fixed oxygen volume percentage at 4.8% and growth temperature varying between 780 and 950 °C, polyangular rods with faceted sidewalls were formed under all growth conditions (Fig. 16.9). The rods have diameters in the range of ~200–700 nm and lengths of ~3–12 μm with 40 min of growth time. As shown in Fig. 16.10, the growth rate is 2.12 μm/h at 780 °C which increases to a maximum of 9.67 μm/h at 870 °C. This can be understood by examining the amount of Ga vapor present as a function of the growth temperature. As the growth temperature increases from 780 to 870 °C, the amount of Ga vapor concurrently

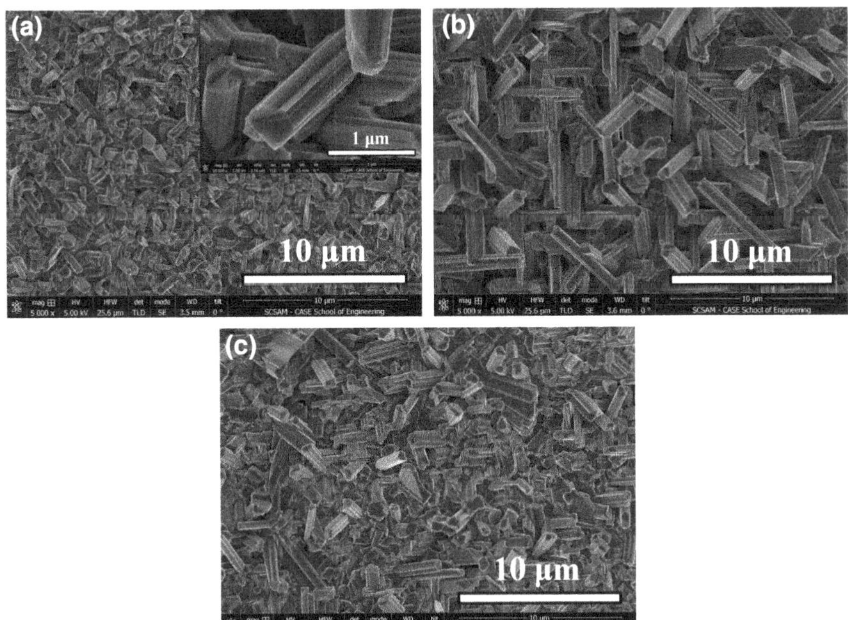

Fig. 16.9 Top-view FESEM images of Ga_2O_3 rods grown at oxygen volume percentage of 4.8% with different growth temperatures. **a** 780 °C, **b** 870 °C, and **c** 900 °C. Inset in (**a**) shows the high-magnification top-view FESEM image of the Ga_2O_3 rods. (Reprinted from [25])

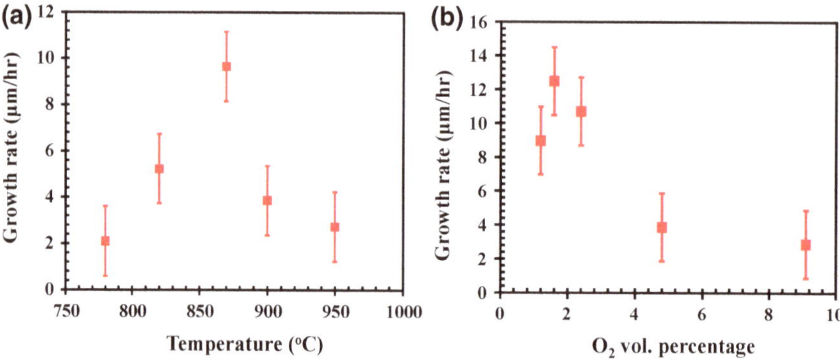

Fig. 16.10 Estimated growth rates of Ga_2O_3 rods as a function of **a** temperature and **b** oxygen volume percentage. (Reprinted from [25])

increases, which leads to an increase in the growth rate of Ga_2O_3. For a fixed oxygen volume percentage, a further increase of the growth temperature might lead to insufficiency of oxygen for reaction of Ga vapor. This could prevent the excess Ga to react with oxygen. Another possibility for the reduced growth rate at higher temperatures is the formation and desorption of Ga_2O which forms at a higher temperature due to the reaction of excess Ga with Ga_2O_3 surface oxides. Such a dependence of growth rate on the deposition temperature has also been observed for the synthesis of β-Ga_2O_3 thin films by PA-MBE [1].

The synthesis of the Ga_2O_3 rods also highly depends on the oxygen volume percentage. With a fixed growth temperature of 900 °C, oxygen flow rate of 5 sccm and Ar flow rate varying between 50 and 400 sccm, the Ga_2O_3 rod growth rate increases when O_2 volume percentage increases from 1.6 to 2% and then decreases as the O_2 volume percentage further increases. With a low flow rate of O_2, the growth rate is mass transport limited. When a sufficient amount of oxygen is introduced in the growth chamber, Ga easily oxidizes to Ga_2O. Ga_2O has a much higher vapor pressure than metallic Ga. Thus, a large amount can be easily evaporated and transported to the substrate by the carrier gas, resulting in a higher growth rate. On the other hand, excessive concentrations of oxygen prevent the Ga vapor from reaching the substrate, which in turn reduces the growth rate.

X-ray diffraction, high-resolution transmission electron microscopy (Fig. 16.11), and Raman spectroscopy measurements were performed for these as-grown β-Ga_2O_3 rods, which revealed a monoclinic phase of β-Ga_2O_3 with single-crystalline microstructure. The selected area electron diffraction pattern recorded on a single β-Ga_2O_3 rod further verified their single-crystalline nature. Because of their high crystalline quality and large surface-area-to-volume ratio, these β-Ga_2O_3 rods have a great potential for surface-related applications such as photocatalysis, chemical sensing, and deep-ultraviolet photodetection.

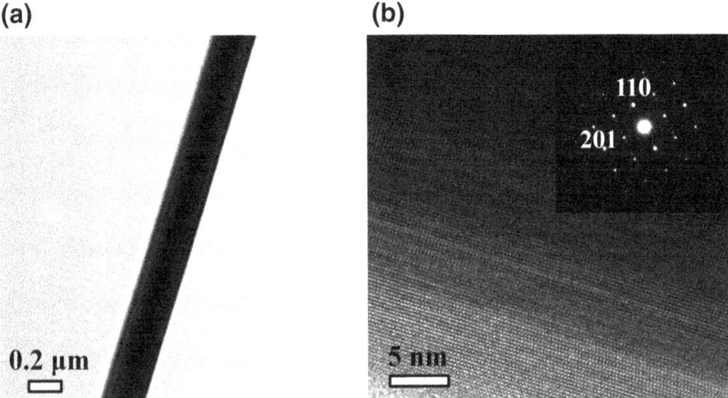

Fig. 16.11 **a** TEM image of a single Ga_2O_3 rod grown at 950 °C and oxygen percentage of 4.8%. **b** HRTEM image of the Ga_2O_3 rod. Inset shows the SAED pattern of the rod taken along the [−112] zone axis. (Reprinted from [25])

16.5 Conclusion

LPCVD growth can produce high-quality β-Ga_2O_3 thin films with controllable doping and wide range tuning of growth rate. High-quality β-Ga_2O_3 films grown on off-axis c-sapphire substrates have been demonstrated with superior charge transport properties. Dislocation-free β-Ga_2O_3 thin films grown on (010) and (001) Ga_2O_3 substrates have also been demonstrated. Vertical Schottky barrier diodes using LPCVD-grown β-Ga_2O_3 exhibited high breakdown electric field.

LPCVD growth parameters such as growth temperature, growth pressure, precursor flow rate, and substrate surface preparation not only affect the film growth rates, but also significantly impact the structural, electrical, and optical properties. The growth mechanisms, the incorporation of impurities, and the generation of native defects have not been fully understood. First-principle theoretical calculations, high-quality epitaxy, advanced material characterization, and device fabrication are important components to facilitate further advancement in this field.

Acknowledgements The work was supported by grants from US National Science Foundation (Drs. T. Paskova and D. Pavlidis) and AFOSR (Dr. A. Sayir).

References

1. H. Okumura, M. Kita, K. Sasaki, A. Kuramata, M. Higashiwaki, J.S. Speck, Appl. Phys. Express **7**, 095501 (2014)
2. G. Wagner, M. Baldini, D. Gogova, M. Schmidbauer, R. Schewski, M. Albrecht, Z. Galazka, D. Klimm, R. Fornari, Phys. Status Solidi A **211**, 27 (2014)

3. H. Murakami, K. Nomura, K. Goto, K. Sasaki, K. Kawara, Q.T. Thieu, R. Togashi, Y. Kumagai, M. Higashiwaki, A. Kuramata, S. Yamakoshi, B. Monemar, A. Koukitu, Appl. Phys. Express **8**, 015503 (2015)
4. S. Rafique, L. Han, H. Zhao, Phys. Status Solidi A **213**, 1002 (2016)
5. S. Rafique, L. Han, A.T. Neal, S. Mou, M.J. Tadjer, R.H. French, H. Zhao, Appl. Phys. Lett. **109**, 132103 (2016)
6. Z. Lin, J. Zhang, S. Xu, Z. Chen, S. Yang, K. Tian, X. Su, X. Shi, Y. Hao, Appl. Phys. Lett. **105**, 082114 (2014)
7. S. Rafique, L. Han, A.T. Neal, S. Mou, H. Zhao, Phys. Status Solidi A **215**, 1700467 (2018)
8. X.Q. Shen, H. Matsuhata, H. Okumura, Appl. Phys. Lett. **86**, 021912 (2005)
9. J.H. Yoo, S. Rafique, A. Lange, H. Zhao, S. Elhadj, APL Mater. **6**, 036105 (2018)
10. S. Rafique, L. Han, M.J. Tadjer, J.A. Freitas Jr., N.A. Mahadik, H. Zhao, Appl. Phys. Lett. **108**, 182105 (2016)
11. P. Vogt, O. Bierwagen, Appl. Phys. Lett. **108**, 072101 (2016)
12. K. Sasaki, A. Kuramata, T. Masui, E.G. Villora, K. Shimamura, S. Yamakoshi, Appl. Phys. Express **5**, 035502 (2012)
13. S. Rafique, M. Rezaul Karim, J.M. Johnson, J. Hwang, H. Zhao, Appl. Phys. Lett. **112**, 052104 (2018)
14. A.B. Yankovich, B. Berkels, W. Dahmen, P. Binev, S.I. Sanchez, S.A. Bradley, A. Li, I. Szlufarska, P.M. Voyles, Nat. Commun. **5**, 4155 (2014)
15. J.M. Johnson, S. Im, W. Windl, J. Hwang, Ultramicroscopy **172**, 17 (2017)
16. J.B. Varley, J.R. Weber, A. Janotti, C.G. Van de Walle, Appl. Phys. Lett. **97**, 142106 (2010)
17. C. Joishi, S. Rafique, Z. Xia, L. Han, S. Krishnamoorthy, Y. Zhang, S. Lodha, H. Zhao, S. Rajan, Appl. Phys. Express **11**, 031101 (2018)
18. J. Yang, F. Ren, R. Khanna, K. Bevlin, D. Geerpuram, L.C. Tung, J. Lin, H. Jiang, J. Lee, E. Flitsiyan, L. Chernyak, S.J. Pearton, A. Kuramata, J. Vac, Sci. Technol B **35**, 051201 (2017)
19. N. Han, F. Wang, Z. Yang, S. Yip, G. Dong, H. Lin, M. Fang, T. Hung, C.J. Ho, Nanoscale Res. Lett. **9**, 347 (2014)
20. D.P. Yu, J.-L. Bubendorff, J.F. Zhou, Y. Leprince-Wang, M. Troyon, Solid State Commun. **124**, 417 (2002)
21. W. Feng, X. Wang, J. Zhang, L. Wang, W. Zheng, P. Hu, W. Cao, B. Yang, J. Mater. Chem. C **2**, 3254 (2014)
22. S. Rafique, L. Han, J. Lee, X.Q. Zheng, C.A. Zorman, P.X.L. Feng, H. Zhao, J. Vac, Sci. Technol. B **35**, 011208 (2017)
23. L. Dai, X.L. Chen, N.X. Zhang, Z.A. Jin, T. Zhou, Q.B. Hu, Z. Zhang, J. Appl. Phys. **92**, 1062 (2002)
24. M. Mitome, S. Kohiki, K. Hori, M. Fukuta, Y. Bando, J. Cryst. Growth **286**, 240 (2006)
25. S. Rafique, L. Han, C.A. Zorman, H. Zhao, Cryst. Growth Des. **16**, 511 (2015)

Part III
Materials Properties

Chapter 17
First-Principles Calculations 1

Electronic and Structural Properties of Ga_2O_3 and Alloys with In_2O_3 and Al_2O_3

Hartwin Peelaers and Chris G. Van de Walle

Abstract In this chapter, we show how hybrid density functional theory can be used to elucidate the basic properties of Ga_2O_3, such as crystal structure, Brillouin zone, and band structure. We also demonstrate how it can predict the properties of Ga_2O_3 alloys and heterojunctions with In_2O_3 or Al_2O_3. The results can guide experimental exploration of materials and design of devices.

17.1 Bulk Properties

17.1.1 Crystal Structure

Ga_2O_3 has a number of polymorphs: α, β, γ, δ, and ε [1]. The most stable polymorph is β-Ga_2O_3, which has a base-centered monoclinic structure with space group 12 ($C2/m$). The structure, as illustrated in Fig. 17.1, contains two symmetry-inequivalent Ga positions, 50% of the Ga positions are tetrahedrally coordinated [Ga(I)] and 50% are octahedrally coordinated [Ga(II)]. There are also three inequivalent O positions: O(I) is three-fold coordinated to 2 Ga(II) and 1 Ga(I) atoms. O(II) is also three-fold coordinated, but to 2 Ga(I) and 1 Ga(II) atom. O(III) is a four-fold coordinated O site. In Fig. 17.1a, the conventional unit cell is shown, which contains 20 atoms; in Fig. 17.1b, we show the primitive unit cell, which contains 10 atoms.

These definitions of conventional and primitive cells and the corresponding crystal lattice vectors are not unique. Here, following [2], we defined the lattice vectors for the conventional cell as:

H. Peelaers (✉)
Department of Physics and Astronomy, University of Kansas, Lawrence, KS 66045, USA
e-mail: peelaers@ku.edu

C. G. Van de Walle
Materials Department, University of California, Santa Barbara, CA 93106-5050, USA
e-mail: vandewalle@mrl.ucsb.edu

M. Higashiwaki and S. Fujita (eds.), *Gallium Oxide*, Springer Series in Materials Science 293, https://doi.org/10.1007/978-3-030-37153-1_17

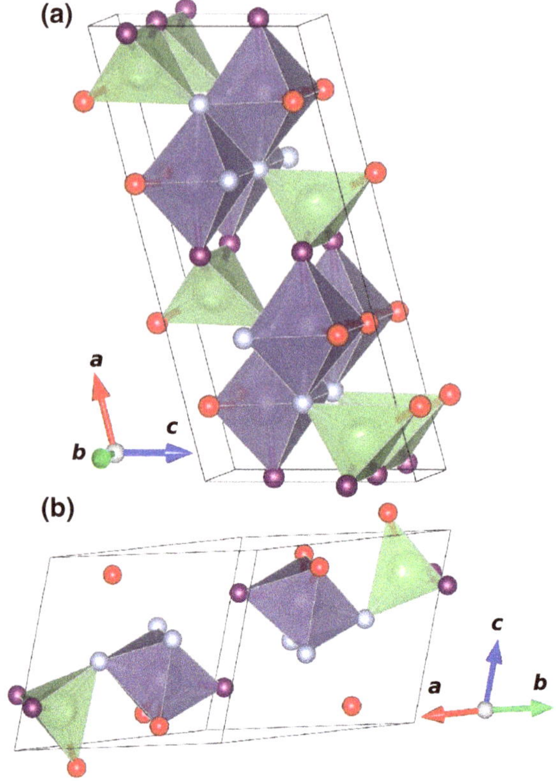

Fig. 17.1 **a** Conventional unit cell of monoclinic β-Ga_2O_3, as defined in the text. The two inequivalent Ga (large spheres) and the three inequivalent O (smaller spheres) sites are indicated with different colors. **b** The transformed primitive unit cell, using the same coloring scheme. The polyhedra indicate the bonding environment of the Ga atoms: 50% are octahedrally coordinated, 50% are tetrahedrally coordinated. Figure reproduced from [2]

$$
\begin{aligned}
\mathbf{a}_1^{\text{conv}} &= a\hat{\mathbf{x}} + 0\hat{\mathbf{y}} + 0\hat{\mathbf{z}} \\
\mathbf{a}_2^{\text{conv}} &= 0\hat{\mathbf{x}} + b\hat{\mathbf{y}} + 0\hat{\mathbf{z}} \\
\mathbf{a}_3^{\text{conv}} &= c\cos\beta\hat{\mathbf{x}} + 0\hat{\mathbf{y}} + c\sin\beta\hat{\mathbf{z}},
\end{aligned}
$$

where $\hat{\mathbf{x}}$, $\hat{\mathbf{y}}$, and $\hat{\mathbf{z}}$ are the cartesian unit vectors. The size of the lattice vectors (a, b, and c) together with the angle β between the a- and c-axes determine the crystal. Note that a different convention can be used, where one orders the lattice vector by increasing length and defines β as being smaller than 90° [3]. A simple coordinate transformation can be employed to go from one convention to the other.

This conventional unit cell can be transformed to the smaller primitive cell (Fig. 17.1b), by defining the following lattice vectors:

$$
\begin{aligned}
\mathbf{a}_1^{\text{prim}} &= (\mathbf{a}_1^{\text{conv}} + \mathbf{a}_2^{\text{conv}})/2 \\
\mathbf{a}_2^{\text{prim}} &= (-\mathbf{a}_1^{\text{conv}} + \mathbf{a}_2^{\text{conv}})/2 \\
\mathbf{a}_3^{\text{prim}} &= \mathbf{a}_3^{\text{conv}}.
\end{aligned}
$$

17.1.2 Brillouin Zone

From the primitive unit cell defined in Sect. 17.1.1, one can obtain the Brillouin zone by transforming to reciprocal space and constructing the Wigner-Seitz cell by the bisection method. Due to the low monoclinic symmetry, the shape of the Brillouin zone is complicated, and shown in Fig. 17.2. We labeled the high-symmetry points; their coordinates can be found in Table 17.1. An alternative labeling scheme (corresponding to a different choice of lattice vectors) is found in [3].

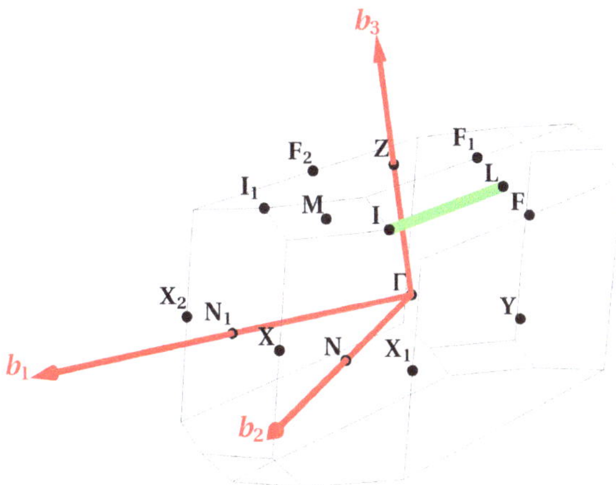

Fig. 17.2 Brillouin zone corresponding to the monoclinic β-Ga_2O_3. Labels indicate high-symmetry points. The lattice vectors of the reciprocal unit cell are also shown. The I-L line is indicated by a (green) line

Table 17.1 Reciprocal-space coordinates of the high-symmetry points for the Brillouin zone of β-Ga_2O_3. Here, $\Psi = \frac{3}{4} - b^2/(4a^2 \sin^2 \beta)$, $\phi = \Psi - (\frac{3}{4} - \Psi)\frac{a}{c}\cos\beta$, $\zeta = (2 + \frac{a}{c}\cos\beta)/4\sin^2\beta$, and $\eta = \frac{1}{2} - 2\zeta\frac{c}{a}\cos\beta$

Label	Coordinates	Label	Coordinates
Γ	$(0, 0, 0)$	Y	$(-\frac{1}{2}, \frac{1}{2}, 0)$
N	$(0, \frac{1}{2}, 0)$	Y_1	$(\frac{1}{2}, -\frac{1}{2}, 0)$
N_1	$(\frac{1}{2}, 0, 0)$	L	$(-\frac{1}{2}, \frac{1}{2}, \frac{1}{2})$
F	$(\zeta - 1, 1 - \zeta, 1 - \eta)$	M	$(0, \frac{1}{2}, \frac{1}{2})$
F_1	$(-\zeta, \zeta, \eta)$	X	$(1 - \Psi, 1 - \Psi, 0)$
F_2	$(\zeta, -\zeta, 1 - \eta)$	X_1	$(\Psi - 1, \Psi, 0)$
I	$(\phi - 1, \phi, \frac{1}{2})$	X_2	$(\Psi, \Psi - 1, 0)$
I_1	$(1 - \phi, 1 - \phi, \frac{1}{2})$	Z	$(0, 0, \frac{1}{2})$

17.1.3 First-Principles Methods

Density functional theory (DFT) is a computational technique to tackle the complex many-body problem inherent in the quantum mechanics of electrons in solids. Instead of using the many-body wavefunction, which depends on $3N$ variables (with N the number of electrons) as the main quantity, it uses the density, which only depends on the three Cartesian coordinates. This tremendously simplifies the problem, allowing a formulation of the Hamiltonian in terms of functionals of the charge density (a functional being a function of a functions). This Nobel-Prize-winning idea was pioneered by Hohenberg and Kohn [4]. However, one would still have to solve a many-body system. Kohn and Sham [5] devised an exact methodology to deal with this complexity: they replaced the many-body system with a system of non-interacting electrons. This system is chosen so that it yields the same ground-state density as the original many-body system. While they proved that such an exact mapping in principle exists, the exact form of the functional to accomplish this is unknown. The crucial information is embedded in the so-called exchange-correlation functional.

Many approximations, such as the local-density approximation (LDA) or the generalized-gradient approximation (GGA), have been proposed and widely used. Unfortunately, these semi-local approximations all suffer from the so-called band-gap problem: they lead to a large underestimation of the band gap, which is related to a self-interaction error. Methods that go beyond DFT and are based on many-body perturbation theory, such as the GW method [6], have been developed. However, these do not allow for a self-consistent determination of the atomic structure, and therefore some dependency on the initial (DFT-determined) structure still exists. Staying within DFT, hybrid functionals offer a pragmatic and accurate approach. Such functionals mix a percentage of Hartree-Fock exchange together with semi-local exchange-correlation (usually at the GGA level). This combination yields very accurate electronic information, while at the same time allowing for a self-consistent determination of structural information. The functional used in the results described in this chapter is the one developed by Heyd, Scuseria, and Ernzerhof (HSE06) [7, 8].

All reported calculations were performed using the VASP code [9], with projected augmented waves [10], a plane-wave basis set with an energy cutoff of 500 eV for all alloy calculations, and 400 eV for all other calculations, and a $8 \times 8 \times 4$ **k**-point grid, or equivalent for larger cells. All structures were fully relaxed. Maximally localized Wannier Functions [11, 12], as implemented in the WANNIER90 package [13], were used to accurately sample the Brillouin zone along the high-symmetry directions (see Fig. 17.2 and Table 17.1) for the shown band structures.

17.1.4 *Structural, Electronic, and Optical Properties of Bulk Ga_2O_3*

Calculated lattice parameters and band gaps are compared with experimental values in Table 17.2. Good agreement is evident. The full band structure of Ga_2O_3 shown in Fig. 17.3; it is plotted along high-symmetry paths in the Brillouin zone, as determined in Sect. 17.1.2. This band structure, obtained using the HSE06 hybrid functional, is in good agreement with calculations using the GW method [14]. The detailed band structure allows determining the exact positions of the band extrema and the nature of the band gap.

Clearly the conduction-band minimum is located at the Γ point. The valence-band maximum, however, is not located at the Γ point, but at a point on the high-symmetry line I-L (see Fig. 17.2), making the band gap indirect. However, the difference between the direct (4.88 eV) and indirect band gap (4.84 eV) is small (0.04 eV) [17–19] so that for practical purposes Ga_2O_3 can be considered a direct-band-gap material [18, 19]. The distinction between the direct and indirect valence-band maximum could be important in applications based on holes in the valence band, but because of the strong tendency for hole-polaron formation [20] this will be irrelevant.

Table 17.2 Calculated and experimental [15, 16] lattice parameters for the conventional unit cell of β-Ga_2O_3, and value of the band gap

	Calculated	Experimental
a (Å)	12.27	12.21
b (Å)	3.05	3.04
c (Å)	5.82	5.80
β	103.82°	103.83°
E_{gap}^{direct} (eV)	4.88	4.76
$E_{gap}^{indirect}$ (eV)	4.84	

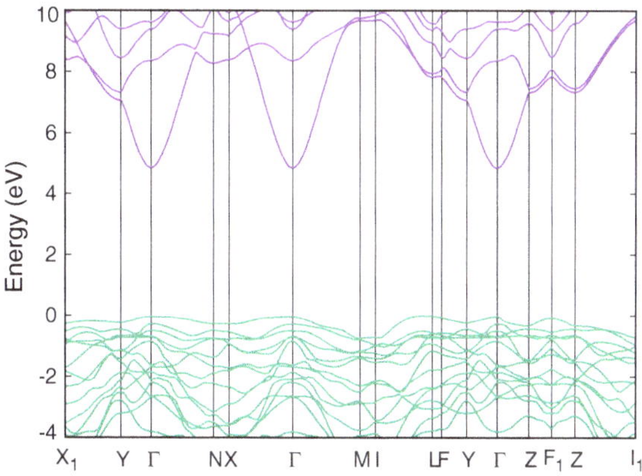

Fig. 17.3 Electronic band structure along high-symmetry paths of the Brillouin zone (see Fig. 17.2). High-symmetry points are labeled according to Table 17.1. The valence and conduction bands are shown in different colors. Reprinted from [2]

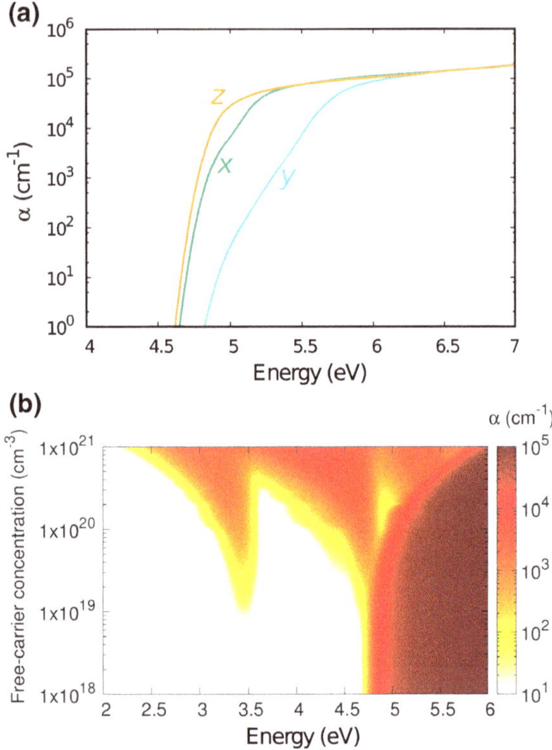

Fig. 17.4 **a** Direct across-the-gap absorption for monoclinic Ga_2O_3 for light polarized along the Cartesian directions. **b** Absorption in Ga_2O_3 for light polarized along the *z* direction as function of photon energy and free-carrier concentration. Figure adapted from [29], with the permission of AIP Publishing

The lowest conduction band is very dispersive, owing to its *s*-orbital character. This also results in low electron effective masses, which are between 0.27 and 0.28 m_e depending on the direction, agreeing well with experimental measurements [17, 21] and with other calculations using hybrid functionals [14, 18, 21, 22]. The valence bands, which consist mainly of O *p* states, are very flat.

Due to the low symmetry of monoclinic Ga_2O_3, the absorption spectra strongly depend on the polarization of the light, which is illustrated in Fig. 17.4a. Experimental absorption measurements also observed a strong anisotropy of the absorption [23–25]. Computationally this has been shown using various levels of theory for dielectric functions [14, 19, 26, 27] and the absorption coefficient [28]. If free carriers are present in the conduction bands, these spectra are modified [29]. A Burstein–Moss shift [30, 31], that is, a shift of the onset of absorption to higher photon energies with increasing free-carrier concentration, will occur. At the same time, the additional carriers can give rise to additional sub-band-gap absorption, as illustrated for light polarized along the *Z* direction in Fig. 17.4b.

Note that the spectra shown in Fig. 17.4 are those in the absence of excitons. Calculated spectra including excitonic effects can be found in [14, 27].

17.2 Alloys

Alloying allows one to modify the basic material properties of Ga_2O_3 described in Sect. 17.1.1. In particular, it allows to change the lattice constants (Sect. 17.2.2), band gaps (Sect. 17.2.3), and the absolute position of the valence and conduction bands (Sect. 17.2.5). Knowledge of how these properties change as a function of alloy concentration is crucial for device design. In the following sections, we will examine the properties of Ga_2O_3 alloyed with either In_2O_3 [32] or Al_2O_3 [33]. These materials all have different ground-state crystal structures: In_2O_3 has the cubic bixbyite structure (see Fig. 17.5a), while Al_2O_3 has a corundum structure (see Fig. 17.5b). The ground-state properties are discussed in Sect. 17.2.1. We then address the dependence on alloy composition of the lattice constant (Sect. 17.2.2) and of the band gap (Sect. 17.2.3). In Sect. 17.2.4, we discuss the stability of the alloys in different crystal structures.

17.2.1 Ground-State Structures

β-Ga_2O_3 has a monoclinic structure, as discussed in Sect. 17.1.4, and shown in Fig. 17.1. For In_2O_3, the ground-state structure is the cubic bixbyite structure, shown in Fig. 17.5a. In this structure, all In atoms are octahedrally coordinated, albeit distorted. There are two symmetry-inequivalent positions, indicated by different colored polyhedra. The ground-state crystal structure of Al_2O_3 is the corundum structure, whose primitive unit cell is depicted in Fig. 17.5b. All Al atoms are octahedrally coordinated. When discussing alloys, we therefore need to consider these two crystal structures, in addition to the monoclinic structure.

Table 17.3 lists the calculated lattice parameters and band gaps for the parent crystal structures,[1] and a comparison with experimental measurements.[2] Figure 17.6 shows the calculated band structures; the band structure for monoclinic Ga_2O_3 is shown in Fig. 17.3.

17.2.2 Lattice Constants

As can be seen from Table 17.3, the lattice constants for monoclinic Al_2O_3 are smaller than those of monoclinic Ga_2O_3, and those of monoclinic In_2O_3 are larger. This

[1]The calculated values for monoclinic Ga_2O_3 reported in Table 17.3 are different from the values reported in Table 17.2 for two reasons: a different mixing is used in the hybrid functional and we included the *d* electrons in the valence. Therefore, the monoclinic Ga_2O_3 results also differ from the results reported in [32].

[2]Note that the experimental values reported for bixbyite Ga_2O_3 might not have been for the bixbyite structure [34].

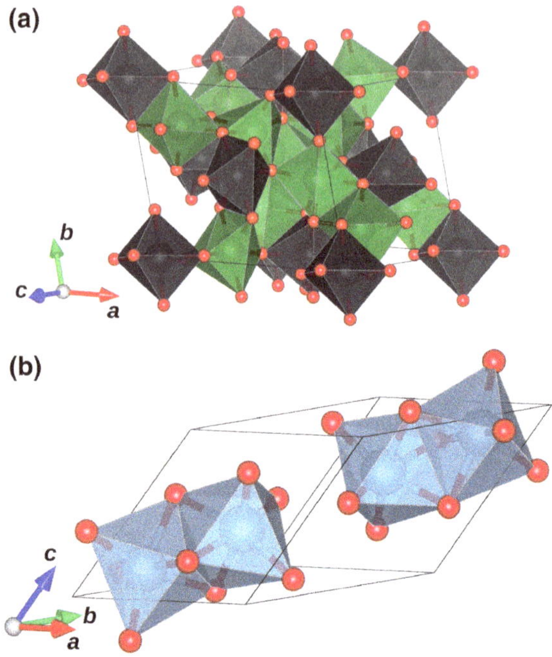

Fig. 17.5 Primitive unit cell of **a** bixbyite In_2O_3 and **b** corundum Al_2O_3. Large spheres depict In atoms in (**a**) and Al atoms in (**b**). Smaller spheres are O atoms. The polyhedra indicate the octahedral environment of the cations

Table 17.3 Calculated and experimental lattice parameters and band gaps of monoclinic Ga_2O_3 [16, 23], corundum Ga_2O_3 [35, 36], bixbyite Ga_2O_3 [1], bixbyite In_2O_3 [37, 38], monoclinic In_2O_3, corundum Al_2O_3 [39, 40], and monoclinic Al_2O_3 [41]

			a	b	c	β	E_{gap}^{direct}	$E_{gap}^{indirect}$
monoclinic	Ga_2O_3	Calc.	12.21	3.03	5.79	103.8	4.87	4.865
		Exp.	12.21	3.04	5.8	103.83	4.76	
	Al_2O_3	Calc.	11.75	2.92	5.57	103.80	7.51	7.24
		Exp.	11.85	2.9	5.62	103.83		
	In_2O_3	Calc.	13.59	3.37	6.45	103.89	2.9	2.78
		Exp.						
corundum	Ga_2O_3	Calc.	4.97	4.97	13.39	–	5.59	5.34
		Exp.	4.98	4.98	13.43	–	5.32	
	Al_2O_3	Calc.	4.74	4.74	12.94	–	8.82	–
		Exp.	4.76	4.76	12.99	–	8.8	–
bixbyite	Ga_2O_3	Calc.	9.25	–	–		5	–
		Exp.	10	–	–	–		–
	In_2O_3	Calc.	10.21	–	–	–	3.04	–
		Exp.	10.12	–	–	–	2.9	–

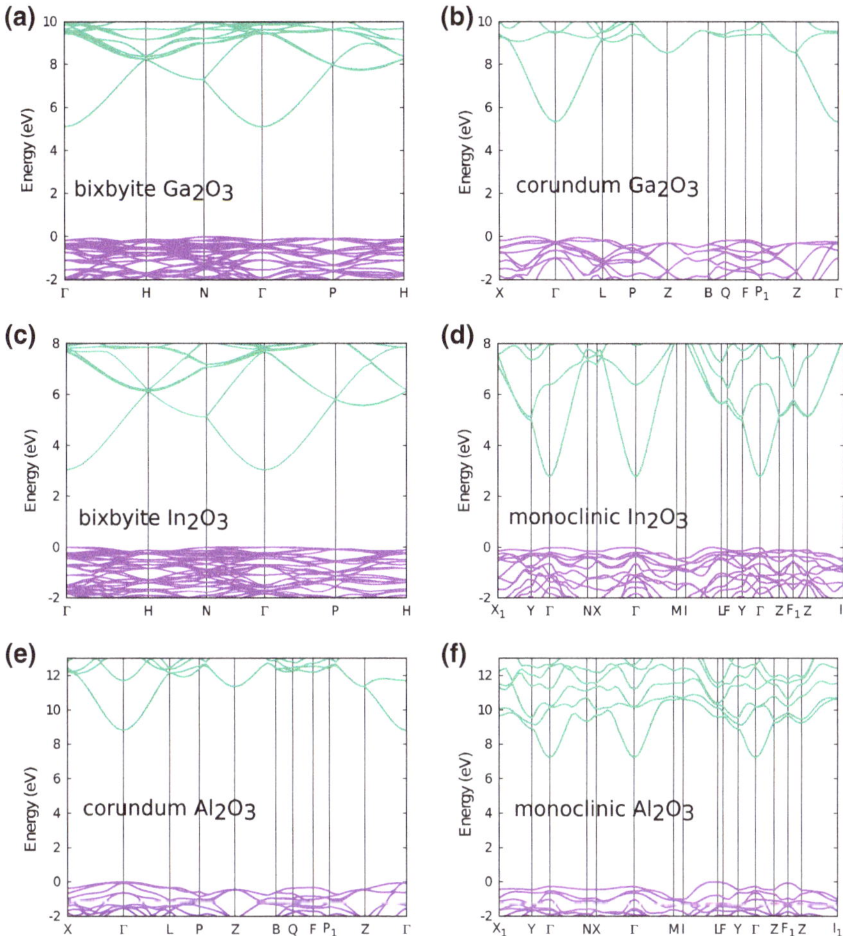

Fig. 17.6 Calculated band structures for **a** bixbyite Ga_2O_3, **b** corundum Ga_2O_3, **c** bixbyite In_2O_3, **d** monoclinic In_2O_3, **e** corundum Al_2O_3, and **f** monoclinic Al_2O_3

suggests that alloying with Al_2O_3 will allow to decrease the lattice constants, and alloying with In_2O_3 increase them. The first-principles results shown in Fig. 17.7 confirm this [32, 33]. Similar results were obtained using the semi-local GGA functional for In_2O_3 in [42] and using hybrid functionals for Al_2O_3 alloys in [43]. In this figure, we plot the pseudocubic lattice constant as function of alloy composition. The pseudocubic lattice constant is defined as the cube root of the volume per formula unit. This measure allows comparing different crystal structures on the same footing. Our calculations show that the increase or decrease in pseudocubic lattice constant is always linear. The results are thus consistent with Vegard's law [44].

The difference in volume between crystal structures can be explained by examining the type of cation positions in these structures: in the monoclinic structure, there

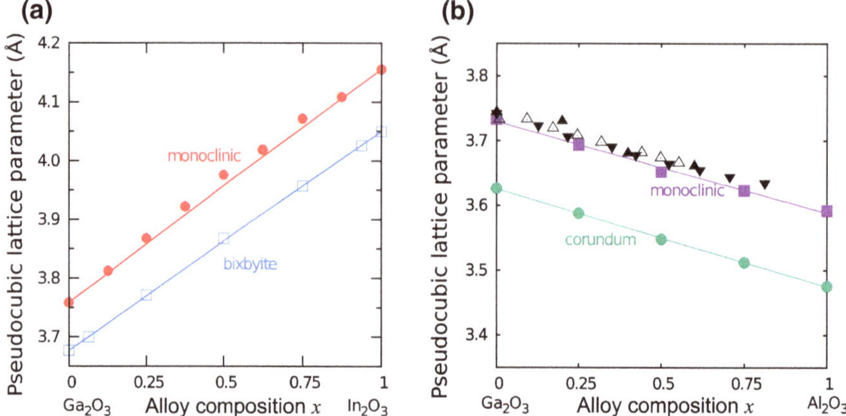

Fig. 17.7 Pseudocubic lattice constant as a function of alloy concentration for alloying of Ga_2O_3 with **a** In_2O_3 and **b** Al_2O_3. Experimental values for monoclinic are shown as solid black triangles [45], open triangles [46], and solid downward triangles [47]. Figure (**a**) adapted with permission from [32]. Copyrighted by the American Physical Society. Figure (**b**) adapted from [33], with the permission of AIP Publishing

are both octahedral and tetrahedral sites, while in corundum or bixbyite there are only octahedral sites. Octahedral sites lead to more compact structures compared to tetrahedral sites, explaining why the volume per cation is always smaller for corundum or bixbyite compared to monoclinic.

17.2.3 Band Gaps and Optical Properties

The bulk band gap of monoclinic Ga_2O_3 is 4.865 eV (indirect), while the direct optical gap is 4.87 eV (see Table 17.3). Corundum Al_2O_3 has a much larger band gap (8.80 eV), while bixbyite In_2O_3 has a smaller band gap of 3.04 eV, but the optical gap is much larger (3.73 eV). The optical gap is larger because transitions from the topmost valence bands to the lowest conduction band are dipole forbidden due to the inversion center of the bixbyite crystal structure. Only transitions originating from states lower in the valence band lead to strong absorption [48]. In absorption spectra, this shows up as a weak onset of absorption, followed by a strong onset [17, 48–51].

Alloying will allow control of the band gap, and accurate knowledge of the dependence of gap on alloy composition, which may deviate from linearity due to bowing, is important. An overview of results is included in Fig. 17.8.

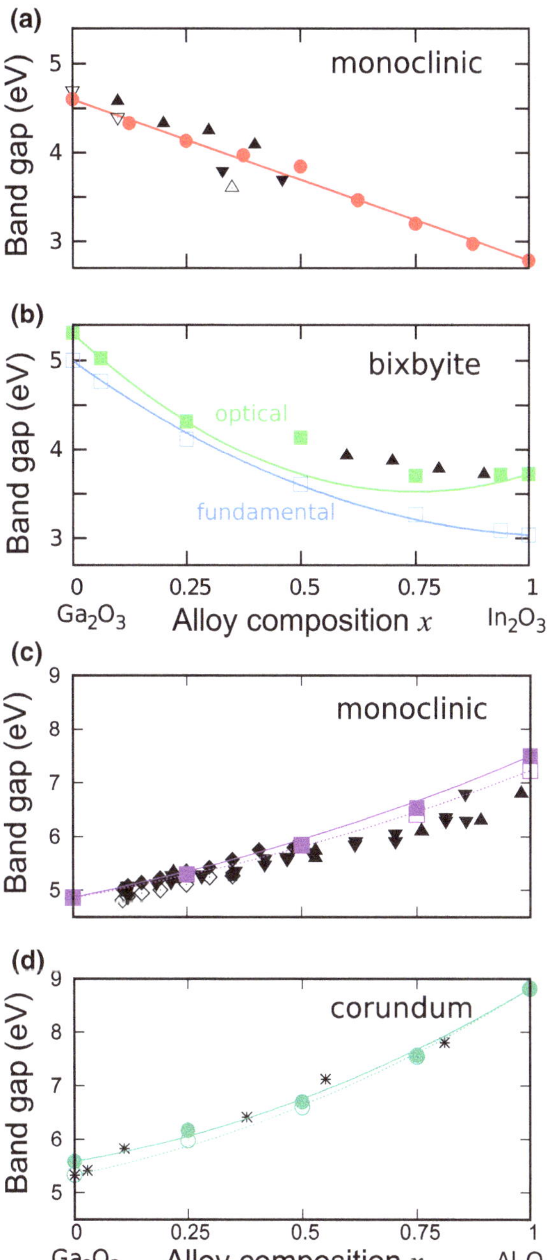

Fig. 17.8 Band gap as function of alloying concentration for alloying of Ga_2O_3 with **a** In_2O_3 in the monoclinic structure, **b** In_2O_3 in the bixbyite structure, **c** Al_2O_3 in the monoclinic structure, and **d** Al_2O_3 in the corundum structure. Available experimental values are shown in (**a**) with triangles: [52] (downward open triangles), [53] (downward filled triangles), [54] (upward open triangle), and [55] (upward filled triangles). In (**b**) the upward filled triangles are from [55]. In (**c**), experimental values are shown with black symbols: solid upward triangles [56], solid downward triangles [47], solid tilted squares [57], and solid pentagons [58]. In (**d**) the experimental data (crosses) are from [36]. Lines are fits to (17.1). In (**c**) and (**d**) two fits are shown: solid lines for the direct band gap, and dashed lines for the indirect band gap. Figure (**a**) and (**b**) adapted with permission from [32]. Copyrighted by the American Physical Society. Figure (**c**) and (**d**) adapted from [33], with the permission of AIP Publishing

Alloying with In_2O_3 leads to a decrease in the magnitude of the band gap. In the case of monoclinic alloys (Fig. 17.8a) bowing is absent and the band gap decreases linearly. The same trend, but with some spread in actual values, was measured experimentally [52–55]. Calculations at the GGA level [42] obtained a similar result. The relationship between band gaps and alloy concentration is more complicated in the bixbyite structure (Fig. 17.8b): in that case, we have to distinguish between the electronic gap (the fundamental band gap) and the optical band gap. While alloying breaks the inversion center of the bixbyite crystal, the absorption is still weak until the energy of the photon is sufficient to reach the lower valence bands. Moreover, the size of the band gap does not vary linearly with alloy concentration. The lines are a fit to the parabolic equation

$$E_g[(In_xGa_{1-x})_2O_3] = (1-x)\,E_g[Ga_2O_3] + x\,E_g[In_2O_3] - bx(1-x), \quad (17.1)$$

where E_g is the band gap, x the alloy concentration, and b the bowing parameter which describes how far the band gap of the alloy deviates from the linear interpolation. We find a bowing parameter of 1.69 eV for the fundamental band gaps, and a bowing parameter of 3.18 eV for the optical band gap. The measured optical band gaps [55] are also shown.

Alloying with Al_2O_3 increase the band gap. In this case, we distinguished between the direct band gap (open colored symbols), which is optically allowed, and the indirect band gap (open colored symbols). In both the monoclinic (Fig. 17.8c) and corundum (Fig. 17.8d), crystal structure significant deviations from linearity are observed. The bowing parameters are 1.37 eV (direct) and 0.93 eV (indirect) for the monoclinic structure, and 1.87 eV (direct) and 1.78 eV (indirect) for the corundum structures. The black symbols in Fig. 17.8c, d denote experimental measurement [36, 47, 56–58]. These follow similar trends, but exhibit substantial scatter.

17.2.4 Crystal Structures

In Sects. 17.2.2 and 17.2.3, we separated the behavior of the different crystal structures. When creating an alloy, the energetics of the different structures will determine what the ground-state structure will be. At 0 K, the enthalpy of formation indicates which structure is energetically favored. This quantity is defined as

$$\begin{aligned}\Delta H[((In/Al)_xGa_{1-x})_2O_3] = {} & E[((In/Al)_xGa_{1-x})_2O_3] - (1-x)\,E[Ga_2O_3] \\ & -x\,E[(In/Al)_2O_3], \end{aligned} \quad (17.2)$$

where monoclinic Ga_2O_3, bixbyite In_2O_3, and corundum Al_2O_3 are used as reference phases. Additionally, if one knows the enthalpies of formation across the different alloy concentrations, one can use a regular solution model to model the enthalpy of mixing, either by using a parabolic dependency:

$$\Delta H(x) = x(1-x)\Omega_0 + (1-x)H_0 + xH_1, \tag{17.3}$$

or by also including an asymmetry (Ω_1):

$$\Delta H(x) = x(1-x)\Omega_0 + \Omega_1(x^2-x)(x-0.5) + (1-x)H_0 + xH_1, \tag{17.4}$$

where H_0 and H_1 are the enthalpies of formation of the end points.

At finite temperature, the relevant quantity is the Gibbs free energy, which is given by

$$\Delta G(x) = \Delta H(x) - T\Delta S(x), \tag{17.5}$$

where $\Delta S(x)$ is the mixing entropy, which can be approximated by a random mixture model. From this, we can also estimate the temperature T at which full miscibility will take place. This is only an estimate, since we are neglecting other entropy contributions such as vibrational entropy and additional configuration interactions.

We will first consider the alloys with In_2O_3, whose enthalpies of formation are shown in Fig. 17.9a. Circles depict monoclinic structures, squares are bixbyite structures. At the end points, we find, as expected, that monoclinic Ga_2O_3 and bixbyite In_2O_3 are the ground states. Note that the energy difference between monoclinic Ga_2O_3 and bixbyite Ga_2O_3 is very small (1.5 meV/cation). For In_2O_3, there is a much larger difference (of 345 meV/cation) between the bixbyite and monoclinic structure. In fact, for alloy concentrations, where $x > 0.5$, the bixbyite structure always has lower energy. Similarly, for $x < 0.5$ the monoclinic alloys are energetically favored. This trend can be explained by considering the type of cation sites present in the different crystal structures: bixbyite only contains octahedrally coordinated positions, while monoclinic has 50% octahedrally and 50% tetrahedrally coordinated positions. In atoms strongly prefer octahedral positions, as can be seen from the energetic preference for bixbyite structures when $x > 0.5$. For smaller In concentrations, there is also a large energy difference between In occupying tetrahedral versus occupying octahedral sites in the monoclinic structure. The open circles indicate alloys where In occupies such unfavorable tetrahedral sites. This also explains the rapid increase of the formation enthalpy for monoclinic structures when $x > 0.5$: in those cases In atoms have to occupy tetrahedral sites, thus, incurring an energy penalty.

The 50/50 monoclinic alloy ($x = 0.5$) has an exceptionally low enthalpy of formation. In this ordered alloy, all In atoms occupy the octahedral sites and all Ga atoms the tetrahedral sites. The bonding environment is such that the In–O bonds closely match the In–O bond lengths in the pristine In_2O_3 bixbyite structure, while the Ga–O bonds have the same length as the Ga–O bonds in the pristine Ga_2O_3 monoclinic structure. Such an arrangement, therefore, minimizes the local strain and its associated energy cost, enabling the ordered alloy to have a very low enthalpy of formation.

For alloys with Al_2O_3, we are comparing the energetics of corundum versus monoclinic alloys (see Fig. 17.9b for the calculated enthalpies of formation, where circles indicate corundum structures and squares monoclinic structures). At the end points, corresponding to the pristine phases, we obtain that monoclinic Ga_2O_3 has

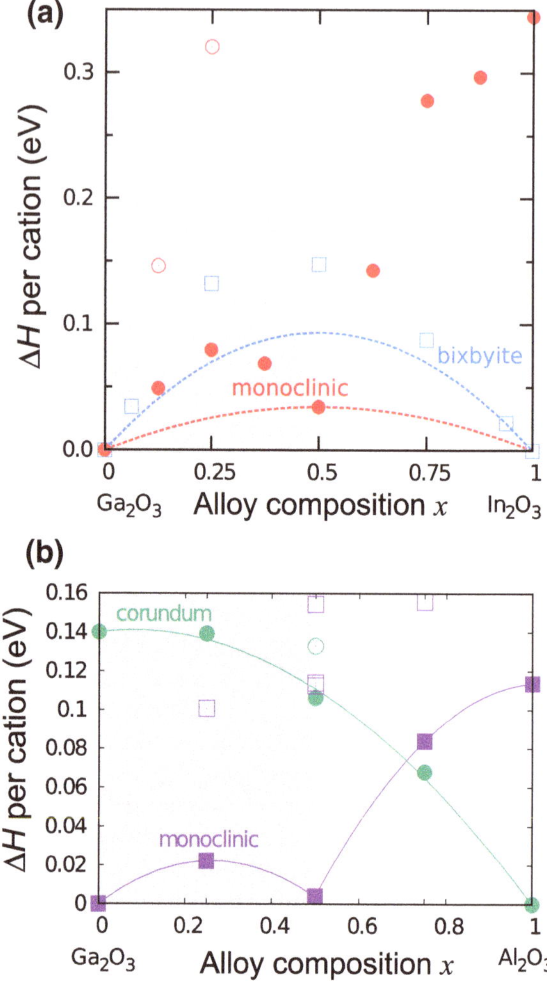

Fig. 17.9 Enthalpy of formation as function of alloy concentration for alloys with **a** In_2O_3 and **b** Al_2O_3. The lines are fits to regular solution models, as described in the text. Figure (**a**) adapted with permission from [32]. Copyrighted by the American Physical Society. Figure (**b**) adapted from [33], with the permission of AIP Publishing

an enthalpy of formation that is 140 meV/cation lower than corundum Ga_2O_3, and that corundum Al_2O_3 has an enthalpy of formation that is 110 meV lower than that of monoclinic Al_2O_3. As mentioned before, the corundum structure only contains octahedral cation positions, while the monoclinic structure has both octahedral and tetrahedral positions. From the energy difference at the Al_2O_3 end point ($x = 1$), we have an indication that Al prefers to occupy octahedral positions, as the corundum crystal structure is much lower in energy. This indication is confirmed when we compare energies of monoclinic alloys, where Al occupying only octahedral positions (the filled squares in Fig. 17.9b) is much lower in energy compared to alloys where Al also occupies tetrahedral positions (open squares). Again, this energetic preference of octahedral sites leads to a strong increase in the enthalpy of formation in the monoclinic crystal structure for alloy compositions with more than 50% Al. Once the alloy composition is above $x = 0.5$ eV, the corundum structure starts to become more favorable. Similar to the case of alloys with In_2O_3, the ordered monoclinic 50/50 alloy, where all octahedral sites are occupied by Al atoms and all tetrahedral sites by Ga atoms, is energetically favored because of the minimization of the strain.

With the information of the enthalpy of formation for all alloy compositions, we can use the regular solution model, combined with an estimate of the mixing entropy, to obtain an estimate of the miscibility temperature. For the In_2O_3 alloys, we used the symmetric (parabolic) expression of (17.3). The lines shown in Fig. 17.9a are fits to this equation ($\Omega_0 = 140$ meV for monoclinic, and 376 meV for bixbyite structures). For the Al_2O_3 alloys, we used the expression including an asymmetric term (17.4) and obtained $\Omega_0 = 164$ meV and $\Omega_1 = 12$ meV for the corundum structure. Since the monoclinic 50/50 alloy has an enthalpy of formation of only 4 meV/cation, we considered it as a stable end point and used 17.3 to fit the $x < 0.5$ and $x > 0.5$ region separately. This yields Ω_0 values of 81 meV and 101 meV. In both types of alloys, the monoclinic structure has lower interaction parameters. Using the common tangent method we determine the miscibility temperature for the monoclinic structures to be 812 K for In_2O_3 alloys, and 472 K for $x < 0.5$ and 584 K for $x > 0.5$ for Al_2O_3 alloys. Note that these temperatures represent an upper bound, as not all entropy contributions are taken into account. The fits also allow to obtain a more accurate determination of the maximum Al composition at which the monoclinic structure is the ground state structure: for x up to 71%, monoclinic structure are preferred. This quantifies the expected experimental range of monoclinic/corundum alloys, in good agreement with observations [36, 46, 47, 56, 59–65]. Note that it is possible to stabilize other alloys (e.g., when growing on specific surfaces), explaining the data points shown in Fig. 17.3d that have Al concentrations larger than 71%. The large spread in experimental band-gap values, in particular, for high Al compositions, can be explained by poorer crystal quality due to the existence of competing crystal phases with similar energies. For In_2O_3, the crossover point is close to $x = 0.5$. Experimental realizations of In_2O_3-Ga_2O_3 alloys indeed observed monoclinic crystals up to $x = 0.5$ [53, 54, 66, 67].

17.2.5 *Band Alignments*

Knowledge of the band alignments is important for device applications in which heterostructures are formed. The alignment of the conduction bands will determine in which material carriers will be confined.

Determining the absolute position of the valence and conduction bands requires a comparison with an absolute energy level. A regular DFT calculation does not provide such an absolute level. There are two methods that can be used to infer an alignment: slab calculations or defect calculations. A slab calculation (within periodic boundary conditions) includes a layer of vacuum, and hence the average electrostatic potential in the interior of the slab can be aligned with the vacuum level.

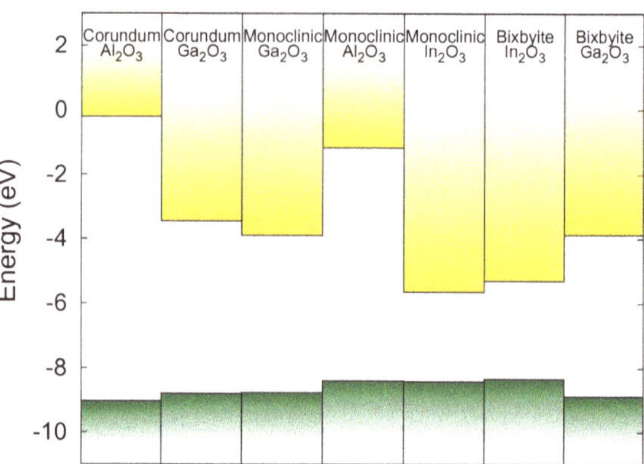

Fig. 17.10 Alignment of the valence-band maximum and conduction-band minimum for the various structures discussed in the text. A combination of slab calculations and hydrogen alignment has been used

If the slab is thick enough, the average electrostatic potential in the interior of the slab can be identified with the average electrostatic potential of the bulk material. In practice, the electrostatic potential assumes its bulk value within a few atomic layers from the surface (in the absence of electrostatic effects due to charging). Referencing the energy of the valence-band maximum or conduction-band minimum to this average electrostatic energy, therefore, allows a determination of their absolute energy position with respect to the vacuum level. In slab calculations, it is important to only consider non-polar surfaces and equivalent surfaces on both sides of the slab, otherwise spurious effects will influence the electrostatic potential.

Alternatively, the alignment of a defect level can be used, and it has been demonstrated that the thermodynamic transition level between the + and − charge states of interstitial hydrogen can be used to obtain a universal alignment [68]. The methodology for defect and impurity calculations is addressed in the chapter by Varley in this volume.

Both approaches have been applied, with good agreement, to obtain the band alignments for the various crystal structures and materials considered so far: monoclinic Ga_2O_3, In_2O_3, and Al_2O_3, bixbyite Ga_2O_3 and In_2O_3, and corundum Ga_2O_3 and Al_2O_3. The specific results shown in Fig. 17.10 are based on the (010) surface for the monoclinic structures and the $(10\bar{1}0)$ surface for corundum structures. The bixbyite structures were aligned using the hydrogen (+/−) level.

The valence bands, which originate mainly from O p states in all structures, have very similar absolute energies. The band offsets occur mainly in the conduction bands: In_2O_3 structures have significantly lower conduction bands, while Al_2O_3 structures have significantly higher conduction band levels compared to Ga_2O_3. We can combine these band alignments with the information about the band gap as a function of alloy composition (Fig. 17.8) to predict the absolute band positions in alloys. For monoclinic $In_xGa_{2-x}O_3$ alloys, the band gap varies linearly with alloy composition, rendering it straightforward to interpolate between the band positions for Ga_2O_3 and In_2O_3. The conduction-band edge is always lower for the alloys, such

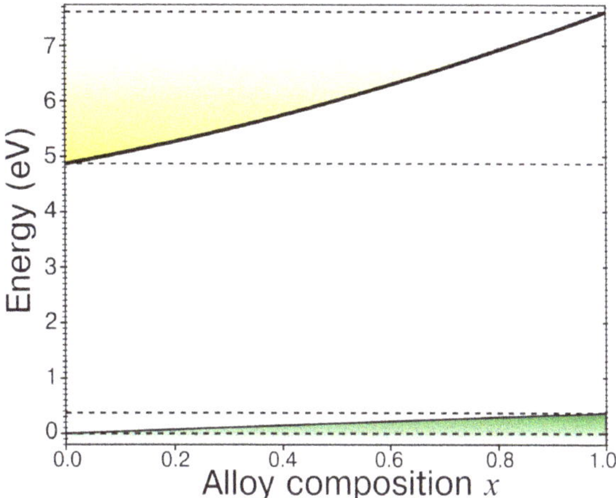

Fig. 17.11 Absolute position of the valence-band maximum and conduction-band minimum as a function of Al concentration in monoclinic Ga_2O_3-Al_2O_3 alloys. Figure reprinted from [33], with the permission of AIP Publishing

that free carriers (electrons) will be confined in the alloy when heterostructures are created.

For monoclinic alloys with Al_2O_3, the band gap exhibits significant bowing. The differences in valence-band position are small; we assume a linear variation [43] for valence-band positions (similar to the case of InGaN alloys [69]), so that all of the bowing occurs in the conduction band. The resulting band energies as a function of alloy concentration are shown in Fig. 17.11. The absolute conduction-band position moves higher with alloying with Al, so that carriers will be confined in Ga_2O_3 in heterostructures with Al_2O_3 alloys.

For heterostructures involving ultrathin layers of Ga_2O_3, it is important to assess the effects of quantum confinement. Recent calculations [70] show that when Ga_2O_3 is embedded in Al_2O_3, quantum confinement occurs. The effective masses of electron remain small ($<0.43\, m_e$), which is promising for devices. Interestingly, unpassivated free-standing Ga_2O_3 layers exhibit a seeming lack of quantum confinement: even for the thinnest possible nanolayers, the band gap remains almost unchanged from the bulk value. This has been attributed to the presence of surface states, whose interaction compensates for the expected quantum confinement [70].

17.3 Conclusions

In this chapter, we briefly discussed the theory behind first-principles calculations, and how hybrid functionals are a powerful computational approach to accurately describe the ground-state properties of Ga_2O_3 and related materials. Structural parameters were obtained, and the exact shape of the Brillouin zone and the corresponding high-symmetry points were derived. These were used to obtain the full band structure of Ga_2O_3. Monoclinic β-Ga_2O_3 is an indirect band-gap material, with the conduction-band minimum at the Γ point and the valence-band maximum on the I-L line. The difference between the indirect and the direct band gap at Γ is small, so that for all intents and purposes, Ga_2O_3 can be considered to be a direct-band-gap material. Due to the low monoclinic symmetry, optical properties are strongly dependent on the polarization of the incoming light, and different polarizations lead to different onsets of absorption. When free carriers are present, the onset of absorption shifts toward higher energies and sub-band-gap absorption, corresponding to transitions of free carriers to higher conduction bands, becomes possible.

Alloying with either In_2O_3 or Al_2O_3 modifies the properties of Ga_2O_3. The ground-state crystal structure of these three materials is different: Ga_2O_3 has the monoclinic crystal structure, In_2O_3 the cubic bixbyite, and Al_2O_3 the corundum structure. Both In and Al strongly prefer octahedrally coordinated positions over tetrahedrally coordinated positions. Since half the cation positions in the monoclinic structure are octahedral and half are tetrahedral, a strong preference exists for alloys to be in the monoclinic structure for concentrations up to 50% In or Al. The ordered monoclinic $AlGaO_3$ and $InGaO_3$ alloys, where all In or Al atoms occupy the octahedral positions is especially favorable, since such configurations minimize local strain. The lattice parameters vary linearly with alloy composition. The band gaps decrease linearly in monoclinic In_2O_3 alloys, or with significant bowing in bixbyite In_2O_3 alloys. The fundamental band gap of the bixbyite alloys is dark, as only transitions originating from lower valence bands have strong dipole matrix elements. Alloys with Al_2O_3 have increasing band gaps with increasing Al concentration, and bowing is present. Heterostructures between Ga_2O_3 and Al_2O_3-Ga_2O_3 alloys can be used to confine electrons in the Ga_2O_3 layer, and band alignments were reported.

Acknowledgements This work was supported by the GAME MURI of the Air Force Office of Scientific Research (FA9550-18-1-0479). Collaborations with A. Janotti, J. S. Speck, and J. B. Varley are gratefully acknowledged.

References

1. R. Roy, V.G. Hill, E.F. Osborn, J. Am. Chem. Soc. **74**(3), 719 (1952)
2. H. Peelaers, C.G. Van de Walle, Phys. Status Solidi B **252**(4), 828 (2015)
3. W. Setyawan, S. Curtarolo, Comput. Mater. Sci. **49**(2), 299 (2010)
4. P. Hohenberg, W. Kohn, Phys. Rev. **136**, B864 (1964)
5. W. Kohn, L.J. Sham, Phys. Rev. **140**(4A), A1133 (1965)
6. L. Hedin, J. Phys.: Condens. Matter **11** (1999)
7. J. Heyd, G.E. Scuseria, M. Ernzerhof, J. Chem. Phys. **118**(18), 8207 (2003)
8. J. Heyd, G.E. Scuseria, M. Ernzerhof, J. Chem. Phys. **124**(21), 219906 (2006)
9. G. Kresse, J. Furthmüller, Phys. Rev. B **54**(16), 11169 (1996)
10. P.E. Blöchl, Phys. Rev. B **50**(24), 17953 (1994)
11. N. Marzari, D. Vanderbilt, Phys. Rev. B **56**(20), 12847 (1997)
12. I. Souza, N. Marzari, D. Vanderbilt, Phys. Rev. B **65**(3), 035109 (2001)
13. A.A. Mostofi, J.R. Yates, Y.S. Lee, I. Souza, D. Vanderbilt, N. Marzari, Comput. Phys. Commun. **178**(9), 685 (2008)
14. J. Furthmüller, F. Bechstedt, Phys. Rev. B **93**(11), 115204 (2016)

15. S. Geller, J. Chem. Phys. **33**(3), 676 (1960)
16. J. Åhman, G. Svensson, J. Albertsson, Acta Crystallogr. C **52**(6), 1336 (1996)
17. C. Janowitz, V. Scherer, M. Mohamed, A. Krapf, H. Dwelk, R. Manzke, Z. Galazka, R. Uecker, K. Irmscher, R. Fornari, M. Michling, D. Schmeißer, J.R. Weber, J.B. Varley, C.G. Van de Walle, New J. Phys. **13**(8), 085014 (2011)
18. H. Peelaers, C.G. Van de Walle, J. Phys.: Condens. Matter **26**(30), 305502 (2014)
19. K.A. Mengle, G. Shi, D. Bayerl, E. Kioupakis, Appl. Phys. Lett. **109**(21), 212104 (2016)
20. J.B. Varley, A. Janotti, C. Franchini, C.G. Van de Walle, Phys. Rev. B **85**(8), 081109(R) (2012)
21. M. Mohamed, C. Janowitz, I. Unger, R. Manzke, Z. Galazka, R. Uecker, R. Fornari, J.R. Weber, J.B. Varley, C.G. Van de Walle, Appl. Phys. Lett. **97**(21), 211903 (2010)
22. Y. Kang, K. Krishnaswamy, H. Peelaers, C.G. Van de Walle, J. Phys.: Condens. Matter **29**(23), 234001 (2017)
23. T. Matsumoto, M. Aoki, A. Kinoshita, T. Aono, Jpn. J. Appl. Phys. **13**(4), 737 (1974)
24. N. Ueda, H. Hosono, R. Waseda, H. Kawazoe, Appl. Phys. Lett. **71**(7), 933 (1997)
25. T. Onuma, S. Saito, K. Sasaki, T. Masui, T. Yamaguchi, T. Honda, M. Higashiwaki, Jpn. J. Appl. Phys. **54**(11), 112601 (2015)
26. K. Yamaguchi, Solid State Commun. **131**(12), 739 (2004)
27. J.B. Varley, A. Schleife, Semicond. Sci. Technol. **30**(2), 024010 (2015)
28. A. Ratnaparkhe, W.R.L. Lambrecht, Appl. Phys. Lett. **110**(13), 132103 (2017)
29. H. Peelaers, C.G. Van de Walle, Appl. Phys. Lett. **111**(18), 182104 (2017)
30. E. Burstein, Phys. Rev. **93**(3), 632 (1954)
31. T.S. Moss, Proc. Phys. Soc., Sect. B **67**(10), 775 (1954)
32. H. Peelaers, D. Steiauf, J.B. Varley, A. Janotti, C.G. Van de Walle, Phys. Rev. B **92**(8), 085206 (2015)
33. H. Peelaers, J.B. Varley, J.S. Speck, C.G. Van de Walle, Appl. Phys. Lett. **112**(24), 242101 (2018); Appl. Phys. Lett. 115, 159901 (2019)
34. H.Y. Playford, A.C. Hannon, E.R. Barney, R.I. Walton, Chem. Eur. J. **19**(8), 2803 (2013)
35. M. Marezio, J.P. Remeika, J. Chem. Phys. **46**(5), 1862 (1967)
36. H. Ito, K. Kaneko, S. Fujita, Jpn. J. Appl. Phys. **51**, 100207 (2012)
37. M. Marezio, Acta Crystallogr. **20**(6), 723 (1966)
38. K. Irmscher, M. Naumann, M. Pietsch, Z. Galazka, R. Uecker, T. Schulz, R. Schewski, M. Albrecht, R. Fornari, Phys. Status Solidi A **211**(1), 54 (2014)
39. R.E. Newnham, Y.M. de Haan, Z. Für Krist. **117**(2–3), 235 (1962)
40. R.H. French, J. Am. Ceram. Soc. **73**(3), 477 (1990)
41. R.S. Zhou, R.L. Snyder, Acta Crystallogr. B **47**(5), 617 (1991)
42. M.B. Maccioni, F. Ricci, V. Fiorentini, J. Phys. Conf. Ser. **566**, 012016 (2014)
43. T. Wang, W. Li, C. Ni, A. Janotti, Phys. Rev. Appl. **10**(1), 011003 (2018)
44. L. Vegard, Z. Phys. **5**(1), 17 (1921)
45. A.L. Jaromin, D.D. Edwards, J. Am. Ceram. Soc. **88**(9), 2573 (2005)
46. C. Kranert, M. Jenderka, J. Lenzner, M. Lorenz, H. von Wenckstern, R. Schmidt-Grund, M. Grundmann, J. Appl. Phys. **117**(12), 125703 (2015)
47. B.W. Krueger, C.S. Dandeneau, E.M. Nelson, S.T. Dunham, F.S. Ohuchi, M.A. Olmstead, J. Am. Ceram. Soc. **99**(7), 2467 (2016)
48. A. Walsh, J. Da Silva, S.H. Wei, C. Körber, A. Klein, L. Piper, A. DeMasi, K. Smith, G. Panaccione, P. Torelli, D. Payne, A. Bourlange, R. Egdell, Phys. Rev. Lett. **100**(16), 167402 (2008)
49. R.L. Weiher, R.P. Ley, J. Appl. Phys. **37**(1), 299 (1966)
50. V. Scherer, C. Janowitz, A. Krapf, H. Dwelk, D. Braun, R. Manzke, Appl. Phys. Lett. **100**, 21 (2012)
51. P.D.C. King, T.D. Veal, F. Fuchs, C.Y. Wang, D.J. Payne, A. Bourlange, H. Zhang, G.R. Bell, V. Cimalla, O. Ambacher, R.G. Egdell, F. Bechstedt, C.F. McConville, Phys. Rev. B **79**(20), 205211 (2009)
52. V.I. Vasyltsiv, Y.I. Rym, Y.M. Zakharko, Phys. Status Solidi B **195**(2), 653 (1996)

53. T. Oshima, T. Okuno, N. Arai, N. Suzuki, S. Ohira, S. Fujita, Appl. Phys. Express **1**(1), 011202 (2008)
54. F. Yang, J. Ma, C. Luan, L. Kong, Z. Zhu, in *2011 Symposium on Photonics and Optoelectronics (SOPO)*, vol. 3 (IEEE, New York, 2011), pp. 1–5
55. A. Kudo, I. Mikami, J. Chem. Soc. Faraday Trans. **94**(19), 2929 (1998)
56. F. Zhang, K. Saito, T. Tanaka, M. Nishio, M. Arita, Q. Guo, Appl. Phys. Lett. **105**(16), 162107 (2014)
57. R. Schmidt-Grund, C. Kranert, H. von Wenckstern, V. Zviagin, M. Lorenz, M. Grundmann, J. Appl. Phys. **117**(16), 165307 (2015)
58. Q. Feng, X. Li, G. Han, L. Huang, F. Li, W. Tang, J. Zhang, Y. Hao, Opt. Mater. Express **7**(4), 1240 (2017)
59. V.G. Hill, R. Roy, E.F. Osborn, J. Am. Ceram. Soc. **35**(6), 135 (1952)
60. M. Mizuno, T. Noguchi, T. Yamada, Bull. Chem. Soc. Jpn. **43**(8), 2614 (1970)
61. T. Oshima, T. Okuno, N. Arai, Y. Kobayashi, S. Fujita, Jpn. J. Appl. Phys. **48**(7), 070202 (2009)
62. X. Wang, Z. Chen, F. Zhang, K. Saito, T. Tanaka, M. Nishio, Q. Guo, AIP Adv. **6**(1), 015111 (2016)
63. R. Stalder, K.H. Nitsch, J. Am. Ceram. Soc. **80**(1), 258 (1997)
64. S.D. Lee, Y. Ito, K. Kaneko, S. Fujita, Jpn. J. Appl. Phys. **54**(3), 030301 (2015)
65. K. Kaneko, K. Suzuki, Y. Ito, S. Fujita, J. Cryst. Growth **436**, 150 (2016)
66. R. Shannon, C. Prewitt, J. Inorg. Nucl. Chem. **30**(6), 1389 (1968)
67. D.D. Edwards, P.E. Folkins, T.O. Mason, J. Am. Ceram. Soc. **80**(1), 253 (1997)
68. C.G. Van de Walle, J. Neugebauer, Nature **423**(6940), 626 (2003)
69. P.G. Moses, C.G. Van de Walle, Appl. Phys. Lett. **96**(2), 021908 (2010)
70. H. Peelaers, C.G. Van de Walle, Phys. Rev. B **96**, 081409(R) (2017)

Chapter 18
First-Principles Calculations 2

Doping and Defects in Ga_2O_3

Joel B. Varley

Abstract Gallium oxide has rapidly developed as a prime candidate for next-generation power electronics owing to the availability of high-quality material (e.g., large single-crystal substrates and epitaxial films) in conjunction with its favorable optical and electronic properties. While a number of advances have been made in improving the crystal quality, improving control over the free carrier and defect concentrations in this wide-band gap semiconductor are critical to realize the full potential of Ga_2O_3-based devices. Identifying the most suitable dopants and the origins of the prominent deep-level defects, as well as possible passivation techniques, is therefore an essential next step in further developments. In this chapter, we use hybrid functional calculations to investigate the behavior of a number of native defects, candidate dopants, and extrinsic impurities in β-Ga_2O_3 and to elucidate how theory has helped in understanding the defect-related properties of this material.

18.1 Introduction

β-Ga_2O_3 has recently become a material of great interest for electronic devices for high-power applications and ultraviolet (UV) optoelectronics due to the appealing combination of a large band gap (~4.7 eV) [1, 2], controllable *n*-type conductivity and availability as large single crystals grown from scalable, melt-growth methods [3, 4]. These properties have already lead to the demonstration of a number of promising devices including solar-blind photodetectors, Schottky barrier diodes, metal-semiconductor, and metal-semiconductor-oxide field-effect transistors both in enhancement and depletion-mode and with champion devices having breakdown-voltages already exceeding several kV [5–8]. Despite possessing a monoclinic crystal structure, β-Ga_2O_3 has also been explored for epitaxial growth of wurtzite GaN and could provide an attractive, electronically conductive, and transparent substrate

J. B. Varley (✉)
Materials Science Division, Lawrence Livermore National Laboratory, Livermore, CA 94551, USA
e-mail: varley2@llnl.gov

M. Higashiwaki and S. Fujita (eds.), *Gallium Oxide*, Springer Series in Materials Science 293, https://doi.org/10.1007/978-3-030-37153-1_18

alternative to sapphire for solid-state lighting applications [3]. A number of other metastable phases have been identified beyond the thermodynamically preferred β phase, adding even more options for developing this material. Additionally, the promise of alloying with other elements like Al and In offers even greater control over the structure, optical, and electrical properties for engineering improved devices [9–12]. While these features make β-Ga_2O_3 extremely promising as the wide-band gap semiconductor of choice for the next generation of power and optoelectronic devices, a detailed understanding of the fundamental and defect-related properties of this material remain fairly limited.

In terms of electrical conductivity, unintentionally doped (UID) material typically exhibits n-type conductivity with free electron concentrations on the order of 10^{18} cm^{-3} [13]. While this has historically been attributed to oxygen vacancies (V_{O}), growing evidence has shown that unintentional impurities are likely responsible for the observed unintentional conductivity, as we discuss in Sects. 18.3.1 and 18.3.2. Doping of Ga_2O_3 has most commonly used Sn as a dopant, while other candidates such as Si are now more routine. The particular choice of optimal dopant(s) for controlling the conductivity is one aspect that has been quite poorly understood. Here, we describe the results of calculations based on hybrid functional calculations to discuss the role of native and extrinsic defects in the observed electrical conductivity in Ga_2O_3 [14–20]. We discuss the properties of known n-type dopants like the group IV elements and hydrogen impurities, as well as the prospects of other dopants like transition metals and halides. We also discuss sources of compensation, such as through native Ga-vacancy (V_{Ga}) related centers in Sect. 18.3.3 and through extrinsic impurities such as Fe in Sect. 18.3.4. Lastly, in Sect. 18.3.5, we comment on efforts at acceptor doping in Ga_2O_3 and how an intrinsic tendency to form small hole polarons is likely to inhibit useful hole mobilities in Ga_2O_3, rendering it limited to unipolar devices.

18.2 Defect Formation Energies and Computational Methodology

The formation energy (E^f) is one of the most important quantities characterizing defects and is a critical ingredient from which many other experimentally observable properties can be derived. When considering the dilute limit, the concentration of a given defect is related to its formation energy via a Boltzmann expression [21]:

$$c = N_0\, e^{-E^f/k_{\mathrm{B}}T}, \tag{18.1}$$

where k_{B} is the Boltzmann constant, and T is the temperature. N_0 represents the number of sites on which the defect can be incorporated, including the number of possible configurations per site, and is on the order of 10^{22-23} sites cm^{-3}. Evaluating the E^f for a representative set of defects (e.g., native defects and impurities) gives insight into which point defects and their associated charge states are expected to be the most prevalent under a given set of conditions. This knowledge can be correlated with the observed electrical and optical properties to identify which defects may have the most impact on the resulting properties of a given material and how to ultimately inform synthesis, processing, and passivation schemes for tailoring desired performance.

The E^f also depends on the chemical potentials relevant to the defect. These chemical potentials reflect the conditions of a chemical environment, i.e., the reservoirs with which atoms or charges are exchanged. While the chemical potentials are fixed for elemental solids, they are variables for non-elemental materials like Ga_2O_3, where the chemical potentials can vary over a range set by the stability conditions of the compound. This is expressed by an equality of the sum of the constituent chemical potentials with the formation enthalpy of the solid. As defects can also exhibit different charge states, the E^f also depends on the energy of the electron reservoir in the material, which is known as the Fermi level (ε_F).

To illustrate the way in which a formation energy is computed in practice, we consider the example of an oxygen vacancy defect using the expression

$$E^f[V_O^q] = E_{tot}[V_O^q] - E_{tot}[Ga_2O_3] + \mu_O + q\varepsilon_F + \Delta^q, \tag{18.2}$$

where $E_{tot}[V_O^q]$ represents the total energy of the supercell containing a single oxygen vacancy in the charge state q, and $E_{tot}[Ga_2O_3]$ is the total energy of a perfect crystal in the same supercell. As previously stated, (18.2) is a function of several chemical potentials, e.g., μ_O for O atoms and which depends on environmental conditions, and ε_F which sets the chemical potential of electrons that can be exchanged with the defect. The typical approach is to consider μ_O that can vary from being in equilibrium with O_2 ($\mu_O = \frac{1}{2}E_{tot}[O_2(g)]$), defining O-rich/Ga-poor conditions, to the opposite extreme of the O-poor/Ga-rich limit where it is additionally limited by the formation enthalpy of Ga_2O_3 (e.g., $\mu_O = \frac{1}{2}E_{tot}[O_2(g)]) + \frac{1}{3}\Delta H[Ga_2O_3]$). The stability condition provides a connected relationship with the chemical potential of Ga, where μ_{Ga} is related to the energy per atom in elemental Ga in the O-poor/Ga-rich limit with an additional contribution of up to $\frac{1}{2}(\Delta H[Ga_2O_3] - 3\mu_O)$ in the O-rich/Ga-poor limit. Similar limits relate the chemical potentials of extrinsic elements (dopants and impurities) that may have their solubilities limited by competing phases that are the most restrictive for a given set of conditions. For example, this includes competing oxides phases in more O-rich conditions (e.g., SnO_2, GeO_2 and SiO_2 for group IV dopants and Fe_2O_3 for Fe impurities) while intermetallic phases can limit the solubility in more Ga-rich conditions (e.g., Ga_3Fe for Fe impurities). The electron chemical potential is defined by the Fermi level position ε_F, which we reference to the valence-band maximum (VBM) and consider up to conduction band minimum (CBM) to evaluate a range spanning the band gap. The Δ^q term defines the finite-size correction for charged defects following the scheme of [22, 23]. This term depends on dielectric screening of the material and details of the simulations and is discussed in Sect. 18.2.2.

In practice, equilibrium conditions may not always apply and one must be very careful when applying such expressions for chemical potential values. For example, relating the chemical potential of an element to the partial pressure of its gas requires that the diffusivity of the species in the solid is sufficient enough to equilibrate with the gas species outside. This can differ considerably for different grown conditions and growth techniques. As a consequence of these issues, calculated formation energies often do not include the additional contributions of temperature and partial pressure to the chemical potentials and thus must be considered with caution when attempting to obtain defect concentrations for direct comparisons with experiment.

18.2.1 Defect Charge-State Transition Levels

Defects almost always introduce levels in the band gap or near the band edges of a semiconductor or insulator. These levels can be either empty or filled, and the transitions between different charge states of the defect are directly related to the occupancy of these levels. We can determine these transition levels from one charge state to another from the E^f. These transitions correspond to the Fermi-level position where charge states q_1 and q_2 have equal energy. Continuing with an explicit example for the V_O in Ga_2O_3, we consider the transition between the 2+ and + charge states:

$$\varepsilon(2+/+) = -\frac{E^f(V_O^{2+}; \varepsilon_F = 0) - E^f(V_O^{+}; \varepsilon_F = 0)}{2-1}. \tag{18.3}$$

Here, $E^f(V_O^q; \varepsilon_F = 0)$ is the formation energy of the V_O in charge state q when the Fermi level is at the VBM. Since the transition level represents the crossing point of the two formation energies as a function of the Fermi level, the V_O^{2+} charge state will be more stable for Fermi-level values below ε(2+/+), while V_O^+ will be more stable for Fermi-level values above ε(2+/+). Also called thermodynamic transition levels, these levels represent the physical situation in which one defect charge state can fully relax to its equilibrium configuration after a charge-state transition. An important feature of (18.3) is that the dependence on chemical potentials cancels, so that concerns expressed above in the choices of particular μ values do not apply. Some defects can also undergo transitions with the preferential trapping of two electrons or holes, with the second bound more strongly than the first. These defects are called negative-U centers based on the fact that a given defect configuration enables a net attraction between otherwise coulombically repelled carriers. Oxygen vacancies in many semiconducting oxides are known to exhibit this type of behavior, where the thermodynamically preferred transition is the ε(2+/0). This is also the case in Ga_2O_3, as we discuss in Sect. 18.3.1.

For defects that introduce states very close to the conduction or valence-band edges, known as shallow centers, these transition levels correspond to the thermal ionization energy of the hydrogenic effective-mass levels. For defects with localized states in the band gap, known as deep centers, these transitions levels are what are experimentally probed in experiments such as deep-level transient spectroscopy (DLTS) measurements. Spectroscopic techniques that rely on optical stimulation, such as photoluminesence measurements, instead probe optical or vertical charge-state transition levels that can be derived from a configuration coordinate diagram analysis [21]. Assignments of defect transition energies associated with optical absorption and emission energies should not be compared with theoretical energies obtained via (18.3), but instead need to incorporate additional relaxation energies associated with the vertical transitions. These relaxation energies are found to be quite large in Ga_2O_3, and can exceed 1 eV for a number of relevant defects like vacancies. These facts make charge-state transition levels important descriptors of a defect's electrical behavior.

18.2.2 Evaluating Formation Energies

The formalism captured in (18.1) and (18.2) is general and can be built upon any level of theory from which to evaluate the total energies of the various defect configurations and reference energies. For our approach, we adopt a methodology based on density functional theory (DFT) to determine the energies that are input into the above expressions to evaluate defect formation energies. Specifically, our calculations adopt the HSE06 screened hybrid functional [25, 26] and projector-augmented wave (PAW) approach [27] as implemented in the VASP code [28].

Hybrid functionals provide advantages over conventional DFT-based methods (e.g., the local density and generalized gradient approximations, LDA, and GGA) that typically suffer from a severe underestimation of the band gap and fundamental errors in their description of charge localization in open-shell systems. The former and the latter points are both critical deficiencies in the study of point defects and their associated transition levels, as both can significantly influence the qualitative and quantitative conclusions of where defect-induced states may fall within the band gap and their associated charge-state transition levels. For example, increasing the fraction of Hartree–Fock exact exchange (α) incorporated in the exchange-correlation potential leads to a larger band gap more comparable to experiment and can begin to correctly predict charge localization (e.g., the formation of small polarons and Jahn–Teller distortions). However, increasing this fraction too much can overshoot experimental band gaps and over-stabilize the localization of charge, making this approach not a true ab initio methodology and sensitive to the parameter choices.

To reproduce the experimental band gaps, the α mixing parameter is adjusted between 32 and 35% for Ga_2O_3, depending on the treatment of the Ga 3*d* electrons [12, 14, 19]. Specifically, this predicts an indirect band gap of 4.86 eV that is nearly equivalent to the direct band gap of 4.87 eV and allows for direct quantitative comparisons with experiments. As summarized in [12, 14], this choice of parameter also leads to an improved description of the lattice constants and bond lengths of Ga_2O_3.

In the defect calculations, we use a supercell approach, where the bulk is typically represented by a periodically repeated unit of 120 to 160 atoms for bulk monoclinic β-Ga_2O_3 [14, 16–19]. All calculations were performed using a $2 \times 2 \times 2$ mesh of Monkhorst-Pack special k-points, and plane-wave basis sets with a cutoff of 400 eV. Owing to placing charged defects in a periodic supercell, corrections for the finite-size effects resulting from the long-range Coulomb interaction of charged defects in neighboring supercells are required. We adopt the approach following the scheme of Freysoldt et al. [21–23], which requires the dielectric screening as an input. To additionally assess the uncertainty in our reported defect levels, we consider the two extremes where the charged defects are corrected using the experimental dielectric constants that reflect the purely electronic contribution (ε_∞) and additionally including ionic screening contributions owing to the response of the lattice to the defect (ε_0) [19]. For these values, we adopt the diagonal elements of the low- and high-frequency anisotropic dielectric tensor as reported by Schubert et al. in [29]. While the Freysoldt approach generally adopts ε_0, recent work has questioned the validity of this choice for the screening in materials that exhibit large differences in the ionic and electronic dielectric constants such as β-Ga_2O_3 and that ε_∞ is more appropriate for the supercell sizes used in the simulations [18]. For the purposes of greater transparency in the reported calculated levels, we include a summary of both approaches to assist in resolving the identification of the defect states experimentally identified in Ga_2O_3. However, as included in [16] and from our own tests for V_{Ga}^{-3} in supercells containing up to 1120 atoms, we find that corrections with ε_0 yield more reliable final energies for evaluating the stability and charge-state transition levels for comparisons with experiment.

18.3 Results and Discussion

The study of point defects in Ga_2O_3 is complicated by the more complex monoclinic β-gallia structure as compared to other conventional semiconductors. As shown in Fig. 18.1a, the β-phase has two crystallographically distinct Ga sites: Ga_I that are tetrahedrally coordinated and Ga_{II} that are octahedrally coordinated. The oxygen sublattice has three distinct sites: O_I and O_{II} that are three-fold coordinated and O_{III} that are tetrahedrally coordinated. The O_I and O_{II} differ in their local bonding environment, with the O_I bonded in a trigonal pyramidal fashion to two Ga_{II} and one Ga_I, while O_{II} are bonded in a trigonal planar fashion to two Ga_I and one Ga_{II}. O_{III} are bound to three Ga_{II} and one Ga_I. Owing to the number of distinct sites in which vacancies may form or impurities may incorporate, the defect physics of Ga_2O_3 is particularly rich. However, we stress that caution must be exercised in comparing the results from theoretical studies, as some works adopt different naming conventions for these sites than those originally reported for β-Ga_2O_3 [14, 16, 17, 30].

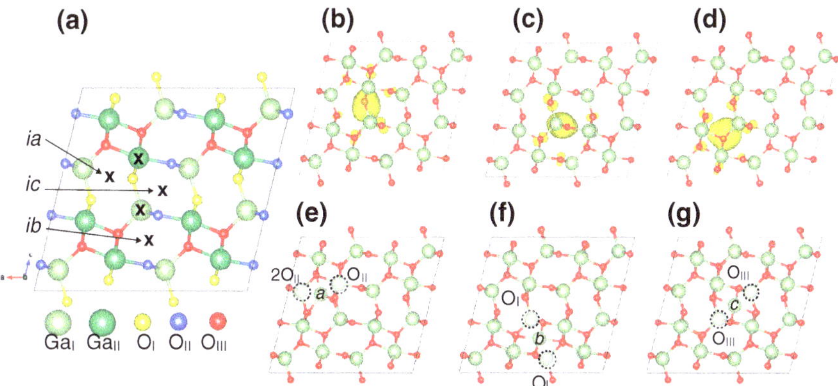

Fig. 18.1 **a** Representation of the β-Ga_2O_3 structure highlighting the crystallographically distinct Ga and O sites with different colors. Configurations of the neutral oxygen vacancies at the O_I (**b**), O_{II} (**c**), and O_{III} (**d**) sites are shown with the charge density isosurfaces of their respective occupied states within the band gap, shown at 10% of their maximum value. Their localized nature supports their behavior as deep, rather than shallow donors. In addition to vacancies forming on the Ga_I and Ga_{II} sites, we highlight possible displaced Ga-vacancy configurations (V_{Ga}^{i}) in (**e**)–(**g**) that correspond to the interstitial sites marked in (**a**). Structures were generated using the VESTA program [24]

18.3.1 Oxygen Vacancies

The calculated formation energies for V_O in Ga_2O_3 are shown in Fig. 18.2 shown for the two extremes of O-rich and O-poor conditions. Owing to the different geometries and local environments of the three O sites, each V_O exhibits distinct formation energies and transition levels. The ε(+2/0) transition levels, denoted by the kinks in the formation energy plots, are all more than 1 eV below the CBM. We summarize our values in Table 18.1, where different theoretical studies have reported slightly different values depending on details of the methods used but give the same general concensus on the behavior of the V_O [14, 16, 18, 20]. For example, the initial reports identified ε(2+/0) transitions of 3.31 eV, 2.70 eV, and 3.57 eV above the VBM for the O_I, O_{II}, and O_{III} sites, respectively [14]. Results in Table 18.1 find that slight changes in the underlying computational details do not lead to significant changes in these values.

These results imply that V_O acts as a *deep* donor and cannot contribute to n-type conductivity, as all three V_O are stable in the neutral charge state for Fermi levels in the upper part of the band gap. This is supported by the localized charge distribution associated with the occupied defect state within the band gap formed for the neutral V_O configurations included in Fig. 18.1b–d. For most values of oxygen chemical potential, the E^f of the V_O are quite high and would suggest a low concentration of V_O vacancies would form. However, in more Ga-rich growth conditions, the E^f can be appreciably low that these defects can form, particularly, at higher-temperatures

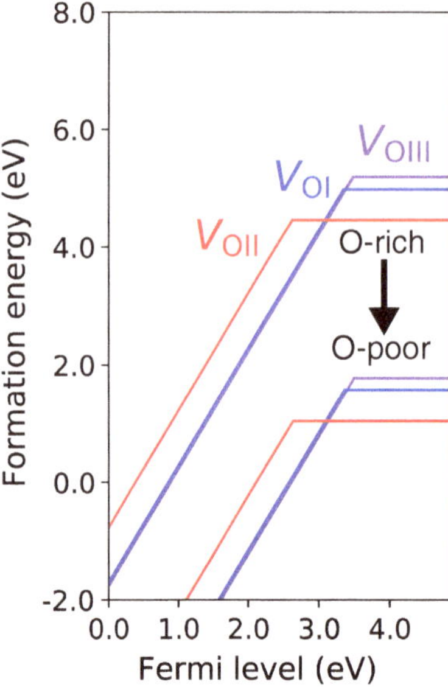

Fig. 18.2 Formation energy diagram for oxygen vacancies in β-Ga_2O_3 shown for both O-rich and O-poor growing conditions. The data is shown as a function of the Fermi level ranging over the entire band gap from the VBM (0 on the x-axis) to the CBM (relevant for n-type conditions typical in Ga_2O_3). Vacancies formed on each of the three crystallographically distinct O sites are included. The slopes of the lines represent the lowest-energy charge state at a given Fermi level (+2 and 0 for the defects shown), and the kinks represent the ε(2+/0) thermodynamic charge-state transition levels

such as those in growth from the melt [4]. This becomes even more pronounced for Fermi levels deeper in the band gap and below the ε(2+/0) transitions, as would be expected via acceptor doping. Therefore, V_O-related defects are expected to be much more relevant in insulating material, while the growth and processing conditions of more n-type Ga_2O_3 may lead to much more variability on whether V_O will be present in significant concentrations. For neutral configurations, V_O^0 is the most favorable on the O_{II} site for n-type conditions to approximately 1.5 eV below the CBM. This likely correlates with the longer O–Ga bond lengths associated with the O_{II} site and may help explain why the (100) is the easy cleavage plane in β-Ga_2O_3, as the O_{II} are oriented in these planes as seen in Fig. 18.1a. For lower Fermi levels, the $V_{O_{III}}^{2+}$ is expected to be the most favorable, with the $V_{O_I}^{2+}$ only slightly higher in energy.

18.3.2 Donor Impurities and Dopants

Because native V_O cannot be the source of the unintentional conductivity in Ga_2O_3, we now turn to impurities. In Fig. 18.3, we include defects associated with hydrogen,

Table 18.1 Calculated energy level positions of the thermodynamic charge-state transition levels shown relative to the conduction band edge as taken from [20]. Values relative to the valence-band edge can be found by adding the band gap energy of 4.85 eV. The values are corrected for spurious finite-size effects in the periodic supercell assuming dielectric screening adopts ionic and electronic contributions (ε_0) and only electronic contributions (ε_∞), which give probable bounds for the levels most relevant for comparison with experiment, although values are expected to skew far closer to the ε_0. We include all transition levels for negative-U defects as indented values, as these may be probed by DLTS measurements. Comparisons with previous theory calculations are included where available, with the naming conventions of each crystallographically distinct site standardized as defined as in this work and [30]

Defect level	Transition level ε	Corrected with ε_0	Corrected with ε_∞
V_{O_I}	(2+/0)	−1.50 (−1.71[a], −1.52[b], −1.72[d])	−1.93 (−2.10[a])
	(2+/+)	−1.30	−1.94
	(+/0)	−1.69	−1.91
$V_{O_{II}}$	(2+/0)	−2.23 (−2.29[a], −2.13[b], −2.42[d])	−2.65 (−2.68[a])
	(2+/+)	−2.00	−2.64
	(+/0)	−2.45	−2.66
$V_{O_{III}}$	(2+/0)	−1.36 (−1.56[a], −1.26[b], −1.52[d])	−1.79 (−1.95[a])
	(2+/+)	−1.14	−1.78
	(+/0)	−1.68	−1.92
V_{Ga_I}	(−2/−3)	−1.76 (−1.64[a], −1.62[c], −1.93[d])	−0.69 (−0.67[a])
	(−1/−2)	−2.32	−1.68
	(0/−1)	−2.62	−2.40
	(+/0)	−3.42	−3.63
$V_{Ga_{II}}$	(−2/−3)	−2.17 (−2.12[a], −1.83[c], −2.27[d])	−1.11 (−1.16[a])
	(−1/−2)	−2.50	−1.85
	(0/−1)	−3.18	−2.97
	(+/0)	−3.65	−3.85
V_{Ga}^{ia}	(−2/−3)	−2.16	−1.07
	(−1/−2)	−2.39	−1.74
	(0/ 1)	−3.16	−2.94
	(+/0)	−3.70	−3.92
V_{Ga}^{ib}	(−2/- 3)	−1.91	−0.87
	(−1/−2)	−2.11	−1.55
	(0/−1)	−3.29	−3.08
	(+/0)	−4.00	−4.21
V_{Ga}^{ic}	(−2/−3)	−2.55	−1.50
	(−1/−2)	−2.82	−2.16
	(0/−1)	−3.23	−3.02
	(+/0)	−3.96	−4.18
V_{Ga}^{ia}−H	(−1/−2)	−2.38	−2.22
V_{Ga}^{ib}−H	(−1/−2)	−2.21	−1.55
V_{Ga}^{ic}−H	(−1/−2)	−2.82	−2.18

(continued)

Table 18.1 (continued)

Defect Level	Transition level ε	Corrected with ε_0	Corrected with ε_∞
V_{Ga}^{ib}−2H	(0/−1)	−3.57	−3.35
	(+/0)	−4.12	−4.34
V_{Ga}^{ic}−2H	(0/−1)	−3.44	−3.26
	(+/0)	−4.18	−4.39
V_{Ga}^{ia}−2H	(0/−1)	−2.83	−2.61
	(+/0)	−3.96	−4.17
Fe_{Ga_I}	(0/−1)	−0.59[e]	−0.36[e]
	(+/0)	−4.29[e]	−4.52[e]
$Fe_{Ga_{II}}$	(0/−)	−0.61[e]	−0.38[e]
	(+/0)	−4.34[e]	−4.57[e]
$Fe_{Ga_{II}}$−H	(+/0)	−1.29	−1.52
Fe_i	(3+/2+)	−2.10	−3.33
	(2+/+)	−0.24	−0.97

[a] [18]
[b] [14]
[c] [15]
[d] [16]
[e] [19]

a ubiquitous impurity that has previously been identified to be electrically active in a number of different semiconducting oxides [15, 31]. Hydrogen can favorably occupy either interstitial (H_i) or substitutional sites (H_O); in both configurations, H acts as shallow donor [14]. Due to the complex crystal structure of β-Ga_2O_3, many configurations exist in which H_i^+ forms a strong bond with an O atom and which are all close in energy. We show the lowest-energy configuration in Fig. 18.3b, where H_i^+ bonds to a lone pair of the three-fold coordinated O_I. From the low E^f of the H_i defects in a large range of conditions, it is expected to be easily incorporated in Ga_2O_3 as an unintentional impurity whenever H is present in the growth or annealing environment. In the acceptor charge state, H_i^-, the H atom preferentially sits near two Ga atoms as shown in Fig. 18.3c, yielding a ε(+/−) transition between the donor and acceptor configurations 4.90 eV above the VBM, which falls just above the CBM [14]. Interstitial hydrogen thus behaves exclusively as a shallow donor for any Fermi level within the band gap of β-Ga_2O_3. This result is consistent with experimental studies on the electrical behavior of muonium in Ga_2O_3, which acts as an electrical analog to H_i [32].

Substitutional hydrogen, H_O, has a low formation energy only under more O-poor conditions, as seen in Fig. 18.3a and discussed in [14]. The overall stability of H_O^+ defects has been predicted to be modest; previous assessments approximated an energy barrier of ∼1.33 eV for the dissociation to a V_{O_I} and mobile H_i^+ species, [14] while explicit calculations identified a barrier of ∼1.22 eV to dissociate the H_O [20]. These barriers suggest that H_O should be readily removable via thermal

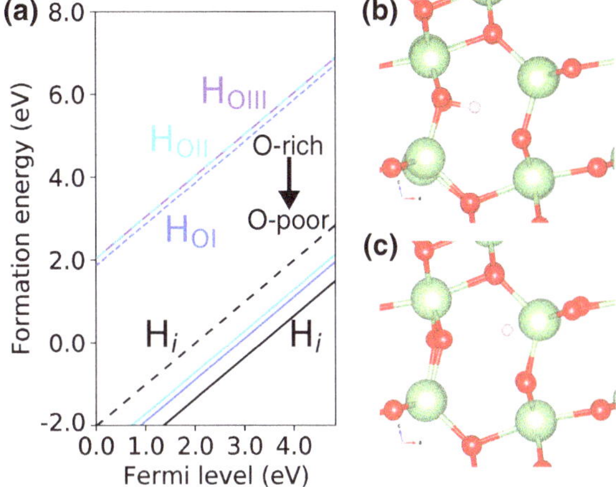

Fig. 18.3 **a** Formation energy diagram for H_i and H_O defects in β-Ga_2O_3 shown for both O-rich and O-poor growing conditions. These defects are found to be stable exclusively in the positive charge state and therefore found to exhibit shallow donor behavior. The dashed lines correspond to O-rich conditions, while the solid lines correspond to O-poor conditions. Structures of the most favorable H_i donor and acceptor configurations are included in (**b**) and (**c**)

annealing, although the induced H_i^+ may also strongly interact with other defects as we discuss in Sect. 18.3.3. For example, barriers associated with H_i^+ migration have been reported to be 0.3–0.4 eV, [14, 33] which suggest that H_i^+ may rapidly diffuse and interact with other defects such as V_{Ga} or other acceptors and facilitate passivation via the formation of highly stable complexes like V_{Ga}−2H species recently observed [20, 34]. Other types of interstitial hydrogen in Ga_2O_3, such as H_2 molecules, were identified to have higher energies than H_i^+ for all conditions but may also act as a source of H_i. These facts suggest the importance of H_i and H_O defects and how understanding their evolution via interactions with other defects requires further study.

In addition to hydrogen, a number of other impurities may also contribute to the observed n-type behavior. In Fig. 18.4, we show the calculated formation energies of Si, Ge, and Sn substituting on the Ga site as in [14]. The results suggest that all of these can behave as favorable shallow donors and may contribute to the n-type conductivity of Ga_2O_3. It has now been established that Si is a dominant background impurity in both high-purity powders (6N) and single crystals that is likely responsible for the unintentional n-type conductivity, [14, 35] and that it and the other group IV elements are effective n-type dopants when incorporated explicitly [2, 36, 37]. Si and Ge prefer the tetrahedral coordination of the Ga_I site, while Sn prefers the octahedral coordination of the Ga_{II} site. For the Si and Sn dopants, there is a strong site preference, with the opposite site energies ∼1 eV less favorable and strongly suggest that these dopants will exclusively incorporate on the tetrahedral and octahedral sites, respectively. For Ge dopants, there is a much smaller energy difference between the two sites, and suggest that Ge may incorporate on both sites. In Fig. 18.4b, we also

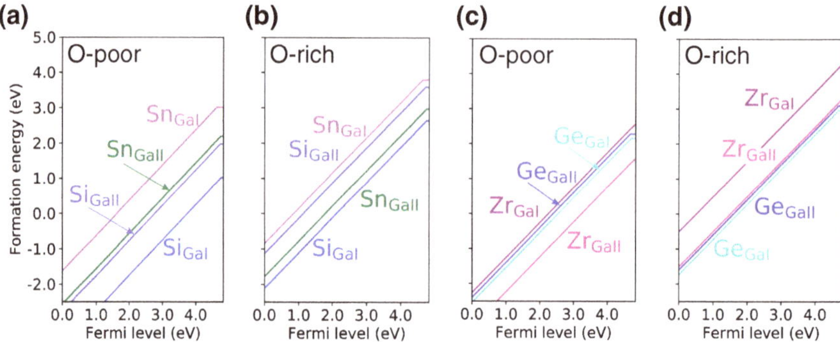

Fig. 18.4 Formation energy diagram for donor impurities like the dominant Si and Sn dopants (**a, b**) and emerging Ge and Zr dopants (**c, d**) in β-Ga_2O_3. The results are shown for both Ga-rich/O-poor conditions in (**a**) and (**c**) and O-rich/Ga-poor conditions in (**b**) and (**d**)

include the formation energy of Zr, which is found to act as a favorable shallow donor that strongly prefers to incorporate on the octahedral site. This behavior is consistent with similarities in the electrical properties of Zr- and Si-doped single crystals [35]. Other transition metal dopants have also been considered, with Nb also predicted to be a favorable shallow donor candidate [38]. Interestingly, the results suggest that Zr is generally expected to be more readily incorporated in Ga_2O_3 than Sn, which has traditionally be the n-type dopant of choice [2, 35]. Additional donors such as the halide incorporation on the O site (F_O and Cl_O) were also predicted to behave as shallow donors, [14], although this behavior has remained to be experimentally verified. Considering these facts, the unintentional n-type conductivity in Ga_2O_3 can be best explained by the unintended incorporation of a number of different elements such as Si and H that act as electrically active donors. Conversely, exploiting this fact with intentional doping has led to high and controllable carrier concentrations necessary for engineering Ga_2O_3-based electronic devices.

18.3.3 Gallium Vacancies and Compensation

Next, we turn to gallium vacancies, which have been identified to be native acceptors that can exhibit low formation energies in n-type conditions [15–18]. Unlike other wide-band gap semiconducting oxides like In_2O_3 or SnO_2, cation vacancies in Ga_2O_3 exhibit low enough formation energies for them to be incorporated in sizeable concentrations during growth, and they act as compensating acceptors, reducing electron conductivity [15].

We plot the formation energy of the V_{Ga} in Fig. 18.5a shown for both the O-rich and O-poor (Ga-rich) limits. We first note that the V_{Ga} are deep acceptors, with the $\varepsilon(-2/-3)$ transition level well below the CBM for both inequivalent sites. As seen from Table 18.1, the theoretically reported values for this transition are believed to

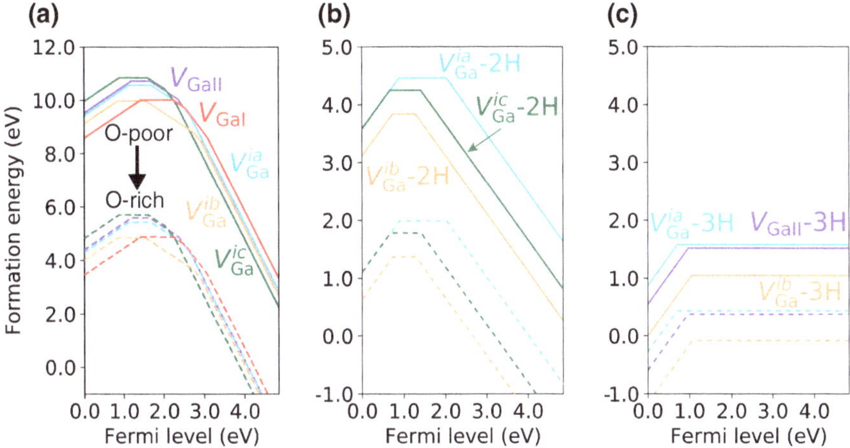

Fig. 18.5 Formation energy diagram for (**a**) isolated V_{Ga} in β-Ga_2O_3 shown for both O-rich and O-poor growing conditions. Vacancies formed on the tetrahedral (Ga_I) and octahedral (Ga_{II}) sites, as well as the three displaced V^i_{Ga} sites shown in Fig. 18.1 are included. V_{Ga} complexes with two H_i are included in (**b**) and three H_i in (**c**). The dashed lines correspond to O-rich conditions, while the solid lines correspond to O-poor conditions

occur ∼1.8 to 2.6 eV below the CBM, depending on the particular V_{Ga} configuration. Such deep transition levels suggest all V_{Ga} will be in the −3 charge state even when the Fermi level is well below the CBM, and not just in *n*-type conditions. We also include the values for the other transition levels that fall within the band gap in Table 18.1, where the V_{Ga} can trap up to 4 holes. We note that particular values of the transition levels can depend on which adjacent O atoms the holes localize in the various charge states. While the most stable configurations are assumed in our summarized results in Table 18.1, higher-energy configurations may exist where holes are trapped on other neighbors. For example, the $V_{Ga_{II}}$ preferentially localizes the first and second holes ($V^{-2}_{Ga_{II}}$ and $V^{-1}_{Ga_{II}}$, respectively) on the two adjacent O_I; a hole trapped instead on an adjacent O_{II} is 0.35 eV higher in energy and would yield correspondingly different charge-state transition levels.

Of the two inequivalent sites, the octahedral $V_{Ga_{II}}$ is predicted to be more favorable than V_{Ga_I} in *n*-type conditions. However, V_{Ga} exhibit a number of stable configurations resulting from a large relaxation of an adjacent Ga to an interstitial octahedral site, as discussed in [15, 17, 20] and illustrated in Fig. 18.1. The relaxation to these interstitial sites, which we denote as V^i_{Ga} configurations *a*, *b*, and *c*, does not occur completely spontaneously, i.e., there exists an energy barrier to go from an "on-site" V_{Ga} local minimum to a lower-energy V^i_{Ga} configuration. The height of these barriers have been calculated to be ∼0.6 eV or less, [15, 17] suggesting they can be easily overcome even at modest temperatures and that the V^i_{Ga} are expected to be the dominant configurations based on their favorable formation energies in *n*-type Ga_2O_3 shown in Fig. 18.5. Of the various configurations, in *n*-type conditions, we find the V_{Ga}^{ic} configuration to be the most favorable, with the V_{Ga}^{ib} slightly higher, and the V_{Ga}^{ia} and $V_{\mathrm{Ga_{II}}}$ slightly higher still and nearly equivalent in energy. Holes in the lower charge states preferentially localize on the O dangling bonds associated with the V_{Ga}^{i} configurations (see Fig. 18.1e, f). For the V_{Ga}^{ib} (V_{Ga}^{ic}), up to 2 holes can be trapped on the O_{I} (O_{III}) dangling bonds before localizing on the Ga_i octahedra. For the V_{Ga}^{ia}, up to 3 holes preferentially localize on the 3 adjacent O_{II} dangling bonds before the fourth localizes on the Ga_i tetrahedra. The first two holes associated with the -1 and -2 charge states of the V_{Ga}^{ia} preferentially localize on the 2 symmetric O_{II} denoted as $2O_{O_{II}}$ in Fig. 18.1e. Owing to the complex bonding environments of the V_{Ga} in β-Ga_2O_3, information on preferential hole-trapping sites may help in the identification of V_{Ga} species via measurement techniques highly sensitive to the local environment such as electron paramagnetic resonance (EPR).

Historically, the changes in conductivity in oxides observed upon annealing in different oxygen atmospheres have traditionally been explained in terms of V_{O} formation. Considering the results for V_{O} in Sect. 18.3.1, the low E^f of V_{Ga} acceptors and their strong dependence on the O chemical potential seen in Fig. 18.5, V_{Ga} are more likely responsible for the reduction in conductivity upon annealing in O-rich environments. Reports on overall migration barriers, including the metastable sites have also been reported in [17], finding barriers ranging from ~0.6 to 2 eV. While these barriers are likely slightly underestimated owing to being calculated within the GGA, [15, 20] they strongly suggest that V_{Ga} are far more mobile than V_{O} and are mobile at lower temperatures compared to other semiconducting oxides like In_2O_3 and SnO_2 [39, 40]. Thus, it is expected that in oxygen-rich environments, the generation of V_{Ga} leads to compensating acceptors and hence decrease the conductivity as observed experimentally [2, 13, 16, 41]. Conversely, in reducing conditions the V_{Ga} are much higher in energy and any V_{Ga} present in the system will tend to diffuse out under these Ga-rich conditions, hence, raising the conductivity. This model of V_{Ga} has been able to explain the experimentally observed oxygen partial pressure dependence on the electrical conductivity in Ga_2O_3 and strongly supports the relevance of V_{Ga}-related defects as sources of compensation [16].

As discussed in Sect. 18.3.2, H_i^+ is also a highly mobile species that may be attracted to V_{Ga} acceptors. Previous studies have highlighted the stability V_{Ga}-H complexes [15, 20], which can exhibit a large number of possible configurations owing to the complexity of the β-Ga_2O_3 lattice. Complexes involving single hydrogens were found to be highly favorable defects in n-type conditions, and predicted to form strong complexes stable to dissociation below temperatures ~700 to 800 °C [15]. Apart from the stability, the electronic consequences of such defects led to passivated acceptors with transition levels roughly equivalent to the transition levels associated with the analogous charge state of the isolated V_{Ga}. For example, a singly passivated V_{Ga}^{ib}-H $\varepsilon(-1/-2)$ falls 2.64 eV above the VBM, whereas the analogous $\varepsilon(-1/-2)$ for the isolated V_{Ga}^{ib} is 2.74 eV. The difference is even smaller for the other V_{Ga}^{i} configurations. Thus, the defect levels can be approximated well from the isolated V_{Ga} for singly hydrogenated complexes.

Experimentally, a highly stable 2H-containing complex formed with the V_{Ga}^{ib} configuration has been found to exist in a number of different samples [20, 34]. We

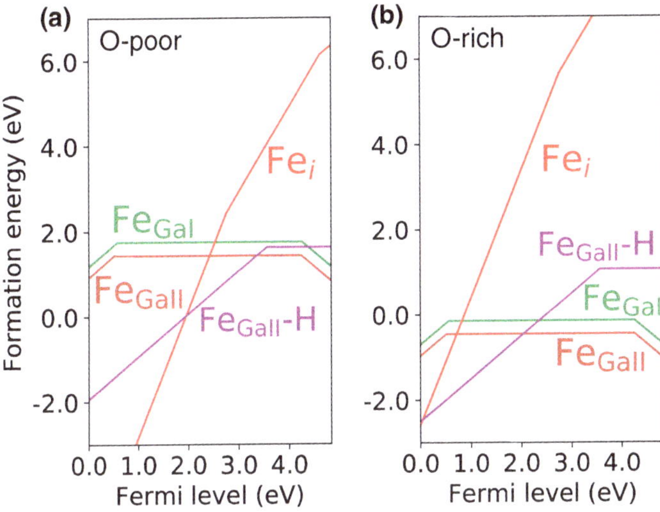

Fig. 18.6 Formation energy diagram for Fe impurities in β-Ga_2O_3 shown for **a** Ga-rich (O-poor) and **b** O-rich (Ga-poor) conditions. The negative formation energies for Fe_{Ga} in O-rich conditions highlight how the theoretically defined O-rich limit is thermodynamically inaccessible

include the formation energies of these complexes in Fig. 18.5b, where the V_{Ga}^{ib}−2H defect is predicted to be very favorable and exhibit an ε(0/−1) deep acceptor level 1.29 eV above the VBM, as well as an ε(+/0) donor level 0.73 eV above the VBM [20]. Comparison with the isolated V_{Ga} species in Table 18.1 suggests that approximation of the levels from the analogous isolated V_{Ga} levels is slightly worse than for singly passivated defects, but still gives reasonable values for comparison with experiment. These complexes are also highly stable, with predicted binding energies of 2.5, 2.0, and 2.3 eV for the second H_i^+ to bind to the $(V_{Ga}^i\text{-H})^{-2}$ species at the b, c, and a configurations. The binding of a third H to form neutral complexes, as seen in Fig. 18.5c, is also found to be stable but with weaker binding than the first two H. For example, the binding energy of the third hydrogen to form a neutral V_{Ga}^{ib}-3H complex is only 0.7 eV relative to the H_i^+ and $(V_{Ga}^{ib}-2\text{H})^{-1}$ constituents, whereas it is 1.55 eV for the V_{Ga}^{ia}-3H complex owing to the three under-coordinated O_{II} dangling bonds associated with this configuration. These are significantly weaker than the binding of the second H within the complexes and suggest the neutral complexes will be more readily dissociated to lead to electrically active centers that may contribute to compensation in n-type material. Owing to the predicted favorability and stability of V_{Ga} and H_i complexes and their role in compensation, [42] these defects are highly relevant in Ga_2O_3 and require more study to determine their overall impact in the performance and longevity of Ga_2O_3-based devices.

18.3.4 Acceptor Impurities

Beyond native acceptors such as V_{Ga} and related complexes, other acceptor impurities or dopants may additionally contribute to compensation of n-type samples or even be utilized to move the Fermi level far away from the conduction band, leading to insulating material. Commercially available substrates from the Tamura Corporation have adopted Fe as a dopant in semi-insulating samples, suggesting its behavior as an acceptor. In Fig. 18.6, we include the formation energy of Fe defects,[19] where it is found to exhibit low formation energies to incorporate as a substitutional dopant on the Ga site and act as an amphoteric impurity. Interstitial configurations and substitution on the O site are found to be significantly higher in energy, particularly, in n-type conditions. The $\varepsilon(0/-)$ transition was calculated to fall 0.61 (0.59) eV below the CBM for the $Fe_{Ga_{II}}$ ($Fe_{Ga_{I}}$), which are in good agreement with experimentally reported levels of ~0.78 to 0.86 eV below the CBM [19, 37]. Additionally, this transition level was strongly correlated with the E2 level identified nearly ubiquitously in bulk single crystals grown from different approaches, [43–47] supporting that Fe_{Ga} is readily incorporated during growth and is the origin of this dominant trap level. In Fig. 18.6, we also include a favorable complex between Fe_{Ga} and H, where H_i binds to an adjacent O_I and yields a deep donor configuration. We find relative to the isolated $Fe^{-}_{Ga_{II}}$ and H_i^{+}, this complex has a binding energy of 0.7 eV and suggests it is much less stable than the V_{Ga}-H complexes. Owing to the influence that various Fe-related defects can have on the electrical properties of Ga_2O_3, its concentration must be controlled in the future development of Ga_2O_3-based devices.

18.3.5 Fundamental Barriers to p-Type Dopability

Lastly, we comment on the prospects of acceptor doping to yield p-type material. Several efforts have focused on Mg doping, as Mg acceptors incorporating on the Ga site in GaN have led to the realization of p-type GaN; however, attempts at Mg doping Ga_2O_3 have shown to produce deep acceptors with ionization energies in excess of 1 eV [37, 48, 49]. Multiple studies have surveyed the prospects of other candidate acceptors in Ga_2O_3 in the hopes of identifying shallow acceptors suitable for yielding p-type material [50–52]. However, these studies identified that other candidates such as substitutional Zn_{Ga}, Be_{Ga}, N_O, and others all produce deep $\varepsilon(0/-)$ levels in excess of 1 eV above the VBM, similar to Mg_{Ga} [50–52]. A number of acceptor candidates were also identified to yield *donor* configurations for Fermi levels even closer to the VBM [50, 52]. This behavior can be rationalized by a strong preference for small hole polaron formation in Ga_2O_3, in which local atomic relaxations lead to a preferential trapping of holes on individual O $2p$ orbitals rather than forming delocalized holes [53]. Previous studies identified that bulk β-Ga_2O_3 exhibits both a very low barrier for polaron formation, as well as a significant energy gain relative to delocalized holes when compared to a number of other wide-band gap oxides [53]. This implies that even if Ga_2O_3 could be suitably doped *p*-type (neglecting difficulties via compensation from a number of favorable donors at these Fermi levels as seen in Figs. 18.2, 18.3, 18.4, 18.5, and 18.6), hole conduction would be mediated by hopping rather than band transport in Ga_2O_3. This behavior has since been corroborated by experimental studies that identify rapid hole polaron formation, which in turn leads to the characteristic UV luminescence bands in β-Ga_2O_3 via recombination with free electrons, rather than band-edge luminescence [48, 53–55]. With an underlying propensity for hole localization in Ga_2O_3, the incorporation of substitutional defects influences the lattice and corresponding energy landscape to further favor hole localization. This can explain the deep acceptor levels observed for a large number of candidate dopants, [50–52] and suggests an intrinsic difficulty in realizing *p*-type Ga_2O_3 through conventional doping approaches.

18.3.6 Conclusions

We have discussed the behavior of a number of native and extrinsic point defects in β-Ga_2O_3 that are believed to dictate the electrical properties in this promising wide-band gap semiconducting oxide. Through the use of atomistic hybrid functional computational studies, we have identified that oxygen vacancies do not contribute to the *n*-type conductivity in typical unintentionally doped samples, whereas a number of impurities (e.g., several group IV elements, transition metals, and hydrogen) are found to be readily incorporated and can act as shallow donors. Considering the relatively high formation energies of native donors such as Ga interstitials and antisites [20], the unintentional conductivity is best attributed to readily incorporated background impurities such as Si and H. This knowledge has been exploited to lead to controllably doped material using dopants like Si, Ge, and Sn, while other candidate dopants like Zr or Nb have not been as widely explored but may also offer promising donor dopant alternatives.

Additionally, other native defects like V_{Ga} are found to be favorable in *n*-type material for a range of conditions, suggesting they can act as problematic sources of compensation. The V_{Ga} are also predicted to form favorable and stable complexes with hydrogen, increasing their solubility but also leading to passivated acceptors relative to the isolated vacancies. These vacancy complexes are found to favorably trap multiple H and are expected to survive annealing temperatures in excess of ~700 °C. Other extrinsic deep acceptors like Fe_{Ga} are also found to be highly soluble and can account for the commonly observed E2 level identified in several independent DLTS studies on a range of different samples. The potentially high solubility and stability of these acceptor defects suggest that they must be considered in further efforts in developing greater control over the electrical properties of Ga_2O_3. A better understanding of the behavior of other transition metal impurities that may be incorporated during bulk growth efforts is also needed.

Lastly, we discuss acceptor doping of Ga_2O_3 and the prospects of achieving *p*-type material. Several studies have identified large ionization energies in excess of 1 eV for a number of shallow acceptor candidates, supporting the difficulty in achieving suitably low Fermi levels for p-type conduction through doping. We explain this tendency for deep acceptor levels as a result of the intrinsic stability of small hole polarons in β-Ga_2O_3, which are found to favorably form even in the absence of substitutional dopants that influence the lattice and local bonding environments. These facts, coupled with favorability of compensating donor species such as V_O^{2+} and extrinsic shallow donors for Fermi levels approaching the VBM, strongly suggest that p-type Ga_2O_3 cannot be realized and that Ga_2O_3 will be limited to unipolar devices. Nonetheless, incorporation of deep acceptor dopants along with control over the other defect populations is a necessary step to finely tuning the carrier concentrations in Ga_2O_3 and realizing the full potential of this material in the future electronic devices.

Acknowledgements This work was partially performed under the auspices of the US DOE by Lawrence Livermore National Laboratory under contract DE-AC52-07NA27344. The work has been supported by the NSF MRSEC Program under award No. DMR05-20415, by Saint-Gobain Research, and by the Critical Materials Institute, an Energy Innovation Hub funded by the U.S. DOE, Office of Energy Efficiency and Renewable Energy, Advanced Manufacturing Office. Collaborations and discussions with J. R. Weber, H. Peelaers, J. L. Lyons, A. Schleife, A. Janotti, C. G. Van de Walle, J. S. Speck, O. Bierwagen, K. Irmscher, Z. Galazka, G. Wagner, M. Baldini, M. Higashiwaki, M. D. McCluskey, A. Perron, V. Lordi, M. E. Ingebrigtsen, and L.Vines are gratefully acknowledged.

References

1. M. Mohamed, C. Janowitz, I. Unger, R. Manzke, Z. Galazka, R. Uecker, R. Fornari, J.R. Weber, J.B. Varley, C.G. Van de Walle, Appl. Phys. Lett. **97**(21), 211903 (2010)
2. M. Orita, H. Ohta, M. Hirano, H. Hosono, Appl. Phys. Lett. **77**(25), 4166 (2000)
3. E.G. Vfllora, K. Shimamura, K. Kitamura, K. Aoki, T. Ujiie, Appl. Phys. Lett. **90**(23), 234102 (2007)
4. Z. Galazka, R. Uecker, K. Irmscher, M. Albrecht, D. Klimm, M. Pietsch, M. Brützam, R. Bertram, S. Ganschow, R. Fornari, Cryst. Res. Technol. **45**(12), 1229 (2010)
5. M. Higashiwaki, K. Sasaki, A. Kuramata, Phys. Status Solidi A **211**(1), 21 (2014)
6. M. Higashiwaki, K. Sasaki, H. Murakami, Y. Kumagai, A. Koukitu, A. Kuramata, T. Masui, S. Yamakoshi, Semicond. Sci. Technol. **31**(3), 034001 (2016)
7. J. Yang, F. Renl, M. Tadjer, S.J. Pearton, A. Kuramata, in *The Proceedings of 76th Device Research Conference* (2018)
8. Z. Hu, K. Nomoto, W. Li, Z. Zhang, N. Tanen, Q.T. Thieu, K. Sasaki, A. Kuramata, T. Nakamura, D. Jena, H.G. Xing, Appl. Phys. Lett. **113**(12), 122103 (2018)
9. H. von Wenckstern, D. Splith, M. Purfürst, Z. Zhang, C. Kranert, S. Müller, M. Lorenz, M. Grundmann, Semicond. Sci. Technol. **30**(2), 024005 (2015)
10. H. Peelaers, D. Steiauf, J.B. Varley, A. Janotti, Phys. Rev. B **92**, 085206 (2015)
11. Y. Zhang, A. Neal, Z. Xia, C. Joishi, J.M. Johnson, Y. Zheng, S. Bajaj, M. Brenner, D. Dorsey, K. Chabak, G. Jessen, J. Hwang, S. Mou, J.P. Heremans, S. Rajan, Appl. Phys. Lett. **112**(17), 173502 (2018)
12. H. Peelaers, J.B. Varley, J.S. Speck, C.G. Van de Walle, Appl. Phys. Lett. **112**(24), 242101 (2018)
13. M.R. Lorenz, J.F. Woods, R.J. Gambino, J. Phys. Chem. Solids **28**(3), 403 (1967)
14. J.B. Varley, J.R. Weber, A. Janotti, C.G. Van de Walle, Appl. Phys. Lett. **97**(14), 142106 (2010)

15. J.B. Varley, H. Peelaers, A. Janotti, C.G. Van de Walle, J. Phys.: Condens. Matter **23**(33), 334212 (2011)
16. T. Zacherle, P.C. Schmidt, M. Martin, Phys. Rev. B **87**(23), 235206 (2013)
17. A. Kyrtsos, M. Matsubara, E. Bellotti, Phys. Rev. B **95**, 245202 (2017)
18. P. Deák, Q. Duy Ho, F. Seemann, B. Aradi, M. Lorke, T. Frauenheim, Phys. Rev. B **95**(7), 35 (2017)
19. M.E. Ingebrigtsen, J.B. Varley, A.Y. Kuznetsov, B.G. Svensson, G. Alfieri, A. Mihaila, U. Badstübner, L. Vines, Appl. Phys. Lett. **112**(4), 042104 (2018)
20. M.E. Ingebrigtsen, A.Y. Kuznetsov, B.G. Svensson, G. Alfieri, A. Mihaila, U. Badstübner, A. Perron, L. Vines, J.B. Varley, APL Mater. **7**, 022510 (2019)
21. C. Freysoldt, B. Grabowski, T. Hickel, J. Neugebauer, G. Kresse, A. Janotti, C.G. Van de Walle, Rev. Mod. Phys. **86**(1), 253 (2014)
22. C. Freysoldt, J. Neugebauer, C.G. Van de Walle, Phys. Rev. Lett. **102**(1), 016402 (2009)
23. Y. Kumagai, F. Oba, Phys. Rev. B **89**(19), 5 (2014)
24. *VESTA: a Three-Dimensional Visualization System for Electronic and Structural Analysis* (2012)
25. J. Heyd, G.E. Scuseria, M. Ernzerhof, J. Chem. Phys. **118**(18), 8207 (2003)
26. J. Heyd, G.E. Scuseria, M. Ernzerhof, J. Chem. Phys. **124**(21), 219906 (2006)
27. P.E. Blöchl, Phys. Rev. B **50**(24), 17953 (1994)
28. G. Kresse, J. Furthmüller, Phys. Rev. B **54**(16), 11169 (1996)
29. M. Schubert, R. Korlacki, S. Knight, T. Hofmann, S. Schoeche, V. Darakchieva, E. Janzén, B. Monemar, D. Gogova, Q.T. Thieu, R. Togashi, H. Murakami, Y. Kumagai, K. Goto, A. Kuramata, S. Yamakoshi, M. Higashiwaki, Phys. Rev. B **93**, 12 (2016)
30. S. Geller, J. Chem. Phys. **33**(3), 676 (1960)
31. A. Janotti, C.G. Van de Walle, Nat. Mater. **6**(1), 44 (2006)
32. P.D.C. King, I. McKenzie, T.D. Veal, Appl. Phys. Lett. **96**(6), 062110 (2010)
33. S. Ahn, F. Ren, E. Patrick, M.E. Law, S.J. Pearton, A. Kuramata, Appl. Phys. Lett. **109**(24), 242108 (2016)
34. P. Weiser, M. Stavola, W.B. Fowler, Y. Qin, S. Pearton, Appl. Phys. Lett. **112**(23), 232104 (2018)
35. T. Harwig, J. Schoonman, J. Solid State Chem. **23**(1), 205 (1978)
36. E.G. Víllora, K. Shimamura, Y. Yoshikawa, T. Ujiie, K. Aoki, Appl. Phys. Lett. **92**(20), 202120 (2008)
37. A.T. Neal, S. Mou, S. Rafique, H. Zhao, E. Ahmadi, J.S. Speck, K.T. Stevens, J.D. Blevins, D.B. Thomson, N. Moser, K.D. Chabak, G.H. Jessen, Appl. Phys. Lett. **113**(6), 062101 (2018)
38. H. Peelaers, C.G. Van de Walle, Phys. Rev. B **94**, 195203 (2016)
39. P. Ágoston, K. Albe, Phys. Rev. B **81**(19), 195205 (2010)
40. W.M.H. Oo, S. Tabatabaei, M.D. McCluskey, J.B. Varley, A. Janotti, C.G. Van de Walle, Phys. Rev. B **82**(19), 193201 (2010)
41. N. Ueda, H. Hosono, R. Waseda, H. Kawazoe, Appl. Phys. Lett. **70**(26), 3561 (1997)
42. E. Korhonen, F. Tuomisto, D. Gogova, G. Wagner, M. Baldini, Z. Galazka, R. Schewski, M. Albrecht, Appl. Phys. Lett. **106**(24), 242103 (2015)
43. K. Irmscher, Z. Galazka, M. Pietsch, R. Uecker, R. Fornari, J. Appl. Phys. **110**(6), 063720 (2011)
44. Z. Zhang, E. Farzana, A.R. Arehart, S.A. Ringel, Appl. Phys. Lett. **108**(5), 052105 (2016)
45. M.E. Ingebrigtsen, L. Vines, G. Alfieri, A. Mihaila, U. Badstübner, B.G. Svensson, A. Kuznetsov, Silicon Carbide and Related Materials, Materials Science Forum, vol. 897 (Trans Tech Publications, 2017). Mater. Sci. Forum **897**, 755–758 (2016)
46. J.F. Mcglone, Z. Xia, Y. Zhang, C. Joishi, S. Lodha, S. Rajan, S.A. Ringel, A.R. Arehart, IEEE Electron Device Lett. **39**, 7 (2018)
47. A.Y. Polyakov, N.B. Smirnov, I.V. Shchemerov, E.B. Yakimov, J. Yang, F. Ren, G. Yang, J. Kim, A. Kuramata, S.J. Pearton, Appl. Phys. Lett. **112**(3), 032107 (2018)
48. T. Onuma, S. Fujioka, T. Yamaguchi, M. Higashiwaki, K. Sasaki, T. Masui, T. Honda, Appl. Phys. Lett. **103**(4), 041910 (2013)

49. J.R. Ritter, J. Huso, P.T. Dickens, J.B. Varley, K.G. Lynn, M.D. McCluskey, Appl. Phys. Lett. **113**(5), 052101 (2018)
50. J.L. Lyons, Semicond. Sci. Technol. **33**(5), 05LT02 (2018)
51. A. Kyrtsos, M. Matsubara, E. Bellotti, Appl. Phys. Lett. **112**(3), 032108 (2018)
52. H. Peelaers, J.L. Lyons, J.B. Varley, C.G. Van de Walle, APL Mater. **7**, 022519 (2019)
53. J.B. Varley, A. Janotti, C. Franchini, C.G. Van de Walle, Phys. Rev. B **85**(8), 081109 (2012)
54. T. Harwig, F. Kellendonk, S. Slappendel, J. Phys. Chem. Solids **39**(6), 675 (1978)
55. S. Yamaoka, Y. Furukawa, M. Nakayama, Phys. Rev. B **95**, 094304 (2017)

Chapter 19
Structural Properties 1

Characterization of Defects in β-Ga_2O_3 Substrates by Transmission Electron Microscopy and Related Techniques

Osamu Ueda

Abstract This chapter describes structural evaluation of β-Ga_2O_3 crystals grown by edge-defined film-fed growth process using etch pitting, focused ion beam scanning ion microscopy, transmission electron microscopy, and related techniques. Three types of defects have been found in the crystals. First, arrays of edge dislocations were observed corresponding to etch pit arrays on a ($\bar{2}01$)-oriented wafer after etching with hot H_3PO_4. In some of the dislocations, the line-shaped appearance is slightly deformed due to an inhomogeneous strain field. Next, platelike nanovoids which correspond to etch pits on the (010) plane were observed. Although the crystallographic configurations of the defects are different from that of nanometer-sized crystalline grooves which have been previously reported, they are both classified as similar type of nanovoids. Finally, twin lamellae as well as regular large twins were observed in the crystal. The twin lamellae correspond to shallow V-grooves revealed after the chemical etching.

19.1 Introduction

β-Ga_2O_3 is an attractive semiconductor material for various optoelectronic devices, having a large band gap energy and good electrical conductivity. Recently, there have been reports on the successful growth of high-quality β-Ga_2O_3 single crystals by edge-defined film-fed growth (EFG) [1–3] and floating zone (FZ) processes [4–8]. Typical features of EFG-grown crystals are as follows:

(a) The crystals are completely twin-free.
(b) Semiconductor substrates of up to 4 in. diameter can be fabricated from the crystals.

O. Ueda (✉)
Meiji University, 1-1-1 Higashimita, Tama-ku, Kawasaki, Kanagawa 214-8571, Japan
e-mail: nanotec4@meiji.ac.jp

M. Higashiwaki and S. Fujita (eds.), *Gallium Oxide*, Springer Series in Materials Science 293, https://doi.org/10.1007/978-3-030-37153-1_19

(c) The main residual impurity is Si and n-type doping is possible with a SnO_2 or SiO_2 source.
(d) The current lowest dislocation density is $1 \times 10^3/cm^3$.

Schottky barrier diodes (SBDs) [9–12] and FETs [13–16] have been successfully developed for power electronics applications. In addition, GaN-related blue LEDs on β-Ga_2O_3 substrates are another promising application [17].

In order to achieve high-performance and high-reliability electron/optical devices, it is crucial to eliminate various defects in bulk β-Ga_2O_3 crystal as well as those in thin films. For this purpose, it is essential to characterize defects in the crystal and to clarify their generation mechanism.

There have been a number of reports on the evaluation of defects in this crystal [18–21]. Nakai et al. have recently found screw dislocations and so-called 'nanopipes' in (010)-oriented β-Ga_2O_3 wafers by etching, X-ray topography, and transmission electron microscopy/energy dispersive X-ray spectrometry (TEM/EDS) [18]. However, the generation mechanism is not clear. We have also reported nanometer-sized crystalline grooves on as-polished (010)-oriented β-Ga_2O_3 wafers [19]. Furthermore, relationship between etch pits and etch patterns revealed on the (010), (001) and the $(\bar{2}01)$ planes, and property of SBDs have been reported in detail [22–24]. Regarding dislocations, Yamaguchi et al. have proposed a slip system composed of four planes: $\{\bar{2}01\}$, $\{101\}$, $\{\bar{3}10\}$, and $\{\bar{3}\bar{1}0\}$ [20].

In this chapter, we describe the characterization of defects in $(\bar{2}01)$ and (010)-oriented β-Ga_2O_3 wafers by chemical etching, focused ion beam scanning ion microscopy (FIB-SIM), transmission electron microscopy, and related techniques.

19.2 Experimental Procedure

In this study, we used $(\bar{2}01)$-and (010)-oriented β-Ga_2O_3 wafers with a thickness of about 400 μm. The β-Ga_2O_3 crystals were grown by the EFG process [3].

A macroscopic observation of crystal defects, such as dislocations and twins, can be performed by observation of the surface of the crystal after chemical etching. Etching was carried out using H_3PO_4 at 130 °C [3]. After the etching, dislocations and twins were detected as etch pits and clear line-shaped patterns, respectively. In this paper, we use the term 'etch pitting' for the process of revealing the dislocations by chemical etching. The etched surface was observed by differential optical microscopy, scanning electron microscopy (SEM), and focused ion beam scanning ion microscopy (FIB-SIM) [25].

TEM observation was carried out using a JEOL-2100F analytical transmission electron microscope operated at 200 kV equipped with an EDS apparatus for compositional analysis of the defects. Thin specimens for cross-sectional TEM (XTEM) were prepared by an FIB thinning method using a Hitachi FB-1200 FIB. FIB-SIM images were also taken using this apparatus.

19.3 Results and Discussion

In this section, we present results of investigation of defects in the crystal by the following procedures:

(1) Macroscopic evaluation of defects in the crystal by chemical etching using typical wafers with a low-index (010) plane and a $(\bar{2}01)$ plane. The (010)-oriented wafers are used for the fabrication of actual electronic devices.
(2) Microscopic evaluation of etch patterns by TEM and related techniques.

Detailed results are described below.

19.3.1 *Arrays of Edge Dislocations*

In this subsection, we describe results of the structural analysis of a $(\bar{2}01)$-oriented β-Ga_2O_3 wafer. First, a typical optical micrograph of the surface of the wafer after etching by H_3PO_4 at 130 °C for 420 min is shown in Fig. 19.1a. Etch pits are observed. They mostly lie in a row along the [010] direction (see the etch pits in the region with red ellipses). The estimated etch pit density is on the order of $10^4/cm^2$. The distance between the etch pits varies in the range 20–100 μm. Apart from these pits, isolated etch pits are also observed. Furthermore, each etch pit does not always precisely lie on a line as shown in Fig. 19.1b. Figure 19.1b also suggests that each etch pit includes four facets, being rectangular with a core, and that its long axis is in the [102] direction (see the schematic diagram of the etch pit denoted by EP). The core of the etch pit is slightly off-centered. In order to determine the Burgers vector of the dislocations, we carried out a TEM contrast experiment. We prepared thin specimens for XTEM observation by the FIB method. The specimens were cut vertically to the substrate surface, where the etch pit was, as shown in Fig. 19.1b.

Figure 19.2a, b are typical bright-field (BF) TEM images corresponding to the core part of a similar etch pit to those shown in Fig. 19.1b. These TEM images were obtained under the $0\bar{2}0$ and $\bar{2}01$ reflections with a positive s ($s > 0$), which is a parameter indicating the deviation from the exact Bragg reflection, respectively. In Fig. 19.2a, we can clearly observe a dislocation just below the core of the etch pit (see the position indicated as 'etch pit'). In this image, only a segment of the dislocation denoted by A–B is observed. This suggests that the direction of the dislocation is not exactly normal to the $(\bar{2}01)$ but slightly tilting toward the [102] direction since the specimens used for the XTEM were cut nearly exactly normal to the wafer as shown in Fig. 19.1b. In contrast, the dislocation is out of contrast in Fig. 19.2b ($\mathbf{g} = \bar{2}01$, $s > 0$). Therefore, according to the $\mathbf{g} \cdot \mathbf{b}$ invisibility criterion, the Burgers vector of the dislocation is perpendicular to the $(\bar{2}01)$, i.e., the [010] or $[0\bar{1}0]$ direction. Therefore, these dislocations are determined to be edge dislocations. Figure 19.3a shows a weak-beam dark-field TEM image of the dislocations obtained under the diffraction condition shown in Fig. 19.3b ($0\bar{6}0$ reflection is strongly

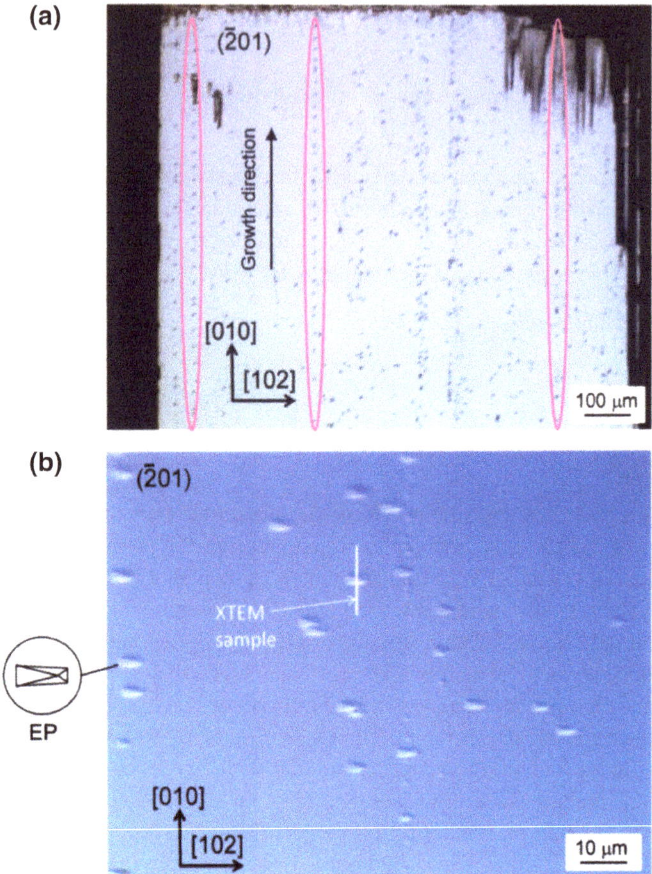

Fig. 19.1 Results of etch pitting analysis on the $(\bar{2}01)$ plane of β-Ga_2O_3 crystal. **a** Optical micrograph of surface of $(\bar{2}01)$ after etch pitting by H_3PO_4 at 130 °C for 420 min. **b** Enlarged image of the etch pits. A XTEM sample is indicated by a white line

excited with a positive *s*). The dislocation is also indicated as *A*–*B*. Although the contrast of the dislocation is weak, very sharp single line contrast is observed. This indicates that a dislocation is not easily dissociated into two partial dislocations in this crystal. This may be due to the relatively high stacking fault energy.

In some cases, comparatively short dislocation segments are observed at region below the etch pit. These dislocations tilt from the direction normal to $(\bar{2}01)$ with a slightly higher angle than that for the dislocation shown in Fig. 19.2a. However, since the dislocation is almost invisible (we call this 'residual contrast') under the $\bar{2}01$ reflection, its Burgers vector is expected to be the same as that for the case shown in Fig. 19.2a. We have also observed dislocations which change their shape slightly during their glide or climb motion due to an inhomogeneous strain field in the crystal growth procedure.

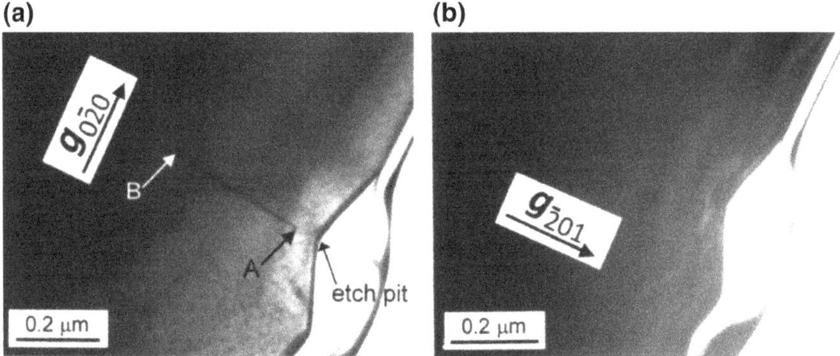

Fig. 19.2 XTEM analysis of defects corresponding to the etch pits shown in Fig. 19.1 as typical examples of dislocations. **a** A bright-field transmission electron microscopy (TEM) image of a dislocation corresponding to an etch pit obtained by $0\bar{2}0$ reflection with a positive s. **b** A bright-field TEM image of the dislocation; $\boldsymbol{g} = \bar{2}01$, $s > 0$

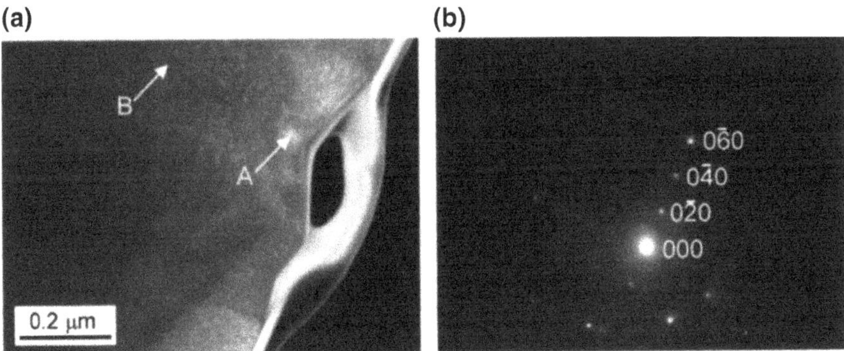

Fig. 19.3 **a** A weak-beam dark-field image of the dislocation shown in Fig. 19.2; $\boldsymbol{g} = 0\bar{2}0$, $s_{3g} > 0$. **b** An electron diffraction pattern for (**a**)

Based on the TEM analysis described above, possible model for generation of the dislocations corresponding to the dislocation array is shown in Fig. 19.4. In this model, the etch pit array corresponds to multiple edge dislocations with a Burgers vector of [010] or $[0\bar{1}0]$. The arrow indicates the direction of dislocation motion. Since these dislocations lie on a plane slightly tilted from the (102) plane, a possible slip plane is the (101) plane, which has been reported as a possible slip plane by Yamaguchi et al. [20]. As previously described, several works on the dislocations in β-Ga_2O_3 crystals have been done: possible slip systems, [20] determination of Burgers vectors of several types of dislocations [18, 26]. However, in order to clarify the generation mechanism for these dislocations, more detailed structural analysis of the crystal is required.

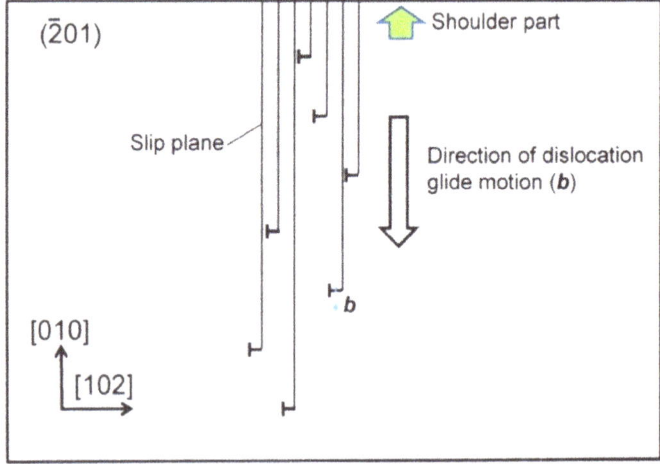

Fig. 19.4 A possible model for generation of the dislocations corresponding to the dislocation dislocation array based on the results shown in Figs. 19.2 and 19.3

19.3.2 Platelike Nanovoids

We also carried out an etch pitting analysis on (010)-oriented β-Ga_2O_3 wafers. Figure 19.5a shows a typical optical micrograph of a surface etched by H_3PO_4 at 130 °C for 120 min. In this plane, arrays of etch pits are observed along the (100) plane (a-plane). Although some other types of etch pits are observed, we only focus on this type of etch pit in this study. A high-magnification SEM image of a part of the etch pit region in Fig. 19.5a is shown in Fig. 19.5b. The etch pits are parallelogram-shaped. However, they consist of rather complicated features:

(1) facets on various crystal planes with brighter contrast,
(2) a parallelogram-shaped region with darker contrast.

In order to clarify the origin of these features, cross-sectional TEM observation was carried out on the etch pits shown in Fig. 19.5b.

Figure 19.6a is an XTEM image of the etch pit region shown in Fig. 19.5b, obtained under a nearly zone-axis condition. The thick tungsten film for protecting the surface of the TEM specimen is indicated by W. In the W-region, we can clearly observe the two facets of the V-groove (see VG in Fig. 19.6a), which are estimated to be ($41\bar{2}$) and ($\bar{4}12$) planes (see also Fig. 19.6a). We can also observe 'platelike nanovoid' region denoted by PNV. The PNV becomes narrower (from W_1 to W_2) with increasing depth. This is assumed to be due to the penetration of the H_3PO_4 down to the deeper PNV region during the etch pitting. Figure 19.6b, c is electron diffraction patterns obtained from the regions denoted by L and R, respectively. In

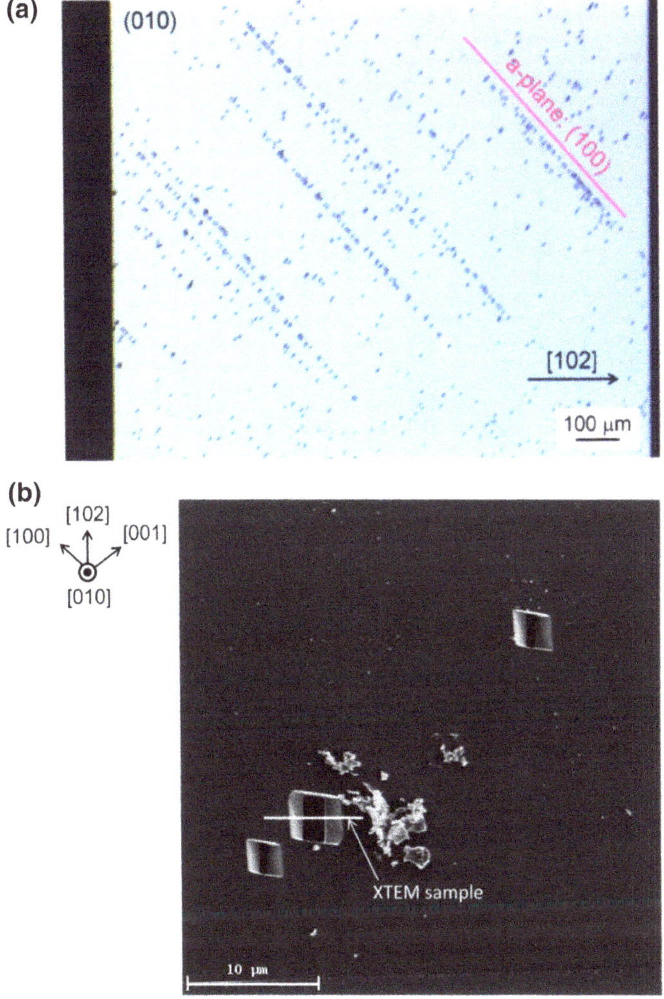

Fig. 19.5 Results of etch pitting on the (010) plane of β-Ga_2O_3 crystal. **a** An optical micrograph of surface of (010) after etching by H_3PO_4 at 130 °C for 120 min. **b** A typical SEM image of etch pits shown in (**a**)

both images, no anomalous features are observed and only the fundamental diffraction spots are observed. In the BF TEM image obtained from the region near the PNV, no defect structures such as dislocations or dislocation loops are observed. Unfortunately, in this experiment, we could not reach the bottom of this defect (whose length was at least 15 μm). Thus, it is not yet clear whether the PNV is hollow or not.

We have previously reported a similar type of defect to the PNVs [19]. Figure 19.7a shows a typical FIB-SIM image of such a defect on the as-polished

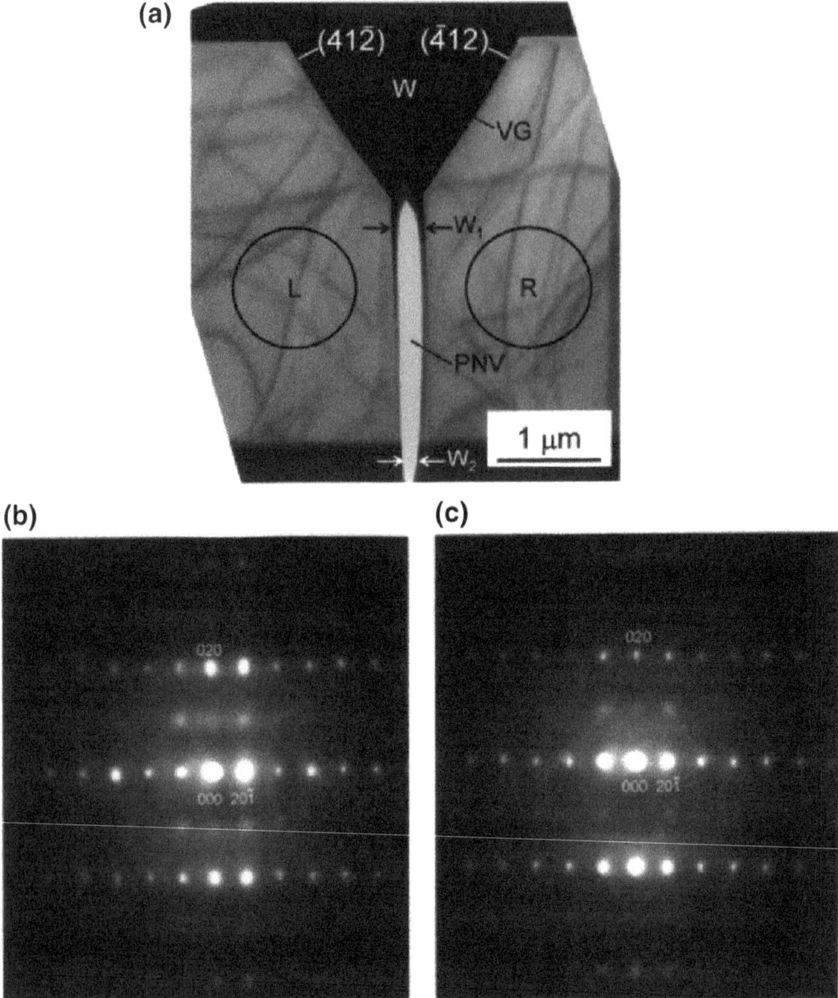

Fig. 19.6 Results of cross-sectional TEM observation of the etch pit region shown in Fig. 19.5b. **a** An XTEM image of the etch pit region. **b** A selected area electron diffraction pattern obtained from the region L. **c** A selected area electron diffraction pattern obtained from the region R

surface of a (010)-oriented β-Ga_2O_3 crystal. As shown in this image, the long axis of the groove is along the [001] direction, which is different from that of the PNV appeared by etching, i.e., the [102] direction. In the previous work, we reported that such defects are observed as nanometer-sized crystalline grooves (NSGs) elongated along the [001] direction. The groove length in the [001] direction is in the range 50–1200 nm and the groove width in the [100] direction is approximately 40 nm. Since most of the grooves disappeared after etching for only 2 min, we concluded that the groove depth is approximately 30 nm. In order to confirm this, we also

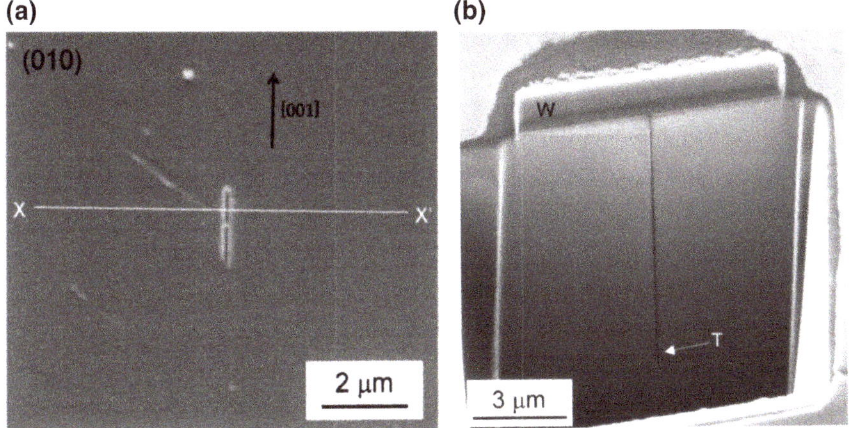

Fig. 19.7 SEM and FIB-SIM observation of PNV. **a** A typical plan-view SEM image of PNV. **b** A cross-sectional FIB-SIM image of the PNV shown in (**a**) (cut by the line x − x′)

performed FIB-SIM observation on the sample shown in Fig. 19.7. Figure 19.7a is a typical plan-view SEM image of the PNV. As shown in Fig. 19.7b, the PNV terminates in the crystal (see point denoted by T). In this case, the depth of the PNV is approximately 7.5 μm. This result strongly indicates that the PNVs are hollow.

Based on the previous work and this work on the PNV and NSGs, it is suggested that both types of defect can be classified as hollow nanovoids although their crystallographic configuration and distribution are different from each other. Regarding the generation mechanisms for these defects, we propose the following candidates:

(1) The condensation of excess oxygen vacancies in the grown crystal generates multiple vacancy-type stacking faults on the (100) plane (as schematically shown in Fig. 19.8 [19]) finally form hollow PNVs.

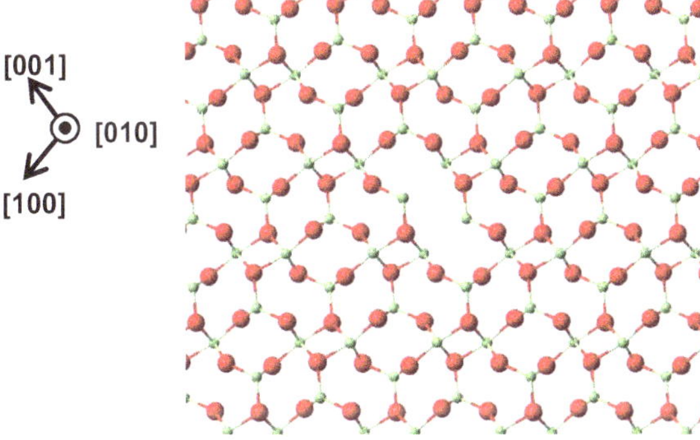

Fig. 19.8 A schematic diagram for a vacancy-type stacking fault loop on the (100) Plane, where gallium and oxygen atoms are painted in green and in red, respectively

(2) The defects described in (1) are formed by the penetration of tiny bubbles at the liquid-solid interface into grain boundaries (proposed in the previous work [19]).
(3) The defects described in (1) are formed by local meltback at some grown-in defects such as screw-type dislocations during growth.

To clarify the generation mechanism of these defects, more detailed examination is required.

19.3.3 Twins and Twin Lamellae

Twins are typical defects that are often generated in the early stage of growth of bulk crystals such as Si, GaAs, and other compound semiconductors. Since twins are assumed to be generated by strain relaxation in the crystal during growth, they can be suppressed by adjusting the growth conditions. Macroscopic evaluations of twins in β-Ga_2O_3 crystals by optical microscopy and polarization microscopy have previously shown that the twin boundary is the (100) plane [3]. An atomic model for the twin boundary has also been given [3]. In this subsection, we describe a detailed microscopic evaluation of twins.

Figure 19.9 shows a typical optical micrograph of $(0\bar{1}0)$ β-Ga_2O_3 crystal after chemical etching with hot H_3PO_4. Since the shapes of the etch pits corresponding to PNVs on both sides of the dotted line have mirror symmetry, the plane corresponding to the dotted line (normal to the $(0\bar{1}0)$ plane) is a (100) twin boundary.

Next, we carried out XTEM analysis on the twin boundary region shown in Fig. 19.9. A specimen for XTEM was prepared from the region denoted by a white line. Since the atomic structures corresponding to the matrix and the twin regions viewed from the [001] direction do not have a mirror symmetry relationship, we selected another pole with a high index by tilting the specimen by a large angle. Figure 19.10a is a bright-field TEM image of a region including the single twin boundary in Fig. 19.9 obtained under zone-axis condition (the corresponding electron diffraction pattern is shown in Fig. 19.10b). The twin boundary is indicated by the dotted line *X*–*X′*. In the region indicated as a white box, we can clearly observe a symmetric strain-related contour. This may be due to the symmetric strain field in the twin boundary region. By comparing the diffraction patterns obtained from the crystals A and B with that shown in Fig. 19.10b, the diffraction spots indexed as hkl^T in Fig. 19.10b are related to twin.

In the case of the $(\bar{2}01)$-oriented wafer, after chemical etching, twins are often observed as areas with line contrast or etched grooves, both along the [010] direction. Figure 19.11a shows a (010) XTEM image of a very shallow etched groove region taken under a many-beam (zone-axis) condition. A ribbon-shaped defect lying on the (100) plane can be observed (see defect denoted by D) that originated from the bottom of the shallow groove. The width of the defect is approximately 60 nm. Figure 19.11b shows an electron diffraction pattern obtained

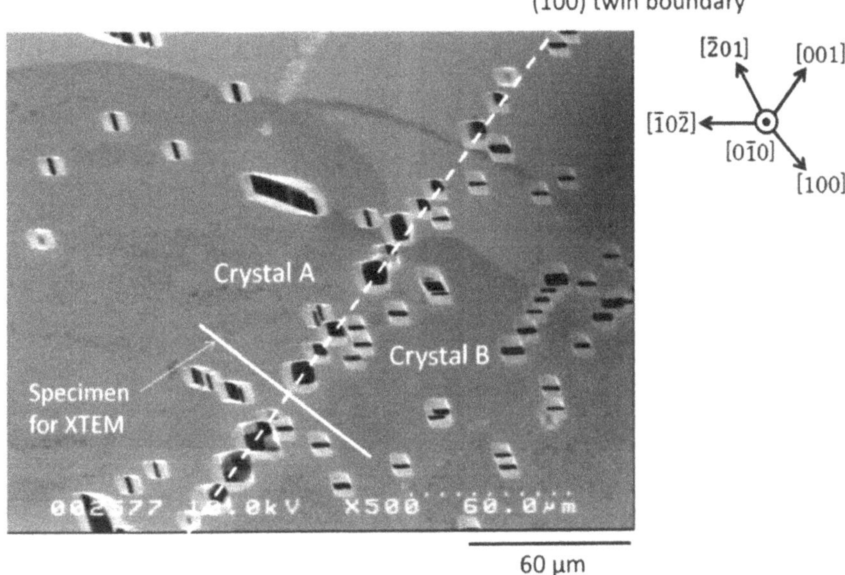

Fig. 19.9 An SEM image of etched surface of (0$\bar{1}$0) plane including a (100) twin boundary

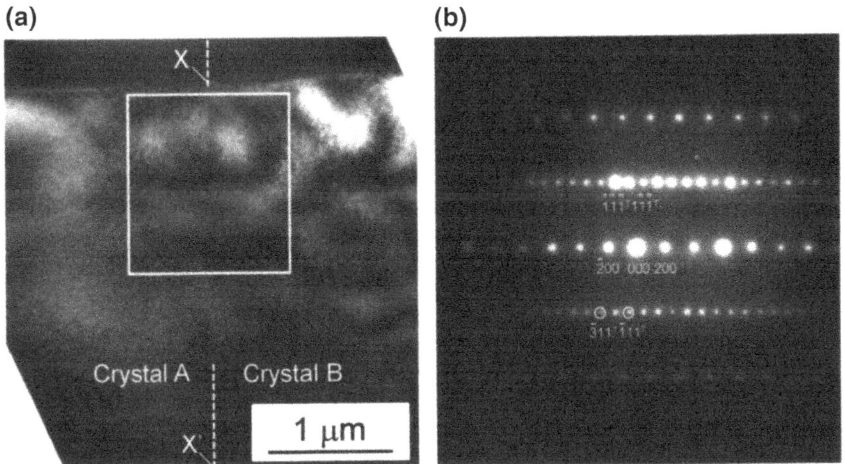

Fig. 19.10 XTEM analysis of single twin boundary shown in Fig. 19.9. **a** A BF-XTEM image of the region indicated as a white line in Fig. 19.9. **b** An electron diffraction pattern obtained from the region indicated by a white box in Fig. 19.10a

from the square region including the defect. We determined that the extra spots denoted by arrows in Fig. 19.11b are associated with twin. The defect region is also observed by high-resolution TEM as shown in Fig. 19.12. The two interfaces

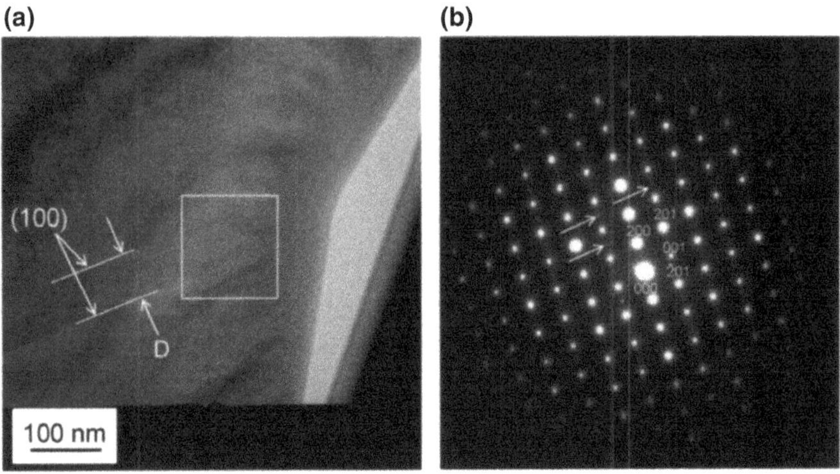

Fig. 19.11 (010) Cross-sectional TEM analysis of a ribbon-shaped defect. **a** A high magnification of the defect near the facet-region obtained under many beam condition. **b** An electron diffraction pattern obtained by the region including the defect (extra spots indicated by arrows are related to twin)

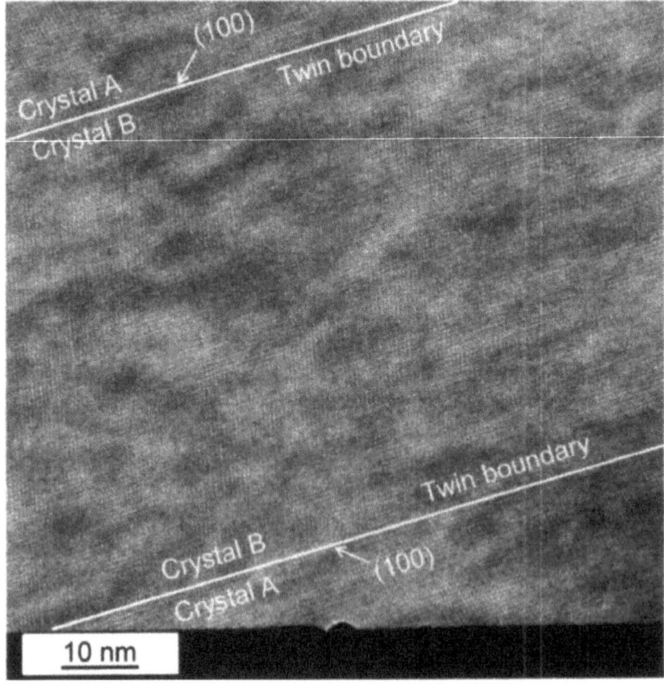

Fig. 19.12 A high-resolution TEM image of the ribbon-shaped defect region

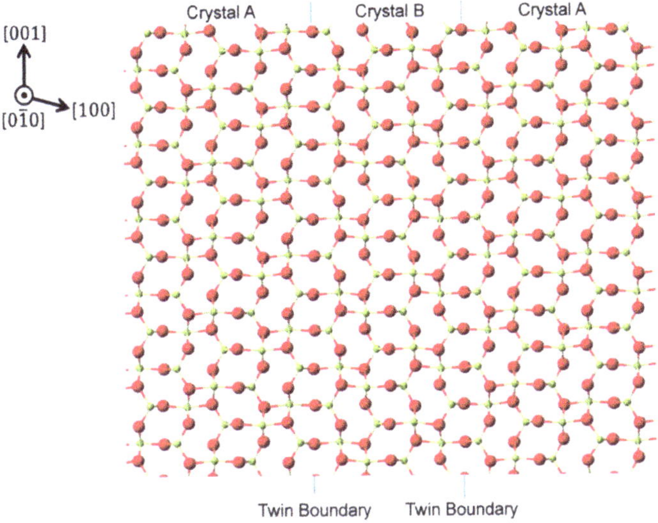

Fig. 19.13 An atomic structure of a region including a twin lamella in β-Ga_2O_3 crystal viewed from the (0$\bar{1}$0) direction

between the matrix and the defect region are clearly observed and are determined to be the (100) plane. Therefore, we can conclude that these interfaces are twin boundaries. In this particular case, we call this twin a 'twin lamella'. A schematic diagram of the atomic structure of the twin lamella is shown in Fig. 19.13. It is assumed that the nucleation of twin lamellae is an initial stage of the twinning process from the surface of the 'shoulder part' of the crystal during its growth. After the twin lamellae are nucleated, their width increases by the relaxation of accumulated strain in the grown crystal. However, the origin of the twin lamellae is not clear at present. Further study is also required.

19.4 Conclusions

We have characterized three types of crystal defects in EFG-grown β-Ga_2O_3 crystals by chemical etching, SEM, FIB-SIM, and TEM/EDS analyses. First, we found arrays of edge dislocations corresponding to etch pit arrays on a ($\bar{2}$01)-oriented wafer after etching with hot H_3PO_4. We also observed dislocations whose line-shaped appearance is slightly deformed due to an inhomogeneous strain field. Next, we found platelike nanovoids (PNVs) corresponding to etch pits on the (010) plane. Although the crystallographic configurations of these defects are different from that of nanometer-sized grooves (NSGs), which we have previously reported, they are both classified as 'hollow nanovoids'. Finally, we found twin lamellae as well as regular large twins in the crystal. The twin lamellae correspond to shallow V-grooves revealed after the chemical etching.

References

1. K. Shimamura, E.G. Villora, K. Matsumura, K. Aoki, M. Nakamura, S. Takekawa, N. Ichinose, K. Kitamura, Nihon Kessho Seicho Gakkaishi **33**, 147 (2006). (in Japanese)
2. H. Aida, K. Nishiguchi, H. Takeda, N. Aota, K. Sunakawa, Y. Yaguchi, Jpn. J. Appl. Phys. **47**, 8506 (2008)
3. A. Kuramata, K. Koshi, S. Watanabe, Y. Yamaoka, T. Masui, S. Yamakoshi, Jpn. J. Appl. Phys. **55**, 1202A2 (2016)
4. M. Saurat, A. Revcolevschi, Rev. Int. Hautes Temper. et Refract. **8**, 291 (1971)
5. N. Ueda, H. Hosono, R. Waseda, H. Kawazoe, Appl. Phys. Lett. **70**, 3561 (1997)
6. E.G. Villora, K. Shimamura, Y. Yoshikawa, K. Aoki, N. Ichinose, J. Cryst. Growth **270**, 420 (2004)
7. J. Zhang, B. Li, C. Xia, G. Pei, Q. Deng, Z. Yang, W. Xu, H. Shi, F. Wu, Y. Wu, J. Xu, J. Phys. Chem. Solids **67**, 2448 (2006)
8. S. Ohira, N. Suzuki, N. Arai, M. Tanaka, T. Sugawara, K. Nakajima, T. Shishido, Thin Solid Films **516**, 5763 (2008)
9. K. Sasaki, A. Kuramata, T. Masui, E.G. Villora, K. Shimamura, S. Yamakoshi, Appl. Phys. Express **5**, 035502 (2012)
10. K. Sasaki, M. Higashiwaki, A. Kuramata, T. Masui, S. Yamakoshi, J. Cryst. Growth **378**, 591 (2013)
11. K. Sasaki, M. Higashiwaki, A. Kuramata, T. Masui, S. Yamakoshi, IEEE Electron Device Lett. **34**, 493 (2013)
12. M. Higashiwaki, K. Koshi, K. Sasaki, K. Goto, K. Nomura, Q.T. Thieu, R. Togashi, H. Murakami, Y. Kumagai, B. Monemar, A. Koukitu, A. Kuramata, S. Yamakoshi, Appl. Phys. Lett. **108**, 133503 (2016)
13. M. Higashiwaki, K. Sasaki, A. Kuramata, T. Masui, S. Yamakoshi, Appl. Phys. Lett. **100**, 013504 (2012)
14. M. Higashiwaki, K. Sasaki, T. Kamimura, M.H. Wong, D. Krishnamurthy, A. Kuramata, T. Masui, S. Yamakoshi, Appl. Phys. Lett. **103**, 123511 (2013)
15. M. Higashiwaki, K. Sasaki, A. Kuramata, T. Masui, S. Yamakoshi, Phys. Status Solidi A **211**, 21 (2014)
16. M.H. Wong, K. Sasaki, A. Kuramata, S. Yamakoshi, M. Higashiwaki, IEEE Electron Device Lett. **37**, 212 (2016)
17. A. Kuramata, K. Iizuka, K. Sasaki, K. Koshi, T. Masui, Y. Morishima, K. Goto, Y. Kumagai, H. Murakami, A. Koukitu, M.H. Wong, T. Kamimura, M. Higashiwaki, S. Yamakoshi, Nihon Kessho Seicho Gakkaishi **42**, 24 (2015). (in Japanese)
18. K. Nakai, T. Nagai, K. Noami, T. Futagi, Jpn. J. Appl. Phys. **54**, 051103 (2015)
19. K. Hanada, T. Moribayashi, T. Uematsu, S. Masuya, K. Koshi, K. Sasaki, A. Kuramata, O. Ueda, M. Kasu, Jpn. J. Appl. Phys. **55**, RC150084 (2016)
20. H. Yamaguchi, A. Kuramata, T. Masui, Superlattices Microstruct. **99**, 99 (2016)
21. K. Hanada, T. Moribayashi, K. Koshi, K. Sasaki, A. Kuramata, O. Ueda, T. Oshima, M. Kasu, Jpn. J. Appl. Phys. **55**, 030303 (2016)
22. M. Kasu, K. Hanada, T. Moribayashi, A. Hashiguchi, T. Oshima, T. Oishi, K. Koshi, K. Sasaki, A. Kuramata, O. Ueda, Jpn. J. Appl. Phys. **55**, 1202BB (2016)
23. T. Oshima, A. Hashiguchi, T. Moribayashi, K. Koshi, K. Sasaki, A. Kuramata, O. Ueda, T. Oishi, M. Kasu, Jpn. J. Appl. Phys. **56**, 086501 (2017)

24. M. Kasu, T. Oshima, K. Hanada, T. Moribayashi, A. Hashiguchi, T. Oishi, K. Koshi, K. Sasaki, A. Kuramata, O. Ueda, Jpn. J. Appl. Phys. **56**, 091101 (2017)
25. T. Ishitani, T. Yamanaka, K. Inai, K. Ohya, Vacuum **84**, 1018 (2010)
26. O. Ueda, N. Ikenaga, K. Koshi, K. Iizuka, A. Kuramata, K. Hanada, T. Moribayashi, S. Yamakoshi, K. Kasu, Jpn. J. Appl. Phys. **55**, 1202BD (2016)

Chapter 20
Structural Properties 2

Crystallographic Defects in β-Ga_2O_3 and X-Ray Topography Analysis

Hirotaka Yamaguchi

Abstract Crystallographic defects, such as dislocations and stacking faults, in β-Ga_2O_3 are analyzed. A slip plane model is proposed, which is derived from the close-packed oxygen subcell. Possible Burgers vectors of dislocations are the translation vectors on the closed packed planes, $(\bar{2}01)$, $(\bar{1}0\bar{1})$, (310), and $(3\bar{1}0)$. Dislocations are observed using synchrotron X-ray topography and are identified based on the proposed slip plane model. Stacking faults on the $(\bar{2}01)$ plane are revealed in the X-ray topographs and are enclosed by a single partial dislocation loop slipping on the plane.

20.1 Introduction

The monoclinic modification of gallium oxide (β-Ga_2O_3) has attracted much attention as a promising material for power electronics applications because of its excellent physical properties, such as a wide energy band gap and high electronic breakdown field [1, 2]. The crystal can be grown by melt-growth techniques, and now wafers for device substrates are commercially available, the ingots for which are grown using the edge-defined film-fed growth (EFG) method. The dislocation density in a high-quality crystal has been measured as on the order of 10^3 cm^{-2} by observing etch pits and dislocations under the pits [3]. Dislocations in substrates affect the performance and reliability of the device fabricated on the substrate. Therefore, the properties and effects on device performance of these substrates have been studied extensively. For typical semiconductor materials, diamond and related structures, such as Si and III–V materials, and hexagonal structures, such as SiC and GaN, the dislocation properties, including characteristics, interactions, movements, behaviors during crystal growth and device process, and effects on the device performance have been thoroughly

H. Yamaguchi (✉)
Advanced Power-electronics Research Center,
National Institute of Advanced Industrial Science and Technology,
1-1-1 Umezono,
Tsukuba, Ibaraki 305-8568, Japan
e-mail: yamaguchi-hr@aist.go.jp

M. Higashiwaki and S. Fujita (eds.), *Gallium Oxide*, Springer Series in Materials Science 293, https://doi.org/10.1007/978-3-030-37153-1_20

investigated. They are understood based on crystal structure, and slip planes in the lattice play an important role [4]. However, β-Ga_2O_3 has a new crystal structure in dislocation theory and the nature of its defects are almost unknown. In this study, we considered the crystal structure to understand dislocations in this crystal, and we characterized the dislocations and stacking faults (SFs) in the crystals grown by the EFG method by using X-ray topography (XRT).

XRT is an X-ray diffraction imaging technique in which defects in a crystal are projected on two-dimensional images that reflect differences in diffraction power [5, 6]. The technique is sensitive to the strain field, has a wide vision, and does not require destructive sample preparation, although the spatial resolution of XRT is lower than that of transmission electron microscopy (TEM) imaging. Thus, XRT and TEM are complementary tools for defect characterization. Synchrotron radiation (SR) gives much more intense parallel X-rays than laboratory X-ray tubes. XRT using SR (SR-XRT) instantaneously projects defect images with high contrast and spatial resolution. Wavelength tunability is another useful feature of SR, which enables optimum diffraction conditions for a specific reciprocal vector. Observing etch pits is another direct observation technique for defects [7] that is convenient for identifying the character of dislocations outcropping at the crystal surface and evaluating the dislocation density.

The first detailed XRT study of defects in β-Ga_2O_3 wafers grown by the EFG method was reported by Nakai et al. [8]. They performed XRT, etch pit analysis, and TEM imaging for (010)-oriented wafers and found two sizes of etch pits. Smaller pits approximately 2 μm in size (type 1) were aligned along the [100] or [102] directions. They exhibited line patterns in X-ray topographs and were edge dislocation arrays with a Burgers vector of $\boldsymbol{b} = \langle 010 \rangle$, which was determined by $\boldsymbol{g} \cdot \boldsymbol{b}$ analysis in TEM imaging. The type 2 etch pits, which had a larger diameter of 10 μm, showed no specific orientational arrangement. TEM observation revealed that a hollow pipe 0.1 μm in diameter extended along the $\langle 010 \rangle$ direction. Because there was no strain contrast observed around the hollow pipes, they were not associated with dislocations like micropipes in SiC, which are induced by the screw dislocation with a large Burgers vector. These hollow pipes extending along the b-axis were observed in crystals grown by the vertical Bridgman method, and their density depended on the growth rate [9]. Nanometer-sized grooves elongated in the $\langle 001 \rangle$ direction were observed on the (010) surface of the as-grown crystal by scanning electron microscopy [10] and belonged to another type of hollow defect.

In this chapter, the analysis of dislocations and SFs by XRT is described [11, 12]. The crystallographic nature of β-Ga_2O_3 is discussed to consider possible dislocations and relevant defects. Defect analysis by SR-XRT is briefly explained, followed by the experimental results. The observation of the SFs is described in detail.

20.2 Crystal Structure and Slip System

The crystal structure of β-Ga_2O_3 is monoclinic (space group C2/m) with lattice constants $a = 1.223$ nm, $b = 0.304$ nm, $c = 0.580$ nm, and $\beta = 103.7°$. Atomic

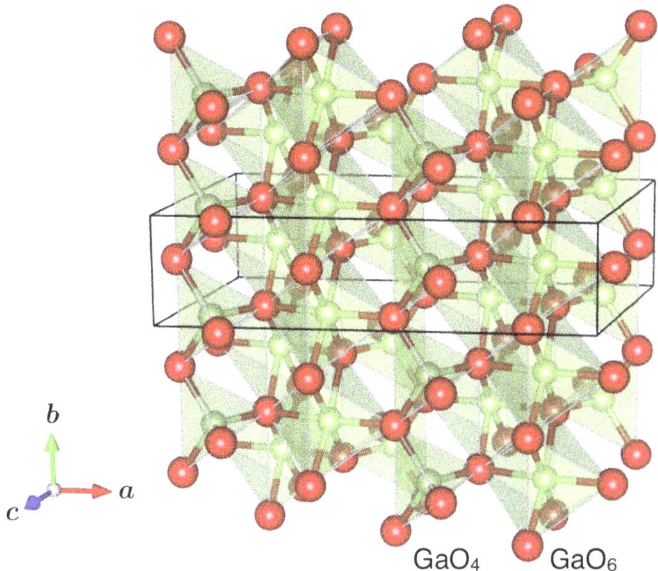

Fig. 20.1 Crystal structure of β-Ga_2O_3, drawn using VESTA [14]. The unit cell is indicated by thin lines

positions were determined by Geller [13], and the crystal structure is as shown in Fig. 20.1. The crystal lattice consists of GaO_4 tetrahedral and GaO_6 octahedral units, which makes it difficult to consider the dislocation structure in this material. Therefore, we start with a simple picture that the O subcell exhibits a three-dimensional close-packed structure, in which Ga ions occupy 2/3 of the interstices. Because the slip plane in typical crystal systems, such as face-centered cubic (FCC) and hexagonal close-packed (HCP) lattices, is generally found in close-packed low-index planes, we first examine the O subcell. In β-Ga_2O_3, a tetrahedron corresponding to {111} of FCC comprises $(\bar{2}01)$, $(\bar{1}0\bar{1})$, (310), and $(3\bar{1}0)$ planes,[1] which are distorted, and the interplanar angles are from 68.9° to 75.0°, whereas they are 70.5° in the regular tetrahedron (Fig. 20.2). The translation vectors in each plane are summarized in Table 20.1 as a model slip system, and if a slip occurs on a plane, one of the translation vectors on the plane is a possible Burgers vector (***b***).

In β-Ga_2O_3, the $(\bar{2}01)$ plane is the unique plane where GaO_4 tetrahedra and GaO_6 octahedra are separately stacked layer-by-layer (Fig. 20.3a). The stacking sequence of O ions is ... *abc* ..., where *a*, *b*, and *c* are the indices denoting in-plane positions.[2] On the other hand, octahedral Ga ions occupy the interstices between the O-layers and tetrahedral Ga ions connect the octahedra. For example, ... $(a\underline{c}b)$ − Ga(T) −

[1] The planes are indexed with vectors outward normal to the tetrahedron.

[2] The real lattice of β-Ga_2O_3 is monoclinic with a distorted close-packed structure. Therefore, the indices of the in-plane positions, *a*, *b*, and *c*, indicate nominal positions, and the direction of stacking is not normal to the $(\bar{2}01)$ plane, but toward the $\langle 10\bar{1}\rangle$ direction.

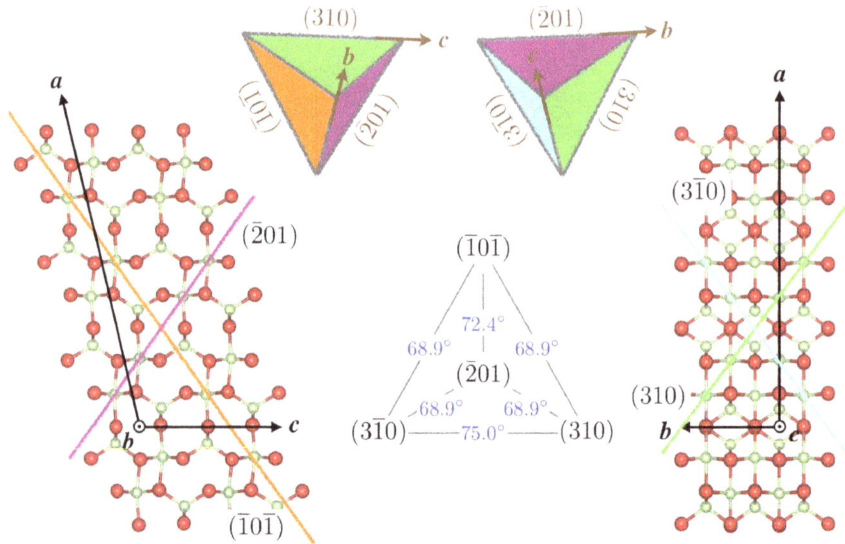

Fig. 20.2 Close-packed planes of β-Ga_2O_3, drawn using VESTA [14] (with permission from [13])

Table 20.1 Translational vectors ***t*** and their norms |***t***| in the close-packed plane in the O subcell

Plane	***t***	|***t***| (nm)
$(\bar{2}01)$	$\langle 010\rangle$	0.304
	$\frac{1}{2}\langle 112\rangle$	0.752
$(\bar{1}0\bar{1})$	$\langle 010\rangle$	0.304
	$\langle 10\bar{1}\rangle$	1.472
(310)	$\langle 001\rangle$	0.580
	$\frac{1}{2}\langle 1\bar{3}0\rangle$	0.763
	$\frac{1}{2}\langle 1\bar{3}2\rangle$	0.866
$(3\bar{1}0)$	$\langle 001\rangle$	0.580
	$\frac{1}{2}\langle 130\rangle$	0.763
	$\frac{1}{2}\langle 132\rangle$	0.866

$(c\underline{b}a)$ – Ga(T) – $(b\underline{a}c)$ – Ga(T) . . ., where the brackets describe octahedral blocks and $\underline{a}$, $\underline{b}$, and $\underline{c}$ denote the indices for the in-plane positions of the octahedral Ga ions, as shown in Fig. 20.3. Ga(T) denotes tetrahedral Ga ions labeled α, β, and γ, which are connected to O(a), O(b), and O(c), respectively. Note that 1/3 of the octahedral interstices in the octahedral Ga layer are vacant. The translation vectors in the $(\bar{2}01)$ plane, $\boldsymbol{b}_1 = \langle 010\rangle$ and $\boldsymbol{b}_2 = \frac{1}{2}\langle 112\rangle$, are identical for the octahedral and tetrahedral blocks.

Here, we consider the in-plane ionic arrangement in an octahedral block. When Ga ions occupy the $\underline{c}$ position (Ga($\underline{c}$)) over the O ions at the a position (O(a)), if some of the Ga ions shift from $\underline{c}$ to $\underline{b}$ positions with their upper section, a partial

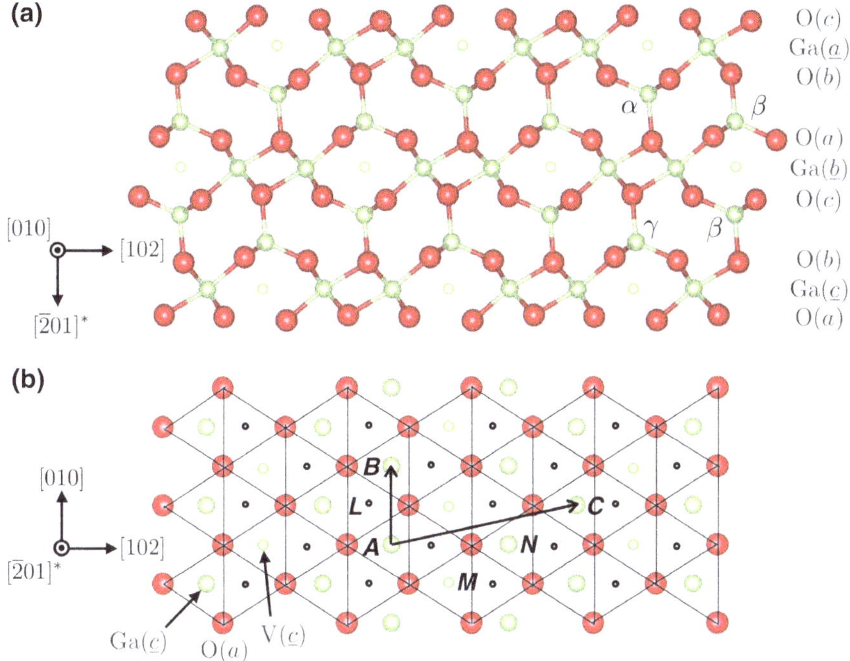

Fig. 20.3 **a** Side view of the $(\bar{2}01)$ stacking structure. Atomic position indices a, b, and c: in-plane positions of O ions and Ga ions in the GaO_6 unit; α, β, and γ: Ga ions in the GaO_4 unit. **b** Plan view. Ga($\underline{c}$) on the O(a) underlayer. Figures drawn using VESTA [14]

dislocation is created on the boundary. Similarly, (Ga($\underline{c}$)) ions can shift from $\underline{c}$ to $\underline{a}$ positions with their downside. These are the same formulae as Shockley partial dislocations in FCC and HCP lattices. As shown in Fig. 20.3b, a perfect dislocation with $\boldsymbol{b}_1 = \overrightarrow{AB}$ dissociates into two partials, $\boldsymbol{b}_{p1} = \overrightarrow{AL}$ and $\boldsymbol{b}_{p2} = \overrightarrow{LB}$:

$$[010] \rightarrow \frac{1}{18}[\bar{1}9\bar{2}] + \frac{1}{18}[192].$$

Because $|\boldsymbol{b}_1|^2 > |\boldsymbol{b}_{p1}|^2 + |\boldsymbol{b}_{p2}|^2$ is satisfied, dissociation is likely to occur.

We consider another type of dissociation, $\boldsymbol{b}_2 = \overrightarrow{AC} = \frac{1}{2}[112]$, the norm of which is larger than that between the neighboring Ga($\underline{c}$) ions. This displacement can be composed of the three shortest $\underline{c} \rightarrow \underline{c}$ displacements, for example,

$$\overrightarrow{AC} \rightarrow \overrightarrow{AM} + \overrightarrow{MN} + \overrightarrow{NC},$$

where

$$\overrightarrow{AM} = \frac{1}{6}[1\bar{3}2], \quad \overrightarrow{MN} = \overrightarrow{NC} = \frac{1}{6}[132].$$

This type of dissociation is analogous to the model proposed by Kronberg [15] for basal slip in sapphire (α-Al_2O_3). The longer period of the Ga ions arises from a superlattice with regularly aligned empty sites on the Ga layer; thus, we refer to this type of dissociation as superlattice partials.

20.3 Dislocations and SFs Observed by XRT

20.3.1 Experiments

Samples were $(\bar{2}01)$-oriented wafers with a diameter of 2 in. and (001)-oriented pieces cut from ingots grown by the EFG method. XRT experiments were performed at the synchrotron radiation facilities, the Photon Factory, Institute of Materials Structure Science, High Energy Accelerator Research Organization (Tsukuba, Japan) and SAGA Light Source (SAGA-LS), Kyushu Synchrotron Light Research Center (Tosu, Japan). The wavelength of X-rays was selected by using a Si (111) double-crystal monochromator. The diffraction geometries were Bragg-case asymmetric diffractions with low incident angles obtained by tuning the wavelength. X-ray topographs were recorded on nuclear emulsion plates (L4, Ilford).

20.3.2 Dislocations

Figure 20.4 compares X-ray topographs taken with different $\boldsymbol{g}$ values from a $(\bar{2}01)$-oriented wafer. Several kinds of dislocations are found in the topographs. We discuss the dislocations in relation to the slip system model proposed in the previous section.

Line and dot images are visible, which are identified as dislocations running in the $(\bar{2}01)$ plane and outcrops of dislocation running on a plane crossing the surface, respectively. The topographs in Fig. 20.4 are taken with $\bar{6}23$ and 006 reflections, $\boldsymbol{g}$ vectors of which are perpendicular to $\langle 130\rangle$ and $\langle 010\rangle$ directions, respectively. Dislocation A is invisible in Fig. 20.4b, indicating that it is a dislocation with the Burgers vector $\boldsymbol{b} = \langle 010\rangle$ slipped on the $(\bar{2}01)$ plane. On the other hand, the dislocations marked B' are visible in both topographs. The Burgers vector of these dislocations cannot be immediately identified from these topographs, but $\boldsymbol{b} = \frac{1}{2}\langle 112\rangle$ is possible on the $(\bar{2}01)$ plane. Dislocations C and D are invisible in the topographs with 006 and $\bar{6}23$ reflections, suggesting that $\boldsymbol{b} = \langle 010\rangle$ on $(\bar{1}0\bar{1})$ and $\boldsymbol{b} = \frac{1}{2}\langle 130\rangle$ on $(3\bar{1}0)$, respectively, in the framework of the slip system model.

Ueda et al. investigated dislocations using TEM technique [16]. They found edge dislocation arrays along the $\langle 010\rangle$ direction on the surface of a $(\bar{2}01)$-oriented wafer. The $\boldsymbol{g} \cdot \boldsymbol{b}$ analysis of the dislocations in the cross section identified the Burgers vector as $\langle 010\rangle$. Because the dislocations lay on a plane slightly tilted from the (102) plane, the $(\bar{1}0\bar{1})$ plane was identified as possible for the dislocations.

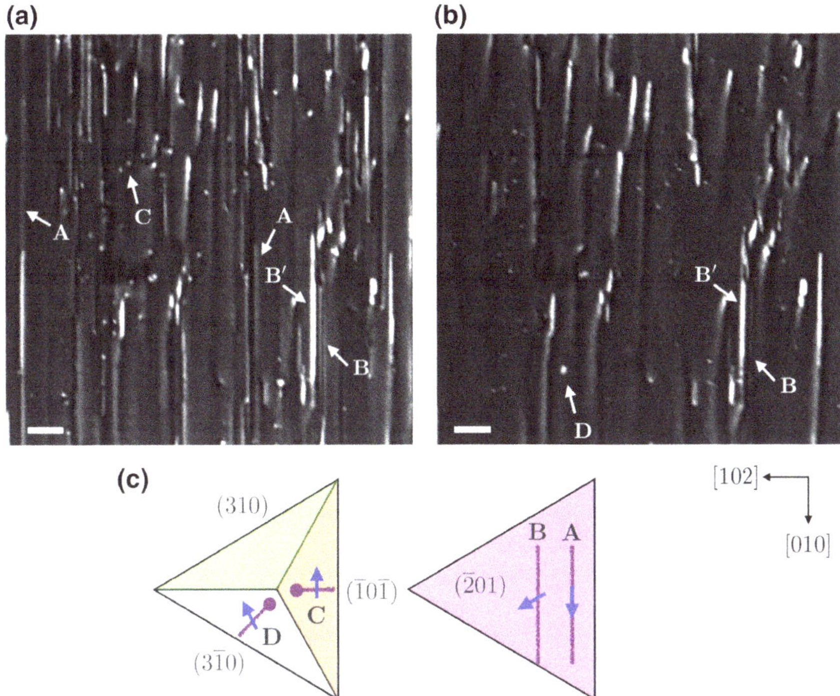

Fig. 20.4 X-ray topographs of a ($\bar{2}01$)-oriented wafer obtained with **a** $\bar{6}23$ and **b** 006 reflections. The scale bars correspond to 100 μm (with permission from [11])

20.3.3 SFs

In the X-ray topographs of the ($\bar{2}01$) wafers, rectangular-shaped planar images were observed (Fig. 20.5). The planar images were visible depending on the $\boldsymbol{g}$ value, suggesting crystallographic defects, such as SFs. The aspect ratios of the rectangles measured with different $\boldsymbol{g}$ values were almost unchanged; thus, the SFs were presumed to lie parallel to the wafer surface ($\bar{2}01$). The SFs were found over the wafer and their density varied by location (typical density of 600 cm^{-2}). The SFs were rectangular with one pair of sides parallel to the *b*-axis with a length of 50–150 μm. The planar defects were isolated, with no connections to any dislocations. Therefore, they were considered to be SFs enclosed by a single partial dislocation loop on the ($\bar{2}01$) plane, indicating that the fault vector of the SF should be identical to the Burgers vector of the partial dislocation loop. As shown in Sect. 20.2, partial dislocation is crystallographically possible in the ($\bar{2}01$) plane as Shockley and superlattice partials.

The contrast of SFs in X-ray topographs is related to the phase shift in the structure factor. The structure factors for $\boldsymbol{g}$ of the two parts divided by the SF with the plane shift of $\boldsymbol{f}$ (fault vector) are given by F_g and $F_g \exp[2\pi i \boldsymbol{g} \cdot \boldsymbol{f}]$. They are different if $\exp[2\pi i \boldsymbol{g} \cdot \boldsymbol{f}] \neq 1$, that is, $\boldsymbol{g} \cdot \boldsymbol{f} \neq$ (integer) [17, 18]. We assume that a SF occurs on

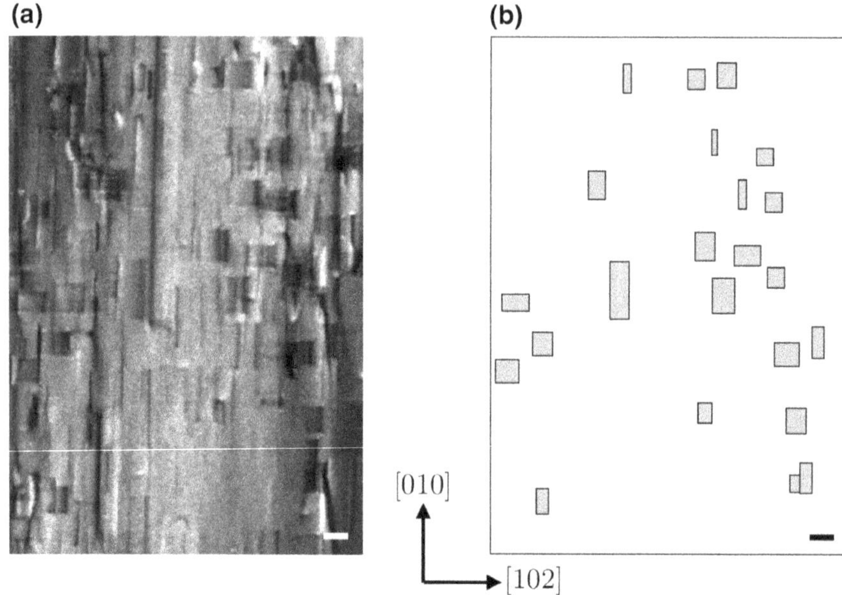

Fig. 20.5 **a** X-ray topographs of a $(\bar{2}01)$-oriented wafer obtained with $\overline{12}01$ reflection. **b** Schematics of some apparent SFs. The scale bars correspond to 100 μm

Table 20.2 Conditions for the appearance of SFs in $(\bar{2}01)$ wafers, $|\boldsymbol{g} \cdot \boldsymbol{f}_1|$ and $|\boldsymbol{g} \cdot \boldsymbol{f}_2|$, for several reflections $\boldsymbol{g}$ used in this study. The fault vectors $\boldsymbol{f}_1$ and $\boldsymbol{f}_2$ correspond to the Shockley and superlattice partials, respectively. The appearance column shows whether the SFs were observed (Yes) or not observed (No)

g	$\lvert g \cdot f_1 \rvert$	$\lvert g \cdot f_2 \rvert$	Appearance
$\overline{12}$ 0 0	2/3	2	Yes
$\bar{6}$ 2 6	4/3	2	Yes
0 0 6	2/3	2	Yes
$\overline{12}$ 0 1	5/9	5/3	Yes
$\bar{5}$ 1 3	5/9	2/3	Yes
$\bar{2}$ 0 6	5/9	5/3	Yes
$\bar{6}$ 2 3	1	1	No
6 0 6	1	3	No

$(\bar{2}01)$, and consider two possible cases, Shockley (fault vector is $\boldsymbol{f}_1$) and superlattice ($\boldsymbol{f}_2$) partials. The $\boldsymbol{g} \cdot \boldsymbol{f}$ values for several selected $\boldsymbol{g}$ values are listed, and compared with the experimental results in Table 20.2. The visible condition of the SFs agrees with the case in which the fault vector is $\boldsymbol{f}_1$. Almost all of the observed SFs showed the same $\boldsymbol{g}$-dependence and no obvious exceptions were found. Therefore, the SFs are identified as SFs enclosed by a single Shockley partial dislocation loop. The other plausible possibility for the SFs is that they are associated with prismatic dislocation loops by the insertion or extraction of partial atomic layers. In this case, the fault vector would contain a component perpendicular to the $(\bar{2}01)$ plane, resulting in different appearance rules from those observed.

The SFs observed in this study are rectangular with a pair of sides elongated along the b-axis. Along the b-axis, there are vacancy arrays in the octahedral Ga layer parallel to the $(\bar{2}01)$ plane (Fig. 20.3). A glide-set slip occurring at the next array to the vacancy would reduce the self-energy of the partial dislocation because the elastic compressive or tensile stress is decreased . In addition, the Ga-O bonds exhibit rotation by 60° and no dangling bonds occur. Our model of SFs is based on this consideration (Fig. 20.6). Figure 20.6a shows the original structure, and Ga ions in the region enclosed by the rectangle are shifted, Ga($\underline{c}$) $\rightarrow$ I($\underline{b}$). The resultant SF is formed by Ga($\underline{b}$), as shown in Fig. 20.6b.

The causes of the SFs are unclear at present, but they are presumed to be generated in the crystal growth process and/or subsequent cooling processes. The SF would be elongated as the crystal grows until it was terminated by elastic instability. We speculate that the typical dimensions of SFs, 50–150 μm, are determined by the critical elastic energy. Because no dislocations associated with SFs were observed, SFs originating from the dissociation of a dislocation are not considered here.

Thus, the SFs are on the $(\bar{2}01)$ plane, which is one of our hypothetical slip planes. This indicates that the SFs may be glissile, potentially leading to enlargement of the SFs by stress or other factors. As demonstrated for SiC diodes, glissile partial dislocations affect degradation in forward-bias operations owing to the recombination-enhanced dislocation glide effect [19]. Therefore, the SFs in β-Ga_2O_3 wafers should be eliminated for electronic device applications.

20.4 Summary

Initially, we described a slip plane model because the slip plane represents the motion of dislocations (slip dislocations) that play a crucial role in the device application of materials. We examined dislocations and SFs by using XRT and proved part of the slip plane model. Research on dislocations in β-Ga_2O_3 has just started, and systematic studies are required to establish the slip system and the dislocation nature. Observations of the dislocation loops and SFs induced by applying stress and annealing should be performed to distinguish these dislocations from grown-in dislocations. Prismatic dislocations originating from atoms in interstitial or vacant sites are also important although they have not been observed yet. Studies combining techniques such as XRT, TEM, and etch pit observations are particularly important.

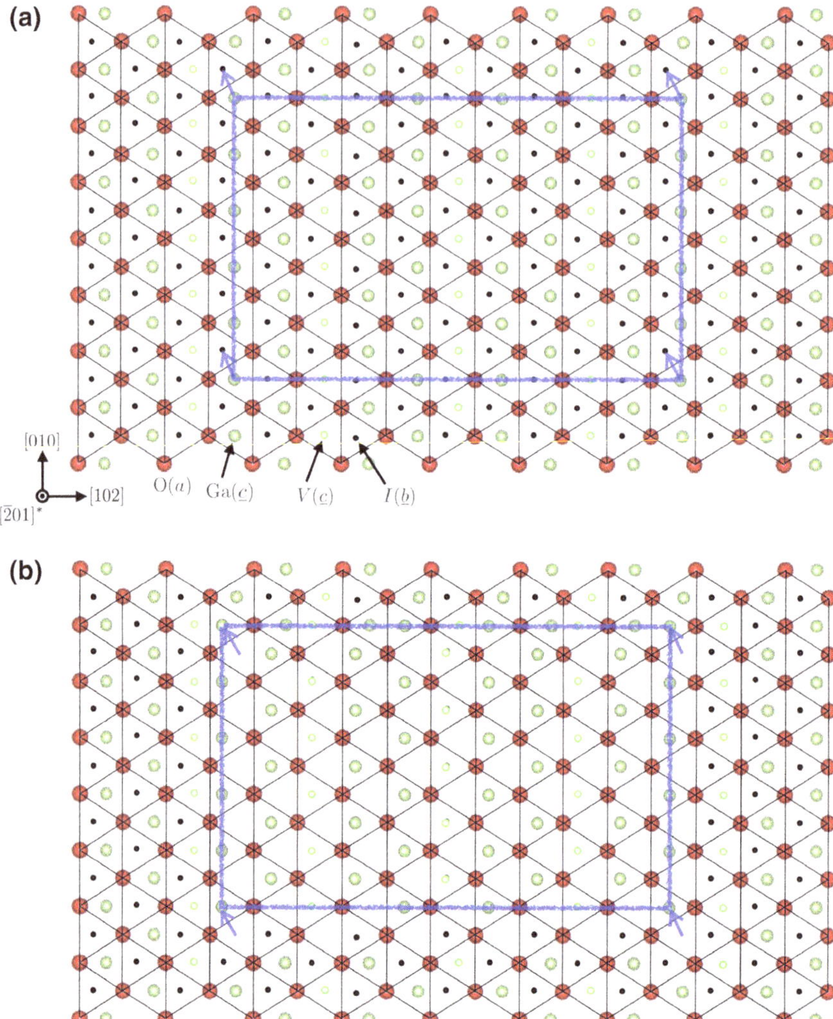

Fig. 20.6 Proposed SF model drawn using VESTA [14]. **a** Original structure. O(a): O ions on the underlayer; Ga($\underline{c}$): Ga ions at $\underline{c}$ position; V($\underline{c}$): vacant sites on the Ga($\underline{c}$) layer; I($\underline{b}$): unoccupied interstice sites on the Ga($\underline{c}$) layer. The rectangle indicates the region that slips from the $\underline{c}$ to $\underline{b}$ positions, as shown by the arrows. **b** After the slip. The region enclosed by the rectangle is an SF

Acknowledgements The XRT experiments were performed at the Photon Factory and SAGA-LS under the approval of the Photon Factory Advisory Committee (Proposal Numbers 2014G142 and 2016G133) and the Kyushu Synchrotron Light Research Center (Proposal Number 1707054G), respectively.

References

1. H.H. Tippins, Phys. Rev. **140**, A316 (1965)
2. M. Higashiwaki, K. Sasaki, A. Kuramata, T. Masui, S. Yamakoshi, Appl. Phys. Lett. **100**, 013504 (2012)
3. A. Kuramata, K. Koshi, S. Watanabe, Y. Yamaoka, T. Masui, S. Yamakoshi, Jpn. J. Appl. Phys. **55**, 1202A2 (2016)
4. J.P. Hirth, J. Lothe, *Theory of Dislocations* (Wiley, New York, 1982), p. 380
5. D.K. Bowen, B.K. Tanner, *High Resolution X-ray Diffractometry and Topography London* (Taylor & Francis, 1998)
6. A. Authier, *Dynamical Theory of X-ray Diffraction* (Oxford University Press, 2001)
7. K. Hanada, T. Moribayashi, K. Koshi, K. Sasaki, A. Kuramata, O. Ueda, M. Kasu, Jpn. J. Appl. Phys. **55**, 1202BB (2016)
8. K. Nakai, T. Nagai, K. Noami, T. Futagi, Jpn. J. Appl. Phys. **54**, 051103 (2015)
9. E. Ohba, T. Kobayashi, M. Kado, K. Hoshikawa, Jpn. J. Appl. Phys. **55**, 1202BF (2016)
10. K. Hanada, T. Moribayashi, T. Uematsu, S. Masuya, K. Koshi, K. Sasaki, A. Kuramata, O. Ueda, M. Kasu, Jpn. J. Appl. Phys. **55**, 030303 (2016)
11. H. Yamaguchi, A. Kuramata, T. Masui, Superlattices Microstruct. **99**, 99 (2016); Corrigendum **130**, 232 (2019)
12. H. Yamaguchi, A. Kuramata, J. Appl. Cryst. **51**, 1372 (2018)
13. S. Geller, J. Chem. Phys. **33**, 676 (1960)
14. K. Momma, F. Izumi, J. Appl. Cryst. **44**, 1272 (2011)
15. M.L. Kronberg, Acta Metall. **5**, 507 (1957)
16. O. Ueda, N. Ikenaga, K. Koshi, K. Iizuka, A. Kuramata, K. Hanada, T. Moribayashi, S. Yamakoshi, M. Kasu, Jpn. J. Appl. Phys. **55**, 1202BD (2016)
17. M.J. Whelan, P.B. Hirth, Phil. Mag. **2**, 1121; ibid. 1303 (1957)
18. K. Kohra, J. Phys. Soc. Jpn. **17**, 1041 (1962)
19. S.I. Maximenko, P. Pirouz, T.S. Sudarshan, Appl. Phys. Lett. **87**, 033503 (2005)

Chapter 21
Structural Properties 3

Vacancy Defects Studied with Positron Annihilation Spectroscopy

Filip Tuomisto

Abstract Positron annihilation spectroscopy has been applied to study vacancy defects in Ga_2O_3. Both positron lifetime and Doppler broadening experiments have been performed, both in bulk single crystals and in thin films. The results show that Ga vacancy defects are efficiently formed in thin film growth and account for the electrical compensation in semi-insulating and highly resistive Ga_2O_3. Their concentrations are very low in n-type material. In $(In_xGa_{1-x})_2O_3$ alloys, the nature and behavior of the cation vacancy defects change from Ga_2O_3-like to In_2O_3-like along with the crystalline phase. Further work is important to elucidate the details of the vacancy formation mechanisms and the origins of the exceptionally strongly anisotropic positron annihilation signals.

21.1 Introduction

Point defects, such as **vacancies**, interstitials and impurity atoms strongly affect—even define completely—the electrical and optical properties of crystalline semiconductors. They are introduced during the materials synthesis and/or post-growth processing, either intentionally or in spite of best efforts to avoid their formation. In addition to point defects, extended defects such as dislocations, stacking faults or aggregates of impurities affect the semiconductor properties. It is not unusual to have important densities of more than one of these kinds of defects, and quite typically, they also affect each other's presence and properties. For example, the formation of stacking faults may induce vacancy defects. In addition to providing a partial answer to the key question in materials science (why is matter the way it is?), in-depth knowledge about the defects in a semiconductor crystal is essential for optimization of the material properties.

F. Tuomisto (✉)
Department of Physics, University of Helsinki, POB 43, 00014 Helsinki, Finland
e-mail: filip.tuomisto@helsinki.fi

M. Higashiwaki and S. Fujita (eds.), *Gallium Oxide*, Springer Series in Materials Science 293, https://doi.org/10.1007/978-3-030-37153-1_21

Numerous experimental and theoretical-computational methods exist for identifying, quantifying and determining the physical properties of defects in semiconductors on the atomic scale. The main advantage of **positron annihilation spectroscopy** is in its selective sensitivity to **vacancy-type defects**. The origin of this sensitivity lies in three key properties of the **positron**: (i) The positron mass is equal to that of the electron, (ii) the positron has a positive charge and (iii) the positron annihilates with an electron producing two or more high-energy photons. Due to the properties (i) and (ii), a free positron in a crystal lattice feels strong repulsion from the positive ion cores and an **open-volume defect** such as vacant lattice site is an attractive center where the positron can get trapped. The time, energy and angular distributions of the positron-electron **annihilation radiation** strongly depend on the spatial and momentum densities of the electrons in the lattice. Positron localization at vacancies leads to an increased **positron lifetime** and to a substantial modification of the shape of the Doppler-broadened 511 keV annihilation line. State-of-the-art theoretical calculations provide important insights into the detailed identification of local atomic structures. Another significant advantage of positron annihilation spectroscopy is that experiments can be performed in bulk crystals and thin film materials of any type of conductivity.

The elementary vacancy defects in Ga_2O_3 are the **Ga vacancy** (missing Ga atom) and the **O vacancy** (missing O atom). When using positron annihilation spectroscopy, it is much more likely to observe Ga vacancies than the O vacancies, thanks to the larger size and non-positive charge states of the former. Unfortunately, the situation is not as simple as in elemental (e.g., Si, Ge, diamond) and binary compound (e.g., GaAs, GaN, ZnO) semiconductors in which the positron methods have brought important successes in defect identification [1–6]. The more complex crystal structure of Ga_2O_3 leads to the fact that there are two non-equivalent Ga sites, that is, two different types of Ga vacancies and three non-equivalent O sites. Based on the earlier work in ZnO [7], it can be expected that V_{Ga} and V_{Ga}–V_O complexes generate positron annihilation signals that are quite similar to each other. Hence, it is likely that eight different defects in Ga_2O_3 produce rather similar signals, a situation that easily leads to smearing out of the "**defect fingerprints**." Only a handful of studies of Ga_2O_3 and related materials have been published [8–10]. The aim of this chapter is to briefly present what has been found so far and in particular to point out what are the further positron steps that could and should be taken in order to gather a more in-depth understanding of the vacancy-type defects. For a full picture of positron annihilation in semiconductors, the reader is referred to review articles [11–13], books and book chapters [14–17], to the conference proceedings of the ICPA (International Conference on Positron Annihilation) and to other references therein.

21.2 Positron Annihilation Spectroscopy

Positrons can be obtained in a variety of ways, out of which the most common in the case of laboratory-scale facilities is using radioactive (β^+) isotopes such as ^{22}Na, where the positron emission is accompanied by a 1.27 MeV photon. The intensity of a ^{22}Na source is limited to about 10^9 positrons/s, but it has a practical half-life of 2.6 years allowing reasonable use of the same source for up to 10 years. These positrons have a wide and continuous energy spectrum with mean energies in the hundreds of keV, and they can be easily used to probe the bulk of a material, up to several hundreds of microns below the surface. In order to study thin films and coatings, the positrons need to be slowed down and monochromated. This is achieved by using crystals with a negative work function for positrons such as W or solid Ne. These slow positrons can be magnetically guided and electrostatically accelerated to form a variable-energy beam allowing for **depth profiling** and near-surface studies, up to a few microns.

The following subsections give a very brief overview of the physics of positrons in solids and of the basic principles of positron lifetime spectroscopy and **Doppler broadening** spectroscopy. These two methods are the most used in studying defects in semiconductors. For thorough reviews on these topics, the reader is referred to [11–17].

21.3 Positrons in Solids: Physical Background

After injection to a crystal, an energetic positron loses its energy through ionization and core electron excitations, and then thermalizes through electron-hole excitations and phonon emission. At room temperature, this whole process takes a few picoseconds. The thermalized positron diffuses behaving similarly to a free carrier in the lattice. In a perfect lattice, the positron state is Bloch-like, and it takes a few hundred picoseconds before the positron annihilates with an electron. In the vast majority of annihilations, and in particular those important from the defect studies point of view, two photons of about 511 keV (equivalent to the rest mass of both the positron and electron) are emitted. During this time, the positron can diffuse up to several hundred nanometers.

A positron diffusing in the lattice may get trapped at a vacancy due to the missing Coulomb repulsion from the atomic cores. This kind of a deep localized positron state has a binding energy of about 1 eV or more. Importantly, in a semiconductor, the vacancy defect needs to be in either a negative or neutral charge state, as the Coulomb repulsion caused by the positive charge state effectively prevents positron trapping. A positron can also get trapped into shallow Rydberg states around a negative ion such as an ionized acceptor impurity, quite analogously to a hole. The positron binding energy to these shallow states is typically 10–100 meV, indicating that positrons are thermally excited from these states at 100–300 K. The hydrogenic positron state around a negative ion has a typical extension of 10–100 Å, and hence, the positron trapped at shallow Rydberg state experiences the same electron environment as in the Bloch-like state in the perfect lattice.

The trapping of a positron at a defect is analogous to carrier capture, and in order to be observed, it needs to be fast enough to compete with annihilation. The **positron trapping rate** κ_D at a defect D (density $[D]$, atomic density of the lattice

N_{at}) is proportional to the defect concentration $c_D = [D]/N_{at}$ through $\kappa_D = \mu_D c_D$. The trapping coefficient μ_D depends on the defect and the host lattice. For a neutral vacancy, it is independent of the temperature, and typical values are in the range $\mu_D^0 = 10^{14}$–10^{15} s^{-1}. In practice, the requirement for detecting defects is $\kappa_D > 0.02\ \lambda_B$, where λ_B is the annihilation rate in the "bulk," i.e., the defect-free lattice. This means that for neutral vacancy defects the detection limit is at roughly 10^{16} cm^{-3}. The positron trapping coefficient at negatively charged vacancies is typically $\mu_D^- = 10^{15}$–10^{16} s^{-1} at room temperature, improving the lower sensitivity limit by an order of magnitude to roughly 10^{15} cm^{-3}. The experimental fingerprint of a negatively charged vacancy is the temperature dependence of the trapping coefficient: $\mu_D^- \sim T^{-1/2}$. The positron trapping rate at the hydrogenic states around negative ions is of the same order of magnitude as that at negatively charged vacancies and behaves in a similar manner as a function of temperature.

A positron state can be experimentally characterized by measuring the positron lifetime. The positron annihilation rate λ, the inverse of the positron lifetime τ, is proportional to the overlap of the electron and positron densities. Typical values of the annihilation rate in the semiconductor lattice are in the range $\lambda_B = 4 \times 10^9$–$1 \times 10^{10}$ s^{-1}, corresponding to $\tau_B = 100$–250 ps. At a vacancy defect, the electron density is locally reduced. This is reflected as a reduction in the annihilation rate and hence a longer positron lifetime, that is, typically 30–80 ps longer than in the perfect lattice. As a consequence, the positron lifetime can be used as a measure of the presence of vacancy defects in a piece of a semiconductor crystal. Both the density of the vacancy defects and their size can be extracted in an experiment, as the trapping at vacancies reduces the effective positron lifetime in the lattice and the annihilation at a vacancy produces a separate signal with a longer lifetime than in the lattice.

More detailed information about the nature of the positron state can be obtained by measuring the energy distribution of the Doppler-broadened 511 keV positron-electron annihilation radiation. The broadening is caused by the center-of-mass motion of the annihilating positron-electron pair, and in practice, the electron momentum dominates over the positron momentum. The energy of the annihilation radiation is shifted in each annihilation event by $\Delta E = c\ p_L/2$, where the longitudinal momentum component p_L is along the measurement direction. A Doppler shift of $\Delta E = 1$ keV corresponds to a momentum value of $p_L = 0.54$ atomic units (a.u.).

When a positron is trapped at a vacancy, the overlap between its wave function and those of the electrons on the atomic core levels is reduced. As the highly localized core electrons have a wider **momentum distribution** than loosely bound or free electrons at reasonable temperatures, the Doppler-broadened 511 keV annihilation line becomes narrower in a typical case. In an area-normalized spectrum, this means an increase in the intensity of the central part of the peak and a decrease in the intensity "wings" of the peak. These two regions of the 511 keV peak also have distinct functional forms as the momentum distribution of the loosely bound and free electrons is very close to the Boltzmann distribution, while the tails of the core electron-momentum distributions decay (sub-)exponentially.

Compared to other atoms in the lattice, the overlap of the vacancy-localized positron wavefunction is significantly stronger with core electrons of the atoms neighboring the vacancy. Hence, the Doppler-broadened signal also carries information about the chemical identity of these atoms, providing, for example, a means for distinguishing "pure" vacancies from vacancy-impurity complexes.

21.3.1 Experimental Methods

The standard technique for bulk crystal studies is sandwiching a low-activity (of the order of 1 MBq) ^{22}Na source between two identical pieces of the sample material. In a typical positron lifetime experiment, the start (1.27 MeV photon) and stop (511 keV photon) signals are detected with collinear scintillation detectors connected to photomultiplier tubes and fast digital acquisition systems. The experiment represents the probability distribution of positron annihilation at time t and is analyzed as a superposition of exponential decay components: $n(t) = \sum_i I_i e^{-t/\tau_i}$ convoluted with the Gaussian resolution function of the spectrometer. A positron in state i annihilates with a specific lifetime τ_i, while the intensity I_i depends on the fraction of positrons annihilating in the state i. About 5–10% of positrons annihilate in the source material, and proper "source corrections" need to be made. Due to the finite time resolution, annihilations in the source materials and random background, typically only 1–3 lifetime components can be reliably resolved in the analysis. The separation of two lifetime components requires their ratio to be $\tau_2/\tau_1 > 1.3$–1.5.

As an example, Fig. 21.1 shows positron lifetime spectra recorded in two diamond samples [5, 13]. One of the samples produces a single-component spectrum with a lifetime of 110 ± 1 ps, and the other a two-component spectrum with $\tau_1 - 125 \pm 5$ ps, $\tau_2 = 420 + 20$ ps and $I_2 = 1 - I_1 = 35 \pm 1\%$. In the latter case, the two lifetime components originate from open core dislocations (τ_1) and large vacancy clusters (τ_1). The average lifetime $\tau_{ave} = \sum_i I_i \tau_i$ is often used as an experimental parameter as it coincides with the center of mass of the spectrum and hence has very high statistical accuracy irrespective of the fit.

The Doppler broadening of the 511 keV positron-electron annihilation radiation is measured using high-purity Ge (HPGe) detectors. In addition to bulk crystal studies with a similar sample-source sandwich as described above, Doppler broadening spectroscopy is typically applied in the variable-energy positron experiments where lifetime spectroscopy is difficult. In thin film studies, the Doppler broadening is often monitored as a function of the beam energy that can be converted to probing depth. Figure 21.2 shows two examples of Doppler-broadened spectra, corresponding to the GaN lattice and the Ga vacancy in GaN. Due to finite **detector resolution** and for the sake of efficiency in measurement time, integrated parameters describing the shape of the peak are conventionally used. The low electron-momentum parameter S is defined as the fraction of the counts in the central region (typically $p_L < 0.4$ a.u.) of the annihilation line, and the

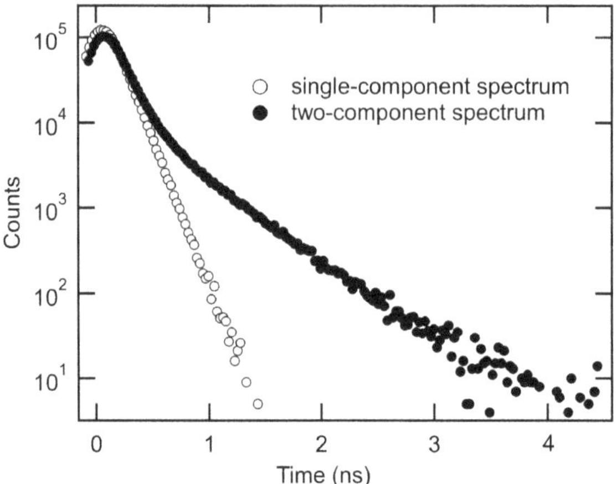

Fig. 21.1 Single- and two-component positron lifetime spectra measured in diamond samples. From [13] with permission

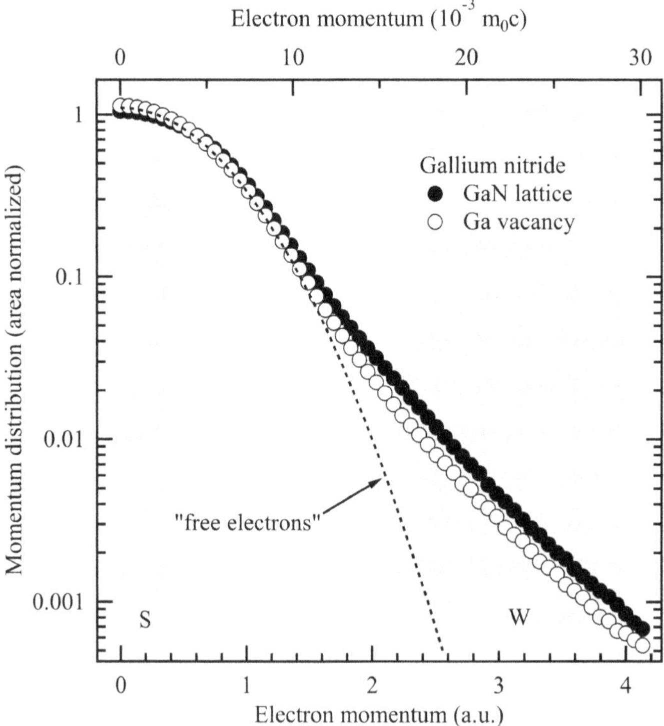

Fig. 21.2 Folded experimental Doppler broadening spectra for the GaN lattice and Ga vacancy. Typical integration windows for the *S* and *W* parameters are shown. From [13] with permission

high electron-momentum parameter W is the fraction of counts in the wing region of the line (typically $p_L > 1.5–2.5$ a.u.).

21.4 Findings in β-Ga_2O_3

21.4.1 Ga_2O_3 Reference Material

Positron annihilation spectroscopy is a comparative methodology in the sense that detailed defect identification and quantification rely on the availability of reference material where positrons annihilate predominantly in the delocalized state in the "defect-free" lattice, that is, no positron trapping at vacancy defects is observed. Hence, for detailed studies of vacancy defects in Ga_2O_3, a "perfect" Ga_2O_3 crystal is needed, often called "bulk" in the positron spectroscopy community. Unfortunately, it is not always straightforward to determine whether the positron annihilation parameters obtained in a given crystal represent the "bulk" or not. Several aspects can and should be considered. In the first approximation, a crystal that produces a spectrum with a single lifetime component is a good candidate. However, a single-component spectrum can also originate from a sample where the concentration of some vacancy defects is so high that all positrons annihilate as trapped in these vacancies (called saturation trapping), and then, the parameters represent the defect instead of the lattice ("bulk"). An additional experiment with a **variable-energy positron beam** can reveal the diffusion length of the positrons that can be used for further justification. If it is long (more than 100 nm), then it is usually assumed that the **defect concentration** in the sample is close to the detection limit. If the trapping is in **saturation** or close to it, the **diffusion length** is strongly reduced. However, this is not a foolproof approach, either, as subsurface defect profiles and surface charge-induced electric fields may affect the apparent diffusion length. By performing experiments in a wide variety of samples, it is usually possible to reach a consensus about the "bulk" parameters. When this is not the case, the shortest center-of-mass lifetime (**average lifetime**) measured in all of the crystals can be assumed to represent the "bulk" state, but this approach needs to be taken with caution.

Early experimental results [8] suggested a **bulk lifetime** of $\tau_B = 175–180$ ps based on experiments in high-temperature sintered Ga_2O_3 samples. A single-component spectrum with $\tau_B = 176$ ps was reported more recently [9] based on experiments in Czochralski-grown single crystals. No further analysis can be found in the literature. With the lack of more detailed knowledge of positron annihilation in Ga_2O_3, it is assumed in the following that $\tau_B = 175$ ps in Ga_2O_3. However, this value and the ensuing conclusion that the Czochralski-grown crystals in question can be assumed to produce annihilation parameters representing the Ga_2O_3 lattice should be taken with caution, as this interpretation may very well evolve with time when more work is published.

21.4.2 Vacancy Defects and Electrical Compensation

Figure 21.3 shows a collection of Doppler broadening **(*S*, *W*) parameters** measured with a variable-energy positron beam in undoped, Si-doped and Sn-doped hetero- and homo-epitaxial Ga_2O_3 thin films grown with metal-organic chemical vapor deposition (MOCVD) and in the Czochralski-grown single crystal discussed above. Some of the data has been published relatively recently [9]. A straight line can be fitted through the data points. This kind of linear behavior can be associated with positrons annihilating in two different states with different state-specific (*S*, *W*) parameters, with the fractions of annihilations in each of the states changing from sample to sample. In the upper left corner of Fig. 21.3, several data points are clustered around the data point obtained in the Czochralski-grown crystal, with no samples producing data with lower *S* (or higher *W*) parameters. This provides further—although circumstantial—evidence that the Czochralski-grown crystal represents the "bulk" for positron annihilation. At this point, it should be noted that the absolute S and W parameters have no physical meaning: They strongly depend

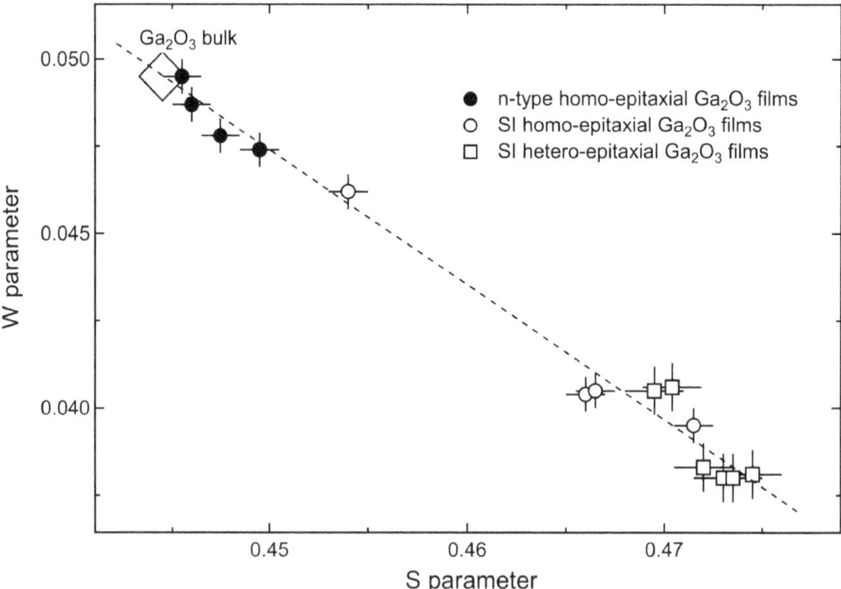

Fig. 21.3 (*S*, *W*) parameters measured in MOCVD-grown hetero- and homo-epitaxial Ga_2O_3 thin films. The large diamond marker represents the data measured in a Czochralski-grown Ga_2O_3 single crystal. Data combined from [9, 18]

on the detector geometry, resolution, calibration, amplifier gain, channel width and in particular in the case of Ga_2O_3 the direction of measurement relative to crystal orientation [18].

If the upper left corner is interpreted as the "bulk" parameters of Ga_2O_3, the simplest interpretation is that the other end represents a V_{Ga}-related defect, as oxygen vacancies are too small to trap positrons [19]. Further, the simplest assumption is that this defect contains a single V_{Ga}, although V_{Ga}–V_O complexes cannot be ruled out. It is likely that the cluster of points in the lower right corner of the figure represents saturation trapping at the observed V_{Ga}-related defects. Comparing the (S, W) values of this vacancy state with those of the "bulk" state gives $S_V/S_B \approx 1.07$ and $W_V/W_B \approx 0.76$. These relative (S, W) values are close to those obtained for V_{Ga} defects in GaN, suggesting that the association of this defect state to V_{Ga} in Ga_2O_3 is reasonable.

With the knowledge of S_B, S_D and τ_B, the concentrations of the V_{Ga}-related defects can in principle be estimated for all the samples in Fig. 21.3 using the so-called trapping model [13]. However, with all the uncertainty in the present values and interpretations, it is cautious not to do so. Instead, as a rule of thumb, we note that data points close to the "bulk" point suggest a vacancy concentration of about 10^{16} cm^{-3} or less, saturation trapping is usually observed when the vacancy concentration exceeds about 5×10^{18} cm^{-3}, and data points halfway between the "bulk" and defect points correspond to a vacancy concentration in the low to mid 10^{17} cm^{-3} range.

An immediate observation can be made from the data in Fig. 21.3. Samples with high resistivity or semi-insulating properties, irrespective of whether doped or undoped, exhibit a V_{Ga}-defect concentration of at least 10^{17} cm^{-3}. In all these samples, the V_{Ga} concentration appears to be higher than the (residual) donor concentration. The other observation is that whenever the samples are n-type conductive, the V_{Ga} concentration appears to be much lower, in the low-to-mid-10^{16} cm^{-3} range. As temperature-dependent experiments have not been performed, we have to rely on theoretical calculations that predict that the V_{Ga} should be in a negative charge state for Fermi levels in the upper half of the band gap [20]. Hence, the observed Ga vacancy concentrations in the Ga_2O_3 films can be interpreted to account for the observed compensation of n-type doping.

While this interpretation of the V_{Ga}-induced **electrical compensation** of n-type conductivity is, in its simplicity, quite appealing, it does not provide a full picture of the processes behind the formation of the compensating centers. In fact, the two concentrations of V_{Ga} in n-type material are somewhat problematic. In addition to predicting that the V_{Ga} should be in a negative charge state, the theoretical calculations [20] also predict that V_{Ga} should be abundant in n-type material and clearly less abundant when the Fermi level is pushed toward the mid-gap, opposite to what is observed and to what one would expect for an acceptor-type defect in general. Further work is required to shed light on this issue and to resolve the possible effects of hydrogen on the apparent V_{Ga} concentrations. One possibility is that the actual V_{Ga} concentrations are much higher than those observed, but a significant fraction is passivated by more than one hydrogen per vacancy [20], in which case they could become much less visible to positrons [6, 7].

21.4.3 Vacancy Formation in $(In_xGa_{1-x})_2O_3$ Alloys

Positron annihilation spectroscopy has also been applied to study the vacancy formation in $(In_xGa_{1-x})_2O_3$ alloys across a wide composition range [10]. This has been possible by studying samples grown by pulsed laser deposition (PLD) in a continuous composition spread (CCS) mode. Several complications arise in analyzing the effects of alloying, the most prominent being the multitude of crystallographic phases that may appear as Ga_2O_3 has several metastable structures and In_2O_3 prefers the cubic bixbyite structure [21–23].

Cation vacancies (V_{Ga} and V_{In}) have slightly different but distinguishable positron signatures in Ga_2O_3 [9] and In_2O_3 [24], respectively. Interestingly, the cation vacancy signatures in the alloys appear to follow the crystalline phase. Alloys with In content up to 20–40% exhibit the standard monoclinic β-phase of Ga_2O_3, and the cation vacancies observed with positrons are V_{Ga}-like. The exact percentage determining the upper bound of this range depends on the overall crystalline quality affected by the substrate for the thin film growth. Alloys with above 70% In content favor the cubic bixbyite structure of In_2O_3, and the cation vacancies exhibit V_{In}-like behavior. In the intermediate composition range, various crystallographic phases (hexagonal $InGaO_3$ II, mixed phase, amorphous) are found and the positron signals are V_{Ga}- or V_{In}-like without clear correlations with the structural phase. Interestingly, the cation vacancy concentrations increase with increasing In content when they are V_{Ga}-like (low In contents) and decrease with increasing In content when they are V_{In}-like (high In contents). This can be interpreted as a manifestation of the tendencies for abundance of V_{Ga} in Ga_2O_3 [9, 20] and for scarcity of V_{In} in In_2O_3 [20, 24].

21.5 Conclusions

This chapter presents a short review of the so far published work performed with positron annihilation spectroscopy with the aim to study vacancy defects in Ga_2O_3. In addition, it introduces the basic concepts and experimental approaches of the methodology. The main finding is that Ga vacancy-related defects appear to be efficient compensating centers in certain cases and can account for semi-insulating or highly resistive behavior even in a material doped with donor impurities. Further, Ga_2O_3 that is n-type after synthesis appears to have very low V_{Ga} contents compared to the carrier concentration. Work on $(In_xGa_{1-x})_2O_3$ alloys has revealed that the crystallographic phase has a strong influence on the Ga_2O_3-like and In_2O_3-like appearance and behavior of the cation (metal) vacancy defects.

Further work is required to shed light on the formation mechanisms of the Ga vacancy defects in Ga_2O_3. Also for improving the quantitative aspects of the positron methods, the determination of the "true" bulk annihilation parameters of Ga_2O_3, equivalent to finding a suitable reference sample for Ga_2O_3 studies, is of high importance. Reaching both of these aims is rendered particularly difficult by the impractical—although very interesting—property of Ga_2O_3: The annihilation parameters are strongly dependent on the measurement direction relative to crystallographic orientation. Colossal differences, of the order of differences found typically only for vacancies compared to "bulk," are observed [18]. A detailed mapping and understanding the exact origin of this property are required.

References

1. K. Saarinen, S. Kuisma, P. Hautojärvi, C. Corbel, C. LeBerre, Phys. Rev. B **49**, 8005 (1994)
2. K. Saarinen et al., Phys. Rev. Lett. **79**, 3030 (1997)
3. K. Saarinen, J. Nissilä, H. Kauppinen, M. Hakala, M.J. Puska, P. Hautojärvi, C. Corbel, Phys. Rev. Lett. **82**, 1883 (1999)
4. F. Tuomisto, V. Ranki, K. Saarinen, D.C. Look, Phys. Rev. Lett. **91**, 205502 (2003)
5. J.-M. Mäki, F. Tuomisto, A. Varpula, D. Fisher, R.U.A. Khan, P.M. Martineau, Phys. Rev. Lett. **107**, 217403 (2011)
6. F. Tuomisto, V. Prozheeva, I. Makkonen, T.H. Myers, M. Bockowski, H. Teisseyre, Phys. Rev. Lett. **119**, 196404 (2017)
7. K.M. Johansen, F. Tuomisto, I. Makkonen, L. Vines, Mater. Sci. Semicond. Process. **69**, 23 (2017)
8. W.-Y. Ting, A.H. Kitai, P. Mascher, Mater. Sci. Eng. **91**, 541 (2002)
9. E. Korhonen, F. Tuomisto, D. Gogova, G. Wagner, M. Baldini, Z. Galazka, R. Schewski, M. Albrecht, Appl. Phys. Lett. **106**, 242103 (2015)
10. V. Prozheeva, R. Hölldobler, H. von Wenckstern, M. Grundmann, F. Tuomisto, J. Appl. Phys. **123**, 125705 (2018)
11. M.J. Puska, R.M. Nieminen, Rev. Mod. Phys. **66**, 841 (1994)
12. P. Asoka-Kumar, K.G. Lynn, D.O. Welch, J. Appl. Phys. **76**, 4935 (1994)
13. F. Tuomisto, I. Makkonen, Rev. Mod. Phys. **85**, 1583 (2013)
14. K. Saarinen, P. Hautojärvi, C. Corbel, Semiconductors and semimetals, in *Identification of Defects in Semiconductors*, vol. 51A, ed. by M. Stavola (Academic Press, New York, 1998), p. 209
15. R. Krause-Rehberg, H.S. Leipner, *Positron Annihilation in Semiconductors* (Springer-Verlag, Berlin, 1998)
16. F. Tuomisto, in *Technology of Gallium Nitride Crystal Growth*, ed. by D. Ehrentraut, E. Meissner, M. Bockowski (Springer, Berlin/Heidelberg, 2010)
17. F. Tuomisto, Semiconductors and Semimetals, in *Oxide Semiconductors*, vol. 88, ed. by B.G. Svensson, S.J. Pearton, C. Jagadish (Elsevier, Oxford, 2013)
18. F. Tuomisto, A. Karjalainen, V. Prozheeva, I. Makkonen, G. Wagner, M. Baldini, Proc. SPIE **10919**, 1091910 (2019)
19. I. Makkonen, E. Korhonen, V. Prozheeva, F. Tuomisto, J. Phys.: Condens. Matter **28**, 224002 (2016)
20. J. Varley, H. Peelaers, A. Janotti, C. Van de Walle, J. Phys.: Condens. Matter **23**, 334212 (2011)
21. H. von Wenckstern, D. Splith, M. Purfürst, Z. Zhang, C. Kranert, S. Müller, M. Lorenz, M. Grundmann, Semicond. Sci. Technol. **30**, 024005 (2015)

22. C. Kranert, J. Lenzner, M. Jenderka, M. Lorenz, H. von Wenckstern, R. Schmidt-Grund, M. Grundmann, J. Appl. Phys. **116**, 013505 (2014)
23. R. Schmidt-Grund, C. Kranert, T. Böntgen, H. von Wenckstern, H. Krauß, M. Grundmann, J. Appl. Phys. **116**, 053510 (2014)
24. E. Korhonen, F. Tuomisto, O. Bierwagen, J.S. Speck, Z. Galazka, Phys. Rev. B **90**, 245307 (2014)

Chapter 22
Electrical Properties 1

Donors and Acceptors

Adam T. Neal and Shin Mou

Abstract Studies of Si, Ge donors and Fe, Mg as acceptors in β-Ga_2O_3 through temperature-dependent van der Pauw and Hall effect measurements of samples grown by a variety of methods are presented in this chapter. Si and Ge are identified as shallow donors with donor energy of 30 meV instead of DX states. Fe is a deep acceptor with its energy level 860 meV below the conduction band edge. Mg-doped samples present an activation energy of 1.1 eV, but the type could not be resolved. Unintentional donors and acceptors are also discussed including intrinsic defects (i.e., vacancies) and extrinsic impurities. Last, an unintentional donor with energy of 110 meV is presented and the impacts of its incomplete ionization to power devices are discussed.

22.1 Introduction

It is key to precisely control the carrier density and resistivity in semiconductors through intentional doping to realize device structures for electronics and opto-electronics. Doping facilitates carrier density control in the channel and drift region of transistors, enables low-contact-resistance Ohmic contacts, controls threshold voltage, forms p-n junctions, generates current blocking layers, and makes semi-insulating substrates. The rapid pace of β-Ga_2O_3 device development can be partly attributed to the effectiveness of the n-type doping as well as compensating acceptor doping. Currently, group IV elements including Si, Ge, and Sn are all choices of n-type dopants [1–7]. On the other hand, Fe, Mg, and N are deep acceptors [1, 8–14] which have been used to form semi-insulating substrates or current blocking layers. Unintentionally, n-type donors and compensating acceptors

A. T. Neal · S. Mou (✉)
Air Force Research Laboratory, Materials and Manufacturing Directorate,
2179 12th Street, Wright-Patterson Air Force Base, OH 45433-7718, USA
e-mail: shin.mou.1@us.af.mil

A. T. Neal
e-mail: adam.neal.3@us.af.mil

M. Higashiwaki and S. Fujita (eds.), *Gallium Oxide*, Springer Series
in Materials Science 293, https://doi.org/10.1007/978-3-030-37153-1_22

are also introduced into β-Ga_2O_3 through extrinsic contamination (e.g., silicon impurities in the melt-grown crystal substrates) and intrinsic defects (e.g., vacancies). Unintentional n-type doping is an important research topic in β-Ga_2O_3 because it sets the lower limit of the residual doping and carrier density, which in turn sets the upper limit on the maximal achievable device breakdown voltages. On the other hand, unintentional compensating acceptors also affect the lower limit for n-type doping, because the target electron concentration ($N_{dn} - N_{ac}$) must be comparable to the compensating acceptor concentration to achieve effective process control of n-type doping. Moreover, the ionized impurity scattering resulting from the unintentional donors and compensating acceptors degrades the mobility of β-Ga_2O_3. Since many of these effects are directly correlated to the electronic transport of β-Ga_2O_3, that is, affecting the carrier density and mobility of the samples, Hall effect measurements are used to characterize β-Ga_2O_3 samples with various dopants (i.e., Si, Ge, Fe, Mg) as a function of temperature. With that approach, critical information is obtained including the donor/acceptor ionization energy, acceptor compensation ratio, and electron scattering mechanisms. All of this knowledge helps us understand the natures of these intentional donors and acceptors, as well as the unintentional dopants, in β-Ga_2O_3. This chapter will be organized in the following way. First, we will discuss the shallow donors, which are appropriate for controllable n-type doping of β-Ga_2O_3. Following that, we will examine intentionally introduced deep acceptors, which can be used to make semi-insulating substrates and current blocking layers. Last, we will discuss unintentional donors and deep acceptors, which often limit the performance of semiconductor electronics. In particular, we will review an unintentional donor with donor energy of 110 meV and discuss its detrimental effects to β-Ga_2O_3.

22.2 Shallow Donors

Ga_2O_3 device development benefits from the effective and controllable n-type doping, which has been achieved using group IV elements including tin (Sn) [2, 3, 5, 6, 15–20], silicon (Si) [6, 7, 21–26], and, more recently, germanium (Ge) [27], consistent with results from DFT calculations [28].

Studies have examined the transport properties of some of these donors in Ga_2O_3; however, there remain some discrepancies regarding the energies of the shallow donors. Estimates of the donor energies range from 7.4 to 60 meV for Sn [15, 29, 30], 16 to 50 meV for Si [31–37], and a 17.5 meV donor energy was previously reported for Ge [4]. Recent electron paramagnetic resonance (EPR) studies report that Si may also exhibit a DX^- state at energy 49 meV [36]; however, other groups have reported no evidence for a DX^- state [38]. Considering theoretical calculations, early on Varley et al. [28] reported Si, Ge, and Sn as shallow donors based on the formation energy calculation using first-principle DFT modeling. More recently, Lany performed detailed DFT modeling including the defect transition energy calculation and concluded that only Si is a "true" shallow donor while Ge and Sn have delocalized impurity states near the conduction band minimum (CBM), which might be close enough for ionization at room temperature but also have the potential to form DX^- states [39]. Given the wide range of donor energies reported in the literature for shallow donors, which is summarized in Fig. 22.1 for Si, and the discrepancy in theoretical modeling, a study to understand the transport properties of Si and Ge is presented in this section in order to clarify the properties of these donors.

To study the properties of shallow donors in Ga_2O_3, temperature-dependent carrier density and mobility of several samples are measured by van der Pauw and Hall effect measurements in a four-terminal configuration. To accurately estimate the donor energies, the carrier density was fit using the charge neutrality equation [40] and the mobility fit using the solution to the Boltzmann transport equation in the relaxation time approximation [41], including the Hall factor [42], where ionized impurity [43], neutral impurity [44], and polar optical phonon [45] scattering mechanisms were included. An effective mass $m^* = 0.3\ m_o$ [46–48], a low-frequency relative dielectric constant $\kappa_s = 10$ [49, 50], a high-frequency relative dielectric constant of $\kappa_\infty = 3.5$ [50–52], an effective phonon energy $\hbar\omega = 44$ meV [34], and an effective number of phonon modes $M = 1.5$ were used in the calculation. Iteration was used to simultaneously and self-consistently fit both the carrier density and mobility. Uncertainty in the compensating acceptor concentration can propagate to the estimated donor energy when fitting the temperature-dependent carrier density alone in moderately doped samples due to the influence of impurity conduction at low temperatures. With this approach, it is possible to independently determine the compensating acceptor concentration from the ionized impurity limited mobility, allowing for more accurate estimation of the donor energies by avoiding said propagation. The Si-doped samples include a bulk CZ sample from Northrop Grumman Synoptics (Sample 3), a bulk EFG sample from Tamura Corporation (Sample 4), and two epitaxial films grown by low-pressure chemical vapor deposition (LPCVD) on c-sapphire substrates with 3.5° (Sample 5) and 6° (Sample 6) offcuts [53]. Glow discharge mass spectrometry (GDMS) and secondary ion mass

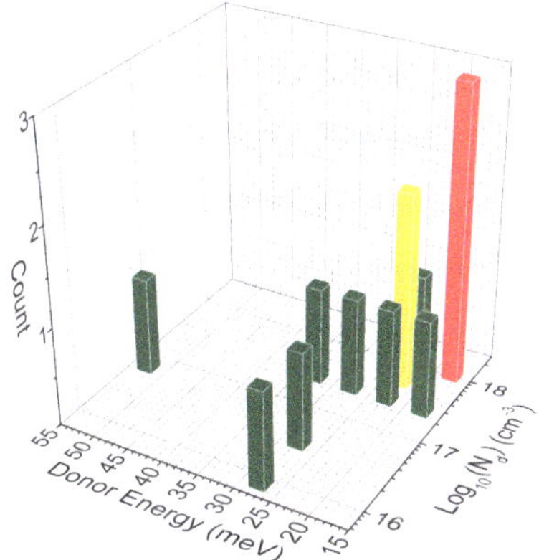

Fig. 22.1 Data point counts of the reported donor energy for Si in the literature [31–37] as a function of log donor density

spectrometry (SIMS) analysis of EFG Ga_2O_3 samples similar to Sample 4 and SIMS analysis of a CZ grown sample similar to Sample 3 confirm that Si is the dominant unintentional donor in the bulk melt-grown Ga_2O_3 used in this study. Our result is consistent with a previous study which determined that the unintentional Si doping comes from the Ga_2O_3 powder used as the source material for bulk melt-growth [21]. The LPCVD films were intentionally doped using $SiCl_4$.

To make ohmic contacts, four 150 nm Ti/500 nm Au contacts were deposited on the sample edges and annealed in a tube furnace under Ar gas flow up to 450 °C. Sample 1 and Sample 2 are Ge-doped MBE-grown Ga_2O_3 epitaxial films on semi-insulating substrates whose fabrication and growth details are published elsewhere [4, 27]. Figures 22.2 and 22.3 show the experimentally measured Hall carrier density and Hall mobility for the samples, along with the temperature-dependent fittings. The mobility due to individual scattering mechanisms is shown for Sample 3 only. Parameters of the temperature-dependent fittings for the samples are shown in Table 22.1. The donor energies for the samples, along with several samples from the literature, are summarized in Fig. 22.4. As the figure shows, the donor energies for samples seem to converge to a value of 30 meV as the donor concentration approaches 1×10^{17} cm^{-3} for both Si and Ge donors. However, as the donor concentration increases above 4×10^{17} cm^{-3}, the donor energy begins to decrease, as is expected for highly doped semiconductors when an impurity band begins to form [54, 55]. Consistent with this hypothesis, the decrease in the donor energy occurs as the donor density approaches $N_{dn} = (0.2/\alpha)^3 = 1.46 \times 10^{18}$ cm^{-3}, where α is the effective Bohr radius for gallium oxide, which is the estimated density at which a Mott metal-insulator transition would occur for the donor level [54, 55].

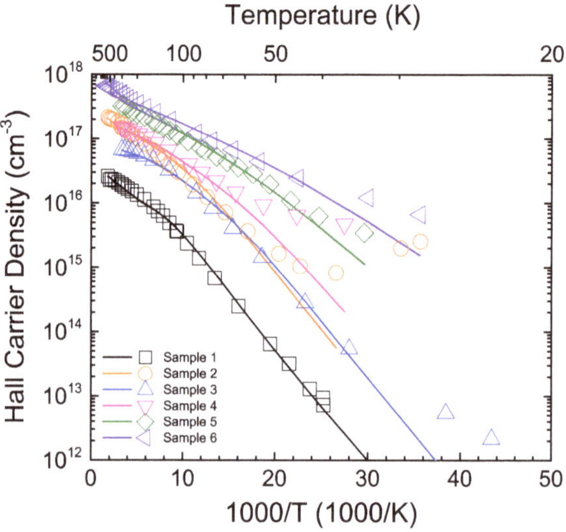

Fig. 22.2 Measured Hall carrier density (symbols) and fittings (solid lines) for Si and Ge doped β-Ga2O3 samples (Reproduced from [1])

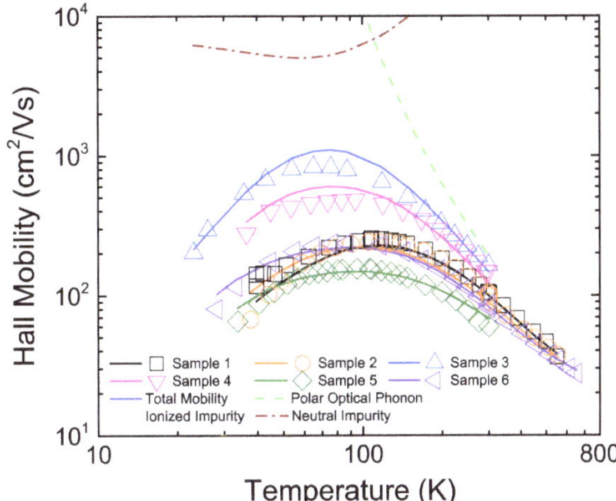

Fig. 22.3 Measured Hall mobility (symbols) and fittings (solid lines) for Si and Ge doped β-Ga_2O_3 samples. Individual components of the mobility are shown for Sample 3, with the scattering mechanisms indicated in the legend (Reproduced from [1])

Table 22.1 Parameters of temperature-dependent carrier density and mobility fitting for n-type β-Ga_2O_3 samples (Reproduced from [1])

	Growth method	Dopant	N_{dn} 10^{16} cm^{-3}	E_{dn}^{a} meV	N_{ac} 10^{16} cm^{-3}	N_{ac}/N_{dn}	$N_{neutral}^{b}$ 10^{16} cm^{-3}	$\hbar\omega^{c}$ meV
Sample 1[d]	MBE	Ge	6.5	28	4.8	0.74	80	44
Sample 2	MBE	Ge	30	29	3.9	0.13	80	44
Sample 3	CZ	Si	13	30	0.91	0.07		44
Sample 4	EFG	Si	30	27	1.5	0.05		44
Sample 5	LPCVD	Si	80	19	5.6	0.07	100	44
Sample 6	LPCVD	Si	100	15	5.0	0.05		44
Fornari et al. #12[e]	CZ	Si	14.3	28.5	4.2			
Fornari et al. #3[e]	CZ	Si	48.3	21.2	14			
Fornari et al. #7[e]	CZ	Si	61.7	24.9	5.4			
Oishi et al.[f]	EFG	Si	14	31	2.7			

[a] Referenced to conduction band
[b] Additional neutral impurities beyond unionized donors
[c] Reference [34]
[d] A 2nd donor with E_{dn2} =100 meV and N_{dn2} = 1.5 × 10^{16} cm^{-3} was also included [35]
[e] References [31, 32]
[f] Reference [33]

The above analysis assumes that the Si and Ge donors behave as typical shallow donors in Ga_2O_3, but recent EPR measurements have suggested that Si may behave as a shallow DX center and theory suggests Ge is potentially a DX center [36, 39]. By contrast, others have reported that they do not see experimental evidence of DX center behavior in Ga_2O_3 [38], and still others predicted Si and Ge to be shallow donors by

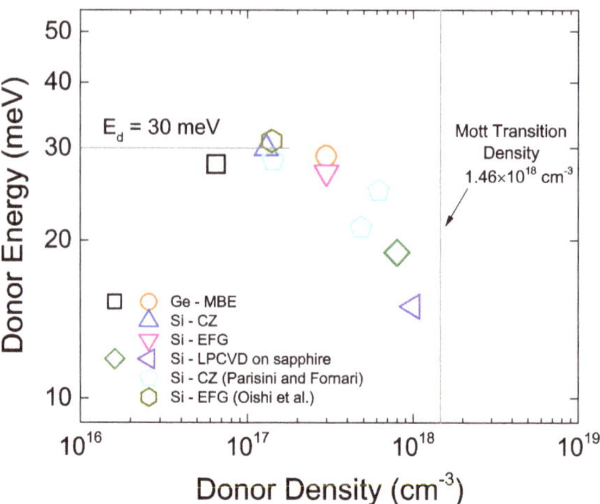

Fig. 22.4 Summary of the donor energies as a function of donor concentration. Data from the literature [32, 33] are also included as indicated in the figure legend (Reproduced from [1])

DFT calculation [28]. To address this open question in light of our Hall effect measurements, let us consider the charge neutrality equation for a DX center [56]

$$N_c \mathcal{F}_{1/2}\left(\frac{E_f - E_c}{kT}\right) + \frac{N_{\mathrm{dn}}}{1 + \exp\left(\frac{(2E_{\mathrm{dn}} + U) - 2E_f}{kT}\right) + 2\,\exp\left(\frac{E_{dn} + U - E_f}{kT}\right)} + N_{\mathrm{ac}} = \frac{N_{\mathrm{dn}}}{1 + \exp\left(\frac{-(2E_{\mathrm{dn}} + U) + 2E_f}{kT}\right) + 2\,\exp\left(\frac{-E_{\mathrm{dn}} + E_f}{kT}\right)} \tag{22.1}$$

where N_c is the conduction band effective density of states, N_{ac} the compensating acceptor concentration, N_{dn} the donor concentration, E_{dn} the donor energy, $\mathcal{F}_{1/2}$ the normalized Fermi-Dirac integral of order one half, E_f the Fermi level, and U the interaction energy for two electrons on a single donor. The interaction energy, U, is the energy difference between two electrons on a single donor (DX^- state) and two non-interacting electrons on two different donors (neutral state). U is negative, which indicates that the DX^- state is the lower energy state, usually due to lattice distortion. For a normal shallow donor, U is a large positive number due to Coulomb repulsion of the electrons, and the conventional charge neutrality equation is recovered. Using (22.1) and the parameters listed in the first two entries of Table 22.2, a comparison of the temperature-dependent carrier density and ionized impurity density for the typical normal shallow donor model and the shallow DX donor model is plotted in Fig. 22.5. For the models in Fig. 22.5, values for the normal donor model were chosen to be representative of the experimentally characterized samples listed in Table 22.1, while the values for the DX donor model were chosen to match the temperature-dependent carrier density of the normal donor model, with $U = -20$ meV as suggested by the recent EPR study [36]. As shown, it is possible to find a set of parameters for the DX donor model that almost match the temperature-dependent carrier density of the normal donor model;

Table 22.2 Parameters for comparing normal donor and DX donor models (Reproduced from [1])

	N_{dn} 10^{16} cm^{-3}	E_{dn} meV	U meV	N_{ac} 10^{16} cm^{-3}	N_{ac}/N_{dn}
Normal Figure 22.5	11	30	+∞	1.1	0.1
DX Figure 22.5	10	16	−20	0	0
Sample 2 DX, Fig. 22.6	27	16	−10	5.4	0.2
Sample 3 DX, Fig. 22.6	13	18	−10	0	0
Sample 4 DX, Fig. 22.6	25	10	−15	0	0

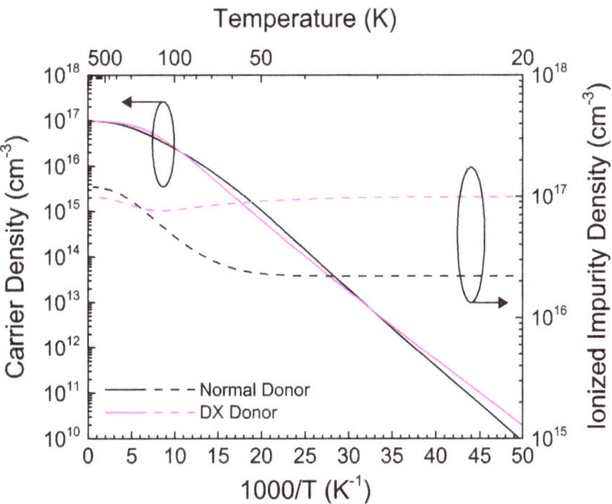

Fig. 22.5 Comparison of carrier density and ionized impurity density versus temperature for a normal donor and a DX center with $N_{dn} - N_{ac}$ of 1×10^{17} cm^{-3}. Other parameters for the two models are shown in Table 22.2. The higher concentration of ionized impurities at low temperature in the DX donor model leads to underestimation of the experimentally measured Hall mobility as shown in Fig. 22.6 (Reproduced from [1])

however, as the dashed lines show, the density of ionized impurities at low temperatures is about a factor of five larger for the DX donor model. Because the density of ionized impurities is so much larger for the DX donor model, it underestimates the experimentally measured mobility of our samples at low temperatures where ionized impurity scattering dominates, as shown in Fig. 22.6, with fitting parameters shown in Table 22.2. This fact means that the DX donor model cannot be used to fit our measured Hall mobility and carrier density data. This analysis assumes that the spatial distribution of donors in the negatively charged DX^- state and those in the typical positively charged ionized state are uncorrelated, so that all donors act as point-charge-like scattering centers. However, there is some evidence that the distributions of DX^- donors and positively charged ionized donors can become correlated, acting as weaker dipole-like scattering centers [57]. While such a correlation may partially account for the discrepancy between our measured mobility data and the DX donor mobility models presented in Fig. 22.6, it does not fully account for the factor of 4–5 discrepancy in low-temperature mobility for Si-doped Sample 3 and Sample 4. Therefore, our measured Hall data for these Si

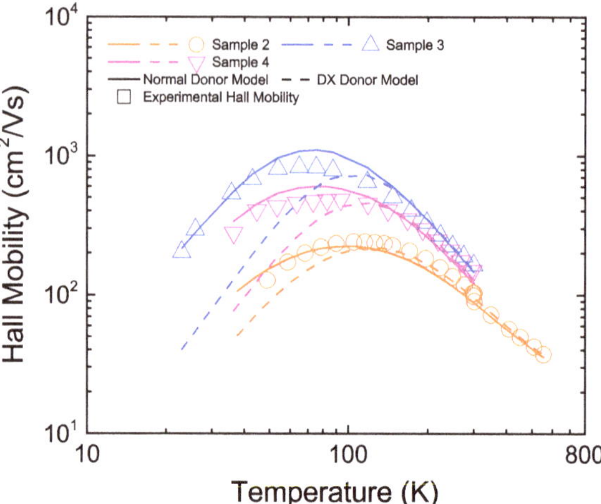

Fig. 22.6 Comparison of mobility fitting using the normal donor model (solid lines) and the DX donor model (dashed lines) to the experimentally measured Hall mobility for Sample 2, Sample 3, and Sample 4. Only the normal donor model can fit the experimental data, as the DX donor model yields too much ionized impurity scattering at low temperatures and underestimates the experimentally measured mobility (Reproduced from [1])

doped samples are consistent with a normal shallow donor and are inconsistent with a shallow DX center. This observation agrees with Irmscher et al. [38] who also reports no evidence of DX behavior. We note, however, that we are only able to rule out DX center behavior in our samples because they have a compensation ratio N_{ac}/N_{dn} less than one-third. For compensation ratios of one-third or greater in the normal donor model, it is always possible to find a set of parameters for the DX donor model that match both the temperature-dependent carrier density and the low-temperature ionized impurity density. This fact means that a normal shallow donor and shallow DX donor can only be distinguished using simultaneous, self-consistent carrier density and mobility fitting for compensation ratios less than one-third, as is the case here. Considering the remarkable consistency in experimental results across a wide variety of growth methods, including EFG, CZ, MBE, and LPCVD, along with the fact that Si and Ge are both group IV elements on the periodic table, we can conclude that Si and Ge act as typical shallow donors in all of the samples presented here.

22.3 Deep Acceptors

It is predicted to be difficult to form shallow acceptors in β-Ga_2O_3 due to the flat heavy-hole dispersion curve at around the Brillouin zone center with very high hole effective mass and the self-trapping nature of holes [58]. Experimentally, only deep acceptors such as Fe, Mg, N, and Zn have been verified [1, 9, 12, 14, 59]. Here, Fe- and Mg-doped samples are characterized using the same four-terminal van der Pauw and Hall effect measurement techniques. Due to the semi-insulating nature of these samples, high-temperature Hall effect measurements were necessary to thermally activate carriers to enable Hall effect measurement. Measurements were performed under N_2 gas flow in a tube furnace with an external silicon carbide heater and electromagnet. Figure 22.7 shows the Hall carrier density for an Fe-doped, EFG grown semi-insulating substrate from Tamura Corporation (Sample 7). Hall effect measurement was possible for temperatures above 400 K, with the sign of the Hall effect indicating that the β-Ga_2O_3 remains weakly n-type at elevated temperatures even with Fe doping. Glow discharge mass spectrometry analysis (GDMS) performed on a similar Fe-doped sample indicates a concentration of about 1×10^{17} cm^{-3} for the unintentional Si donor and 8×10^{17} cm^{-3} for the intentional Fe acceptor. Therefore, we ascribe the observed n-type behavior to thermal activation of electrons on ionized Fe acceptors into the conduction band. Fitting with the charge neutrality equation yields an acceptor energy $E_c - E_{ac}$ of 860 meV. Note that this acceptor energy is referenced to the conduction band edge due to the n-type behavior. This acceptor energy near the conduction band is consistent with preliminary DFT calculations of Fe doped Ga_2O_3 which indicate that Fe induces midgap states closer to the conduction band [8]. The acceptor energy is somewhat higher than that recently observed via DLTS for Fe impurities in n-type bulk substrates [60]; however, this difference can be explained by the broadening of the Fe acceptor energy level at the much higher Fe concentration in this semi-insulating substrate. Because the carrier density does not saturate at higher temperatures, it is not possible to estimate the absolute value of the Fe and Si concentrations in this sample. However, the ratio of Fe acceptors to Si donors, N_{ac}/N_{dn}, is uniquely determined to be 1.65 by fitting the temperature-dependent carrier density using the charge neutrality equation. We note that this ratio is smaller than one would expect based on the GDMS results, which could suggest that not all of the Fe dopants in the sample are electrically active. Finally, the inset of Fig. 22.7 shows the temperature-dependent conductivity measured by the van der Pauw method for an Mg-doped sample grown by Northrop Grumman Synoptics using the

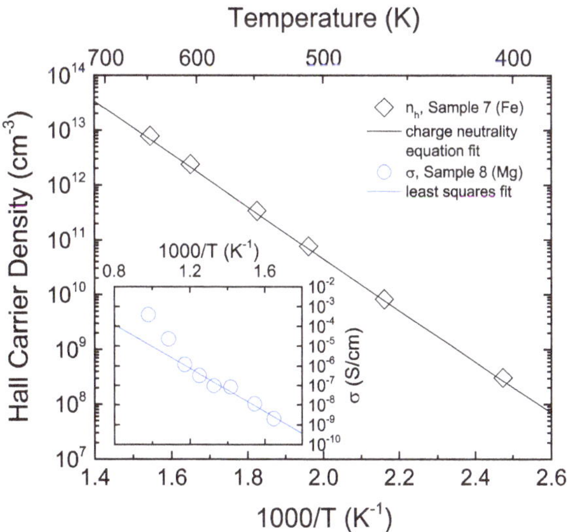

Fig. 22.7 High temperature Hall carrier density for Sample 7, an Fe doped semi-insulating β-Ga_2O_3 sample. The sign of the Hall effect indicates that the sample is weakly n-type at elevated temperatures. Inset: Temperature-dependent conductivity of Sample 8, an Mg doped semi-insulating β-Ga_2O_3 sample (Reproduce from [1])

CZ method (Sample 8). The melt from which the sample was pulled contained 0.1 mol% of MgO. Hall effect measurement was attempted, but it was not possible to resolve the Hall voltage due to low signal to noise ratio as a result of the very low conductivity for the sample. Least squares fitting of the temperature-dependent conductivity indicates an empirical activation energy of 1.1 eV, which is consistent with a previous report on the activation energy of highly Mg-doped Ga_2O_3 [10]. Because Si contamination is present in these CZ-grown samples, it is expected that the activation energy of 1.1 eV is approximately equal to the acceptor energy for Mg. However, without Hall effect measurement, it is not possible to determine the carrier type of the Mg-doped sample. This fact means that we are unable to determine if this 1.1 eV activation energy is referenced to the conduction band or valence band edge of Ga_2O_3 from experiment, although recent DFT studies indicate the Mg acceptor level is closer to the valence band [61]. In conclusion, while Fe and Mg are the easy choices for deep compensation acceptors in the melt for β-Ga_2O_3 semi-insulating substrate growth, it warrants more study to not only understand the natures of all kinds of compensating acceptors but also find the best choice for the formation of current blocking layers.

22.4 Unintentional Donors and Acceptors

Unintentional doping is an important research topic for semiconductors because it sets the lower limit of the residual doping and carrier density, which in turn sets the upper limit on device breakdown voltages. To illustrate this fact, the dependence of breakdown voltage for p-n single-sided junction and Schottky junction devices, normalized by material parameters, is plotted in Fig. 22.8. The black line indicates the theoretical limit, assuming the full-depletion approximation and one-dimensional electrostatics. By normalizing the breakdown voltages in this way, one can isolate the effect of doping on device breakdown voltage to allow comparisons of devices made from different materials. Comparing early Ga_2O_3 devices to those of more mature semiconductor materials gives insight into the future potential of Ga_2O_3 devices. The plotted experimental data are among the best reported for vertical devices of each material with large breakdown voltage and low doping of the device drift layer. While edge effects prevent real devices of any material from matching the simplified theoretical limit exactly, the N_d^{-1} trend is nevertheless observed considering the best performing devices of the different material systems. Comparing Ga_2O_3 to the more mature materials, it is clear that some improvement can be expected from the optimization of device geometries to mitigate edge effects, as shown by the blue arrow. However, much greater improvement in breakdown voltage is possible from reducing drift layer doping in Ga_2O_3 devices, as indicated by the orange arrow. For instance, if the drift layer doping of a Ga_2O_3 device can be reduced to about 10^{14} cm^{-3} like the 21.7 kV SiC device indicated in Fig. 22.8, then the Ga_2O_3 device will have a breakdown voltage

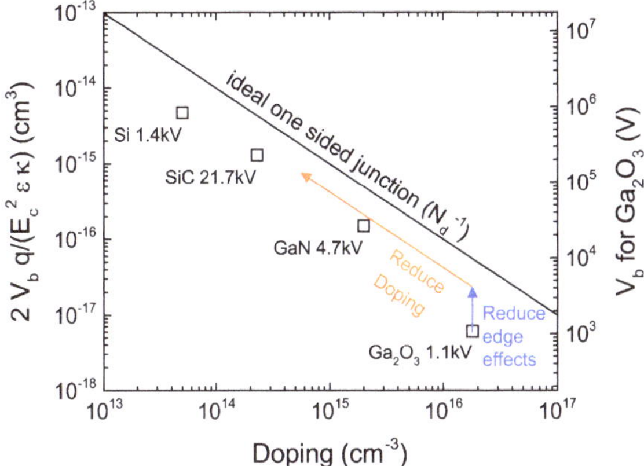

Fig. 22.8 Normalized breakdown voltage (left axis) versus doping concentration for single-sided junction power devices. The ideal relationship assuming the full-depletion approximation and 1D electrostatics is shown by the black line. The right axis indicates the projected absolute breakdown voltage for Ga_2O_3 devices for comparison. Square symbols indicate experimental measurements of devices with breakdown voltages among the highest reported for each material: Si [69], SiC [70], GaN [71], Ga_2O_3 [62] (Reproduced from [35])

over 100 kV as indicated by the right axis of Fig. 22.8. Therefore, understanding and mitigating unintentional doping in Ga_2O_3 is key to increasing the maximum achievable breakdown voltages in Ga_2O_3 beyond the recently demonstrated 1 kV devices [62]. While unintentional donors obviously set a lower limit on the doping that can be achieved in Ga_2O_3, compensating acceptors also set a lower limit on controlled n-type doping when considering process control. The target electron concentration ($N_{dn} - N_{ac}$) must be comparable to the compensating acceptor concentration in order to achieve effective process control of n-type doping in Ga_2O_3 or any semiconductor. In other words, it is not practical to perfectly control the intentional donor concentration N_{dn} to achieve a difference $N_{dn} - N_{ac}$ which is much less than the unintentional compensating acceptor concentration N_{dn}. Therefore, from the standpoint of a practical technology, both unintentional donors and acceptors set a lower limit on the doping and upper limit on breakdown voltage of practical device. Additionally, unintentional doping is also a source of additional scattering centers which degrade the carrier mobility.

In melt-grown substrates, Si is the main unintentional donor resulting from SiO_2 impurities in the Ga_2O_3 powder precursor [21]. In epitaxial samples grown by, for example, molecular beam epitaxy (MBE) or metal-organic chemical vapor deposition (MOCVD), the residual unintentional dopants are not very well studied except for the impurities out diffused from the substrate during the growth such as Fe and Si [9, 63]. For doping resulting from intrinsic defects, first-principle calculation indicates that the Ga vacancy is a deep compensating acceptor [64, 65] while the O vacancy is a deep donor, which can be a neutral impurity scattering center [28]. A positron annihilation study did confirm that the Ga vacancy is a deep acceptor in β-Ga_2O_3 [66]. Also, the Ga vacancy is energetically more favorable than the oxygen vacancy even in a Ga-rich growth environment [65]. Therefore, it is expected Ga vacancies would form during the growth to compensate the donors to lower the total energy of the β-Ga_2O_3 crystal. One observation is that most of the samples with non-degenerate doping in Table I have similar compensating acceptor densities in the 10^{16} cm^{-3} range. While more studies are needed to understand the source of the unintentional compensating acceptors in β-Ga_2O_3, the Ga vacancy theory is consistent with this observation. On the other hand, oxygen vacancy-related defects were identified using catholuminescence (CL) as deep levels [67]. This observation is consistent with the aforementioned theoretical calculation. Since oxygen vacancies are deep donors, they do not provide ionized-free electrons at room temperature that can be measured by transport measurements. Other possible unintentional residual impurities include hydrogen, carbon, and chlorine. Hydrogen is hard to remove in MBE and can come from the metal-organic precursors in MOCVD. It can act as a shallow donor in β-Ga_2O_3 either as an interstitial or incorporated on the oxygen site [28], or it can passivate the Ga vacancies based on ab initio calculations [64]. Carbon, like hydrogen, can be a residual in the films due to similar reasons.

Based on DFT calculation, carbon is a DX center in β-Ga_2O_3 [39], which is a deep level donor. Chlorine is a shallow donor substituting the oxygen sites based on first-principle calculation [28], which can potentially be a residual in halide vapor phase epitaxiay that uses a chlorine-based vapor precursor [25].

Last, there are unintentional donors with energy about 100 to 200 meV reported by Neal et al. [35] and Farzana et al. [68] whose material origin remains unknown. Figure 22.9 shows the temperature-dependent carrier density, normalized to room temperature, of two EFG grown samples measured via high temperature Hall effect measurement up to T = 1000 K. The increase in carrier density above room temperature indicates the existence of these unintentional donors and cannot be explained by Si donors alone. Fitting of the temperature-dependent carrier density estimates a donor energy of 110 meV for these samples in addition to the Si shallow donor. Admittance spectroscopy measurements also confirm the existence of this unintentional donor [35]. Because the doping of the drift layer cannot be reduced below the level of unintentional doping, the density of unintentional donors sets the lower limit on electric field at the metal-semiconductor interface and the upper limit on the depletion width of the Ga_2O_3 drift layer for a given applied voltage. Therefore, a limit is set on the trade-off which allows higher breakdown voltage to be achieved at the expense of larger on-resistance. Moreover, incomplete ionization of the 110 meV donor poses an additional challenge to maximizing breakdown voltage and minimizing on-resistance due to the four times larger donor energy as compared to regular shallow donors such as Si and Ge. The 110 meV donor is shallow enough in energy that it becomes fully ionized when the Schottky diode is reverse biased, reducing the depletion width, increasing the electric field at the metal-semiconductor interface, and reducing the breakdown voltage in the same way as the shallow donor. However, the 110 meV donor energy is large enough that incomplete ionization can be significant, meaning that only a fraction of the

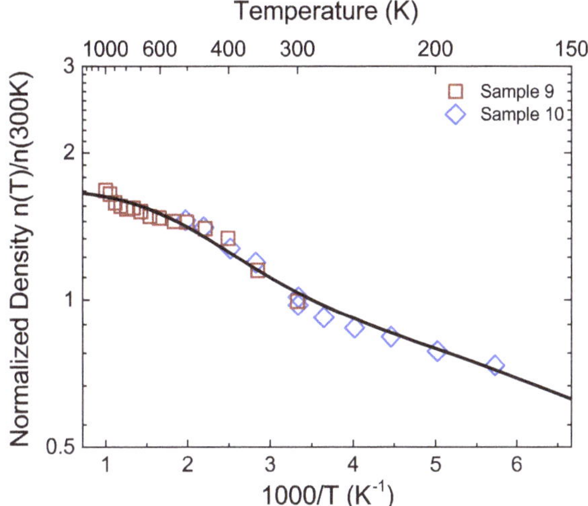

Fig. 22.9 Hall carrier density (log scale) versus 1000/T of Ga_2O_3 for two samples measured in the square geometry. The data are normalized to the Hall carrier density at 300 K. The symbols are the measured data and the black line a fit. Room temperature electron densities are 1.71×10^{17} cm^{-3} for Sample 9 (red square) and 1.21×10^{17} cm^{-3} for Sample 10 (blue diamond) (Reproduced from [35])

110 meV donors are ionized when the Schottky diode is forward biased, reducing the carrier concentration available to conduct the on-current. To quantitatively examine those effects, specific on-resistance (R_{onsp}) versus breakdown voltage characteristics have been calculated for Ga_2O_3 Schottky diode devices including the incomplete ionization effect, with the results plotted in Fig. 22.10. As Fig. 22.10 shows, the on-resistance versus breakdown voltage characteristics are degraded where the maximum breakdown voltage decreases and the on-resistance increases as the concentration of the 110 meV donor increases. To determine the maximum acceptable concentration of 110 meV donors for a particular target breakdown voltage, the percent increase in R_{onsp} is calculated for a Schottky diode containing both 110 meV donors and shallow donors versus a Schottky diode containing only

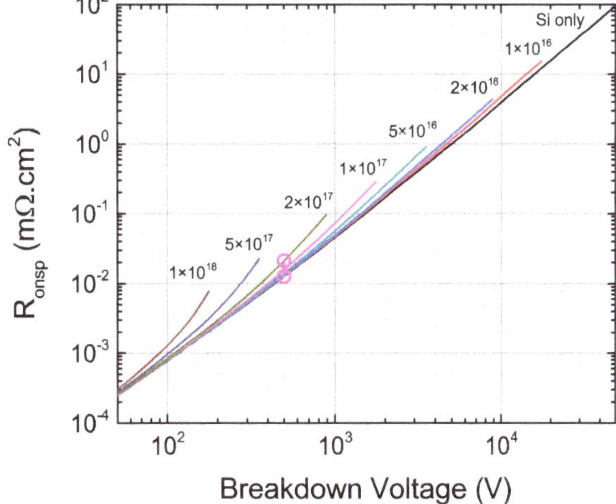

Fig. 22.10 Analytical calculation of specific on-resistance (R_{onsp}) versus breakdown voltage for Ga_2O_3 based Schottky junction devices with 110 meV donors including the effects of incomplete ionization. The label for each curve indicates the fixed concentration of 110 meV donors in units of cm^{-3}. The pink circles and line segment in this figure illustrate the percent increase marked as a pink circle in Fig. 22.11 (Reproduced from Neal [35])

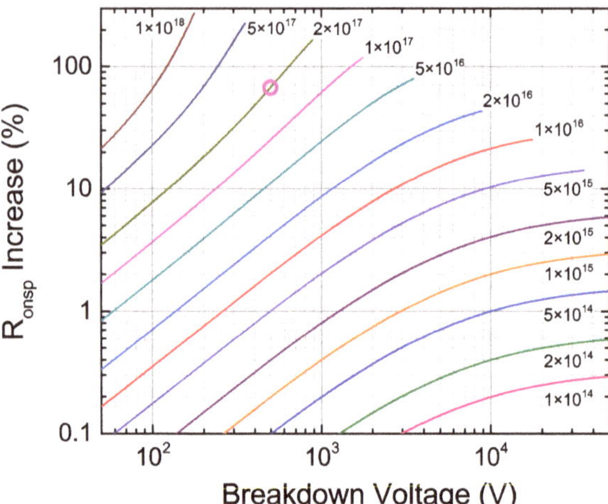

Fig. 22.11 Percent increase in specific on-resistance (R_{onsp}) due to incomplete ionization as a function of breakdown voltage comparing Ga_2O_3 Schottky diodes with 110 meV donors to those without. The labels indicate the fixed concentration of 110 meV donors in cm^{-3} for each curve. The pink circle in this figure marks the percent increase illustrated by the pink circles and line segment in Fig. 22.10 (Reproduced from [35])

shallow donors designed for the same breakdown voltage. The results of the calculation are plotted in Fig. 22.11. With Fig. 22.11, we can estimate the maximum concentration of the 110 meV donors acceptable for 10 kV operation of Ga_2O_3 Schottky diode devices. To limit the increase in R_{onsp} to one percent, the concentration of 110 meV donors must be less than 5×10^{14} cm^{-3}. The origin of this donor is still largely unknown, but, due to the resulting impact, it warrants further studies to identify its origin and reduce its density.

22.5 Conclusion

We have disclosed the nature of Si, Ge, Fe, and Mg doping in β-Ga_2O_3. Si and Ge are believed to be true shallow donors with energy of 30 meV, which can be explained by the hydrogenic model. We could not observe the DX center behavior from Si and Ge as suggested by others [36, 39]. Therefore, both Si and Ge are practically useful n-type dopants for β-Ga_2O_3. Fe and Mg are both deep acceptors which can be used to form semi-insulating substrates and current blocking layers. Last, unintentional donors and acceptors of various kinds were discussed including extrinsic impurities and intrinsic defects. In general, the study of unintentional dopants in Ga_2O_3 is still in its infancy, where some are only predicted by theoretical calculation without strong experimental confirmation, while others were observed but their origins unknown, like the 110 meV donor we discussed. More experiments are needed to confirm the properties, material origin, and impacts of these unintentional donors.

References

1. A.T. Neal, S. Mou, S. Rafique, H. Zhao, E. Ahmadi, J.S. Speck, K.T. Stevens, J.D. Blevins, D.B. Thomson, N. Moser, K.D. Chabak, G.H. Jessen, Appl. Phys. Lett. **113**, 062101 (2018)
2. M. Higashiwaki, K. Sasaki, A. Kuramata, T. Masui, S. Yamakoshi, Appl. Phys. Lett. **100**, 013504 (2012)
3. W. Mi, X. Du, C. Luan, H. Xiao, J. Ma, RSC Adv. **4**, 30579 (2014)
4. N. Moser, J. McCandless, A. Crespo, K. Leedy, A. Green, A. Neal, S. Mou, E. Ahmadi, J. Speck, K. Chabak, N. Peixoto, G. Jessen, IEEE Electron Device Lett. **38**, 775 (2017)
5. M. Baldini, M. Albrecht, A. Fiedler, K. Irmscher, D. Klimm, R. Schewski, G. Wagner, J. Mater. Sci. **51**, 3650 (2016)
6. M. Baldini, M. Albrecht, A. Fiedler, K. Irmscher, R. Schewski, G. Wagner, ECS J. Solid State Sci. Technol. **6**, Q3040 (2017)
7. E.G. Víllora, K. Shimamura, Y. Yoshikawa, T. Ujiie, K. Aoki, Appl. Phys. Lett. **92**, 202120 (2008)
8. H. He, W. Li, H.Z. Xing, E.J. Liang, Adv. Mat. Res. **535–537**, 36 (2012)
9. M.H. Wong, K. Sasaki, A. Kuramata, S. Yamakoshi, M. Higashiwaki. Appl. Phys. Lett. **106**, 032105 (2015)
10. T. Harwig, J. Schoonman, J. Solid State Chem. **23**, 205 (1978)
11. M.H. Wong, K. Goto, A. Kuramata, S. Yamakoshi, H. Murakami, Y. Kumagai, M. Higashiwaki, in *Proceedings of the 75th Device Research Conference*, South Bend, IN, USA (2017)
12. T. Onuma, S. Fujioka, T. Yamaguchi, M. Higashiwaki, K. Sasaki, T. Masui, T. Honda, Appl. Phys. Lett. **103**, 041910 (2013)
13. C. Tang, J. Sun, N. Lin, Z. Jia, W. Mu, X. Tao, X. Zhao, RSC Adv. **6**, 78322 (2016)
14. M.H. Wong, C.-H. Lin, A. Kuramata, S. Yamakoshi, H. Murakami, Y. Kumagai, M. Higashiwaki, Appl. Phys. Lett. **113**, 102103 (2018)
15. M. Higashiwaki, K. Sasaki, T. Kamimura, M.H. Wong, D. Krishnamurthy, A. Kuramata, T. Masui, S. Yamakoshi, Appl. Phys. Lett. **103**, 123511 (2013)
16. N. Ueda, H. Hosono, R. Waseda, H. Kawazoe, Appl. Phys. Lett. **70**, 3561 (1997)
17. J. Zhang, C. Xia, Q. Deng, W. Xu, H. Shi, F. Wu, J. Xu, J. Phys. Chem. Solids **67**, 1656 (2006)
18. N. Suzuki, S. Ohira, M. Tanaka, T. Sugawara, K. Nakajima, T. Shishido, Phys. Status Solidi C **4**, 2310 (2007)
19. K. Sasaki, A. Kuramata, T. Masui, E.G. Víllora, K. Shimamura, S. Yamakoshi, Appl. Phys. Express **5**, 035502 (2012)
20. D. Gogova, M. Schmidbauer, A. Kwasniewski, Cryst. Eng. Comm. **17**, 6744 (2015)
21. A. Kuramata, K. Koshi, S. Watanabe, Y. Yamaoka, T. Masui, S. Yamakoshi, Jpn. J. Appl. Phys. **55**, 1202A2 (2016)
22. D. Gogova, G. Wagner, M. Baldini, M. Schmidbauer, K. Irmscher, R. Schewski, Z. Galazka, J. Cryst. Growth **401**, 665 (2014)
23. S. Rafique, L. Han, A.T. Neal, S. Mou, M.J. Tadjer, R.H. French, H. Zhao, Appl. Phys. Lett. **109**, 132103 (2016)
24. M. Higashiwaki, K. Sasaki, K. Goto, K. Nomura, Q.T. Thieu, R. Togashi, H. Murakami, Y. Kumagai, B. Monemar, A. Koukitu, A. Kuramata, S. Yamakoshi, in *Proceedings of the 73rd Annual Device Research Conference (DRC)*, Columbus, Ohio, 21 June 2015, pp. 29–30
25. M. Higashiwaki, K. Konishi, K. Sasaki, K. Goto, K. Nomura, Q.T. Thieu, R. Togashi, H. Murakami, Y. Kumagai, B. Monemar, A. Koukitu, A. Kuramata, S. Yamakoshi, Appl. Phys. Lett. **108**, 133503 (2016)
26. S. Krishnamoorthy, Z. Xia, S. Bajaj, M. Brenner, S. Rajan, Appl. Phys. Express **10**, 051102 (2017)
27. E. Ahmadi, O.S. Koksaldi, S.W. Kaun, Y. Oshima, D.B. Short, U.K. Mishra, J.S. Speck, Appl. Phys. Express **10**, 041102 (2017)

28. J.B. Varley, J.R. Weber, A. Janotti, C.G. Van de Walle, Appl. Phys. Lett. **97**, 142106 (2010)
29. M. Orita, H. Ohta, M. Hirano, Appl. Phys. Lett. **77**, 4166 (2000)
30. T. Oishi, K. Harada, Y. Koga, M. Kasu, Jpn. J. Appl. Phys. **55**, 030305 (2016)
31. K. Irmscher, Z. Galazka, M. Pietsch, R. Uecker, R. Fornari, J. Appl. Phys. **110**, 063720 (2011)
32. A. Parisini, R. Fornari, Semicond. Sci. Technol **31**, 035023 (2016)
33. T. Oishi, Y. Koga, K. Harada, M. Kasu, Appl. Phys. Express **8**, 031101 (2015)
34. N. Ma, N. Tanen, A. Verma, Z. Guo, T. Luo, H.G. Xing, D. Jenna, Appl. Phys. Lett. **109**, 212101 (2016)
35. A.T. Neal, S. Mou, R. Lopez, J.V. Li, D.B. Thomson, K.D. Chabak, G.H. Jessen, Sci. Rep. **7**, 13218 (2017)
36. N.T. Son, K. Goto, K. Nomura, Q.T. Thieu, R. Togashi, H. Murakami, Y. Kumagai, A. Kuramata, M. Higashiwaki, A. Koukitu, S. Yamakoshi, B. Monemar, E. Janzén, J. Appl. Phys. **120**, 235703 (2016)
37. M. Higashiwaki, A. Kuramata, H. Murakami, Y. Kumagai, J. Phys. D: Appl. Phys. **50**, 333002 (2017)
38. K. Irmscher, in *Abstract of the 2nd International Workshop on Ga_2O_3 and Related Materials*, Parma, Italy, 2017, p. I5
39. S. Lany, APL Mater. **6**, 046103 (2018)
40. S.M. Sze and K.K. Ng, Physics of Semiconductor Devices, 3rd ed. (John Wiley & Sons, 2007) pp. 17–27
41. M. Lundstrom, *Fundamentals of Carrier Transport*, 2nd edn. (Cambridge University Press, New York, 2000), pp. 136–137
42. M. Lundstrom, *Fundamentals of Carrier Transport*, 2nd edn. (Cambridge University Press, New York, 2000), p. 171
43. M. Lundstrom, *Fundamentals of Carrier Transport*, 2nd edn. (Cambridge University Press, New York, 2000), p. 70
44. C. Erginsoy, Phys. Rev. **79**, 1013 (1950)
45. M. Lundstrom, *Fundamentals of Carrier Transport*, 2nd edn. (Cambridge University Press, New York, 2000), p. 86
46. H. He, R. Orlando, M.A. Blanco, R. Pandey, E. Amzallag, I. Baraille, M. Rérat, Phys. Rev. B **74**, 195123 (2006)
47. H. Peelaers, C.G. Van de Walle, Phys. Status Solidi B **252**, 828 (2015)
48. J. Furthmüller, F. Bechstedt, Phys. Rev. B **93**, 115204 (2016)
49. B. Hoeneisen, C.A. Mead, M.A. Nicolet, Solid-State Electron. **14**, 1057 (1971)
50. M. Schubert, R. Korlacki, S. Knight, T. Hofmann, S. Schöche, V. Darakchieva, E. Janzén, B. Monemar, D. Gogova, Q.-T. Thieu, R. Togashi, H. Murakami, Y. Kumagai, K. Goto, A. Kuramata, S. Yamakoshi, M. Higashiwaki, Phys. Rev. B **93**, 125209 (2016)
51. M. Rebien, W. Henrion, M. Hong, J.P. Mannaerts, M. Fleischer, Appl. Phys. Lett. **81**, 250 (2002)
52. C. Sturm, J. Furthmuller, F. Bechstedt, R. Schmidt-Grund, M. Grundmann, APL Mater. **3**, 106106 (2015)
53. S. Rafique, L. Han, A.T. Neal, S. Mou, J. Boeckl, H. Zhao, Phys. Status Solidi A **215**, 1700467 (2018)
54. E.M. Conwell, Phys. Rev. **103**, 51 (1956)
55. N.F. Mott, Rev. Mod. Phys. **40**, 677 (1968)
56. D.C. Look, Phys. Rev. B **24**, 5852 (1981)
57. E.P. O'Reilly, Appl. Phys. Lett. **55**, 1409 (1989)
58. J.B. Varley, A. Janotti, C. Franchini, C.G.V.d. Walle, Phys. Rev. B **85**, 081109 (2012)
59. Q. Feng, J. Liu, Y. Yang, D. Pan, Y. Xing, X. Shi, X. Xia, H. Liang, J. Alloys Compd. **687**, 964 (2016)
60. M.E. Ingebrigtsen, J.B. Varley, A.Y. Kuznetsov, B.G. Svensson, G. Alfieri, A. Mihaila, U. Badstübner, L. Vines, Appl. Phys. Lett. **112**, 042104 (2018)
61. A. Kyrtsos, M. Matsubara, E. Bellotti, Appl. Phys. Lett. **112**, 032108 (2018)

62. K. Konishi, K. Goto, H. Murakami, Y. Kumagai, A. Kuramata, S. Yamakoshi, M. Higashiwaki, Appl. Phys. Lett. **110**, 103506 (2017)
63. C. Joishi, Z. Xia, J. McGlone, Y. Zhang, A.R. Arehart, S. Ringel, S. Lodha, S. Rajan, Appl. Phys. Lett. **113**, 123501 (2018)
64. J.B. Varley, H. Peelaers, A. Janotti, C.G.V. de Walle, J. Phys.: Condens. Matter **23**, 334212 (2011)
65. T. Zacherle, P.C. Schmidt, M. Martin, Phys. Rev. B **87**, 235206 (2013)
66. E. Korhonen, F. Tuomisto, D. Gogova, G. Wagner, M. Baldini, Z. Galazka, R. Schewski, M. Albrecht, Appl. Phys. Lett. **106**, 242103 (2015)
67. S.M. Hantian Gao, N. Pronin, M.R. Karim, S.M. White, T. Asel, G. Foster, S. Krishnamoorthy, S. Rajan, L.R. Cao, M. Higashiwaki, H.v. Wenckstern, M. Grundmann, H. Zhao, D.C. Look, L.J. Brillson, Appl. Phys. Lett. **112**, 242102 (2018)
68. E. Farzana, E. Ahmadi, J.S. Speck, A.R. Arehart, S.A. Ringel, J. Appl. Phys. **123**, 161410 (2018)
69. H. Yilmaz, I.E.E.E. Trans, Electron Devices **38**, 1666 (1991)
70. H. Niwa, G. Feng, J. Suda, T. Kimoto, in *Proceedings of the 24th International Symposium on Power Semiconductor Devices and ICs*, pp. 381–384 (2012)
71. H. Ohta, N. Kaneda, F. Horikiri, Y. Narita, T. Yoshida, T. Mishima, T. Nakamura, IEEE Electron Device Lett. **36**, 1180 (2015)

Chapter 23
Electrical Properties 2

Electron Transport Studies in β-Ga_2O_3

Krishnendu Ghosh, Avinash Kumar and Uttam Singisetti

Abstract This chapter discusses the electron transport properties and their implications on power devices under very high electric fields (>1 MV/cm). A combination of first-principles calculation of microscopic interactions, Monte Carlo simulation of electron transport, and TCAD simulation on power transistors manifests a bottom-up view of the hot electron dynamics in β-Ga_2O_3 devices. While the focus of the chapter is on impact ionization and resulting avalanche breakdown, a brief overview on low-field and moderately high-field transport are also presented. The role of crystal orientation on breakdown is discussed in detail. Compact model parameters for breakdown phenomena are provided. Estimates of breakdown field are shown through simple analytical calculations and also through a TCAD simulation.

23.1 Introduction

β-Ga_2O_3 has recently emerged as a promising ultra-wide band gap material for applications in power switching electronics and optoelectronics [1]. Experimental demonstrations of power transistors with record high breakdown voltages [2–7], Schottky diodes [8–10], and deep ultraviolet photodetectors [11–13] have been remarkable. Matured crystal growth techniques and conventional device processing techniques [14–16] make the promise of the material further strong. High accuracy of n-type doping and difficulty in achieving p-type doping make electrons the

K. Ghosh · A. Kumar · U. Singisetti (✉)
Electrical Engineering Department, University at Buffalo, Buffalo, NY 14260, USA
e-mail: uttamsin@buffalo.edu

K. Ghosh
e-mail: krisg@umich.edu

A. Kumar
e-mail: a42@buffalo.edu

K. Ghosh
Mechanical Engineering Department, University of Michigan, Ann Arbor, MI 48109, USA

M. Higashiwaki and S. Fujita (eds.), *Gallium Oxide*, Springer Series in Materials Science 293, https://doi.org/10.1007/978-3-030-37153-1_23

primary carrier of transport. An important measure for the performance of a power semiconductor is the Baliga's figure of merit [17] which is a joint effect of the breakdown electric field and the electron mobility both of which are dependent upon the transport properties of the carriers. Hence, understanding the electron transport mechanism in β-Ga_2O_3 is extremely crucial for efficient design and engineering of devices. However, a low crystal symmetry and a large primitive cell make the calculation of transport properties challenging in this material. Nevertheless, low-field transport properties have been studied and reported a few times in recent years both theoretically and experimentally [18–22]. The temperature and doping dependence along with anisotropy have been investigated [23]. Still, there remains a lot to investigate in the low-field transport properties which are the mobility in two-dimensional electron gas (2-DEG) in hetero structures, and effect of alloy-scatterings on 2-DEGs. On the other hand, high-field transport properties have not been explored except a few recent reports [24, 25]. The high-field transport regime can be broadly classified into two categories—moderately high-field and very high-field. Under moderately high-field, the primary physics of interest are the velocity overshoot and velocity saturation. While under very high electric field impact ionization, Zener tunneling breakdown mechanism, and high-frequency Bloch oscillations are expected. The electric field ranges for velocity saturation and impact ionization vary drastically with material systems. Typically, as a rule of thumb, wider band gap materials show a shift toward higher values of electric field. This chapter focuses on very high-field transport properties, particularly the impact ionization mechanism, in β-Ga_2O_3 and the resulting breakdown limits.

When large electric bias is applied in power devices in the OFF state, sharply peaked electric field profiles are observed in the gate–drain junction. In narrow-band gap, semiconductor such as InAs, such a high electric field immediately results to interband tunneling leading to breakdown limit of the device. However, in wide band gap semiconductors such as β-Ga_2O_3 such interband tunneling is unlikely and the dominant breakdown mechanism is from impact ionization which is introduced in details later in the chapter. Besides, there could be potential extrinsic breakdown mechanisms in β-Ga_2O_3 devices resulting from the breakdown of the gate dielectric due to hot carrier injection. However, in this chapter, the focus will be on impact ionization only. Before discussing impact ionization, we provide a brief overview on low and moderately high-field transport in β-Ga_2O_3 based on the recent work by the authors [18, 23, 24] and other colleagues, [19, 20] since mobility and velocity-field curves are also important figure of merits for design of power devices. Next, we discuss the impact ionization mechanism, ionization rates in β-Ga_2O_3, calculation of the ionization coefficients using a full-band Monte Carlo simulation, estimation of Chynoweth parameters, and also some important device simulations to predict breakdown voltages in β-Ga_2O_3 devices respectively.

23.2 Overview on Mobility and Velocity-Field Curves in β-Ga_2O_3

β-Ga_2O_3 has a base-centered monoclinic crystal with a 10 atom primitive cell. The conventional unit cell is shown in Fig. 23.1a along with the Cartesian direction convention followed throughout this chapter. The numerous experimental reports of room temperature bulk mobility range from about 100–150 cm^2/V s based on the direction of the transport and on the purity of the sample. The dominant intrinsic mobility limiting mechanism is the polar optical phonon (POP) scattering. There are 12 IR active modes that can scatter electrons via polar interactions. It is found that the low-energy modes B_u^1, B_u^2, and A_u^2 contribute the most to the scattering rate at room temperature for low-energy electrons. Like the conventional semiconductors, β-Ga_2O_3 also shows an initial increase in mobility as temperature is lowered from room temperature (see Fig. 23.1b) because of the reduced POP scattering at lower temperatures. The mobility peaks and starts to decrease again due to the dominance of ionized impurity scattering. Usually, in conventional semiconductors, anisotropy in mobility is a result of the anisotropy in the conduction band minima. However, in β-Ga_2O_3, although the bottom of the conduction band is isotropic, a ~10–15% anisotropy in mobility is experimentally observed [22]. This anisotropy in mobility is a result of the anisotropy in polar electron–phonon coupling, which is a consequence of the low-symmetry of the crystal. Hall measurements of electron mobility have been performed several times and Hall factor correction schemes have been proposed [21]. It is difficult to use relaxation time approximation due to the presence of many active phonon modes. Also, the coupling of the bulk plasmon mode with that of longitudinal optical phonon modes is investigated rigorously under dynamic screening [23]. This reveals an interesting trend of the electron mobility with increasing electron concentration. In short, as the electron concentration is increased the plasmon energy increases and the plasmon mode couples with the multiple LO modes differently. As a consequence, based on the plasmon energy, some of the LO modes get screened, while others get anti-screened. This results to a non-monotonic variation of the electron mobility with electron concentration as shown in Fig. 23.1c.

Apart from bulk mobility, 2DEG mobility is also estimated incorporating appropriate screening and LO-plasmon coupling. The essential idea in realizing high mobility in 2DEG is to spatially separate out the dopants from the channel. An enhanced remote doping increases the electron concentration in the channel providing a good screening of the LO modes, while the dopant remains far from the channel resulting to a reduced impurity scattering. However, based on the distance of the dopant layer from the channel, the mobility varies significantly due to varying scattering rates caused by the remote potential of the charged dopant layer. Figure 23.1d shows the variation of the 2DEG mobility with electron concentration in the channel for different distance of separation from the dopant layer. This way the mobility can theoretically improve up to 1000 cm^2/V s from its bulk counterpart of about 120 cm^2/V s. In practice, there are other scattering mechanisms that

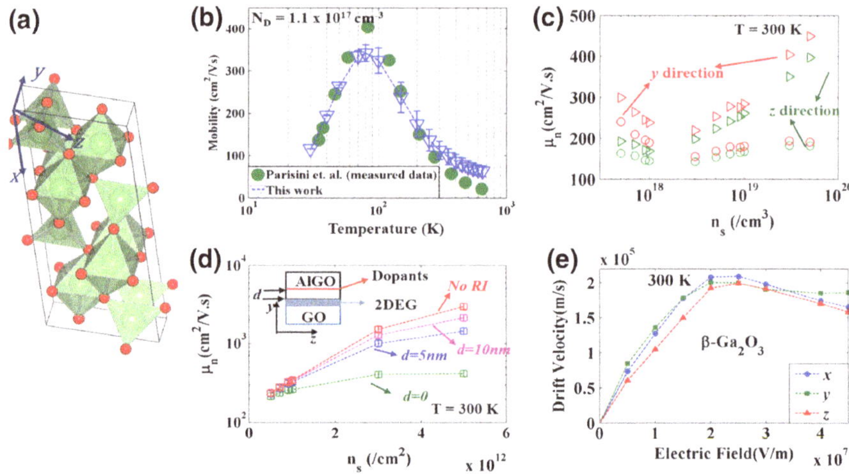

Fig. 23.1 Overview of mobility and velocity [18, 23, 24]. **a** The conventional unit cell of β-Ga_2O_3 with the Cartesian direction convention followed in this chapter. **b** Dependence of electron mobility with temperature and comparison with experimentally measured Hall mobility. **c** Dependence of electron mobility on electron concentration for two different directions. Triangles indicate only phonon scattering limited mobility, while circles represent the mobility including both phonon and impurity scatterings. **d** Electron mobility in 2-DEG formed at hetero-junctions shown for different distance of the dopant layer from the channel. **e** The velocity-field curves in β-Ga_2O_3 for three different directions at room temperature. **b** is reproduced from [18] with permissions from the AIP, **c–d** are reproduce from [23] with permissions from the Cambridge University Press, and **e** is reproduced from [24] with permissions from the AIP

will reduce the 2-DEG mobility including roughness and surface phonon scatterings.

There has also been a recent report [26] in modeling parameters for analytical scattering rate models in β-Ga_2O_3. The effort is motivated by a previous work by Chang and James [27] where a spherical average of the angle-dependent POP coupling strengths is used to compute the scattering rates. Multiplicative constants are used to account for the fact that the POP scattering in low-symmetry crystals are highly angle-dependent. These multiplicative constants are obtained for β-Ga_2O_3 by fitting the first-principles computed scattering rates to the Chang-James equation proposed in [27]. The fitted constants and the phonon modes are shown in Table 23.1. It is to be noted that the fitted phonon energies do not represent the existence of any actual phonon mode at those energies. With these fitted constants and phonon energies in the Chang-James analytical model, one can determine the mobility in β-Ga_2O_3, thereby eliminating the need for first-principles computation of the scattering rates.

Figure 23.1e which shows computed velocity-field curves in β-Ga_2O_3 using a combination of first-principles calculation and full-band Monte Carlo simulation. We discuss here some key aspects of the velocity-field profile without going into the technical details of the computation. With increasing electric field, the velocity

Table 23.1 Multiplicative coupling constants and phonon energies for analytical modeling of POP scattering in β-Ga_2O_3 [26]

Mode	ω_j(eV)(fitted)	C_j' (eV/Å^2)
B_u (1)	0.025	2.0
B_u (2)	0.029	1.6
B_u (3)	0.035	0.5
B_u (4)	0.043	1.15
B_u (5)	0.050	3.2
B_u (6)	0.070	4.5
B_u (7)	0.080	3.1
B_u (8)	0.090	3.2
A_u (1)	0.014	0.15
A_u (2)	0.037	2.0
A_u (3)	0.060	3.8
A_u (4)	0.081	3.0

keeps increasing up to 200 kV/cm at which a peak velocity of about 2×10^7 cm/s is estimated and beyond which the velocity gets saturated to a slightly lower value. For comparison, the saturation velocity is slightly lower than that of wurtzite GaN [28]. The initial anisotropy in the velocity profile is due to POP phonons as discussed in the case of mobility, while the anisotropy in the saturated portion is from the different slope of the conduction bands along y-direction compared to that along the other two directions. The slight negative differential conductivity in β-Ga_2O_3, unlike GaN and GaAs, is a result of strong intra-band scattering within the Γ valley and the decrease in group velocity due to non-parabolicity of the conduction band.

23.3 Impact Ionization Rates

Impact ionization is a two-electron process where an electron in the valence band interacts with an electron in the conduction band creating an additional electron in the conduction band and a hole in the valence band. So this creates secondary charge carriers and the generation of these secondary carriers can eventually become self-sustaining leading to the breakdown of the device. Although this phenomenon is utilized in several high-frequency devices like IMPATT diodes and ionization field-effect transistors, it is not desired in a power device. Figure 23.2a shows a real-space schematic of the impact ionization mechanism described here, while Fig. 23.2b shows the same in the reciprocal space. To clarify the energy and momentum conservation in the process, we note that the hot electron in the $|m\boldsymbol{k}$ interacts with the valence band electron at state $|n'\boldsymbol{k}'$ and then while the hot electron ends up at state $|n\boldsymbol{k}+\boldsymbol{q}$ in the conduction band, the valence band electron gets promoted to the conduction band at state $|m'\boldsymbol{k}'-\boldsymbol{q}$. The corresponding Feynman diagrams are shown in Fig. 23.2b for completeness. The ionization rates are computed from the matrix elements of the screened Coulomb operator.

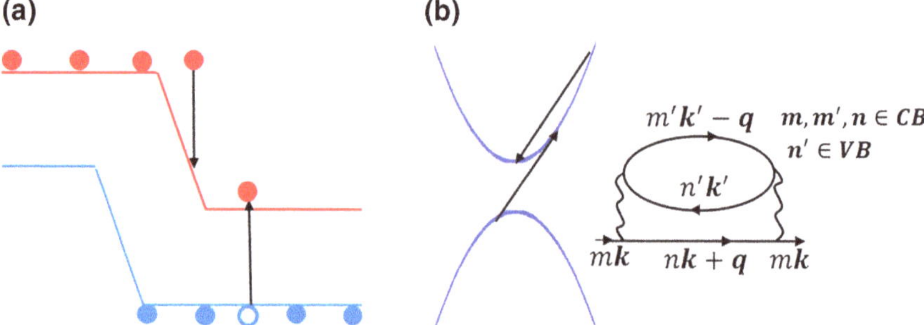

Fig. 23.2 Impact ionization mechanism [25]. **a** Real-space schematic of impact ionization process. The red dots show hot electrons on the conduction band and blue ones are electron in the valence band. As the hot electron loses energy, a valence electron gets promoted to the conduction band. **b** Reciprocal space schematic of impact ionization. The Feynman diagram depicts conservation and momentum and all the internal degrees of freedom, see text for details

Computation of the ionization rates requires summation over the internal degrees of freedom on the Feynman diagram which implies integrations over $\boldsymbol{k}'$ and $\boldsymbol{q}$, while summations over m', n', and n. The electronic bands computed based on density functional theory (DFT) calculations are shown on Fig. 23.3a for β-Ga_2O_3. It is well known that DFT underestimates the optical gap. However, for the ionization rate computation, the band gap needs to be accurate. The experimental band gap in β-Ga_2O_3 is reported to be in the range of 4.5–4.9 eV. So the conduction band manifold is scissor shifted to match the experimental band gap. One can do a more rigorous job by carrying out a G_0W_0 calculation; however, in the current chapter, we keep ourselves to the simple scissor shifted bands only. On the other hand, the DFT-computed wave-functions provide a good description of both valence band and conduction band wave-functions and can be directly used for ionization rate calculation.

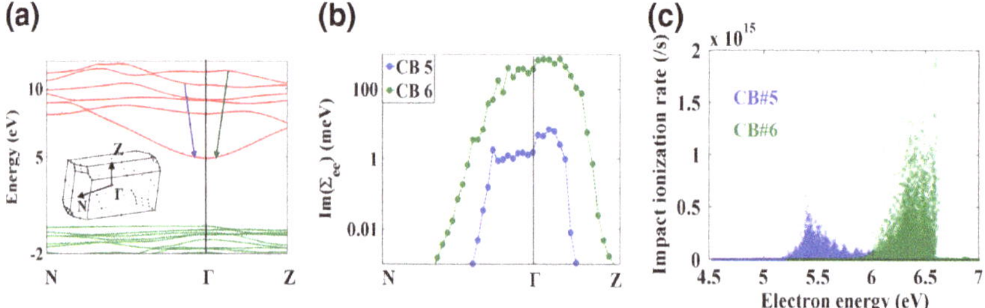

Fig. 23.3 Impact ionization rates [25]. **a** Electronic bands in β-Ga_2O_3 computed under DFT. The conduction bands are scissor shifted to match the experimental band gap. **b** The ionization rates along the two high-symmetry lines. **c** The ionization rates with respect to energy for all the *k*-points in the Brillouin zone. **a–c** are reproduced from [25] with permissions from the AIP

Next, we turn to discuss some key features of the IIR in β-Ga_2O_3. Like the phonon-mediated scatterings, IIR is also dependent upon the density of final available states for both the ionizing and the ionized electrons. β-Ga_2O_3 has an isolated conduction band minimum that is about ~2.5 eV below the satellite valleys. During the onset of ionization, the final state for both ionizing and ionized electrons is on the Γ valley. Moreover, the high-energy remote valleys occur near the Brillouin zone edge as can be seen on Fig. 23.3a. It is very unlikely that a hot electron near the zone edge can cause ionization. This is because the Coulomb interaction responsible for causing the ionization is long-ranged meaning that the interaction cannot cause large changes in the electron momentum. Hence, the zone-edge electron fails to find a final state that can satisfy both energy and momentum conservation. This effect can be observed on Fig. 23.3b where the IIR is much higher near the zone center compared to that near the zone edges. This has crucial consequences on the IIC and the overall high-field transport properties. This makes the possibility of Bloch oscillations and interband tunneling higher near the zone edges.

Ionization cannot be caused by the electrons present in the conduction bands (CB) 1–3 since they do not possess enough energy (>band gap) to promote the valence band electrons to the conduction band. While CB 4 does have some energy states that have higher energy than the band gap, those states lie near the zone-edge and fail to cause ionization due to reasons mentioned on the previous paragraph. Hence, the major ionization contributions come from CB 5 and beyond. The ionization contribution from all over the BZ is shown on Fig. 23.3c as a function of energy. The anisotropy in the IIR can be clearly observed by the range of the IIR values spanned by different *k*-points on the BZ at a given energy. The reason behind such anisotropy is the flatness of the higher conduction bands. Near the onset of ionization, the IIR follows a power law as is observed in other semiconductors [29]. The abrupt drop in the IIR around 6.5 eV is due to the same reason of not having final density of states after ionization. In the present chapter, we only consider up to CB 6. The best practice to determine the number of CB that need to be considered in the transport calculation is a posteriori study of the convergence of the IIC. However, a couple of important approximation and limitation need to be clarified here. First, the hot electron distribution coming from the Monte Carlo simulation (discussed in next section) reveals that the distribution dies off beyond 6.5 eV and hence considering beyond 6 bands is not likely to change the computed IIC. Secondly, due to the presence of many phonons, considering more bands make the computation extremely expensive and intractable. So, we limit our discussion here up to 6 bands.

23.4 Transport Under Very High Field

Electrons gain energy from the externally applied electric field and lose the same though electron–phonon and electron–electron interactions. In the current context, electron–electron interactions imply ionizing events. IIC is the reciprocal of the distance traversed by an electron between two successive ionizing events. The IIC can be computed from the generation rate of secondary electrons and the ensemble average of the electron drift velocity. The generation rate, G at a given electric field F is related to the drift velocity v_d and IIC α by the relation, $G(F) = \alpha(F)v_d(F)$. The generation rate and the velocity of electrons both follow from a full-band Monte Carlo simulation. It is important to mention here that both G and v_d are to be extracted from the FBMC simulation after a steady state is reached. The FBMC simulation takes into account the electronic structure and lattice dynamics, electron–phonon interaction elements, electron–electron interactions elements, and the different types of scattering rates including polar and nonpolar phonon scatterings and ionization events. The FBMC scheme is a time-propagation manifold where an ensemble of electrons are thrown under an electric field. The trajectory of the electrons is recorded as the electrons drift under the field and gets scattered by scattering events described before. The time between two scattering events, the scattering mechanism, and the final state of the electron after the scattering are obtained stochastically within the FBMC. Every time an ionization event occurs, an additional electron is added to the ensemble of electrons. It is important to note here that the valence electron distribution remains unaffected by the applied electric field which implies that the mean field generated by the valence electrons and hence the ionization rates are independent of the applied field. However, the generation rates do increase with increasing the electric field since at a higher field there are more number of hot electrons present to provide the required energy barrier to the valence electrons. So, while the ionization rates are completely deterministic prior to the FBMC simulation, the generation rates are a stochastic result of the FBMC simulation.

Figure 23.4a shows the transient dynamics of the electrons under an electric field of 2 MV/cm. The oscillations, known as Bloch oscillations [30], occur as the electrons hit the surface of the Brillouin zone. The group velocity of the electron reverses direction at the BZ surface. In conventional semiconductors, there are primarily two reasons that prohibit the observation of Bloch oscillation. Firstly, as

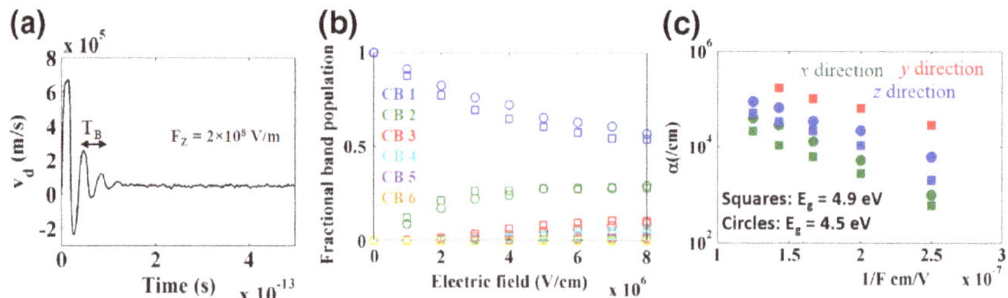

Fig. 23.4 Transport under high field [25]. **a** Transient results from Monte Carlo transport simulation shows Bloch oscillation, see text for details. **b** The fractional population of electrons in different bands as a function of electric field. **c** Impact ionization coefficients as a function of electric for two different band gaps, see text for details. **a–c** are reproduced from [25] with permissions from the AIP

electrons try to reach the zone surface, they acquire high energy and hence face high scattering rates randomizing their momentum. Broadly but quantitatively speaking, if the order of the average scattering time for electrons is τ_S and the time taken by a ballistic electron to reach the zone surface is τ_B, then the necessary criterion for the electron to undergo 'reflection' at the BZ surface is $\tau_S \gg \tau_B$. However, satisfying this criterion requires an extremely high electric field and hence, this inhibits most of the electrons from reaching the surface of the BZ except a few 'lucky' electrons. The ensemble average of the velocity remains unaffected as the percentage of 'lucky' electrons is usually negligible. Secondly, another important aspect here is that near the zone surface instead of remaining on the same band and reversing the group velocity, an electron might tunnel to a higher band. This also preserves the direction of the group velocity and hence prohibits Bloch oscillations. In the FBMC study of β-Ga_2O_3, it was found that the first criterion $\tau_S \gg \tau_B$ can be satisfied with the reason being the absence of satellite valleys until the zone edges. The intervalley scattering which is usually much higher in magnitude than intravalley scatterings potentially slow down the electrons from reaching the BZ surface, but in β-Ga_2O_3, the onset of such intervalley scattering occurs in almost at the same time when the electrons reach the BZ surface allowing several electrons to satisfy the criterion of $\tau_S \gg \tau_B$. However, the second issue, interband tunneling, is not yet studied in β-Ga_2O_3 although the interband tunneling effects can also be included in the FBMC simulation. So, to conclude on the possibility of Bloch oscillations, further FBMC simulations including interband tunneling and importantly experimental efforts are required.

Figure 23.4b shows the relative population of the electrons in different conduction bands with respect to the applied electric field. The transition of electrons from one band to the other occurs only through nonpolar phonon scatterings and ionization events as no tunneling is considered. At a glance, it might seem that the electron population at CB5 and CB6 are negligible and one might speculate that since ionization events are also negligible since they are mediated through electrons located beyond CB4 as discussed in Sect. 3. However, it is to be noted that ionization can become cumulative and even a small number of electrons can lead to Avalanche breakdown. The population on Fig. 23.4b is shown for the field applied along two different directions. Figure 23.4c shows the computed IIC along three different directions. There is a strong anisotropy in the IIC which is attributed to the anisotropy of the ionization rates as shown in Fig. 23.3b. This anisotropy in turn results from the anisotropy of the higher conduction bands even near the zone center as can be seen on Fig. 23.3a. There is some uncertainty on the experimentally observed band gap in β-Ga_2O_3 which ranges from 4.5 to 4.9 eV. Hence, the IIC is shown on Fig. 23.4c for the two limits of the band gap. As expected, the IIC is lower for a higher band gap. For comparison, we can see that the electron IIC is lower than that compared to GaN [28].

23.5 Avalanche Breakdown and Implication on Devices

Now, we focus on using the computed IIC to predict breakdown electric field in power devices. The first step is to extract compact model parameters that can be used in TCAD simulation. A well-known model for device simulation including impact ionization is the Chynoweth model [31] that relates the electric field (F) to IIC (α) as $\alpha(F) = ae^{-\frac{b}{F}}$, where a and b are the fitting parameters that either come from experimental breakdown measurement data or from macroscopic transport calculations. Here, the fitted Chynoweth parameters to the FBMC-computed IIC are shown on Table 23.1 for all the three Cartesian directions. The IIC along x direction is the lowest and hence is likely to show a higher breakdown field. We explore more on that through some analytical calculation and some TCAD simulation in the following paragraphs.

Avalanche breakdown occurs only when impact ionization becomes self-sustainable. This means the denominator of the multiplication factor vanishes. The denominator [32] has a form $1 - \int_0^W \alpha_n e^{-\int_x^W (\alpha_n - \alpha_p) dx'} dx$ where W is the length of the space charge region, α_n and α_p are the IIC of electrons and holes as a function of space. In the current context, the IIC of the electrons are only known, while hole IIC of β-Ga_2O_3 is never reported in literature. Hence, here, we discuss two limiting cases—(i) $\alpha_n \approx \alpha_p$ and (ii) $\alpha_n \gg \alpha_p$. To estimate the breakdown field, we consider a simple triangular electric field profile (which approximates the case of a p-n junction) with a peak electric field of 8 MV/cm and a width of 1 μm. For the first case, the denominator vanishes if the ionization integral $\int_0^W \alpha_n dx$ becomes 1. Table 23.2 shows the critical electric field limits in each direction at which the denominator vanishes. As can be seen the breakdown is less likely along x compared to the other two directions. In fact for electric field along y direction, breakdown occurs for both limiting cases. As observed experimentally and also calculated theoretically, the electron mobility is also superior along x compared to the other two directions. So one can infer that x direction is the most suitable direction in terms of Baliga's figure of merit. In the following, a more realistic device operation is shown using the estimated Chynoweth parameters in TCAD simulation.

Table 23.2 Chynoweth parameters and ionization integral [25]

	a (/cm)	b (V/cm)	$I_\alpha \big\vert_{\alpha_n \approx \alpha_p,\ E_p = 8\text{MV/cm}}$ <!endarray>	$I_\alpha \big\vert_{\alpha_n \gg \alpha_p,\ E_p = 8\text{ MV/cm}}$	$E_c\big\vert_{\alpha_n \approx \alpha_p}$ (MV/cm)
x	0.79×10^6	2.92×10^7	0.38	0.32	10.2
y	2.16×10^6	1.77×10^7	Breakdown	Breakdown	4.8
z	0.706×10^6	2.10×10^7	Breakdown	0.70	7.6

A field-plated gallium oxide lateral FET is designed and simulated [33, 34] in the TCAD tool Silvaco ATLAS [35] to demonstrate the effect of impact ionization on breakdown voltage. An electric field peaking at almost 7 MV cm^{-1} near the field plate edges is observed, when the breakdown occurs, in the channel region. A very high breakdown voltage of around 1.78 kV is estimated using the ionization integral method. As shown in Fig. 23.5a, the simulated device has a typical lateral MOSFET structure with an additional 1.5 μm long field plate formed over a 0.5 μm thick field plate oxide. The source–drain and the channel regions are doped with n-type dopants with concentration of 9×10^{19} cm^{-3} and 1×10^{18} cm^{-3}, respectively. A 2 μm metal gate with a work function of 5.93 eV is formed over a 20 nm thick gate oxide. There is a separation of 4 μm maintained between the metal gate and the drain region. Material parameters used in the simulation are taken from the experiments. For example, the band gap energy and the electron affinity for gallium oxide are assumed to be 4.8 eV and 4.0 eV, respectively. The effective density of states in the conduction band is calculated using an electron effective mass of 0.3 times the free electron mass. At room temperature, electron mobility is taken to be 118 cm^2 V^{-1} s^{-1}. The Chynoweth parameters in the current direction (x in this case) are taken from Table 23.2. It has been observed that the electric field gets stronger near the field plate and the metal gate as the drain voltage is ramped up and reaches to almost 7 MV cm^{-1}, as shown in Fig. 23.5b, near the field plate edge when the breakdown happens at 1.78 kV. The advantage of having a field plate is the field spread which leads a higher breakdown voltage. Figure 23.5c shows the variation of ionization coefficient with position in x direction which peaks at the

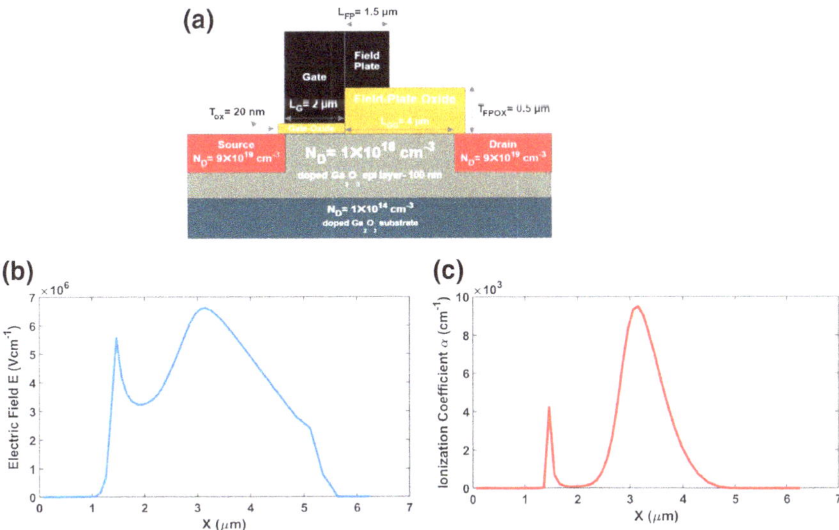

Fig. 23.5 Device operation [33, 34]. **a** Cross section of the field-plated β-Ga_2O_3 MOSFET. **b** Electric field profile along the channel direction (x). **c** Ionization coefficient profile along the channel direction (x). **a**–**b** are reproduced from [34] with permissions from the IEEE

same position as electric field. For simplicity, ionization coefficients of electrons and holes are assumed to be the same. The length of the field plate and the thickness of the field plate oxide are adjusted in order to get the breakdown as high as possible. If the doping in the channel region is changed to get more current, above two parameters can be altered accordingly in order to keep the breakdown voltage same.

23.6 Conclusions

This chapter provides a description of some early stage research of hot electron transport in the novel ultra-wide band gap material β-Ga_2O_3. Recent reports on first-principles based impact ionization calculation are discussed in detail with some important physical insights. The large energy isolation of the conduction band minima from other remote valleys help in reducing ionization rates. This same isolation is responsible for reducing phonon-mediated intervalley scattering rates and providing a path for Bloch oscillations. IIC in β-Ga_2O_3 is lower than GaN at any given electric field and has a significant anisotropy. The x direction of electron transport enjoys a superior Baliga's figure of merit compared to other directions. TCAD simulation using field-plated MOSFET device shows a breakdown voltage of about 1.8 kV. There is a huge room for the future research on high-field transport in β-Ga_2O_3 spanning from exploring transport in two-dimensional electron gas at hetero-junctions, interband tunneling phenomena, estimating IIC of holes and many more. In the coming years, integration of microscopic interactions, full-band transport calculations, and TCAD simulation will play a key role in probing and engineering novel devices based on β-Ga_2O_3.

Acknowledgements The first principles calculations are carried out using Quantum ESPRESSO [36], Wannier90 [37] and a modified version of the EPW code [38, 39]. This work is supported by National Science Foundation (NSF) grant (ECCS 1607833) monitored by Dr. Dimitris Pavilidis. The authors sincerely acknowledge the high-performance computing clusters provided by the Center for Computational Research at the University at Buffalo.

References

1. M. Higashiwaki, K. Sasaki, H. Murakami, Y. Kumagai, A. Koukitu, A. Kuramata, T. Masui, S. Yamakoshi, Semicond. Sci. Technol. **31**, 034001 (2016)
2. M. Higashiwaki, K. Sasaki, T. Kamimura, M.H. Wong, D. Krishnamurthy, A. Kuramata, T. Masui, S. Yamakoshi, Appl. Phys. Lett. **103**, 123511 (2013)
3. M. Higashiwaki, K. Sasaki, A. Kuramata, T. Masui, S. Yamakoshi, Appl. Phys. Lett. **100**, 013504 (2012)
4. K. Zeng, J.S. Wallace, C. Heimburger, K. Sasaki, A. Kuramata, T. Matsui, J.A. Gardella, U. Singisetti, IEEE Electron Device Lett. **38**, 513 (2017)

5. A. Green, K. Chabak, E. Heller, R. Fitch, M. Baldini, A. Fiedler, K. Irmscher, G. Wagner, Z. Galazka, S. Tetlak, A. Crespo, K. Leedy, G. Jessen, IEEE Electron Device Lett. **37**, 902 (2016)
6. K.D. Chabak, N. Moser, A.J. Green, D.E. Walker Jr., S.E. Tetlak, E. Heller, A. Crespo, R. Fitch, J.P. McCandless, K. Leedy, M. Baldini, G. Wagner, Z. Galazka, X. Li, G. Jessen, Appl. Phys. Lett. **109**, 213501 (2016)
7. M.H. Wong, K. Sasaki, A. Kuramata, S. Tamakoshi, M. Higashiwaki, IEEE Electron Device Lett. **37**, 212 (2016)
8. T. Oishi, Y. Koga, K. Harada, M. Kasu, Appl. Phys. Express **8**, 031101 (2015)
9. M. Higashiwaki, K. Sasaki, K. Goto, K. Nomura, Q. T. Thieu, R. Togashi, H. Murakami, Y. Kumagai, B. Monemar, A. Koukitu, A. Kuramata, S. Yamakoshi, in *Proceedings of the 73rd Device Research Conference*, Columbus, pp. 29–30 (2015)
10. K. Sasaki, M. Higashiwaki, A. Kuramata, T. Masui, S. Yamakoshi, IEEE Electron Device Lett. **34**, 493 (2013)
11. T. Oshima, T. Okuno, N. Arai, S. Suzuki, S. Ohira, S. Fujita, Appl. Phys. Express **1**, 011202 (2008)
12. D.Y. Guo, Z.P. Wu, P.G. Li, Y.H. An, H. Liu, X.C. Guo, H. Yan, G.F. Wang, C.L. Sun, L.H. Li, W.H. Tang, Opt. Mater. Express **4**, 1067 (2014)
13. Q. Feng, L. Huang, G. Han, F. Li, X. Li, L. Fang, X. Xing, J. Zhang, W. Mu, Z. Jia, D. Guo, W. Tang, X. Tao, Y. Hao, IEEE Trans. Electron Devices **63**, 3578 (2016)
14. Z. Galazka, R. Uecker, K. Irmscher, M. Albrecht, D. Klimm, M. Pietsch, M. Brützam, R. Bertram, S. Ganschow, R. Fornari, Cryst. Res. Technol. **45**, 1229 (2010)
15. K. Irmscher, Z. Galazka, M. Pietsch, R. Uecker, R. Fornari, J. Appl. Phys. **110**, 063720 (2011)
16. Y. Tomm, P. Reiche, D. Klimm, T. Fukuda, J. Cryst. Growth **220**, 510 (2000)
17. B.J. Baliga, Semicond. Sci. Technol. **28**(7), 074011 (2013)
18. K. Ghosh, U. Singisetti, Appl. Phys. Lett. **109**, 072102 (2016)
19. Y. Kang, K. Krishnaswamy, H. Peelaers, C.G. Van de Walle, J. Phys.: Condens. Matter **29**, 234001 (2017)
20. N. Ma, N. Tanen, A. Verma, Z. Guo, T. Luo, H.G. Xing, D. Jena, Appl. Phys. Lett. **109**, 212101 (2016)
21. A. Parisini, R. Fornari, Semicond. Sci. Technol. **31**, 035023 (2016)
22. M.H. Wong, K. Sasaki, A. Kuramata, S. Yamakoshi, M. Higashiwaki, Jpn. J. Appl. Phys. **55**, 1202B9 (2016)
23. K. Ghosh, U. Singisetti, J. Mater. Res. **32**(22), 4142 (2017)
24. K. Ghosh, U. Singisetti, J. Appl. Phys. **122**, 035702 (2017)
25. K. Ghosh, U. Singisetti, J. Appl. Phys. **124**, 085707 (2018)
26. A. Parisini, K. Ghosh, U. Singisetti, R. Fornari, Semicond. Sci. Technol. **33**, 105008 (2018)
27. Y.-C. Chang, R.B. James, Phys. Rev. B **53**, 14200 (1996)
28. F. Bertazzi, M. Moresco, E. Bellotti, J. Appl. Phys. **106**(6), 063718 (2009)
29. M. Bernardi, D. Vigil-Fowler, C.S. Ong, J.B. Neaton, S.G. Louie, PNAS **112**, 5291 (2015)
30. T. Dekorsy, R. Ott, H. Kurz, K. Köhler, Phys. Rev. B **51**, 17275 (1995)
31. A.G. Chynoweth, Phys. Rev. **109**, 1537 (1958)
32. R.J.E. Hueting, A. Heringa, B.K. Boksteen, S. Dutta, A. Ferrara, V. Agarwal, A.J. Annema, IEEE Trans. Electron Devices **64**(1), 264 (2017)
33. I. Lee, A. Kumar, K. Zeng, U. Singisetti, X. Yao, in *Proceedings of IEEE Energy Conversion Congress and Expo (ECCE)*, Cincinnati, pp. 4377–4382 (2017)
34. I. Lee, A. Kumar, K. Zeng, U. Singisetti, X. Yao, in *Proceedings of the IEEE 5th Workshop on Wide Bandgap Power Devices and Applications (WiPDA)*, Albuquerque, pp. 185–189 (2017)
35. D.S. Atlas, *Atlas User's Manual* (Silvaco International Software, Santa Clara, CA, 2014)
36. P. Giannozzi, S. Baroni, N. Bonini, M. Calandra, R. Car, C. Cavazzoni, D. Ceresoli, G.L. Chiarotti, M. Cococcioni, I. Dabo, A. Dal Corso, S. de Gironcoli, S. Fabris, G. Fratesi, R. Gebauer, U. Gerstmann, C. Gougoussis, A. Kokalj, M. Lazzeri, L. Martin-Samos, N. Marzari,

F. Mauri, R. Mazzarello, S. Paolini, A. Pasquarello, L. Paulatto, C. Sbraccia, S. Scandolo, G. Sclauzero, A.P. Seitsonen, A. Smogunov, P. Umari, R.M. Wentzcovitch, J. Phys.: Condens. Matter **21**, 395502 (2009)
37. A.A. Mostofi, J.R. Yates, Y.-S. Lee, I. Souza, D. Vanderbilt, N. Marzari, Comput. Phys. Commun. **178**, 685 (2008)
38. S. Poncé, E. Margine, C. Verdi, F. Giustino, Comput. Phys. Commun. **209**, 116 (2016)
39. C. Verdi, F. Giustino, Phys. Rev. Lett. **115**, 176401 (2015)

Chapter 24
Electrical Properties 3

Traps in β-Ga_2O_3: From Materials to Transistors

Aaron R. Arehart and Steven A. Ringel

Abstract Traps in ultra-wide bandgap semiconductors (UWBG) are problematic for devices due to the very wide range of performance degradation phenomena they can cause, from high leakage currents, dynamic resistance and voltage dispersion in transistors, to high dark currents, low quantum efficiencies and low responsivities in various optoelectronic devices. For β-Ga_2O_3, the early stage of development for this promising UWBG semiconductor and associated lack of knowledge of many basic materials properties add more challenges as many of the trap states are unknown and their physical sources are poorly understood at present. This chapter summarizes the state of knowledge concerning deep level defects in β-Ga_2O_3 materials and early stage transistors. Deep level transient and optical spectroscopies (DLTS/DLOS) are the primary characterization methods being focused on here. DLTS/DLOS measurements made on β-Ga_2O_3 materials prepared by several growth methods and irradiated by high energy particles are discussed. Defect spectroscopy measurements made directly on β-Ga_2O_3 transistors are also described. Several traps that are in common across the range of materials and devices are revealed, several unique traps are identified, and by comparing with theory and other physical characterization results, the potential physical sources for several traps are considered. Finally, this chapter attempts to correlate defect levels found in β-Ga_2O_3 transistors with the fundamental materials studies, leading toward possible identification of specific defects as primary sources for transistor instabilities such as threshold voltage shifts.

A. R. Arehart (✉) · S. A. Ringel
Department of Electrical and Computer Engineering, The Ohio State University, Columbus, OH 43210, USA
e-mail: arehart.5@osu.edu

M. Higashiwaki and S. Fujita (eds.), *Gallium Oxide*, Springer Series in Materials Science 293, https://doi.org/10.1007/978-3-030-37153-1_24

24.1 Introduction

Beta-phase gallium oxide (β-Ga_2O_3) is a promising material for applications in next generation power and RF electronics due to a unique combination of advantageous properties that include an ultrawide bandgap (UWBG) of ~4.5–4.9 eV, a high predicted breakdown field, the ability to form heterostructures with $(Al,Ga)_2O_3$, ease of n-type doping, and high peak electron velocity [1–3]. However, the fact that β-Ga_2O_3 can be synthesized by melt-based growth methods makes these properties potentially transformational, since homoepitaxial UWBG devices are feasible, implying that dislocation-induced degradation in future high-power RF devices are possible. Numerous groups are working to develop and optimize β-Ga_2O_3 epitaxy, improve substrate quality, expand the range of doping while maximizing transport properties, develop heterostructures with β-$(Al,Ga)_2O_3$ barrier layers, and advance device performance toward theoretically predicted values. To achieve this requires identifying and attempting to eliminate crystalline defects that affect transport properties, and if they cannot be eliminated fully, then to determine their electronic properties so that device engineering can mitigate possible deleterious effects. As with any relatively immature semiconductor technology, understanding the properties of electrically active defects, translating that into quantifiable parameters that can be used to guide device optimization, and understanding the modes by which defect influence terminal characteristics, are central. Thus, this chapter focuses on the characterization of electrically active defects in β-Ga_2O_3 materials and devices to establish the current state of knowledge of deep level defects in this promising, but immature UWBG semiconductor system. We focus on deep level defect characterization of β-Ga_2O_3 material with an emphasis on identifying traps in β-Ga_2O_3 and how their concentrations depend on growth methods. The impact of high energy particle radiation is reviewed since this enables differentiation between native and extrinsic defect sources as a first step in experimental evaluation of physical sources. Then the impact of individual traps on the electrical characteristics of modulation-doped β-Ga_2O_3 metal semiconductor field effect transistors (MESFETs) is revealed and the specific traps responsible are identified using defect spectroscopy methods applied directly on transistors.

24.2 Materials Characterization Approach for Deep Level Defects in β-Ga_2O_3

Two techniques, deep level transient and optical spectroscopies (DLTS/DLOS) are reviewed and are the primary techniques discussed in this chapter. The advantages of DLTS and DLOS are the ability to quantitatively determine the trap energies with respect to a band edge and quantify each trap's concentration where the trap detection limit is $\sim 10^{-4} N_D$, which translates to an extremely sensitive, and yet quantitative detection limit of 10^{12}–10^{14} cm^{-3} for most studies. Both techniques are applied on diode test structures and rely on modulating trap occupation by applying a voltage, which controls $E_C - E_F$. If the trap is below the Fermi level, it will fill with an electron, and if above the Fermi level and given enough time, it will emit its trapped electron by thermal or optical stimulation. The trap occupation changes the charge in the depletion region, which results in a change in the depletion depth. Electron trap emission results in a decrease in the depletion depth for n-type semiconductors, which is measured by monitoring the junction capacitance versus time. In β-Ga_2O_3, there are no identified shallow acceptors, the hole effective mass is extremely large, and small polarons are predicted to trap free holes [4, 5]. Therefore, the rest of the defect spectroscopy discussion is with n-type semiconductors in mind. For DLTS measurements, traps are filled with a positive bias (fill pulse), then the diode is brought to reverse bias to monitor the trap emission, which is strongly temperature dependent. The trap emission time constant τ is [6]

$$\tau = \frac{1}{\sigma_n \langle v_{th} \rangle N_C} \exp\left(\frac{E_C - E_T}{kT}\right) \tag{24.1}$$

where $\langle v_{th} \rangle$ is the average thermal velocity of the electrons, N_C is the conduction band density of states, k is Boltzmann's constant, and T is the temperature. Then the fill pulse, reverse bias, and time constant extraction is repeated at each temperature where typically a double boxcar or Fourier analysis method is used to determine τ versus T. By plotting the data on an Arrhenius plot ($1/kT$ versus $\ln(\tau T^2)$) and fitting a line, the trap energy and capture cross section for each trap can be extracted. The trap concentrations N_T are extracted from the magnitude of the capacitance transient ΔC, using [6]

$$N_T \approx \frac{2N_D \Delta C}{C_\infty} \tag{24.2}$$

where N_D is the doping and C_∞ is the equilibrium capacitance at the applied reverse bias. There are several corrections to this equation to obtain more accurate individual trap concentrations since this simple relation can lead to underestimated values in some cases. The most common corrections address the fact that the sampling times in the double boxcar method only detect a fraction of the total ΔC (the correction results in ~3× increase in concentration over the uncorrected value) and the "lambda" correction. The "lambda" correction refers to the case where a trap is not modulated throughout the entire depletion depth but only where the Fermi level crosses the particular trap level at the fill and reverse biases, and this correction can increase the obtained trap concentration by as much as 10× over the uncorrected value. When the lambda effect correction is important depends on the positions of the trap energy level relative to the Fermi level. Full details of the technique and analysis methods are in [6, 7].

DLTS is usually limited to detecting traps within ~1.0 eV of the conduction band edge in β-Ga_2O_3 because of the practical thermal limits of the measurement stage and the stability of the metal contacts on the devices at very high temperatures that are otherwise needed to thermally stimulate trapped carriers with larger ionization energies, such as expected for any wide bandgap semiconductor. DLOS avoids this limitation and allows the rest of the bandgap to be characterized. DLOS uses a spectrally-resolved monochromatic light source to photo-emit trapped carriers to either the conduction or valence band. If the incident photons have energy equal to or greater than $E_C - E_T$ for a majority carrier trap in an n-type semiconductor, trapped electrons within the diode depletion region will be stimulated out of the trap and into the conduction band, which causes the capacitance to increase. Plotting the change in capacitance, ΔC, versus the incident photon energy will show onsets (changes in slope) in ΔC at the trap energy. Since photocapacitance onsets can be broadened by local lattice relaxation effects, to more accurately determine the energy level and extract any energy values associated with lattice relaxation (i.e. the Franck-Condon energy), the optical cross section is extracted by analyzing the dC/dt spectrum at time $t = 0$ [6, 8, 9]. The extracted optical cross section data can then be numerically fitted using the Lucovsky model (no Franck-Condon energy), or the Pässler and Bois-Chantre models (both include the Franck-Condon energy) [8–10]. For β-Ga_2O_3, the most appropriate capture cross section models were not known, so Farzana et al. examined the different models and determined that the Lucovsky and Pässler models produced the best fits for the various traps along with the most realistic fitting parameters (i.e. the trap energy and Franck-Condon energy) [7].

24.3 Deep Levels in β-Ga_2O_3

There are relatively few reports of the impact of defects in early β-Ga_2O_3 transistors, and these have focused on dielectric/β-Ga_2O_3 interfaces due to their importance in device characteristics. However, defect states within bulk epitaxial and substrate materials are also expected to be significant issues and work on characterizing traps within the β-Ga_2O_3 material itself has been ramping considerably. Efforts are well underway toward establishing a baseline of traps present in β-Ga_2O_3 prepared using different growth methods and conditions, and initial efforts to identify the sources of these traps have begun. As early as 2010, Varley et al. were predicting the traps states and energies that would exist in β-Ga_2O_3 using DFT calculation methods [11], and this preceded much of the experimental work to actually characterize traps in this material. The earliest published experimental DLTS work was that of Irmscher et al., done on Czochralski bulk β-Ga_2O_3, in which three traps were identified, at $E_C - 0.55$ eV, $E_C - 0.74$ eV, and $E_C - 1.04$ eV where the $E_C - 0.74$ eV trap, labeled E2 by the authors, had the highest concentration, which in most cases were in the range of low- to mid-10^{16} cm^{-3} [12]. Trap concentrations this high will undoubtedly be problematic in many devices. In 2016, Zhang et al. was the first to investigate the trap spectrum throughout the entire bandgap, which was achieved using both DLTS and DLOS. This work was done on unintentionally-doped (010) edge-defined film-fed grown

(EFG) β-Ga_2O_3 substrates from the Tamura Corporation [13]. A summary of the traps for the Tamura substrates from that effort is shown in Fig. 24.1 where the $E_C - 0.8$ eV and 4.4 eV traps dominate the total trap concentration, which was $\sim 6 \times 10^{16}$ cm^{-3} [13]. Note we do not reference the 4.4 eV energy level to the conduction band because there is a possibility this is related to self-trapped holes (STHs), which will be discussed shortly. Subsequent work showed the $E_C - 0.8$ eV trap here to be the same E2 trap identified by Irmscher [12], which will be explained below. Even with advances in the growth techniques over the years from the initial work on Czochralski crystals, the trap concentrations from the EFG-grown material for the $E_C - 0.8$ eV trap (E2) remained in the mid-10^{16} cm^{-3} range and dominated the trap spectrum. At the time, this indicated significant work is still needed to optimize EFG materials. Importantly, the work by Zhang revealed several other defect states not seen prior, including defect levels at $E_C - 0.6$, $E_C - 1.0$, $E_C - 2.0$, and 4.4 eV, where the 4.4 eV trap concentration was of comparable concentration to the $E_C - 0.8$ eV (E2) trap [3]. While this initial work baselined the trap spectrum for EFG material, to explore their physical sources and, ultimately, their impact on devices, additional research was necessary.

To begin identifying the physical sources of these traps, high-energy particle radiation was employed as it is an ideal method to separate native point defects from extrinsic point defects or extended defects. Proton, neutron, electron, etc. radiation with sufficient energy will create intrinsic defects including vacancies, interstitials, and possibly point defect complexes, while extrinsic defects and most traps resulting from extended defects will not change in concentration due to radiation effects. Farzana et al. have explored the effects of neutron irradiation on bulk β-Ga_2O_3 grown by EFG [14]. Figure 24.2 shows the DLTS spectra before and after 8.5×10^{14} cm^{-2} equivalent fluence of 1 MeV neutrons where the $E_C - 0.6$, $E_C - 0.8$, and $E_C - 1.0$ eV trap concentrations remained unchanged after the neutron irradiation suggesting that none of these traps are simple intrinsic point defects. The behavior of the DLOS spectra, shown in Fig. 24.3, however, is very

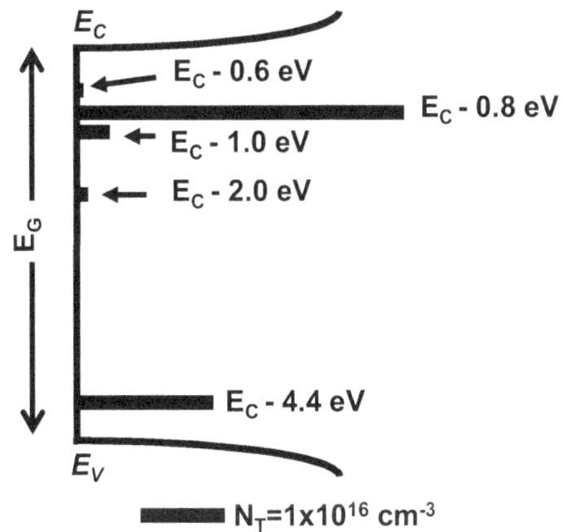

Fig. 24.1 Summary of trap energies and concentrations in unintentionally-doped (010) EFG- grown β-Ga_2O_3 from Tamura Corporation

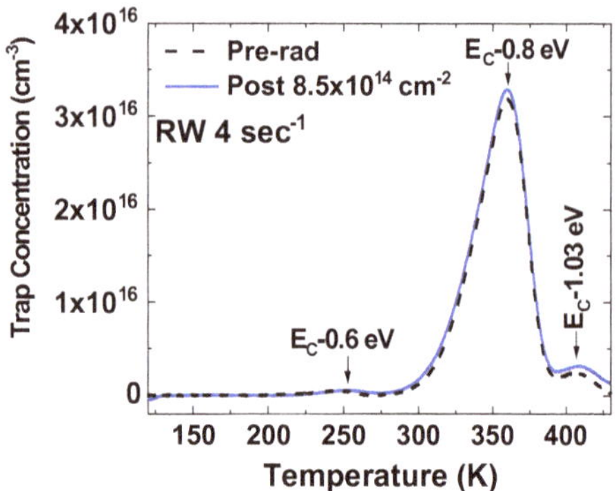

Fig. 24.2 DLTS spectra of EFG-grown β-Ga_2O_3 before and after an 8.5×10^{14} cm^{-2} fluence of fast neutrons at the OSU Nuclear Reactor Laboratory with energies ranging from 1 eV to ~2 MeV. After [14]

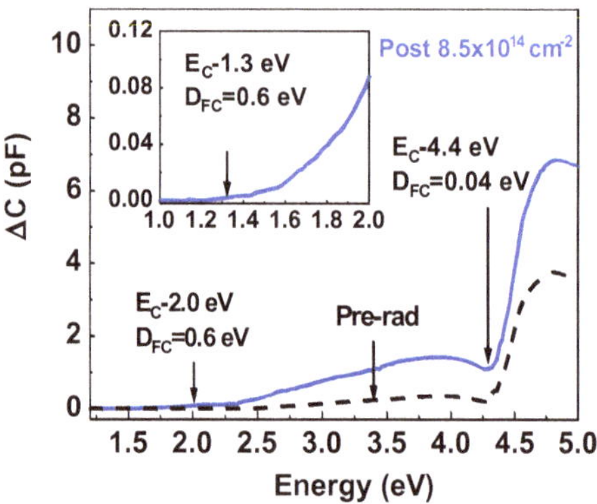

Fig. 24.3 Steady-state DLOS spectra of EFG-grown β-Ga_2O_3 before and after an 8.5×10^{14} cm^{-2} fluence of fast neutrons at the OSU Nuclear Reactor Laboratory with energies ranging from 1 eV to ~2 MeV. After [14]

different than the DLTS results. Prior to irradiation, the DLOS measurements revealed deep levels at $E_C - 2.0$ and at a photon energy of 4.4 eV. After irradiation, it is clearly seen that while the 4.4 eV state did not change in its concentration, there was a strong increase in concentration observed for the $E_C - 2.0$ eV state and a new state at $E_C - 1.3$ eV appeared. Subsequent work found that both the $E_C - 1.3$ and $E_C - 2.0$ eV trap concentrations strongly depend on the magnitude of radiation fluence, and Farzana et al. demonstrated through a series of lighted C-V (LCV) experiments that these traps are the primary source of strong carrier compensation that was observed for these samples after irradiation [14]. The dependence on radiation fluence suggests that the states at $E_C - 1.3$ and $E_C - 2.0$ eV likely are related to the formation of native defects, whereas the $E_C - 0.6$ and $E_C - 0.8$ eV traps appear to be of extrinsic or extended defect origin with the possible exception of the 4.4 eV feature, as will now be discussed.

The 4.4 eV feature has been the source of much controversy. This state is ubiquitous having been observed by all DLOS experiments carried out to date on β-Ga_2O_3 material grown by multiple methods [7, 13, 14]. Not only has its energy position and DLOS spectral features been virtually identical in all these studies, but the trap concentration is very consistent, varying by only a factor of approximately two with concentrations ranging from $\sim$2–5 × 10^{16} cm^{-3}. This is despite investigating β-Ga_2O_3 using different dopants, bulk crystals versus epitaxial layers, and different crystal orientations. This concentration invariance has been confounding since it implies a physical source that is also unchanged with respect to all materials variables. One possibility, first suggested by Armstrong, was the tentative association of this DLOS feature with the presence of self-trapped holes, or STHs [15], which are theoretically expected to form in β-Ga_2O_3 [4, 16] and have been observed experimentally [17, 18]. In fact, STHs resulting from small polarons are predicted in a wide-range of semiconducting oxides [19]. If the DLOS feature at 4.4 eV is somehow related to STHs, this would be consistent with the invariance of its concentration versus all independent materials variables and radiation damage as just noted. Intense studies to find a correlation of this strong DLOS feature to its source is ongoing and currently an open question.

The above discussion focuses on discerning possible physical sources for DLTS and DLOS features in bulk β-Ga_2O_3 substrates. Next, we summarize results from a series of investigations of traps in MBE-grown β-Ga_2O_3 both in the as-grown state and after high energy particle irradiation. This opens a broader range of material variables with which physical sources of traps in gallium oxide could be determined, as well as the important goal of determining the critical trap states in MBE material, since MBE growth is currently a prevalent epitaxial deposition process used to fabricate β-Ga_2O_3 transistors.

Figure 24.4 shows the summary of the trap spectrum obtained from a lightly Ge-doped (n $\sim$3 × 10^{16} cm^{-3}) β-Ga_2O_3 layer that is representative for materials grown using plasma-assisted MBE (PAMBE). A total of nine distinct trap states are present. However, the concentrations of the traps within 1 eV of the conduction band, detected by DLTS, are more than one order of magnitude lower than the trap concentrations typically observed for EFG bulk material. While it is possible that the EFG material contains all the same trap states, its higher n-type background doping (1 × 10^{17} cm^{-3}) decreases DLTS sensitivity for very low trap concentrations, and together with the high concentrations of the E_C − 0.8 eV (E2) trap in the

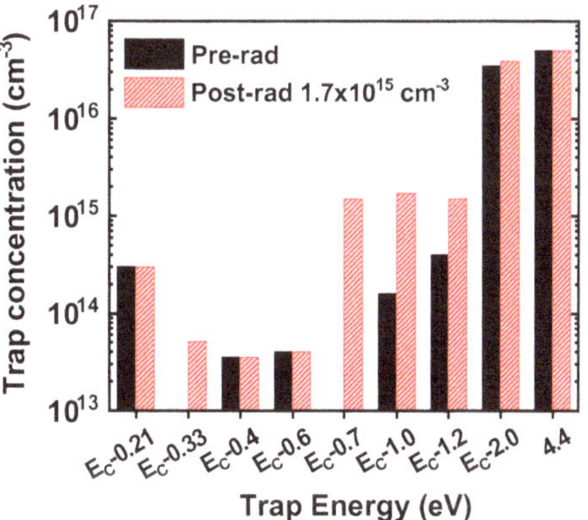

Fig. 24.4 Bar chart showing the trap concentrations in MBE-grown Ge-doped β-Ga_2O_3 before and after neutron irradiation with a fluence of 1.7 × 10^{15} cm^{-2} in the OSU Nuclear Reactor Laboratory. After [20]

EFG material can obscure the observation of very low DLTS trap concentrations. Regarding the traps detected with DLOS (i.e. those with energies greater than 1 eV from the conduction band), the trap spectrum obtained from the Ge-doped PAMBE material reveals a high concentration of the $E_C - 2.0$ eV state ($\sim 3 \times 10^{16}$ cm^{-3}) along with the ubiquitous 4.4 eV feature. With the PAMBE material now baselined, these samples were then exposed to an equivalent 1 MeV neutron fluence of 1.5×10^{15} cm^{-2}, identical to that which was used for the EFG studies. No new states were observed by DLTS, even at this higher level of sensitivity afforded by the lower background doping; however, there was significantly different dependencies of the pre-existing traps on irradiation, which are now described. The concentration of the $E_C - 0.21$, $E_C - 0.4$, and $E_C - 0.6$, and 4.4 eV traps were found not to depend on neutron irradiation, whereas the $E_C - 0.33$ eV trap showed very weak dependence. The other traps at $E_C - 0.7$ eV, $E_C - 1.0$, $E_C - 1.2$ and $E_C - 2.0$ eV all show stronger concentration dependence on the neutron irradiation, which is consistent with the behavior expected for intrinsic point defect sources for these states. The $E_C - 0.7$ eV trap likely existed in the neutron-irradiated EFG material in Fig. 24.2, but as this trap occurs at nearly the same temperature as the $E_C - 0.8$ eV trap it was likely obscured by the high-concentration radiation-independent $E_C - 0.8$ eV trap. Like the neutron-irradiated β-Ga_2O_3 grown by EFG, the 4.4 eV feature was found to be independent of radiation and its 5×10^{16} cm^{-3} concentration in the PAMBE material ($\sim 2\times$ higher than for EFG) is the highest concentration observed. On the other hand, the $E_C - 2.0$ eV trap concentration, like in the EFG material, increased after irradiation with neutrons, again suggesting a native defect. Deák et al. performed DFT calculations that suggested this state could be the V_O native point defect [16]. However, this is contradicted by Peelaers et al., who instead predicted the V_O defect to be at $\sim E_C - 3.5$ eV, with defect states associated with O_i interstitial and the V_{Ga} defects to be close to $E_C - 2.0$ eV [21]. The exact source of this defect other than its native defect character is still under investigation. However, one thing to note is the DFT results suggest a low formation energy (~ 0.5 eV) for the O_i defect in oxygen-rich growth conditions used in the MBE [21], so this is possibly the most likely candidate.

The various dependencies of all the defect states on growth method and high energy particle radiation provide some strong indicators regarding the physical sources for the traps. This discussion will be provided toward the end of this chapter, once the trap measurement results taken directly from gallium oxide transistors are reviewed, since those studies were found to be profoundly useful in identifying the physical sources of the observed traps within the constituent materials. However, at this point, we summarize the DLTS and DLOS results presented thus far. Figure 24.5 shows the Arrhenius behavior for each DLTS trap, which includes data overlaid from several groups. The similarities in this comparison are revealing, as this suggests that the physical sources are fairly common across a broad range of studies. However, the Arrhenius behavior also provides the correct way to discern whether traps that are closely spaced in energy are indeed

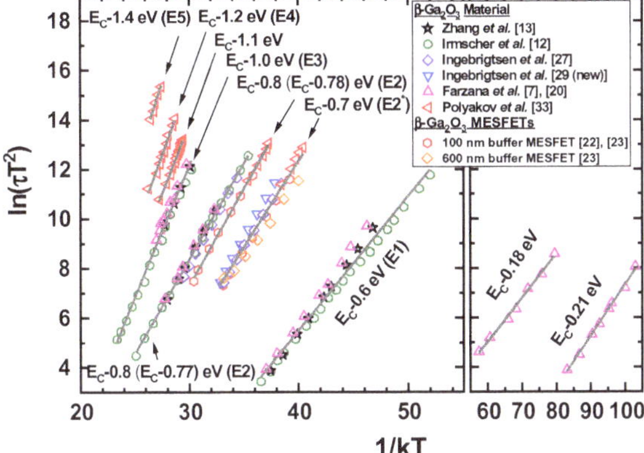

Fig. 24.5 Arrhenius data of β-Ga_2O_3 traps over a wide range of DLTS measurements from several groups, several growth methods, different doping, and irradiation. The lines represent fitting a line to all of the related Arrhenius trap data. There are closely spaced levels at $E_C - 0.77$ and $E_C - 0.78$ eV. These traps may be one and the same but fitting all the $E_C - 0.77$ eV and $E_C - 0.78$ eV Arrhenius data together gives an erroneously low trap energy. Therefore, these are fit separately until it can be determined if they are the same trap or not

different defects or are statistically spread but are the same defect. This type of critical and detailed comparison will become essential in understanding the transistor-based DLTS results described in the next section.

24.4 Trap Identification in β-Ga_2O_3 Transistors and Their Influence on Device Instabilities

The impact of traps in β-Ga_2O_3 devices is just beginning to be fully explored. Groups have seen a number of classic trap related problems in β-Ga_2O_3 transistors like dispersion and hysteresis [22–25]. To commercialize β-Ga_2O_3 devices, optimizing growth and device designs will be necessary, which must include accounting for defect-driven behaviors. The previous section focused on trap characterization at the materials level using specialized test structures. In this section, that work is extended to the direct characterization of traps within actual β-Ga_2O_3 MESFETs, specifically investigating the linkage between individual deep levels and transistor instabilities. As will be seen, several of the traps observed in the materials-level studies have significant roles in defining the MESFET instabilities.

24.4.1 MESFET Experimental Details and Terminal Characteristics

The basic device used for this study was a Si delta-doped MESFET, with the device structure shown in Fig. 24.6. The device layers were grown by PAMBE on Tamura (010) EFG, Fe-doped semi-insulating substrates. Layers were grown under oxygen-rich conditions at a temperature of 700 °C, with details provided elsewhere [22, 26]. Two device structures were compared, one with a 100 nm thick, PAMBE-grown UID buffer layer between the channel and the Fe-doped substrate and one with a 600 nm thick buffer. Comparing DLTS results on these two structures was chosen due to our observations of significant differences in threshold voltage behavior as a function of buffer layer thickness, which motivated exploring possible links between individual trap states and MESFET terminal characteristics, as will be described below.

Device fabrication was completed by using regrown *n*+ contact layers to achieve low Ohmic contact resistance as discussed in [22, 26]. Ti/Au/Ni source and drain Ohmic contacts were formed with e-beam evaporation with a one minute anneal at 470 °C in an N_2 atmosphere. After dry etching for mesa isolation, the Ni/Au/Ni Schottky gates were patterned through a lift-off process. These samples had Hall mobilities between 70 and 110 cm^2/Vs and charge densities between 5×10^{12} and 1×10^{13} cm^{-2} indicating these devices are of high quality. Despite this high quality, these MESFETs displayed significant threshold voltage instabilities of up to ~ 1 V after biasing the drain-source voltage V_{DS} at 15 V while pinched-off.

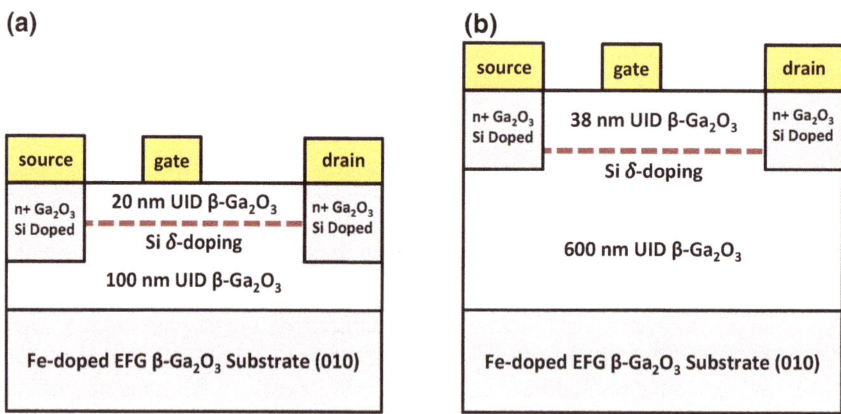

Fig. 24.6 Schematic cross-sectional diagrams of the MBE-grown MESFETs. The dashed line represents the silicon δ-doping channel. The Ohmic contacts utilize an MBE regrown heavily doped region to reduce contact resistance to the channel and metal. The 100 nm buffer MESFET in **a** utilizes a 100 nm buffer with a gate-channel spacing of 20 nm while the 600 nm buffer MESFET in **b** utilizes a 600 nm buffer sample with a 38 nm spacer with both on Tamura Fe-doped EFG β-Ga_2O_3 substrates. After [23]

Figure 24.7 shows double-pulsed output I-V and transconductance curves of MBE-grown Si δ-doped β-Ga_2O_3 MESFETs with the two buffer thicknesses with a maximum drain current of over 100 mA/mm at zero gate voltage, consistent with high material and device quality [23]. The double pulsed I-V allows for a fixed quiescent bias, which enables both the self-heating and trap occupancy to be controlled. In the zero-bias quiescent condition, the quiescent condition is 0 V for the gate-source voltage V_{GS} and 0 V for the drain-source voltage V_{DS}. In the high-V_{DS} quiescent condition ($V_{DS,q} = 15$ V and $V_{GS,q} = -5$ V), there is a threshold voltage V_T shift of 0.78 V for the 100 nm buffer MESFET and 1.18 V for the 600 nm buffer MESFET. Both quiescent conditions are nominally zero power dissipation states, so the differences in the V_T shift are not due to self-heating and must be due to traps under the gate that shift V_T. To understand the trapping behavior, first for the zero-bias quiescent condition, the traps are empty of electrons because V_T is more negative relative to the high-V_{DS} quiescent condition. This

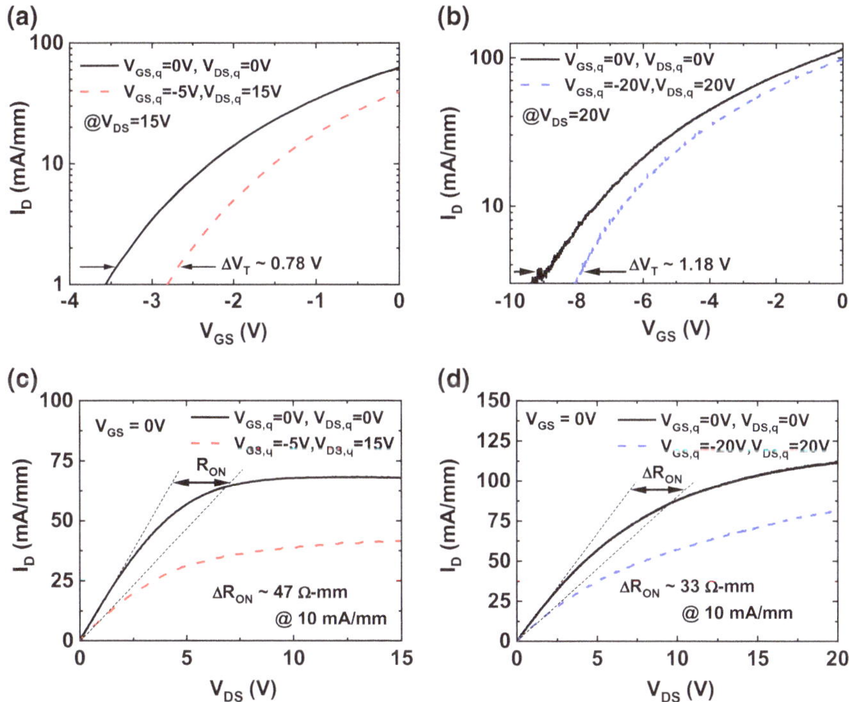

Fig. 24.7 Double-pulsed 50 μs pulsed transconductance curves for the **a** 100 nm buffer thickness and **b** 600 nm buffer thickness MESFETs with quiescent conditions shown in the figures. The double-pulsed output curves are also plotted in **c** and **d** for the 100 nm and 600 nm buffer, respectively. The zero-bias quiescent condition (black) represents the non-trapped state while the high V_{DS} pinch-off quiescent condition (red) exacerbates the trapping effect. Both devices show dispersion between the two conditions indicating an electron trapping mechanism below the channel is responsible (change to transconductance). After [23]

indicates the traps are above the Fermi level or the Fermi level is pinned and the traps are partially filled. In the high-V_{DS} quiescent condition, the traps are now filled with electrons, causing the positive V_T shift, where the likely filling mechanism is the high gate leakage current (~ 100 μA/mm). This then moves the electron quasi-Fermi level closer to the conduction band, which allows the traps to fill.

24.4.2 Isothermal Constant Drain Current Deep Level Transient Spectroscopy

Even though the trap location and filling mechanism are understood, the trap energies and time constants of the traps are difficult to understand by just using pulsed I-V measurements. However, much like DLTS, if the time constants of trap emission can be determined as a function of temperature then the trap energies and concentrations can be determined. Isothermal gate-controlled constant drain current DLTS (GC CI_D-DLTS) was employed for this purpose. GC CI_D-DLTS is similar to DLTS except it is performed directly on transistors and is based on dynamically adjusting the gate voltage (V_{GS}) to maintain a constant drain current after voltage pulsing. In this mode, the trap-induced threshold voltage instability is tracked by recording the gate voltage transient in saturation where $\Delta V_T \approx \Delta V_{GS}$. Also, in saturation, the drain-to-source current (I_{DS}) is insensitive to drain resistance (R_D) changes if the output conductance is low, and by measuring at low I_D, this measurement is therefore insensitive to changes in source resistance R_S (i.e. $I_{DS} * \Delta R_S \ll \Delta V_T$). In GC CI_D-DLTS, the transistor is biased to $V_{GS} \geq 0$ V to empty traps in this case, then biased into saturation with a low I_{DS} where the gate (threshold) voltage transient is recorded. This is repeated at several temperatures to extract the trap energy and the electron capture cross section by using a standard Arrhenius relationship. Just like DLTS, the trap concentration is extracted from the height of the peak in the double boxcar analysis except in the case of CI_D-DLTS, majority carrier traps are negative peaks while for DLTS they are positive. In the isothermal mode, a gate voltage transient at a particular temperature is analyzed by plotting ΔV_{GS} versus τ, which are given by [6]

$$\tau = \frac{t_2 - t_1}{\ln(t_2/t_1)} \tag{24.3}$$

and

$$\Delta V_{GS} = V_{GS}(t_2) - V_{GS}(t_1) \tag{24.4}$$

where t_2 and t_1 are times where $t = 0$ is at the end of the fill pulse. The t_2/t_1 ratio is kept constant, and we choose use $t_2 = 2.5t_1$. Here, ΔV_{GS} peaks at the τ value that corresponds to the time constant of the transient, and if multiple time constants are present then multiple peaks will be observed in the isothermal analysis. From the height of the ΔV_{GS} peak, the trap concentration $N_{T,sheet}$ (i.e. the change in the channel charge due to the trap) can be estimated using

$$N_{T,sheet} = \frac{\varepsilon_s}{qt_s}|\Delta V_T| \quad (24.5)$$

where ε_s is the β-Ga_2O_3 permittivity, t_s is the channel to gate distance, and q is the elementary charge.

24.4.3 Identifying the Traps Responsible for Threshold Voltage Instability

The isothermal GC CI_D-DLTS results obtained from the MESFETs with the two different buffer thicknesses are shown in Fig. 24.8. The 100 nm buffer MESFET reveals two traps where their Arrhenius characteristics are plotted in Fig. 24.5 alongside the conventional DLTS results obtained from both substrate and epitaxial layer materials. As seen, the two traps observed in the MESFETs clearly correspond to the $E_C - 0.7$ eV and $E_C - 0.8$ eV traps, which Fig. 24.5 shows are the same traps others have referred to as the $E2^*$ and E2 traps, respectively [12, 27]. For the 100 nm thick buffer MESFET, the $E_C - 0.7$ eV trap dominates and is responsible for at least 0.5 V of the total threshold voltage instability, while the $E_C - 0.8$ eV (E2) peak is responsible for less than ~0.2 V. For the 600 nm buffer MESFET, Fig. 24.5 shows only the trap at $E_C - 0.7$ eV is present. In both cases, the GC CI_D-DLTS trap magnitudes underestimate the V_T instability from the pulsed I-V. This difference is due to different pulsing conditions between the two measurements that

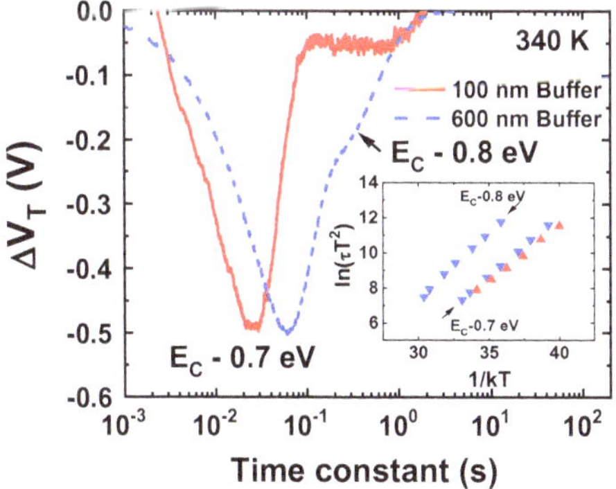

Fig. 24.8 Isothermal gate-controlled CI_D-DLTS analysis the 100 and 600 nm buffer MESFETs. Two trap levels are revealed at $E_C - 0.8$ eV (shoulder of 100 nm buffer spectrum) and $E_C - 0.7$ eV from the inset Arrhenius plot. The small non-distinct peak in the 100 nm buffer MESFET near the $E_C - 0.8$ eV trap was not temperature dependent, so this is not the $E_C - 0.8$ eV trap. After [23]

leads to reduced trap filling and V_T instability magnitude in the GC CI_D-DLTS measurements. It should be noted that the V_T shift caused by the $E_C - 0.7$ eV trap is similar regardless of buffer thickness. Even though the 600 nm buffer MESFET V_T instability is larger than the 100 nm buffer MESFETs, the 600 nm buffer MESFET has a larger (38 nm) spacer, and overall the $E_C - 0.7$ eV trap density calculated using (24.5) ends up being slightly lower for the 600 nm buffer sample than the 100 nm sample, which are 2.2×10^{12} cm^{-2} and 1.8×10^{12} cm^{-2}, respectively. The $E_C - 0.8$ eV trap concentration in the 100 nm buffer MESFET is <3×10^{11} cm^{-2}, so it is not the dominant defect even in the 100 nm buffer MESFET. These results immediately imply that the $E_C - 0.7$ eV (E2*) trap is probably present in the PAMBE-grown buffer and that the $E_C - 0.8$ eV (E2) trap could be related to a diffusing impurity from the Fe-doped substrate, or the substrate itself. Indeed, SIMS results shown in Fig. 24.9 reveal a non-negligible tailing of Fe into the buffer region of the MESFET, and as seen the Fe concentration does not reduce below detection limits until 200 nm of buffer layer growth. This suggests that the $E_C - 0.8$ eV (E2) trap correlates with the presence of residual Fe impurities that diffuse from the Fe-doped substrate. This connection is supported by DFT calculations that predict energy levels relatively close to this value for Fe_{Ga} point defects [27]. Moreover, the assignment of this defect state to an extrinsic source is consistent with the lack of sensitivity of this defect state to high energy particle irradiation discussed earlier in this chapter and previously published [14, 27].

While the $E_C - 0.8$ eV (E2) state seems to correlate with Fe_{Ga} defect, it is not the trap that dominates the V_T instability. As noted earlier, the $E_C - 0.7$ eV (E2*) state accounts for most of the V_T instability, and its possible physical source can also be understood from the materials studies as well. As noted earlier, the concentration of this trap strongly responds to high energy particle radiation, not only for neutrons [14] but also for protons [28]. Such an observation supports the assertion that its source is intrinsic in nature, and based on comparisons with the most current DFT calculations, aligns well with V_{Ga}-related defects [16, 29]. Since the same growth conditions were used for the PAMBE buffer in the MESFET devices and since the concentration of this trap is very similar independent of the

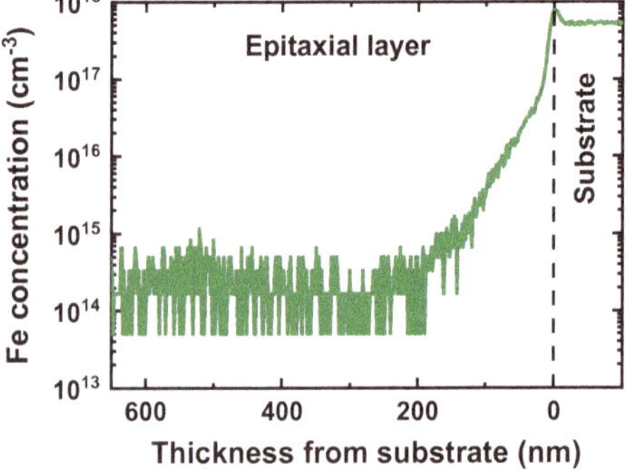

Fig. 24.9 SIMS profile of the MESFET structures indicating the Fe concentration where zero indicates the epitaxy/substrate interface. The 100 nm buffer MESFET total epilayer thickness is 120 nm while the 600 nm buffer MESFET total epilayer thickness is 638 nm. After [23]

buffer thickness, this explanation seems plausible. The ramifications of both physical source assignments for these traps are now discussed.

Traps in transistors can have many effects, including increasing the subthreshold swing, reducing the transconductance, causing time-dependent buffer leakage, sourcing system-level problems like reduced and time-dependent linearity, and more. First, we consider the presumed native point defect at E_C − 0.7 eV (E2*). This trap dominates the isothermal CI_D-DLTS signal in both cases, which indicates that improved growth to minimize intrinsic defects in the buffer is necessary for improved V_T stability. It also indicates the Fe can have some influence of the V_T stability, but in these devices it is not the dominant effect. Second, with the trap energies and capture cross sections known, the time constant of the V_T instability can be calculated at any temperature using (24.1) or the Arrhenius plot. For example, the 600 nm buffer MESFET with its E_C − 0.7 eV trap, will have an emission (i.e. the recovery after going from high V_{DS} to low V_{DS}) time constant of ~3 s at 300 K and 1 ms at 400 K. The capture time constant when the device is biased to a high V_{DS} from a low V_{DS} has not be characterized but will generally be very different than the emission time constant and would typically be non-exponential [30]. This is because the capture time constant is dictated by the number of empty traps, the number of free electrons in the conduction band, and the trap capture cross section. However, none of these terms depend on the $E_C - E_T$ energy. In fact, the capture time constant should only be weakly temperature dependent. This is true unless there is a capture barrier, in which case it will have a stronger temperature dependence. The emission time constant does depend on $E_C - E_T$ and is given by (24.1) and is strongly temperature dependent. Therefore, care is needed to understand if emission or capture is occurring during device operation so both the temperature dependence and time constant can be predicted.

24.5 Summary of Traps and Potential Sources

In the previous example of the δ-doped β-Ga_2O_3 MESFETs, it was shown that two traps cause the V_T instability and that these traps are most likely due to Fe_{Ga} and a native point defect, which is potentially V_{Ga}-related. However, the materials studies are beginning to reveal the sources of many other deep levels that will be important in the future for maturing β-Ga_2O_3 devices such as MESFETs, modulation-doped FETs (MODFETs), UV photonic devices, and high voltage and RF devices in general. The deep levels observed by DLTS and DLOS are compiled in Table 24.1 along with the current understanding of their physical sources. It is important to note that these states are identified by their communication to the conduction band (either thermally or optically stimulated electron emission) in all cases. Because the hole effective mass in the valence band is very large, and self-trapped holes are potential charging sources, the ability to change the fixed charge and capacitance due to hole emission to the valence band is not well-defined and will require careful study. Beyond density functional theory predictions of energy levels for different

Table 24.1 Summary of traps detected by DLTS and DLOS in β-Ga_2O_3

Label (eV)	Alternate names	Avg. trap energy (eV)[a]	Avg. thermal cross section (cm^2)[a]	Growth methods observed in[b]	Likely physical identity	References
DLTS						
$E_C - 0.18$		$E_C - 0.18$	3×10^{-19}	Ge-doped MBE	Unknown	[7]
$E_C - 0.21$	E6	$E_C - 0.21$	6×10^{-16}	Ge-doped MBE, MOCVD	Likely extrinsic point defect or extended defect, possibly Ge_{Ga}	[7, 31]
$E_C - 0.3$	E7	$E_C - 0.27$ to $E_C - 0.29$	6×10^{-18}	HVPE	Unknown	[32]
$E_C - 0.33$		$E_C - 0.33$	2×10^{-15}	Ge-doped MBE	Native point defect	[20]
$E_C - 0.4$		$E_C - 0.42$	1×10^{-15}	Ge-doped MBE	Extrinsic or extended (non-native)	[20]
$E_C - 0.6$	E1	$E_C - 0.54$	1×10^{-14}	MBE, HVPE, EFG, CZ	Extrinsic or extended (non-native)	[12–14, 28]
$E_C - 0.7$	E2*	$E_C - 0.70$	5×10^{-15}	MBE, HVPE, EFG	Native point defect, possibly V_{Ga}-related [16, 28]	[20, 22, 27, 28]
$E_C - 0.8$	E2	$E_C - 0.77$ and $E_C - 0.78$	$5–7 \times 10^{-14}$	CZ, MBE, HVPE, EFG	Fe_{Ga}	[7, 12–14, 22, 27, 28]
$E_C - 1.0$	E3	$E_C - 1.05$	2×10^{-13}	CZ, EFG	Native point defect	[12–14, 27, 28]
$E_C - 1.1$	E3	$E_C - 1.13$	5×10^{-13}	EFG	Unknown	[33]
$E_C - 1.2$	E4	$E_C - 1.22$	1×10^{-12}	EFG	Native point defect, Possibly V_O [35]	[34]
$E_C - 1.4$	E5	$E_C - 1.38$	6×10^{-12}	CZ	Unknown	[33]

(continued)

Table 24.1 (continued)

Label (eV)	Alternate names	Avg. trap energy (eV)[a]	Avg. thermal cross section (cm^2)[a]	Growth methods observed in[b]	Likely physical identity	References
DLOS/LCV						
$E_C - 1.3$	–	$E_C - 1.27$ to $E_C - 1.29$ $D_{FC} \sim 0.5$	–	Ge-doped MBE, EFG, HVPE	Native point defect, compensating center	[7, 14, 28]
$E_C - 2.0$	–	$E_C - 2.0$ to $E_C - 2.2$ $D_{FC} \sim 0.5$	–	MBE, EFG, HVPE, MOCVD	Native point defect, compensating center, Trap conc. proportional to [Sn], possibly V_O, O_i, or V_{Ga} complex [14, 16, 21]	[7, 13, 14, 28, 31]
$E_C - 3.2$	–	$E_C - 3.25$ $D_{FC} \sim 0.3$	–	Ge-doped MBE	Unknown [3]	[7]
4.4[c]	–	4.37–4.48 $D_{FC} < 0.1$	–	MBE, EFG	Possibly STH related or non-native defect	[7, 13–15, 20, 36]

[a]Average trap energy and cross section are extracted from linear fits to the Arrhenius data in Fig. 24.5. DFC is the Franck-Condon energy
[b]CZ is Czochralski, HPVE is hydride vapor phase epitaxy, MBE is molecular beam epitaxy, MOCVD is metal-organic chemical vapor deposition, and EFG is edge-defined film-fed grown
[c]Trap is possibly $E_c - 4.4$ eV, $E_v + 4.4$ eV, and/or STH related

intrinsic and extrinsic point defects, the use of high-energy particle radiation has been a helpful tool to separate intrinsic point defects from other defects. In summary, the $E_C - 0.21$ eV, $E_C - 0.4$, $E_C - 0.8$, and 4.4 eV traps do not respond to irradiation suggesting they are not related to intrinsic point defects. The $E_C - 0.21$ eV trap has been suggested to be Ge_{Ga} while the $E_C - 0.8$ eV is likely to be Fe_{Ga} as previously discussed [7, 22, 27]. Regarding the 4.4 eV trap, as previously discussed, not only is it observed in all samples investigated so far, regardless of growth methods and conditions, it also does not respond to particle irradiation. There are several possibilities and competing theories at the moment that could explain this. First, its source could be an extrinsic or extended defect at $E_C - 4.4$ eV that is mostly invariant of the materials preparation. Second, it could be evidence for a state that exists near $E_V + 4.4$ eV, which upon illumination with 4.4 eV photons would create STHs and free electrons where the STHs act to store positive charge, thus adding a photocapacitance that creates a causes a DLOS feature appear at 4.4 eV, or third and least likely, the state could be related to direct creation of STHs.

While there are traps whose concentrations do not respond to radiation, there are many that strongly do, which suggests they are related to native point defects. The removal of intrinsic defects can be very challenging as this requires careful optimization of growth conditions that also must be balanced by conditions to optimize doping, the role of the Fermi level on defect formation energies, and other factors. The $E_C - 1.0$ and $E_C - 1.2$ eV trap concentrations increase under neutron irradiation suggesting they are related to native point defects, and along with the $E_C - 2.0$ eV state discussed earlier, are the primary sources of free carrier compensation as seen using LCV measurements [14, 31]. The $E_C - 2.0$ eV trap is particularly interesting. Not only is it observed in samples grown by multiple methods [7, 13, 20, 28, 31] with strongly varying concentrations, but it also responds to high-energy particle radiation, which implies a native defect source [7, 13, 32]. In recent investigations, the concentration of this state also has been found to strongly depend on the concentration of n-type doping using Sn in EFG material [35] but has also been seen in relatively high concentration for lightly Ge-doped PAMBE material [7]. Taken together, these observations point toward a source that involves intrinsic point defects, the formation of which strongly depend on growth condition and on the Fermi level, with the relative degree of sensitivity to either being unclear at the moment. While it is difficult to resolve the specific physical source of the $E_C - 2.0$ eV trap, we note that theoretical calculations have predicted energy levels close to this trap for O_i, V_O, and V_{Ga} defects, all of which would be sensitive to high energy particle radiation and likely to growth conditions [11, 16, 21]. Work is continuing to make this identification at present.

Other than the above discussion, there are several other traps at various energy levels, whose source identification is too early or too confusing to be useful at this point in time. This includes the traps seen at $E_C - 0.18$, $E_C - 0.3$, $E_C - 1.1$, $E_C - 1.4$, and $E_C - 3.2$ eV. Each of these states demonstrate varying degrees of sensitivity to growth conditions, doping, and in some cases, high energy particle irradiation, but none have received enough systematic evaluation to initiate the process of source identification with any degree of confidence at this point, and more work is needed.

One of the challenges of assigning physical sources to deep levels in general for β-Ga_2O_3 is related to the β-gallia structure of β-Ga_2O_3 itself, due to its low crystal symmetry along with Ga and O having two bonding configurations. This means that DFT calculations will result in many possible bandgap states for vacancies, interstitials, antisites, and impurities. At the time of this writing, extensive efforts are ongoing to continue to unravel the source identification for observed defect states in β-Ga_2O_3 and the implications of specific defects on device characteristics. These efforts are largely based on expanding interactions between theoretical calculations, broader ranges of growth studies, and involving many research groups worldwide.

24.6 Conclusions

Identification of the physical sources of defects is an ongoing process, but great strides have been made by the community in the past several years. Not only have trap states been identified in a wide range of β-Ga_2O_3 prepared by many methods, but many of the same defect states are being seen by many groups over a wide range of materials sources. This has increased the confidence in assigning possible physical sources in some cases, and it has helped to refine and validate theoretical predictions. In other cases, many of the findings have opened up fundamental questions about potential physical sources for observed traps. The work thus far is pointing to the need for continued improvement in material quality and optimization of growth. This need will only grow as the range of epitaxial growth methods expands and matures, with more groups using methods such as metal organic chemical vapor deposition (CVD), low-pressure CVD (LPCVD) and plasma-enhanced CVD to grow β-Ga_2O_3, and due to the need to characterize defects within β-Ga_2O_3 based heterostructures and β-$(Al,Ga)_2O_3$ alloys. β-Ga_2O_3 devices have been achieving new records in metrics like maximum current density, low gate leakage, and breakdown voltage, but these are still far from theoretically-predicted and practically-achievable values. Therefore, there are many optimization strategies needed here and many revolve around device design and material improvements. As was shown, issues like threshold voltage instability due to traps are still a problem, but other concerns like dynamic on-resistance, also due to traps, are problems as well. Optimization strategies related to the semi-insulating β-Ga_2O_3 substrate must be considered for the mitigation of trapping effects, including not only adjustment of the space between the active device and the substrate, but perhaps even the choice of impurity to be used to achieve the semi-insulating property of an ideal transistor substrate. As larger substrates become available, growth uniformity over large will be required, and this will become a challenge for defect control. Thus, the characterization of defects, their association with physical sources, and their direct connections to device characteristics will continue to be an important part of advancing β-Ga_2O_3 from an exciting area of semiconductor research to a bona fide device technology.

Acknowledgements The authors gratefully acknowledge financial support from the following sources: AFOSR via the GAME MURI Program, Grant No. FA9550-18-1-0479, and Grant No. FA9550-18-1-0059, both managed by Ali Sayir; and DTRA Grant No. HDTRA1-17-10034, managed by Jacob Calkins. The authors also acknowledge the students and postdoctoral researchers at The Ohio State University for their work, James S. Speck's group at the University of California, Santa Barbara for providing samples, Siddharth Rajan's group at The Ohio State University for providing the transistors in this study, and Gregg Jessen's team at the Air Force Research Laboratory, WPAFB, OH for measurements and substrates.

References

1. M. Orita, H. Ohta, M. Hirano, H. Hosono, Appl. Phys. Lett. **77**, 4166 (2000)
2. T. Onuma, S. Saito, K. Sasaki, T. Masui, T. Yamaguchi, T. Honda, M. Higashiwaki, Jpn. J. Appl. Phys. **54**, 112601 (2015)
3. E.G. Víllora, K. Shimamura, K. Kitamura, K. Aoki, Appl. Phys. Lett. **88**, 031105 (2006)
4. J.B. Varley, A. Janotti, C. Franchini, C.G. Van de Walle, Phys. Rev. B **85**, 081109 (2012)
5. H. He, R. Orlando, M.A. Blanco, R. Pandey, E. Amzallag, I. Baraille, M. Rérat, Phys. Rev. B **74**, 195123 (2006)
6. P. Blood, J.W. Orton, *The Electrical Characterization of Semiconductors: Majority Carriers and Electron States* (Academic Press, San Diego, 1992)
7. E. Farzana, E. Ahmadi, J.S. Speck, A.R. Arehart, S.A. Ringel, J. Appl. Phys. **123**, 161410 (2018)
8. A. Chantre, G. Vincent, D. Bois, Phys. Rev. B **23**, 5335 (1981)
9. R. Passler, J. Appl. Phys. **96**, 715 (2004)
10. G. Lucovsky, Solid State Commun. **3**, 299 (1965)
11. J.B. Varley, J.R. Weber, A. Janotti, C.G. Van de Walle, Appl. Phys. Lett. **97**, 142106 (2010)
12. K. Irmscher, Z. Galazka, M. Pietsch, R. Uecker, R. Fornari, J. Appl. Phys. **110**, 063720 (2011)
13. Z. Zhang, E. Farzana, A.R. Arehart, S.A. Ringel, Appl. Phys. Lett. **108**, 052105 (2016)
14. E. Farzana, M.F. Chaiken, T.E. Blue, A.R. Arehart, S.A. Ringel, APL Mater. **7**, 022502 (2018)
15. A.M. Armstrong, M.H. Crawford, A. Jayawardena, A. Ahyi, S. Dhar, J. Appl. Phys. **119**, 103102 (2016)
16. P. Deák, Q. Duy Ho, F. Seemann, B. Aradi, M. Lorke, T. Frauenheim, Phys. Rev. B **95**, 075208 (2017)
17. S. Yamaoka, M. Nakayama, Phys. Status Solidi C **13**, 93 (2016)
18. S. Yamaoka, Y. Furukawa, M. Nakayama, Phys. Rev. B **95**, 094304 (2017)
19. O.F. Schirmer, J. Phys.: Condens. Matter **18**, R667 (2006)
20. E. Farzana, A. Mauze, J.B. Varley, T.E. Blue, J.S. Speck, A.R. Arehart, S.A. Ringel, APL Mater. **7**, 121102 (2019)
21. H. Peelaers, J.L. Lyons, J.B. Varley, C.G. Van de Walle, APL Mater. **7**, 022519 (2019)
22. J.F. McGlone, Z. Xia, Y. Zhang, C. Joishi, S. Lodha, S. Rajan, S.A. Ringel, A.R. Arehart, IEEE Electron Device Lett. **39**, 1042 (2018)
23. J.F. McGlone, Z. Xia, C. Joshi, S. Lodha, S. Rajan, S.A. Ringel, A.R. Arehart, Appl. Phys. Lett.

24. N. Moser, J. McCandless, A. Crespo, K. Leedy, A. Green, A. Neal, S. Mou, E. Ahmadi, J. Speck, K. Chabak, N. Peixoto, G. Jessen, IEEE Electron Device Lett. **38**, 775 (2017)
25. M.H. Wong, A. Takeyama, T. Makino, T. Ohshima, K. Sasaki, A. Kuramata, S. Yamakoshi, M. Higashiwaki, Appl. Phys. Lett. **112**, 023503 (2018)
26. Z. Xia, C. Joishi, S. Krishnamoorthy, S. Bajaj, Y. Zhang, M. Brenner, S. Lodha, S. Rajan, IEEE Electron Device Lett. **39**, 568 (2018)
27. M.E. Ingebrigtsen, J.B. Varley, A.Y. Kuznetsov, B.G. Svensson, G. Alfieri, A. Mihaila, U. Badstübner, L. Vines, Appl. Phys. Lett. **112**, 042104 (2018)
28. A.Y. Polyakov, N.B. Smirnov, I.V. Shchemerov, E.B. Yakimov, J. Yang, F. Ren, G. Yang, J. Kim, A. Kuramata, S.J. Pearton, Appl. Phys. Lett. **112**, 032107 (2018)
29. M.E. Ingebrigtsen, A.Y. Kuznetsov, B.G. Svensson, G. Alfieri, A. Mihaila, U. Badstübner, A. Perron, L. Vines, J.B. Varley, APL Mater. **7**, 022510 (2018)
30. W. Sun, J. Joh, S. Krishnan, S. Pendharkar, C.M. Jackson, S.A. Ringel, A.R. Arehart, IEEE Trans. Electron Devices **66**, 890 (2019)
31. A.Y. Polyakov, N.B. Smirnov, I.V. Shchemerov, D. Gogova, S.A. Tarelkin, S.J. Pearton, J. Appl. Phys. **123**, 115702 (2018)
32. J. Kim, S.J. Pearton, C. Fares, J. Yang, F. Ren, S. Kim, A.Y. Polyakov, J. Mater. Chem. C **7**, 10 (2019)
33. A.Y. Polyakov, N.B. Smirnov, I.V. Shchemerov, E.B. Yakimov, S.J. Pearton, F. Ren, A.V. Chernykh, D. Gogova, A.I. Kochkova, ECS J. Solid State Sci. Technol. **8**, Q3019 (2019)
34. A.Y. Polyakov, N.B. Smirnov, I.V. Shchemerov, S.J. Pearton, F. Ren, A.V. Chernykh, P.B. Lagov, T.V. Kulevoy, APL Mater. **6**, 096102 (2018)
35. J.M. Johnson, Z. Chen, J.B. Varley, C.M. Jackson, E. Farzana, Z. Zhang, A.R. Arehart, H.-L. Huang, A. Gene, S.A. Ringel, C.G. Van de Walle, D.A. Muller, J. Hwang, Phys. Rev. **X**, 9, 041027 (2019)
36. H. Ghadi, J.F. McGlone, C.M. Jackson, E. Farzana, Z. Feng, A.F.M. Bhuiyan, H. Zhao, A. R. Arehart, and S.A. Ringel, APL Mater. **8**, 021111 (2020)

Chapter 25
Electrical Properties 4

Band Offsets and Interface State Density Characterization of Dielectric/Ga_2O_3 Interfaces

Marko J. Tadjer, Virginia D. Wheeler and David I. Shahin

Abstract This chapter reviews recent literature on band offsets of dielectrics and semiconductors to gallium oxide and its ternary alloy, aluminum gallium oxide. Band diagram principles are reviewed, and an X-Ray photoelectron spectroscopy method for accurate band offset measurement is discussed. Interface state density measurements of Ga_2O_3 MOS capacitors are reviewed, and Terman method results for the HfO_2/β-Ga_2O_3 and ZrO_2/β-Ga_2O_3 are presented. Fowler-Nordheim tunneling model was used to fit the leakage current for the HfO_2/β-Ga_2O_3 MOS structure. A need for new dielectrics with both wide bandgap and high dielectric constant is noted, and possible new directions for research are presented.

25.1 Introduction

The monoclinic, β-phase of Ga_2O_3 is an attractive next generation semiconductor for high power and high temperature electronics due to its ultra-wide bandgap of 4.85 eV, high theoretical critical electric field of 8 MV/cm, and the availability of large-area, inexpensive bulk substrates grown from the melt [1, 2]. These fundamental properties, combined with recent major advances in substrate and epitaxial growth, have motivated intensive research of novel ultra-wide-bandgap Ga_2O_3 Schottky barrier diodes and transistor devices [3–11]. The latter in particular requires high-quality dielectric/Ga_2O_3 interfaces to maximize capacitive coupling with Ga_2O_3 without significant off-state gate leakage and achieve optimum device performance, particularly in the presence of high electric fields. The problem of

M. J. Tadjer (✉) · V. D. Wheeler
Electronics Science and Technology Division, Power Electronics Branch, U.S. Naval Research Laboratory, Washington, DC 20375, USA
e-mail: marko.tadjer@nrl.navy.mil

D. I. Shahin
Department of Materials Science and Engineering, University of Maryland, College Park, MD 20742, USA

M. Higashiwaki and S. Fujita (eds.), *Gallium Oxide*, Springer Series in Materials Science 293, https://doi.org/10.1007/978-3-030-37153-1_25

finding a suitable gate dielectric for emerging ultra-wide-bandgap materials becomes more challenging as the bandgap of the semiconductor starts to approach that of available dielectrics. The bandgap of SiO_2, 8.9 eV, is the largest among that of dielectrics considered viable for devices, as shown in Fig. 25.1 [12]. However, its low dielectric constant of 3.9 limits its use in scaled electronics due to increased tunneling current through the oxide [13]. For this reason, much of the Si industry has transitioned away from its native oxide toward high-*k* dielectrics such as HfO_2, where tunneling currents are reduced at low gate bias and Fowler-Nordheim tunneling is not prevalent [14]. From Fig. 25.1, it is evident that a trade-off exists between the dielectric constant and bandgap of dielectrics, which limits the options for Ga_2O_3 power devices to only a few viable materials. This limitation is an excellent opportunity for research into new dielectrics, such as composite nanolaminates, which could potentially combine high bandgap and high dielectric constant to address the gate dielectric needs for ultra-wide-bandgap devices.

Characterization of a dielectric/semiconductor interface is typically performed by complementary methods designed to assess the quality of the deposited dielectric and the characteristics of the interface. A variety of electrical characterization methods have been developed over the years based on current and capacitance measurements as a function of voltage, frequency, and temperature [15]. The operation of semiconductor devices, including semiconductor-semiconductor or oxide-semiconductor junctions, is often predicted by their band diagram, in which the energy levels in the components (E_G, E_C, E_V, E_F, etc.) are plotted as a function of physical space. Band diagrams help explain the influence of material properties, such as conductivity type and doping, on the transport characteristics of the particular device. Therefore, the ability to directly measure the band offsets in a junction is particularly useful. A generalized band diagram is shown in Fig. 25.2 for the case of a metal-semiconductor and a semiconductor-semiconductor junction with band bending ignored as it is typically only qualitatively drawn to reflect the doping level

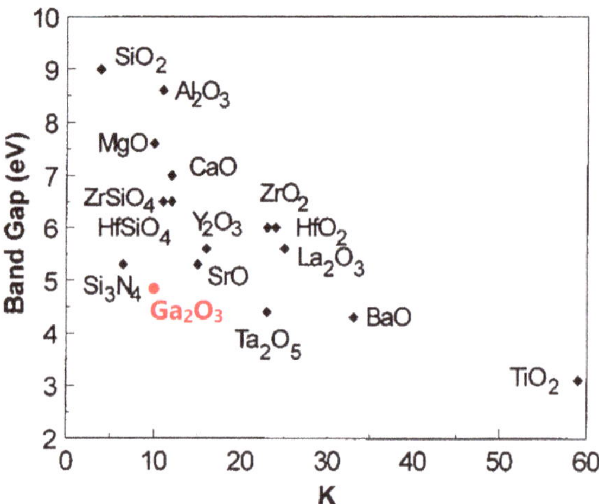

Fig. 25.1 Correlation of bandgap and dielectric constant of candidate dielectrics for β-Ga_2O_3. Reprinted with permission from Peacock and Robertson, J. Appl. Phys. 92, 4712 (2012), with the permission of AIP Publishing [12]

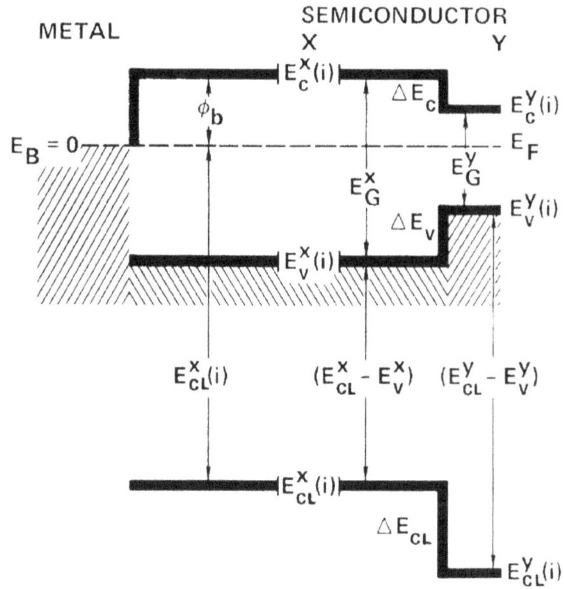

Fig. 25.2 A schematic of a flat-band diagram at a metal-semiconductor (left) or heterojunction (right) interface. Reprinted Fig. 25.1b with permission from [16]. Copyright 1980 by the American Physical Society

in the material and does not consider the effects of interface states and Fermi-level pinning, among others [16].

Experimentally, the band diagram can be determined by X-ray photoelectron spectroscopy (XPS), which measures the binding energy of core electrons ejected when a sample is irradiated with monoenergetic soft x-rays. The interaction between the X-rays and atoms throughout a limited penetration depth causes electrons to be emitted by the photoelectric effect. The specific binding energy of the emitted photoelectrons is unique to the particular material; thus, XPS is widely used in chemical analysis of various materials. Because XPS spectra also give information about differences in energy levels of the constituent atoms, information such as stoichiometry, bandgap, and band offsets is also possible to extract using XPS. At this time, interfaces of Ga_2O_3 with many dielectrics, as well as heterojunctions with other wide-bandgap semiconductors such as SiC and the binary III-nitrides, have been characterized for their band offsets and electrical properties when deposited on Ga_2O_3. Table 25.1 summarizes these results [17–37]. The XPS method to determine heterojunction band offsets was originally proposed by Kraut and coworkers in 1980 and has since become the de facto standard due to its high precision [16]. The next section in this chapter will explain Kraut's method using the HfO_2/β-Ga_2O_3 as an example; however, the same method was used for all band offset measurements summarized in Table 25.1. Of these materials, only a small subset has been employed as gate dielectrics in recent device reports of Ga_2O_3 MOSFETs and MOS capacitors.

Band diagrams are also quite useful when illustrating the presence and influence of traps on the device, e.g., either in the oxide or in the heterojunction interface.

Table 25.1 Reported values for band offsets for different materials on Ga_2O_3

Dielectric material (crystalline nature of Ga_2O_3)	Synthesis method	ΔE_C (eV)	ΔE_V (eV)	Alignment type	References
SiO_2 (single crystal)	PECVD	3.1 (±0.2)	1.0 (±0.2)	I	Konishi et al. [22]
SiO_2 (single crystal)	ALD	3.63–3.76	0.3–0.43	I	Jia et al. [23]
SiO_2 (single crystal)	ALD	2.9 (±0.7)	1.2 (±0.2)	I	Carey et al. [21]
Al_2O_3	ALD	1.5–1.6 (± 0.2)	0.7 (±0.2)	I	Kamimura et al. [24]
Γ-Al_2O_3 (s.c.)	PLD	1.9	0.5	I	Hattori et al. [25]
Al_2O_3 (single crystal)	ALD	2.23 (±0.2)	0.07 (±0.2)	I	Carey et al. [20]
Al_2O_3 (single crystal)	PVD	3.16 (±0.2)	−0.86 (±0.2)	II	
$LaAl_2O_3$ (single crystal)	PVD	2.01 (±0.60)	−0.21 (±0. 02)	II	Carey et al. [19]
Si (amorphous)	PLD	−0.2 (±0.1)	−3.5 (±0.1)	I	Chen et al. [26]
GaN (polycrystalline)	oxidation	−0.1 (±0.08)	−1.4 (±0.08)	I	Wei et al. [27]
6H-SiC (amorphous)	PVD	0.89 (±0.1)	−2.8 (±0.1)	II	Chang et al. [28]
ZrO_2 (single crystal)	ALD	1.2	−0.3 (±0.04)	II	Wheeler et al. [29]
HfO_2 (single crystal)	ALD	1.3	−0.5 (±0.04)	II	
$HfSiO_4$ (single crystal)	ALD	2.38 (±0.5)	0.02 (±0.003)	I	Carey et al. [21]
AZO (single crystal)	PVD	−0.79 (±0.34)	−0.61 (±0.023)	I	Carey et al. [18]
ITO (single crystal)	PVD	−0.32 (±0.34)	−0.78 (±0.34)	I	Carey et al. [17]
IZGO (single crystal)	Sputter	0.71 (±0.1)	0.49 (±0.05)	I	Huan et al. [34]
poly-In_2O_3	ALD	0.35	0.45	I	Sun et al. [30]
a-In_2O_3	ALD	0.39	0.79	I	
AlN	PEALD	1.39 (±0.1)	−0.09 (±0.1)	II	Chen et al. [31]
	T-ALD	0.58 (±0.1)	0.72 (±0.1)	I	
AlN	PLD of β-Ga_2O_3	−1.75 (±0.05)	−0.55 (±0.05)	II	Sun et al. [37]

(continued)

Table 25.1 (continued)

CuI-Ga_2O_3	Iodinated sputtered Cu	1.25 (±0.25)	−0.25 (±0.04)	II	Fares et al. [32]
CuI-$(Al_{0.14}Ga_{0.86})_2O_3$		1.85 (±0.35)	−0.05 (±0.01)	I	
InN	ALEp	3.35	0.55	I	Fares et al. [33]
$(Al_{0.27}Ga_{0.73})_2O_3/Ga_2O_3$	PLD	0.37 (±0.08)	0.52 (±0.08)	I	Wakabayashi et al. [35]
$(Al_{0.37}Ga_{0.63})_2O_3/Ga_2O_3$	PLD	0.02 (±0.07)	0.13 (±0.07)	I	
SiO_2-$(Al_{0.14}Ga_{0.86})_2O_3$	ALD	2.1 (±0.08)	1.6 (±0.40)	I	Fares et al. [36]

Reprinted with permission from [2], with the permission of AIP publishing, and extended with results published since then [2]

Thus, band offset measurements of novel MOS devices are typically followed by an assessment of the density of defects (a.k.a., traps) using a method applicable to the type of defect present. Historically, the initial ideas that lead to the concept of interface states date back to Bardeen's 1947 paper in which he proposed the existence of energy levels in the forbidden zone (the gap) at the surface of a semiconductor [38]. Studies of the GeO_2-Ge and SiO_2-Si interfaces in the following years provided the experimental test vehicle for quantifying surface states in semiconductors [39]. Improved understanding of these instabilities in silicon devices eventually lead to their elimination and the production of commercial quality thermal SiO_2 that lead to the foundation of the modern information age [40]. This vast body of research has been summarized in several works and is beyond the scope of this chapter [15, 41]. However, many of the methods initially developed for Si have found application in wide-bandgap semiconductors, yet often in revised form due to the larger bandgap.

While only SiO_2, Al_2O_3, and HfO_2 from the materials listed in Table 25.1 have been evaluated for their electrical performance (e.g., density of interface states, D_{it}) on β-Ga_2O_3, we note that Ga_2O_3 itself has been also studied as a dielectric on GaN [42]. Consideration of Ga_2O_3 as a dielectric is beyond the scope of this chapter. Two-dimensional semiconductors such as h-BN (E_G, E_C = 5.2 eV) and WSe_2 (*p*-type, E_G, E_C ~1 eV) have also been demonstrated as gate materials for nanomembrane Ga_2O_3 devices [43, 44]. However, the small area of these exfoliated materials has prevented their study using the traditional surface science methods. Thus, Table 25.2 summarizes results only from the aforementioned relevant dielectrics, along with additional details of dielectric deposition, surface pre-treatment, and post-deposition annealing [45–53]. Furthermore, in this chapter we report for the first time the electrical analysis of atomic-layer deposited (ALD) ZrO_2 on $(\bar{2}01)$ β-Ga_2O_3 substrates using sister samples from the previously reported band offset study [29]. Similar ALD-grown ZrO_2 thin films have been previously demonstrated

Table 25.2 Reported values for interface state density using different dielectrics deposited on β-Ga_2O_3

Dielectric	Dep. method	Thickness (nm)	Ga_2O_3	Clean	Post-dep. anneal	D_{it} (cm^{-2} eV^{-1})	References
Al_2O_3	ALD	20	($\bar{2}01$) or (010) Si-doped EFG substrate	Solvent clean	None, but 100-300 °C films were grown on (010) substrates	$2–20 \times 10^{11}$ for ($\bar{2}01$); $0.59–9.3 \times 10^{11}$ for (010); lowest D_{it} for 300 °C Al_2O_3 on (010)	Kamimura et al. [45]
SiO_2	ALD	20	Sn-doped EFG substrate	HCl or HF	Not reported	6×10^{11} for untreated 2×10^{12} for HF or HCl	Zeng et al. [46]
Al_2O_3	ALD	15	($\bar{2}01$) Sn-doped EFG sub.	H_2SO_4: H_2O_2, 1 min.	Not reported	2.3×10^{11}	Zhou et al. [47]
Al_2O_3	ALD	40	Si-doped	Acetone, H_2SO_4: H_2O_2, HF	500 °C, 2 min., O_2	7×10^{12}	Su et al. [48]
SiO_2	Thermal on Si	270	Sn-doped Ga_2O_3 nanomembrane FET	Solvent clean	N/A	$0.5–5.5 \times 10^{11}$ (donor) $2.1–11 \times 10^{11}$ (acceptor)	Bae et al. [49]
SiO_2	ALD	20	Si-doped MBE Ga_2O_3 on (010) Fe-doped Ga_2O_3 sub.	Not reported	Not reported	$<2.2 \times 10^{12}$	Chabak et al. [50]
Al_2O_3 on HfO_2	ALD	10/10	(100) EFG Ga_2O_3 substrate	Not reported	Not reported	$2.2–80 \times 10^{11}$	Dong et al. [51]
HfO_2 on Al_2O_3	ALD	10/10	(100) EFG Ga_2O_3 substrate	Not reported	Not reported	$1.0–80 \times 10^{11}$	Dong et al. [51]
SiO_2	ALD	20	MBE Sn-doped channel, UID buffer, Fe-doped (010) Ga_2O_3 sub.	Not reported	450 °C, 1 min.	$\sim 2.0 \times 10^{13}$	Zeng et al. [52]
HfO_2	ALD	40	($\bar{2}01$) EFG Ga_2O_3 substrate	UV-O_3, Buffered HF	None	1.3×10^{11}	Shahin et al. [53]

to enable enhancement mode operation in AlGaN/GaN HEMTs, which motivated its investigation as a potential gate dielectric for Ga_2O_3 devices [54].

25.2 Characterization of Dielectric-Semiconductor Interfaces

25.2.1 Band Offsets

As mentioned above, several methods are typically used for determining band offsets in newly reported heterojunctions. For example, capacitance-voltage methods allow for straightforward extraction of barrier height (x-intercept in $1/C^2$-V plot). For isotype heterojunctions (*n–n* or *p–p*), Kroemer's method is applicable to both abrupt and linearly graded junctions [55]. However, capacitance-voltage methods have been known to exhibit significant error due to their sensitivity to carrier traps [56]. Additional inaccuracies arise from errors in the positional determination of the metallurgical junction, often assuming an abrupt junction. For the same reasons, current-voltage rectification and deep-level transient spectroscopy (DLTS) face accuracy challenges to determining the band offsets in a heterojunction as well. Photocurrent measurements of a device as a function of incident energy (a. k.a., internal photoemission) have also been reported [57, 58]. Absorption and photoreflectance spectroscopy methods developed by Dingle and Shanabrook, respectively, are widely used to quantify intra-band transitions in quantum wells such as AlGaAs-GaAs [59, 60]. However, it was not until Kraut and coworkers proposed photoelectron spectroscopy as a highly precise method for determining band offsets in a heterojunction that a reliable method to address this important problem emerged [16]. Prior to that technique, band offset determinations had to be performed on devices, whereas XPS allowed for a nondestructive approach through the direct measurement of core energy level difference in the constituent materials.

Following a similar nomenclature as that in Fig. 25.2, the energy positions of core levels in each material in the heterostructure (E^X_{CL} and E^Y_{CL}), as well as the valence band maxima (E^X_{VBM} and E^Y_{VBM}), are directly measured with XPS. The valence band maxima (VBM) are determined from the intersection of the background energy and the linearly fitted valence band leading edge. The valence and conduction band offsets are then determined using the expressions below, where prior knowledge of the bandgap (ΔE_g) of each material is required.

$$\Delta E_V = \left(E^X_{core} - E^X_{VBM}\right) - \left(E^Y_{core} - E^Y_{VBM}\right) - \left(E^X_{core} - E^Y_{core}\right) \tag{25.1}$$

$$\Delta E_C = \Delta E^X_G - \Delta E^Y_G - \Delta E_V \tag{25.2}$$

Thus, XPS alone provides sufficient information to determine the valence band offsets, and the conduction band offsets can be easily determined once bandgaps are measured independently. Using this approach, we have experimentally determined the band offsets for the ZrO_2/Ga_2O_3 and the HfO_2/Ga_2O_3 heterojunctions [29]. Figure 25.3 shows the XPS data used for determining the valence band offsets for the HfO_2/Ga_2O_3 heterostructure (see also Table 25.1), and Fig. 25.4 shows the resulting band diagram. Further details of this experiment have been elucidated by Wheeler et al. [29]. Similarly, Konishi and Kamimura have determined the band offsets and the resulting band diagrams using Kraut's XPS experimental method for the SiO_2/Ga_2O_3 and Al_2O_3/Ga_2O_3 interfaces [22, 24]. Kamimura has gone a step further and has solved the Poisson equation for the MOS capacitor structure shown in Fig. 25.5 using an Au contact to the Al_2O_3 film. Interface state density (D_{it}) analysis was also performed in Kamimura's report, showing that the formation of crystalline Al_2O_3 near the interface upon a 300 °C anneal suppressed traps at the interface (Table 25.2). These experiments demonstrate that this XPS technique,

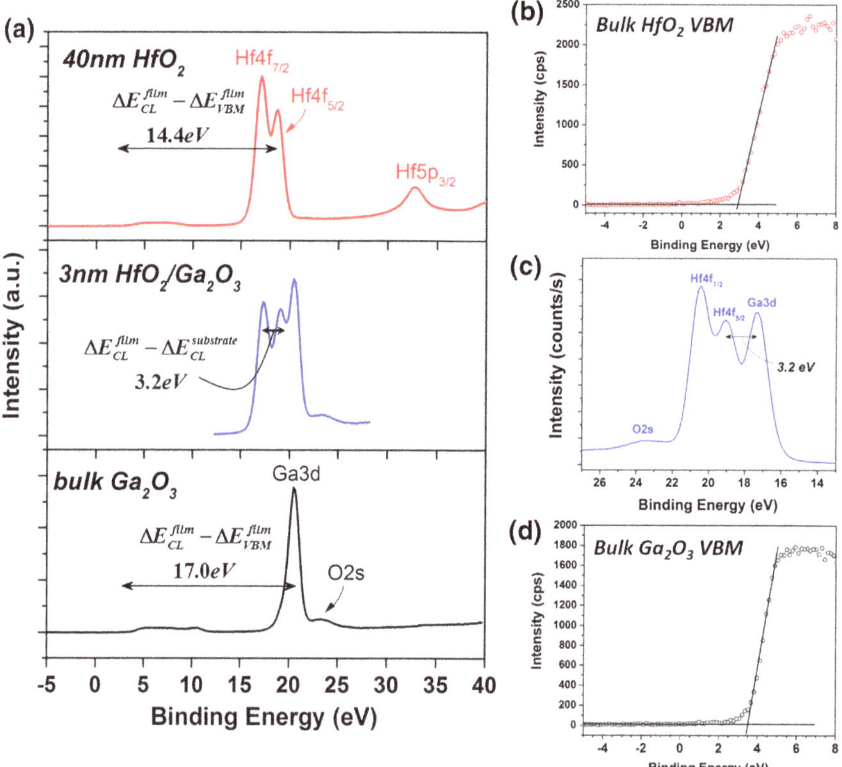

Fig. 25.3 XPS spectra used to calculate the VBO of the HfO_2/Ga_2O_3 heterojunction using **a** compiled spectra from the 3 nm HfO_2, 40 nm HfO_2, and bulk Ga_2O_3 measurements, **b** linear extrapolation for the VBM of 40 nm HfO_2 film, **c** zoomed-in region showing the HfO_2 4f/Ga_2O_3 3d core level peak split, and **d** the linear VBM extrapolation for the bare Ga_2O_3 substrate. Reprinted with permission from [29]. Copyright 2017, The Electrochemical Society [29]

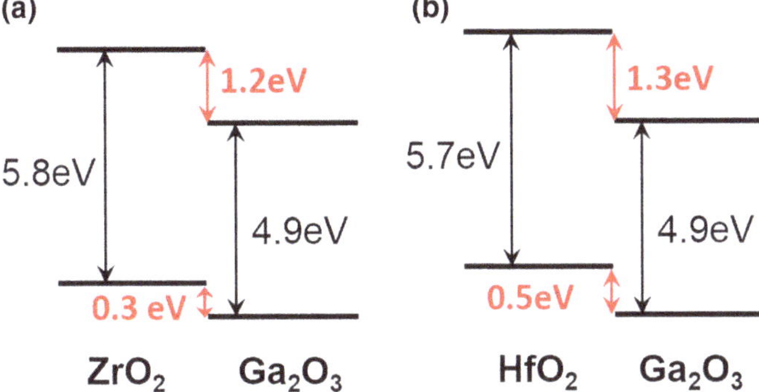

Fig. 25.4 Schematic of the band alignments for the heterojunctions between β-Ga_2O_3 and (**a**) ALD ZrO_2 or (**b**) ALD HfO_2 dielectrics. Reprinted with permission from [29]. Copyright 2017, The Electrochemical Society [29]

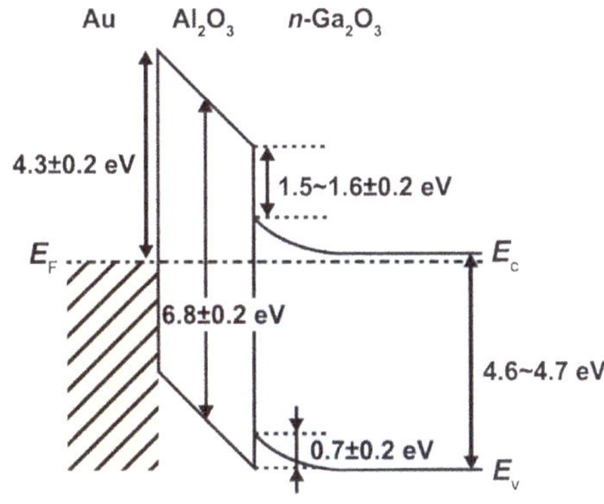

Fig. 25.5 Band diagram of $Au/Al_2O_3/n\text{-}Ga_2O_3$ calculated by Poisson simulation using the extracted band offsets and a typical bandgap of 4.6–4.7 eV for Ga_2O_3. Reprinted with permission from [24], with the permission of AIP Publishing [24]

coupled with current and capacitance-based electrical characterization methods, forms a powerful set of methods for the understanding of heterojunction devices.

25.2.2 Capacitance-Voltage Characterization of HfO_2/Ga_2O_3 and ZrO_2/Ga_2O_3 Interfaces

A number of electrical studies characterizing dielectrics deposited on Ga_2O_3 have been published recently, as shown in Table 25.2. For these experiments, the fabrication of MOS capacitors, after various oxide depositions, was performed in a nearly identical manner among all groups, using Ti/Au contacts in almost all cases. Similarly to these previous reports, Shahin et al. also used a top-side contact on the HfO_2 dielectric and a Ti/Au backside Ohmic contact to the unintentionally Si-doped $(\bar{2}01)$ Ga_2O_3 substrate [53]. The top-side Au contact was chosen without the typical

Ti adhesion layer in order to minimize any undesirable reaction between the reactive Ti with O atoms from the HfO_2, thus changing its stoichiometry and charge characteristics. To preserve the as-deposited HfO_2/Ga_2O_3 interface, no subsequent annealing was performed and a large-area Ti/Au backside Ohmic contact was employed to ensure sufficiently low contact resistance of the as-deposited metal stack. A schematic of the fabricated structure is shown in Fig. 25.6 [53]. Similar MOS capacitors were also fabricated on ZrO_2/Ga_2O_3 structures.

Capacitance-voltage measurements were performed between the two contacts of the identically processed HfO_2/Ga_2O_3 and ZrO_2/Ga_2O_3 MOS capacitors at room temperature using a 10, 100, and 1000 kHz measurement frequency. The C–V and $1/C^2$–V characteristics are shown in Figs. 25.7 and 25.8, respectively. In both cases, almost no hysteresis was observed independent of measurement frequency; however, significant frequency dispersion in the accumulation capacitance was measured for the ZrO_2/Ga_2O_3 sample. Comparison with calculated theoretical depletion capacitance (25.3) revealed about 1 V positive shift for the threshold voltage, which was consistent with the presence of negative charge in the oxide estimated to have a density of about 3×10^{-7} C/cm^2 [61].

$$C_D = \frac{\varepsilon_{Ga_2O_3}}{\sqrt{2}L_D}\left[\frac{\exp\left(\frac{q\varphi_s}{k_BT}\right) - 1 - \left(\frac{n_i}{N_D}\right)^2 \exp\left(\frac{-q\varphi_s}{k_BT}\right)}{\left[-\frac{q\varphi_s}{k_BT} + \exp\left(\frac{q\varphi_s}{k_BT}\right) - 1 + \left(\frac{n_i}{N_D}\right)^2\left[\exp\left(\frac{-q\varphi_s}{k_BT}\right) - 1\right]\right]^{\frac{1}{2}}}\right] \tag{25.3}$$

In (25.3), ε is the permittivity of the HfO_2 (or ZrO_2), φ_s is the surface potential, n_i is the intrinsic carrier concentration of Ga_2O_3, N_D is the doping density of the Ga_2O_3 substrate, and L_D is the Debye length. Detailed derivation of (25.3) is given in [61]. The surface potential φ_s was obtained by solving the following equation

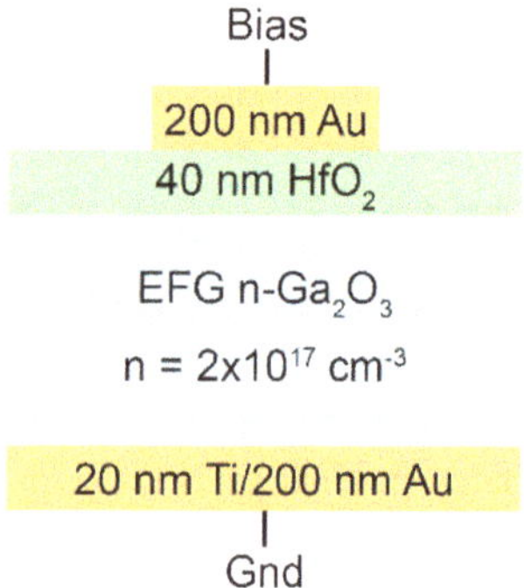

Fig. 25.6 HfO_2/Ga_2O_3 MOS capacitor structure employed in this work. Reprinted with permission from [53], with the permission of AIP Publishing [53]

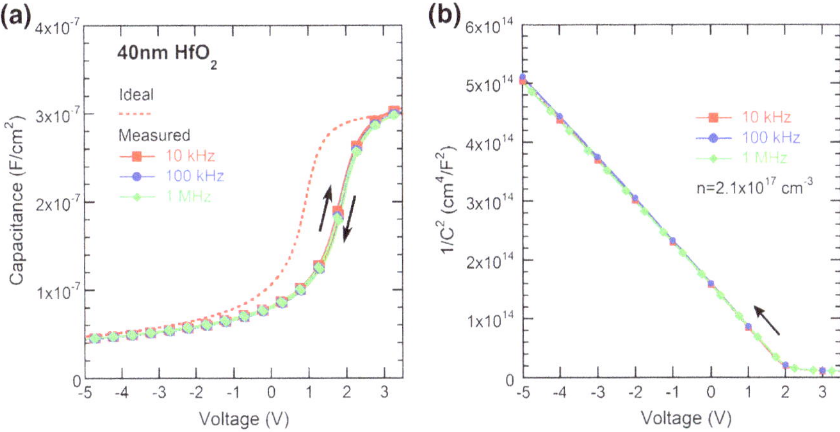

Fig. 25.7 **a** Measured and calculated ideal C–V curves for 40 nm HfO_2 on $(\bar{2}01)$ β-Ga_2O_3, indicating positive shifts in the C–V curves and extremely low hysteresis and stretch-out. **b** The corresponding $1/C^2$–V plot for C–V sweeps from 3.5 to −5 V, yielding an apparent carrier density of 2.1×10^{17} cm^{-3}. Reprinted with permission from [53], with the permission of AIP Publishing [53]

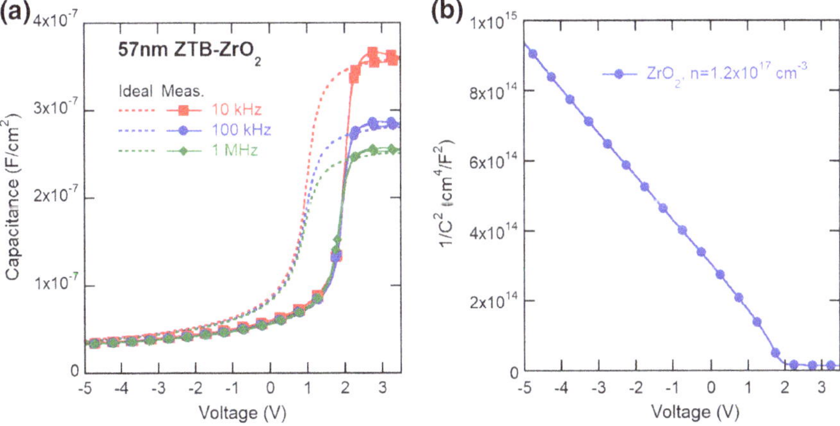

Fig. 25.8 **a** Measured and ideal C–V curves for 57 nm ZrO_2 on $(\bar{2}01)$ β-Ga_2O_3, indicating positive shifts in the C–V curves and extremely low hysteresis. **b** $1/C^2$–V plots and average apparent carrier densities for ZrO_2/Ga_2O_3 MOS capacitor. Nonuniform surface depletion effects for the ZrO_2/Ga_2O_3 can be observed from the deviations from linearity between +1 V and +2 V

$$V - V_{fb} = \pm\frac{1}{C_{ox}}\left[\sqrt{2}\varepsilon_{Ga_2O_3}\left(\frac{k_BT}{qL_D}\right)\left[-\frac{q\varphi_s}{k_BT} + \exp\left(\frac{q\varphi_s}{k_BT}\right) - 1 + \left(\frac{n_i}{N_D}\right)^2\left[\exp\left(\frac{-q\varphi_s}{k_BT}\right) - 1\right]\right]^{\frac{1}{2}}\right] + q\varphi_s \quad (25.4)$$

where V_{fb} is the theoretical value for the flat-band voltage, and C_{OX} is the oxide capacitance. The first term in (25.4) represents the voltage drop across the oxide and the second term ($q\varphi_s$) the voltage drop across the semiconductor (i.e., surface potential in a series capacitance model).

25.2.3 Interface State Density Analysis

As discussed earlier, a number of methods are available for interface state density analysis in MOS capacitor structures. Similarly to [53], this work uses the Terman method to report characterization of the ZrO_2/Ga_2O_3 interface since it enables analysis using high-frequency C–V measurements at room temperature. The disadvantage is that it is only sensitive to shallow states within about 0.2–0.6 eV below the conduction band, where their density is typically highest, which can lead to overlooking of significant midgap states, particularly in an ultra-wide-bandgap material such as Ga_2O_3. D_{it} was calculated using the following equations for the interface state density (25.5) and trap energy (25.6),

$$D_{it} = \frac{C_{OX}}{q^2}\left(\frac{dV_g}{d\varphi_s} - 1\right) - \frac{C_s}{q} = \frac{C_{OX}}{q^2}\frac{d\Delta V_g}{d\varphi_s} \quad (25.5)$$

$$E_{trap} = \frac{E_g}{2} + \varphi_s + \left(\frac{k_BT}{q}\right)\ln\left(\frac{N_D}{n_i}\right) \quad (25.6)$$

where C_{OX} is the oxide capacitance, C_S is the depletion capacitance, $\Delta V_G = V_{G,measured} - V_{G,ideal}$, φ_S is the surface potential, and N_D is the substrate carrier concentration. Figure 25.9 shows D_{it} as a function of E_{TRAP} for the HfO_2/Ga_2O_3 and the ZrO_2/Ga_2O_3 MOS capacitors. We note that the significant increase in D_{it} for the ZrO_2 compared to HfO_2 was due to the particular ALD precursor used in that growth that promoted an oxygen-rich film [29]. The HfO_2 films, however, yielded similar D_{it} compared to reports with the wider bandgap SiO_2 or Al_2O_3 as the gate dielectric.

The leakage current in the HfO_2/Ga_2O_3 sample was fitted using the Fowler-Nordheim tunneling model (25.7) in Fig. 25.10,

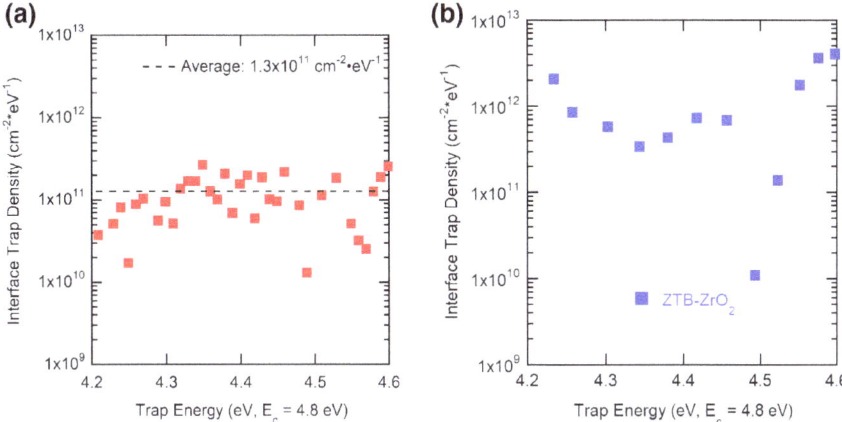

Fig. 25.9 Interface trap density (D_{it}) estimations using the Terman method for **a** HfO_2/Ga_2O_3 (Reprinted with permission from [53], with the permission of AIP Publishing [53].) and **b** ZrO_2/Ga_2O_3 MOS capacitors, with an average D_{it} value of 1.3×10^{11} cm^{-2} eV^{-1} for E_c−0.6 V $\leq E_{trap} \leq E_C$−0.2 V

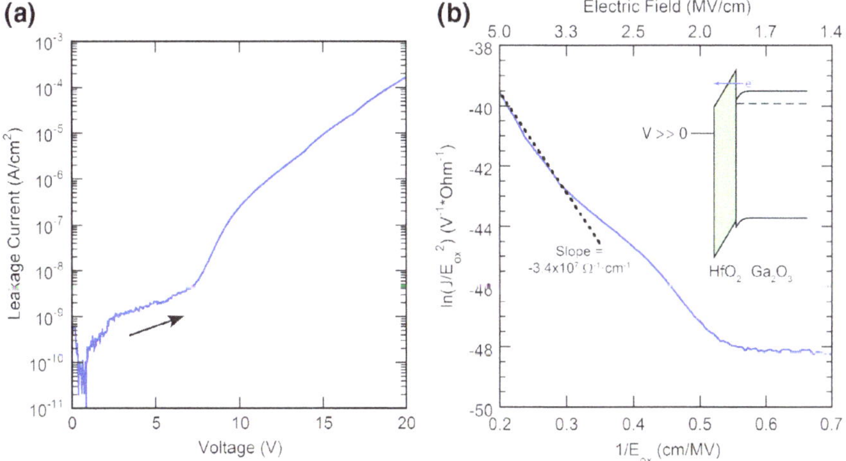

Fig. 25.10 **a** Forward bias leakage current measurement and **b** corresponding Fowler-Nordheim plot for 40 nm HfO_2 on Ga_2O_3. The slope of the $\ln(J/E_{ox}^2)-1/E_{OX}$ plot was used to extract a 1.3 eV CBO for the HfO_2/Ga_2O_3 that closely matches the CBO determined from XPS. This slope was extracted at the highest field (5 MV/cm) to ensure that the leakage current was due to true Fowler-Nordheim tunneling through a triangular potential barrier. Reprinted with permission from [53], with the permission of AIP Publishing [53]

$$J = q^2E_{\mathrm{ox}}^2/16\pi^2\hbar\phi_{\mathrm{ox}}\exp\left[-4\sqrt{2m^*}(q\phi_{\mathrm{ox}})^{3/2}/3\hbar qE_{\mathrm{ox}}\right] \quad (25.7)$$

where E_{OX} is the electric field in the oxide, m^* is the electron effective mass in HfO_2 taken as $0.11 \times m_0$, and ϕ_{OX} is the barrier height between oxide and the Ga_2O_3 substrate (equivalent to ΔE_C in the absence of band bending) [51]. The barrier height can be extracted from the slope of the $\ln(J/E_{OX}^2)$ versus $1/E$ox plot in Fig. 25.10b and in this case was found to be 1.3 eV, which agreed well with the conduction band offset measured by XPS (Fig. 25.4). Temperature-dependent I–V measurements were additionally performed for the case of a 4 nm thick HfO_2 deposited on the Ga_2O_3 sample. Scaled to the thickness of the HfO_2, it was possible to plot the I–V data for the 40 nm and 4 nm thick HfO_2 films as a function of electric field (Fig. 25.11a) and obtain very good agreement at room temperature. The temperature-dependent data for the 4-nm-thick film was then used to extract the barrier height in the 30–100 °C range (shown in Fig. 25.11b). The linear fit of the experimental barrier height data in Fig. 25.11b yields 1.3 eV at room temperature, as expected. An additional leakage mechanism was observed at low electric field, particularly at high temperature, the origin of which is still a matter of investigation but is likely to be assisted by tunneling through interfacial traps and the thin oxide barrier. This effect has been researched widely, particularly in the field of deeply scaled Si microelectronics, where any tunneling through the ultra-thin high-*k* gate dielectrics must be minimized to maintain reliable operation of the digital transistor circuits.

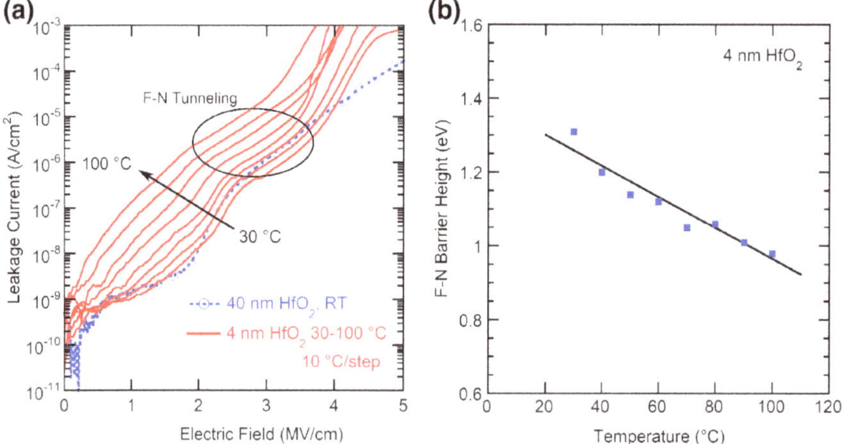

Fig. 25.11 **a** Forward leakage current comparisons between MOS capacitors with 40 nm (dashed line) and 3 nm (solid lines) HfO_2, normalized to applied electric field. A temperature dependence of the 3 nm film leakage between 30 and 100 °C is also shown to be especially prevalent at lower fields (< 2MV/cm). At intermediate fields around 3 MV/cm, a linear I–V region, circled in (**a**), was observed that fits well with a temperature-dependent Fowler-Nordheim tunneling current (**b**) and allows extrapolation of the expected 1.3 eV barrier height at room temperature. Reprinted with permission [53], with the permission of AIP Publishing [53]

25.3 Summary

This chapter introduced recent results on characterization of dielectric interfaces with β-Ga_2O_3. We have compiled a summary of the published literature to date (as of November 2018) on the band offsets of semiconductors and dielectrics deposited on β-Ga_2O_3. While Kraut's XPS method has yielded band offset results with excellent precision, interface state density characterization methods for Ga_2O_3 remains an area of active investigation. Deep levels in the 0.62−1 eV range below the conduction band, which are the topic of the next chapter, could prove a significant source of interface traps and thus might necessitate the use of complementary methods such as deep level transient or optical spectroscopy (DLTS/DLOS), photoionization spectroscopy, and cathodoluminescence which are able to probe electrically active states deeper into the gap than previously required.

A subset of dielectrics, namely SiO_2, Al_2O_3, and HfO_2, is considered as viable candidates for future Ga_2O_3 device applications as they offer favorable band offsets, low leakage current, and acceptable interface state densities. SiO_2 and Al_2O_3 are the widest bandgap dielectrics known and, therefore, their interfaces to Ga_2O_3 will be most relevant as passivation and field spreading layers for vertical β-Ga_2O_3 power devices. However, the trade-off between energy gap and dielectric constant presents a unique challenge for finding high-*k* dielectrics with sufficiently large band offsets for ultra-wide-bandgap Ga_2O_3 devices. We predict that high-*k* dielectrics such as HfO_2 and ZrO_2 will become more viable in scaled devices such as $(Al_xGa_{1-x})_2O_3$/Ga_2O_3 ternary heterostructures developed for RF power applications. Devices with a combination of a low-D_{it} HfO_2 or Al_2O_3 gate dielectric and a thick SiO_2 passivation layer are likely to be explored in the near future as well. Finally, high-quality nanolaminates combining the benefits of materials such as Al_2O_3 and TiO_2, or even a ternary alloy, could offer a path forward if appropriately engineered [62, 63]. It is therefore quite possible that the best dielectric for Ga_2O_3 is yet to be reported.

Acknowledgements The authors thank the NRL Institute for Nanoscience (Dr. Anthony Boyd, Walter Spratt, and Dean St. Amand) for processing equipment and support, as well as Dr. Chip Eddy (NRL) for proofreading this manuscript. Research at NRL was supported by the Office of Naval Research.

References

1. H. Aida, K. Nishigchi, H. Takeda, N. Aota, K. Sunakawa, Y. Yaguchi, Jpn. J. Appl. Phys. **47**, 8506 (2008)
2. S.J. Pearton, J. Yang, P.H. Cary IV, F. Ren, J. Kim, M.J. Tadjer, M.A. Mastro, Appl. Phys. Rev. **5**, 011301 (2018)
3. H. Zhang, R. Jia, Y. Lei, X. Tang, Y. Zhang, Y. Zhang, J. Phys. D Appl. Phys. **51**, 075104 (2018)
4. A. Jayawardena, R.P. Ramamurthy, A.C. Ahyi, D. Morisette, S. Dhar, Appl. Phys. Lett. **112**, 192108 (2018)

5. M.A. Bhuiyan, H. Zhou, R. Jiang, E.X. Zhang, D.M. Fleetwood, P.D. Ye, T.-P. Ma, IEEE Electron Device Lett. **39**(7), 1022 (2018)
6. K. Zeng, U. Singisetti, Appl. Phys. Lett. **111**, 122108 (2017)
7. H. Zhou, S. Alghmadi, M. Si, G. Qiu, P.D. Ye, IEEE Electron Device Lett. **37**(11), 1411 (2016)
8. L. Yuan, H. Zhang, R. Jia, L. Guo, Y. Zhang, Y. Zhang, Appl. Surf. Sci. **433**, 530 (2018)
9. H. Dong, W. Mu, Y. Hu, Q. He, B. Fu, H. Xue, Y. Qin, G. Jian, Y. Zhang, S. Long, Z. Jia, H. Lv, Q. Liu, X. Tao, M. Liu, AIP Adv. **8**, 065215 (2018)
10. Z. Hu, K. Nomoto, W. Li, N. Tanen, K. Sasaki, A. Kuramata, T. Nakamura, D. Jena, H.G. Xing, IEEE Electron Device Lett. **39**(6), 869 (2018)
11. M.H. Wong, Y. Nakata, A. Kuramata, S. Yamakoshi, M. Higashiwaki, Appl. Phys. Express **10**, 041101 (2017)
12. P.W. Peacock, J. Robertson, J. Appl. Phys. **92**, 4712 (2012)
13. J. Cai, C.-T. Sah, J. Appl. Phys. **89**, 2272 (2001)
14. W.J. Zhu, T.-P. Ma, T. Tamagawa, J. Kim, Y. Di, IEEE Electron Device Lett. **23**(2), 97 (2002)
15. D.K. Schroder, *Semiconductor Material and Device Characterization*, 3rd edn. (Wiley, New Jersey, 2006), pp. 342–363
16. E.A. Kraut, R.W. Grant, J.R. Waldrop, S.P. Kowalczyk, Phys. Rev. Lett. **44**, 1620 (1980)
17. P.H. Carey, F. Ren, D.C. Hays, B.P. Gila, S.J. Pearton, S. Jang, A. Kuramata, Appl. Surf. Sci. **422**, 179 (2017)
18. P.H. Carey IV, F. Ren, David C. Hays, B.P. Gila, S.J. Pearton, S. Jang, A. Kuramata, Vacuum **141**, 103 (2017)
19. P.H. Carey, F. Ren, D.C. Hays, B.P. Gila, S.J. Pearton, S. Jang, A. Kuramata, J. Vac. Sci. Technol., B **35**, 041201 (2017)
20. P. Carey, F. Ren, D.C. Hays, B.P. Gila, S.J. Pearton, S. Jang, A. Kuramata, Vacuum **142**, 52 (2017)
21. P. Carey, F. Ren, D.C. Hays, B.P. Gila, S.J. Pearton, S. Jang, A. Kuramata, Jpn. J. Appl. Phys. **56**, 071101 (2017)
22. K. Konishi, T. Kamimura, M.H. Wong, K. Sasaki, A. Kuramata, S. Yamakoshi, M. Higashiwaki, Phys. Status Solidi B **253**, 623 (2016)
23. Y. Jia, K. Zheng, J.S. Wallace, J.A. Gardella, U. Singisetti, Appl. Phys. Lett. **106**, 102107 (2016)
24. T. Kamimura, K. Sasaki, M.H. Wong, D. Krishnamurthy, A. Kuramata, T. Masui, S. Yamakoshi, M. Higashiwaki, Appl. Phys. Lett. **104**, 192104 (2014)
25. M. Hattori, T. Oshima, R. Wakabayashi, K. Yoshimatsu, K. Sasaki, T. Masui, A. Kuramata, S. Yamakoshi, K. Horiba, H. Kumigashira, A. Ohtomo, Jpn. J. Appl. Phys. **55**, 1202B6 (2016)
26. Z. Chen, K. Hishihagi, X. Wang, K. Saito, T. Tanaka, M. Nishio, M. Arita, Q. Guo, Appl. Phys. Lett. **109**, 102106 (2016)
27. W. Wei, Z. Qin, S. Fan, Z. Li, K. Shi, Q. Zhu, G. Zhang, Nanoscale Res. Lett. **7**, 562 (2012)
28. S.H. Chang, Z.Z. Chen, W. Huang, X.C. Liu, B.Y. Chen, Z.Z. Li, E.W. Shi, Chin. Phys. B **20**, 116101 (2011)
29. Virginia D. Wheeler, David I. Shahin, Marko J. Tadjer, Charles R. Eddy, Jr., ECS J. Solid State Sci. Technol. **6**, Q3052 (2017)
30. S.M. Sun, W.J. Liu, Y.P. Wang, Y.W. Huan, Q. Ma, B. Zhu, S.D. Wu, W.J. Yu, R.H. Horng, C.T. Xia, Q.Q. Sun, S.J. Ding, D.W. Zhang, Appl. Phys. Lett. **113**, 031603 (2018)
31. J.X. Chen, J.J. Tao, H.P. Ma, H. Zhang, J.J. Feng, W.J. Liu, C. Xia, H.L. Lu, D.W. Zhang, Appl. Phys. Lett. **112**, 261602 (2018)
32. C. Fares, F. Ren, D.C. Hays, B.P. Gila, M.J. Tadjer, K.D. Hobart, S.J. Pearton, Appl. Phys. Lett. **113**, 182101 (2018)
33. C. Fares, F. Ren, J. Woodward, N. Nepal, M.J. Tadjer, C.R. Eddy, Jr., S.J. Pearton, Appl. Phys. Lett., in preparation

34. Y.-W. Huan, X.-L. Wang, W.-J. Liu, H. Dong, S.-B. Long, S.-M. Sun, J.-G. Yang, S.-D. Wu, W.-J. Yu, R.-H. Horng, C.-T. Xia, H.-Y. Yu, H.-L. Lu, Q.-Q. Sun, S.-J. Ding, D.W. Zhang, Jpn. J. Appl. Phys. **57**, 100312 (2018)
35. R. Wakabayashi, M. Hattori, K. Yoshimatsu, K. Horiba, H. Kumigashira, A. Ohtomo, Appl. Phys. Lett. **112**, 232103 (2018)
36. C. Fares, F. Ren, E. Lambers, D.C. Hays, B.P. Gila, S.J. Pearton, J. Vac. Sci. Technol., B **36**, 061207 (2018)
37. H. Sun, C.G. Torres Castanedo, K. Liu, K.-H. Li, W. Guo, R. Lin, X. Liu, J. Li, X. Li, Appl. Phys. Lett. **111**, 162105 (2017)
38. J. Bardeen, Phys. Rev. **71**, 717 (1947)
39. H. Statz, L. Davis, Jr., G.A. deMars, Phys. Rev. **98**, 540 (1955)
40. B.E. Deal, U.S. Patent No. 3,426,422, 11 Feb 1969
41. S.M. Sze, *Physics of Semiconductor Devices*, 2nd edn. (Wiley, New York, 1981), pp. 379–390
42. H. Altuntas, I. Donmez, C. Ozgit-Akgun, N. Biyikli, J. Alloys Compd. **593**, 190 (2014)
43. J. Kim, M.A. Mastro, M.J. Tadjer, J. Kim, ACS Appl Mater. Interfaces **9**, 21322 (2017)
44. J. Kim, M.A. Mastro, M.J. Tadjer, J. Kim, ACS Appl Mater. Interfaces **10**, 29724 (2018)
45. T. Kamimura, D. Krishnamurthy, A. Kuramata, S. Yamakoshi, M. Higashiwaki, Jpn. J. Appl. Phys. **55**, 1202B5 (2016)
46. K. Zeng, Y. Jia, U. Singisetti, IEEE Electron Device Lett. **37**, 906 (2016)
47. H. Zhou, S. Alghamdi, M. Si, G. Qiu, P.D. Ye, IEEE Electron Device Lett. **37**, 1411 (2016)
48. C.Y. Su, T. Hoshii, I. Muneta, H. Wakabayashi, K. Tsutsui, H. Iwai, K. Kakushima, ECS Trans. **85**(7), 27 (2018)
49. H. Bae, J. Noh, S. Alghamdi, M. Si, P.D. Ye, IEEE Electron Device Lett. **39**(11), 1708 (2018)
50. K.D. Chabak, J.P. McCandless, N.A. Moser, A.J. Green, K. Mahalingam, A. Crespo, N. Hendricks, B.M. Howe, S.E. Tetlak, K. Leedy, R.C. Fitch, D. Wakimoto, K. Sasaki, A. Kuramata, G. Jessen, IEEE Electron Device Lett. **39**(1), 67 (2018)
51. H. Dong, W. Mu, Y. Hu, Q. He, B. Fu, H. Xue, Y. Qin, G. Jian, Y. Zhang, S. Long, Z. Jia, H. Lv, Q. Liu., X. Tao, M. Liu, AIP Adv. **8**, 065215 (2018)
52. K. Zeng, A. Vaidya, U. Singisetti, IEEE Electron Device Lett. **39**(9), 1385 (2018)
53. D.I. Shahin, M.J. Tadjer, V.D. Wheeler, A.D. Koehler, T.J. Anderson, C.R. Eddy Jr., A. Cristou, Appl. Phys. Lett. **112**, 042107 (2018)
54. T.J. Anderson, V.D. Wheeler, D.I. Shahin, M.J. Tadjer, A.D. Koehler, K.D. Hobart, A. Christou, F.J. Kub, C.R. Eddy Jr., Appl. Phys. Express **9**, 071003 (2016)
55. H. Kroemer, W.-Y. Chen, J.S. Harris Jr., D.D. Edwall, Appl. Phys. Lett. **36**, 295 (1980)
56. M.A. Rao, E.J. Caine, H. Kroemer, S.I. Long, D.I. Babic, J. Appl. Phys. **61**, 643 (1987)
57. G. Abstreiter, U. Prechtel, G. Weimann, W. Schlapp, Surf. Sci. **174**, 312 (1986)
58. M.A. Haase, M.J. Hafich, G.Y. Robinson, Appl. Phys. Lett. **58**, 616 (1991)
59. R. Dingle, in *Festkörperprobleme XV—Advances in Physics*, ed. by H.J. Queisser (Pergamon, Münster, Germany, 1975), p. 21
60. B.V. Shanabrook, O.J. Glembocki, W.T. Beard, Phys. Rev. B **35**(5), 2540 (1987)
61. H.C. Casey, *Devices for Integrated Circuits: Silicon and III-V Compound Semiconductors* (Wiley, New York, 1999), pp. 331–341
62. L.H. Kim, K. Kim, S. Park, Y.J. Jeong, H. Kim, D.S. Chung, S.H. Kim, C.E. Park, ACS Appl Mater. Interfaces **6**, 6731 (2014)
63. K.-T. Lee, C.-F. Huang, J. Gong, B.-H. Liou, IEEE Electron Device Lett. **30**(9), 907 (2009)

Chapter 26
Electrical Properties 5

Defects in β-Ga_2O_3 Crystals and Their Influence on Schottky Barrier Diode Characteristics

Makoto Kasu

Abstract β-Gallium oxide is promising for use in semiconductor power devices. First, the type and character of crystal defects, such as dislocations and voids, are described. Next, I describe the fabrication and measurement of Schottky barrier diodes (SBD) on the entire surface and investigate the relation between the leakage current and defects, as revealed mainly by the etch-pit method. The dislocations that appeared on the (010) surface became SBD leakage paths, whereas the dislocations on the $(\bar{2}01)$ and (001) surfaces apparently had no relation with the SBD leakage current. Voids that extend in [010] direction and appeared on all surface orientations did not affect the SBD leakage current.

26.1 Introduction

Gallium oxide (Ga_2O_3) is a wide-bandgap semiconductor with a bandgap of 4.7–5.2 eV [1–6] and is expected to be a next-generation power semiconductor material. Ga_2O_3 shows five crystalline structural phases: α, β, γ, δ, and ε. Among them, the β-phase is the most thermodynamically stable. β-Ga_2O_3 bulk single crystals can be grown by melt growth methods, including the floating-zone (FZ) method [7–13], edge-defined film-fed growth (EFG) method [14–16], Czochralski (CZ) method [17–19], and the vertical Bridgman (VB) method [20]. Therefore, a high growth rate is possible. The electron concentration in *n*-type Ga_2O_3 can be controlled in a wide range. These properties are advantages over other wide-bandgap semiconductors and have attracted much attention to Ga_2O_3.

In fact, Ga_2O_3-based Schottky barrier diodes (SBDs) and field-effect transistors have been demonstrated at the application level. Therefore, from a device viewpoint, studies of defect properties and their influence on the reliability of Ga_2O_3

M. Kasu (✉)
Department of Electrical and Electronic Engineering, Saga University,
1 Honjo-Machi, Saga 840-8502, Japan
e-mail: kasu@cc.saga-u.ac.jp

M. Higashiwaki and S. Fujita (eds.), *Gallium Oxide*, Springer Series
in Materials Science 293, https://doi.org/10.1007/978-3-030-37153-1_26

devices are essential. Therefore, in this chapter, I describe crystal defects such as dislocations and stacking faults, along with residual impurities and their relation to defect properties.

26.2 β-Ga_2O_3 Crystal Structure

β-Ga_2O_3 exhibits a monoclinic structure. Figure 26.1 shows the crystal structure viewed from the (010) face. The lattice constants are a = 1.223 nm, b = 0.304 nm, and c = 0.580 nm, and the angle between the a and c axes is β = 103.7° [21, 22]. The β-Ga_2O_3 cleave planes are (100) and (001). A Ga(1) atom bonds with one O(1) atom, two O(2) atoms, and one O(3) atom, resulting in four total bonds. A Ga(2) atom bonds with two O(1) atoms, one O(2) atom, and three O(3) atoms, totaling six bonds. However, an O(1) atom bonds with one Ga(1) atom and two Ga(2) atoms for a total of three bonds. An O(2) atom bonds with two Ga(1) atoms and one Ga(2) atom for a total of three bonds, and an O(3) atom bonds with one Ga(1) atom and three Ga(2) atoms for a total four bonds.

If the crystal structure in Fig. 26.1 is cut parallel to the [100] axis, the side plane is (001). If the crystal structure in the figure is cut parallel to the [001] axis, the side plane is (100). Notably, the [100] axis and the [001] axis are not perpendicular; they have an angle of 103.7°. Therefore, a direction perpendicular to the (001) plane is not [001] and a direction perpendicular to the (100) plane is not [100].

The crystal structure in Fig. 26.1 is cut perpendicular along [102], and the side plane is ($\bar{2}$01). Figure 26.2 shows the ($\bar{2}$01) crystal structure. On ($\bar{2}$01), along [010], Ga atoms and O atoms are stacked sequentially. When viewed from the surface, the

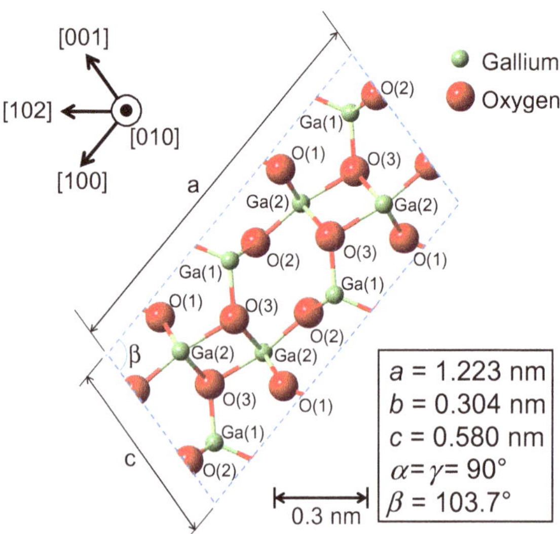

Fig. 26.1 Crystal structure of β-Ga_2O_3 viewed from the (010) surface

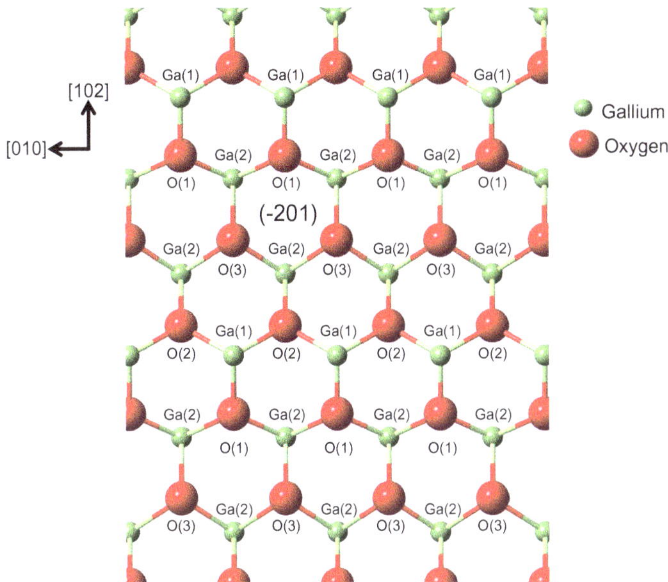

Fig. 26.2 Crystal structure of β-Ga_2O_3 viewed from the ($\bar{2}01$) surface

crystal structure is similar to (0001) and the lattice spacing between β-Ga_2O_3 and GaN is similar ($\bar{2}01$); Ga_2O_3 has been used as a substrate for a GaN epitaxial layer.

26.3 Experimental Procedure

β-Ga_2O_3 was grown by the EFG method at Novel Crystal Technology [16]. The melt was supplied through a die via capillary action to the seed crystal. The pulling direction was [010], and the as-grown crystal had a plane shape with a main plane of ($\bar{2}01$). In this work, the crystal was not intentionally doped. The grown crystal was cut along the (010), (001), or ($\bar{2}01$) plane, and the surface was treated by chemical mechanical polishing (CMP).

Figure 26.3 shows the Schottky barrier diode (SBD) structure fabricated on the (0$\bar{1}$0) surface. First, on the entire backside of the (0$\bar{1}$0) surface, i.e., the (010) face, Si-ion implantation was carried out and annealing was subsequently performed for 30 min at 950 °C. Then, Ti (100 nm thick)/Au (100 nm thick) were evaporated and ohmic contacts were formed. Next, Ti (10 nm thick)/Au (30 nm thick) were evaporated onto the (010) surface through a metal mask. SBD contacts 350 µm in diameter were formed in pixels on the entire surface. The distance between the nearest two SBD contacts is 650 µm

SBDs were first fabricated and measured; after etching of the SBD contacts, surface was then etched at 140 °C in H_3PO_4. The resultant etch pits and patterns

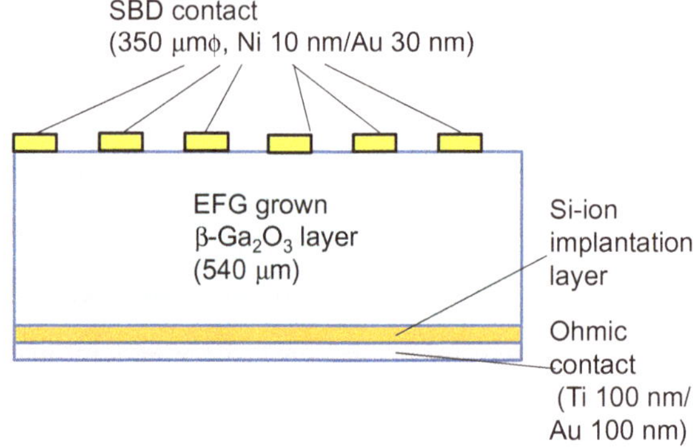

Fig. 26.3 Schottky barrier diode structure of β-Ga_2O_3

were observed by differential interference contrast microscopy (DICM), scanning electron microscopy (SEM), and atomic force microscopy (AFM).

26.4 Defect Observations

26.4.1 Unetched Pits on (010)

The (010) face was the high-growth-rate plane in EFG, and [010] was the pulling direction of EFG growth. Here, the surface before etching (unetched surface) is observed, and nanoscale groove-shaped unetched pits, denoted as type-G pits, are observed (Fig. 26.4) [23]. The type-G pits are longer in the [001] direction, and the length along [001] is in the range from 50 nm to 1.2 μm. The width in the [100] direction was approximately the same as 40 nm. Thus, the main side facet is the (100) plane.

The properties of type-G pits were subsequently investigated. The type-G pits extend in the [010] direction, perpendicular to the image, like a pipe. When the surface is etched further, the type-G pits are eliminated. The type-G pits were not observed at the same position on the backside. From these results, the type-G pits

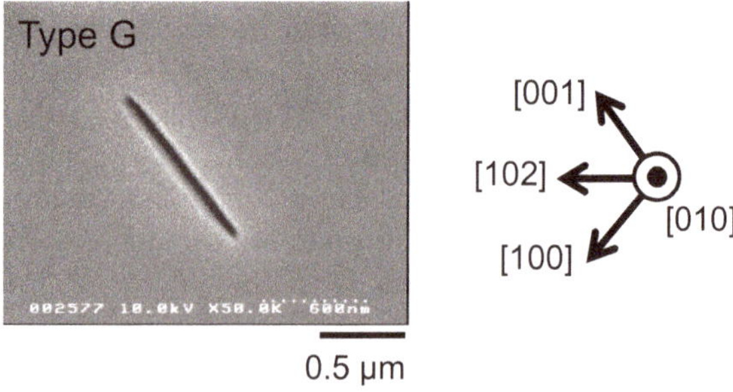

Fig. 26.4 Unetched pits observed on the (010) surface of as-received β-Ga_2O_3 (SEM image)

were concluded to be like a pipe that extends in the [010] direction (the growth direction) and terminates at both ends inside crystal; thus, the type-G pits are void defects, not like micropipes that appeared in SiC.

The type-G pit defects also exhibit an array-like character. In most cases, an array of type-G pits aligned in the [001] direction was observed, although some were aligned in the [100] direction. In the view of the cross section of (010) plane of EFG as-grown crystal, type-G voids are found to concentrate in the circumference of the as-grown crystal.

With further etching, the type-G pits changed to type-A etch pits and, as described in the next section, the type-G pits were eventually eliminated. Nakai et al. observed nanopipes on the (010) EFG-grown surface by transmission electron microscopy (TEM) and X-ray topography. Their nanopipes appear to be type-G pits [24]. Ueda et al. observed edge dislocations with **b** = [010] and void-originated type-G defects on the $(\bar{2}01)$ plane, along with microtwins with a twin plane of (100) [25].

Concerning the origin of the type-G pits, during EFG growth, bubbles in the melt are speculated to be solidarized at the growth front, i.e., the liquid/solid interface [23]. The shape of the type-G pits is determined by the slowest growth rate of specific surface orientation. The [010] direction is the fastest growth direction, and the void is therefore longer in the [010] direction. The [100] direction is the slowest growth direction; thus, the side facet of (100) mainly appears.

26.4.2 Void-Related and Dislocation-Related Etch Pits on the (010) Plane

Type-G pits are observed on the unetched surface; therefore, type-G pits are considered "unetch pits." In addition to type-G pits, various shapes of etch pits are observed on the etched surface. The etch pits are categorized into types A–F, as shown in Fig. 26.5 [26]. The etching times for the images in Fig. 26.5b–g differ.

Type-A etch pits contain a void at their center. In the direction perpendicular to the image (i.e., the [010] direction), each type-A pit consists of a two-stair structure: The lower structure consists of the (010) bottom plane and $(\bar{2}01)$ and $(20\bar{1})$ vertical planes, whereas the upper structure consists of $(41\bar{2})$ and $(\bar{4}12)$ facet planes. These orientated surfaces are considered thermodynamically stable.

Type-*B–D* etch pits have the shape of a rectangle; the ridgeline is one side and is not straight but slightly wandering. Type-B etch pits have two ridgelines; among four distorted facets, two are $(21\bar{1})$ and $(\bar{2}11)$.

Upon etching, the type-G pits change into type-A pits. With further etching, the pits transform as $A \rightarrow B \rightarrow C \rightarrow D$ [23]. These results show that the origin of type-*A–D* pits is the same as that of type-G pits, i.e., void defects. With etching, the thermally stable facets appear in sequence.

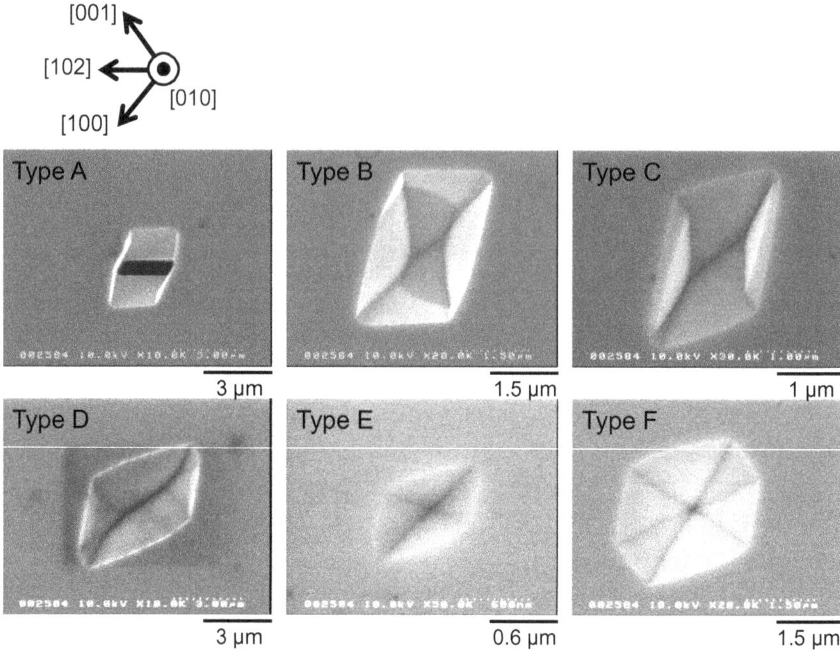

Fig. 26.5 Etch pits observed on the (010) β-Ga_2O_3 surface (SEM images). The etch pits were sorted into six groups (types *A*, *B*, *C*, *D*, *E*, and *F*) according to the shapes of the etch pits. The etching time was varied

Type-E etch pits exhibit a rectangular shape. They appear similar to type-*D* pits. However, the type-D pits have only one slightly wandering ridgeline, whereas type-E pits have two straight ridgelines. The four facets of the type-E pits are (111), ($\bar{1}1\bar{1}$), ($21\bar{1}$), and ($2\bar{1}\bar{1}$). Type-F is hexagonal-shaped and has three straight ridgelines. Its two facets are (310) and ($\bar{3}01$).

As etching proceeds, the shape of both type-E and type-F pits remains unchanged, although their size increases. This characteristic of types *E* and *F* is different from that of types *A*–*D*. Thus, type-E and type-F pits possess a core in their center, i.e., a threading dislocation. Thus, type-*E* and type-*F* pits originate from dislocation(s).

26.4.3 ($\bar{2}01$) Orientation Etch Pits and Pattern

On the ($\bar{2}01$) surface, line-shaped etch patterns, arrow-shaped etch pits, and gourd-shaped etch pits are observed (Figs. 26.6 and 26.7) [27]. Line-shaped etch patterns extend in the [010] direction; they are therefore etch pits that originate from voids. Thus, these pits share a common origin as type-G unetch pits (Fig. 26.4) and type-A–D etch pits on (010) (Fig. 26.5).

As shown in Fig. 26.6, the arrow of arrow-shaped etch pits points in the [102] direction. The arrow-shaped etch pits consist of slightly deep arrow-shaped facets; on both sides of it are two declined facets. AFM observations show that the angles between the respective declined facets and the surface are 13.2° and 5.4°; therefore, the declined facets are very shallow.

Figure 26.7 shows gourd-shaped etch pits, which point in the [102] direction. The cross section obtained by AFM shows that the slope angle is only 1.3°. Thus, the gourd-shaped etch pits are very shallow and therefore difficult to observe by SEM. Compared with the arrow-shaped etch pits, the gourd-shaped etch pits exhibit a rounder shape at their points.

26.4.4 The (001) Surface

On the (001) surface, line-shaped etch pits and arrow (bullet)-shaped etch pits are observed [28]. As shown in Fig. 26.8, the line-shaped etch pits extend in the [010] direction but their lengths vary; they were therefore concluded to have originated at the voids observed on (010). As the etching proceeds, the width in the [100] direction widens. In contrast, the arrow (bullet)-shaped etch pits are shown in the red square in Fig. 26.8, point in the [$\bar{1}$00] direction. The arrow-shaped etch pits would be due to the defects.

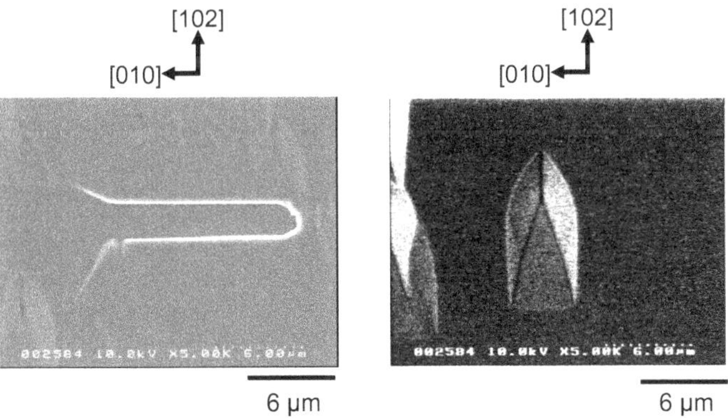

Fig. 26.6 Line-shaped etch pattern (left) and arrow-shaped etch pit (right) observed on the ($\bar{2}$01) β-Ga_2O_3 surface (SEM images)

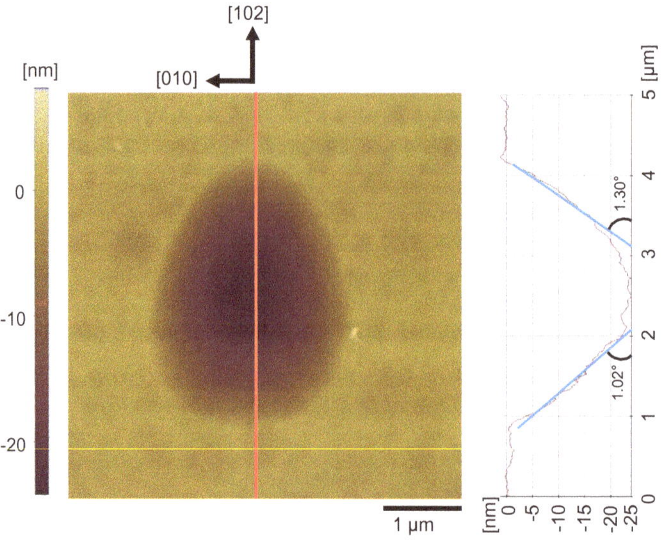

Fig. 26.7 Gourd-shaped Etch pits observed on the pits observed on the ($\bar{2}01$) β-Ga_2O_3 surface (AFM image and its cross section)

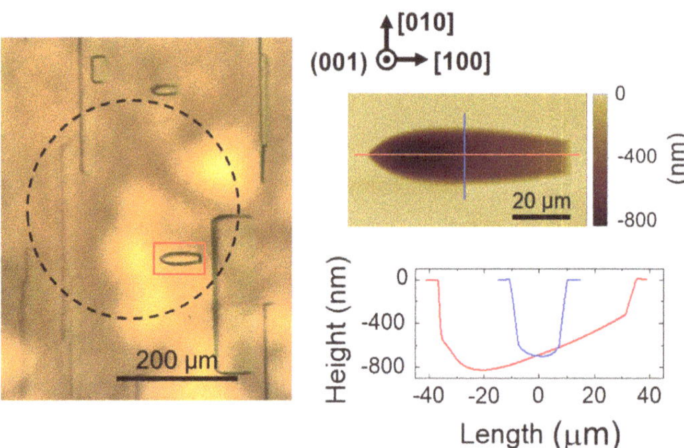

Fig. 26.8 Etch pits observed on the (001) β-Ga_2O_3 surface. The line-shaped etch pit originates from a void defect. The red square shows arrow (bullet)-shaped etch pits, which originate from defects

26.4.5 The (100) Surface Orientation

Ohba et al. reported the characteristics of a VB-grown (100) surface [29]. Their pulling direction of VB growth was perpendicular to the (100) face. The (100) face of the VB-grown crystal exhibits a low etch-pit density. The (100) surface is thermodynamically stable, and step-flow growth therefore occurs.

26.5 Relation with SBD Characteristics

26.5.1 The (010) Surface

A pixel of SBDs with the same separation was fabricated and their current–voltage (*I–V*) and capacitance–voltage (*C–V*) characteristics were measured [30]. The obtained forward *I–V* characteristics were fitted assuming a thermal field emission (TFE) model, and the ideality factor (n_s) and the energy barrier height (ϕ_B) were obtained:

$$I = SA^{*}T^{2}\exp\left(\frac{-q\phi_B}{k_BT}\right)\left\{\exp\left[\frac{q(V-IR_S)}{n_sk_BT}\right]-1\right\} \tag{26.1}$$

where *I*, *S*, *A**, and *V* are the current, SBD contact area, Richardson constant, and the applied voltage, respectively; *A** is assumed to be 41.1 A/cm^2. Parameter q is the electron charge. The ideality factor (n_s), Schottky barrier height (ϕ_B), and series resistance (R_S) determined by least-squares fitting are shown in the graph. Figure 26.9a shows the *I–V* characteristics of the SBD contact, which contains six type-F dislocation-related etch pits (density: 6×10^3 cm^{-2}). This figure shows the real *I–V* curve, where the ideality factor n_s is 1.08 in the forward direction and the reverse current is less than 1×10^{-10} A at a reverse bias of −15 V. Figure 26.9b shows the *I–V* characteristics of SBD with 40 dislocation-related etch pits (density: 4×10^2 cm^{-2}), the intrinsic diode component with a ϕ_B of 1.08 eV and an n_s of 1.10 and, on the lower-voltage side, a defect-related leakage current component with an ϕ_B of 0.83 eV and an n_s of 1.17. In the reverse direction, at −15 V, the reverse current was as high as 10^{-7} A. I therefore concluded that a relation exists between dislocations and the reverse leakage current.

Figure 26.10 presents an SBD map that shows the relation between the reverse current level at −15 V and the number of dislocation-related type-F etch pits. The diameter (size) of the circle represents the number of dislocation-related etch pits, and the color represents the reverse current level. The SBDs shown in the upper-left

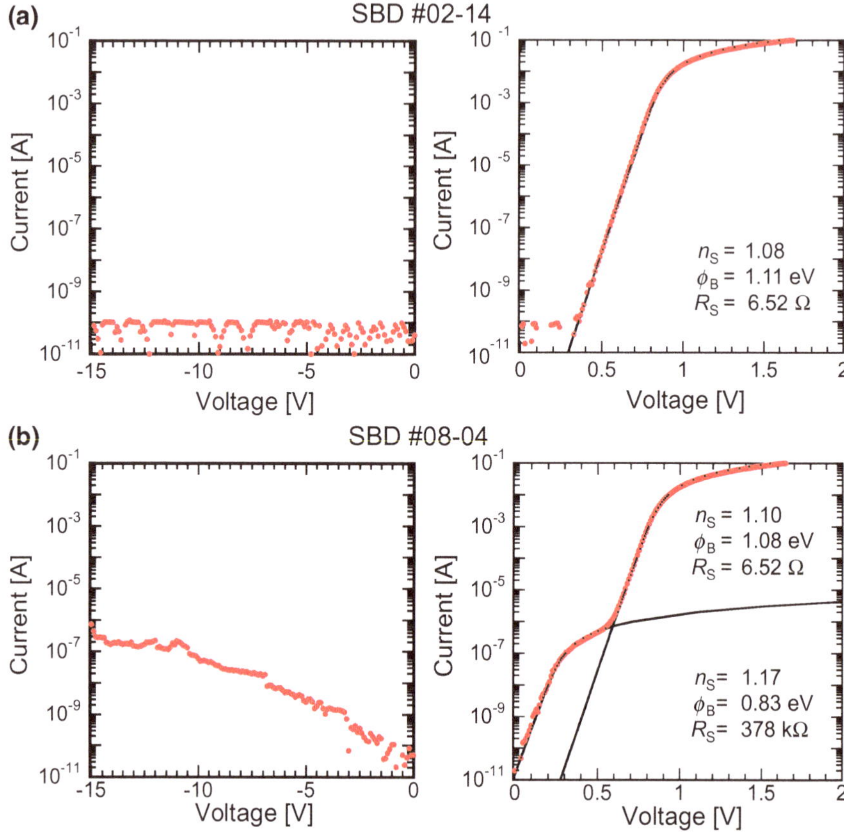

Fig. 26.9 Current–voltage characteristics of Schottky barrier diodes on the (010) β-Ga_2O_3 surface: **a** with no reverse and forward leakage currents (#02–14); **b** with high reverse and forward leakage currents (#08–04). Six and forty dislocation-related (type-F) etch pits are observed inside the Schottky electrode for (**a**) and (**b**), respectively. The forward *I–V* curve was fitted on the basis of the thermal field emission theory; the best-fitted parameters, including the ideality factor (n_S), Schottky barrier height (ϕ_B), and series resistance (R_S), are shown

corner of Fig. 26.10 exhibit a high dislocation density; the SBDs show a high reverse current level. Therefore, the relation between the dislocation-related etch pits and the reverse current level was confirmed.

On the other hand, the relation between the void-related etch pits and the reverse current level was similarly investigated, but no clear relation between them was confirmed.

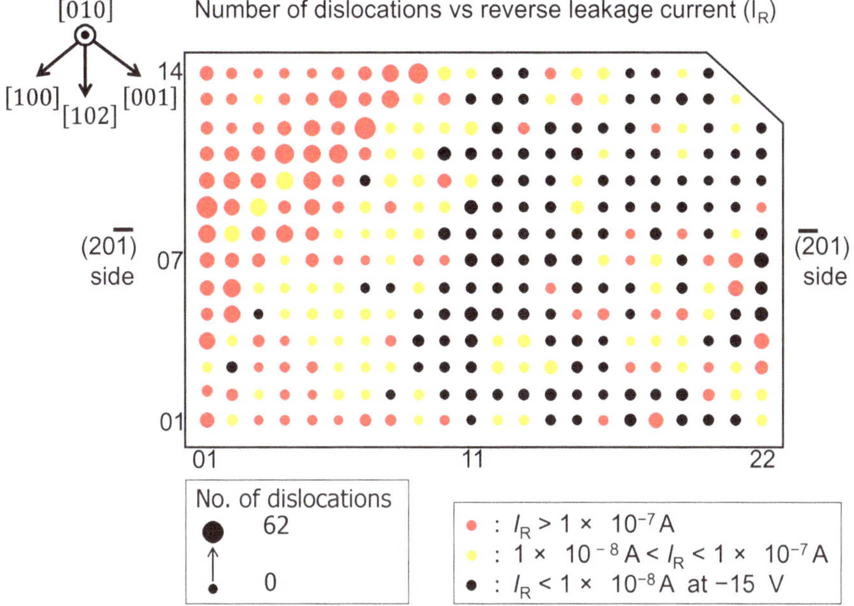

Fig. 26.10 On the (010) β-Ga_2O_3 surface, distributions of the number of dislocations and reverse leakage current (I_R) in SBDs

26.5.2 *The ($\bar{2}$01) and (001) Orientations*

In these surface orientations, no relation between the etch pits and leakage current is observed. In EFG Ga_2O_3, dislocations extend in the [010] direction, which is the growth direction. On surface orientations other than (010), these dislocations are not a leakage current path.

26.6 Summary

All results are summarized in Table 26.1. In EFG-grown β-Ga_2O_3, dislocation-related and void-related etch pits are observed, and only SBDs on the (010) surface show leakage current through dislocations.

Table 26.1 Properties of pits observed on the β-Ga_2O_3 surface

Surface orientation	Type	Origin	SBD leak current	Comment
(010)	A	Void	No	Etch pit
	B	Void	No	Etch pit
	C	Void	No	Etch pit
	D	Void	No	Etch pit
	E	Dislocation	Probably yes	Etch pit
	F	Dislocation	Probably yes	Etch pit
	G	Void	No	Observed on as-grown surface
($\bar{2}$01)	Line-shaped	Void	No	Etch pit
	Arrow-shaped	Defect	No	Etch pit
	Gourd-shaped	Defect	No	Etch pit
(001)	Line-shaped	Void	No	Etch pit
	Arrow (bullet)-shaped	Defect	No	Etch pit

Acknowledgements I appreciate Drs. Kenji Hanada (Saga University, currently, Aichi Synchrotron), Takayoshi Oshima (Saga University, currently, Flosfia), Drs. Kohei Sasaki, Akito Kuramata (Novel Crystal Tech.), and Prof. Osamu Ueda (Meiji University) for their collaborations.

References

1. H.H. Tippins, Phys. Rev. **140**, A316 (1965)
2. M. Orita, H. Ohta, M. Hirano, H. Hosono, Appl. Phys. Lett. **77**, 4166 (2000)
3. W.S. Hwang, A. Verma, H. Peelaers, V. Protasenko, S. Rouvimov, H.G. Xing, A. Seabaugh, W. Haensch, C. Van de Walle, Z. Galazka, M. Albrecht, R. Fornari, D. Jena, Appl. Phys. Lett. **104**, 203111 (2014)
4. T. Onuma, S. Saito, K. Sasaki, T. Masui, T. Yamaguchi, T. Honda, M. Higashiwaki, Jpn. J. Appl. Phys. **54**, 112601 (2015)
5. M. Higashiwaki, H. Murakami, Y. Kumagai, A. Kuramata, Jpn. J. Appl. Phys. **55**, 1202A1 (2016)
6. S. Fujita, M. Oda, K. Kaneko, T. Hitora, Jpn. J. Appl. Phys. **55**, 1202A3 (2016)
7. E.G. Víllora, K. Shimamura, Y. Yoshikawa, K. Aoki, N. Ichinose, J. Cryst. Growth **270**, 420 (2004)
8. M. Saurat, A. Revcolevschi, Rev. Int. Hautes Temp. Refract. **8**, 291 (1971)
9. N. Ueda, H. Hosono, R. Waseda, H. Kawazoe, Appl. Phys. Lett. **70**, 3561 (1997)
10. J. Zhang, B. Li, C. Xia, G. Pei, Q. Deng, Z. Yang, W. Xu, H. Shi, F. Wu, Y. Wu, J. Xu, J. Phys. Chem. Solids **67**, 2448 (2006)
11. S. Ohira, N. Suzuki, N. Arai, M. Tanaka, T. Sugawara, K. Nakajima, T. Shishido, Thin Solid Films **516**, 5763 (2008)
12. V.I. Vasyltsiv, Y.I. Rym, Y.M. Zakharko, Phys. Status Solidi B **195**, 653 (1996)
13. Y. Tomm, J.M. Ko, A. Yoshikawa, T. Fukuda, Sol. Energy Mater. Sol. Cells **66**, 369 (2001)
14. K. Shimamura, E.G. Villora, K. Matsumura, K. Aoki, M. Nakamura, S. Takekawa, N. Ichinose, K. Kitamura, Nihon Kessho Seicho Gakkaishi **33**, 147 (2006). [in Japanese]

15. H. Aida, K. Nishiguchi, H. Takeda, N. Aota, K. Sunakawa, Y. Yaguchi, Jpn. J. Appl. Phys. **47**, 8506 (2008)
16. A. Kuramata, K. Koshi, S. Watanabe, Y. Yamaoka, T. Masui, S. Yamakoshi, Jpn. J. Appl. Phys. **55**, 1202A2 (2016)
17. Z. Galazka, K. Irmscher, R. Uecker, R. Bertram, M. Pietsch, A. Kwasniewski, M. Naumann, T. Schulz, R. Schewski, D. Klimm, M. Bickermann, J. Cryst. Growth **404**, 184 (2014)
18. Y. Tomm, P. Reiche, D. Klimm, T. Fukuda, J. Cryst. Growth **220**, 510 (2000)
19. K. Irmscher, Z. Galazka, M. Pietsch, R. Uecker, R. Fornari, J. Appl. Phys. **110**, 063720 (2011)
20. K. Hoshikawa, E. Ohba, T. Kobayashi, J. Yanagisawa, C. Miyagawa, Y. Nakamura, J. Cryst. Growth **447**, 36 (2016)
21. J. Åhman, G. Svensson, J. Albertsson, Acta Crystallogr. Sect. C **52**, 1336 (1996)
22. S. Geller, J. Chem. Phys. **33**, 676 (1960)
23. K. Hanada, T. Moribayashi, T. Uematsu, S. Masuya, K. Koshi, K. Sasaki, A. Kuramata, O. Ueda, M. Kasu, Jpn. J. Appl. Phys. **55**, 030303 (2016)
24. K. Nakai, T. Nagai, K. Noami, T. Futagi, Jpn. J. Appl. Phys. **54**, 051103 (2015)
25. O. Ueda N. Ikenaga, K. Koshi, K. Iizuka, A. Kuramata, K. Hanada, T. Moribayashi, S. Yamakoshi, M. Kasu, Jpn. J. Appl. Phys. **55**, 1202BD (2016)
26. K Hanada, T. Moribayashi, K. Koshi, K. Sasaki, A. Kuramata, O. Ueda, M. Kasu, Jpn. J. Appl. Phys. **55**, 1202BG (2016)
27. M. Kasu, T. Oshima, K. Hanada, T. Moribayashi, A. Hashiguchi, T. Oishi, K. Koshi, K. Sasaki, A. Kuramata, O. Ueda, Jpn. J. Appl. Phys. **56**, 091101 (2017)
28. T. Oshima, A. Hashiguchi, T. Moribayashi, K. Koshi, K. Sasaki, A. Kuramata, O. Ueda, T. Oishi, M. Kasu, Jpn. J. Appl. Phys. **56**, 086501 (2017)
29. E. Ohba, T. Kobayashi, M. Kado, K. Hoshikawa, Jpn. J. Appl. Phys. **55,** 1202BF (2016)
30. M. Kasu, K. Hanada, T. Moribayashi, A. Hashiguchi, T. Oshima, T. Oishi, K. Koshi, K. Sasaki, A. Kuramata, O. Ueda, Jpn. J. Appl. Phys. **55,** 1202BB (2016)

Chapter 27
Optical Properties

Fundamental Absorption Edge and Emission Properties of β-Ga_2O_3

Takeyoshi Onuma

Abstract This chapter provides fundamental optical properties of β-Ga_2O_3. Valence band ordering was investigated by polarized transmittance and reflectance measurements. Anisotropic optical properties were also investigated by spectroscopic ellipsometry measurements. The optical anisotropy in a biaxial crystal as well as the gradual increase in the absorption coefficient were recognized as origins of the scattering in optical bandgap energies in a range 4.5–5.0 eV. Temperature-dependent exciton resonance energies were studied by using polarized reflectance measurement. The large changes in the exciton resonance energies with temperature were found to be originated from the exciton—longitudinal optical phonon interaction. Correlation between blue luminescence (BL) intensity and formation energy of oxygen vacancy (V_O) was found by measuring temperature-dependent cathodoluminescence spectra. Suppression of the BL band in the heavily nitrogen-doped epitaxial films was shown as an evidence for the decrease in the V_O concentration by N-doping and resultant high resistivity in the N-doped epitaxial films.

27.1 Introduction

Monoclinic β-Ga_2O_3 is now recognized as one of the attractive wide bandgap semiconductors including SiC, II-VI, and III–V nitride semiconductors. However, special care must be taken due to its characteristic monoclinic structure. According to Geller [1], β-Ga_2O_3 has a base-centered monoclinic structure with the space group symmetry of C_{2h}^3 (C_2/m). As shown in Fig. 27.1a, the conventional unit cell contains four formula units. Projections of the unit cell along the surface normal b- and c^*-axes of the (010) and (001) planes are shown in Fig. 27.1b, c. The a^*-axis

T. Onuma (✉)
Department of Applied Physics, School of Advanced Engineering
and Department of Electrical Engineering and Electronics, Kogakuin University,
2665-1, 192-0015 Hachioji, Tokyo, Japan
e-mail: onuma@cc.kogakuin.ac.jp

M. Higashiwaki and S. Fujita (eds.), *Gallium Oxide*, Springer Series
in Materials Science 293, https://doi.org/10.1007/978-3-030-37153-1_27

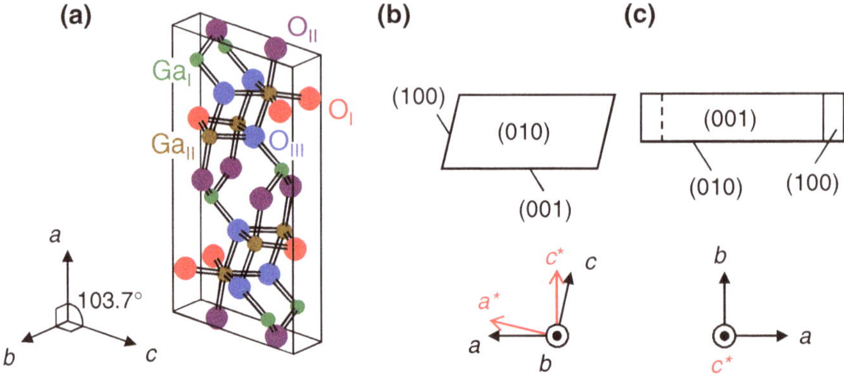

Fig. 27.1 **a** Crystal structure of β-Ga_2O_3 with a unit cell, in which two types of gallium ions (Ga_I and Ga_{II}) and three types of oxygen ions (O_I, O_{II}, and O_{III}) are composed. Projections of the unit cell along the surface normal **b** *b*- and **c** *c**-axes of the (010) and (001) planes

is perpendicular to the (100) plane and is angled at −13.7° from the *a*-axis. The c^*-axis is perpendicular to the (001) plane and is angled at 13.7° from the *c*-axis.

Even though the bandgap energy (E_g) is one of the important material parameters, reported energies of absorption edge (E_{edge}) [2–8] are widely distributed from 4.4 to 5.0 eV at room temperature (RT) as summarized in Table 27.1. The pioneering study was conducted by Tippins, who investigated the optical absorption on the (100) primary cleavage plane of single crystals grown by the Verneuil method [2]. Value of E_{edge} was found to be 4.60 eV at RT. The optical [3, 4] and electrical [4, 9] properties were found to be different in the directions parallel and perpendicular to the *b*-axis. Polarized optical absorption and reflectance measurements were made on single crystals grown by halide vapor phase epitaxy [3]. The energies of the absorption edge were found to be 4.54 eV for the electric field vector of incidence light *E* parallel to the *c*-axis (*E*//*c*), 4.56 eV for *E* parallel to the *a**-axis (*E*//*a**), and 4.90 eV for *E* parallel to the *b*-axis (*E*//*b*). Ueda and coworkers investigated the anisotropy of the optical properties of the (100) plane of single crystals grown by the floating zone (FZ) growth method [4]. The energies of the absorption edge were determined to be 4.52 eV for *E*//*c* and 4.79 eV for *E*//*b*. Our group investigated the optical anisotropy using the polarized optical transmittance and reflectance measurements [10]. Details will be shown in Sect. 27.3. Since the reported experimental E_{edge} values are widely distributed from 4.4 to 5.30 eV, the energies can be divided into five ranges, labeled *E*(2)–*E*(6), in order of energy, as shown in Table 27.1. Obviously, E_{edge} in the films [6–8] (4.75–5.0 eV) are larger than those in bulk crystals [2–4] (4.52–4.60 eV). Recently, the optical properties were further investigated using the Mueller matrix ellipsometry [11–13], where the excitonic effects were taken into account to develop the model dielectric function. The excitonic transition energies were respectively determined to be 4.8–4.92 eV and 5.10–5.17 eV for the lowest and second lowest ones in the *ac*-plane, and

Table 27.1 Experimental energies (eV) of the absorption edge in β-Ga_2O_3 at RT. Calculated energies at 0 K are also shown for comparison. *E*(1)–*E*(6) are assigned to the transitions from the valence bands to the CBM according to the selection rules

	E(1)	*E*(2)	*E*(3)	*E*(4)	*E*(5)	*E*(6)
Experiment (RT)						
Tippins[a]	–	4.60	–	–	–	–
Matsumoto et al.[b]	–	4.54 (*E*//*c*)	4.56 (*E*//*a**)	–	4.90 (*E*//*b*)	–
		4.63[c] (*E*//*c*)			5.06[c] (*E*//*b*)	5.30[c]
Ueda et al.[d]	–	4.52 (*E*//*c*)	–	4.79 (*E*//*b*)	–	–
Passlack et al.[e]	–	4.4	–	–	–	–
Orita et al.[f]	–	–	–	–	4.9	–
Víllora et al.[g,h]	–	–	–	4.75[g]	4.87[h]	5.21[g]
Oshima et al.[i]	–	–	–	–	5.0	–
Sturm et al.[j]	–	4.8[k] (*ac*-plane)	–	5.3[k] (*E*//*b*)	–	–
Sturm et al.[l]	–	4.88[k] (*ac*-plane)	5.10[k] (*ac*-plane)	5.41[k] (*E*//*b*)	–	–
Mock et al.[m]	–	4.92[k] (*ac*-plane)	5.17[k] (*ac*-plane)	5.46[k] (*E*//*b*)	–	–
	E(1)	*E*(2)	*E*(3)	*E*(4)	*E*(5)	*E*(6)
Calculation (0 K)	Indirect	Direct	Direct	Direct	Direct	Direct
Yamaguchi[n]	2.190	2.307	2.478	2.975	3.130	3.864
He et al.[o]	4.37	4.40	–	–	–	–
He et al.[p]	4.66	4.69	–	5.29	–	–
Varley et al.[q]	4.83	4.87	–	–	–	–
Varley and Schleife[r]	–	4.5 (*E*//c^*)	4.7 (*E*//*a*)	5.0 (*E*//*b*)	5.3 (*E*//*b*)	–
Peelaers and Van de Walle[s]	4.84	4.88	–	–	–	–
Furthmüller and Bechstedt[t]	–	4.65 (*E*//c^*)	4.90 (*E*//*a*)	5.50 (*E*//*b*)	–	–
Mengle et al.[u]	4.240	4.269 (*E*//*c*)	4.48 (*E*//*a**)	4.79 (*E*//*b*)	5.12 (*E*//*b*)	–
Onuma et al.[v] (RT)						
Valence band	I-L line	$\Gamma_2^-(1)$	Γ_2^- (2)	Γ_1^- (1)	Γ_1^- (2)	Γ_{1or2}^-
(010) Mg-doped	4.40	4.48 (*E*//*c*)	4.57 (*E*//*a**)	–	–	–
		4.48[c] (*E*//*c*)	4.58[c] (*E*//*a**)			5.29[c]

(continued)

Table 27.1 (continued)

	E(1)	E(2)	E(3)	E(4)	E(5)	E(6)
(001) undoped	4.46	–	4.55 (*E*//*a*)	4.70 (*E*// *b*)	–	–
			4.54[c] (*E*//*a*)	4.73[c] (*E*//*b*)	4.88[c] (*E*//*b*)	

[a][2]. 86-μm-thick (100) bulk. [b][3]. (100), (010), and (001) bulk. [c]Reflectance measurements. [d][4]. (100) bulk. [e][5]. Polycrystalline film. [f][6]. Polycrystalline film. [g][7]. Single crystalline (100) film. [h][7]. Single crystalline ($\bar{2}$01) film. [i][8]. Single crystalline ($\bar{2}$01) film. [j][11]. ($\bar{2}$01) and (010) bulk. [k]Mueller matrix ellipsometry measurement. [l][12]. ($\bar{2}$01) and (010) bulk. [m][13]. ($\bar{2}$01) and (010) bulk. [n][14]. LDA-DFT. [o][15]. GGA-DFT. [p][16]. B3LYP-DFT. [q][17]. HSE06-DFT. [r][18]. PBE-DFT including excitonic and local-field effects. Values correspond to excitonic peak positions of imaginary part of the dielectric function. [s][19]. HSE06-DFT [t][20]. HSE + GW-DFT including excitonic effect. [u][21]. LDA + GW-DFT [v][10]

5.3–5.46 eV for *E*//*b*, though the values were larger than the E_{edge} values. Moreover, the measurements suggested significant contributions of excitonic effects with relatively large exciton binding energies of 0.12–0.27 eV. Theoretically calculated values [14–21] are also shown in Table 27.1 for comparison. The calculations [14–19, 21] have predicted that the indirect bandgap energy ($E_{g,indir}$) is slightly smaller than the direct bandgap energy ($E_{g,dir}$) with the energy difference in the range of 0.03–0.04 eV [15–17, 19], though the dipole matrix element of the indirect transition is an order of magnitude smaller than those of the direct Γ-Γ transitions [17]. The conduction band minimum (CBM) locates at the Γ point, and the valence band maximum (VBM) locates on the E line in the reciprocal lattice of the conventional unit cell [14]. Here, the theoretical study [19] has pointed out that the E line is not a high-symmetry line; the VBM locates on the I–L line in the proper Brillouin zone [19].

Ueda and coworkers evaluated the carrier concentration dependence of the absorption edge [4]. It is well known that the doping leads to an effective shrinkage of E_g, and further increase in the doping density induces the Burstein–Moss (BM) shift. The BM shift is given by

$$\Delta E_g^{BM} = \frac{\hbar^2}{2m_r^*}\left(3\pi^2 n\right)^{2/3}, \tag{27.1}$$

where n is the carrier density. The reduced effective mass m_r^* is given by the effective hole mass m_h^* and the effective electron mass m_e^* as $1/m_r^* = m_h^* + 1/m_e^*$.

By applying m_r^* = 1.24 m_0 for *E*//*c* and m_r^* = 0.44 m_0 for *E*//*b*, the BM shifts of the absorption edge are drawn in Fig. 27.2. Here, m_0 is the rest mass of electron. It is found that negligibly small BM shift of 1 meV is expected for the substrates with $n < 10^{18}$ cm^{-3}.

Luminescence measurement is recognized as a very useful tool for detection and identification of impurity or intrinsic point defects through the recombination of generated excess carriers in the defect levels. The pioneering studies on the emission properties were performed on the single crystals grown by the Verneuil method

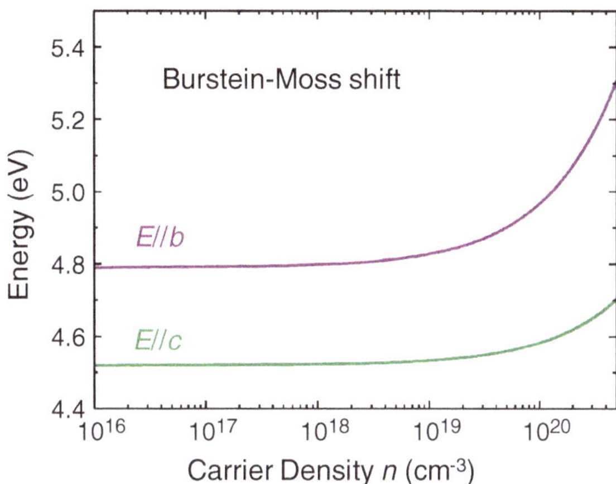

Fig. 27.2 Direct absorption edge of FZ-grown (100) β-Ga_2O_3 crystal as a function of n at RT

[22–24]. They have been followed by the emission measurements on the single crystals grown by the FZ growth method [25–27]. Typical emission spectra do not exhibit near-band-edge (NBE) emissions. The emission spectra alternatively show UV luminescence (UVL) [22, 26], blue luminescence (BL) [23, 24, 26, 27], and green luminescence (GL) bands [25]. The UVL band is traditionally recognized as being impurity independent [22], and experimental [22, 27–29] and theoretical [30] studies have assigned it to recombination of free electrons and self-trapped holes (STHs) or self-trapped excitons (STEs). Note that STHs were originally discovered in alkali halides [31, 32]. The STHs trap holes in a local electric field induced by small displacements of oxygen atoms. Recently, we have shown a significant influence of the donor-acceptor-pair (DAP) recombination involving Si_{Ga} donor on the UVL band in the Si-doped crystals [33]. The BL band has been attributed to the DAP transitions involving deep donors and acceptors [23, 24]. Possible donors are intrinsic point defects such as oxygen vacancies (V_O) and interstitial Ga (Ga_i) [23, 24]. The ionization energy of the V_O donor was calculated to be 1 eV [17], which is close to the photoemission energy of the deep-level defect located 0.8–0.82 eV below the conduction band probed by the deep-level optical spectroscopy [34, 35]. Possible acceptors for the BL band are Ga vacancies (V_{Ga}), V_O–V_{Ga} complexes [23, 24], Mg acceptors (Mg_{Ga}) for Mg-doped crystals, and N acceptors (N_O) for N-doped crystals.

In this chapter, we show the polarized transmittance and reflectance spectra of β-Ga_2O_3 crystals to clarify origins of the difference in the reported E_g values (4.4–5.0 eV) [10]. The anisotropic optical properties are further clarified by spectroscopic ellipsometry (SE) measurements [36]. Temperature-dependent polarized reflectance spectra are shown to discuss transition energy and the broadening parameter in terms of lattice vibrations [37]. Then, temperature-dependent cathodoluminescence (CL) spectra of β-Ga_2O_3 single crystals and epitaxial films are comprehensively shown to investigate their electronic structure and defect states [27, 33]. Correlation between the BL intensity and formation energy of V_O will be shown. For the N-doped films, suppression of the BL band in the heavily N-doped epitaxial films will be shown as an evidence for the decrease in the V_O concentration by N-doping and resultant high resistivity in the N-doped epitaxial films.

27.2 Experimental Procedure

Three series of β-Ga_2O_3 crystals and epitaxial films were prepared as summarized in Table 27.2. First series consists of single crystals having a resistivity ρ in a range of 1.4×10^{-2}–6×10^{11} Ω cm. A (001) unintentionally doped (undoped) substrate was grown with the FZ growth method [38]. The substrate exhibited n-type conductivity due to unintentional Si doping. The Hall-effect measurement yielded the carrier density $n = 7.7 \times 10^{16}$ cm^{-3} and the Hall electron mobility $\mu_H = 172$ cm^2/(Vs) at room temperature (RT). The resistivity ρ of $4.7{\times}10^{-1}$ Ω cm is given by $\rho = 1/(qn\mu_H)$, where q is the electron charge. Another (001) undoped substrate was also grown by the edge-defined film-fed growth (EFG) method [39]. The Hall-effect measurement yielded a carrier density of $n < 10^{17}$ cm^{-3} at RT. A (010) Mg-doped substrate grown by the FZ growth method exhibited a semi-insulating behavior, and a ρ of 6×10^{11} Ω cm was obtained with a Mg concentration of 4×10^{18}–2×10^{19} cm^{-3}. An (100) Si-doped substrate was grown using the EFG method. Values of $n = 4.9 \times 10^{18}$ cm^{-3}, $\mu_H = 93$ cm^2/(Vs), and $\rho = 1.4{\times}10^{-2}$ Ω cm are obtained at RT. Second series consists of ($\bar{2}$01) Si-doped crystals grown by the EFG method [39]. A full width at half maximum (FWHM) value of the X-ray rocking curve of ($\bar{4}$02) diffraction peak is as narrow as 17 arcsec and the crystals are free from the twin boundaries. Effective donor concentration (N_d–N_a) was estimated by the electrochemical capacitance–voltage measurements to be 3.0×10^{17}, 2.1×10^{18}, 4.0×10^{18}, and 7.3×10^{18} cm^{-3}. Third series consists of N-doped epitaxial films grown on (010) Sn-doped β-Ga_2O_3 substrates at 630 °C by radio-frequency plasma-assisted MBE (RF-MBE) [40]. The Ga beam equivalent pressure was fixed with $2.4{\times}10^{-5}$ Pa, and the growth rate increased from 0.03 to 0.38 μm/h with increasing the oxygen gas flow rates from 0.5 to 3 sccm. The film thicknesses are 750, 690, 490, and 100 nm for the films grown with the oxygen gas flow rates of 3, 2, 1, and 0.5 sccm, respectively.

A clear streak pattern was maintained throughout all the growths of N-doped epitaxial films. The resultant N-doped films have smooth surfaces with the arithmetical mean roughness (Ra) in the range of 0.35–2.0 nm, and similar FWHM values for the X-ray rocking curves of (020) diffraction peak are obtained for the films with the substrates. The N atoms incorporated during the growth using the 99.99995% pure oxygen gas, in which 0.2 ppm of nitrogen gas was incorporated. The N concentrations are analyzed by the secondary ion mass spectrometry to be [N] = 7×10^{16}, 2×10^{17}, 1×10^{18}, and 1×10^{18} cm^{-3} for the films grown with the oxygen gas flow rates of 3, 2, 1, and 0.5 sccm, respectively. Note that the residual Si impurity concentrations are [Si] = 6×10^{16}, 6×10^{16}, 8×10^{16}, and 5×10^{17} cm^{-3} for the films with the O_2 gas flow rates of 3, 2, 1, and 0.5 sccm,

Table 27.2 List of β-Ga_2O_3 single crystal substrates and epitaxial films

First series	Growth	n (cm^{-3})	μ_H [$cm^2/(Vs)$]		ρ ($\Omega \cdot cm$)
(001) undoped	FZ	7.7×10^{16}	172		4.7×10^{-1}
(001) undoped	EFG	$<10^{17}$	–		–
(010) Mg-doped	FZ	–	–		6×10^{11}
(100) Si-doped	EFG	4.9×10^{18}	93		1.4×10^{-2}
Second series	Growth	N_d–N_a (cm^{-3})			
($\bar{2}$01) Si-doped	EFG	3.0×10^{17}			
		2.1×10^{18}			
		4.0×10^{18}			
		7.3×10^{18}			
Third series	Growth	O_2 flow (sccm)	N (cm^{-3})	Thickness (nm)	ρ (Ω·cm)
(010) N-doped	RF-MBE	3	7×10^{16}	750	10^{11}–10^{12}
		2	2×10^{17}	690	
		1	1×10^{18}	490	
		0.5	1×10^{18}	100	

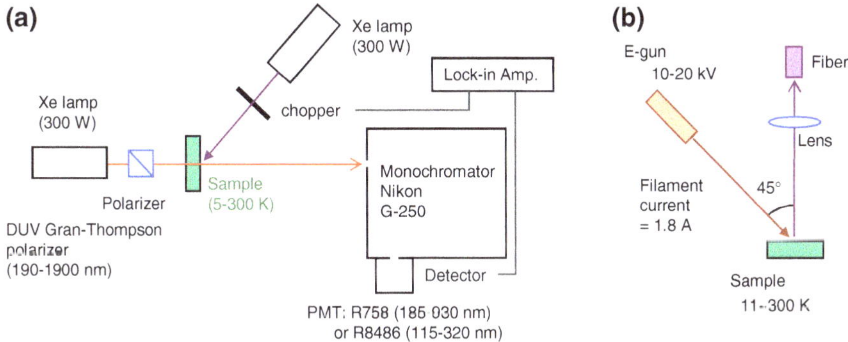

Fig. 27.3 Schematic diagrams of **a** polarized optical transmittance and reflectance and **b** CL setup

respectively. The N-doped films typically exhibit the increase in ρ on the order of 10^{11}–10^{12} Ω cm [41, 42], and N-implanted layer works well as a current blocking layer in vertical metal-oxide-semiconductor FETs [43].

As shown in Fig. 27.3a, polarized optical transmittance and reflectance spectra were measured at RT using a 300 W Xe lamp and a Glan-Thompson polarizer. The polarizer has a high transparency in a wavelength range of 190–1900 nm. The incidence angle was 30°. The transmitted and reflected lights were dispersed by a 25 cm focal-length Czerny-Turner monochromator equipped with a 1200 groove/mm grating. The dispersed light was detected by a photomultiplier tube using the lock-in detection technique. The spectral resolution was 0.03 nm (nearly 0.6 meV at the wavelength of 250 nm). Off-normal transmission SE spectra were measured for the (010) Mg-doped substrate using a J.A. Woolam RC2 system. The infrared spectroscopic ellipsometry (IRSE) spectra were measured for the EFG-grown (001) undoped substrate using a J.A. Woolam IR-VASE system at RT for wavenumbers from 250 to 1000 cm^{-1} with a step of 1 cm^{-1} and incidence angles of 50, 60, and 70°. As shown in Fig. 27.3b, CL was excited by an electron beam operated at 15–20 kV for the crystals and 10 kV for the N-doped epitaxial films with a filament current of 1.8 A. A typical probe current at the sample was 1 μA. The Monte Carlo simulation [44, 45] suggests that the electron energy loss profile has a peak around 120 nm with an acceleration voltage of 10 kV. Thus, the CL mostly probed the epitaxial layers, and influence of the underlying Sn-doped substrate was suppressed in the CL spectra. The CL was detected using an optical fiber and a Hamamatsu C7473 multichannel analyzer. Back-thinned charge-coupled device image sensors were used to make simultaneous measurements of wavelengths from 200 to 950 nm with a wavelength resolution of 2 nm. Details of the CL setup can be found elsewhere [27, 46].

27.3 Valence Band Ordering of β-Ga_2O_3

The band structure of β-Ga_2O_3 is schematically shown in Fig. 27.4. The valence band ordering was determined from our experiments with the aid of the theoretical calculation [14]. According to theoretical calculations [14–21], the CBM is isotropic with the irreducible representation of Γ_1^+ and the effective electron mass near

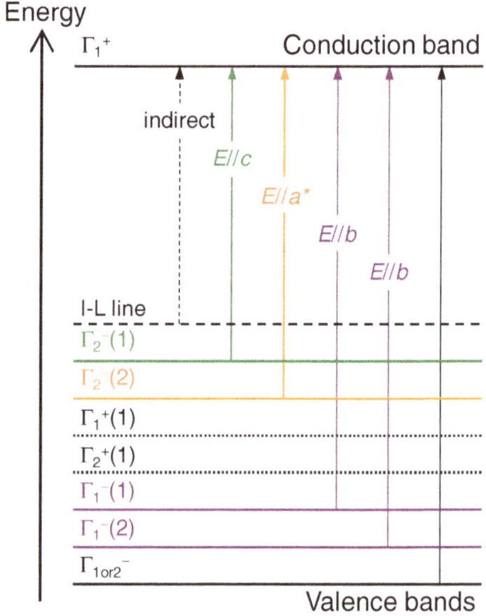

Fig. 27.4 Band structure and irreducible representations in β-Ga_2O_3. The valence band ordering was determined from our experiments with the aid of the theoretical calculation [14]. The dashed line represents the valence band on the I–L line in the Brillouin zone [19]. Solid and dotted lines represent the valence bands for the dipole allowed and forbidden transitions at Γ, respectively. *E*//*a**, *E*//*b*, and *E*//*c* represent electric field vectors *E* parallel to the *a**-, *b*-, and *c*-axes, respectively [10]

it is extremely light. On the other hand, the valence bands are almost flat, and the effective mass of holes in them is heavy. The C_{2h} point group is a subgroup of T_d for zinc-blend crystals [47]. Without the spin-orbit interaction, the VBM in the T_d is the threefold Γ_5 [48]. When the crystal symmetry is reduced from T_d to C_{2h}, the threefold Γ_5 band splits into $2\Gamma_1^+ + \Gamma_2^+$ or $\Gamma_1^- + 2\Gamma_2^-$ [47]. Theoretical calculations suggest that the VBM at the Γ point is composed of oxygen 2p states [14–21]. The primitive unit cell consists of two formula units. The six O atoms construct 18 bands including six Γ_1^+, 3 Γ_2^+, 3 Γ_1^-, and 6 Γ_2^- bands: the transitions from the Γ_1^+ and Γ_2^+ bands to the Γ_1^+ band of the CBM are dipole forbidden, the transition from the Γ_1^- band to the Γ_1^+ band is dipole allowed for *E*//*b*, and the transition from the Γ_2^- band to the Γ_1^+ band is dipole allowed for *E*//*a** or *E*//*c* (axes in the (010) plane).

As shown by the bold solid lines in Fig. 27.5, the unpolarized transmittance spectra at RT from the substrates exhibited distinct shoulders at 4.5–4.7 eV. Polarized transmittance spectra are shown by the thin solid lines in the figure. Obviously, both the substrates exhibited optical anisotropy. The absorption coefficient α was calculated by $\alpha = \ln[(1-R)^2/T]/d$, where R, T, and d represent the reflectance, transmittance, and thickness of the substrates, respectively. As shown in Fig. 27.6, linear regression on the $(\alpha h\nu)^2$ versus photon energy $h\nu$ plots gives the energies of the direct absorption edge. The observed energies are shown by the dotted lines in the figure, and the values fall into three of the ranges listed in

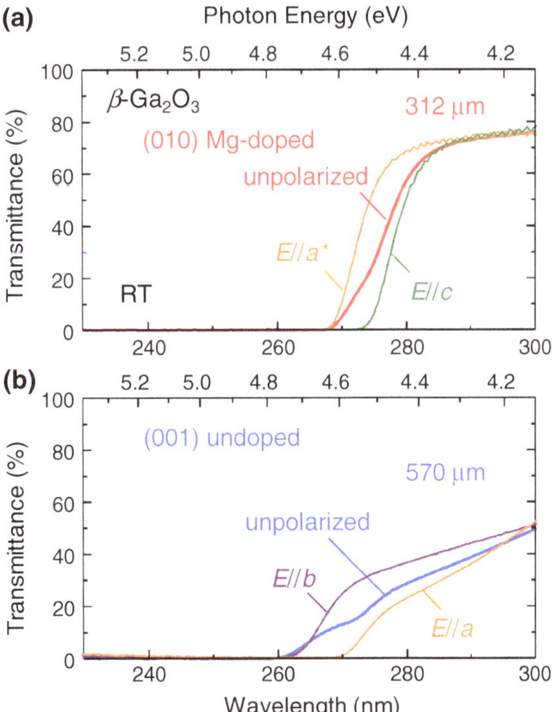

Fig. 27.5 Unpolarized (bold solid lines) and polarized (thin solid lines) transmittance spectra of **a** (010) Mg-doped and **b** (001) undoped β-Ga_2O_3 substrates at RT [10]

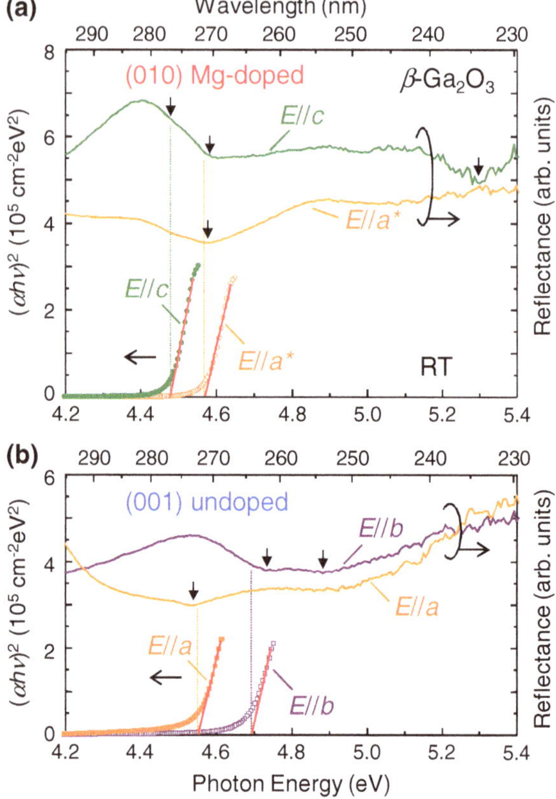

Fig. 27.6 $(\alpha h\nu)^2$ versus $h\nu$ plots for **a** (010) Mg-doped substrate for *E*//*c* (closed circles) and *E*//*a** (open circles) at RT. **b** Data set of (001) undoped substrate for *E*//*a* (closed squares) and *E*//*b* (open squares). Dotted lines represent the energies of the direct absorption edge. Partially polarized reflectance spectra at RT are shown in the upper part of the figures. Energies of the dips and the shoulder are indicated by the vertical arrows [10]

Table 27.1, i.e., *E*(2)–*E*(4). Note that the values include an experimental error of 0.03 eV. The variation in energies [*E*(2)–*E*(4)] may be intrinsically attributed to the various transitions from the valence bands to the CBM or more or less extrinsically to the Urbach tail state induced by the crystallinity, strain, impurity, etc. [48]. In order to determine the origin, partially polarized reflectance spectra were measured, as shown in the upper part of Figs. 27.6a, b. Here, the spectra were measured at the incidence angle of 30° without the Glan-Thompson polarizer, where the reflectivity for *s*-polarized light was higher than that for *p*-polarized light. As shown by the vertical arrows, distinct dips and a shoulder occur in the spectra. Some of the energies nearly match the direct absorption edge, e.g., the absorption edge at 4.48 eV (4.57 eV) for *E*//*c* (*E*//*a**) appears as a shoulder (a dip) in the reflectance spectrum for *E*//*c* (*E*//*a**) in the (010) Mg-doped substrate, and the one at 4.55 eV (4.70 eV) for *E*//*a* (*E*//*b*) appears as a dip in the reflectance spectrum for *E*//*a* (*E*//*b*) in the (001) undoped substrate. Note that a dip at 4.58 eV in the reflectance spectrum for *E*//*c* in the (010) Mg-doped substrate may be attributed to overlapping of the dip for *E*//*a**. Furthermore, additional dips were observed at 4.88 eV for *E*//*b* in the (001) undoped substrate and 5.29 eV for *E*//*c* in the (010) Mg-doped substrate, and the values fall into two of the ranges listed in

Table 27.1, i.e., $E(5)$ and $E(6)$. Because the reflectance anomalies appear as a consequence of band-to-band or excitonic transitions [48], these results imply the existence of the various transitions from the valence bands to the CBM. According to the selection rules, $E(2)$–$E(5)$ correspond to the transitions from the $\Gamma_2^-(1)$, $\Gamma_2^-(2)$, Γ_1^- (1), and Γ_1^- (2) bands to the Γ_1^+ band of the CBM. Here, the numbers in the parentheses are labeled in order of energy. $E(6)$ can be tentatively assigned as Γ^-_{1or2}, since the reflectance anomaly at 5.29 eV was not reproducibly observed. The valence band ordering is almost the same as that of the theoretical calculation by Yamaguchi [14] (see Fig. 27.4). $E_{g,dir}$ is thus determined to be 4.48±0.03 eV with *E*//*c*. As shown in Table 27.1, the value is close to the previous values reported for the bulk crystals. In contrast, E_g in the films [6–8] (4.75–5.0 eV) correspond to the transitions from the Γ_1^- (1) or Γ_1^- (2) bands to the CBM. The optical anisotropy is found to be one of the causes of the difference in the reported E_g of the thin films and the bulk crystals. Indeed, the energies of the direct absorption edge have been evaluated by the unpolarized transmittance measurements for the optically anisotropic (100) or ($\bar{2}$01) films [6–8], and the E_g values have been preferentially determined for *E*//*b*.

$(\alpha h\nu)^{0.5}$ is plotted as a function of $h\nu$ in Fig. 27.7 to evaluate the indirect absorption edge. Here, α was calculated using the transmittance in the unpolarized measurements. The energies of the indirect absorption edge were determined to be the points of intersection of straight lines, which represent guidelines for the background and the low-energy tails. As shown by the vertical arrows in Fig. 27.5, these values are respectively 4.40 and 4.46 eV for (010) Mg-doped and (001) undoped substrates. $E_{g,indir}$ is thus determined to be 4.43±0.03 eV. $E_{g,indir}$ is smaller than $E_{g,dir}$ by 0.05 eV, and the energy difference nearly matches the theoretically calculated values of 0.03–0.04 eV [15–17, 19].

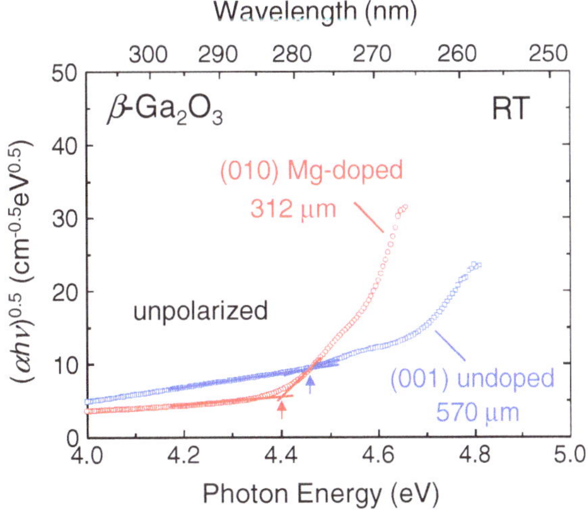

Fig. 27.7 $(\alpha h\nu)^{0.5}$ versus $h\nu$ plots for (010) Mg-doped (circles) and (001) undoped (squares) substrates at RT. Straight lines represent guidelines for the background and the low-energy tails. Vertical arrows indicate the energies of the indirect absorption edge [10]

Anisotropic optical properties are further investigated on the (010) Mg-doped substrate by transmission SE measurements. β-Ga_2O_3 is a biaxial crystal, whose optic axis is parallel to the *b*-axis. In particular, the angle $\beta = 103.7°$ between the *a*- and *c*-axes [1] leaves the off-diagonal element in the dielectric tensor nonvanishing as

$$\boldsymbol{\varepsilon} = \begin{pmatrix} \varepsilon_{xx} & \varepsilon_{xy} & \\ \varepsilon_{xy} & \varepsilon_{yy} & \\ & & \varepsilon_{zz} \end{pmatrix}. \tag{27.2}$$

Here, the optic *b*-axis is parallel to the observed *z*-axis, and *c*- and *a**-axes are parallel to the observed *x*- and *y*-axes, respectively. The off-diagonal element ε_{xy} induces the cross-coupling of the *p*- and *s*-polarized lights through the color-dependent rotation of the dielectric ellipsoid in the *ac*-plane around the *b*-axis as shown in Fig. 27.8 [49]. The real and imaginary parts of the dielectric function have different rotation angles [49–51]. According to the Mueller matrix ellipsometry study on single crystals [11], ε_{xy} is small ($|\varepsilon_{xy}| < 0.02$) in the transparent spectral range, but it reaches values up to 0.30 in the absorbing spectral range and the cross-coupling of the polarized lights cannot be neglected [see Fig. 27.9]. In this study, the optical anisotropy was evaluated using the biaxial layer mode in WVASE32 [52], where the spectral dispersion of the optical constant along the *z*-axis (*b*-axis) is first determined, and those along the *x*-axis (*c*-axis) and *y*-axis (*a**-axis) are given by the differences. Indeed, the data reflect the dielectric constants that projected the dielectric ellipsoid on the observed *x*-, *y*-, and *z*-axes. Although the information acquired is less than that obtained by the Mueller matrix ellipsometry [11–13, 53–56] where the projected components and ε_{xy} are simultaneously determined, the present measurements still give reasonable sets of optical constants.

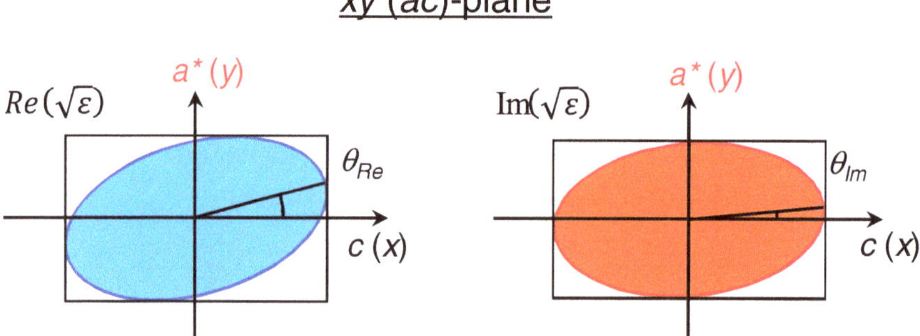

Fig. 27.8 Dielectric ellipsoid in the *ac*-plane of β-Ga_2O_3. Real and imaginary parts of the dielectric function have different rotation angles

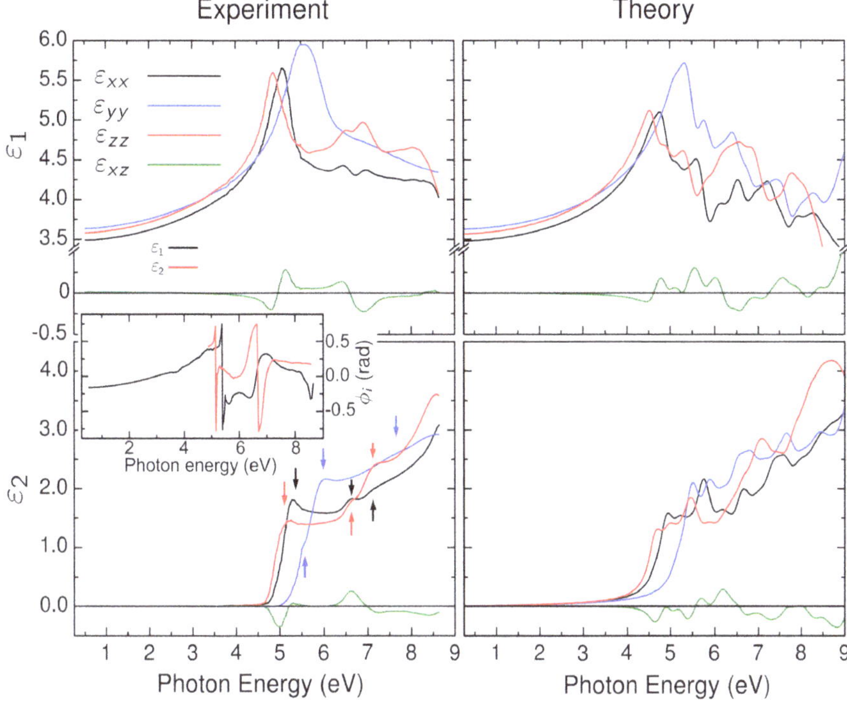

Fig. 27.9 Experimentally determined (left column) and theoretically derived (right column) tensor components of the dielectric function. Inset: Rotation angle of the dielectric axes around the *y*-axis for real (black) part and imaginary (red) part. Here, the optic *b*-axis is defined as being parallel to the observed *y*-axis, and *c**- and *a*-axes are parallel to the observed *z*- and *x*-axes, respectively. Figure is reproduced from Sturm et al. [11]

As shown in Fig. 27.10a, the refractive indexes increase along the a^*- (n_{a*}), c- (n_c), and b- (n_b) axes in order in the transparent spectral range, and the order is changed in the absorbing spectral range. The absorption coefficients α along the c- (α_c), a^*- (α_{a*}), and b- (α_b) axes are shown in Fig. 27.11. Here, the weak absorption at around 300 nm in the substrate is reflected in the spectrum along the *c*-axis. As shown by the vertical broken lines, E_g was estimated to be 4.458, 4.605, and 4.744 eV along the *c*-, *a**-, and *b*-axes, respectively, which are in good agreement with those determined by the polarized optical transmittance and reflectance measurements [10]. As shown, α for the transition from the uppermost valence band at Γ to the CBM is small on the order of 10^3 cm^{-1}, and it gradually increases up to 10^6 cm^{-1} at around 5.2 eV. The characteristic optical anisotropy as well as the gradual increase in α is eventually recognized as origins of the scattering in the reported E_g values [see Table 27.1].

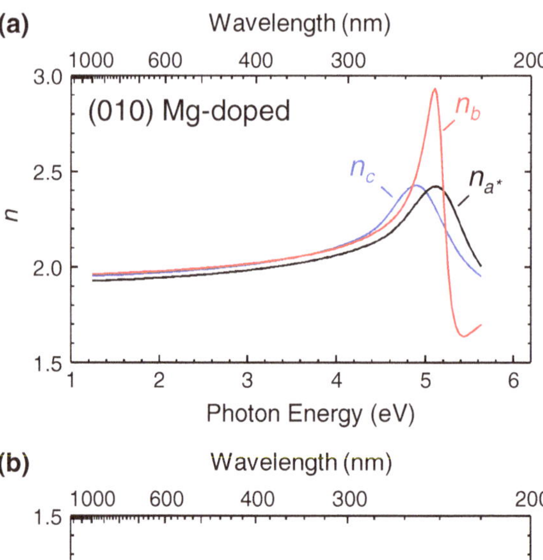

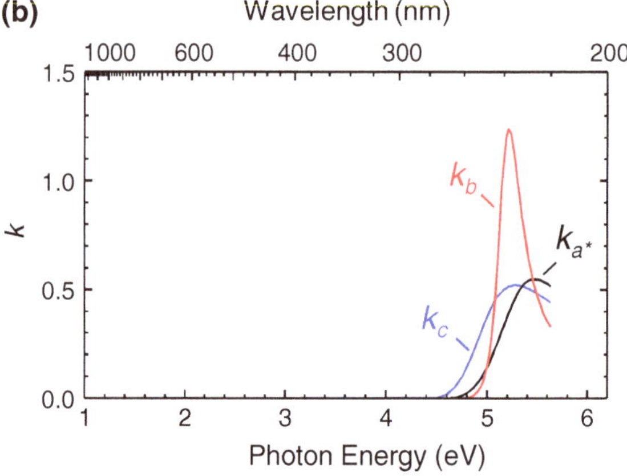

Fig. 27.10 Spectra for **a** refractive index n and **b** extinction coefficient k of (010) Mg-doped substrate [36]

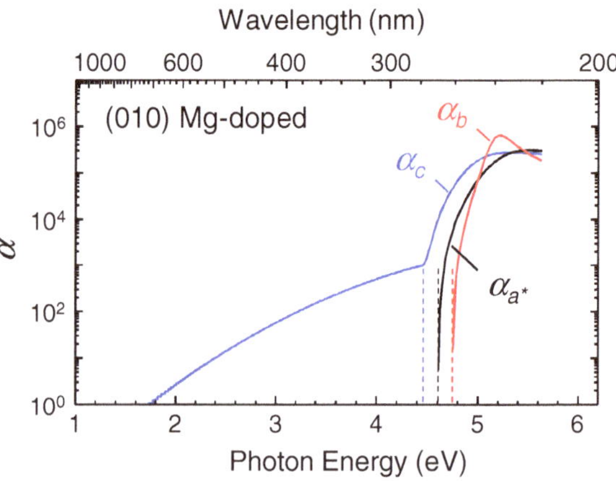

Fig. 27.11 Spectra for absorption coefficient α of (010) Mg-doped substrate. E_g values are indicated by the vertical broken lines [36]

27.4 Temperature-Dependent Exciton Resonance Energies and Their Correlation with IR-Active Optical Phonon Modes

Polarized reflectance spectra for *E*//*c* and *E*//*a** of (010) Mg-doped substrate at RT are shown in Fig. 27.12. The spectra exhibited distinct anomalies owing to the Γ-Γ direct transitions, and they showed energy shifts by reflecting the variation of the valence bands in the transitions. The spectra were analyzed using a well-known lineshape function [57, 58]

$$I(E) = I_0 + bE + I_1 e^{i\phi}(E - E_0 + i\Gamma_{\text{broad}})^{-n}, \quad (27.3)$$

where $I_0 + bE$ is a background, I_1 is an amplitude, E_0 is a transition energy, Γ_{broad} is a broadening parameter, and ϕ is a phase factor. The three-dimensional ($n = 1/2$) and exciton ($n = 1$) lineshape functions were used to analyze the present spectra for the single crystals as shown in Fig. 27.12. The spectra are well fitted by the exciton lineshape function, indicating that the exciton transition dominants the Γ-Γ direct transitions in β-Ga_2O_3. The significant parameters obtained from the fitting are E_0 = 4.45 eV, Γ_{broad} = 108 meV, and ϕ = 1.14π for *E*//*c* and E_0 = 4.54 eV, Γ_{broad} = 105 meV, and ϕ = 1.11π for *E*//*a**. E_0 represents the exciton resonance energy E_{exciton}, and the values are slightly smaller than E_{edge} by about 30 meV.

Temperature dependences of polarized reflectance spectra for *E*//*c*, *E*//*a** and *E*//*b* are shown in Fig. 27.13. The spectra for *E*//*c* and *E*//*a** were measured for the (010) Mg-doped substrate, and those for *E*//*b* were measured for the FZ-grown (001) undoped substrate. Distinct reflectance anomalies were observed for all measured temperatures, and E_{exciton} values obtained from the fitting are plotted as a

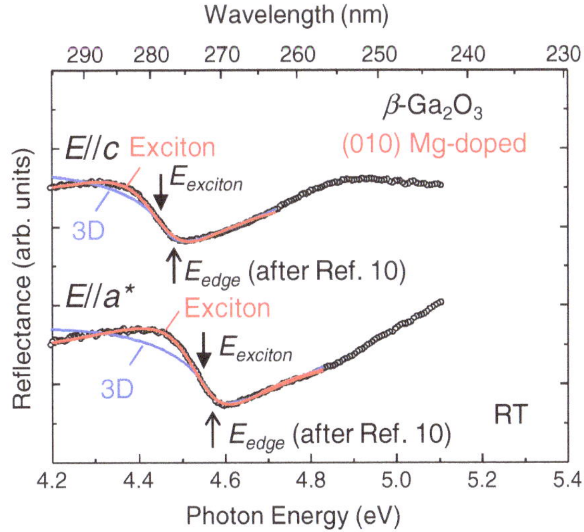

Fig. 27.12 Polarized reflectance spectra for *E*//*c* and *E*//*a** of (010) Mg-doped substrate at RT (open circles). Solid lines represent fitting curves using three-dimensional (blue lines) and exciton (red lines) lineshape functions [37]

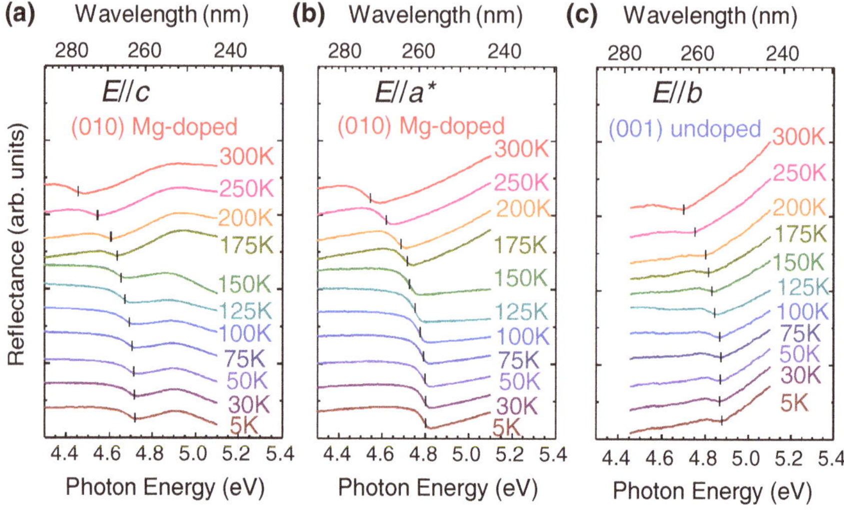

Fig. 27.13 Temperature dependences of polarized reflectance spectra for **a** *E*//*c* and **b** *E*//*a** of (010) Mg-doped substrate, and that for **c** *E*//*b* of (001) undoped substrate. Energy positions of E_{exciton} are marked on the spectra [37]

function of temperature in Fig. 27.14a. Valence band ordering does not change from low temperature to room temperature. The data were fitted using the Varshni's empirical equation [59]

$$E_{\text{exciton}}(T) = E_{\text{exciton}}(0) + \frac{\alpha_V T^2}{T + \beta_V}, \tag{27.4}$$

where $E_{\text{exciton}}(0)$ is E_{exciton} at 0 K and α_V and β_V are fitting parameters. Since β_V is related to the Debye temperature, the value was fixed with the reported value of 738 K [60]. As shown in Fig. 27.14a, the experimental data are well reproduced by the fitting curves with the parameters in Table 27.3. The energy differences in E_{exciton} between 5 and 300 K are respectively 268, 257, and 179 meV for *E*//*c*, *E*//*a**, and *E*//*b*, and the values are found to be larger than those in the other materials [61].

Table 27.3 Significant parameters obtained from the fitting for (27.4)

	$E_{\text{exciton}}(0)$ (eV)	α_V (meV/K)	β_V (K)	*E*(5)-*E*(300) (meV)
E//*c*	4.73	3	738	268
E//*a**	4.81	3	738	257
E//*b*	4.88	2	738	179

In order to clarify the origin of the large E_{exciton} changes with temperature, temperature dependences of Γ_{broad} values obtained from the fitting were investigated as plotted in Fig. 27.14b. The data were fitted using an equation [62]

$$\Gamma_{\text{broad}}(T) = \Gamma_{\text{broad}}(0) + \gamma T + \frac{\Gamma_{\text{LO}}}{\exp\left(\frac{\hbar\omega_{\text{LO}}}{k_{\text{B}}T} - 1\right)}, \tag{27.5}$$

where $\Gamma_{\text{broad}}(0)$ is Γ_{broad} at 0 K, γ and Γ_{LO} are fitting parameters, $\hbar\omega_{LO}$ is the longitudinal optical (LO) phonon energy, k_{B} is the Boltzmann constant, and the second and third terms on the right-hand side arise from exciton interactions with acoustic and longitudinal optical phonons, respectively. The equation is traditionally used to analyze temperature dependence of exciton linewidths in II–VI and III–V semiconductors [62]. According to the factor group analysis at the Γ point [47], the irreducible representations for acoustical and optical zone center modes are $\Gamma_{\text{aco}} = A_u + 2B_u$ and $\Gamma_{\text{opt}} = 10A_g + 5B_g + 4A_u + 8B_u$, respectively. For the optical modes, A_g and B_g modes are Raman-active [63, 64], while A_u and B_u modes are IR-active. According to the polarization selection rules [47], the A_u modes are allowed for *E*//*b*, and the B_u modes are allowed for *E*//*a* and *E*//*c*. To separate the A_u

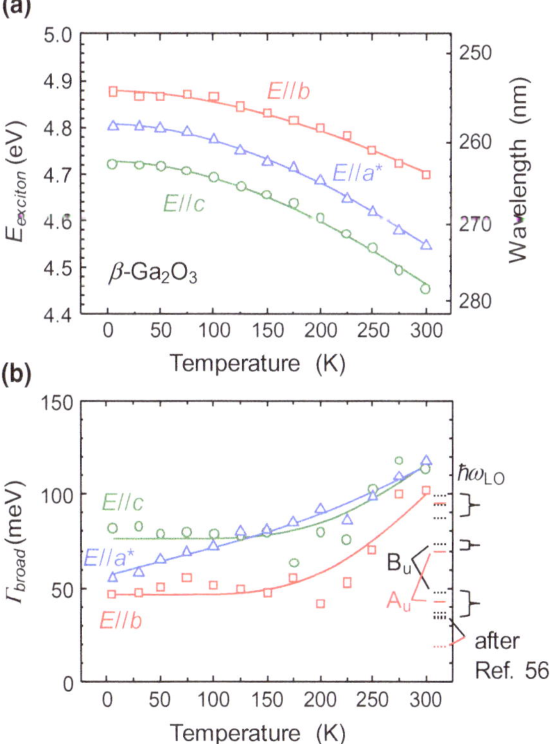

Fig. 27.14 Temperature dependences of **a** exciton resonance energy E_{exciton} and **b** broadening parameter Γ_{broad} for *E*//*c* (open circles), *E*//*a** (open triangles), and *E*//*b* (open squares). Solid lines represent the fitting curves. Three ranges of $\hbar\omega_{LO}$ are indicated on the right axis of (**b**) [37]

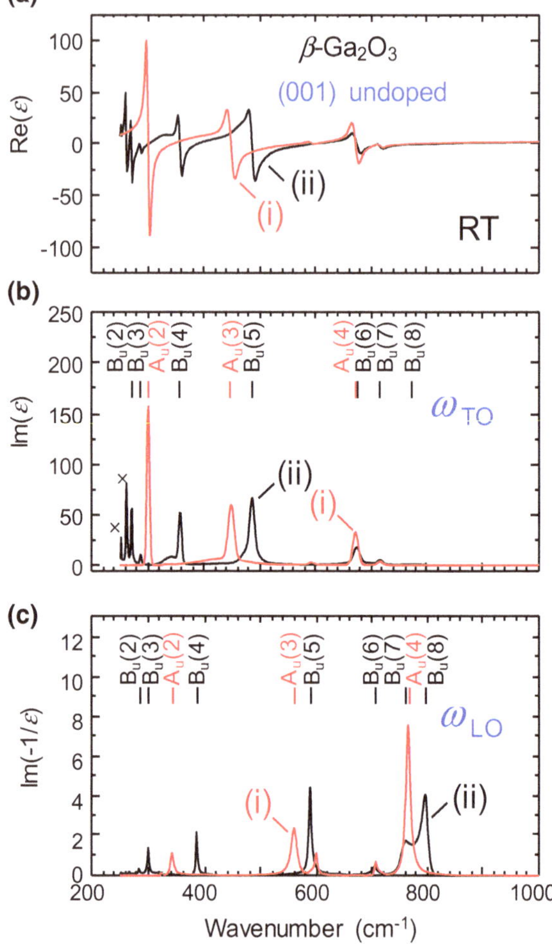

Fig. 27.15 Spectra for **a** real [Re(ε)] and **b** imaginary part [Im(ε)] of the dielectric function, and those for **c** imaginary part of the energy loss function [Im(-1/ε)] of (001) undoped substrate at RT. Spectra were measured for two configurations: (i) *x*//*b*, *y*//*a*, and *z*//*c** for the A_u modes and (ii) *x*//*a*, *y*//*b*, and *z*//*c** for the B_u modes [37]

and B_u mode components and to determine the phonon frequencies independently, the standard IRSE spectra of the EFG-grown (001) undoped substrate were measured for two configurations (i) *x*//*b*, *y*//*a*, and *z*//*c** for the A_u modes and (ii) *x*//*a*, *y*//*b*, and *z*//*c** for the B_u modes. Spectra for the real [Re(ε)] and imaginary part [Im(ε)] of the dielectric function and for the imaginary part of the energy loss function [Im(-1/ε)] are shown as a function of wavenumber in Fig. 27.15. The transverse optical (TO) phonon frequencies ω_{TO} were determined by the peak positions in the Im(ε) spectra, and the LO phonon frequencies ω_{LO} were determined by the peak positions in the Im(-1/ε) spectra. The A_u and B_u modes were well separated according to the polarization selection rules. The ω_{LO} values can be divided into three ranges, i.e., 284–387, 561–590, and 707–798 cm^{-1} ($\hbar\omega_{LO}$= 35–48, 70–73, and 88–99 meV). The $\hbar\omega_{\mathrm{LO}}$ values and the three ranges are

Table 27.4 Significant parameters obtained from the fitting for (27.5)

	$\Gamma_{\text{broad}}(0)$ (meV)	γ (meV/K)	Γ_{LO} (meV)	$\hbar\omega_{\text{LO}}$ (meV)
E//*c*	76	0	683	75
E//*a**	56	0.15	228	75
E//*b*	47	0	929	75

indicated on the right axis of Fig. 27.14b. Note that the lowest ω_{LO} values of 156 cm^{-1} for A_u mode and 269 cm^{-1} for B_u mode are shown by referring the report by Schubert et al. [see Chap. 28]. The Γ_{broad} values at low temperature correspond to the lower two ranges of $\hbar\omega_{LO}$, implying that the exciton-LO-phonon interaction dominates the Γ_{broad} even at low temperature. The second term on the right-hand side of the fitting functions in Fig. 27.14b is interpreted as indication of transformation of LO phonon energies instead of the acoustic phonon. Here, the existence of the STHs has been pointed out in the experimental [22, 27–29] and theoretical [30] studies. The UV luminescence band at around 3.2–3.6 eV in the emission spectra at low temperature has been ascribed to the evidence of the STHs or STEs as described in Sect. 27.5 [22, 27]. Stability of the STE has been confirmed by temperature dependence of Urbach tails of the absorption spectra [28]. In principle, there are few real LO phonons at low temperature. But the hole induces virtual LO phonons through the polaron behavior [32]. The hole dressed with the virtual phonons is called as the STH. The STE will be formed if the free electron is trapped by the STH. Thus, the Γ_{broad} values at low temperature might be ascribed to the exciton-LO phonon interaction induced by the STE formation process. Similar phenomenon has been recognized in alkali halides [31, 32]. The significant parameters obtained from the fitting are summarized in Table 27.4. The moderate

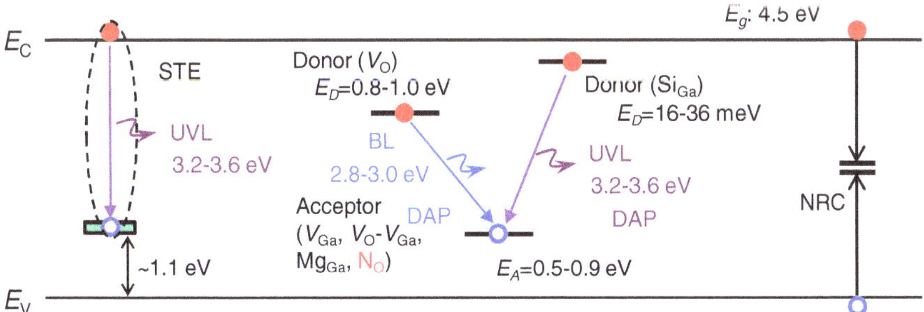

Fig. 27.16 Schematic band diagram and emission models in β-Ga_2O_3. The STE state is drawn as a radiative recombination center for simplicity. The BL band is attributed to DAP transitions involving V_O and deep acceptors, such as V_{Ga}, V_O–V_{Ga} complexes, Mg_{Ga}, and N_O. The UVL band is also attributed to DAP transitions involving Si_{Ga} and the same deep acceptors [33]

increases in Γ_{broad} above 150 K are well reproduced by fixing the $\hbar\omega_{LO}$ as 75 meV. The large E_{exciton} changes with temperature in β-Ga_2O_3 were found to be originated from the exciton-LO-phonon interaction for all measured temperatures.

27.5 Emission Properties of β-Ga_2O_3

As mentioned in Sect. 27.1, luminescence measurement is recognized as a very useful tool for detection and identification of impurity or intrinsic point defects through the recombination of generated excess carriers in the defect levels. β-Ga_2O_3 is known to exhibit no NBE emissions, and they typically exhibit UVL, BL, and GL bands as schematically shown in Fig. 27.16. Note that the GL band has been reported to correlate with the oxygen partial pressure during the FZ growth though its emission process is unknown at present [25]. Indeed, temperature-dependent CL spectra of the FZ-grown (001) undoped, EFG-grown (100) Si-doped, and FZ-grown (010) Mg-doped β-Ga_2O_3 substrates show the luminescence bands at 3.2–3.6, 2.8–3.0, and 2.4 eV for the UV, BL, and GL bands, respectively [see Fig. 27.17]. Note here that the multiple-peak structure in the CL spectra originates from unavoidable multiple internal reflections in the measurement system. These are measurement artifacts. The STE state is drawn as a radiative recombination center in Fig. 27.16 for simplicity though the formation processes of self-trapped holes and self-trapped excitons should be phenomenologically drawn using a configuration coordinate diagram [30, 32]. The UV bands were impurity independent, as has been observed in the spectra for Verneuil-grown crystals [22]. The results are evidence of the proposed emission mechanism, i.e., recombination of free electrons and STHs or STEs [22, 30]. In contrast, the BL intensity exhibited an impurity dependency at 300 K: The Si-doped substrate was the lowest intensity, the undoped substrate was in the middle, and the Mg-doped substrate had the highest intensity. The BL band has been attributed to a DAP transition involving deep donors and acceptors [23, 24]. Possible donors are V_O and Ga_i, and possible acceptors are V_{Ga} and/or the V_O-V_{Ga} complex Mg_{Ga} for the Mg-doped substrate, and N_O for N-doped crystals. It has been theoretically shown [17] that V_O acts as a deep donor with an ionization energy of more than 1 eV. The formation energies of charged defects and impurities depend on the Fermi level (E_F) position, where E_F is referenced with respect to the VBM, i.e., $E_F = 0$ at the VBM and $E_F = E_g$ at the CBM. In the case of V_O, the formation energy decreases as E_F shifts toward the VMB [17]. Since the Mg-doped substrate exhibited a high ρ of 6×10^{11} Ω cm, the Fermi level more or less shifts toward the VBM. In contrast, the formation energy would increase for the Si-doped substrate. Therefore, the increase in ρ induces a simultaneous increase in the V_O concentration and BL intensity. The results clearly indicate strong correlation between the BL intensity and formation energy of V_O.

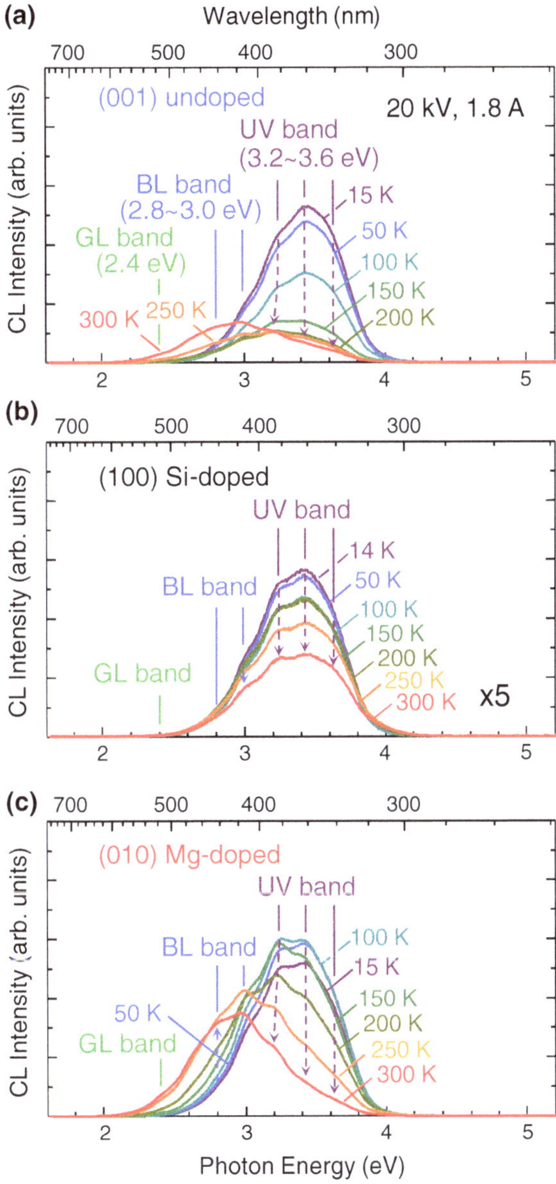

Fig. 27.17 Temperature-dependent CL spectra of **a** (001) undoped, **b** (100) Si-doped, and **c** (010) Mg-doped β-Ga_2O_3 substrates. Spectral intensities for the (100) Si-doped substrate were multiplied by 5 for ease of viewing [27]

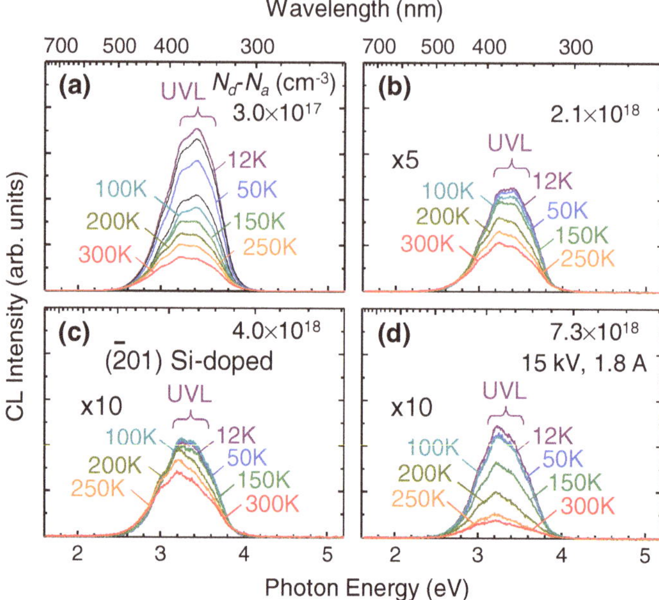

Fig. 27.18 **a** Temperature-dependent CL spectra of ($\bar{2}01$) Si-doped β-Ga_2O_3 single crystals grown by the EFG method with N_d–N_a of **a** 3.0 × 10^{17}, **b** 2.1 × 10^{18}, **c** 4.0 × 10^{18}, and **d** 7.3 × 10^{18} cm^{-3}. Spectral intensities for (**b**) are multiplied by 5 and those for (**c**) and (**d**) are multiplied by 10 for ease of viewing [33]

Temperature-dependent CL spectra of the EFG-grown ($\bar{2}01$) Si-doped β-Ga_2O_3 single crystals are shown in Fig. 27.18. The spectra predominately exhibit the UVL band at 3.2–3.6 eV, which is attributed to recombination of STEs. The BL band at 2.8–3.0 eV is suppressed in the *n*-type crystals. The UVL intensities decrease with increasing N_d–N_a even at low temperature (LT). The decrease of the STE emission at LT in heavily Si-doped crystals implies the Debye-Hückel screening [65] of STEs with the critical charge density $N_{\rm crit}$ larger than 2 × 10^{18} cm^{-3}. The Debye-Hückel screening length is calculated from

$$\lambda_D = \sqrt{\frac{\varepsilon_s \varepsilon_0 k_{\rm B} T}{N_{\rm crit} e^2}} \tag{27.6}$$

to be $\lambda_D = 2.7$ nm with $\varepsilon_s = 10.2$ [66]. Here, ε_s is the static dielectric permittivity, ε_0 is the permittivity in vacuum, T is a temperature, and e is the electron charge. Consequently, the UVL band in the heavily Si-doped crystals may be ascribed to the DAP emission involving Si_{Ga} and V_{Ga} and/or V_O–V_{Ga}. Note that the ionization energies of Si_{Ga} donor [67–69] in a range E_D = 16–36 meV and V_{Ga} and/or V_O–V_{Ga} acceptors [23, 24] in a range E_A= 0.5–0.9 eV subtracted from E_g= 4.5 eV give transition energies at around 3.5 eV as shown in Fig. 27.16.

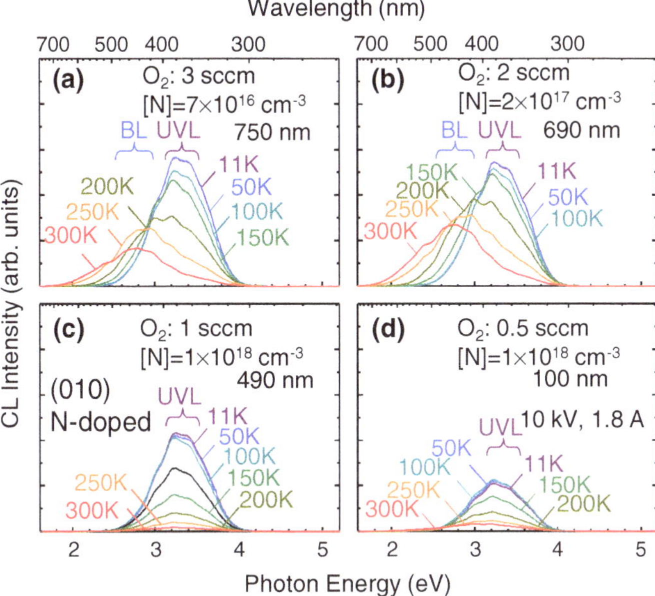

Fig. 27.19 Temperature-dependent CL spectra of (010) N-doped β-Ga_2O_3 epitaxial films grown by RF-MBE with nitrogen concentrations [N] of **a** 7×10^{16}, **b** 2×10^{17}, **c** 1×10^{18}, and **d** 1×10^{18} cm^{-3} with the oxygen gas flow rates of 3, 2, 1, and 0.5 sccm, respectively [33]

As shown in Fig. 27.19, CL spectra at LT for the lightly N-doped films with [N] = 7×10^{16}–2×10^{17} cm^{-3} exhibited the UVL band. The UVL intensities decrease as the temperature increased, and the BL band is predominately observed at 300 K. On the contrary, BL band is suppressed for the heavily N-doped films with [N] = 1×10^{18} cm^{-3} at all measured temperatures. Note that the multiple-peak structure in the CL spectra originates from unavoidable multiple internal reflections in the measurement system. These are measurement artifacts. Since the BL band correlates with V_O [27], the result may imply the decrease in V_O concentration by N-doping. It has been proposed that V_O compensates V_{Ga} or related deep acceptors, contributing to activate Si impurities in n-type β-Ga_2O_3 [70]. The decrease in V_O concentration by N-doping is thus considered to induce the compensation of residual Si donors by deep acceptors. Indeed, the N-doped films typically exhibit increase in ρ on the order of 10^{11}–10^{12} Ω cm [41, 42]. The similarity in the energy position of the UVL band in the N-doped films and the Si-doped crystals gives an expectation for an ionization energy of N_O acceptor as the same with those in V_{Ga} and related deep acceptors.

27.6 Summary

Fundamental optical properties of β-Ga_2O_3 are reviewed by focusing on its fundamental absorption edge and emission properties. Polarized transmittance and reflectance spectra of β-Ga_2O_3 crystals were measured to clarify origins of the difference in the reported E_g values (4.4–5.0 eV). The anisotropic optical properties were further clarified by the SE measurements. The characteristic optical anisotropy as well as the gradual increase in α are eventually recognized as origins of the scattering in the reported E_g values. Temperature-dependent polarized reflectance spectra were shown to discuss exciton resonance energy E_{exciton} and the broadening parameter Γ_{broad} in terms of lattice vibrations. The Γ_{broad} corresponds to the lower two ranges of LO phonon energies $\hbar\omega_{LO}$ at LT and 75 meV above 150 K. The large E_{exciton} changes with temperature in β-Ga_2O_3 were found to be originated from the exciton-LO-phonon interaction. Temperature-dependent CL spectra of β-Ga_2O_3 single crystals and epitaxial films were comprehensively shown to investigate their electronic structure and defect states. The BL intensity depended on ρ in the crystals, and the results clearly indicate strong correlation between the BL intensity and formation energy of V_O. For the N-doped films, suppression of the BL band in the heavily N-doped epitaxial films was shown as an evidence for the decrease in the V_O concentration by N-doping and resultant high resistivity in the N-doped epitaxial films.

Acknowledgements The author would like to acknowledge Dr. M. Higashiwaki for introducing and continuously giving me the opportunity to work on the β-Ga_2O_3. The author would like to thank Dr. K. Sasaki and Dr. A. Kuramata of Novel Crystal Technology Inc. and Dr. K. Goto, Dr. T. Masui, and Dr. S. Yamakoshi of Tamura Corporation for preparation of β-Ga_2O_3 crystals grown by the FZ and EFG methods. The author is grateful to Dr. Y. Nakata and Dr. M. Higashiwaki for growths of the series of N-doped epitaxial films by RF-MBE. The author would like to thank Dr. T. Sato of J. A. Woollam Japan Co., Inc., for technical support on the IRSE measurements. The author is truly grateful to Prof. T. Yamaguchi, Prof. T. Honda, Prof. Y. Itoh, Prof. H. Nagai, and Prof. M. Sato of Kogakuin University for their fruitful discussion and continuous support on the experiment. The author would like to acknowledge Prof. M. Sugiyama of Tokyo University of Science and Prof. S.F. Chichibu of Tohoku University for their continuous encouragements.

This work was supported in part by Grants-in-Aid for Scientific Research Nos. 25390071, 25289093, 25420341, and 25706020 from the Ministry of Education, Culture, Sports, Science and Technology, Japan (MEXT).

References

1. S. Geller, J. Chem. Phys. **33**, 676 (1960)
2. H.H. Tippins, Phys. Rev. **140**, A316 (1965)
3. T. Matsumoto, M. Aoki, A. Kinoshita, T. Aono, Jpn. J. Appl. Phys. **13**, 1578 (1974)
4. N. Ueda, H. Hosono, R. Waseda, H. Kawazoe, Appl. Phys. Lett. **71**, 933 (1997)
5. M. Passlack, E.F. Schubert, W.S. Hobson, M. Hong, N. Moriya, S.N.G. Chu, K. Konstadinidis, J.P. Mannaerts, M.L. Schnoes, G.J. Zydzik, J. Appl. Phys. **77**, 686 (1995)

6. M. Orita, H. Ohta, M. Hirano, H. Hosono, Appl. Phys. Lett. **77**, 4166 (2000)
7. E.G. Víllora, K. Shimamura, K. Kitamura, K. Aoki, Appl. Phys. Lett. **88**, 031105 (2006)
8. T. Oshima, T. Okuno, S. Fujita, Jpn. J. Appl. Phys. **46**, 7217 (2007)
9. L. Binet, D. Gourier, C. Minot, J. Solid State Chem. **113**, 420 (1994)
10. T. Onuma, S. Saito, K. Sasaki, T. Masui, T. Yamaguchi, T. Honda, M. Higashiwaki, Jpn. J. Appl. Phys. Part 1 **54**, 112601 (2015)
11. C. Sturm, J. Furthműller, F. Bechstedt, R. Schmidt-Grund, M. Grundmann, APL Mater. **3**, 106106 (2015)
12. C. Sturm, R. Schmidt-Grund, C. Kranert, J. Furthműller, F. Bechstedt, M. Grundmann, Phys. Rev. B **94**, 035148 (2016)
13. A. Mock, R. Korlacki, C. Briley, V. Darakchieva, B. Monemar, Y. Kumagai, K. Goto, M. Higashiwaki, M. Schubert, Phys. Rev. B **96**, 245205 (2017)
14. K. Yamaguchi, Solid State Commun. **131**, 739 (2004)
15. H. He, M.A. Blanco, R. Pandey, Appl. Phys. Lett. **88**, 261904 (2006)
16. H. He, R. Orlando, M.A. Blanco, R. Pandey, E. Amzallag, I. Baraille, M. Rérat, Phys. Rev. B **74**, 195123 (2006)
17. J.B. Varley, J.R. Weber, A. Janotti, C.G. Van de Walle, Appl. Phys. Lett. **97**, 142106 (2010) [**108**, 039901 (2016)]
18. J.B. Varley, A. Schleife, Semicond. Sci. Technol. **30**, 024010 (2015)
19. H. Peelaers, C.G. Van de Walle, Phys. Status Solidi B **252**, 828 (2015)
20. J. Furthműller, F. Bechstedt, Phys. Rev. B **93**, 115204 (2016)
21. K.A. Mengle, G. Shi, D. Bayerl, E. Kioupakis, Appl. Phys. Lett. **109**, 212104 (2016)
22. T. Harwig, F. Kellendonk, S. Slappendel, J. Phys. Chem. Solids **39**, 675 (1978)
23. T. Harwig, F. Kellendonk, J. Solid State Chem. **24**, 255 (1978)
24. L. Binet, J. Gourier, J. Phys. Chem. Solids **59**, 1241 (1998)
25. E.G. Víllora, T. Atou, T. Sekiguchi, T. Sugawara, M. Kikuchi, T. Fukuda, Solid State Commun. **120**, 455 (2001)
26. K. Shimamura, E.G. Víllora, T. Ujiie, K. Aoki, Appl. Phys. Lett. **92**, 201914 (2008)
27. T. Onuma, S. Fujioka, T. Yamaguchi, M. Higashiwaki, K. Sasaki, T. Masui, T. Honda, Appl. Phys. Lett. **103**, 041910 (2013)
28. S. Yamaoka, M. Nakayama, Phys. Status Solidi C **13**, 93 (2016)
29. S. Yamaoka, Y. Furukawa, M. Nakayama, Phys. Rev. B **95**, 094304 (2017)
30. J.B. Verley, A. Janotti, C. Franchini, C.G. Van de Walle, Phys. Rev. B **85**, 081109 (2012)
31. T.G. Castner, W. Känzig, J. Phys. Chem. Solids **3**, 178 (1957)
32. Y. Toyozawa, *Optical Process in Solids* (Cambridge University Press, Cambridge, 2003)
33. T. Onuma, Y. Nakata, K. Sasaki, T. Masui, T. Yamaguchi, T. Honda, A. Kuramata, S. Yamakoshi, M. Higashiwaki, J. Appl. Phys. **124**, 075103 (2018)
34. Z. Zhang, E. Farzana, A.R. Arehart, S.A. Ringel, Appl. Phys. Lett. **108**, 052105 (2016)
35. Y. Nakano, ECS J. Solid State Sci. Technol. **6**, P615 (2017)
36. T. Onuma, S. Saito, K. Sasaki, T. Masui, T. Yamaguchi, T. Honda, A. Kuramata, M. Higashiwaki, Jpn. J. Appl. Phys. **55**, 1202B2 (2016)
37. T. Onuma, S. Saito, K. Sasaki, K. Goto, T. Masui, T. Yamaguchi, T. Honda, A. Kuramata, M. Higashiwaki, Appl. Phys. Lett. **108**, 101904 (2016)
38. E.G. Víllora, K. Shimamura, Y. Yoshikawa, K. Aoki, N. Ichinose, J. Cryst. Growth **270**, 420 (2004)
39. A. Kuramata, K. Koshi, S. Watanabe, Y. Yamaoka, T. Masui, S. Yamakoshi, Jpn. J. Appl. Phys. **55** 1202A2 (2016)
40. Y. Nakata, T. Kamimura, A. Kuramata, S. Yamakoshi, M. Higashiwaki, in *65th The Japan Society of Applied Physics Spring Meeting*, 2018, No. 19p-P11-18 [in Japanese]
41. T. Kamimura, Y. Nakata, A. Kuramata, S. Yamakoshi, M. Higashiwaki, *2nd International Workshop on Gallium Oxide and Related Materials*, Parma, Italy, September 13 (2017), No. O2
42. M.H. Wong, C.-H. Lin, A. Kuramata, S. Yamakoshi, H. Murakami, Y. Kumagai, M. Higashiwaki, Appl. Phys. Lett. **113**, 102103 (2018)

43. M.H. Wong, K. Goto, H. Murakami, Y. Kumagai, M. Higashiwaki, IEEE Electron Device Lett. **40**, 431 (2019)
44. P. Hovington, D. Drouin, R. Gauvin, Scanning **19**, 1 (1997)
45. D. Drouin, A.R. Couture, D. Joly, X. Tastet, V. Aimez, R. Gauvin, Scanning **29**, 92 (2007)
46. T. Onuma, T. Yamaguchi, T. Honda, Phys. Status Solidi C **10**, 869 (2013)
47. F.A. Cotton, *Chemical Applications of Group Theory,* 2nd edn. (Wiley, New York, 1971)
48. C.F. Klingshirn, *Semiconductor Optics,* 4th edn. (Springer, Heidelberg, 2012)
49. M. Born, E. Wolf, *Principles of Optics: Electromagnetic Theory of Propagation, Interference and Diffraction of Light,* 7th expanded edn. (Cambridge University Press, Cambridge, U.K., 1999)
50. M. Dressel, B. Gompf, D. Faltermeier, A.K. Tripathi, J. Pflaum, M. Schubert, Opt. Express **16**, 19770 (2008)
51. G.E. Jellison Jr., M.A. McGuire, L.A. Boatner, J.D. Budai, E.D. Specht, D.J. Singh, Phys. Rev. B **84**, 195439 (2011)
52. J.A. Woollam Co., Inc., Guide to Using WVASE32
53. H. Fujiwara, *Spectroscopic Ellipsometry,* 2nd edn. (Maruzen, Tokyo, 2011)
54. M. Schubert, Phys. Rev. B **53**, 4265 (1996)
55. M. Schubert, *Infrared Ellipsometry on Semiconductor Layer Structures* (Springer, Heidelberg, 2004)
56. M. Schubert, R. Korlacki, S. Knight, T. Hofmann, S. Schöche, V. Darakchieva, E. Janzén, B. Monemar, D. Gogova, Q.-T. Thieu, R. Togashi, H. Murakami, Y. Kumagai, K. Goto, A. Kuramata, S. Yamakoshi, M. Higashiwaki, Phys. Rev. B **93**, 125209 (2016)
57. M. Cardona, *Modulation Spectroscopy* (Academic, New York, 1969)
58. D.E. Aspnes, Surf. Sci. **37**, 418 (1973)
59. Y.P. Varshni, Physica **34**, 149 (1967)
60. Z. Guo, A. Verma, X. Wu, F. Sun, A. Hickman, T. Masui, A. Kuramata, M. Higashiwaki, D. Jena, T. Luo, Appl. Phys. Lett. **106**, 111909 (2015)
61. K.P. O'Donnell, X. Chen, Appl. Phys. Lett. **58**, 2924 (1991)
62. S. Rudin, T.L. Reinecke, B. Segall, Phys. Rev. B **42**, 11218 (1990)
63. D. Dohy, G. Lucazeau, A. Revcolevschi, J. Solid State Chem. **45**, 180 (1982)
64. T. Onuma, S. Fujioka, T. Yamaguchi, Y. Itoh, M. Higashiwaki, K. Sasaki, T. Masui, T. Honda, J. Cryst. Growth **401**, 330 (2014)
65. H. Haug, S.W. Koch, *Quantum Theory of The Optical and Electronic Properties of Semiconductors*, 5th edn. (World Scientific, 2009)
66. M. Passlack, N.E.J. Hunt, E.F. Schubert, G.J. Zydzik, M. Hong, J.P. Mannaerts, R.L. Opila, R.J. Fischer, Appl. Phys. Lett. **64**, 2715 (1994)
67. K. Irmscher, Z. Galazka, M. Pietsch, R. Uecker, R. Fornari, J. Appl. Phys. **110**, 063720 (2011)
68. T. Oishi, Y. Koga, K. Harada, M. Kasu, Appl. Phys. Express **8**, 031101 (2015)
69. N. Ma, N. Tanen, A. Verma, Z. Guo, T. Luo, H. Xing, D. Jena, Appl. Phys. Lett. **109**, 212101 (2016)
70. E. Korhonen, F. Tuomisto, D. Gogova, G. Wagner, M. Baldini, Z. Galazka, R. Schewski, M. Albrecht, Appl. Phys. Lett. **106**, 242103 (2015)

Chapter 28
Phonon Properties

Phonon and Free Charge Carrier Properties in Monoclinic-Symmetry β-Ga_2O_3

Mathias Schubert, Alyssa Mock, Rafał Korlacki, Sean Knight, Bo Monemar, Ken Goto, Yoshinao Kumagai, Akito Kuramata, Zbigniew Galazka, Günther Wagner, Marko J. Tadjer, Virginia D. Wheeler, Masataka Higashiwaki and Vanya Darakchieva

Abstract We present and discuss the complete set of infrared-active phonon modes in monoclinic-symmetry crystal modification gallium oxide (gallia, β-Ga_2O_3). The phonon mode set is obtained from a comprehensive analysis of generalized spectroscopic ellipsometry data in the farinfrared and infrared spectral regions investigating

---

M. Schubert (✉) · R. Korlacki · S. Knight
Department of Electrical and Computer Engineering, University of Nebraska - Lincoln, Lincoln, NE 68588, USA
e-mail: schubert@engr.unl.edu

M. Schubert
Leibniz-Institut für Polymerforschung, 01069 Dresden, Germany

M. Schubert · B. Monemar · V. Darakchieva
Department of Physics, Chemistry and Biology (IFM), Linköpings Universitet, 58183 Linköping, Sweden
e-mail: bomon@ifm.liu.se

V. Darakchieva
e-mail: vanya@ifm.liu.se

A. Mock
Electronics Science and Technology Division, U.S. Naval Research Laboratory, Washington, DC 20375, USA
e-mail: alyssa.lynn.mock@gmail.com

K. Goto
Tokyo University of Agriculture and Technology, Koganei, Tokyo 184-8588, Japan
e-mail: gotoken@go.tuat.ac.jp

Y. Kumagai
Institute of Global Innovation Research, Tokyo University of Agriculture and Technology, Koganei, Tokyo 184-8588, Japan
e-mail: 4470kuma@cc.tuat.ac.jp

A. Kuramata
Novel Crystal Technology, Inc., Sayama, Saitama 350-1305, Japan
e-mail: akito.kuramata@tamura-ss.co.jp

M. Higashiwaki and S. Fujita (eds.), *Gallium Oxide*, Springer Series in Materials Science 293, https://doi.org/10.1007/978-3-030-37153-1_28

various n-type electrically conductive single crystal samples with different free electron volume density parameters cut under different orientations. The analysis of the ellipsometry data is performed using an eigendielectric displacement vector summation (EDVS) approach. In this approach, the effect of the free charge carriers onto the lattice modes of intrinsic β-Ga_2O_3 are removed by calculation. Density functional theory calculations are performed in the general gradient approximation and all phonon modes at the Brillouin-center and their displacement direction dependencies are obtained. Transverse and longitudinal optical phonon mode parameters polarized within the monoclinic plane as well as perpendicular to the monoclinic plane agree excellently between experiment and theory. We also present and discuss the directional limiting frequency parameters within the monoclinic plane, the shape and anisotropy of the reststrahlen band, and the order of the phonon modes in semiconductors with polar phonon modes and monoclinic crystal structure. We further present and discuss the effect of coupling of longitudinal optical phonons with free charge carriers, leading to longitudinal-phonon-plasmon coupled modes. We reveal that the coupled modes, which affect electric and thermal transport, change amplitude, frequency, and direction within the monoclinic plane as a function of free electron concentration. Finally, we show optical Hall effect measurements, and provide experimentally determined effective electron mass parameters in β-Ga_2O_3 for moderately-doped n-type samples.

28.1 Introduction

This chapter attempts to summarize the current state of knowledge about the optical phonon mode properties and their coupling with free charge carriers in gallium oxide with monoclinic crystal structure—β-Ga_2O_3 (Fig. 28.1). Up to the very recent point in time when this low-symmetry crystal modification of an ultra-wide band gap metal oxide semiconductor became of interest to the communities of semiconductor scien-

Z. Galazka · G. Wagner
Leibniz-Institut für Kristallzüchtung, 12489 Berlin, Germany
e-mail: zbigniew.galazka@ikz-berlin.de

G. Wagner
e-mail: guenter.wagner@ikz-berlin.de

M. J. Tadjer · V. D. Wheeler
U.S. Naval Research Laboratory, Washington, DC 20375, USA
e-mail: marko.tadjer@nrl.navy.mil

V. D. Wheeler
e-mail: virginia.wheeler@nrl.navy.mil

M. Higashiwaki
National Institute of Information and Communications Technology, Koganei, Tokyo 184-8795, Japan
e-mail: mhigashi@nict.go.jp

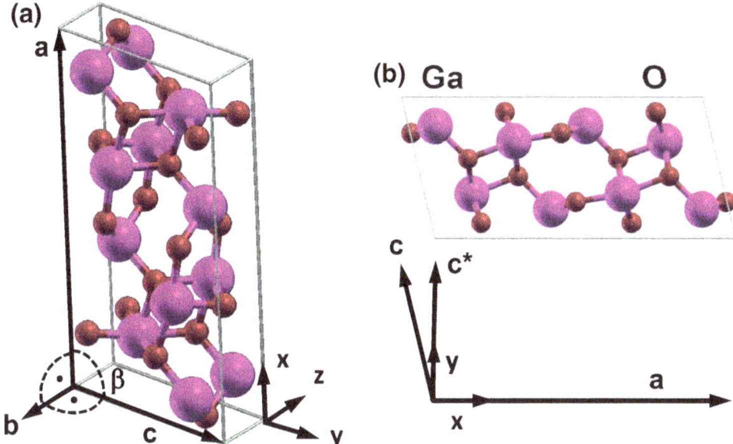

Fig. 28.1 **a** Definition of Cartesian laboratory coordinate system (x, y, z), and unit cell of β-Ga_2O_3 with monoclinic angle β, and crystal unit axes **a**, **b**, **c**. **b** Monoclinic plane **a**–**c** viewed along axis **b**. (**b** points into the plane.) Vector $\mathbf{c}^{\star}$ parallel to axis y is used for convenience. Reprinted from [1] with copyright permission by American Physical Society

tists and engineers, [2–4] very little was known about the fundamental properties of phonons and free charge carriers in semiconductors with low-symmetry crystal structures. In particular, due to the low symmetry, within the monoclinic plane, phonons are polarized in different directions, and their frequency order as well as their coupling behavior with free charge carriers deviate from the traditionally observed behaviors of polar lattice vibrations in semiconductors with higher symmetries such as in cubic, tetragonal, trigonal, hexagonal, and orthorhombic crystal structures. Thereby, β-Ga_2O_3 has become a model system for elucidating the effects of symmetry removal within the monoclinic lattice plane onto phonon order, phonon eigenpolarizations, coupled phonon-plasmon eigenpolarization rotations, and directional mode dispersions. While gallium oxide exists in multiple crystal modifications, the monoclinic (gallia) crystal modification is thermodynamically stable. This chapter focuses on the monoclinic crystal modification only.

Phonon properties of β-Ga_2O_3 have been studied by experiment and by theory. Experimental data of complete phonon mode determination is available from Schubert et al. [1]. Sturm et al. [5] performed similar experiments within a narrower spectral range and obtained a limited phonon mode set. Also, Schubert et al. observed, considered, and explained the effects of free charge carriers onto longitudinal phonon modes [1]. Knight et al. performed infrared optical Hall effect measurements and reported the conduction band effective mass parameters [6]. Liu, Mu, and Liu studied the lattice dynamical properties using the generalized gradient approximation (GGA) density functional theory (DFT) [7]. They provided frequencies of modes with transverse optical (TO) and longitudinal optical (LO) character, but only the TO modes match experimentally observed modes [1]. Schubert et al. calculated phonon mode parameters (including the displacement patterns) for all long-wavelength active phonon modes using local density approximation (LDA) DFT [1]. Both Liu, Mu, and Liu and Schubert et al. used a plane-wave pseudopotentials basis. Sturm et al. [5] performed DFT calculations using a localized Gaussian-type functions basis set, which allowed them to use a more sophisticated hybrid functional, and which is otherwise computationally expensive with plane waves. The momentum scattering time for electron-phonon interaction in β-Ga_2O_3 was derived within the relaxation time approximation considering all long-wavelength active optical phonon modes [8]. Electron-phonon coupling and its effect on the electron mobility was studied theoretically by Ghosh and Singisetti [9] and Kang et al. [10]. Ghosh and Singisetti investigated electron transport in β-Ga_2O_3 using a combination of perturbation DFT and coupling of lattice modes with collective plasmon oscillations, and Boltzmann theory-based transport calculations [11]. The same authors studied high-field transport in β-Ga_2O_3 using a combination of ab-initio calculations and full band Monte Carlo (FBMC) simulation considering the complete wave-vector (both electron and phonon) and crystal direction dependent electron phonon interactions [12].

A recent review of the present theoretical understanding of low-field and high-field electron transport properties in β-Ga_2O_3 was published by Krishnendu Ghosh and Uttam Singisetti [13]. The electron transport is governed by the electronic structure, the lattice dynamics, and the electron-phonon interactions, which are treated within a computational framework of DFT and Boltzmann transport calculations [13]. Of particular interest are the temperature and carrier concentration dependence of the electron mobility, and the influence of plasmon-phonon coupling and dynamic screening [13]. A contemporary subject of interest is the question whether the electron transport in monoclinic-symmetry modification gallium oxide is anisotropic. Presently, no experimental studies are available investigating phonon and phonon plasmon coupling at elevated temperatures.

In this chapter we summarize results from our DFT calculations to obtain the polar (transverse optical: TO; longitudinal optical: LO) phonon mode frequencies, direction and amplitude parameters at the center of the Brillouin zone (Fig. 28.6), and their dispersion along a high symmetry path through the Brillouin zone (Fig. 28.7). An eigendielectric polarization summation approach is shown which suffices to express the frequency dependence of the dielectric function tensor for long wavelengths of monoclinic-symmetry semiconductors with polar lattice vibrations. We use this approach to derive phonon and free charge carrier properties from experimental (ellipsometry) investigations. We show the relationships between TO and LO modes and the dielectric function tensor. Both TO and LO modes possess eigenpolarization and eigenpolarization loss displacement vectors (directions), respectively, parameterized within or perpendicular to the monoclinic lattice plane. We show and use coordinate-invariant generalizations of the dielectric response functions for analysis of experimental data. A generalization of the Lyddane-Sachs-Teller relation is shown for materials with monoclinic and triclinic symmetries. We introduce the optical Hall effect for materials with monoclinic symmetry. We discuss mode multiplicities, and phonon mode orders. Due to the monoclinic symmetry, the phonon mode order differs from all previously observed mode orders (frequency sequence, polarization directions) in semiconductors with higher symmetry, and we provide the explanation and classification into inner and outer phonon modes. The bands of total reflection (a.k.a. reststrahlen band) follow different rules than those usually observed in solid state physics text books for materials with higher symmetries, and are explained for β-Ga_2O_3 as an example for monoclinic symmetry semiconductors with polar lattice vibrations. We discuss the effects of coupling with free charge carriers - the longitudinal optical phonon plasmon (LPP) coupled mode frequency, amplitude, and direction parameters, as a function of free charge carrier (electron) density. We use generalized infrared and farinfrared spectroscopic Mueller matrix ellipsometry on a large set of single crystal substrates (for example: Fig. 28.2: (010) surface; Fig. 28.3: $(\bar{2}01)$ surface; Fig. 28.4: (100) surface) to determine the dielectric function tensor using a wavelength-by-wavelength method, that is, without physical model line shape function assumptions (Fig. 28.5). The thereby obtained spectra are then analyzed by our eigendielectric polarization summation approach, and we determine all TO (Fig. 28.8) and LO modes (Fig. 28.9), their frequencies, amplitude, and orientation parameters, and the mode broadening parameters. We show the directional dispersion of the so-called limiting modes within the monoclinic plane (Fig. 28.10), and we derive and discuss the reststrahlen bands (Fig. 28.11). We find excellent agreement with all parameters obtained by DFT and by ellipsometry for the lattice without free charge carriers. We augment into our dielectric function tensor model the effects of free charge carriers, and we show the behaviors of the LPP mode frequencies (Fig. 28.12), orientations (Fig. 28.13), and amplitude parameters (Fig. 28.14), in comparison with data obtained from our experiment. Finally, we show experimental and best-match model calculated optical Hall effect data (Fig. 28.18) and we determine the tensor of the conduction band effective electron mass at the Brillouin zone center.

28.2 Methods

28.2.1 Density Functional Theory

Lattice vibrations are among the best established crystal properties to be computed using density functional theory (DFT). The linear response of the electron density allows the calculation of the dynamical matrix (i.e., the matrix of interatomic force constants) at arbitrary points of the Brillouin zone using density functional perturbation theory [14, 15]. In short, eigenvalues and eigenvectors of the diagonalized dynamical matrix provide frequencies and atomic displacement patterns for all phonon modes; eigenvectors multiplied by Born effective charges for polar phonons provide vectors of infrared transition dipoles for each mode (for phonons that are not long-wavelength active, these vectors have all components equal to zero as required by symmetry); and squares of these vectors are infrared absorption intensities.

The dynamical matrix calculated at the Brillouin zone center ($\mathbf{q} = 0$) does not include contributions from the macroscopic electric fields (see section on limiting frequencies below) that are direction-dependent and constitute the so called non-analytical term of the dynamical matrix [14, 15]. Eigenvectors and eigenvalues of the dynamical matrix without the macroscopic electric fields correspond to the TO modes [16]. Including the nonanalytical term into the dynamical matrix allows obtaining parameters of the LO modes, but it requires specifying a direction of approaching the Brillouin zone center ($\mathbf{q} \rightarrow 0$) parallel to the particular phonon eigenvector. This poses a problem in the case of B_u modes, for which the modes forming the TO-LO pairs do not need to be parallel to each other. Probing the entire monoclinic plane with a small step produces a set of curves corresponding to the limiting frequencies and oscillating between various TO and LO frequencies (see Fig. 28.10), but the assignment and identification of the actual LO modes requires a careful analysis as presented in [17], and which is discussed further below.

28.2.2 Generalized Ellipsometry

Generalized ellipsometry has been successfully used previously to investigate anisotropic materials including biaxial, uniaxial, and multilayered materials [18–33]. Recently, it has been applied to materials with monoclinic crystal structure [1, 5, 34–38]. Application to phonon mode characterization in materials with orthorhombic crystal symmetry was reported recently by Schubert et. al for single crystalline stibnite (Sb_2S_3) [39] and by Mock et. al for single crystalline centrosymmetric neodymium gallate ($NdGaO_3$) [40]. Application to phonon mode characterization in monoclinic symmetry materials was also reported (β-Ga_2O_3: [1, 5]; $CdWO_4$: [41]; Y_2SiO_5: [42]). Often, the Mueller matrix formalism is used to describe the interaction of electromagnetic plane waves with anisotropic samples. Real-valued 4×4 Mueller matrix elements are measured, which connect the Stokes vector components before and after interaction with the sample,

$$\begin{pmatrix} S_0 \\ S_1 \\ S_2 \\ S_3 \end{pmatrix}_{\text{output}} = \begin{pmatrix} M_{11} & M_{12} & M_{13} & M_{14} \\ M_{21} & M_{22} & M_{23} & M_{24} \\ M_{31} & M_{32} & M_{33} & M_{34} \\ M_{41} & M_{42} & M_{43} & M_{44} \end{pmatrix} \begin{pmatrix} S_0 \\ S_1 \\ S_2 \\ S_3 \end{pmatrix}_{\text{input}}, \quad (28.1)$$

with the Stokes vector components defined by $S_0 = I_p + I_s$, $S_1 = I_p - I_s$, $S_2 = I_{45} - I_{-45}$, $S_3 = I_{\sigma+} - I_{\sigma-}$. Here, I_p, I_s, I_{45}, I_{-45}, $I_{\sigma+}$, and $I_{\sigma-}$denote the intensities for the p-, s-, +45°, −45°, right handed, and left handed circularly polarized light components, respectively [43]. Model calculations are performed where experimental Mueller matrix data are best matched against model calculated data. The model calculated data require the dielectric function tensor, ε, as input. The dielectric function tensor elements are considered as model parameters in the so-termed point-by-point tensor function approach. In the model dielectric function tensor approach, the spectral and structural properties of the dielectric function tensor are rendered by physical model lineshape functions. The lineshape model parameters are used as

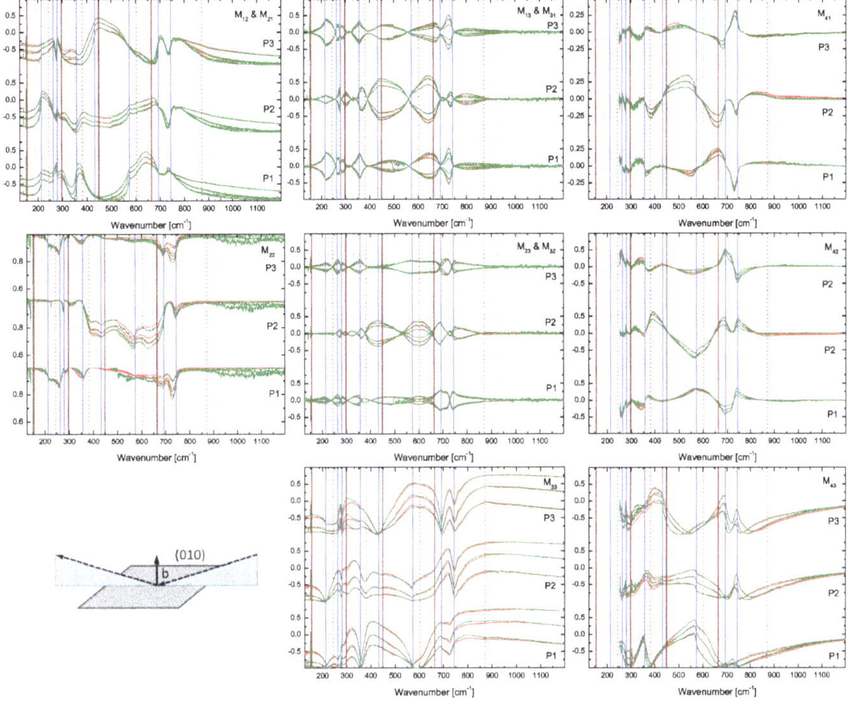

Fig. 28.2 Experimental (dotted, green lines) and best match model calculated (solid, red lines) Mueller matrix data obtained from a (010) surface at three different sample azimuth orientations. (P1: $\varphi = 62.5(4)°$, P2: $\varphi = 107.5(4)°$, P3: $\varphi = 152.5(4)°$). Data were taken at three angles of incidence ($\Phi_a = 50°, 60°, 70°$). Equal Mueller matrix data, symmetric in their indices, are plotted within the same panels for convenience. Vertical lines indicate wavenumbers of TO (solid lines) and LPP modes (dotted lines) with B_u symmetry (blue) and A_u symmetry (brown). Fourth column elements are only available from the IR instrument. Note that all elements are normalized to M_{11}. The remaining Euler angle parameters are $\theta = 0.4(2)$ and $\psi = 0.0(1)$ consistent with the crystallographic orientation of the (010) surface. Data from sample ESN11 3-19 described in Tables 28.3 and 28.4. Reprinted from [1] with copyright permission from American Physical Society

model parameters in the best-match model calculation (data regression) algorithm. The Euler angles describing the orientation of the crystal axes are free parameters in both approaches. Details of generalized ellipsometry data analysis procedures for monoclinic crystals are discussed in [1, 41, 42].

Example data for measured and best-match model calculated ellipsometry data are shown for selected angles of incidence, and for selected in-plane azimuth orientations (sample rotations with a sample edge relative to the plane of incidence) for samples with (010) (Fig. 28.2), ($\bar{2}01$) (Fig. 28.3), and (100) surface orientation (Fig. 28.4). Vertical lines indicate frequencies of TO and LPP modes (see further below) with A_u and B_u symmetry.

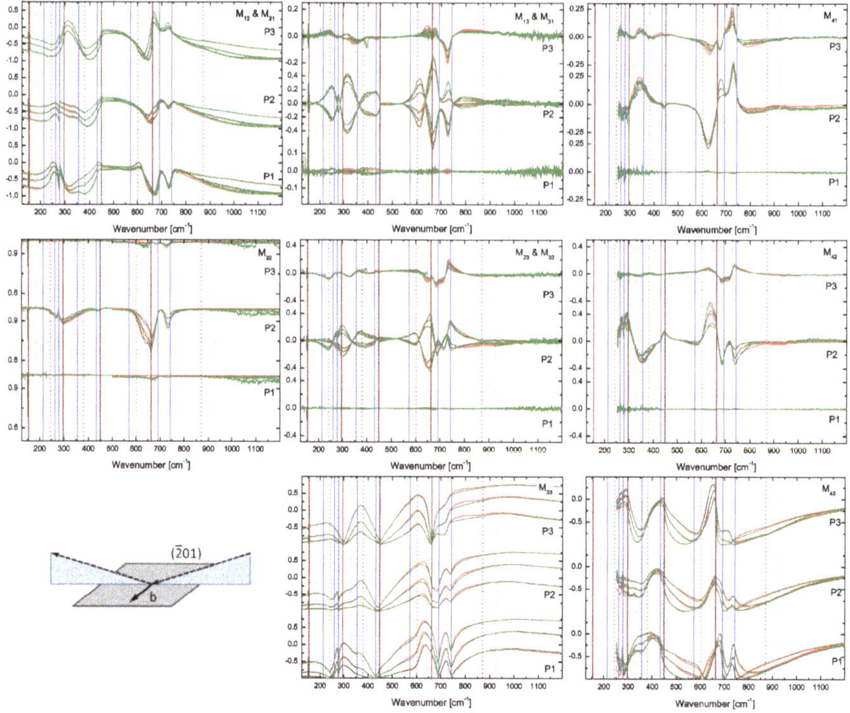

Fig. 28.3 Same as Fig. 28.2 for a $(\bar{2}01)$ sample at azimuth orientation P1: $\varphi = 179.(3)°$, P2: $\varphi = 224.(3)°$, P3: $\varphi = 269.(3)°$. $\theta = 90.(5)$ and $\psi = -28(1)$, consistent with the crystallographic orientation of a $(\bar{2}01)$ surface. Data from T1730-5 described in Tables 28.3 and 28.4. Reprinted from [1] with copyright permission from American Physical Society

Data for a complete set of wavelength-by-wavelength (point-by-point) obtained and best-match model lineshape calculated dielectric function tensor element spectra are shown in Fig. 28.5. In our choice of coordinates, modes with B_u symmetry shape the dielectric function spectra within the $(x - y)$ plane, and modes with A_u affect ε_{zz} only. Note that ε is symmetric. Note that $\varepsilon_{xz} = \varepsilon_{yz} = 0$. Note also that the appearance of the spectra within the $(x - y)$ plane depend on a rotation around the z axis. Details of coordinate system transformations are discussed in [1].

28.3 Phonon Mode Properties in β-Ga_2O_3

28.3.1 Unit Cell and Brillouin Zone Center Modes

Figure 28.1 depicts the unit cell of β-Ga_2O_3, the definition of the unit cell axes, and Cartesian axes used in this work. The primitive cell of β-Ga_2O_3 contains 10 atoms

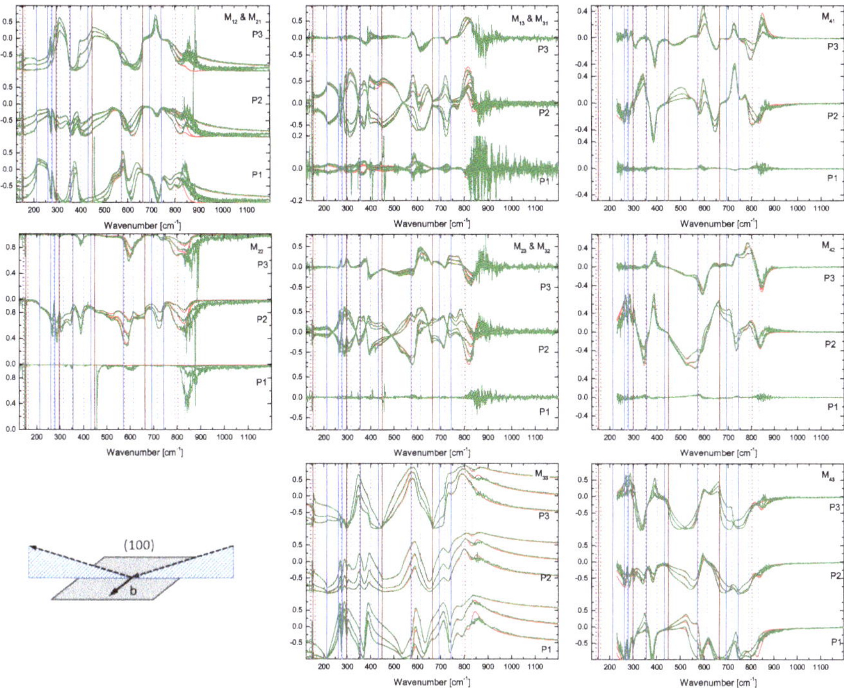

Fig. 28.4 Same as Fig. 28.2 for a (100) sample at azimuth orientation P1: $\varphi = -1.13(1)^\circ$, P2: $\varphi = 43.8(1)^\circ$, P3: $\varphi = 88.6(1)^\circ$. $\theta = 89.8(1)$ and $\psi = 99.0(1)$, consistent with the crystallographic orientation of a (100) surface. The free electron density parameter is lower than in the previous figures, hence the LPP modes (dotted vertical lines) appear at lower frequency than in Figs. 28.2 and 28.3. Data from GAO-37-23 described in Tables 28.3 and 28.4

(6 oxygen and 4 gallium), which results in 30 phonon modes: 3 acoustic and 27 optical modes. At the Brillouin zone center (Γ) the optical modes belong to the irreducible representation:

$$\Gamma^{\mathrm{opt}} = 10A_{\mathrm{g}} + 5B_{\mathrm{g}} + 4A_{\mathrm{u}} + 8B_{\mathrm{u}}.$$

A_{g} and B_{g} modes are Raman active, and A_{u} and B_{u} modes are long wavelength (farinfrared (FIR) and infrared (IR)) active. The long wavelength active phonon modes have their transition dipoles oriented parallel (A_{u} modes) and perpendicular (B_{u} modes) to the main symmetry axis. Thus, the B_{u} modes are distributed within the monoclinic plane (**a**–**c** plane) without additional symmetry constraints.

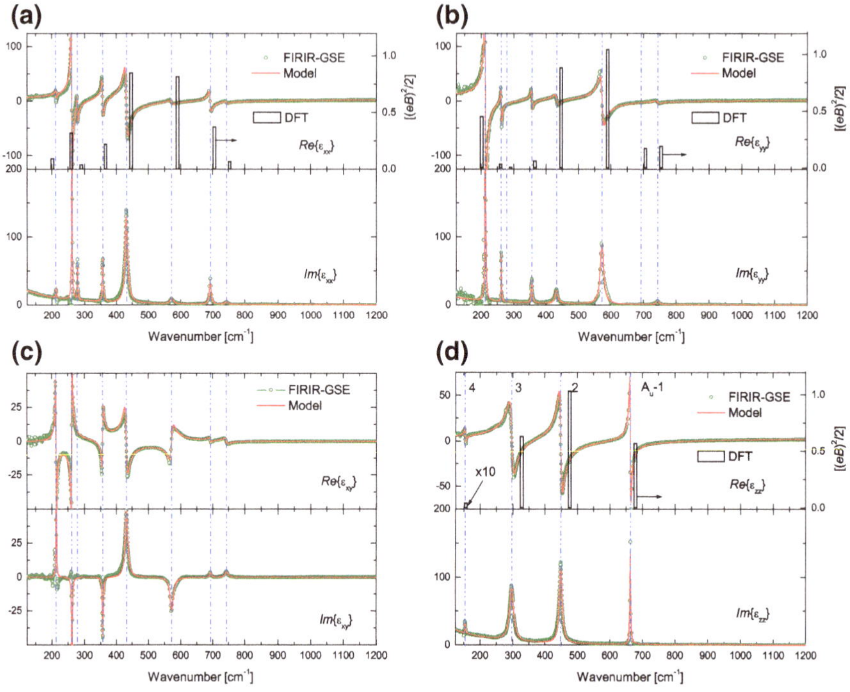

Fig. 28.5 Dielectric function tensor elements ε_{xx} (**a**), ε_{yy} (**b**), ε_{xy} (**c**), and ε_{zz} (**d**). Lines indicate results from wavelength by wavelength best match model calculation to experimental Mueller matrix data (dotted; green) and best match model lineshape analysis (solid; red). Vertical lines indicate B_u mode TO frequencies. Vertical bars indicate relative displacement amplitudes of phonon mode from DFT calculations projected onto direction **x**. Reprinted from [1] with copyright permission from American Physical Society

28.3.2 TO and LO Mode Frequency and Orientation Parameters

28.3.2.1 Phonon Properties Calculated by DFT

Table 28.1 lists parameters [frequencies, amplitudes/intensities (if available), and orientations of the transition dipoles for B_u phonons within the monoclinic **a**–**c** plane (if available)] of long wavelength active phonon modes in β-Ga_2O_3 from selected literature. Figure 28.6 shows renderings of TO phonon modes with A_u and B_u symmetries. We observed that DFT at the local density approximation (LDA) level, despite overall better agreement of phonon mode frequency parameters with experimentally determined frequency parameters fails to reproduce our experimentally observed sequence of TO-LO pairs (order of TO-LO modes, see further below) [17]. Therefore, we include in Table 28.1 a complete set of parameters of TO and LO phonon modes from the calculation presented in [17], which were performed by

Table 28.1 Phonon mode parameters for A_u and B_u modes obtained from DFT calculations. Renderings of atomic displacement patterns for TO modes are shown in Fig. 28.6

$X = B_u$		$k=1$	2	3	4	5	6	7	8
$(A_k^X)^2$ ((D/Å)^2/amu)	a	4.65	9.48	30.57	28.1	5.37	0.89	7.33	10.43
$\omega_{TO,k}$)(cm^{-1})	a	753.76	705.78	589.86	446.83	365.84	289.71	260.4	202.4
$\alpha_{TO,k}$ (°)	a	70.9	25.0	128	46.1	165	7.5	175	101
$(A_k^X)^2$ ((D/Å)^2/amu)	b	3.76	8.24	30.81	29.20	5.34	1.03	8.93	12.65
$\omega_{TO,k}$ (cm^{-1})	b	710.52	661.63	539.90	413.11	344.48	267.29	250.17	200.08
$\alpha_{TO,k}$ (°)	b	74.2	28.2	128.7	46.7	171.1	21.0	179.4	104.6
$(A_k^X)^2$ ((D/Å)^2/amu)	b	30.02	34.18	14.62	15.76	2.099	1.559	0.6185	1.107
$\omega_{LO,k}$ (cm^{-1})	b	759.30	725.07	672.60	550.27	370.05	290.67	269.79	254.38
$\alpha_{LO,k}$ (°)	b	66.7	158.9	112.0	38.4	147.3	147.2	94.3	85.5
$(A_k^X)^2/(A_3^X)^2$	c	0.13	0.23	1	0.96	0.16	0.03	0.33	0.41
$\omega_{TO,k}$ (cm^{-1})	c	742.5	692.5	560.8	434.2	361.0	281.9	267.3	224.3
$\alpha_{TO,k}$ (°)	c	76	30	130	47	173	39	176	101
$X = A_u$		$k=1$	2	3	4				
$(A_k^X)^2$ ((D/Å)^2/amu)	a	12.76	23.24	14.34	0.07				
$\omega_{TO,k}$ (cm^{-1})	a	678.39	475.69	327.45	155.69				
$(A_k^X)^2$ ((D/Å)^2/amu)	b	13.19	24.81	15.18	0.1968				
$\omega_{TO,k}$ (cm^{-1})	b	633.52	432.26	294.65	148.52				
$(A_k^X)^2$ ((D/Å)^2/amu)	b	40.99	10.84	1.528	0.0208				
$\omega_{LO,k}$ (cm^{-1})	b	732.04	537.08	335.97	149.58				
$(A_k^X)^2/(A_{B_u-3}^X)^2$	c	0.43	0.78	0.50	0.01				
$\omega_{TO,k}$ (cm^{-1})	c	665.8	447.0	300.5	160.7				

[a]LDA-DFT (PZ), [1]
[b]GGA-DFT (PBE), [17]
[c]HF-DFT (B3LYP), [5]

DFT with generalized gradient approximation (GGA). The TO mode parameters in this set are slightly different than the TO modes presented by Schubert et al. [1]. Figure 28.7 depicts the dispersion of all 30 phonon mode branches, including acoustic and Raman active modes, along a high symmetry path across the Brillouin zone using our GGA-DFT results. The definition of the high symmetry path is given in [36].

28.3.2.2 Phonon Modes Obtained from Dielectric Function Tensor

The frequency dependence of the dielectric function tensor, $\varepsilon(\omega)$, and the dielectric loss function tensor, $\varepsilon^{-1}(\omega)$, are determined by two sets of parameters describing the effects of characteristic eigenmodes. The eigenmode sets are determined by the TO and LO phonon mode properties, except for constant contributions from higher energy electronic polarizations (ε_∞). TO modes occur at frequencies at which

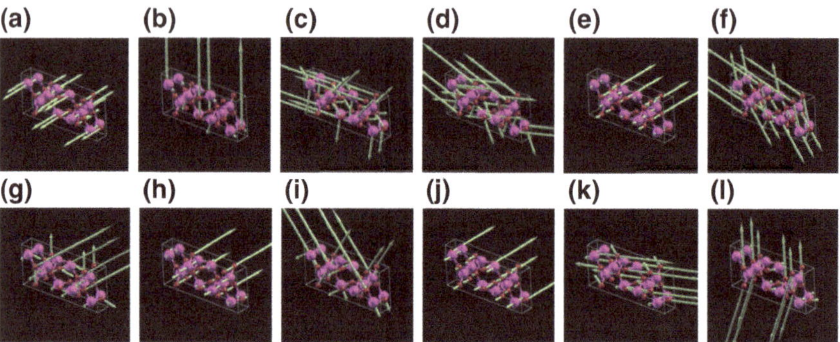

Fig. 28.6 Renderings of TO phonon modes in β-Ga_2O_3 with A_u (**a** $A_u(4)$, **e** $A_u(3)$, **h** $A_u(2)$, **j** $A_u(1)$) and B_u symmetry (**b** $B_u(8)$, **c** $B_u(7)$, **d** $B_u(6)$, **f** $B_u(5)$, **g** $B_u(4)$, **i** $B_u(3)$, **k** $B_u(2)$, **l** $B_u(1)$). Reprinted from [1] with copyright permission from American Physical Society

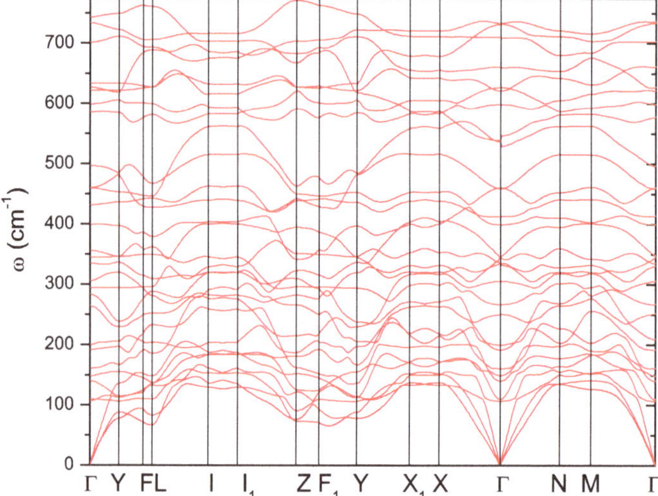

Fig. 28.7 Phonon mode band dispersion obtained from GGA-DFT calculations on a 8 × 8 × 8 regular grid over the first Brillouin zone. The definition of the high symmetry path and the coordinates of the high symmetry points are given in [36]

dielectric resonance occurs for electric fields along $\hat{\mathbf{e}}_l$ with eigendielectric displacement unit vectors then defined as $\hat{\mathbf{e}}_l = \hat{\mathbf{e}}_{\mathrm{TO},l}$. Similarly, LO modes occur when the dielectric loss approaches infinity for electric fields along $\hat{\mathbf{e}}_l$ with eigendielectric displacement unit vectors then defined as $\hat{\mathbf{e}}_l = \hat{\mathbf{e}}_{\mathrm{LO},l}$. This can be written as:

$$|\det\{\varepsilon(\omega = \omega_{\mathrm{TO},l})\}| \to \infty, \tag{28.2a}$$

$$|\det\{\varepsilon^{-1}(\omega = \omega_{\mathrm{LO},l})\}| \to \infty, \tag{28.2b}$$

$$\varepsilon^{-1}(\omega = \omega_{\mathrm{TO},l})\hat{\mathbf{e}}_{\mathrm{TO},l} = 0, \tag{28.2c}$$

$$\varepsilon(\omega = \omega_{\mathrm{LO},l})\hat{\mathbf{e}}_{\mathrm{LO},l} = 0, \tag{28.2d}$$

where l is an index for multiple frequencies in the sets [35]. We state here without further proof that in a given crystal regardless of crystal structure symmetry the total number of all TO modes must always equal the total number of all LO modes.

28.3.3 The Eigendielectric Summation Approaches

28.3.3.1 The Eigendielectric Displacement Vector Summation

In the eigendielectric displacement vector summation (EDVS) approach [1, 35, 41] tensor ε is composed of sums of dyadics, $(\hat{\mathbf{e}}_{\mathrm{TO},l} \otimes \hat{\mathbf{e}}_{\mathrm{TO},l})$, scaled with frequency-dependent complex-valued response functions, $\varrho_{\mathrm{TO},l}$

$$\varepsilon = \varepsilon_\infty + \sum_{l=1}^{N} \varrho_{\mathrm{TO},l}(\hat{\mathbf{e}}_{\mathrm{TO},l} \otimes \hat{\mathbf{e}}_{\mathrm{TO},l}). \tag{28.3}$$

In this formulation of ε, parameters in functions $\varrho_{\mathrm{TO},l}$ and directions $\hat{\mathbf{e}}_{\mathrm{TO},l}$ are directly accessible by comparison to a tensor ε determined from experiment. Contributions to the dielectric polarizability from electronic excitations at much shorter wavelength are compounded into the constant tensor ε_∞. Note that ε_∞ may be written as the sum of three frequency-independent dyadics[1]

$$\varepsilon_\infty = \varepsilon_{\infty,\mathrm{a}}(\hat{\mathbf{e}}_\mathrm{a} \otimes \hat{\mathbf{e}}_\mathrm{a}) + \varepsilon_{\infty,\mathrm{b}}(\hat{\mathbf{e}}_\mathrm{b} \otimes \hat{\mathbf{e}}_\mathrm{b}) + \varepsilon_{\infty,\mathrm{c}}(\hat{\mathbf{e}}_\mathrm{c} \otimes \hat{\mathbf{e}}_\mathrm{c}), \tag{28.4}$$

where $\varepsilon_{\infty,\mathrm{a}}$, $\varepsilon_{\infty,\mathrm{b}}$, $\varepsilon_{\infty,\mathrm{c}}$ are real-valued parameters, and $\hat{\mathbf{e}}_\mathrm{a}$, $\hat{\mathbf{e}}_\mathrm{b}$, $\hat{\mathbf{e}}_\mathrm{c}$ are unit vectors along the crystal axes **a**, **b**, **c**, respectively. Data for a complete set of wavelength by wavelength obtained and best-match model lineshape calculated dielectric function tensor element spectra are shown in Fig. 28.5.

28.3.3.2 The Eigendielectric Displacement Loss Vector Summation

In full analogy to the EDVS approach, the inverse of the dielectric function can be modeled by an eigendielectric displacement loss vector summation (EDLVS) approach

[1] In such a presentation, the tensor elements are over determined, meaning, an unambiguous set of dyadics cannot be obtained from knowledge of tensor ε_∞.

$$\varepsilon^{-1} = \varepsilon_{\infty}^{-1} - \sum_{l=1}^{N} \varrho_{\mathrm{LO},l}(\hat{\mathbf{e}}_{\mathrm{LO},l} \otimes \hat{\mathbf{e}}_{\mathrm{LO},l}). \tag{28.5}$$

Note that the minus sign in front of the summation is chosen to ensure positive real-valued amplitude parameters for LO modes defined further below. In this formulation of ε^{-1}, parameters in functions $\varrho_{\mathrm{LO},l}$ and directions $\hat{\mathbf{e}}_{\mathrm{LO},l}$ are directly accessible by comparison to the numerically obtained inverse tensor ε^{-1} of tensor ε determined from experiment. Contributions to the dielectric polarizability loss from electronic excitations at much shorter wavelength are compounded into the constant tensor $\varepsilon_{\infty}^{-1}$. Note that $\varepsilon_{\infty}^{-1}$ can be written as the sum of three frequency-independent dyadics

$$\varepsilon_{\infty}^{-1} = \varepsilon_{\infty,\mathrm{a}}^{-1}(\hat{\mathbf{e}}_{\mathrm{a}}^{\dagger} \otimes \hat{\mathbf{e}}_{\mathrm{a}}^{\dagger}) + \varepsilon_{\infty,\mathrm{b}}^{-1}(\hat{\mathbf{e}}_{\mathrm{b}}^{\dagger} \otimes \hat{\mathbf{e}}_{\mathrm{b}}^{\dagger}) + \varepsilon_{\infty,\mathrm{c}}^{-1}(\hat{\mathbf{e}}_{\mathrm{c}}^{\dagger} \otimes \hat{\mathbf{e}}_{\mathrm{c}}^{\dagger}), \tag{28.6}$$

where $\varepsilon_{\infty,\mathrm{a}}^{-1}$, $\varepsilon_{\infty,\mathrm{b}}^{-1}$, $\varepsilon_{\infty,\mathrm{c}}^{-1}$ are real-valued parameters, and $\hat{\mathbf{e}}_{\mathrm{a}}^{\dagger}$, $\hat{\mathbf{e}}_{\mathrm{b}}^{\dagger}$, $\hat{\mathbf{e}}_{\mathrm{c}}^{\dagger}$ are the reciprocal space unit vectors of β-Ga_2O_3.

28.3.3.3 Phonon Response Model Functions

Anharmonic broadened Lorentzian oscillator functions can be used to describe functions ϱ in (28.3) and (28.5).

$$\varrho_{k,l}(\omega) = \frac{A_{k,l}^2 - i\Gamma_{k,l}\omega}{\omega_{k,l}^2 - \omega^2 - i\omega\gamma_{k,l}}. \tag{28.7}$$

Here, $A_{k,l}$, $\omega_{k,l}$, $\gamma_{k,l}$, and $\Gamma_{k,l}$ denote amplitude, resonance frequency, harmonic broadening, and anharmonic broadening parameter for TO (k = "TO") or LO (k = "LO") mode l, respectively, and ω is the frequency of the driving electromagnetic field. Note that $A_{k,l}$, $\omega_{k,l}$, $\gamma_{k,l}$, and $\Gamma_{k,l}$ are energy dimension parameters, and given in units of inverse centimeter.[2]

28.3.3.4 Coordinate-Invariant Generalized Dielectric Function

A scalar form of the dielectric response of β-Ga_2O_3 can be expressed from the determinant of ε [1, 35]

$$\det\{\varepsilon(\omega)\} = \det\{\varepsilon_{\infty}\} \prod_{l=1}^{N} \frac{\omega_{\mathrm{LO},l}^2 - \omega^2 - i\omega\gamma_{\mathrm{LO},l}}{\omega_{\mathrm{TO},l}^2 - \omega^2 - i\omega\gamma_{\mathrm{TO},l}}. \tag{28.8}$$

Parameters $\gamma_{\mathrm{LO},l}$ and $\gamma_{\mathrm{TO},l}$ account for lifetime broadenings of TO and LO modes [42].

---

[2] We note a misprint in [1] where the square on the amplitude parameter in (28.10) was erroneously omitted.

28.3.3.5 Coordinate-Invariant Generalized Dielectric Loss Function

A function analogous to (28.8) can be obtained for the dielectric loss response, and which has the following form [42]:

$$\det\{\varepsilon^{-1}(\omega)\} = \det\{\varepsilon_\infty^{-1}\}\prod_{l=1}^{N}\frac{\omega_{\mathrm{TO},l}^2 - \omega^2 - i\omega\gamma_{\mathrm{TO},l}}{\omega_{\mathrm{LO},l}^2 - \omega^2 - i\omega\gamma_{\mathrm{LO},l}}. \tag{28.9}$$

28.3.3.6 Coordinate-Invariant Generalized Lyddane-Sachs-Teller (Schubert-LST) Relation

While omitting derivation here for the coordinate-invariant functions above, an immediate consequence can be expressed for $\omega = 0$, which leads to a generalization of the well-known Lyddane-Sachs-Teller relation [35, 44]

$$\det\{\varepsilon(\omega = 0)\} = \det\{\varepsilon_\infty\}\prod_{l=1}^{N}\frac{\omega_{\mathrm{TO},l}^2}{\omega_{\mathrm{LO},l}^2}. \tag{28.10}$$

This expression holds true for all materials with polar lattice vibrations and for all crystal structure symmetries. The product runs over all mode pairs. It can be shown that all modes of type TO and type LO can be arranged into pairs of exactly one TO and exactly one LO, where every mode occurs only once. This is discussed below for monoclinic symmetry structures.[3] The product expands over all such pairs.

28.3.4 Mode Multiplicity in β-Ga_2O_3 Without Free Charge Carriers

B_u modes:
In β-Ga_2O_3, 8 TO modes and 8 associated LO modes (B_u symmetry) are oriented (polarized) within the monoclinic **a–c** plane. Note that none of the eigenvectors, $\hat{\mathbf{e}}_{\mathrm{TO},l}$, $\hat{\mathbf{e}}_{\mathrm{LO},l}$ are parallel or perpendicular to each other.

A_u modes:
In β-Ga_2O_3, 4 TO modes and 4 associated LO modes (A_u symmetry) are directed perpendicular to the monoclinic plane. All eigenvectors, $\hat{\mathbf{e}}_{\mathrm{TO},l}$, $\hat{\mathbf{e}}_{\mathrm{LO},l}$ are parallel to axis **b**.

---

[3]In materials with higher symmetries, for example, in a hexagonal lattice, multiple modes may have the same frequency. In a hexagonal lattice, there may be one TO mode frequency observed experimentally, but actually 3 different TO modes with equal amplitude parameters but polarized in different directions exist, where each direction is $\frac{\pi}{3}$ rotated from its neighboring mode within the hexagonal plane. Then there are also exactly 3 corresponding LO modes with the same frequency. See also Schubert [35].

28.3.4.1 Comparison Between DFT and GSE Results

Figure 28.8 depicts a graphic rendering of the amplitude (length), frequency (color), and direction (polar angle) of all TO modes within the monoclinic plane. Figure 28.8a depicts data obtained by DFT, Fig. 28.8b depicts data obtained by generalized spectroscopic ellipsometry (GSE). Figure 28.9 depicts a graphic rendering of the amplitude (length), frequency (color), and direction (polar angle) of all LO modes within the monoclinic plane. Figure 28.9a depicts data obtained by DFT, Fig. 28.9b depicts data obtained by GSE. Data for polarization along axis **b** are not plotted here, and given below in Tables 28.1 (DFT) and 28.2 in [17] (GSE). Table 28.2 summarizes all TO and LO mode parameters determined this far from analysis of multiple samples. LO mode parameters are derived from the best-match TO mode parameter set, ignoring broadening. Hence, no broadening parameters are shown, which differ for different

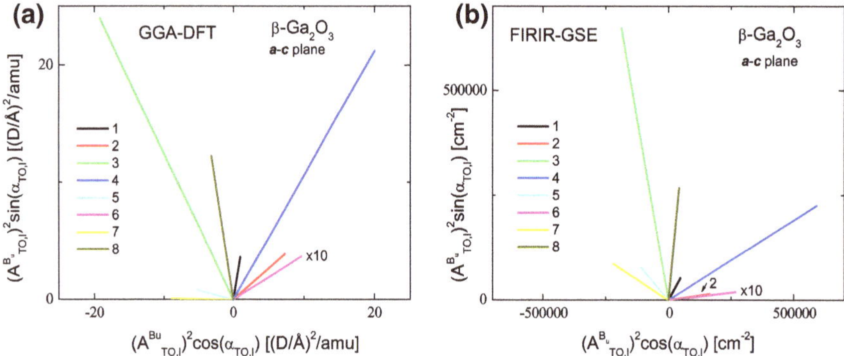

Fig. 28.8 Monoclinic plane TO modes obtained from GGA-DFT calculations (**a**) and long wavelength GSE data analysis (**b**). In (**a**), the abscissa is parallel to **a** and the ordinate is parallel to **c*** defined in Fig. 28.1b. In (**b**), the results are plotted as obtained from the ellipsometry analysis. See text for explanation of the rotational shift between (**a**) and (**b**)

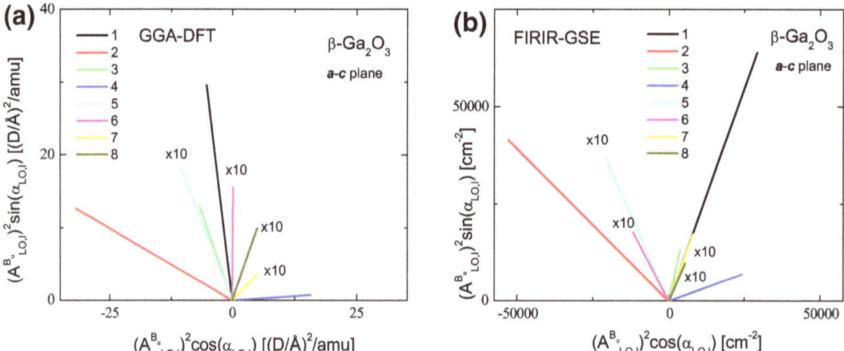

Fig. 28.9 Same as Fig. 28.8 for LO modes

Table 28.2 Phonon mode parameters for A_u and B_u modes obtained from GSE data analyses. Parameters are obtained from multiple samples with different free electron densities. The LO mode parameters are derived from the TO mode parameters assuming zero broadening. The broadening parameters vary among different samples, and are omitted here for the TO modes. For LO modes, no broadening parameters have been measured yet since no sample without free charge carriers has been investigated yet. ($\varepsilon_{\infty,xx} = 3.749$, $\varepsilon_{\infty,xy} = 0.0799$, and $\varepsilon_{\infty,yy} = 3.205$, and $\varepsilon_{\infty,zz} = 3.714$)

$X = B_u$		$k = 1$	2	3	4	5	6	7	8
$A_{TO,l}$ (cm^{-1})	a	266.21	406.55	821.90	795.71	365.87	164.073	485.71	520.70
$\omega_{TO,l}$ (cm^{-1})		743.48	692.44	572.52	432.58	356.79	279.16	262.34	213.79
$\alpha_{TO,l}$ (°)		47.87	5.14	106.24	20.90	144.81	4.28	158.36	80.92
$A_{LO,l}$ (cm^{-1})		265.18	259.28	117.83	158.72	65.13	46.10	43.59	33.25
$\omega_{LO,l}$ (cm^{-1})		806.04	774.36	715.59	585.67	391.04	305.73	287.31	270.06
$\alpha_{LO,l}$ (°)		65.45	−38.15	74.21	15.92	−60.47	−56.30	65.16	61.10
$X = A_u$		$k = 1$	2	3	4				
$A_{TO,l}$ (cm^{-1})	a	544.90	727.05	592.07	77.89				
$\omega_{TO,l}$ (cm^{-1})		663.17	448.66	296.63	154.84				
$A_{LO,l}$ (cm^{-1})		261.113	123.082	48.131	6.010				
$\omega_{LO,l}$ (cm^{-1})		781.26	562.86	345.97	156.35				

[a]Note the misprint in [1], where the square was omitted accidentally in (28.10)

samples. TO mode and LPP mode broadening parameters are shown in [1] for a specific set of n-type conductive samples. Systematic analyses of broadening as a function of doping and/or lattice strain has not been performed yet. Note further that in ellipsometry analysis the absolute orientation of the TO modes relative to the crystal axes cannot be determined, but must be input from structural analysis, e.g., from X-ray investigations. Hence, there is an approximate 20° rotation shift between the DFT and GSE results, which is due to an accidental rotational shift in the ellipsometry sample analysis of what was thought to be the direction of the axis **a** and what was actually the orientation of axis **a**. We note that all A_u modes are parallel to each other, and perpendicular to all B_u modes. Note further that for all A_u modes, the TO-LO rule is fulfilled.[4] Note that the TO-LO rule is not fulfilled for modes of B_u symmetry, which will be explained in the next section. We also note very close agreement between DFT and GSE results. Note further that these modes are the result for the case of insulating β-Ga_2O_3.

[4]The TO-LO rule is observed from the fact that in materials with symmetries higher than monoclinic and triclinic, along major, orthogonal polarization directions, with ascending frequency, all frequencies of TO and LO modes polarized parallel to such major direction fall into pairs TO-LO, TO-LO, etc., where every TO mode is succeeded directly by one LO mode.

28.3.5 *Inner and Outer Phonon Modes and Phonon Mode Order in Monoclinic Plane of* β-Ga_2O_3

28.3.5.1 Phonon Mode Order

The order of phonon modes can be understood from an eigenpolarization reflectance analysis for the monoclinic surface [17]. Electric fiels $\mathbf{E}_{\pm}$ and corresponding wave propagation constants, $n_{\pm}$ (indices of refraction), follow from Maxwell's postulates

$$\left[\varepsilon_{ij} - n^2\left(\delta_{ij} - k_i k_j\right)\right] E_j = 0, \tag{28.11}$$

where δ is the Kronecker symbol, $\mathbf{k} = k_0(k_x, k_y, k_z)$ is the wave vector, $k_0 = \frac{\omega}{c}$, and c is the speed of light. At normal incidence to the **a**–**c** plane ($k_x = k_y = 0$), $n_{\pm} = \sqrt{p \pm q}$, with $p = \varepsilon_{xx} + \varepsilon_{yy}$, and $q = \sqrt{(\varepsilon_{xx} - \varepsilon_{yy})^2 + 4\varepsilon_{xy}^2}$. When $n_{\pm} \rightarrow 0$ it follows that $\det(\varepsilon_{ij}) \rightarrow 0$, and hence $\omega \rightarrow \omega_{\mathrm{LO},l}$. Consequently, 2 types of LO modes exist, one for $p - q = 0$, and one for $p + q = 0$. We refer to those as LO$_-$ and LO$_+$, respectively. The reflectance is $r_{\pm} = \frac{n_{\pm}-1}{n_{\pm}+1}$ [45]. The two eigenmodes lead to two different types of bands of total reflection

$$\sqrt{r_- r_-^{\star}} = 1 \Leftrightarrow p - q < 0, \tag{28.12}$$

and

$$\sqrt{r_+ r_+^{\star}} = 1 \Leftrightarrow p + q < 0, \tag{28.13}$$

thereby defining so-called inner (index +) and outer (index −) modes [17]. At normal incidence, for ascending frequency, total reflection in r_- occurs and forms a band, which stretches from a TO$_-$ to a LO$_-$. We refer to these modes as "outer" modes, which form pairs: [TO$_{j,-}$, LO$_{j',-}$], where j and j' index the occurrence of TO and LO modes, for example, with ascending frequency. In the second case, at normal incidence, bands of total reflection in r_+ stretch between pairs of frequencies of TO$_+$ and LO$_+$ modes, for ascending frequencies. We refer to these modes as "inner" modes, which form pairs: [TO$_{j',+}$, LO$_{j',+}$]. Pairs [TO$_+$, LO$_+$] fall within pairs of [TO$_-$, LO$_-$] [17].

28.3.5.2 Limiting Frequencies (Directional LO Modes in β-Ga_2O_3)

In the Born and Huang approach, [46] the lattice dynamic properties in crystals with arbitrary symmetries are categorized under different electric field **E** and dielectric displacement **D** conditions [16]. $\mathbf{E} = 0$ and $\mathbf{D} = 0$ defines the transverse optical (TO) modes, $\omega_{\mathrm{TO},l}$, associated with dipole moment. $\mathbf{E} \neq 0$ but $\mathbf{D} = 0$ defines the longitudinal optical (LO) modes, $\omega_{\mathrm{LO},l}$, identical with the definitions through the dielectric tensor above. $\mathbf{E} \neq 0$ and $\mathbf{D} \neq 0$ defines the so-called limiting frequencies

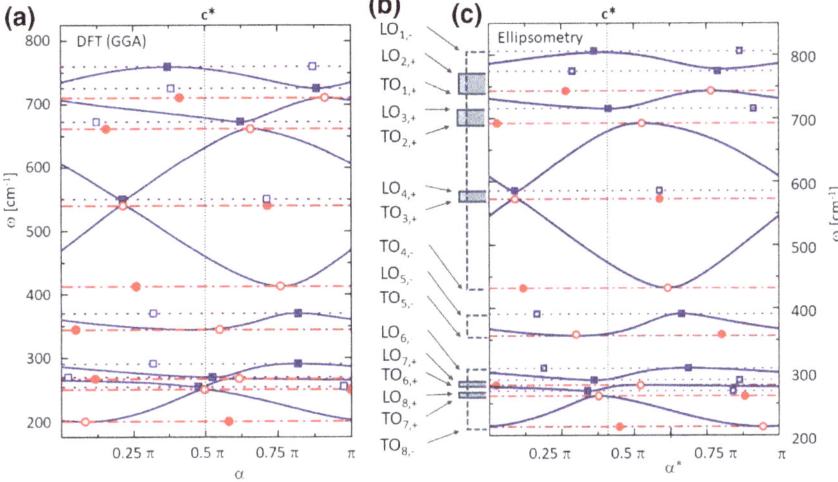

Fig. 28.10 Limiting long-wavelength mode frequencies $\omega(\hat{\alpha})_l$ (solid lines) of monoclinic symmetry β-Ga_2O_3 as a function of unit direction $\hat{\alpha} = \cos\alpha\hat{\mathbf{x}} + \sin\alpha\hat{\mathbf{y}}$ in the **a**–**c** plane obtained from **a** density functional theory calculations, and (**c**) generalized ellipsometry investigations. $\alpha = 0 : \hat{\alpha} \parallel \mathbf{a}$, $\alpha = \pi/2 : \hat{\alpha} \parallel \mathbf{c}^\star$; $\mathbf{c}^\star$ is perpendicular to **a** within the **a**–**c** plane, indicated by the vertical dashed line. The angular parameter α was offset by approximately −20° during experiment, $\alpha^\star = \alpha - 20°$. Indicated are frequencies and eigenvectors (closed symbols) and their normal directions (open symbols) of TO (red circles), and LO modes (purple squares). Horizontal dash-dotted and dotted lines indicate frequencies of TO and LO modes, respectively. **b** 3 pairs of outer modes, and 5 pairs of inner modes occur in β-Ga_2O_3, forming 3 phonon bands with phonon mode order matching the previously observed order of B_u-symmetry TO and LO frequencies in [1]. Parameters in (**a**) are calculated in this work, parameters in (**c**) are taken from experiment described in [1]. Reprinted from [17] with copyright permission from American Physical Society

$\omega(\hat{\alpha})_l$. The limiting frequencies, $\omega(\hat{\alpha})_l$, depend on the direction of a unit vector within the $\mathbf{a} - \mathbf{c}$ plane, $\hat{\alpha} = \cos\alpha\hat{\mathbf{x}} + \sin\alpha\hat{\mathbf{y}}$. No solution for $\omega(\hat{\alpha})_l$ exists within bands of total reflection, and the allowed frequency regions are confined to frequency regions $[\mathrm{TO}_-, \mathrm{LO}_-] \cap [\mathrm{TO}_+, \mathrm{LO}_+]$. Hence, minimum - maximum bounds for $\omega(\hat{\alpha})_l$ can be (i) $\omega_{\mathrm{TO}_{j,-}}$ - $\omega_{\mathrm{TO}_{j',+}}$, (ii) $\omega_{\mathrm{LO}_{j,+}}$ - $\omega_{\mathrm{TO}_{j',+}}$, (iii) $\omega_{\mathrm{LO}_{j,+}}$ - $\omega_{\mathrm{LO}_{j',-}}$, and (iv) $\omega_{\mathrm{TO}_{j,-}}$ - $\omega_{\mathrm{LO}_{j',-}}$ [17]. The limiting mode frequencies for β-Ga_2O_3 are shown in Fig. 28.10, in comparison with those obtained by our DFT calculations, and those calculated from the model dielectric function approach detailed above, assuming that all broadening parameters are zero. A detailed discussion is given in [17].

28.3.5.3 Bands of Total Reflection (a.k.a. Reststrahlen Band)

Figure 28.11 depicts calculated linearly polarized reststrahlen ($R_{\alpha^\star}$) bands of β-Ga_2O_3 in the monoclinic plane (Fig. 28.11a) using the EDVS approach, and a schematic of the associated phonon bands (Fig. 28.11b). The influence of broadening

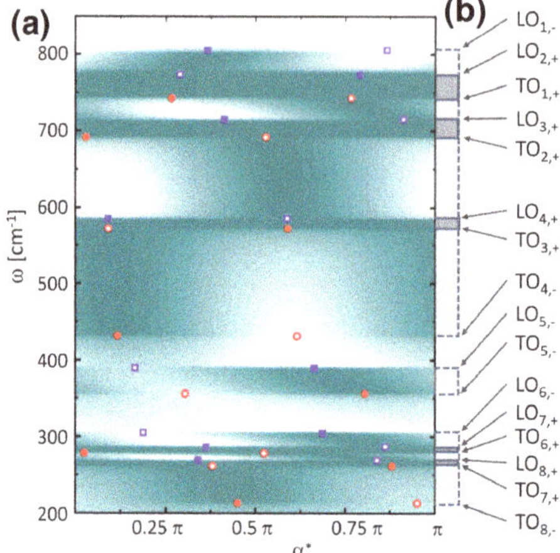

Fig. 28.11 **a** Reflectance color-density plot (10 equally spaced density scales between dark olive for $R_{\alpha^\star} = 1$ and white for $R_\alpha = 0$) rendering the linearly polarized reststrahlen bands for the β-Ga_2O_3 **a–c** plane as a function of $\alpha^\star$ and symbols as defined in Fig. 28.10. The incident polarization is parallel to $\mathbf{a}^\star$. **b** 3 pairs of outer modes, and 5 pairs of inner modes occur in β-Ga_2O_3, forming 3 phonon bands with phonon mode order matching the previously observed order of B_u-symmetry TO and LO frequencies in [1]. Reprinted from [17] with copyright permission from American Physical Society

is ignored here for clarity, and all broadening parameters are set to zero. A detailed discussion of the restrahlen band is given in [17]. The reststrahlen bands in the monoclinic plane are polarization dependent. For linearly polarized incident light, the reststrahlen bands are characterized by unpolarized bands of total reflection, polarized bands of total reflection, and intermediate bands with no total reflection. The latter occur outside bands of outer modes. Within inner mode pairs, $[\mathrm{TO}_+, \mathrm{LO}_+]$, bands of unpolarized total reflection occur, since $r_+ = r_- = 1$. Hence, at normal incidence the **a–c** plane of a monoclinic crystal is totally reflective within all inner bands, regardless of light polarization. Within spectral regions inside outer mode pairs not overlaid by inner mode pairs, $[\mathrm{TO}_-, \mathrm{LO}_-] \cap [\mathrm{TO}_+, \mathrm{LO}_+]$, total reflection occurs only for one linear polarization state. The angle depends on the wavelength, hence, we refer to these bands as polarization (or angular) dependent bands. Narrow lines of total reflection form, which begin and end at frequencies and directions of the modes in $[\mathrm{TO}_-, \mathrm{LO}_-] \cap [\mathrm{TO}_+, \mathrm{LO}_+]$. The angles at which the bands begin and end can be directly read from the directions of the eigenvectors, $\varphi_\pm = \tan^{-1}(E_{\pm,y}/E_{\pm,x})$ at begin and end of the polarization dependent bands. Bands begin and end at outer mode frequencies, and mark the direction of the phonon modes. Bands connecting to inner mode frequencies identify directions perpendicular to the phonon mode directions [17].

28.4 Phonon and Free Charge Carrier Properties in β-Ga_2O_3

The presence of free charge carriers, or charge carrier plasmas, leads to coupling with LO modes. This phenomenon is well understood for semiconductors with polar phonon modes and orthorhombic and higher crystal symmetries.[5]

28.4.1 LO-Phonon-Plasmon Coupling in β-Ga_2O_3

In monoclinic symmetry semiconductors, such as β-Ga_2O_3, coupling with free charge carriers also causes LPP modes, however, LPP mode frequencies, amplitude, and eigenvectors behave differently with increasing free charge carrier density than in materials with higher crystal symmetry. LPP modes which originate from LO modes within the monoclinic plane remain polarized within the monoclinic plane but their eigenpolarization directions change continuously with increasing charge carrier density. This is perhaps the most consequential difference of free charge carrier coupling in semiconductors with monoclinic symmetry and polar phonon modes in comparison to high-symmetry materials. Also, some of the LPP modes can cross the frequency of a TO mode. This will be discussed in more detail further below. The contributions of free charge carriers to ε can simply be expressed by the Drude model for free charge carriers. In order to account for the three dimensional nature of the motion of free charge carriers, and their directional anisotropy for transport properties (optical mobility parameters), free charge carrier contributions are augmented by 3 additional terms to (28.3)

$$\varepsilon_{\mathrm{LPP}} = \varepsilon + \sum_{l=1}^{3} \varrho_{\mathrm{TO}=0,l}(\hat{\mathbf{e}}_{\mathrm{TO}=0,l} \otimes \hat{\mathbf{e}}_{\mathrm{TO}=0,l}). \tag{28.14}$$

Functions $\varrho_{\mathrm{TO}=0,l}$ are identical with those in (28.7) while here $\omega_{\mathrm{TO}=0,l} = 0$. Directions $\hat{\mathbf{e}}_{\mathrm{TO}=0,l}$ must not necessarily coincide with Cartesian axes directions. Note that the ideal (if defect/dopant induced strain effects are ignored) addition of free charge carriers does not affect the frequencies of the lattice TO modes in β-Ga_2O_3. All amplitude parameters in (28.3) remain the same. For the inverse of the dielectric function, in order to express the effect of the addition of free charge carriers,

[5] In materials with orthorhombic and higher crystal symmetries, the coupling leads to so-called coupled LO-phonon-plasmon (LPP) modes, which shift upward in frequency as a function of increased carrier density. All LPP modes maintain the same polarization direction as their originating LO modes. All polarization directions coincide with high-symmetry directions of the crystal. All LPP modes common to one set of polarization are bound by associated TO modes, except for the LPP mode with the highest frequency, which approaches infinity when the plasma density approaches infinity. See, e.g., [47–49].

3 additional terms must be added and all other frequency and broadening parameters must be modified in (28.5)

$$\varepsilon_{\mathrm{LPP}}^{-1} = \varepsilon_{\infty}^{-1} - \sum_{l=1}^{N+3} \varrho_{\mathrm{LPP},l}(\hat{\mathbf{e}}_{\mathrm{LPP},l} \otimes \hat{\mathbf{e}}_{\mathrm{LPP},l}). \tag{28.15}$$

If Cartesian axes (x, y, z) are chosen for the description of the free charge carrier response, functions $\varrho_{\mathrm{TO}=0,l}$ can be expressed as follows

$$\varrho_{\mathrm{TO}=0,(x,y,z)} = -\frac{e^2 N}{\tilde{\varepsilon}_0 m_{\mathrm{eff},(x,y,z)} m_{\mathrm{e}} \omega(\omega + i\gamma_{\mathrm{p},(x,y,z)})}, \tag{28.16}$$

where N is the free charge carrier volume density parameter, e is the electronic charge, m_{e} is the free electron mass, $m_{\mathrm{eff},(x,y,z)}$ are the three directional effective mass parameters, and the directional plasma broadening parameters $\gamma_{\mathrm{p},(x,y,z)}$ are connected with the directional optical mobility parameters

$$\mu_{(x,y,z)} = \frac{e}{m_{\mathrm{eff},(x,y,z)} m_{\mathrm{e}} \gamma_{\mathrm{p},(x,y,z)}}. \tag{28.17}$$

The LPP mode frequencies and eigenvectors are then obtained from the dielectric function tensor, analogous to (28.18):

$$|\det\{\varepsilon^{-1}(\omega = \omega_{\mathrm{LPP},l})\}| \rightarrow \infty, \tag{28.18a}$$

$$\varepsilon(\omega = \omega_{\mathrm{LPP},l})\hat{\mathbf{e}}_{\mathrm{LPP},l} = 0, \tag{28.18b}$$

28.4.2 Mode Multiplicity in β-Ga_2O_3 with Free Charge Carriers

B_{u} modes:
For a single species carrier (such as single band holes, or single band electrons) 2 TO modes with zero frequency $\omega_{\mathrm{TO}} = 0$ are added, hence, 10 TO modes and 10 associated LPP modes occur, oriented (polarized) within the monoclinic **a**–**c** plane. Note that none of the eigenvectors, $\hat{\mathbf{e}}_{\mathrm{TO},l}$, $\hat{\mathbf{e}}_{\mathrm{LPP},l}$ are parallel to each other. However, as will be shown below, certain free charge carrier density values exist where certain LPP modes are exactly polarized perpendicular to certain TO modes.

A_{u} modes:
For a single species carrier (such as single band holes, or single band electrons) one TO mode with zero frequency $\omega_{\mathrm{TO}} = 0$ is added, hence, the displacement directions of 5 TO modes and 5 associated LPP modes (A_{u} symmetry) are oriented perpendicular to the monoclinic plane. Thus all eigenvectors, $\hat{\mathbf{e}}_{\mathrm{TO},l}$, $\hat{\mathbf{e}}_{\mathrm{LPP},l}$ are parallel to axis **b**.

28.4.3 LPP Mode Parameters as a Function of the Plasma Frequency Parameter

The plasma frequency parameter is commonly used to express the density of free charge carriers provided that the effective mass parameter of a given semiconductor is known

$$\omega_{\mathrm{p}} = \sqrt{\frac{N}{m_{\mathrm{eff}}}}\frac{|e|}{\sqrt{\varepsilon_0 m_{\mathrm{e}}}}, \tag{28.19}$$

where N is the volume density of free charge carriers (electrons/holes), $|e|$ is the magnitude of the electron charge, ε_0 is the vacuum dielectric constant, and m_{e} and m_{eff} are the free electron and effective carrier mass parameters, respectively. We show further below that our current experiment reveals an isotropic effective mass behavior for the conduction band electron states near the Brillouin zone center, hence, the effective mass is here approximated as a scalar. Then, the plasma frequency can be considered as a scalar as well, and one can evaluate the effect of the LPP mode coupling onto the LPP mode parameters as a function of the scalar parameter ω_{p}. This is discussed in the following section.

28.4.3.1 LPP Mode Frequencies

Figure 28.12 depicts the dependencies of the LPP mode frequency parameters as a function of the plasma frequency parameter. At small ω_{p}, 2 additional branches emerge from zero, polarized within the monoclinic plane. With increasing ω_{p}, the LPP modes shift away from their associated LO mode frequencies at $\omega_{\mathrm{p}} = 0$, and change the phonon mode order. The two highest frequency branches approach infinity for $\omega_{\mathrm{p}} \rightarrow \infty$. Note that some LPP modes cross TO frequencies. Inserted in Fig. 28.12 are data obtained from analysis of single crystal samples with slightly different concentrations of free electrons.

28.4.3.2 LPP Mode Orientation Angles

Figure 28.13 depicts the dependencies of the LPP mode eigenpolarization vector orientation parameters within the monoclinic plane as a function of the plasma frequency parameter. Most notably, all LPP modes change their polarization direction as a function of ω_{p}. Note further that all LPP modes merge to be oriented parallel to a TO mode orientation for very large free charge carrier concentration. Inserted in Fig. 28.13 are data obtained from analysis of single crystal samples with slightly different concentrations of free electrons.

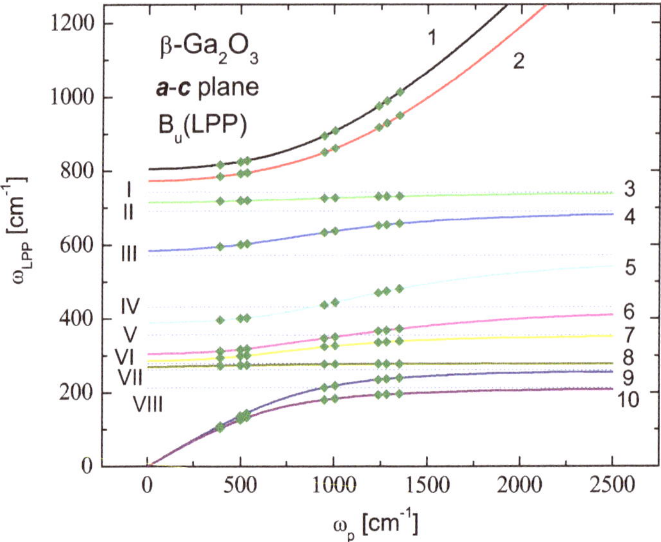

Fig. 28.12 **a** Frequencies of coupled LPP modes polarized within the **a**–**c** plane as a function of isotropic plasma frequency parameter ω_p. The horizontal lines and Roman numerals indicate frequencies of the B_u symmetry TO modes. Note that in addition to the deviation from the so called TO-LO rule at $\omega_p = 0$, some LPP modes can cross the frequencies of TO modes. Symbols (diamonds) indicate experimental observations for different bulk crystals listed in Tables 28.3 and 28.4. Reprinted from [50] with copyright permission from American Institute of Physics Publishing

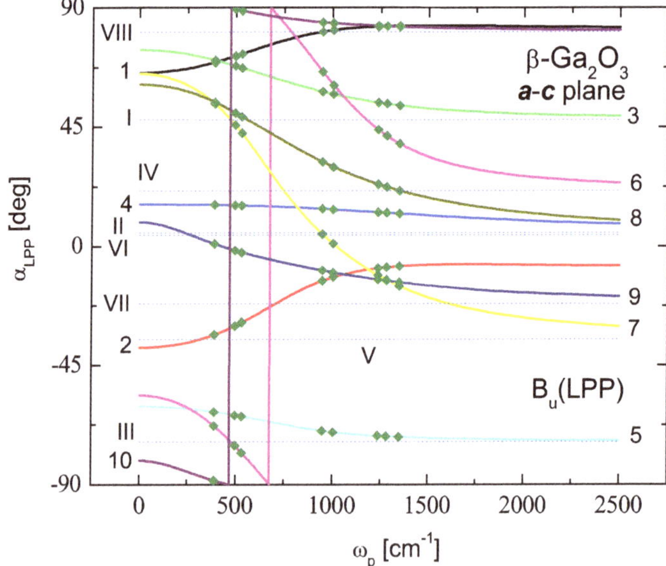

Fig. 28.13 Same as Fig. 28.12 for coupled LPP mode unit eigendisplacement vectors. Reprinted from [50] with copyright permission from American Institute of Physics Publishing

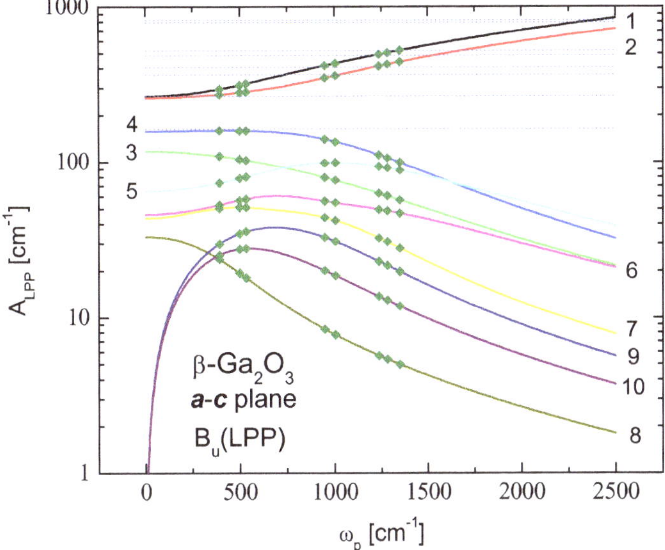

Fig. 28.14 Same as Fig. 28.12 for coupled LPP mode amplitude parameters. Horizontal dotted lines indicate the TO mode amplitude parameters for comparison. Two additional modes (plasma-like modes) emerge for small plasma frequencies, which couple with the LO modes. For $\omega_p \rightarrow \infty$, the plasma-like mode amplitude parameters approach ∞ while all other LPP modes cease. Reprinted from [50] with copyright permission from American Institute of Physics Publishing

28.4.3.3 LPP Mode Amplitudes

Figure 28.14 depicts the dependencies of the LPP mode amplitude parameters as a function of the plasma frequency parameter. At small ω_p, 2 additional branches emerge polarized within the monoclinic plane whose amplitude parameters increase. With increasing ω_p, only the two highest frequency LPP mode branches remain and with increasing amplitude parameters, while all other LPP mode amplitude parameters approach zero for $\omega_p \rightarrow \infty$. Note that some LPP modes cross TO frequencies. Inserted in Fig. 28.14 are data obtained from analysis of single crystal samples with slightly different concentrations of free electrons.

28.4.3.4 Limiting Frequencies for Single-Species Free Charge Carriers (Directional LPP Modes in β-Ga_2O_3)

In analogy to the so-called limiting frequencies, $\omega(\hat{\alpha})_l$, in the presence of free charge carriers the limiting frequencies, $\omega(\hat{\alpha})_{\mathrm{LPP},l}$, depend on the direction of a unit vector within the **a**–**c** plane, $\hat{\alpha} = \cos\alpha\hat{\mathbf{x}} + \sin\alpha\hat{\mathbf{y}}$. No solution for $\omega(\hat{\alpha})_{\mathrm{LPP},l}$ exists within bands of total reflection, and the allowed frequency regions are confined to frequency regions $[\mathrm{TO}_-, \mathrm{LPP}_-] \cap [\mathrm{TO}_+, \mathrm{LPP}_+]$. Note that 2 new TO modes have appeared at

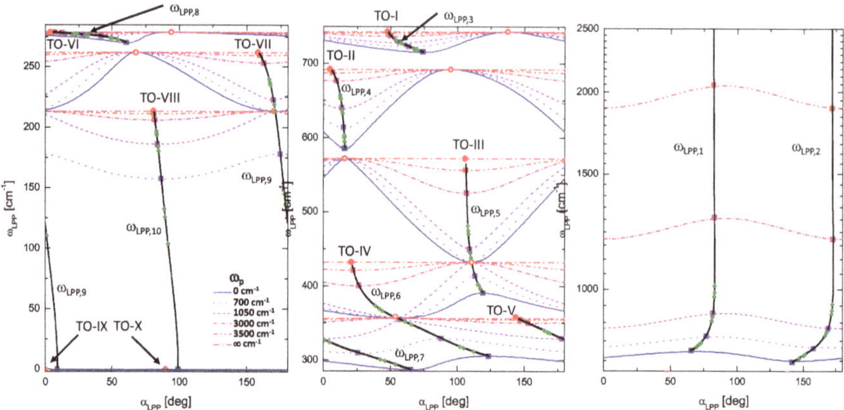

Fig. 28.15 Colored lines with different styles: B_u-symmetry directional limiting frequencies, $\omega(\alpha)_{LPP,l}$, for β-Ga_2O_3 as a function of displacement direction, α, relative to axis a within the monoclinic plane, for selected plasma frequency parameters (see inset for labels). Black solid lines indicate the evolution of LPP modes as a continuous function of ω_p. Square symbols indicate B_u-symmetry LPP modes. Left panel: Modes $\omega(\alpha)_{LPP,8}$-$\omega(\alpha)_{LPP,10}$, middle panel: modes $\omega(\alpha)_{LPP,3}$-$\omega(\alpha)_{LPP,7}$, right panel: modes $\omega(\alpha)_{LPP,1}$ and $\omega(\alpha)_{LPP,2}$. Note the different y axis scales. Full red circles and numerals indicate B_u-symmetry TO modes. Open red circles indicate the TO frequencies at orientations perpendicular to the lattice TO mode polarization. Green diamonds indicate the experimental data observed in [50]. Reprinted from [50] with copyright permission from American Institute of Physics Publishing

$\omega = 0$ due to the free charge carrier contributions. Note further that frequencies $\omega_{LPP,l}$ increase with increasing ω_p. The minimum–maximum bounds for $\omega(\hat{\alpha})_{LPP,l}$ can be (i) $\omega_{TO_{j,-}} - \omega_{TO_{j',+}}$, (ii) $\omega_{LPP_{j,+}} - \omega_{TO_{j',+}}$, (iii) $\omega_{LPP_{j,+}} - \omega_{LPP_{j',-}}$, and (iv) $\omega_{TO_{j,-}} - \omega_{LPP_{j',-}}$ [17]. The limiting mode frequencies for β-Ga_2O_3 are shown in Fig. 28.15. As can be seen, some LPP modes can bypass a TO mode, and when the frequency of a LPP mode equals said TO mode, then both modes are polarized perpendicular to each other. For example, $\omega_{LPP,9}$ passes TO-VIII at exactly 90° difference in α. Same for $\omega_{LPP,6}$ passes TO-V, $\omega_{LPP,5}$ passes TO-IV. Note how the directional dependencies reduce with increasing carrier density, and how at the same time the spectral intervals for total reflection increase. Note that for $\omega_p \rightarrow \infty$ all limiting frequencies merge with TO modes and loose their directional dependencies. Because of the two highest frequency LPP modes approaching infinity also (plasma-like modes), the monoclinic plane becomes totally reflective due to the screening of the electromagnetic waves by the free charge carriers. See also [50].

28.4.4 LPP Modes for Polarization Along b

28.4.4.1 LPP Mode Frequencies

Figure 28.16 depicts the dependencies of the LPP mode frequency parameters as a function of the plasma frequency parameter. At small ω_p, 1 additional branch emerges from zero, polarized perpendicular to the monoclinic plane. With increasing ω_p, the LPP modes shift away from their associated LO mode frequencies at $\omega_p = 0$, but do not change the phonon mode order. The highest frequency mode approaches infinity for $\omega_p \to \infty$. Note that no LPP mode crosses TO frequencies. Inserted in Fig. 28.16 are data obtained from analysis of single crystal samples with slightly different concentrations of free electrons listed in Tables 28.3 and 28.4.

28.4.4.2 LPP Mode Frequencies

Figure 28.17 depicts the dependencies of the LPP mode amplitude parameters as a function of the plasma frequency parameter. At small ω_p, 1 additional branch emerges polarized perpendicular to the monoclinic plane whose amplitude parameter

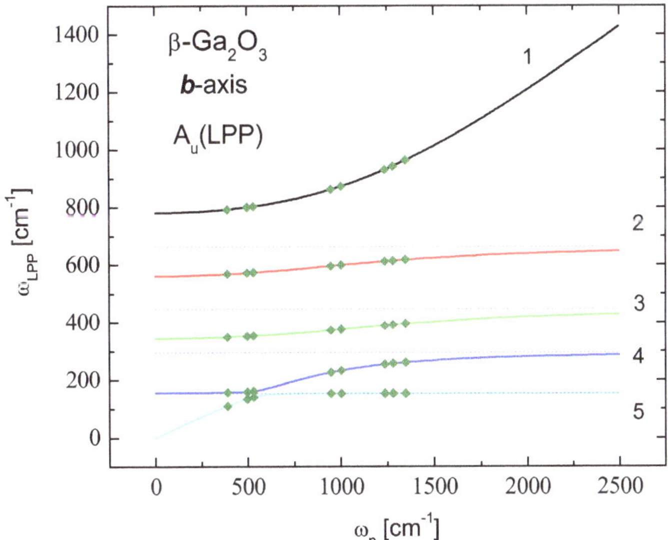

Fig. 28.16 **a** Frequencies of coupled LPP modes polarized parallel to axis **b** as a function of isotropic plasma frequency parameter ω_p. The horizontal lines indicate frequencies of the A_u-symmetry TO modes. Note that the so called TO-LO rule is fulfilled for all ω_p. Symbols (diamonds) indicate experimental observations for different bulk crystals listed in Tables 28.3 and 28.4. The mode behavior is that seen previously in all semiconductors with orthorhombic or lesser symmetry. Reprinted from [50] with copyright permission from American Institute of Physics Publishing

Table 28.3 Sample parameters for all samples investigated in this work with nominal free electron density, $N_D - N_A$

Sample name	Surface	Dopant	Source	$N_D - N_A$ (cm^{-3})
T3151	$(\bar{2}01)$	Sn	NCT[a]	7×10^{18}
ESN05-3 21-1	(001)	Sn	NCT[a]	3×10^{18}
T1730-5	$(\bar{2}01)$	Sn	NCT[a]	5.9×10^{18}
ESN11 3-19x	(010)	Sn	NCT[a]	3×10^{18}
ESN21a1-1	(100)	Sn	NCT[a]	5×10^{18}
GAO-37-23	(100)	X[b]	IKZ[c]	5×10^{17}
T2456	$(\bar{2}01)$	Sn	NCT[a]	1.1×10^{18}
EUND28a	$(\bar{2}01)$	X[a]	NCT[a]	4.3×10^{17}

[a]Novel Crystal Technology, Inc., Japan
[b]Unintentionally doped
[c]Leibniz-Institut für Kristallzüchtung e.V., Berlin, Germany

Table 28.4 Best-model free charge carrier parameter results for all samples investigated in this work. The free charge carrier density parameter (N) and optical mobility parameters ($\mu_{\mathbf{a}}$, $\mu_{\mathbf{c}^\star}$, $\mu_{\mathbf{b}}$) were calculated from the plasma broadening and plasma frequency parameters assuming a constant and isotropic conduction band effective mass parameter determined by optical Hall effect [6] previously of $m^\star = 0.28m_e$, with m_e as the inertial mass of the free electron

Sample name	N (cm^{-3})	ω_p (cm^{-1})	$\mu_{\mathbf{a}}$ (cm^2/Vs)	$\mu_{\mathbf{c}^\star}$ (cm^2/Vs)	$\mu_{\mathbf{b}}$ (cm^2/Vs)
T3151	5.68×10^{18}	1350	57	50	56
ESN05-3 21-1	5.1×10^{18}	1282	47	41	48
T1730-5	4.7×10^{18}	1238	48	42	48
ESN11 3-19x	3.1×10^{18}	1006	50	50	44
ESN21a1-1	2.8×10^{18}	950	41	57	62
GAO-37-23	8.9×10^{17}	533	37	37	37
T2456	0.78×10^{18}	499	41	37	51
EUND28a	4.78×10^{17}	392	24	19	35

increases with ω_p. With increasing ω_p, only the highest frequency LPP mode remains with increasing amplitude parameters, while all other LPP mode amplitude parameters approach zero for $\omega_p \rightarrow \infty$. Inserted in Fig. 28.17 are data obtained from analysis of single crystal samples with slightly different concentrations of free electrons, listed in Tables 28.3 and 28.4.

28.4.5 *The Optical Hall Effect in β-Ga_2O_3*

The optical Hall effect (OHE) is a physical phenomenon that describes the occurrence of magnetic-field-induced dielectric displacement at optical wavelengths, transverse and longitudinal to the incident electric field, and analogous to the static electrical

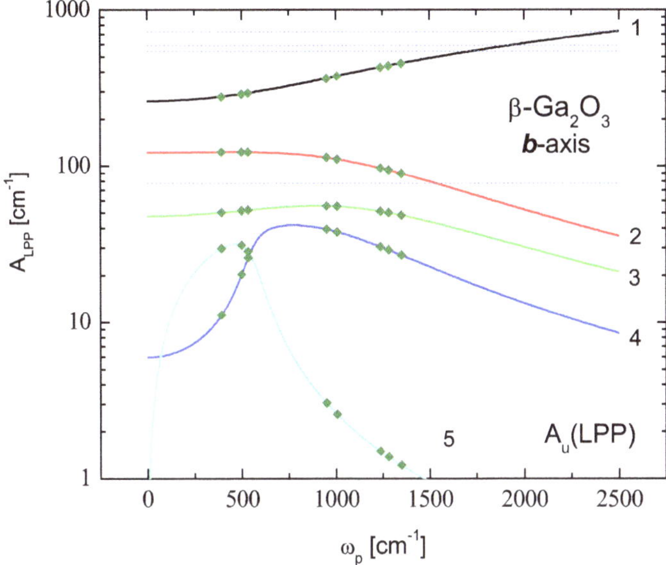

Fig. 28.17 Same as Fig. 28.16 for coupled LPP mode amplitude parameters. Horizontal dotted lines indicate the TO mode amplitude parameters for comparison. One additional mode (plasma-like mode) emerges for small plasma frequencies, which couples with the LO modes. For $\omega_p \rightarrow \infty$, the plasma-like mode amplitude parameter approaches ∞ while all other LPP modes cease. The mode behavior is that seen previously in all semiconductors with orthorhombic or lesser symmetry. Reprinted from [50] with copyright permission from American Institute of Physics Publishing

Hall effect [51]. The OHE can be measured in semiconductors with free charge carriers and can be used to differentiate between charge carrier type (electron or hole), structure (one dimensional, two dimensional, three dimensional density), effective mass and mobility parameters including their anisotropy, and number density of the free charge carriers. Hence, the OHE is a non-contact, non-destructive optical means to measure the effective mass parameters in semiconductor heterostructures with plane parallel interfaces, and also for single bulk crystals.

An exact model equation system for the OHE in semiconductor materials with monoclinic and triclinic symmetries has not been reported yet. The response of free charge carriers in β-Ga_2O_3 is approximated here by a system of free charge carriers in a material with orthorhombic symmetry, where the phonon response is described by the monoclinic model tensor described above. Thereby it is assumed that the effective mass parameters and the optical mobility parameters can be mapped within tensors of common major orthogonal axes, (x, y, z), $(\mathbf{m})_{ij}$ and $(\mu)_{ij}$, respectively

$$\varepsilon_{\text{FCC-OHE},ik} = \varepsilon_{ik} + \frac{Nq^2}{\varepsilon_0}\left[m_{ik}(-\omega^2\delta_{ij} - i\omega\gamma_{\text{p},ik}) - i\omega\varepsilon_{ijk}qB_j\right]^{-1}, \quad (28.20)$$

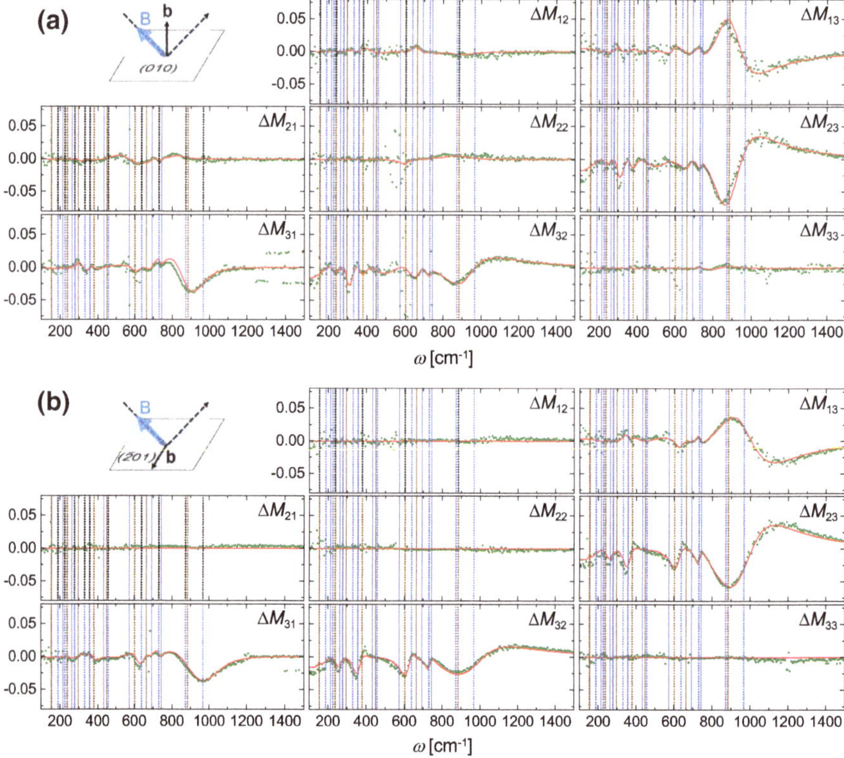

Fig. 28.18 **a** Experimental (green dotted lines) and best-match model calculated (solid red lines) FIRIR OHE spectra ($\Delta M_{ij} = M_{ij}(+\mathbf{B}\ \mathrm{T}) - M_{ij}(-\mathbf{B}\ \mathrm{T})$) for a (010) sample at nominal azimuth angle ϕ = 115°. For the FIR-OHE measurements **B** = 7 T (100 cm^{-1} to 555 cm^{-1}), and for the IR-OHE measurements **B** = 6 T (555 cm^{-1} to 1500 cm^{-1}). All measurements are performed at temperature T = 300 K, and at angle of incidence Φ_a = 45°. The magnetic field **B** is parallel to the incoming infrared beam. Vertical lines indicate wavenumbers of TO (solid lines) and LPP modes (dotted lines) with B_u symmetry (blue) and A_u symmetry (brown). **b** Same as (**a**) for a ($\bar{2}01$) sample at nominal azimuth angle ϕ = 180°. Samples from Novel Crystal Technology, Inc., Japan

where B_j are the components of the magnetic field, and δ_{ij} is the Kronecker symbol [51].

Exemplary data for measured and best-match model calculated OHE data are shown for samples with (010) (Fig. 28.18a), and ($\bar{2}01$) (Fig. 28.18b), for one angle of incidence, and one in-plane azimuth orientation. Vertical lines indicate frequencies of TO and LPP modes with A_u and B_u symmetry. Details of a combined analysis of OHE data obtained in the mid-infrared spectral range on a (010) sample and a ($\bar{2}01$) sample are presented by Knight et al. [6]. No detectable anisotropy of the effective mass parameter was found, and an isotropic effective mass parameter of $(0.284 \pm 0.013)m_e$ was obtained. Our experimental OHE result agrees well with our calculations of the conduction band structure using Gaussian-attenuation Perdew-

Burke-Ernzerhof hybrid density functionals, which predict an isotropically average effective mass parameter of $0.267m_e$ [6]. Our OHE investigations up to this point did not indicate detectable anisotropy in the electron mobility parameters. Further work is needed to explore free electron and free hole parameters in differently doped materials.

28.5 Outlook

Further analysis is needed for the effects of defects (vacancies, interstitials, site exchanges, etc.) onto the phonon and free charge carrier properties in β-Ga_2O_3. For example, the group of Michael Stavola et al. reported on hydrogen playing a key role in the conductivity of Ga_2O_3 by passivating deep defects and acting as a shallow donor. Vibrational spectroscopy experiments found a dominant hydrogen center with a polarized O-H line at 3437 cm^{-1} [52, 53]. DFT analysis identified a defect consisting of two equivalent H atoms trapped at a relaxed Ga vacancy. Other possible species may be associated with defects and vacancies, and much work remains to be performed by combination with experiment and theory to identify signatures and origins of vacancies, interstitials, and site exchanges, for example. The effects of dopants onto the phonon and free carrier properties is another emerging topic as more and more high quality material emerges with different dopants [54–56]. The search for acceptors will continue, [57] and issues of acceptor diffusivity and defects will be of interest [58]. An emerging topic with the availability of epitaxial material is the quest to determine the influence of strain and stress onto the phonon and free charge carrier properties. Of further importance is the lack of knowledge about the influence of alloying with, for example, aluminum and/or indium, as well as other elements. Critical discussion and comparisons will be needed to differentiate the existence and structural properties of crystal modifications of different crystal structure and/or quality. The understanding of phonon and free charge carrier properties in low-symmetry materials has just begun, and much remains yet to be discovered.

28.6 Summary

We presented the current state of knowledge about the long-wavelength phonon and free charge carrier properties of monoclinic symmetry gallium oxide. We presented results obtained by experiment (ellipsometry and optical Hall effect) from multiple samples of single crystalline material, and we provided for comparison our results from density functional theory. Beta-crystal modification gallium oxide is a model system for understanding of phonon and free charge carrier properties in monoclinic crystals with polar lattice vibrations. We developed an eigendielectric polarization model, which can successfully be exploited to shed light on the peculiarities of phonon mode properties and phonon-plasmon coupling. Consequential findings are the non-parallelism of all phonons within the monoclinic plane, the change of LPP mode directions with plasma frequency, and an unusual separation of the polar lattice modes within the monoclinic planes into inner and outer phonon mode pairs. Numerous questions remain to be addressed such as the effects of defects, strain, and alloying, for example.

28.7 Note Added in Proof

It may be of importance to note further that there is no distinguished direction independent of wavelength within the a-c plane which aligns with a simply measurable physical property. Such property could be, for example, an optical axis, or a polarization direction for which total reflection is recorded. While the latter directions exist, none aligns with any of the high-symmetry crystallographic directions. This is perhaps another interesting observation about polar materials with monoclinic symmetry, where the paradigm is broken that high-packing density directions along which X-ray diffraction occur align with eigendirections of other physical properties. A direct consequence is that the here reported value for the off-diagonal element of the dielectric constant at zero frequency is meaningless, or more accurately formulated, fully dependent on the choice of coordinates made here for the in-plane orientations of the phonon modes. A rotation can be found such that the off-diagonal DC value will vanish. However, the high-frequency dielectric tensor off diagonal component will not vanish then and will reveal a distinct value. This value has not been reported yet and appears to be therefore another intrinsic property of beta Gallium Oxide. Consequently, as will be shown in a forthcoming work, the so-called Superconvergence and Sum Rules for the optical constants derived by Altarelli et al. in Phys. Rev. 6, 4502 (1972) appear to require revisions for monoclinic and triclinic symmetry materials.

Acknowledgements This work was supported in part by the National Science Foundation under award DMR 1808715, by Air Force Office of Scientific Research under award FA9550-18-1-0360, by the Nebraska Materials Research Science and Engineering Center under award DMR 1420645, and by the Knut and Alice Wallenbergs Foundation grant 2018-2023 "Wide-bandgap semiconductors for next generation quantum components". We acknowledge support from the Swedish Energy Agency under award P45396-1, the Swedish Research Council VR under award No. 2016-00889, the Swedish Foundation for Strategic Research under Grant Nos. FL12-0181, RIF14-055, and EM16-0024, and the Swedish Government Strategic Research Area in Materials Science on Functional Materials at Linköping University, Faculty Grant SFO Mat LiU No. 2009-00971. M. S. acknowledges support by the University of Nebraska Foundation and the J. A. Woollam Foundation.

References

1. M. Schubert, R. Korlacki, S. Knight, T. Hofmann, S. Schöche, V. Darakchieva, E. Janzén, B. Monemar, D. Gogova, Q.T. Thieu, R. Togashi, H. Murakami, Y. Kumagai, K. Goto, A.

Kuramata, S. Yamakoshi, M. Higashiwaki, Phys. Rev. B **93**, 125209 (2016)
2. M. Higashiwaki, K. Sasaki, A. Kuramata, T. Masui, S. Yamakoshi, Phys. Stat. Sol. (A) **211**, 2126 (2014)
3. M. Higashiwaki, G.H. Jessen, Appl. Phys. Lett. **112**, 060401 (2018)
4. Z. Galazka, Semicond. Sci. Technol. **33**, 113001 (2018)
5. C. Sturm, R. Schmidt-Grund, C. Kranert, J. Furthmüller, F. Bechstedt, M. Grundmann, Phys. Rev. B **94**, 035148 (2016)
6. S. Knight, A. Mock, R. Korlacki, V. Darakchieva, B. Monemar, Y. Kumagai, K. Goto, M. Higashiwaki, M. Schubert, Appl. Phys. Lett. **112**, 012103 (2018)
7. B. Liu, M. Gu, X. Liu, App. Phys. Lett. **91**, 172102 (2007)
8. A. Parisini, K. Ghosh, U. Singisetti, R. Fornari, Sem. Sci. Tech. **33**, 105008 (2018)
9. K. Gosh, U. Singisetti, Appl. Phys. Lett. **109**, 072102 (2016)
10. Y. Kang, K. Krishnaswamy, H. Peelaers, C.G.V. de Walle, J. Phys.: Cond. Matt. **29**, 234001 (2017)
11. K. Gosh, U. Singisetti, J. Mat. Res. **32**, 4142 (2018)
12. K. Gosh, U. Singisetti, J. Appl. Phys. **122**, 035702 (2017)
13. K. Ghosh, U. Singisetti, *Low-Field and High-Field Transport in* β-Ga_2O_3 (Elsevier, 2019)
14. X. Gonze, C. Lee, Phys. Rev. B **55**, 10355 (1997)
15. S. Baroni, S. de Gironcoli, A.D. Corso, S. Baroni, S. de Gironcoli, P. Giannozzi, Rev. Mod. Phys. **73**, 515 (2001)
16. G. Venkataraman, L.A. Feldkamp, V.C. Sahni, *Dynamics of Perfect Crystals* (The MIT Press, Cambridge, MA, 1975)
17. M. Schubert, A. Mock, R. Korlacki, V. Darakchieva, Phys. Rev. B **99**, 041201(R) (2019)
18. M. Schubert, T.E. Tiwald, C.M. Herzinger, Phys. Rev. B **61**(12), 8187 (2000)
19. S. Schöche, T. Hofmann, R. Korlacki, T.E. Tiwald, M. Schubert, J. Appl. Phys. **113**, 111906 (2013)
20. M. Dressel, B. Gompf, D. Faltermeier, A.K. Tripathi, J. Pflaum, M. Schubert, Opt. Exp. **16**, 19770 (2008)
21. M. Schubert, C. Bundesmann, G. Jakopic, H. Arwin, Appl. Phys. Lett. **84**, 200 (2004)
22. N. Ashkenov, B.N. Mbenkum, C. Bundesmann, V. Riede, M. Lorenz, E.M. Kaidashev, A. Kasic, M. Schubert, M. Grundmann, G. Wagner, H. Neumann, J. Appl. Phys. **93**, 126 (2003)
23. A. Kasic, M. Schubert, S. Einfeldt, D. Hommel, T.E. Tiwald, Phys. Rev. B **62**(11), 7365 (2000)
24. A. Kasic, M. Schubert, Y. Saito, Y. Nanishi, G. Wagner, Phys. Rev. B **65**(11), 115206 (2002)
25. V. Darakchieva, P.P. Paskov, E. Valcheva, T. Paskova, B. Monemar, M. Schubert, H. Lu, W.J. Schaff, Appl. Phys. Lett. **84**, 3636 (2004)
26. V. Darakchieva, J. Birch, M. Schubert, T. Paskova, S. Tungasmita, G. Wagner, A. Kasic, B. Monemar, Phys. Rev. B **70**, 045411 (2004)
27. V. Darakchieva, E. Valcheva, P.P. Paskov, M. Schubert, T. Paskova, B. Monemar, H. Amano, I. Akasaki, Phys. Rev. B **71**, 115329 (2005)
28. V. Darakchieva, T. Paskova, M. Schubert, H. Arwin, P. Paskov, B. Monemar, D. Hommel, M. Heuken, J. Off, F. Scholz, B. Haskell, P. Fini, J. Speck, S. Nakamura, Phys. Rev. B **75**, 195217 (2007)
29. V. Darakchieva, M. Schubert, T. Hofmann, B. Monemar, Y. Takagi, Y. Nanishi, Appl. Phys. Lett. **95**, 202103 (2009)
30. V. Darakchieva, T. Hofmann, M. Schubert, B.E. Sernelius, B. Monemar, P.O.A. Persson, F. Giuliani, E. Alves, H. Lu, W.J. Schaff, Appl. Phys. Lett. **94**, 022109 (2009)
31. V. Darakchieva, K. Lorenz, N. Barradas, E. Alves, B. Monemar, M. Schubert, N. Franco, C. Hsiao, L. Chen, W. Schaff, L. Tu, T. Yamaguchi, Y. Nanishi, Appl. Phys. Lett. **96**, 081907 (2010)
32. M.Y. Xie, M. Schubert, J. Lu, P.O.A. Persson, V. Stanishev, C.L. Hsiao, L.C. Chen, W.J. Schaff, V. Darakchieva, Phys. Rev. B **90**, 195306 (2014)
33. M.Y. Xie, N.B. Sedrine, S. Schöche, T. Hofmann, M. Schubert, L. Hong, B. Monemar, X. Wang, A. Yoshikawa, K. Wang, T. Araki, Y. Nanishi, V. Darakchieva, J. Appl. Phys. **115**, 163504 (2014)

34. G.E. Jellison, M.A. McGuire, L.A. Boatner, J.D. Budai, E.D. Specht, D.J. Singh, Phys. Rev. B **84**, 195439 (2011)
35. M. Schubert, Phys. Rev. Lett. **117**, 215502 (2016)
36. A. Mock, R. Korlacki, C. Briley, V. Darakchieva, B. Monemar, Y. Kumagai, K. Goto, M. Higashiwaki, M. Schubert, Phys. Rev. B **96**, 245205 (2017)
37. C. Sturm, J. Furthmüller, F. Bechstedt, R. Schmidt-Grund, M. Grundmann, APL Mater. **3**(10), 106106 (2015)
38. C. Sturm, R. Schmidt-Grund, V. Zviagin, M. Grundmann, Appl. Phys. Lett. **111**(8), 082102 (2017)
39. M. Schubert, T. Hofmann, C.M. Herzinger, W. Dollase, Thin Solid Films **455–456**, 619 (2004)
40. A. Mock, R. Korlacki, S. Knight, M. Stokey, A. Fritz, V. Darakchieva, M. Schubert, Phys. Rev. B (2019)
41. A. Mock, R. Korlacki, S. Knight, M. Schubert, Phys. Rev. B **95**, 165202 (2017)
42. A. Mock, R. Korlacki, S. Knight, M. Schubert, Phys. Rev. B **97**, 165203 (2018)
43. H. Fujiwara, *Spectroscopic Ellipsometry* (Wiley, New York, 2007)
44. R.H. Lyddane, R. Sachs, E. Teller, Phys. Rev. **59**, 613 (1941)
45. A.B. Kuzmenko, E.A. Tishchenko, V.G. Orlov, J. Phys.: Condens. Matter **8**, 6199 (1996)
46. M. Born, K. Huang, *Dynamical Theory of Crystal Lattices* (Clarendon, Oxford, 1954)
47. P. Yu, M. Cardona, *Fundamentals of Semiconductors* (Springer, Berlin, 1999)
48. C. Klingshirn, *Semiconductor Optics* (Springer, Berlin, 1995)
49. C. Kittel, *Introduction To Solid State Physics* (Wiley India Pvt. Ltd., 2009)
50. M. Schubert, R. Korlacki, A. Mock, Y. Kumagai, K. Goto, A. Kuramata, Z. Galazka, G. Wagner, V. Darakchieva, Appl. Phys. Lett. **114**, 102102 (2019)
51. M. Schubert, P. Kuehne, V. Darakchieva, T. Hofmann, J. Opt. Soc. Am. A **33**, 1553 (2016)
52. P. Weiser, M. Stavola, W.B. Fowler, Y. Qin, S. Pearton, Appl. Phys. Lett. **112**(23), 232104 (2018)
53. Y. Qin, M. Stavola, W.B. Fowler, P. Weiser, S.J. Pearton, ECS, J. Sol. Stat. Sci. Tech. **8**, Q3103 (2019)
54. Z. Galazka, S. Ganschow, A. Fiedler, R. Bertram, D. Klimm, K. Irmscher, R. Schewski, M. Pietsch, M. Albrecht, M. Bickermann, J. Cryst. Growth **486**, 82 (2018)
55. S.H. Han, A. Mauze, E. Ahmadi, T. Mates, Y. Oshima, J.S. Speck, Semicond. Sci. Technol. **33**(4), 045001 (2018)
56. M.N. Fireman, G. L'Heureux, F. Wu, T. Mates, E.C. Young, J.S. Speck, J. Cryst. Growth **508**, 19 (2019)
57. A.T. Neal, S. Mou, S. Rafique, H. Zhao, E. Ahmadi, J.S. Speck, K.T. Stevens, J.D. Blevins, D.B. Thomson, N. Moser, K.D. Chabak, G.H. Jessen, Appl. Phys. Lett. **113**(6), 062101 (2018)
58. H. Peelaers, J.L. Lyons, J.B. Varley, C.G. Van de Walle, APL Mat. **7**(2), 022519 (2019)

Chapter 29
Thermal Properties

Zeyu Liu and Tengfei Luo

Abstract In this chapter, an overview of the current research progress on the thermal properties of beta-gallium oxide (β-Ga_2O_3) is provided. Thermal properties of β-Ga_2O_3 are of great significance to the device reliability and performance in its potential applications. Previous research through both computational and experimental studies on β-Ga_2O_3 using various methods is reviewed. The most notable findings are the relatively low and highly anisotropic thermal conductivity. At room temperature, the [010] direction has the highest thermal conductivity of around 25 W/mK, while that in the [100] direction is measured to be the lowest, which is around 13 W/mK. We also make comparison between β-Ga_2O_3 and GaN, another widely used semiconductor for power electronics. The relatively low thermal conductivity of β-Ga_2O_3 compared to GaN may present a major challenge for its potential applications. Another important thermal property, heat capacity, of β-Ga_2O_3 at room temperature is measured to be 18.7 J/mol K. On the other hand, the effective thermal conductivity in β-Ga_2O_3 thin film is shown to be larger than other gate oxides, providing a possibility of using it as gate dielectrics in GaN device contacts. The thermal properties discussed in this chapter might be useful for thermal management and design of β-Ga_2O_3 devices.

Thermal conductivity is a physical property describing the capability of conducting heat from a high-temperature region to a low-temperature region in materials or structures [1]. The classical Fourier's law states that the heat flux is proportional to the temperature gradient by the thermal conductivity:

Z. Liu
Department of Aerospace and Mechanical Engineering, University of Notre Dame, Notre Dame, IN 46615, USA
e-mail: zliu10@nd.edu

T. Luo (✉)
Department of Aerospace and Mechanical Engineering, Department of Chemical and Biomolecular Engineering and Center for Sustainable Energy of Notre Dame (ND Energy), University of Notre Dame, Notre Dame, IN 46615, USA
e-mail: tluo@nd.edu

M. Higashiwaki and S. Fujita (eds.), *Gallium Oxide*, Springer Series in Materials Science 293, https://doi.org/10.1007/978-3-030-37153-1_29

$$\boldsymbol{q} = -\kappa \nabla T \tag{29.1}$$

where $\boldsymbol{q}$ is the heat flux (W/m^2), κ is the thermal conductivity of the material (W/mK) and ∇T is the temperature gradient (K/m).

As a wide bandgap (4.8 eV) [2, 3] semiconductor with a high breakdown field, β-Ga_2O_3 is attractive in high voltage device applications as both substrate and device materials. In high voltage devices, however, most of the power is dissipated in the channel due to the Joule heating, and the channel temperature can be tens or even over one hundred degrees higher than the ambient temperature if heat is not transferred efficiently out of the near junction region. The high temperature can be detrimental to the device reliability and can further lead to electron transport performance degradation due to the stronger electron-phonon scattering. Thermal properties of β-Ga_2O_3 are also critical for applications as substrate materials. For example, the control of substrate temperature and its uniformity during epitaxial growth is very important, and the thermal conductivity of the substrate is also critical to the device packaging design [4]. Therefore, the knowledge of thermal properties of β-Ga_2O_3 is of great importance for device design and thermal management.

For semiconductors like β-Ga_2O_3, the dominant contribution to thermal transport is from the lattice vibration [5]. The picture of phonon as the quanta of lattice vibration in crystals has been adopted to study the thermal transport properties of β-Ga_2O_3. The transport of phonon in defect-free bulk crystals is mainly restricted by the three-phonon scattering process, which needs the description of interatomic interactions beyond the harmonic approximation. Recent progress in first-principles calculations enables the accurate prediction of interatomic force constants (IFCs) in both the harmonic and anharmonic levels with no adjusting parameters. These first-principles IFCs can then be fed into numerical schemes to solve phonon Boltzmann transport equation (BTE) to calculate the thermal conductivity. Lattice thermal conductivity computed in this approach for various materials can be highly accurate [6–10]. Such a method with predictive power also offers a means to deeply understand the phonon scattering physics and provides critical guidelines for future material and device design [11].

From the harmonic IFCs, a dynamical matrix can be constructed and the phonon eigenfrequency for different wavevectors (i.e., dispersion relation) can be calculated. The phonon dispersion can provide us with information on phonon group velocity, $\boldsymbol{v}_g$, as well as specific heat, c_v. Phonon lifetime can be solved either iteratively or using relaxation time approximation (RTA) in a linearized phonon BTE, in which usually up to three-phonon scattering processes are considered in Fermi's golden rule [12], where absorption or emission process of phonons happens only when energy conservation and momentum (or pseudo-momentum) conservation are achieved [13]. A simple diagram describing all possible three phonon scattering processes is presented in Fig. 29.1. With all these results, lattice thermal conductivity of a bulk crystal can be computed.

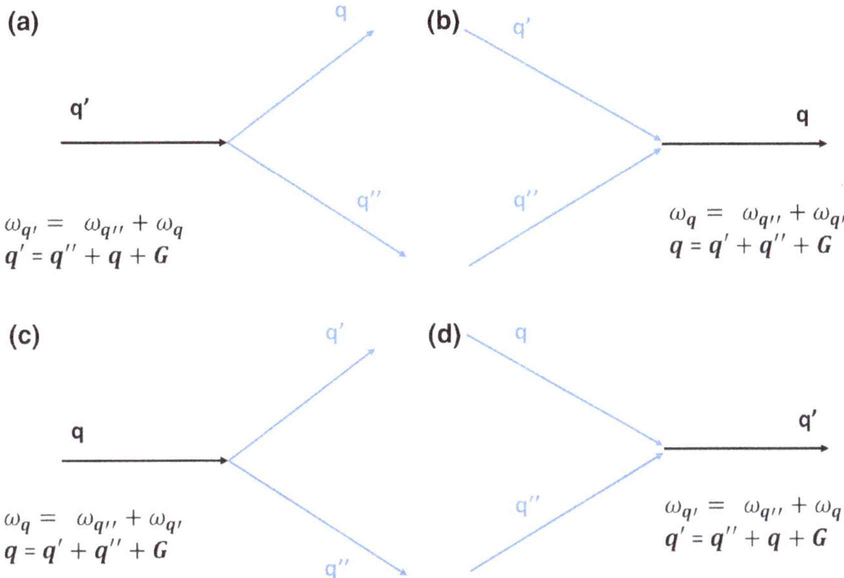

Fig. 29.1 Simple diagram describing three-phonon scattering process. The scattering processes creating a phonon with wavevector $\boldsymbol{q}$ are shown in (**a**) and (**b**), while (**c**) and (**d**) are the destruction processes of a phonon. In three-phonon scattering, energy conservation and pseudo-momentum conservation, where an extra reciprocal lattice vector $\boldsymbol{G}$ is involved, are satisfied

The first-principles study of the lattice thermal conductivity of β-Ga_2O_3 was first performed by Santia et al. [4]. In their study, a real space finite difference method was applied to calculate the IFCs in *Phonopy* [14] (for harmonic IFCs) and *ShengBTE* [15] (for anharmonic IFCs) using the first-principles package *Quantum Espresso* [16]. The DFT calculation was conducted using a plane-wave pseudo-potential technique using a projector augmented wave generalized gradient approximation (GGA) exchange-correlation functional [17, 18]. Lattice constants obtained after fully relaxing the structure were found to be a = 12.214 Å, b = 3.041 Å, and c = 5.647 Å with β = 103.72° compared with an experimental result of a = 12.276 Å, b = 3.0371 Å, and c = 5.7981 Å with β = 103.83° [19]. Based on the relaxed structures, 2 × 2 × 2 supercell size was used for both harmonic and anharmonic IFCs calculations. Using a similar method, we calculate the phonon dispersion relation in Fig. 29.2, and the results agree well with that from Santia et al. Considering the low symmetry and high anisotropy in β-Ga_2O_3, the phonon dispersion is much more complex compared to other wide bandgap materials like GaN and ZnO [20].

A 10 × 10 × 10 $\boldsymbol{q}$-point mesh in the first Brillouin zone was sampled in the work by Santia et al. [4] using the Methfessel-Paxton scheme [21] to solve the linearized phonon BTE in ShengBTE. The converged anisotropic thermal conductivity projected to different crystallographic directions is $\kappa_{[100]}$ = 16.06 W/mK,

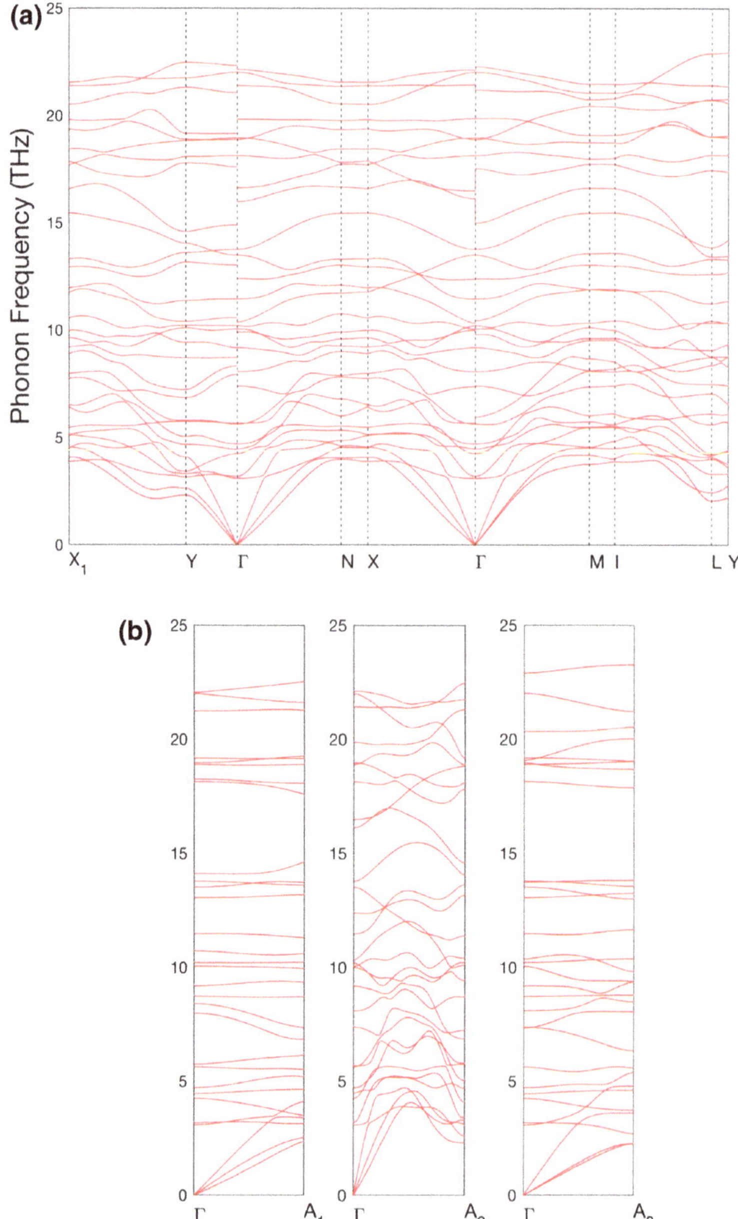

Fig. 29.2 **a** Phonon dispersion relation of β-Ga_2O_3 along selected high symmetry lines. The dispersion relation agrees well with the phonon dispersion in Figure S2 [4]. **b** Phonon dispersion along three crystallographic directions, denoted by $\Gamma \rightarrow A_1$ ([100]), $\Gamma \rightarrow A_2$ ([010]), and $\Gamma \rightarrow A_3$ ([001]). It can be seen that there is no phonon band gap observed in the $\Gamma \rightarrow A_2$ ([010]) direction. This lack of phonon dispersion gap will usually lead to larger opportunity of phonon scattering and smaller thermal conductivity

Table 29.1 Lattice thermal conductivity of bulk β-Ga_2O_3 from different experimental and computational methods at 300 K

	First-principles [4]	TDTR [22]	3ω [23]	3ω [24]	2ω [25]	Laser flash [26]
$\kappa_{[100]}$	16.06	10.9		13	11	13.0
$\kappa_{[010]}$	21.54	27.0	29.21		29	
$\kappa_{[001]}$	21.15	13.7			21	

Units are W/mK

$\kappa_{[010]} = 21.54$ W/mK, and $\kappa_{[001]} = 21.15$ W/mK as listed in Table 29.1. The direction and a conventional cell of β-Ga_2O_3 are presented in Fig. 29.3. The three crystallographic directions are along the $\overrightarrow{OA}$, $\overrightarrow{OB}$, and $\overrightarrow{OC}$ vectors in the conventional cell. In the high-temperature regime, a $\kappa \propto T^{-m}$ trend, where m was close to 1, was observed, indicating a Umklapp dominating scattering nature in bulk β-Ga_2O_3 [5, 13]. It should be noted that the relatively small $10 \times 10 \times 10$ $\boldsymbol{q}$-point mesh used in [4] due to the computational resources limitation may not be large enough to capture the full phonon transport properties in bulk gallium oxide, especially at low temperature where most heat is transferred by acoustic phonons near the Brillouin zone center and thus a large density of $\boldsymbol{q}$-point sampling is usually required. However, low-temperature thermal conductivity values may be of less interest to this community because the application of β-Ga_2O_3 is anticipated to be mainly at room or elevated temperatures.

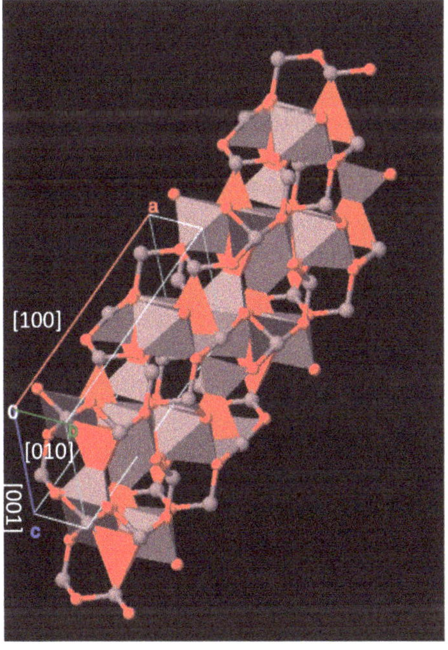

Fig. 29.3 A conventional cell of the β-Ga_2O_3 crystal and three crystallographic directions. [100], [010], and [001] are the three principle crystallographic directions along the $\overrightarrow{OA}$, $\overrightarrow{OB}$, and $\overrightarrow{OC}$ directions

The thermal transport of phonon can usually be understood in the relaxation time approximation (RTA) model. RTA can lead to error for high thermal conductivity crystals (e.g., graphene, diamond). However, for low thermal conductivity materials like β-Ga_2O_3, RTA is usually sufficient compared to the more sophisticated iterative BTE solutions [4, 7, 27–29]. It should be noted that they used iterative scheme to solve the BTE to obtain thermal conductivity, but the thermal conductivity calculated from iterative solutions was found close to the result using RTA [4]. In RTA, we have

$$\kappa \propto \sum_{q,n} C_{qn} \boldsymbol{v}_{qn}^{2} \tau_{qn} \tag{29.2}$$

where C_{qn} is the volumetric heat capacity for phonon mode with wavevector $\boldsymbol{q}$ and polarization index n. $v_{qn} = \frac{\partial \omega_{qn}}{\partial q}$ denotes the group velocity, and τ_{qn} is the phonon relaxation time [5, 30]. The summation is over phonon modes with wavevector $\boldsymbol{q}$ and polarization n in the first Brillouin zone. Since the anisotropy in lattice thermal conductivity comes solely from the vector term, the orientation-dependent phonon group velocity, and thus the scalar relaxation time and specific heat can be viewed as weighting factors, it is of great importance to study the group velocity profile in different propagation direction. Although the [010] crystallographic direction has the largest thermal conductivity, its acoustic phonon dispersion slope ($\Gamma \rightarrow A_2$) is only slightly larger than that of the [100] ($\Gamma \rightarrow A_1$) direction (Fig. 29.2b) (group velocity comparison in these three directions is shown in Fig. 29.4a), meaning the difference in thermal conductivity is not entirely from the difference in acoustic phonon group velocity. Further decomposition of thermal conductivity to different phonon polarizations by Santia et al. [4] shows that the thermal conductivity from acoustic phonon for [010] is actually the smallest among three crystallographic directions (12.3 W/mK in [010] vs. 13.92 W/mK in [100] and 15.75 W/mK in [001]) meaning that the unusual large optical phonon contribution in the [010] direction, contributing about 44% to the total thermal conductivity at 300 K, is the main reason for the anisotropic thermal transport. The absence of a phonon band gap in [010] compared with other directions may work as the reason for the relatively small acoustic phonon contribution, since the phonon band gap will reduce the opportunity for satisfying the energy conservation condition in three-phonon scattering process [31]. The large contribution of optical phonons has also been observed in other complex materials [32].

The thermal transport properties of bulk β-Ga_2O_3 calculated from first-principles simulations were also compared with that of GaN, another widely used semiconductor for power electronic applications [22]. GaN was reported to have a relatively large thermal conductivity of 130–230 W/mK at room temperature [33–36]. Considering the fact that their volumetric heat capacities differ by only ~20% and the acoustic mode group velocities are in the same orders of magnitude (a detailed phonon group velocity comparison of GaN and β-Ga_2O_3 is presented in Fig. 29.4b), their difference in phonon relaxation time must be the main reason for

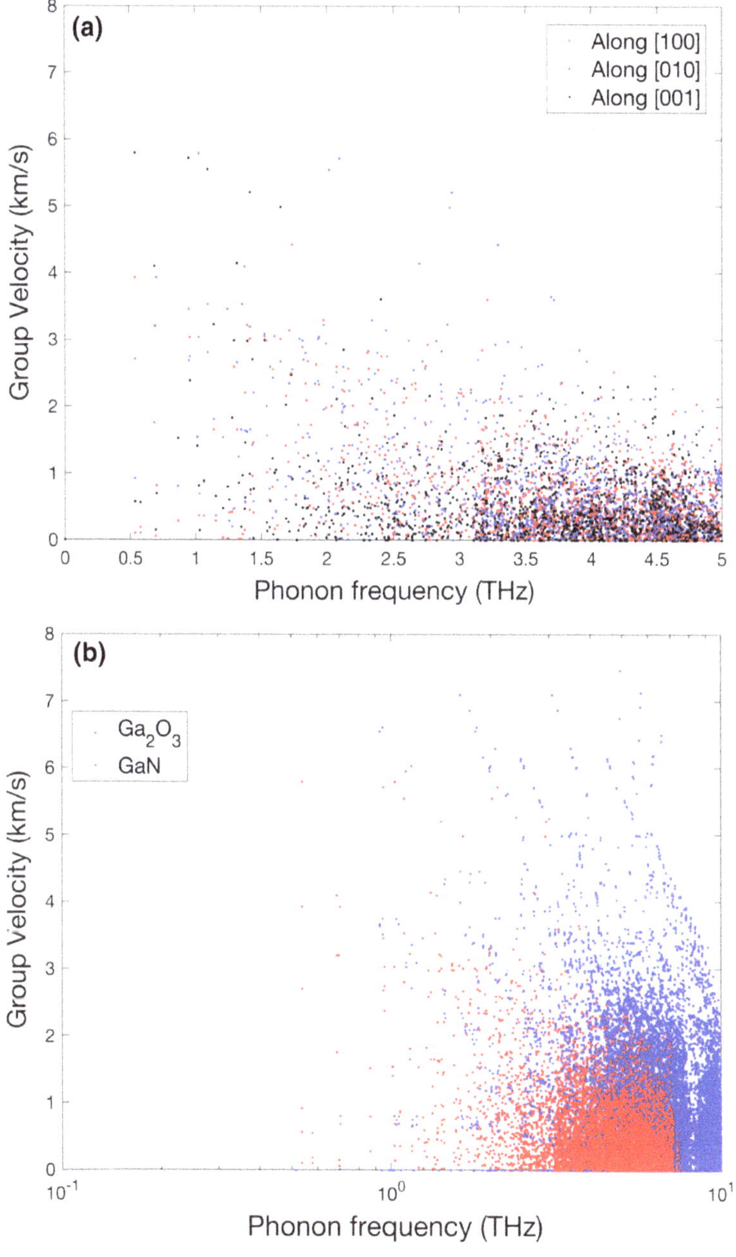
(a)
Along [100]
Along [010]
Along [001]
Group Velocity (km/s)
Phonon frequency (THz)
(b)
Ga_2O_3
GaN
Group Velocity (km/s)
Phonon frequency (THz)

◀ **Fig. 29.4** **a** Comparison of group velocities of low-frequency phonons (0–5 THz) along the [100] ($\Gamma \rightarrow A_1$), [010] ($\Gamma \rightarrow A_2$), and [001] ($\Gamma \rightarrow A_3$) directions in β-Ga_2O_3. The group velocities in the [010] direction are only slightly larger than those in the other two directions. **b** Comparison of phonon group velocities of β-Ga_2O_3 and GaN for the acoustic phonon regime. The phonon group velocity of GaN is only slightly larger than that of β-Ga_2O_3. Considering the orders of magnitude difference in thermal conductivity (130–230 W/mK for GaN vs. 10–30 W/mK for β-Ga_2O_3), their difference in phonon relaxation time must be the main reason for the thermal conductivity difference

the thermal conductivity difference. In three-phonon scattering, both the number of possible scattering channels and the strength of scattering (anharmonicity) can directly impact the relaxation time. These can be characterized by the three-phonon scattering phase space (scattering channel numbers) and Grüneisen parameters (anharmonicity), respectively. It was found that the Grüneisen parameters of the acoustic polarizations in β-Ga_2O_3 are comparable to that in GaN, while the scattering phase space in β-Ga_2O_3 is orders of magnitude larger, meaning that three-phonon scattering opportunity is much larger than that in GaN. Detailed data are presented in Table 29.2. This large scattering possibility can be attributed to the absence of a large phonon band gap [36] and a more complex phonon dispersion relation originated from the low symmetry in β-Ga_2O_3.

Besides first-principles simulations, experimental studies of lattice thermal conductivity for bulk β-Ga_2O_3 are also reported using different methods. The thermal conductivity at a wide temperature range from 80 to 495 K is measured by Guo et al. [22] for single crystal β-Ga_2O_3 using the time-domain thermoreflectance (TDTR) technique. Their samples are grown via an edge-defined film-fed growth (EFG) method [37] by Tamura Corporation (Japan), doped with Sn during the growth process, resulting in a room temperature carrier concentration of around 5–6 × 10^{18} cm^{-3}. It is noted that this doping level should not lead to significant contribution from charges to thermal conductivity.

TDTR is an optical pump-probe technique used widely to measure the thermal conductivity of both bulk and thin film materials [30, 38]. In such measurements, a metal layer working as the photo-to-thermal transducer is deposited on the sample surface. The transducer absorbs the pump laser energy and converts it into heat. The heat spreads into the sample, and consequently, the transducer layer is cooled down.

Table 29.2 Comparison of Grüneisen parameters and scattering phase space for β-Ga_2O_3 and GaN for acoustic phonons

	Grüneisen parameters of different branches	Normalized three-phonon scattering phase space
β-Ga_2O_3	−1.06 (TA1), −0.30 (TA2), 0.83 (LA)	1200
GaN	−0.86 (TA1), −0.86 (TA2), 0.51 (LA)	38

Here, TA1 and TA2 denote the different transverse acoustic phonon modes, and LA denotes the longitudinal acoustic phonon modes [22]

The surface reflectance of the transducer changes linearly with temperature when the temperature rise is sufficiently small, and thus, its reflectance probed by the probe laser is indicating the temperature evolution of the transducer layer. This recovered temperature profile can be fitted using appropriate heat equations, and thus, the thermal diffusivity ($\kappa/\rho c$, thermal conductivity κ, density ρ, and specific heat c) is obtained, from which thermal conductivity is calculated. More details of this technique can be found in the literature [38–40].

To obtain the thermal conductivity of bulk β-Ga_2O_3, temperature-dependent heat capacity needs to be measured first since fitting TDTR signal yields the thermal diffusivity ($\alpha = \kappa/\rho c$). The specific heat can be measured using a differential scanning calorimeter and was found well fitted in the Debye model with an Debye temperature of 738 K, which is close to a first-principles result of 872 K [41]. The room temperature heat capacity is measured to be 18.7 J/mol K. With the help of measured specific heat, the anisotropic thermal conductivities are shown in Table 29.1. It should be noted that both the first-principles result and the TDTR result show that [010] is the optimal direction for heat transfer and [100] is the direction with the weakest thermal transport capability. The temperature-dependent thermal conductivity in this TDTR measurement can also be fitted using a $\kappa \propto T^{-m}$ relation where m is close to 1 above 200 K, indicating that the phonon transport is still Umklapp scattering limited, in spite of the Sn doping. It is likely that the Sn doping concentration is so low that it does not have significant effect on phonon scattering. A comparison of the fitting parameters compared with first-principles result is shown in Table 29.3 as well as Fig. 29.5.

Besides this experiment, thermal conductivity of bulk β-Ga_2O_3 has also been measured using different methods at different conditions [23, 24, 26, 42]. A thermal conductivity of 13 W/mK in the [100] direction of bulk β-Ga_2O_3 at room temperature was measured by Víllora et al. [26] using the laser flash technique. Later, a thermal conductivity of 21 W/mK was measured by Galazka et al. [42] in the [010] direction for sample grown using the Czochralski method. Using a 3ω method [43], thermal conductivity of undoped semiconducting and Mg-doped insulating β-Ga_2O_3 grown using a floating-zone method was confirmed to be 13 W/mK in the [100] direction by Handwerg et al. [24]. These authors further studied the anisotropic thermal conductivity using a 2ω method on the same sample [25], and the anisotropic thermal conductivity in the [100], [010], and [001] directions are 11, 29,

Table 29.3 Comparison of the fitted temperature-dependent lattice thermal conductivity for TDTR and first-principles results at high-temperature regime (above 200 K)

	A_{TDTR}	m_{TDTR}	A_{FP}	m_{FP}
$\kappa_{[100]}$	1.06×10^4	1.21	1.99×10^4	1.27
$\kappa_{[010]}$	3.28×10^4	1.27	2.47×10^4	1.21
$\kappa_{[001]}$	8.14×10^3	1.12	9.71×10^3	1.39

A function form of $\kappa = A \times T^{-m}$ is used, where A_{TDTR}, m_{TDTR}, A_{FP}, and m_{FP} are the corresponding fitting parameters for data from the TDTR experiment [22] and the first-principles calculation [4]

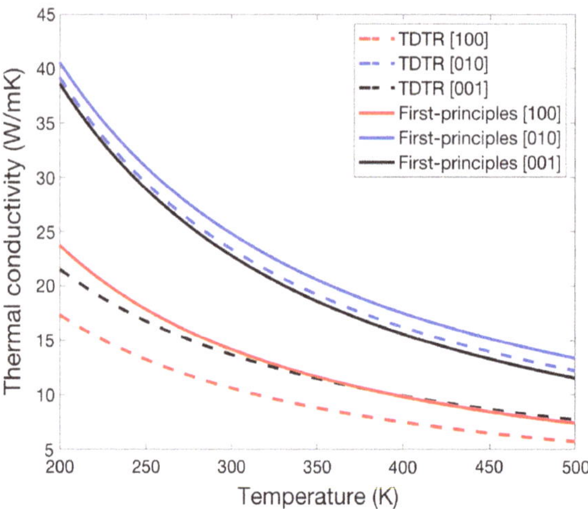

Fig. 29.5 Fitted temperature-dependent anisotropic lattice thermal conductivity of β-Ga_2O_3 for the high-temperature regime. The dashed lines denote the measured thermal conductivity from the TDTR experiments [22] and the solid lines denote the first-principles calculation results [4]

and 21 W/mK, respectively. A more detailed 3ω measurement was performed by Slomski et al. [23] for the thermal conductivity in the temperature range of 295–410 K in the [010] and [102] directions of undoped, Sn-doped, and Fe-doped bulk β-Ga_2O_3 grown using the EFG method. A 29.21 W/mK thermal conductivity in the [010] direction was measured, close to the TDTR result [22], and the thermal conductivity along the [102] direction they obtained was 18.39 W/mK. In this experiment, a $m < 1$ fitting parameter using the $\kappa = A \times T^{-m}$ relation indicates that a relatively large phonon-point-defect scattering and phonon-free-carrier scattering exist in the measured samples, regardless of the doping conditions or orientations. This was explained by the relatively high oxygen vacancy concentration in the crystal growth process, but the larger thermal conductivity than first-principles result makes this explanation questionable. Of course, first-principles calculations of such complicated lattice structures like the one of β-Ga_2O_3 are also non-trivial and can have uncertainties. For example, all the measured thermal conductivities in the [010] directions are higher than that of the first-principles result (Table 29.1). The type of doping (i.e., Sn-doping and Fe-doping) was found to affect little the thermal conductivity. This property is ideal for power electronic devices applications since we can modulate electron properties easily while holding thermal transport properties unsacrificed.

β-Ga_2O_3 may also be used as a component of gate oxides in high-frequency devices, like the GaN MOSFET [44, 45]. Due to the small size of the devices and large number of interfaces, the interfacial thermal conductance, a quantity describing efficiency of transmitting energy across the interface, is of great importance [5]. Interfacial thermal conductance ($Wm^{-2}K^{-1}$) is defined as

$$h = q/\Delta T \tag{29.3}$$

where q is the heat flux (W/m^2), and ΔT is the temperature drop across the interface (K) [46]. Considering this, study of thermal conductance of the interface between β-Ga_2O_3 and GaN, as well as the effective thermal conductivity of the Ga_2O_3 thin films, is of interest. Szwejkowski et al. [47] studied the thermal transport in β-Ga_2O_3 thin film and across the Ga_2O_3/GaN interface using the TDTR technique. In this study, phase pure and polycrystalline β-Ga_2O_3 thin films with 10–1000 nm thicknesses were first grown via annealing the (0001) GaN surfaces in open atmosphere. To analyze the TDTR data, a three-layer model considering the metal transducer/Ga_2O_3/GaN junction was applied. In this study, the interfacial thermal resistance at Ga_2O_3/GaN was first neglected to study an "effective" thickness-dependent thermal conductivity. A strong thickness dependence was observed in the effective thermal conductivity, where the effective thermal conductivity of Ga_2O_3 increased from ~0.3 W/mK for a 10-nm-thick thin film to close to bulk thermal conductivity of ~10 W/mK in a 1000-nm-thick film. This trend, different from other gate dielectrics like amorphous SiO_2 and SiN_x thin film, was explained by a longer phonon mean free path in Ga_2O_3. Different metal transducers were also applied and the same trend was observed. This thickness dependence trend was also confirmed by a recent 3ω method measurement by Blumenschein et al. [48]. Detailed analysis in the measurement from [47] indicated that the Ga_2O_3/GaN interfacial thermal boundary conductance was 31.2 ± 8.1 MW $m^{-2}K^{-1}$, which is in the range of typical thermal boundary conductance for non-metal/non-metal interfaces. An intrinsic thermal conductivity for the bulk polycrystalline β-Ga_2O_3 was measured to be 8.8 W/mK, lower than that of single crystal β-Ga_2O_3. This study on Ga_2O_3/GaN interface sheds light on the application of gallium oxide as potential gate dielectrics in GaN devices.

In summary, the thermal properties of β-Ga_2O_3 are reviewed. Using various experimental and computational methods, the thermal conductivity of bulk β-Ga_2O_3 is proven to be highly anisotropic. The largest thermal conductivity at room temperature is around 25 W/mK in β-Ga_2O_3, and this relatively small thermal conductivity compared with GaN might be a major challenge for potential applications in power electronics. These thermal properties may be useful for thermal management and design of β-Ga_2O_3 devices.

References

1. T.L. Bergman, F.P. Incropera, D.P. DeWitt, A.S. Lavine, *Fundamentals of Heat and Mass Transfer* (Wiley, 2011)
2. K. Sasaki, M. Higashiwaki, A. Kuramata, T. Masui, S. Yamakoshi, MBE grown Ga2O3 and its power device applications. J. Cryst. Growth **378**, 591 (2013)
3. H. Peelaers, C.G. Van de Walle, Brillouin zone and band structure of β-Ga2O3. Phys. Status Solidi B **252**, 828 (2015)
4. M.D. Santia, N. Tandon, J. Albrecht, Lattice thermal conductivity in β−Ga2O3 from first principles. Appl. Phys. Lett. **107**, 041907 (2015)

5. G. Chen, *Nanoscale Energy Transport and Conversion: A Parallel Treatment of Electrons, Molecules, Phonons, and Photons* (Oxford University Press, 2005)
6. D. Broido, M. Malorny, G. Birner, N. Mingo, D. Stewart, Intrinsic lattice thermal conductivity of semiconductors from first principles. Appl. Phys. Lett. **91**, 231922 (2007)
7. A. Ward, D. Broido, D.A. Stewart, G. Deinzer, Ab initio theory of the lattice thermal conductivity in diamond. Phys. Rev. B **80**, 125203 (2009)
8. K. Esfarjani, G. Chen, H.T. Stokes, Heat transport in silicon from first-principles calculations. Phys. Rev. B **84**, 085204 (2011)
9. J. Garg, N. Bonini, B. Kozinsky, N. Marzari, Role of disorder and anharmonicity in the thermal conductivity of silicon-germanium alloys: a first-principles study. Phys. Rev. Lett. **106**, 045901 (2011)
10. T. Luo, J. Garg, J. Shiomi, K. Esfarjani, G. Chen, Gallium arsenide thermal conductivity and optical phonon relaxation times from first-principles calculations. EPL (Europhys. Lett.) **101**, 16001 (2013)
11. Z. Liu, X. Wu, V. Varshney, J. Lee, G. Qin, M. Hu, A.K. Roy, T. Luo, Bond saturation significantly enhances thermal energy transport in two-dimensional pentagonal materials. Nano Energy **45**, 1 (2018)
12. A. Maradudin, A. Fein, Scattering of neutrons by an anharmonic crystal. Phys. Rev. **128**, 2589 (1962)
13. G.P. Srivastava, *The Physics of Phonons* (CRC Press, 1990)
14. A. Togo, I. Tanaka, First principles phonon calculations in materials science. Scr. Mater. **108**, 1 (2015)
15. W. Li, J. Carrete, N.A. Katcho, N. Mingo, ShengBTE: a solver of the Boltzmann transport equation for phonons. Comput. Phys. Commun. **185**, 1747 (2014)
16. P. Giannozzi et al., QUANTUM ESPRESSO: a modular and open-source software project for quantum simulations of materials. J. Phys.: Condens. Matter **21**, 395502 (2009)
17. J.P. Perdew, W. Yue, Accurate and simple density functional for the electronic exchange energy: generalized gradient approximation. Phys. Rev. B **33**, 8800 (1986)
18. J.P. Perdew, K. Burke, M. Ernzerhof, Generalized gradient approximation made simple. Phys. Rev. Lett. **77**, 3865 (1996)
19. J. Åhman, G. Svensson, J. Albertsson, A reinvestigation of [beta]-gallium oxide. Acta Crystallogr. C **52**, 1336 (1996)
20. X. Wu, J. Lee, V. Varshney, J.L. Wohlwend, A.K. Roy, T. Luo, Thermal conductivity of wurtzite zinc-oxide from first-principles lattice dynamics–a comparative study with gallium nitride. Sci. Rep. **6**, 22504 (2016)
21. M. Methfessel, A. Paxton, High-precision sampling for Brillouin-zone integration in metals. Phys. Rev. B **40**, 3616 (1989)
22. Z. Guo et al., Anisotropic thermal conductivity in single crystal β-gallium oxide. Appl. Phys. Lett. **106**, 111909 (2015)
23. M. Slomski, N. Blumenschein, P. Paskov, Anisotropic thermal conductivity of β-Ga2O3 at elevated temperatures: Effect of Sn and Fe dopants. J. Muth, T. Paskova, J. Appl. Phys. **121**, 235104 (2017)
24. M. Handwerg, R. Mitdank, Z. Galazka, S.F. Fischer, Temperature-dependent thermal conductivity in Mg-doped and undoped β-Ga2O3 bulk-crystals. Semicond. Sci. Technol. **30**, 024006 (2015)
25. M. Handwerg, R. Mitdank, Z. Galazka, S.F. Fischer, Temperature-dependent thermal conductivity and diffusivity of a Mg-doped insulating β-Ga2O3 single crystal along [100], [010] and [001]. Semicond. Sci. Technol. **31**, 125006 (2016)
26. E.G. Víllora, K. Shimamura, T. Ujiie, K. Aoki, Electrical conductivity and lattice expansion of β-Ga2O3 below room temperature. Appl. Phys. Lett. **92**, 202118 (2008)
27. A. Ward, D.A. Broido, Intrinsic phonon relaxation times from first-principles studies of the thermal conductivities of Si and Ge. Phys. Rev. B **81**, 085205 (2010)
28. L. Lindsay, D. Broido, N. Mingo, Flexural phonons and thermal transport in graphene. Phys. Rev. B **82**, 115427 (2010)

29. L. Lindsay, D. Broido, T. Reinecke, Ab initio thermal transport in compound semiconductors. Phys. Rev. B **87**, 165201 (2013)
30. T. Luo, G. Chen, Nanoscale heat transfer – from computation to experiment. Phys. Chem. Chem. Phys. **15**, 3389 (2013)
31. L. Lindsay, D. Broido, Three-phonon phase space and lattice thermal conductivity in semiconductors. J. Phys.: Condens. Matter **20**, 165209 (2008)
32. R. Guo, X. Wang, B. Huang, Thermal conductivity of skutterudite CoSb3 from first principles: substitution and nanoengineering effects. Sci. Rep. **5**, 7806 (2015)
33. E. Sichel, Thermal conductivity of GaN, 25-360 K. J. Phys. Chem. Solids **38**, 330 (1977)
34. A. Jeżowski, B.A. Danilchenko, M. Boćkowski, I. Grzegory, S. Krukowski, T. Suski, T. Paszkiewicz, Thermal conductivity of GaN crystals in 4.2–300 K range. Solid State Commun. **128**, 69 (2003)
35. G.A. Slack, L.J. Schowalter, D. Morelli, J.A. Freitas, 'Some effects of oxygen impurities on AlN and GaN. J. Cryst. Growth **246**, 287 (2002)
36. L. Lindsay, D.A. Broido, T.L. Reinecke, Thermal conductivity and large isotope effect in GaN from first principles. Phys. Rev. Lett. **109**, 095901 (2012)
37. A. Kuramata, K. Koshi, S. Watanabe, Y. Yamaoka, T. Masui, S. Yamakoshi, High-quality β-Ga2O3 single crystals grown by edge-defined film-fed growth. Jpn. J. Appl. Phys. **55**, 1202A2 (2016)
38. P. Jiang, X. Qian, R. Yang, Time-domain thermoreflectance (TDTR) measurements of anisotropic thermal conductivity using a variable spot size approach. Rev. Sci. Instrum. **88**, 074901 (2017)
39. D.G. Cahill, Analysis of heat flow in layered structures for time-domain thermoreflectance. Rev. Sci. Instrum. **75**, 5119 (2004)
40. A.J. Schmidt, Optical characterization of thermal transport from the nanoscale to the macroscale. Ph.D. Dissertation, Massachusetts Institute of Technology, 2008
41. H. He, M.A. Blanco, R. Pandey, Electronic and thermodynamic properties of β-Ga2O3. Appl. Phys. Lett. **88**, 261904 (2006)
42. Z. Galazka et al., On the bulk β-Ga2O3 single crystals grown by the Czochralski method. J. Cryst. Growth **404**, 184 (2014)
43. D.G. Cahill, Thermal conductivity measurement from 30 to 750 K: the 3ω method. Rev. Sci. Instrum. **61**, 802 (1990)
44. T. Lin, H. Chiu, P. Chang, L. Tung, C. Chen, M. Hong, J. Kwo, W. Tsai, Y. Wang, High-performance self-aligned inversion-channel In0.53Ga0.47As metal-oxide-semiconductor field-effect-transistor with Al2O3/Ga2O3(Gd2O3) as gate dielectrics. Appl. Phys. Lett. **93**, 033516 (2008)
45. J. Johnson et al., Gd2O3/GaN metal-oxide-semiconductor field-effect transistor. Appl. Phys. Lett. **77**, 3230 (2000)
46. E.T. Swartz, R.O. Pohl, Thermal boundary resistance. Rev. Mod. Phys. **61**, 605 (1989)
47. C.J. Szwejkowski, N.C. Creange, K. Sun, A. Giri, B.F. Donovan, C. Constantin, P.E. Hopkins, Size effects in the thermal conductivity of gallium oxide (β-Ga2O3) films grown via open-atmosphere annealing of gallium nitride. J. Appl. Phys. **117**, 084308 (2015)
48. N. Blumenschein, M. Slomski, P. Paskov, F. Kaess, M. Breckenridge, in *Proceedings of SPIE 10533*, pp. 105332G, 2018

Chapter 30
Scintillation Properties

Takayuki Yanagida, Go Okada and Noriaki Kawaguchi

Abstract Scintillators are key materials for radiation detectors for a wide range of applications such as medicine, security, well-logging, environmental monitoring and high energy physics. In recent years, Ga_2O_3 has attracted much attention as a scintillator. In this chapter, we introduce some backgrounds of scintillators and scintillation detectors, which are followed by recent R&D of Ga_2O_3 for scintillation detectors. In this chapter, in order to improve the scintillation properties, we have examined some dopants for Ga_2O_3. The dopants are rare earth, ns^2 and alkali earth ions. In addition to the investigations of dopants, we have also investigated the transparent ceramic form. These results are introduced in the present chapter.

30.1 Scintillators and Scintillation Detectors

Scintillators are a type of luminescent materials, and they have a function to absorb ionizing radiations and convert to low energy (1–6 eV) photons immediately [1]. The photo emission due to absorption of ionizing radiations is called scintillation, and radiation detectors which is equipped with scintillators are called scintillation detectors. Up to now, continuous efforts have been made to develop new scintillators since the properties, especially light yield and decay time, of scintillators directly affect the detector properties. If bright and fast scintillators are developed, the spatial resolution and throughput of non-destructive imaging detectors can be improved. Recently, semiconductor scintillators including Ga_2O_3 have attracted much attention [2], and in this chapter, we introduce recent progress of the related topics.

Today, most of existing scintillators consist of insulator as host material as well as a fraction of emission center included. The scintillation light yield is empirically

T. Yanagida (✉) · N. Kawaguchi
Nara Institute of Science and Technology, Ikoma, Nara 630-0192, Japan
e-mail: t-yanagida@ms.naist.jp

G. Okada
Kanazawa Institute of Technology, Nonoichi, Ishikawa 921-8501, Japan

M. Higashiwaki and S. Fujita (eds.), *Gallium Oxide*, Springer Series in Materials Science 293, https://doi.org/10.1007/978-3-030-37153-1_30

expressed as $LY_{sc} \propto E/(\beta E_g) \times S \times Q$ [3–6], where LY_{sc} is scintillation light yield, E is absorbed energy of ionizing radiation, β is an empirical parameter (= 2–3), E_g is band-gap energy, S is energy transportation efficiency from the host to emission center and Q is quantum efficiency of the localized emission centers, which is internal photoluminescence (PL) quantum efficiency. For βE_g, we generally understand as a mean energy required to generate a single electron–hole pair in (e.g., in Si photodiode, $\beta E_g \sim 3.6$ eV). In semiconductor scintillators, potentially high LY_{sc} is expected since $S = 1$, which means that the host matrix itself has a function of emission center if the material is a direct transition semiconductor. If the material is not a direct transition semiconductor, generated carriers can smoothly move to localized emission centers than typical insulators. In addition, for a direct transition semiconductor, a fast decay time is expected. This is a brief understanding why researchers investigate semiconductor scintillators. Commonly known semiconductor scintillators are ZnO [7], GaN [8], CdS [9] and ZnSe [10]. Following the studies of these materials, we have recently found that Ga_2O_3 has very promising scintillation properties such as high light yield and fast decay time [11]. In the latter study, we used a commercial Ga_2O_3 crystal substrate of Tamura Corp., and any R&D for applications as scintillator were not conducted at all. Therefore, we think there remains a large room for studying and developing Ga_2O_3-based semiconductor scintillators.

30.2 Methodology of R&D of Ga_2O_3 Scintillators

In our research, Ga_2O_3 bulk crystal was grown by the floating zone (FZ) method. Before the crystal growth, polycrystalline rods of doped Ga_2O_3 were prepared by the conventional solid-state reaction. Then, the polycrystalline rod was loaded in the FZ furnace (Canon Machinery, FZD0192). During the crystal growth, the rotation rate was 20 rpm, and the pull-down rate was 2.5–5 mm/h. After the crystal growth, the obtained crystal samples were cut to a typical size of 3–5 mmϕ × 1–2 mmt, and the wide surfaces were mechanically polished. Characterizations for applications as scintillator were performed as follows.

As described above, the scintillation light yield is expressed as $LY_{sc} \propto E/(\beta E_g) \times S \times Q$ [3–6], so the internal quantum efficiency is one of the important properties for scintillators. Firstly, we generally measure the PL quantum efficiency by using Quantaurus-QY (Hamamatsu, C11347). In addition to the PL quantum efficiency, we measured PL decay time profiles by using Quantaurus-τ (Hamamatsu, C11367). Although PL properties are sometimes very different from scintillation, they are still important to consider the emission origins and physics of scintillation.

As a fundamental scintillation property, X-ray-induced scintillation spectrum was measured by using our original setup [12]. The X-ray generator (Spellman, XRB80N100/CB) was equipped with an X-ray tube having a tungsten anode target and beryllium window. The X-ray tube was operated by applying the bias DC voltage of 80 kV and tube current of 1.2 mA. The scintillation photons from the samples were guided to a spectrometer unit consisting of a CCD (Andor, DU-420-BU2) and monochromator (Shamrock, SR163) through a 2-m optical fiber to measure the spectrum. The CCD was cooled to 188 K by a Peltier module to reduce the thermal noise. Furthermore, scintillation decay time profiles were measured using an afterglow characterization system equipped with a pulse X-ray tube [13]. The system allows measurements by the time-correlated single-photon counting (TCSPC) technique. The photomultiplier tube (PMT) used in this measurement covers the spectral range from 160 to 650 nm. For scintillation decay time profiles, the decay time constant is deduced by the least-squares fitting with a multi-exponential decay function.

Scintillation light yield was evaluated by measuring pulse height spectrum. In this measurement, a sample piece was firmly placed on the window of PMT (Hamamatsu, R7600-200) with a silicon grease (OKEN, 6262A), and the sample was covered by several layers of Teflon reflectors to guide the scintillation photons to the PMT. The scintillation light signal is converted to an electrical signal by the PMT, which is then processed by the pre-amplifier (ORTEC, 113), shaping amplifier (ORTEC, 570) and a multichannel analyzer (Amptek, Pocket MCA 8000A). Eventually, the pulse height spectrum was stored on the computer. Here, the shaping time was typically 2 μs. In this measurement, we used ^{137}Cs 662 keV γ-ray and ^{241}Am 5.5 MeV α-ray sources. Although this measurement is not common in other fields related to luminescent materials, pulse height spectroscopy is a common technique to selectively detect a type of ionizing radiations and also estimate a scintillation light yield, which is a number of photons emitted per unit energy of incident ionizing radiation.

30.3 Results

In this section, we show some selected results of our recent R&D on Ga_2O_3. In particular, we have tested effects of doping with some impurity ions. The dopant concentrations described in this section are nominal value, and the actual dopant concentrations are 1–10% of the nominal values due to the segregation (confirmed by X-ray fluorescence measurement). As it is common in the crystal growth science, the segregation coefficient depends on the crystal growth methodology and materials. Figure 30.1 shows a photograph of as-grown undoped Ga_2O_3 bulk single crystal. By using the FZ method, we can typically grow a sample as shown in Fig. 30.1. The crystal looks transparent, and we typically cut the sample perpendicularly to the growth direction. It must be mentioned that Ga_2O_3 crystals grown by the FZ method typically have the β-phase, although it is known that Ga_2O_3 can be in the α, β, γ, δ and ε phases.

In the scintillation emission spectrum (in other words, radioluminescence spectrum) of undoped Ga_2O_3 under X-rays, the peak emission wavelength was typically ~380 nm (blue) with a shoulder ~340 nm (UV). In addition, the blue

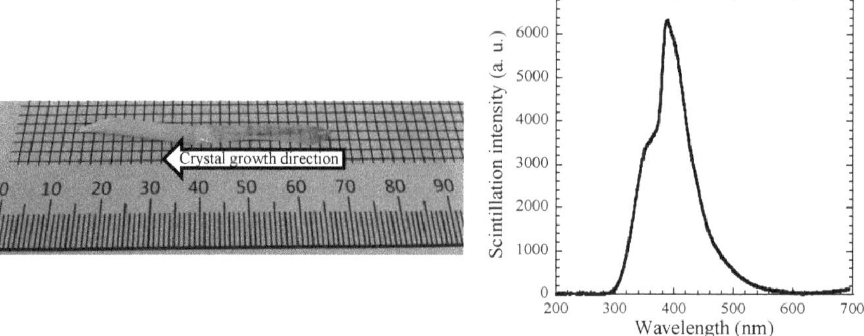

Fig. 30.1 As-grown Ga_2O_3 single crystal (left) and scintillation emission spectrum under X-ray irradiation (right)

emission band has a long tail to longer wavelengths up to 500–600 nm. Ga and oxygen vacancies play important roles to the luminescence properties. The spectral range of this emission is well-agreed with the spectral response of most conventional PMTs, which is a strong advantage for practical applications. PMTs have a high multiplication gain up to 10^{6-7}, so PMTs are the most common photodetector for scintillation detectors. In addition to PMTs, Si-based photodetectors such as a photodiode (PD) are also common. The wavelength sensitivity of PDs is high at red and near-infrared wavelengths, thus the signal read out by PDs do not suit for Ga_2O_3-based scintillation detectors. Apart from the practical application view point, some groups have discussed the origin of these emission bands, and a typical interpretation is that the origin of the UV emission is due to recombination of free electrons and self-trapped holes [14]. On the other hand, the donor–acceptor recombination involving the oxygen vacancy, interstitial Ga and Ga vacancies is considered to be the origin of the blue–green emission [15, 16]. Therefore, the emission spectral feature of undoped Ga_2O_3 resembles to that in PL or electroluminescence (EL).

In the scintillator field, it is common to dope with rare earth ions as emission center. Particularly, Eu^{2+} and Ce^{3+} are common doping ions, so we examined these for Ga_2O_3. Eu^{2+} and Ce^{3+} can show the luminescence due to the electron transitions from 5d to 4f levels (5d–4f transition). The 5d–4f transition is spin and parity allowed transition so a high luminescence intensity can be expected. In addition, emission wavelength based on 5d–4f transition of Eu^{2+} and Ce^{3+} typically appear blue–green wavelengths, and such an emission wavelength is preferable for conventional PMT readout.

In general, it is challenging to design and control the valence state of Eu ion, and sometimes Eu^{2+} can appear if Eu ions are substituted to different valence cations. Therefore, experimental approach is the most reliable way to find materials with Eu^{2+}. Figure 30.2 shows Eu- and Ce-doped Ga_2O_3. For the 10% Eu-doped sample, the color looks red and it is transparent. X-ray induced scintillation spectra of these samples are also shown in Fig. 30.2. In some samples, sharp emission lines due to Eu^{3+}:4f–4f transitions were observed while we could not confirm 5d–4f transition of Eu^{2+} by PL and scintillation decay times. The emission around 420 nm would be ascribed to the host emission of Ga_2O_3. Ce-doping was also investigated, and we could grow Ce-doped crystals successfully (Fig. 30.2 right). Although we could not distinguish the host emission of Ga_2O_3 and emission due to Ce^{3+}: 5d–4f transition in the spectral domain, we confirmed the presence of emission due to the 5d–4f transitions of Ce^{3+} by checking the PL and scintillation decay curves [17]. When we evaluated the scintillation detector properties (pulse height spectroscopy), these rare earth doped Ga_2O_3 did not show preferable properties.

Figure 30.3 represents Ga_2O_3 crystals doped with other rare earth ions. The sample colors represent characteristics of each dopant ions (e.g., purple = Nd^{3+}, green = Pr^{3+}), so these rare earth ions are activated. However, their scintillation properties were not promising.

In addition to rare earth ions, doping with ns^2 ions is also common in scintillators as Tl-doped NaI [18] and $Bi_4Ge_3O_{12}$ [19] are known to be standard scintillators. In order to investigate the effects of ns^2-ion doping, we synthesized In, Sn, Sb, Tl, Pb and Bi 1% doped Ga_2O_3 single crystals. Figure 30.4 shows the appearance of grown crystals (left) and X-ray induced scintillation emission spectra (right). As seen in the picture, we grew ns^2-ion-doped single crystals successfully, and all of them showed higher scintillation intensity than that of the undoped Ga_2O_3. Generally, emissions due to ns^2-ions appear around 400–600 nm, and the emissions due to ns^2-ions and Ga_2O_3 host overlap. Actually, several μs decay

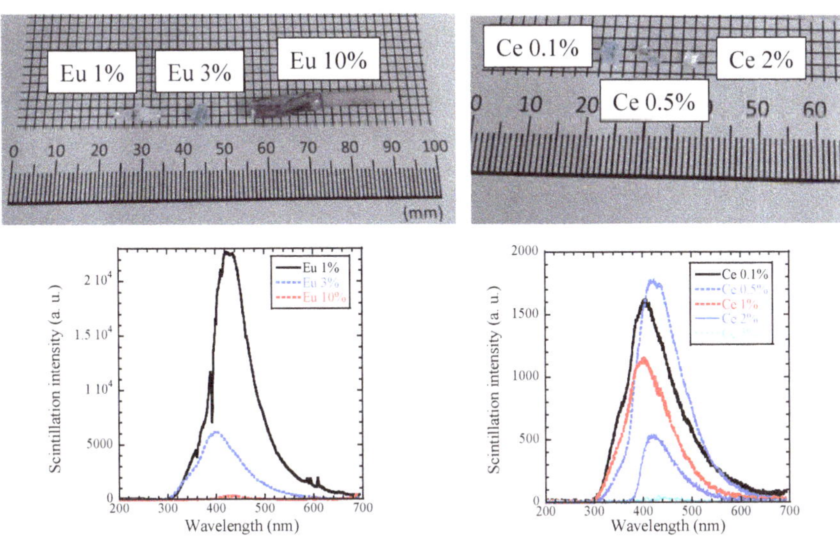

Fig. 30.2 Eu-doped (top left) and Ce-doped (top right) Ga_2O_3 single crystals and scintillation emission spectrum under X-ray irradiation of Eu-doped (bottom left) and Ce-doped (bottom right) ones

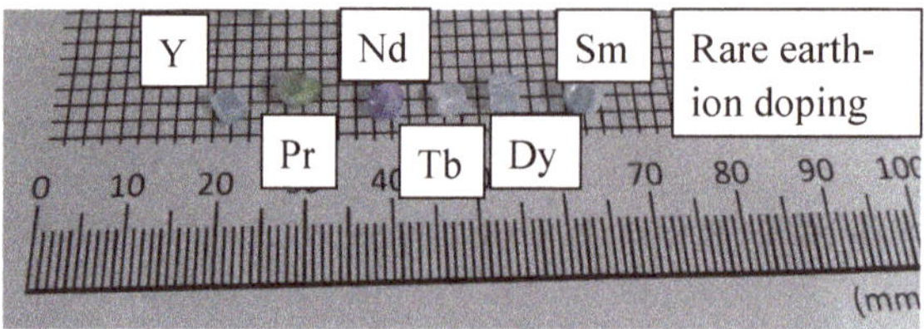

Fig. 30.3 From left to right, Y-, Pr-, Nd-, Tb-, Dy- and Sm-doped Ga_2O_3 bulk single crystals

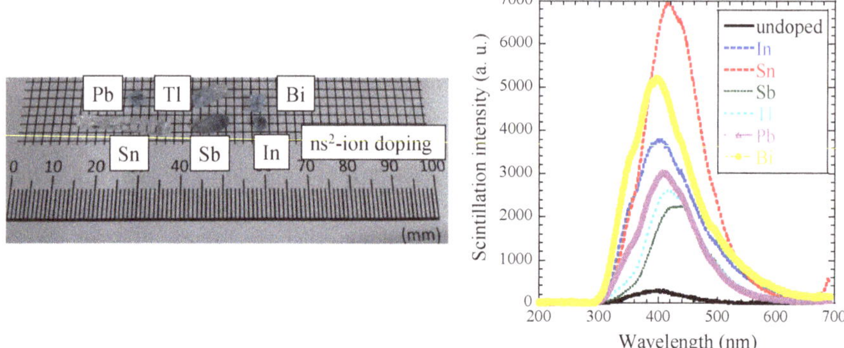

Fig. 30.4 Appearance of the Ga_2O_3 crystals doped with ns^2-ions (left) and X-ray induced scintillation emission spectra of these crystals (right)

component is detected in PL and scintillation decay curves, and the origin of this μs order component is nsnp–ns^2 transitions of dopant ions. Especially, Sn- and Bi-doped Ga_2O_3 show relatively higher emission intensities [20], and we further investigated with these dopant ions of different concentrations.

Figure 30.5 represents results of Ga_2O_3 single crystals doped with different concentrations of Sn. The dopant concentrations of Sn are ranged from 1 to 10%. As shown in the figure, we could successfully grow crystal samples. When we irradiated with X-rays, all of the samples showed similar emission spectra, and the Sn 2% doped sample exhibited the highest scintillation intensity among the present samples. As a detector property, we evaluated pulse height spectra. As a result, the Sn 1, 2 and 5% doped samples exhibited similar scintillation light yields [21]. We think Sn doping is one of the promising ways to improve the scintillation properties of Ga_2O_3.

Results on Bi-doped Ga_2O_3 crystals are shown in Fig. 30.6, and the appearance is similar to the Sn-doped ones. In X-ray induced scintillation spectra, an emission peak shift occurred when the Bi concentration increased. For the Bi 0.1% doped sample, the emission peak was around 380 nm, while the peak wavelength in the other samples was around 420 nm. When we investigated the scintillation detector properties, scintillation light yield was found to be inferior to those of the Sn-doped crystals. In ns^2-ion-doped Ga_2O_3, the existence of emission due to ns^2 ions was

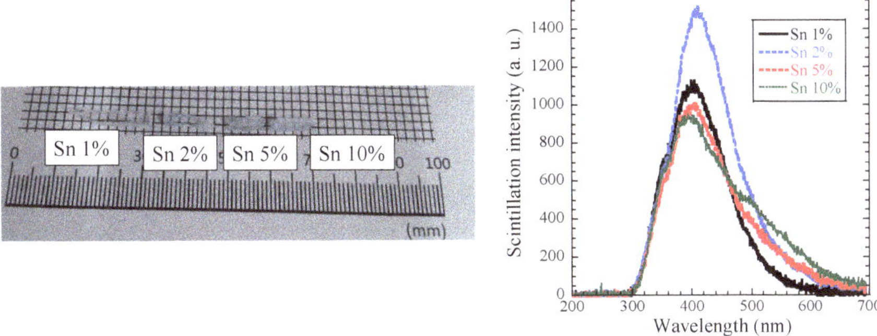

Fig. 30.5 Appearance of the Ga_2O_3 crystals doped with Sn-ions (left) and X-ray induced scintillation emission spectra of these crystals (right)

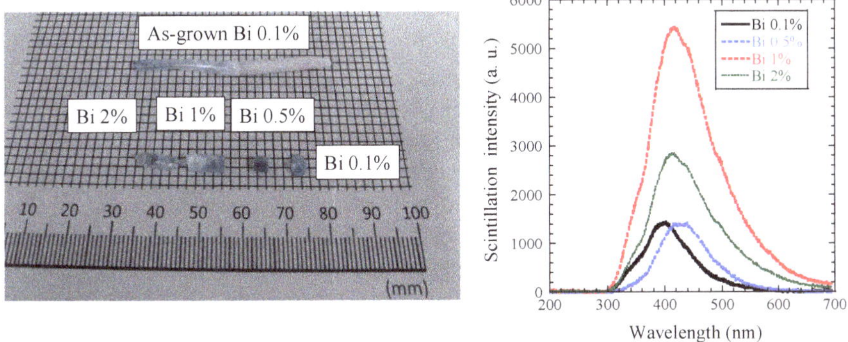

Fig. 30.6 Appearance of the Ga_2O_3 crystals doped with Bi-ions (left) and X-ray induced scintillation spectra of these crystals (right)

confirmed by X-ray induced scintillation and PL decay time measurements, but the relative intensity of such emissions was not high. Thus, the dominant emission origin was the Ga_2O_3 host itself, and the doping with ns^2 ion worked to increase the luminescence intensity of defect emissions. Overall, among the ns^2-ion-doped Ga_2O_3 tested here, the Sn-doping gave the most promising contribution to Ga_2O_3 as scintillator.

Ga_2O_3 has been studied for next-generation semiconductor devices. For semiconductor applications, one of the important aspects for R&D is to study p-type and n-type doping. Actually, such investigations on Ga_2O_3 have been conducted [22, 23]. Although effects of p- and n-type doping are not clear in scintillator materials, we investigated p-type (Be, Mg, Ca, Sr and Ba) and n-type (Si, Ge and Sn) doping on Ga_2O_3. It must be noted that we were not interested in electrical properties, so we did not investigate them at all, and we only focused to study scintillation properties.

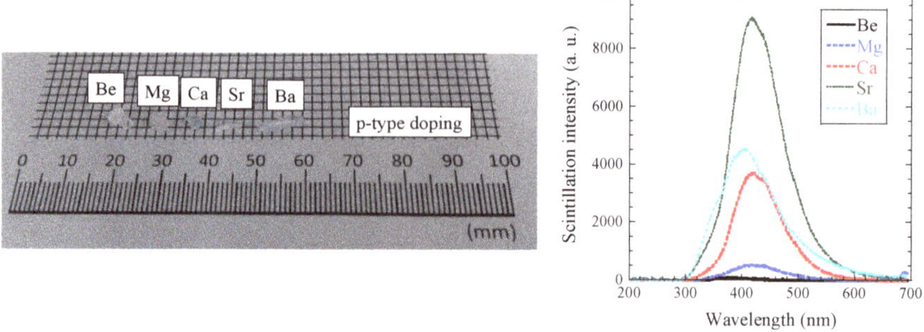

Fig. 30.7 Appearance of the Be-, Mg-, Ca-, Sr- and Ba-doped Ga_2O_3 crystals (left) and X-ray-induced scintillation emission spectra of these crystals (right)

Figure 30.7 (left) shows crystals of Be, Mg, Ca, Sr and Ba 1% doped Ga_2O_3 bulk single crystals. In the viewpoint of crystal growth, we could easily obtain transparent and crack-free bulk crystals. Figure 30.6 (right) exhibits X-ray induced scintillation emission spectra of p-type-doped Ga_2O_3. The Be-doped one had an emission peak around 380 nm which is close to the undoped sample. The Mg-, Ca- and Sr-doped samples had an emission peak around 420 nm, and the Ba-doped one had an emission peak around 400 nm with a shoulder around 340 nm. In contrast to above-mentioned rare earth ions and ns^2-ions, the largest difference of these dopants is that the Be–Ba ions do not have electronic transition levels to show luminescence. Therefore, it was considered that the effects of generating vacancies were different for each alkali metal ion, confirmed from the scintillation emission spectra.

Figure 30.8 exhibits Si-, Ge- and Sn-doped (n-type doping) Ga_2O_3 crystals. The emission peak of all the samples appeared around 420 nm, and unlike the case of

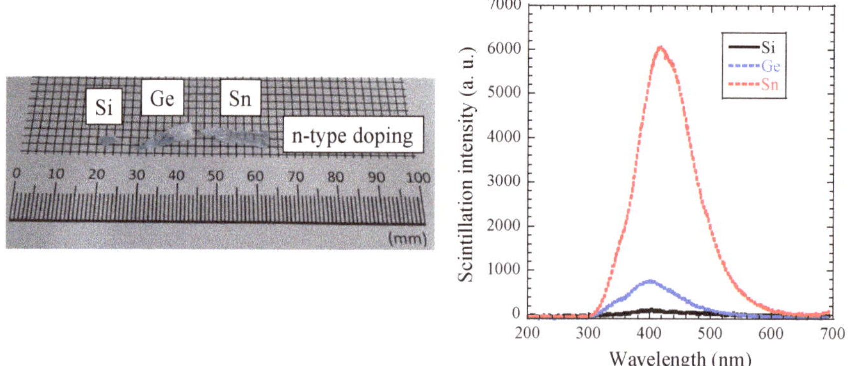

Fig. 30.8 Appearance of the Si-, Ge and Sn-doped Ga_2O_3 crystals (left) and X-ray induced scintillation emission spectra of these crystals (right)

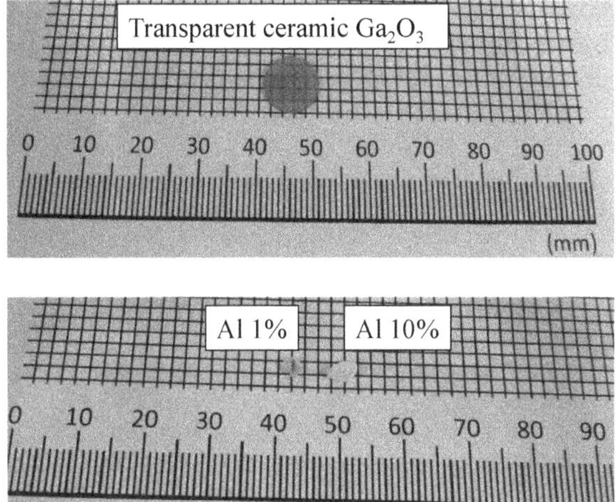

Fig. 30.9 Transparent ceramic of Ga_2O_3 produced by the SPS method (top) and Al-doped Ga_2O_3 crystals (bottom)

p-type doping, spectral shape is similar regardless of the difference of doping ions. The emission intensity varied as Sn > Ge > Si, and it could be understood that the large discrepancy of the ionic radius with Ga made more defect centers which act as emission center of Ga_2O_3. In the pulse height spectrum analysis under ^{137}Cs γ-ray and ^{241}Am α-ray irradiations, the Sn- and Ge-doped samples exhibited similarly high scintillation light yield, while that of Si-doped one was inferior to these two samples.

Figure 30.9 shows other approaches to develop Ga_2O_3-based scintillators. Figure 30.8 (top) shows a transparent ceramic of Ga_2O_3 synthesized by the spark plasma sintering (SPS) method [24]. The sample had a thickness of ~ 1 mm and the in-line transmittance was around 20%. The bottom panel in Fig. 30.8 exhibits Al-substituted Ga_2O_3 crystals grown by the FZ method. The band-gap engineering between Al and Ga can be expected as for the case of AlN, GaN and InN. These new approaches have just started, and we will continue further R&D on Ga_2O_3 for the next-generation semiconductor scintillator since Ga_2O_3 has better scintillation light yield than the other semiconductor scintillators such as GaN, ZnO and CdS.

30.4 Summary and Conclusion

We investigated Ga_2O_3 single crystals various ions doped, which were synthesized by the FZ method, for scintillator applications. In this chapter, we introduced effects of rare earth, ns^2-ion, p-type and n-type dopings. For the rare-earth doping, although the observed scintillation properties were interesting, no positive effects on scintillation properties were confirmed. For the ns^2-ion doping, Sn- and Bi-doped Ga_2O_3 showed promising scintillation properties compared with the rare-earth doped and undoped samples. The p-type (Be, Mg, Ca, Sr and Ba) doping showed interesting emission features that the emission peak position differed for each dopant, while the n-type (Si, Ge and Sn) doping did not affect the spectral shape.

References

1. T. Yanagida, Inorganic scintillating materials and scintillation detectors. Proc. Jpn. Acad. B **94**, 75 (2018)
2. E.D. Stephen, E. Bourret-Courshesne, B. Gregory, C. Andrew, Bright and ultra-fast scintillation from a semiconductor? Nucl. Instrum. Methods Phys. Res., Sect. A **805**, 36 (2016)
3. D.J. Robbins, On predicting the maximum efficiency of phosphor systems excited by ionizing radiation. J. Electrochem. Soc. **127**, 2694 (1980)
4. A. Lempicki. A. Wojtowicz, and A.J.E. Berman, Fundamental limits of scintillator performance. Nucl. Instrum. Methods Phys. Res., Sect. A **333**, 304 (1993)
5. P.A. Rodnyi, P. Dorenbos, C.W.E. van Eijk, Energy loss in inorganic scintillators. Phys. Status Solidi "C" **187**, 15 (1995)
6. P. Dorenbos, Light output and energy resolution of Ce^{3+}-doped scintillators. Nucl. Instrum. Methods Phys. Res., Sect. A **486**, 208 (2002)
7. T. Yanagida, Y. Fujimoto, M. Miyamoto, H. Sekiwa, Optical and scintillation properties of Cd doped ZnO film. Jpn. J. Appl. Phys. **53**, 02BC13 (2014)
8. T. Yanagida, Y. Fujimoto, M. Koshimizu, Evaluation of scintillation properties of GaN. e-J. Surf. Sci. Nanotechnol. **12**, 396 (2014)
9. T. Yanagida, M. Koshimizu, G. Okada, Scintillation properties of undoped CdS for ionizing radiation detectors. Jpn. J. Appl. Phys. **55**, 02BC03 (2016)
10. P. Shootanus, P. Dorenbos, V.D. Ryzhikov, Detection of CdS(Te) and ZnSe(Te) scintillation light with silicon photodiodes. I.E.E.E. Trans, Nucl. Sci. **39**, 546 (1992)
11. T. Yanagida, G. Okada, T. Kato, D. Nakauchi, S. Yanagida, Fast and high light yield scintillation in Ga_2O_3 semiconductor material. Appl. Phys. Express **9**, 042601 (2016)
12. T. Yanagida, K. Kamada, Y. Fujimoto, H. Yagi, T. Yanagitani, Comparative study of ceramic and single crystal Ce:GAGG scintillator. Opt. Mat. **35**, 2480 (2013)
13. T. Yanagida, Y. Fujimoto, T. Ito, K. Uchiyama, K. Mori, Development of X-ray induced afterglow characterization system. Appl. Phys. Express **7**, 062401 (2014)
14. T. Harwig, F. Kellendonk, S. Slappendel, The ultraviolet luminescence of β-galliumsesquioxide. J. Phys. Chem. Solids **39**, 675 (1978)
15. T. Harwig, F. Kellendonk, Some observations on the photoluminescence of doped β-galliumsesquioxide. J. Solid State Chem. **24**, 255 (1978)
16. L. Binet, J. Gourier, Origin of the blue luminescence of β-Ga_2O_3. J. Phys. Chem. Solids **59**, 1241 (1998)
17. Y. Usui, T. Oya, G. Okada, N. Kawaguchi, T. Yanagida, Development of Ga_2O_3 single crystalline semiconductor scintillator doped with Ce. Optik **143**, 150 (2017)
18. D.E. Persyk, T.E. Moi, State-of-the-art photomultipliers for anger cameras. IEEE Trans. Nucl. Sci. **25**, 615 (1978)
19. M.J. Weber, R.R. Monchamp, Luminescence of $Bi_4Ge_3O_{12}$: spectral and decay properties. J. Appl. Phys. **44**, 5495 (1973)
20. Y. Usui, T. Oya, G. Okada, N. Kawaguchi, T. Yanagida, Comparative study of scintillation and optical properties of Ga_2O_3 doped with ns^2 ions. Mater. Res. Bull. **90**, 266 (2017)
21. Y. Usui, D. Nakauchi, N. Kawano, G. Okada, N. Kawaguchi, T. Yanagida, Scintillation and optical properties of Sn-doped Ga_2O_3 single crystals. J. Phys. Chem. Solids **117**, 36 (2018)

22. A. Kyrtsos, M. Matsubara, E. Bellotti, On the feasibility of p-type Ga_2O_3. Appl. Phys. Lett. **112**, 032108 (2018)
23. S. Rafique, L. Han, A.T. Neal, S. Mou, M.J. Tadjer, R.H. French, H. Zhao, Heteroepitaxy of N-type β-Ga_2O_3 thin films on sapphire substrate by low pressure chemical vapor deposition. Appl. Phys. Lett. **109**, 132103 (2016)
24. Y. Usui, T. Kato, G. Okada, N. Kawaguchi, T. Yanagida, Comparative study of scintillation properties of Ga_2O_3 single crystals and ceramics. J. Lumin. **200**, 81 (2018)

Part IV
Devices

Chapter 31
Field-Effect Transistors 1

Introduction of Early Transistor Developments for Power Switching and RF Applications

Neil Moser, Andrew Green, Kelson Chabak, Eric Heller and Gregg Jessen

Abstract In this chapter, we describe the early development of gallium oxide transistors that led to the most common gallium oxide device: n-type, depletion-mode, metal insulator semiconductor field effect transistor. We relate the enticing material properties of gallium oxide described in earlier chapters to the requirements in design and fabrication of marketable devices for low-loss power switching and radio frequency applications. A route for device optimization is shown by maximizing electric field in the drift region while reducing parasitic resistance. We analyze the developments that will be required to overcome device-related technical barriers such as achieving low contact resistance, reducing effects of self-heating, and increasing device gain. The chapter concludes with recent research and goals for gallium oxide transistors moving forward with subsequent chapters detailing these topics.

N. Moser · A. Green · K. Chabak · G. Jessen (✉)
Air Force Research Laboratory, Sensors Directorate, 2241 Avionics Circle, Wright-Patterson Air Force Base, OH 45433, USA
e-mail: gregg.h.jessen@ieee.org

N. Moser
e-mail: neil.moser@us.af.mil

A. Green
e-mail: andrew.green.25@us.af.mil

K. Chabak
e-mail: kelson.chabak.1@us.af.mil

E. Heller
Air Force Research Laboratory, Materials and Manufacturing Directorate, 2241 Avionics Circle, Wright-Patterson Air Force Base, OH 45433, USA
e-mail: eric.heller.2@us.af.mil

M. Higashiwaki and S. Fujita (eds.), *Gallium Oxide*, Springer Series in Materials Science 293, https://doi.org/10.1007/978-3-030-37153-1_31

31.1 Introduction

Early motivation for gallium oxide (Ga_2O_3) transistor development stemmed from the cubic dependence of DC conduction losses on critical electric field strength (E_C), the n-type dopant concentration control over a large range, and wafer scalability of melt-grown substrates. Besides a few nanowire devices [1], the first FET results were reported in 2006 to characterize conductivity control in Ga_2O_3 using an applied electric field [2]. Ga_2O_3 device research then surged significantly following the 2012 MESFET [3] and 2013 MOSFET [4] by the National Institute of Information and Communications Technology in Japan. The relative ease and rate of progress in which these devices achieved multi-hundred volt operation were impressive and inspired rapid development in the community. As presented in previous chapters, this rapid development was realized in all research areas including exploration of substrate and epitaxial growth and doping techniques, physics-based modeling for better understanding of material properties, and updated characterization of measurable material properties to improve data originally taken in the mid-1900s [5].

Device research was also spurred with engineers focusing first on realization of the material promise for high E_C and the associated low on-resistance and, second, on mitigation of limitations expected from the low thermal conductivity and lack of p-type conduction. Accomplishment of these device research steps moves Ga_2O_3 further toward high-performance low-loss power switches and integrated circuits combining both radio frequency and high-efficiency power conversion. A timeline of modern β-Ga_2O_3 device milestones is shown in Fig. 31.1, depicting the recent surge in research and summarizing the milestones that will be discussed in the chapters in this section.

31.2 Exploiting Critical Electric Field Strength for Low-Loss Devices

The promise of achieving $\sim$8 MV/cm critical field strength in Ga_2O_3 devices [3, 6, 7] leads to an ideal combination of high three-terminal breakdown voltage, V_{BK}, and low on-resistance, R_{ON}, which we discuss here in terms of high-voltage, low-loss switching devices. Equation (31.1) shows the power dissipation of a power switch device assuming that the off-state and gate leakage currents are negligible. I_{rms} is the root-mean-square drain current, V_G is the applied gate voltage, f is the switching frequency, and A is the device area. $R_{ON,sp}$ and $C_{IN,sp}$ are the specific on-resistance and input capacitance, respectively, normalized for the device area [8].

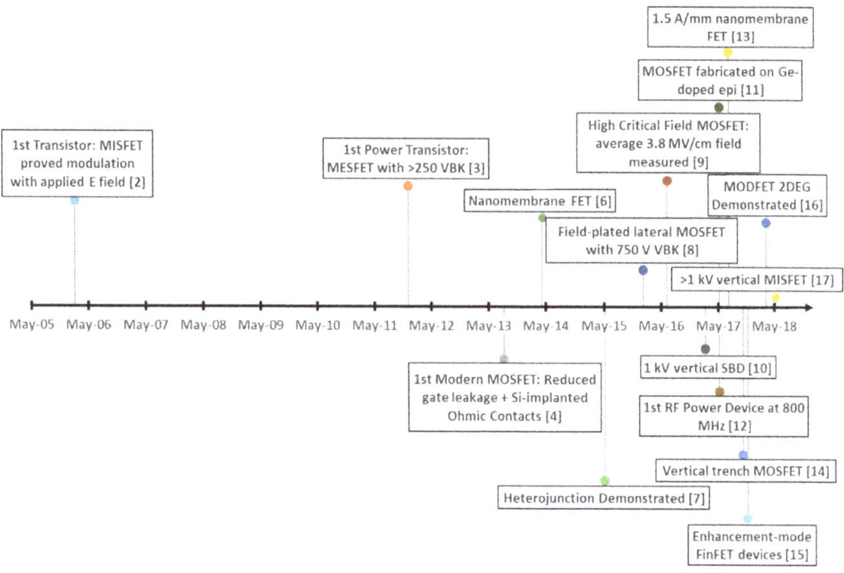

Fig. 31.1 A timeline of modern β-Ga_2O_3 devices with select milestones highlighted. The dramatic increase in research is evident after the demonstration of the first high-voltage devices in 2012 and 2013 [2–4, 6–17]

$$P_{\text{loss}} = \underbrace{I_{\text{rms}}^2 \frac{R_{\text{ON,sp}}}{A}}_{\text{Conduction loss}} + \underbrace{C_{\text{IN,sp}} A V_G^2 f}_{\text{Dynamic Switching loss}} \tag{31.1}$$

For a low-frequency switch, f is small and the first term dominates power loss. Since the power handling capability is determined by the current and voltage product, it is desirable to minimize R_{ON} for a given V_{BK}, to minimize the power loss.

As shown in (31.2), Baliga used a simple one-sided junction approximation to derive a high-frequency switch loss figure of merit (BHFFOM) that incorporates both conduction losses and dynamic switching losses by relating R_{ON} at maximum operating V_B to fundamental material properties. In this equation, the fundamental material parameters are the critical field strength, E_C, the effective mobility of the carriers, μ, and the dielectric constant of the semiconductor ϵ_S. Device operating parameters are defined by the breakdown voltage required, V_B, the gate voltage drive, V_G, specific on-resistance required for the desired efficiency, $R_{\text{ON,sp}}$, and specific input capacitance $C_{\text{IN,sp}}$ [8]. A higher FOM indicates reduced power losses. Highlighted on the left side of (31.1), this metric reduces to Baliga's figure of merit (BFOM) for DC conduction losses at low frequency. While BHFFOM and BFOM are idealized expressions, they provide an excellent basis for comparing material and device requirements. BFOM indicates that materials with a high E_C provide a substantial advantage for DC power loss due to the cubic dependence on electric field strength. Separate reference material which discusses other techniques to generate a FOM for dynamic switch loss can be found in work by Huang [9]. Huang uses input voltage instead of input capacitance in his derivation ultimately leading his FOM to scale with E_c and $\mu^{0.5}$.

$$\mathrm{BHFFOM} = \frac{1}{R_{\mathrm{ON,sp}} C_{\mathrm{IN,sp}}} = \underbrace{\frac{\in_s E_C^3 \mu}{4V_B^2}}_{\text{BFOM for DC}} \cdot \frac{2\sqrt{V_G V_B}}{\in_s E_C} = \frac{E_C^2 \mu V_G^{0.5}}{2V_B^{1.5}} \tag{31.2}$$

In real devices, design steps can be used to achieve the maximum BFOM for a material by assessing V_{BK} versus the doping concentration and corresponding drift region depletion distance. Additionally, it is necessary to remove all parasitic resistance, i.e., any resistance not necessary to achieving V_{BK}, which contributes to R_{ON}. From (31.2) and [9], it becomes clear that drift region scaling and channel scaling enabled by high E_C maintain an advantage even at higher frequency operation. So, while the primary motivation for Ga_2O_3 is for low-loss power switching, there is also promise for low-loss high-frequency switches. Moreover, the high E_C potentially allows for aggressive device scaling to reduce carrier transit time for use of the material in RF devices. This is represented in Johnson's Figure of Merit [10] as shown in (31.3).

$$\mathrm{JFOM} = f_T V_B = \frac{E_C v_{\mathrm{sat}}}{2\pi} \tag{31.3}$$

The JFOM for an arbitrary semiconductor material relates the critical field strength, E_C, and the saturation velocity, v_{sat}, to the cutoff frequency, f_T, and breakdown voltage, V_B, product. From this, the maximum RF power that can be delivered to a load at a given frequency in a given material system can be estimated. For Ga_2O_3, this value exceeds that of GaN [7]. Ga_2O_3 is thus expected to have potential as an integrated platform for both low-loss, high-frequency power switching and high-efficiency, high-voltage RF power amplifiers.

It can be seen from these metrics that a material with high E_C can achieve lower loss in both DC and dynamic switching devices and large gain bandwidth products in RF devices. These figures of merit help guide device design toward next-generation electronics. In the following sections, we will outline routes toward maximizing the potential for this material.

31.2.1 Minimizing Channel Resistance for a Target Breakdown Voltage

Previously, we noted that Ga_2O_3 research surged because of the high breakdown voltage easily achieved by early devices. These devices exhibited high breakdown voltages of ~250 V and ~370 V for the MESFET and MOSFET, respectively. However, the electric fields achieved as estimated from (31.2) were 0.64 and 0.87 MV/cm and were far short of the maximum predicted for Ga_2O_3. More compelling device results for the high E_C of Ga_2O_3, however, would soon follow. Before discussing these results, further understanding of Baliga's method for realized and reported devices is required.

Figure 31.2 illustrates the calculation of W_B from a lateral or vertical transistor biased at the breakdown voltage. This is derived from a 1D Poisson's equation with an applied voltage difference between two planes, V_B, dropped across a constant concentration of ionized donors, N_B, creating a triangular shape to the electric field with a peak value, E_C, at the higher potential plane. As shown by the realized Ga_2O_3 transistor in Fig. 31.2, the planes are approximately at the gate and drain contact edges. Thus, V_B, which is analogous to V_{GD}, the doping concentration, equivalent to N_B, and gate-drain spacing, L_{GD}, can be simultaneously optimized for a desired breakdown voltage. This, in turn, minimizes the total resistance of the depletion region in a lateral device or the drift region in a vertical device as shown in (31.4) for total resistance of a channel of a semiconductor material. In (31.4), R_{drift} is the total resistance of the drift region of a semiconductor device. The value is based on the length, ℓ, surface area, A, and resistivity, ρ, of the drift region.

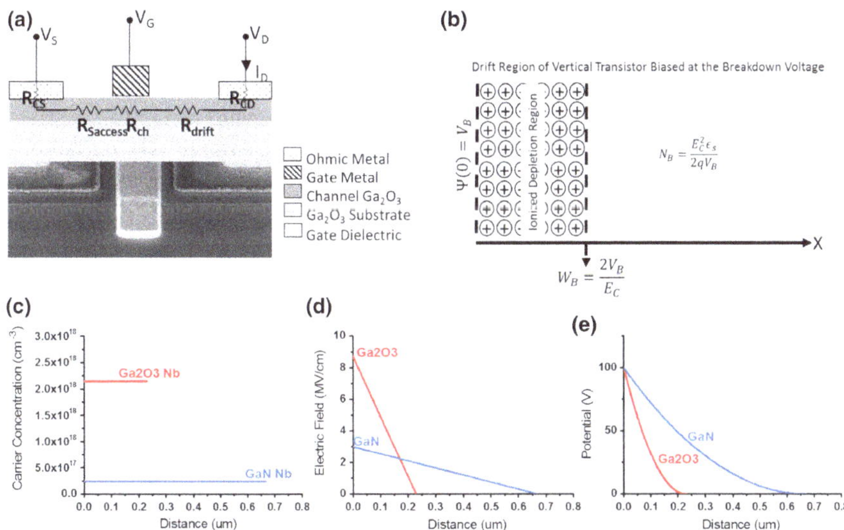

Fig. 31.2 **a** Schematic cross section and angled scanning electron microscope image of a Ga_2O_3 MOSFET. **b** One-sided junction approximation of the drift region of a MOSFET as proposed by Baliga [8]. **c** W_B calculated for Ga_2O_3 at high N_B and GaN at low N_B. 1D solution of Poisson's equation on the drain side of the gate showing **d** electric field and **e** potential as a function of position for a GaN and Ga_2O_3 MOSFET biased at a breakdown voltage, $V_{BK} = 100$ V. The higher critical field strength of Ga_2O_3 allows for larger N_B and smaller W_B compared to the GaN device. This results in reduced on resistance

$$R_{\text{drift}} = \rho\frac{\ell}{A} = \frac{1}{\sigma}\frac{\ell}{A} = \frac{1}{nq\mu}\frac{\ell}{A} = \frac{1}{N_B q\mu}\frac{W_B}{A} \tag{31.4}$$

In a device, these values relate to the doping concentration, N_B, the depletion width necessary for a given breakdown voltage, W_B, and the mobility of carriers, μ, as shown. What is important to realize is that the device blocking voltage will be a significant part of device design, and if a larger blocking voltage is required, an increase in series resistance is unavoidable. More specifics will be discussed in the next section. For now, we will look at the progress toward achieving the maximum E_C in realized devices.

What had not been demonstrated by the early devices were scaled drift regions. As seen in other technologies, high-voltage devices in materials such as Si, SiC, or GaN require significant field management to support high blocking voltages. This changed in 2016 when Green et al. demonstrated the first high-voltage Ga_2O_3 device with a submicron drift region [11]. Here, because the drift region is very small, the expected W_B is approximately that of the fabricated L_{GD}. This can be seen in Fig. 31.3f, where [11] is very close to the theoretical Baliga line.

Figure 31.3 shows a cross section and family of current-voltage curves (drain current, I_D, versus drain-source voltage, V_{DS}, with gate-source voltage, V_{GS}, as a parameter) with three-terminal V_{BK} shown in the off-state for a lateral Ga_2O_3 transistor with a gate contact to drain contact separation, L_{GD}, of 0.6 μm [11]. The device achieved a minimum peak electric field, E_C, based on averaging over L_{GD} of 3.8 MV/cm which is higher than the theoretical values for GaN and SiC [21]. Using TCAD modeling, the authors also estimated a peak electric field of 5.3 MV/cm, and using the depletion approximation of Baliga, the value may be as high as 7.6 MV/cm which is near the estimated critical field strength of Ga_2O_3.

In Fig. 31.3e, f, V_{BK} for several devices from the literature is plotted versus doping concentration, N_B, and gate to drain spacing, L_{GD}, to illustrate progress in reaching the maximum E_C for Ga_2O_3 in a device. The theoretical line is also plotted for E_C = 8 MV/cm for comparison. To achieve minimum R_{ON} with maximum V_{BK}, engineered devices must be close to both of the theoretical lines in Fig. 31.3. From the equations in Fig. 31.2, the minimum L_{GD} required to support the maximum achievable V_{BK} can be estimated by the depletion width, W_B. Values of L_{GD} smaller than W_B result in premature breakdown, and values larger than W_B increase parasitic access resistance. Reducing the peak electric fields across the drift region will be a major body of work with device performance trade-offs. It should be noted that BFOM was originally derived for a vertical junction where R_{ON} is normalized to the area of the depletion region viewed in the direction of the applied field—in many cases, this is simply the area of the anode contact. For a lateral device, the anode (gate) and the direction of the depletion toward the drain are normal to one another. For a uniformly doped channel, the R_{ON} should be normalized to $d_{CH} \cdot W_G$ which is the area of the depletion region looking into the applied field along the channel.

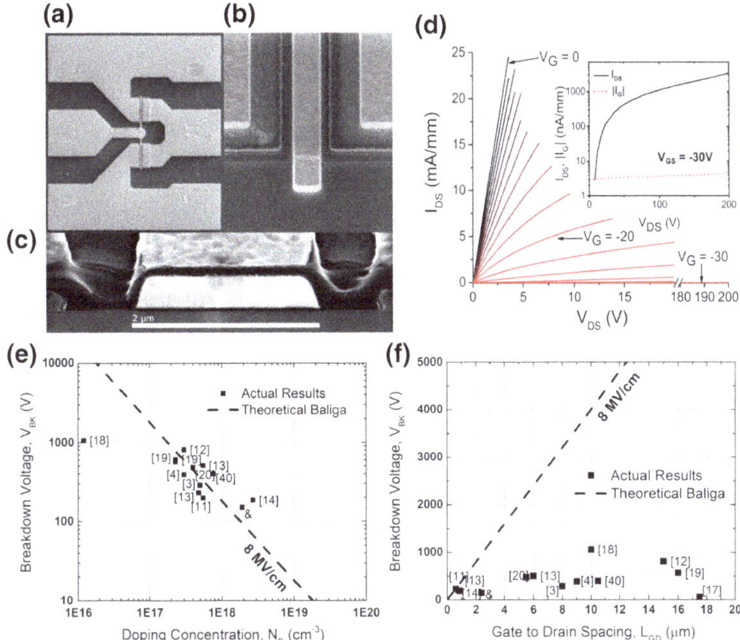

Fig. 31.3 **a** Top-down SEM image of a 2 × 50 μm β-Ga_2O_3 MOSFET with $L_G = 2$ μm, $L_{GS} = 0.8$ μm, and $L_{GD} = 0.6$ μm. **b** Angled view SEM image of the gate finger. **c** Cross-sectional SEM view of the source-drain space and gate finger. **d** Family of curves for the same lateral Ga_2O_3 transistor. The device achieved a total V_{GD} of 230 V before failure in the off-state ($V_G = -30$ V) as shown by the bottom curve. Breakdown voltage versus doping concentration (**e**) and gate-drain spacing (**f**) for several real β-Ga_2O_3 MOSFETs compared to the prediction by Baliga for a material with 8 MV/cm critical field strength. Note that devices marked [12], [13], and & have at least a partial field plate. The device labeled with & is an unreported recent result. Device [14] is a back-gated device using a 2-terminal breakdown voltage [3, 4, 11 20]. The '&' represents unpublished devices fabricated at the Air Force Research Laboratory

For heterostructure devices forming a 2DEG, the normalized R_{ON} is usually overestimated using $W_G \cdot L_{SD}$ since d_{CH} is negligible [22]. As technology matures and variation from targeted doping levels in delivered samples is reduced, engineers can move closer to the theoretical line instead of using the conservative L_{GD} values shown currently in Fig. 31.3f.

To illustrate the trade between device scaling on parasitic access resistance and the electric field profile, a lateral power device is modeled in commercial TCAD software with $L_{GD} = 500$ nm, 75-nm-thick passivation with a dielectric constant of 7.5, and channel thickness of ~70 nm with 1.5×10^{17} cm^{-3} and 4.5×10^{17} cm^{-3} net donor concentration. A dielectric constant of 10.0 is used for the Ga_2O_3 channel and buffer. The epi characteristics are very similar to [11] but with a more tightly scaled gate. The contour plots are for $V_D = 40$ V applied at an off-state. The gate rests on top of the channel and does not penetrate through it, and the materials

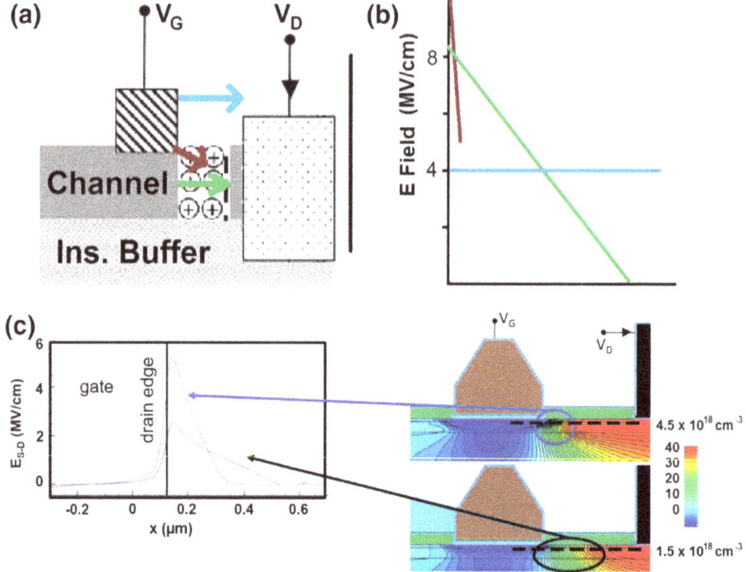

Fig. 31.4 **a** Schematic for a gated doped Ga_2O_3 channel on an insulating buffer. **b** Corresponding electric field profiles for the vectors shown in (**a**). **c** TCAD model of the schematic showing good agreement with the single-side abrupt junction assumption

above and below the channel make the field profile approximately two-dimensional. We assume the gate width is many times the channel dimensions as is common for GaN power handling FETs and consider the average channel properties for this discussion; otherwise, the profile becomes fully three-dimensional. We see that the lateral electric field at the top of the channel (black dotted line) does in fact approximate the one-sided channel well enough for comparison purposes for the device modeled as seen in the plot. Figure 31.4 shows a schematic for a lateral device and the expected field profile for multiple directions away from the gate.

Once the channel material is selected and the properties of the semiconductor, the dielectric, and metals are known, field management engineering becomes the next technological hurdle. This engineering is too advanced for the level of maturity of the materials and beyond the scope of this discussion, but optimization of dopant placement and field plate dimensions will be necessary to maximize the potential for Ga_2O_3.

31.2.2 *Removing Parasitic Resistance for Minimal* R_{ON}

As mentioned in the previous section, to achieve the minimum R_{ON} for a given V_{BK}, the material conductivity, $N_B \cdot q \cdot \mu$ where q is the electron charge and μ the electron mobility, must be maximized and L_{GD} must be minimized. The design of the drift/depletion region can be estimated, therefore, using the simple methods above and the equations in Fig. 31.2. Increasing μ is primarily achieved by optimizing material growth as was presented in the preceding chapters. This will also be affected by heterostructure design as described in a future chapter. N_B is achieved by precise control of the doping concentration during growth and is limited by the design parameters and the operating voltage required by the application. Device engineers are thus limited in design changes that may reduce the resistance in the access region, and the verification of E_C presented above is comprehensive of the methods needed. This section will consider strategies for removing parasitic resistors in the device.

From source to drain, the device R_{ON} is determined by the source contact resistance, R_{CS}, the gate-source access region resistance, $R_{saccess}$, the channel resistance with the device turned on, R_{ch}, the drift/depletion region resistance, R_{drift}, and finally, the drain contact resistance, R_{CD}. As mentioned above, R_{drift} is limited by the material parameters and design needs. Similarly, R_{ch} is limited by the requirements for gate length discussed later in this chapter and the material conductivity. R_{Cx} and $R_{saccess}$ are the parasitic resistances that are the primary focus of device design engineers and will be discussed in this section.

Eliminating access resistance is incredibly important for Ga_2O_3 transistors since even the upper bound for mobility is relatively low. Theoretically, $R_{saccess}$ can be eliminated using a self-aligned source contact with no spacing between the source and gate. Degenerately doping the source contact region through implant doping techniques is the most common method to create a self-aligned structure. Wong et al. have successfully implanted Ga_2O_3 with silicon to create degenerate doping levels, $N_D = 5 \times 10^{19}$ cm^{-3}, between the gate and source contact [23], which can be effective for creating a self-aligned source. This method was used to create an enhancement-mode device with a measured sheet resistance of 84 Ω/sq in the degenerately doped region.

Implant doping under the source and drain contacts is also an effective way to reduce R_{Cx} in Ga_2O_3 as shown by Higashiwaki et al. [15]. Here the authors have obtained specific contact resistance of 8.1×10^{-6} Ω cm^2 which is comparable to

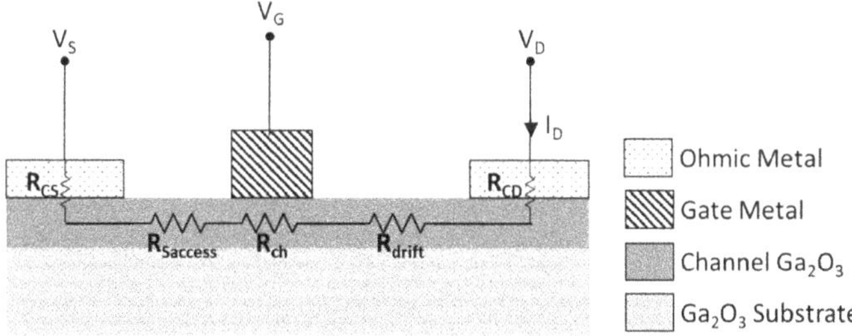

Fig. 31.5 Schematic cross section of a Ga_2O_3 lateral transistor showing the resistors from source to drain when the device is in the on-state

the highly optimized contacts made to the mature AlGaN/GaN heterostructure system. Besides implant doping, R_{Cx} can be effectively reduced by employing a highly doped ohmic cap layer under the contacts that is subsequently removed under the gate. While this process does not lend as easily to creating a self-aligned source, it avoids the implant processing step, the high temperature implant activation steps (>900 °C in [15, 23]), and allows for other dopant species since, to date, only Si has been implanted as a donor. A third method for reducing R_{Cx} is through selective regrowth of a highly doped layer under the area planned for ohmic contacts. Again this method does not lend directly to a self-aligned source and adds a complicated regrowth step to the fabrication process. However, it does avoid etch damage to the channel area unlike the highly doped cap layer technique. An example regrowth process using MBE with ohmic contact resistance of 1.5 Ω mm can be found in [24]. All three of these methods are a result of reducing the barrier thickness between metal and semiconductor as the donor concentration beneath the metal is increased.

31.3 Unipolar Operation

Current state-of-the-art Ga_2O_3 lateral and vertical devices, which follow the example in [15], operate entirely in the depletion mode. P-type conductivity is not available and an acceptor with low activation energy has not been identified. Furthermore, valence band structure is known to be relatively flat and holes are self-trapped as mentioned in previous chapters. Consequently complementary operation is not expected. Additionally, it is unclear in the device literature if accumulation mode operation is achievable, with most devices that operate with the gate forward biased never reaching the expected flat-band capacitance due to the position of the conduction band. Further, due to the extremely wide band gap of Ga_2O_3, few dielectrics are known to have a conduction band offset greater than 1.5 eV.

Because of the exclusive depletion-mode operation, it will be important to frame our physical understanding of the MOSFET with those constraints. The gate voltage swing, ΔV_G, required from the fully on-state to the off-state is determined by the flat-band voltage, V_{FB}, and the voltage required to completely deplete the channel thickness. The off-state voltage, (31.5), is easily determined by the depletion approximation and can be measured from the electrical characteristics.

$$V_{\text{off}} = V_{\text{FB}} - dqN_d\left(\frac{d}{2\,\epsilon_s\,\varepsilon_0} + \frac{1}{C_{\text{OX}}}\right) \tag{31.5}$$

The condition is reached when the gate voltage equals the off-state voltage, $V_G = V_{\text{off}}$, where V_{off} is the point when the entire active channel is depleted, $x_{\text{dep}} = d$, at the source side of the gate. V_{FB} is the flat-band voltage, d is the thickness of the channel, N_d is the active carrier concentration, q is the electron charge, C_{ox} is the gate oxide capacitance per unit area, and $\varepsilon_s \cdot \varepsilon_0$ is the permittivity of β-Ga_2O_3. The V_{FB} can also be determined from the gate metal work function and the density of traps or measured from several methods using capacitance-voltage curves. R_{ON} can also be estimated from (31.6).

$$R_{ON} = \left(\frac{dI_{DS}}{dV_{DS}}\right)^{-1} = \frac{L}{q\mu N_d W}\left\{\left(d + \frac{\varepsilon_G}{C_{OX}}\right) - \sqrt{\frac{\varepsilon_G^2}{C_{OX}^2} - \frac{2\varepsilon_G}{qN_d}(V_{GS} - V_{FB} - V_{DS})}\right\}^{-1} \quad (31.6)$$

Equation (31.6) shows R_{ON} in the linear region of a β-Ga_2O_3 MOSFET. I_{DS} is the drain current, V_{DS} is the drain-source voltage, L is the gate length, W is the gate width, q is the electron charge, N_d is the active carrier concentration, μ is the effective mobility of electrons in the channel, d is the thickness of the channel, V_{FB} is the flat-band voltage, C_{OX} is the oxide capacitance per unit area, and $\varepsilon_G = \varepsilon_S\varepsilon_0$ is the dielectric constant of Ga_2O_3. From (31.6), or a simple sheet resistance calculation, it is clear that to reduce R_{ON} it is desirable to increase the channel thickness, d, or increase the doping, N_d, with the assumption that maximum mobility, μ, has already been achieved through optimized epitaxial growth. From (31.5), it is clear that increasing N_d is preferred compared to increasing d to maintain the lowest R_{ON} simultaneously with the lowest ΔV_G (i.e., V_{off} closest to V_{FB}). Thus, device engineers are targeting thin, highly doped channels and two-dimensional electron gas (2DEG) heterostructure devices which present the lowest sheet resistance for a given channel thickness. Unlike GaN, β-Ga_2O_3 does not offer a spontaneous and piezoelectric field for creating a 2DEG, and thus, modulation-doped FETs (MODFETs) similar to those used in the GaAs material system are the primary device type in development [17, 25, 26]. Many device engineering parallels can be drawn between the unipolar MODFET and HEMT systems. For these devices, reducing ohmic contact resistance and controlling interface traps at the surface, substrate, and heterojunction are paramount. The methods for reducing ohmic contact resistance mentioned above also apply here although the contact is often made to an $(Al_xGa_{1-x})_2O_3$ layer. Numerous reports also exist for characterizing each interface [27–31]. The dominant scattering mechanism limiting mobility of confined carriers in a 2DEG is phonon scattering. However, achieved mobility values are still below the modeled polar optical phonon scattering limit. This suggests that mobility can improve significantly in a highly optimized doped channel. To date, Zhang et al. [32] have demonstrated a mobility of 180 $cm^2/(V\ s)$ which is 1.6 times the highest reported value in a Ga_2O_3 MOSFET device while achieving the highest transconductance reported at $G_m = 39$ mS/mm [20]. Heterostructure epitaxy and design will be discussed further in the chapters that follow.

31.4 State-of-the-Art RF Devices

In the following section, RF results for Ga_2O_3 devices are discussed. As mentioned earlier, motivation to look at Ga_2O_3 as a new semiconductor system was largely motivated by the prospect of scaling the device dimension for superior conduction and dynamic loss performance. As Johnson pointed out, device scaling has implications on radio frequency performance also due to the benefit of transit time being reduced. Currently, ab initio modeling of v_{SAT} suggests values as may be as high as 2.0×10^7 cm/s [7]. The combination of high v_{SAT} and electric field strength greater than 2.5 times that of GaN suggests the potential for good RF performance in Ga_2O_3.

Class-A operation is used as a simple example to illustrate the impact of high V_{BK} on power output and efficiency. Figure 31.6 shows the trade space for maximizing I_{DS} and the difference between V_{BK} and V_{knee} to maximize RF output power. Figure 31.6 also shows the impact of high V_{BK} to V_{knee} ratio and high transconductance, G_M, on efficient RF operation. For Ga_2O_3, the V_{BK} to V_{knee} ratio is addressed by the preceding discussions for reducing R_{ON} at high V_{BK}. G_M and $I_{DS,MAX}$ can be increased by aggressive scaling of the gate length as shown in Fig. 31.6 relating to achieving the Johnson's Figure of Merit. As device development matures, the operation space is expected to be RF amplifiers with conduction angle <180° pushing well past Class-B into switch-mode operation where the high breakdown voltage achievable is advantageous for high-efficiency design. There is substantial synergy between fast switching power devices and switch-mode RF amplification. This section shows the latest status in achieving Ga_2O_3 material system goals for RF operation.

In Fig. 31.7, a comparison of the small signal gain (h_{21}^2) at 1 GHz for GaN and Ga_2O_3 is shown versus gate length using the saturated carrier velocity and transit

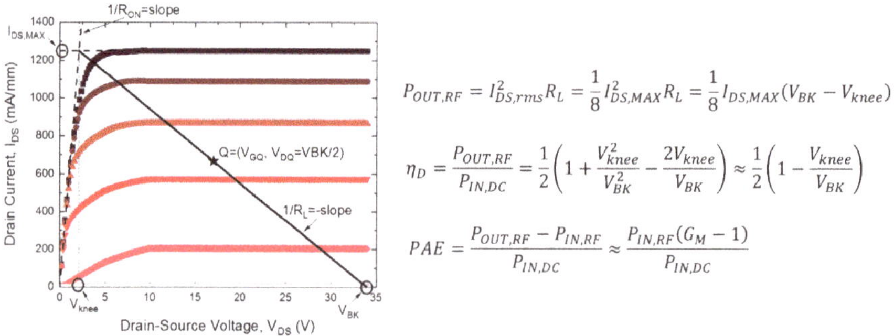

Fig. 31.6 Basic calculations for an example RF transistor acting as a power amplifier with Class-A operation. The family of curves has been extended to the breakdown voltage of the device (34 V) for clarity. The RF output power, P_{OUT}, drain efficiency, η_D, and power added efficiency, PAE, can be estimated from the device performance from the parameters shown on the drain IV curves. The significance of the V_{knee}/V_{BK} ratio and the device gain, G_M, is evident

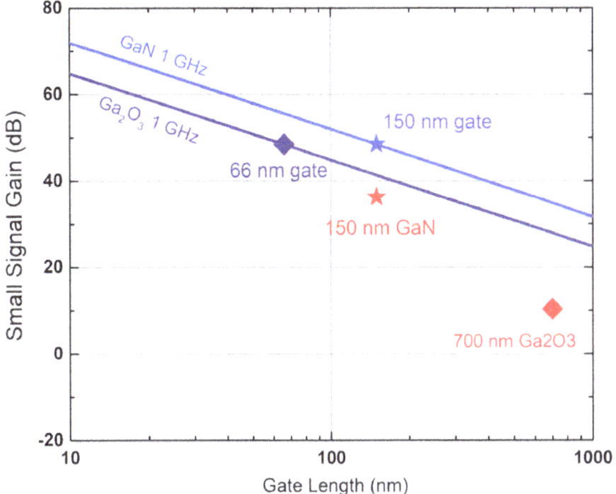

Fig. 31.7 A comparison of the small signal gain versus gate length for GaN and Ga_2O_3 technologies. At 1 GHz, the Ga_2O_3 device requires a gate length of 66 nm (navy diamond) to operate at the same small signal gain as the GaN device with gate length of 150 nm (blue star). Real device results are shown for a 150 nm GaN device operating at 1 GHz (red star) and a 700 nm Ga_2O_3 device operating at 1 GHz (red diamond)

time through the gate region. A conservative saturation velocity of 1.1×10^7 cm/s was used for Ga_2O_3, while 2.5×10^7 cm/s was used for GaN. This shows that for equivalent gain with respect to a GaN device with 150 nm gate length, a gate length of around 66 nm is required for Ga_2O_3. A measured GaN and a Ga_2O_3 device are also included in this plot showing the relative departure from the theoretical line for the respective gate lengths. Early studies indicate little difference in these saturation velocities [7] suggesting potential for RF operation of Ga_2O_3 transistors. Still, the saturation velocity and electric field required to reach saturation must be verified for Ga_2O_3 in real devices as the material system matures.

Promising RF results for lateral Ga_2O_3 transistors have already been achieved as shown in Fig. 31.7. Green et al. reported the first RF performance in a device with a highly doped ohmic cap layer and a 0.7 μm-long recessed gate to achieve a f_t = 3.3 GHz, a f_{max} = 12.9 GHz, and an output power, P_{out}, of 0.23 W/mm and gain, G_P, of 5.1 dB at 800 MHz [33]. These early results, like the results in the previous sections, were limited by material quality and high extrinsic contact resistances including R_{Cx} and $R_{Saccess}$. For the device in Fig. 31.8, the extracted Hall mobility was 96 cm^2/(V s) corresponding to a sheet resistance, R_{sheet}, of 4850 Ω/Sq, and the process achieved a measured R_{Cx} of 3.3 Ohm-mm. Improvement in all of these values has been subsequently achieved. Gain, power output, and efficiency have been limited by self-heating in the device due to the low thermal conductivity of Ga_2O_3. Early simulation work toward understanding the heat transport during RF application will be discussed in the next section.

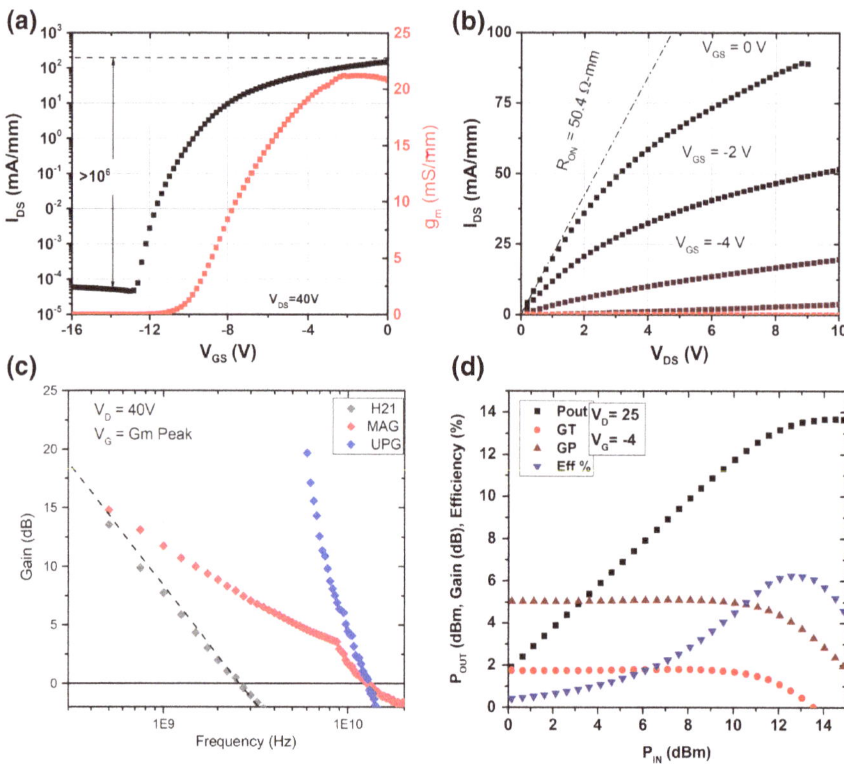

Fig. 31.8 **a** Transfer and **b** output characteristics for an L_{GD} = 0.7 um device. **c** Extrinsic small signal RF gain performance recorded at $V_{GS} = -3.5$ V (peak g_m) and V_{DS} = 40 V for a β-Ga_2O_3 gate-recessed MOSFET. A gain decay of −20 dB/dec is plotted with the dashed line. **d** 800 MHz Class-A power sweep of the same β-Ga_2O_3 gate-recessed MOSFET

Because of the immaturity of the material system, very few scaled Ga_2O_3 transistors exist for RF comparison, with most of the reported results being on relatively long gate-length devices in the 2–4 μm range. Even the previously mentioned RF result [33] had a gate length of 0.7 μm which is quite long compared with current GaN technologies which are commonly in the 0.25–0.10 μm range for RF [34]. Even using a long gate device; however, Singh et al. achieved G_P = 4.8 dB and 0.13 W/mm at a frequency of 1 GHz. The authors avoided self-heating effects by conducting pulsed load-pull measurements and published the highest PAE of 12% for a Ga_2O_3 device. Shortly after, small signal RF data was also reported for a MODFET with a 0.7 μm gate [32]. The device achieved similar results to [33] with f_T = 3.1 GHz and $f_{\max}$ = 13.1 GHz at a supply voltage of 10 V.

A device using a 0.14 μm T-shaped gate [35] is shown in Fig. 31.9. In spite of the aggressive gate-length scaling, the high contact resistance, R_{Cx} = ~15 Ω-mm, allowed for only a moderate increase in small signal performance with f_T = 5.1 GHz and $f_{\max}$ = 17.1 GHz. A similar device was tested using pulsed

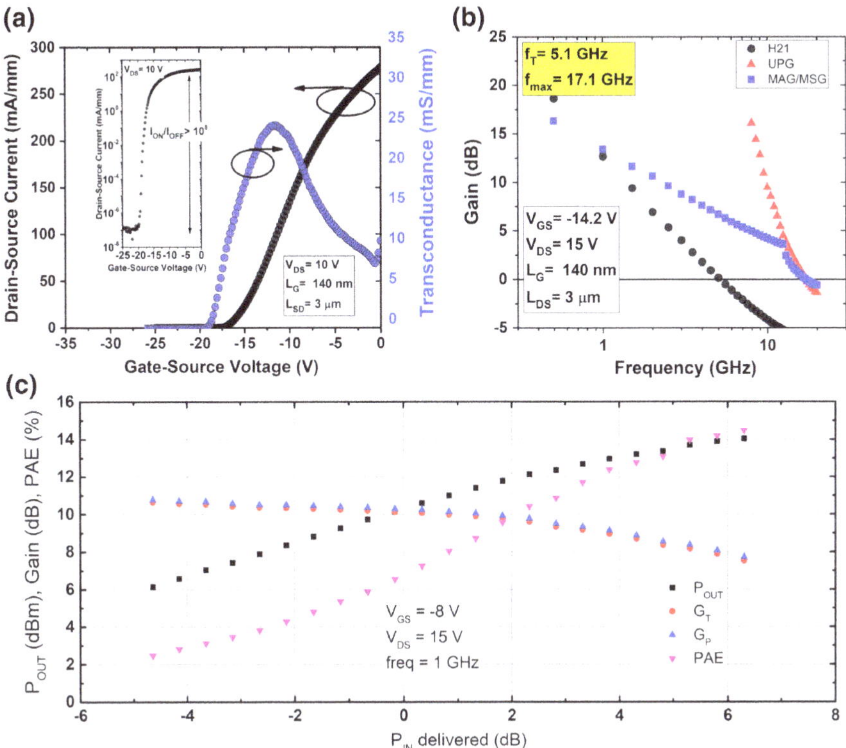

Fig. 31.9 Device characteristics of a thin-channel RF Ga_2O_3 MOSFET with T-gate and L_G = 0.14 μm. **a** Transfer characteristics of the thin-channel Ga_2O_3 MOSFET with T-gate. The G_M reaches 25 mS/mm with current density over 275 mA/mm, and the inset shows a good I_{ON}/I_{OFF} ratio greater than 10^8. **b** Small signal gain at V_{DS} = 15 V of the thin-channel Ga_2O_3 MOSFET with f_t/f_{max} = 5.1/17.1 GHz. **c** 5-μs pulsed power sweep at 1 GHz for a 2 × 50 μm Ga_2O_3 transistor with L_G = 0.14 μm. The device threshold voltage was ∼−11 V, so the operation is Class AB. Source and load impedance were tuned for maximum gain

load-pull at 1 GHz as shown in Fig. 31.9. With a 5 μs pulse, the device achieved a maximum *PAE* = 14% simultaneously with an output power of P_{out} = 0.25 W/mm. The maximum output power obtained was 0.5 W/mm (not shown) and the maximum gain was G_P = 11 dB. Results were significantly degraded under continuous-wave measurements with G_P = 7.5 dB and maximum P_{out} = 0.03 W/mm. The results are encouraging since contact and thermal resistance can be reduced using the methods in the next section.

Although early demonstrations for radio frequency performance are encouraging, there is still significant development required. Two tracks are currently being pursued for lateral RF devices. The first track involves scaling MOSFET channel thickness and exploiting the field strength of Ga_2O_3 to achieve higher doping concentration and short depletion regions. A MOSFET with properly scaled drift region and self-aligned gate-source contacts has not yet been realized. The second track involves using the vertical scaling advantage of a MODFET structure allowing deep submicron gate lengths. In doing so, engineering the heterostructure with higher sheet charge concentration must be realized to reduce sheet resistance beyond what is achievable with thin-channel Ga_2O_3 MOSFETs.

31.5 Thermal Analysis and Solutions

Self-heating occurs in Ga_2O_3 devices even at low power due to the low thermal conductivity of the material [12]. Fortunately, GaN and GaAs have paved the way for power semiconductor device thermal analysis, including the characterization of thermal resistance and channel temperature [36, 37], and thermal solutions including top-side thermal-shunt solutions [38, 39]. In the early development of Ga_2O_3 devices, interface quality makes it difficult to differentiate thermal effects from interface trapping effects in electrical measurements. In Fig. 31.10, pulsed-IV data is shown for a Ga_2O_3 transistor operating under different environmental temperatures [40]. This device has a sharp increase in saturated drain current with temperature at moderate base plate temperatures followed by a decrease in saturated drain current for temps above 200 °C. The authors attribute the increase to a decrease in negative trapped charge with temperature followed by mobility degradation as the temp continues to increase. A model using temperature-dependent Hall effect measurements increases the authors' confidence in this assumption, though it is apparent a complicated relationship exists between the trap occupancy and the self-heating effects. Device engineers will need innovative

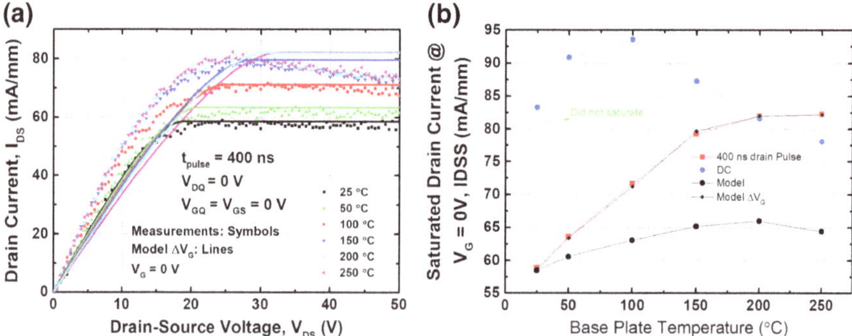

Fig. 31.10 **a** $V_{GS} = 0$ V pulsed-IV curve at various temperatures for a Ge-doped β-Ga_2O_3 MOSFET. **b** Extracted saturated drain current from the same MOSFET versus base plate temperature under both DC and 400 ns pulsed conditions. The device shows an increase in saturated current with temperature until the mobility begins to degrade. A model with actual temperature-dependent Hall data is used to confirm this effect, but the magnitude of the current for the model only matches the real device after incorporating reduction in trap occupancy with increasing temperature

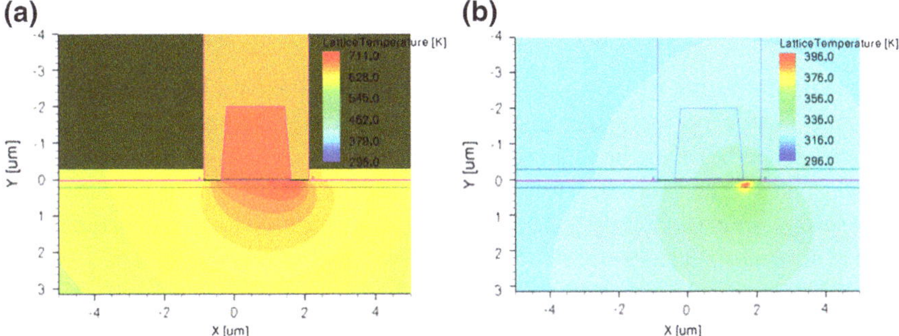

Fig. 31.11 **a** Thermal model of the gate region of a Ga_2O_3 transistor showing a large increase in temperature at the drain side of the gate. The calculated thermal resistance is 200 K/W. **b** The same gate region with an incorporated top side high thermal conductivity material showing reduced increase in temperature and a calculated thermal resistance of 22 K/W

methods to differentiate these effects, as well as potential device degradation under stress, to create reliable characterization methods for thermal solutions. Complexity is increased if, as described in [41], self-heating effects manifest even in the linear region of the device ruling out some of the methods used previously for GaN devices.

In Fig. 31.11, modeling results are shown for a top-side thermal solution for Ga_2O_3 devices and specifically consider very tightly scaled devices of interest for lower voltage applications (e.g., 200 V breakdown), where the hot spot will be localized in the channel. For very high voltage, larger depletion zone devices (e.g., tens of kV and tens of μm depletion), less attention to the nanoscale thermal environment and more attention to the microscale thermal environment are proper. Larger depletion zone devices will benefit from more conventional approaches such as backside thinning. The model shows that using a high thermal conductivity material mounted on the top side of a Ga_2O_3 chip and thermally contacting the channel between gate and drain can reduce the thermal resistance by an order of magnitude from 200 to 22 K/W assuming low thermal boundary resistance from the Ga_2O_3 and into the thermal spreader. This dramatic improvement comes about because the thermal path length through Ga_2O_3 from hot spot to the thermal spreader is only ~100 nm. Sinking heat through the ohmic contacts without this high thermal conductivity material generates a significant benefit but does not approach 22 W/m/K. This shows that even with the poor thermal conductivity Ga_2O_3, engineering methods can be used to reduce the thermal resistance of the device to a level comparable to GaN devices. Other operational modes are also a more straightforward way to avoid generating heat. Pulsed mode operation as seen in Fig. 31.9 is one example of this.

31.6 Summary and Conclusion

To ensure the future of Ga_2O_3 as a material system for ultra-low conduction and dynamic switch power losses and high-efficiency RF amplifiers, the promising material properties must be realized in engineered devices. As demonstrated in this chapter and the chapters that follow in this section, device engineers have made tremendous progress in only a short time. Further progress, however, is required to realize all material achievements (high mobility, doping control, thickness control) and fabrication achievements (low ohmic contact resistance, short gate length, trap-free gate oxide interface) in a single device. Additionally, engineers must address thermal challenges and concerns associated with exclusively depletion-mode operation. Although not addressed here, engineers must continue to develop testing solutions to verify the low power loss properties Ga_2O_3 offers with its high E_C as was done for GaN and SiC in the past. With the rapid accession of Ga_2O_3 as a power semiconductor material in the research community, it is likely that these breakthroughs will come quickly from a community driven to realize the potential of this material system.

References

1. P.C. Chang, Z. Fan, W.-T. Tseng, A. Ragagopal, J.G. Lu, Appl. Phys. Lett. **87**, 222102 (2005)
2. K. Matsuzaki, H. Yanagi, T. Kamiya, H. Hiramatsu, K. Nomura, M. Hirano, H. Hosono, Appl. Phys. Lett. **88**, 092106 (2006)
3. M. Higashiwaki, K. Sasaki, A. Kuramata, T. Masui, S. Yamakoshi, Appl. Phys. Lett. **100**, 013504 (2012)
4. M. Higashiwaki, K. Sasaki, T. Kamimura, M.H. Wong, D. Krishnamurthy, A. Kuramata, T. Masui, S. Yamakoshi, Appl. Phys. Lett. **103**, 123511 (2013)
5. R. Roy, V.G. Hill, E.F. Osborn, J. Am. Chem. Soc. **74**, 719 (1952)
6. J.L. Hudgins, G.S. Simin, E. Santi, M.A. Khan, IEEE Trans. Power Electron. **18**, 907 (2003)
7. K. Ghosh, U. Singisetti, J. Appl. Phys. **122**, 035702 (2017)
8. B.J. Baliga, IEEE Electron Device Lett. **10**, 455 (1989)
9. A.Q. Huang, IEEE Electron Device Lett. **25**, 298 (2004)
10. E.O. Johnson, RCA Rev. **26**, 163 (1965)
11. A.J. Green, K.D. Chabak, E.R. Heller, R.C. Fitch, M. Baldini, A. Fiedler, K. Irmscher, G. Wagner, Z. Galazka, S.E. Tetlak, A. Crespo, K. Leedy, G.H. Jessen, IEEE Electron Device Lett. **37**, 902 (2016)
12. M.H. Wong, K. Sasaki, A. Kuramata, S. Yamakoshi, M. Higashiwaki, IEEE Electron Device Lett. **37**, 212 (2016)
13. K.D. Chabak, J.P. McCandless, N.A. Moser, A.J. Green, K. Mahalingam, A. Crespo, N. Hendricks, B.M. Howe, S.E. Tetlak, K. Leedy, R.C. Fitch, D. Wakimoto, K. Sasaki, A. Kuramata, G.H. Jessen, IEEE Electron Device Lett. **39**, 67 (2018)
14. H. Zhou, M.W. Si, S. Alghamdi, G. Qiu, L.M. Yang, P.D. Ye, IEEE Electron Device Lett. **38**, 103 (2017)
15. M. Higashiwaki, K. Sasaki, M.H. Wong, T. Kamimura, D. Krishnamurthy, A. Kuramata, T. Masui, S. Yamakoshi, in *Technical Digest of the IEEE International Electron Devices Meeting*, 2013

16. W.S. Hwang, A. Verma, H. Peelaers, V. Protasenko, S. Rouvimov, H.G. Xing, A. Seabaugh, W. Haensch, C. Van de Walle, Z. Galazka, M. Albrecht, R. Fornari, D. Jena, Appl. Phys. Lett. **104**, 203111 (2014)
17. S. Krishnamoorthy, Z. Xia, C. Joishi, Y. Zhang, J. McGlone, J. Johnson, M. Brenner, A.R. Arehart, J. Hwang, S. Lodha, S. Rajan, Appl. Phys. Lett. **111**, 023502 (2017)
18. Z.Y. Hu, K. Nomoto, W.S. Li, Z.X. Zhang, N. Tanen, Q.T. Thieu, K. Sasaki, A. Kuramata, T. Nakamura, D. Jena, H.G. Xing, Appl. Phys. Lett. **113**, 122103 (2018)
19. K.D. Chabak, N. Moser, A.J. Green, D.E. Walker, S.E. Tetlak, E. Heller, A. Crespo, R. Fitch, J.P. McCandless, K. Leedy, M. Baldini, G. Wagner, Z. Galazka, X. Li, G. Jessen, Appl. Phys. Lett. **109**, 213501 (2016)
20. N. Moser, J. McCandless, A. Crespo, K. Leedy, A. Green, A. Neal, S. Mou, E. Ahmadi, J. Speck, K. Chabak, N. Peixoto, G. Jessen, IEEE Electron Device Lett. **38**, 775 (2017)
21. M. Östling, Sci. China Inform. Sci. **54**, 1087 (2011)
22. M. Amato and V. Rumennik, in *Technical Digest of the IEEE International Electron Devices Meeting*, 1985
23. M.H. Wong, Y. Nakata, A. Kuramata, S. Yamakoshi, M. Higashiwaki, Appl. Phys. Express **10**, 041101 (2017)
24. Z. Xia, C. Joishi, S. Krishnamoorthy, S. Bajaj, Y. Zhang, M. Brenner, S. Lodha, S. Rajan, IEEE Electron Device Lett. **39**, 568 (2018)
25. E. Ahmadi, O.S. Koksaldi, X. Zheng, T. Mates, Y. Oshima, U.K. Mishra, J.S. Speck, Appl. Phys. Express **10**, 071101 (2017)
26. Y. Oshima, E. Ahmadi, S.C. Badescu, F. Wu, J.S. Speck, Appl. Phys. Express **9**, 061102 (2016)
27. M.A. Bhuiyan, H. Zhou, R. Jiang, E.X. Zhang, D.M. Fleetwood, P.D. Ye, T.-P. Ma, IEEE Electron Device Lett. **39**, 1022 (2018)
28. J.F. McGlone, Z. Xia, Y. Zhang, C. Joishi, S. Lodha, S. Rajan, S.A. Ringel, A.R. Arehart, IEEE Electron Device Lett. **39**, 1042 (2018)
29. T. Kamimura, D. Krishnamurthy, A. Kuramata, S. Yamakoshi, M. Higashiwaki, Jpn. J. Appl. Phys. **55**, 1202B5 (2016)
30. S. Muller, H. von Wenckstern, F. Schmidt, D. Splith, F.L. Schein, H. Frenzel, M. Grundmann, Appl. Phys. Express **8**, 121102 (2015)
31. M.H. Wong, K. Sasaki, A. Kuramata, S. Yamakoshi, M. Higashiwaki, Appl. Phys. Lett. **106**, 032105 (2015)
32. Y. Zhang, A. Neal, Z. Xia, C. Joishi, J.M. Johnson, Y. Zheng, S. Bajaj, M. Brenner, D. Dorsey, K. Chabak, G. Jessen, J. Hwang, S. Mou, J.P. Heremans, S. Rajan, Appl. Phys. Lett. **112**, 173502 (2018)
33. A.J. Green, K.D. Chabak, M. Baldini, N. Moser, R. Gilbert, R.C. Fitch, G. Wagner, Z. Galazka, J. McCandless, A. Crespo, K. Leedy, G.H. Jessen, IEEE Electron Device Lett. **38**, 790 (2017)
34. G.H. Jessen, R.C. Fitch, J.K. Gillespie, G. Via, A. Crespo, D. Langley, D.J. Denninghoff, M. Trejo, E.R. Heller, IEEE Trans. Electron Devices **54**, 2589 (2007)
35. K.D. Chabak, D.E. Walker, A.J. Green, A. Crespo, M. Lindquist, K. Leedy, S. Tetlak, R. Gilbert, N.A. Moser, G. Jessen, in *Proceedings of the IEEE MTT-S International Microwave Workshop Series on Advanced Materials and Processes for RF and THz Applications*, 2018
36. J. Joh, J.A. del Alamo, U. Chowdhury, T.M. Chou, H.Q. Tserng, J.L. Jimenez, IEEE Trans. Electron Devices **56**, 2895 (2009)
37. S. Martin-Horcajo, A. Wang, M.F. Romero, M.J. Tadjer, F. Calle, IEEE Trans. Electron Devices **60**, 4105 (2013)
38. J. Sewell, L.L. Liou, D. Barlage, J. Barrette, C. Bozada, R. Dettmer, R. Fitch, T. Jenkins, R. Lee, M. Mack, G. Trombley, P. Watson, IEEE Electron Device Lett. **17**, 19 (1996)

39. M.J. Tadjer, M.A. Mastro, N.A. Mahadik, M. Currie, V.D. Wheeler, J.A. Freitas, J.D. Greenlee, J.K. Hite, K.D. Hobart, C.R. Eddy, F.J. Kub, J. Electron. Mater. **45**, 2031 (2016)
40. N.A. Moser, J.P. McCandless, A. Crespo, K.D. Leedy, A.J. Green, E.R. Heller, K.D. Chabak, N. Peixoto, G.H. Jessen, Appl. Phys. Lett. **110**, 143505 (2017)
41. M.H. Wong, Y. Morikawa, K. Sasaki, A. Kuramata, S. Yamakoshi, M. Higashiwaki, Appl. Phys. Lett. **109**, 193503 (2016)

Chapter 32
Field-Effect Transistors 2

Ga_2O_3 Field-Effect Transistors for Power Switching and Radiation-Hard Electronics

Man Hoi Wong and Masataka Higashiwaki

Abstract Ga_2O_3 has exploded onto the semiconductor landscape for next-generation power electronics because its enticing material properties, most notably a large critical field strength stemming from its ultrawide bandgap, promise miniaturized circuits and systems with high conversion efficiency. Intense pursuit of Ga_2O_3 power devices galvanized by the demonstration of a high-voltage Ga_2O_3 field-effect transistor (FET) in 2012 has brought about tremendous advancements in this new technology, whose strong radiation tolerance and high thermal stability also befit harsh-environment applications that impose stringent reliability requirements. This chapter reviews the designs and properties of various types of depletion- and enhancement-mode Ga_2O_3 FETs—which have predominantly been lateral devices—for power switching and radiation-hard electronics. The development of vertical Ga_2O_3 transistors based on a low-cost, highly manufacturable ion implantation doping process will also be presented.

32.1 Introduction

Many of humanity's most pressing needs and exciting advances today are critically dependent on the effective use of electrical energy, including in motor drives, electrified transportation, data centers, and the grid. These applications are impacted by the performance of power converters, which consist of rectifiers (diodes) and switches (transistors) as the core device components. Today, Si power devices are the mainstream but they are rapidly approaching fundamental performance limitations, rendering commercial power systems bulky and inefficient. These challenges are

M. H. Wong (✉) · M. Higashiwaki
National Institute of Information and Communications Technology, Koganei, Tokyo 184-8795, Japan
e-mail: mhwong@nict.go.jp

M. Higashiwaki
e-mail: mhigashi@nict.go.jp

M. Higashiwaki and S. Fujita (eds.), *Gallium Oxide*, Springer Series in Materials Science 293, https://doi.org/10.1007/978-3-030-37153-1_32

converging with technology opportunities created by wide-bandgap semiconductors to define a new design landscape for power conversion.

Recent years have seen an upsurge in research activities on beta-phase gallium oxide (β-Ga_2O_3, referred to hereinafter as Ga_2O_3) for power electronics [1]. Boasting a large bandgap of 4.5–4.9 eV [2–4] and resultant experimental critical fields of 5.1–5.9 MV/cm that surpass the theoretical bulk values for GaN and SiC [5–9], Ga_2O_3 promises dramatic reductions in the size, weight, and energy consumption of power systems by simultaneously increasing power density and power conversion efficiency at the device level. Ga_2O_3 is also thermally stable and radiation hard [10], thus rendering it particularly suitable for building devices and systems for aerospace, detector, survey, and nuclear technologies that demand stable operations in harsh environments [11, 12]. The impact of these emerging Ga_2O_3 technologies is bolstered by the availability of melt-grown native substrates (floating zone [13–15], Czochralski [16–19], edge-defined film-fed growth [20, 21], vertical Bridgman [22]), whose high quality and scalability offer Ga_2O_3 a unique platform that carries a compelling cost advantage over the incumbent wide-bandgap technologies. This chapter summarizes the progress made and lessons learnt on both lateral and vertical Ga_2O_3 field-effect transistors (FETs) for power switching and radiation-hard applications, with a particular focus on device technologies developed in our group.

32.2 Lateral Depletion-Mode Ga_2O_3 FETs

32.2.1 MESFET

The first single-crystal, high-voltage Ga_2O_3 transistor is a depletion-mode (D-mode) Ga_2O_3 metal–semiconductor FET (MESFET) consisting of a 300-nm-thick Sn-doped *n*-type Ga_2O_3 channel layer (Fig. 32.1) [23]. The device, which is grown by ozone molecular beam epitaxy (MBE) [24, 25] on a Mg-doped semi-insulating Ga_2O_3 (010) substrate, demonstrates successful transistor action with sharp drain current (I_D) pinch-off (Fig. 32.2a). A three-terminal off-state breakdown voltage (V_{br}) of 257 V is obtained for a gate–drain spacing (L_{GD}) of 8 μm (Fig. 32.2a), which is notable considering that this is the first attempt on a high-voltage Ga_2O_3 transistor. However, the I_D on/off ratio (I_{on}/I_{off}) is limited to about 10^4 by a small leakage current through the unpassivated Ga_2O_3 surface (Fig. 32.2b). The ohmic contact resistance also leaves much room for improvement.

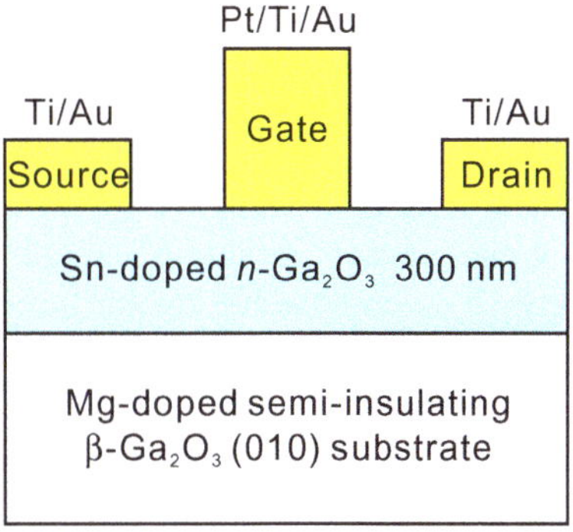

Fig. 32.1 Cross-sectional schematic of Ga_2O_3 MESEFT. Reprinted from [23], with the permission of AIP Publishing

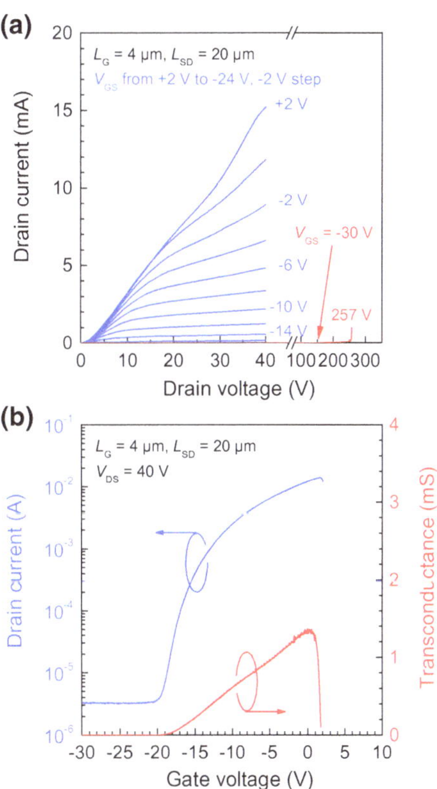

Fig. 32.2 DC characteristics of Ga_2O_3 MESFET with gate length (L_G) of 4 μm and source–drain spacing (L_{SD}) of 20 μm: **a** I_D–V_D and **b** I_D–V_G. Reprinted from [23], with the permission of AIP Publishing

32.2.2 MOSFET with Si-Ion-Implanted Source/Drain Contacts

To address the two major shortcomings of the MESFET, a metal–oxide–semiconductor FET (MOSFET) structure engineered with an Al_2O_3 dielectric grown by plasma-assisted atomic layer deposition (ALD) and n^+ source/drain contact layers formed by Si-ion (Si^+) implantation doping is fabricated on an Fe-doped semi-insulating Ga_2O_3 (010) substrate (Fig. 32.3) [26, 27]. The Al_2O_3 dielectric suppresses I_{off} by functioning as both a surface passivation layer and a gate insulator, leading to a dramatic increase in I_{on}/I_{off} to over 10^{10}. Degenerate doping of the ohmic contacts results in a small specific contact resistivity of 8.1×10^{-6} Ω cm^2, which is comparable to the typical resistivity values of optimized Ti/Al-based contacts made to n-GaN and AlGaN/GaN heterostructures. The MOSFET, which has a Sn-doped channel layer nominally the same as that of the MESFET introduced in Sect. 32.2.1, delivers a maximum I_D of 39 mA/mm and an off-state V_{br} of 370 V (Fig. 32.4a). High-temperature operation at 250 °C is demonstrated without permanent degradation in electrical characteristics, albeit with a lower I_{on}/I_{off} of 10^4 (Fig. 32.4b).

32.2.3 MOSFET with Si-Ion-Implanted Channel and Contacts

The Ga_2O_3 MESFET and MOSFET with a Sn-doped channel have shown great promise as an emerging wide-bandgap device technology, yet further advancement is hampered by a narrow MBE growth temperature window required to promote Sn incorporation into the Ga_2O_3 epitaxial layer while maintaining high crystal quality [25]. Delayed Sn incorporation due to surface segregation of the dopant leads to poor control of channel thickness and non-uniform distribution of in-plane carrier concentration. These challenges are circumvented by applying Si^+ implantation doping to form the channel in an unintentionally doped (UID) Ga_2O_3 epilayer (Fig. 32.5) [28], which allows higher growth temperatures to ensure optimal crystal quality without concern for irreproducible dopant profiles. The MOSFET with a Si^+-implanted channel delivers comparable performance to its Sn-doped counterpart, including a maximum I_D of 65 mA/mm (Fig. 32.6), an off-state V_{br} of 415 V, and an I_{on}/I_{off} of over 10^{10}.

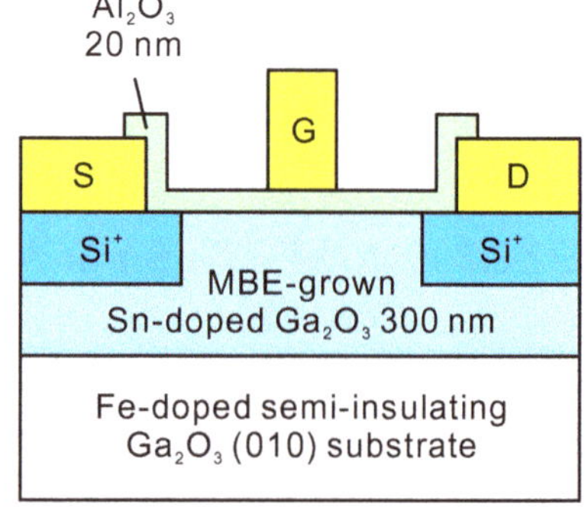

Fig. 32.3 Cross-sectional schematic of Ga_2O_3 MOSFET with Si^+-implanted source/drain ohmic contacts. Reprinted from [26], with the permission of AIP Publishing

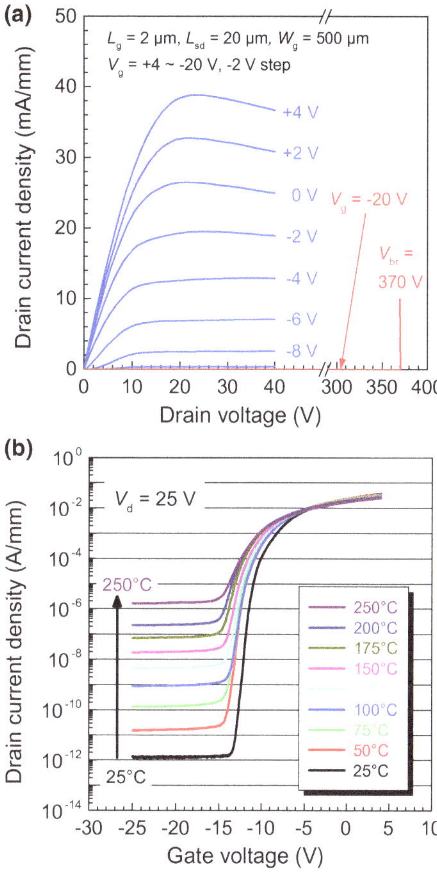

Fig. 32.4 DC characteristics of Ga_2O_3 MOSFET with Si^+-implanted source/drain ohmic contacts: **a** I_D–V_D and **b** temperature-dependent I_D–V_G. Reprinted from [26], with the permission of AIP Publishing

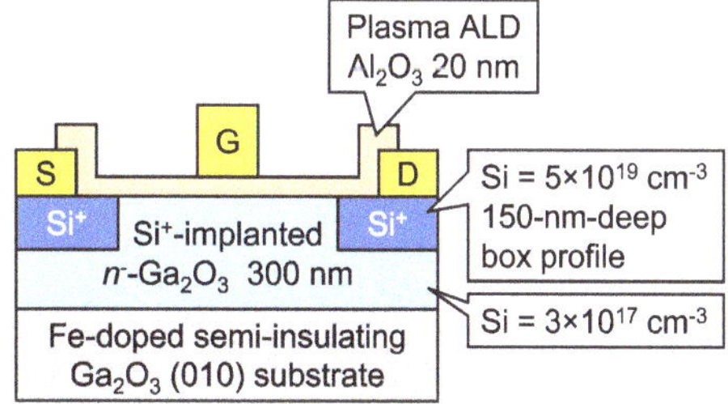

Fig. 32.5 Cross-sectional schematic of Ga_2O_3 MOSFET with Si^+-implanted channel and ohmic contacts. © 2013 IEEE. Reprinted, with permission, from [28]

32.2.4 Field-Plated MOSFETs

Field-plating is an edge termination technique commonly employed in FETs to enhance V_{br} by splitting a single large electric field peak at the gate metal edge into two or more smaller peaks located at the edges of the gate and field plate(s).

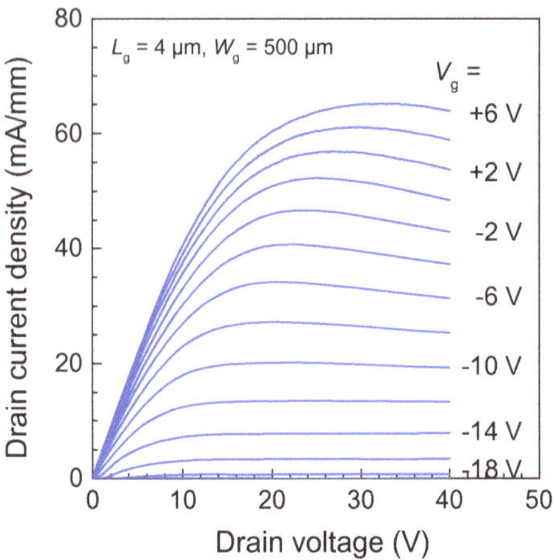

Fig. 32.6 DC I_D–V_D characteristics of Ga_2O_3 MOSFET with Si^+-implanted channel and ohmic contacts. © 2013 IEEE. Reprinted, with permission, from [28]

Accordingly, a gate-connected field plate (FP) is introduced into the Si^+-implanted MOSFET discussed in Sect. 32.2.3 for a target V_{br} of over 1 kV [29]. The field-plated MOSFET (FP-MOSFET), illustrated schematically in Fig. 32.7, incorporates a 0.9-µm-thick UID Ga_2O_3 buffer layer under its Si^+-implanted channel to protect the substrate from implantation damage, thereby preventing defect-assisted outdiffusion of Fe toward the channel upon implant activation annealing and the consequent charge compensation [30, 31]. Proximity of Fe doping to the electron channel has also been ascribed to mobility degradation, DC–RF dispersion, and threshold voltage (V_T) instability in Ga_2O_3 MOSFETs with non-ion-implanted channels [32, 33]. At room temperature, the FP-MOSFET delivers a maximum I_D of 78 mA/mm and an I_{on}/I_{off} of over 10^9 (Fig. 32.8). Stability against thermal stress is sustained up to at least 300 °C, where an I_{on}/I_{off} higher than 10^3 is maintained (Fig. 32.8b). Gate- and drain-lag measurements

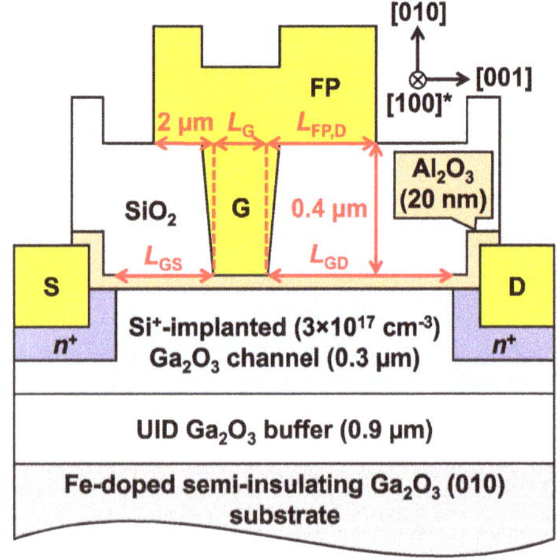

Fig. 32.7 Cross-sectional schematic of Ga_2O_3 FP-MOSFET with gate-connected FP of length $L_{FP,D}$ extended toward the drain

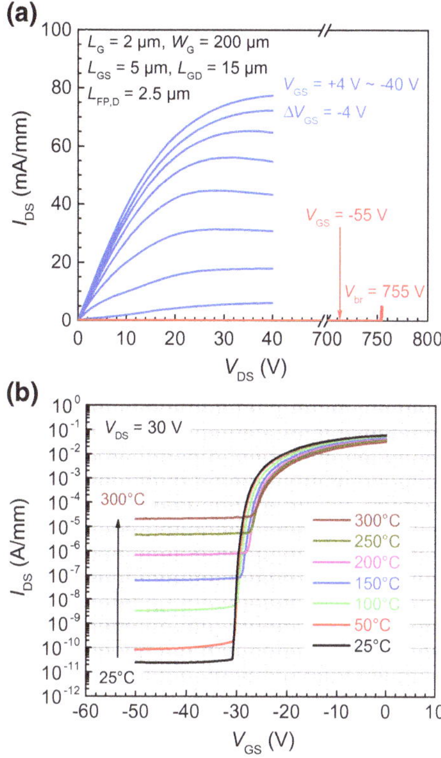

Fig. 32.8 DC characteristics of Ga_2O_3 FP-MOSFET with 2.5-μm-long FP: **a** I_D–V_D and **b** temperature-dependent I_D–V_G. © 2016 IEEE. Reprinted, with permission, from [29]

reveal negligible bulk trapping and a well-passivated surface (Fig. 32.9) [29, 34]. Compared with the non-field-plated baseline device, a 2.5-μm-long FP contributes to an 80% increase in off-state V_{br} from 415 to 755 V (Fig. 32.8a); however, hard breakdown of this device is limited by a parasitic source–drain punch-through

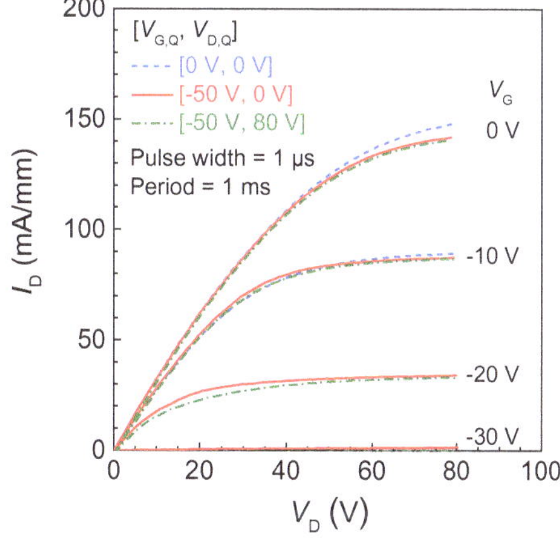

Fig. 32.9 Pulsed I_D–V_D characteristics of Ga_2O_3 FP-MOSFET measured at quiescent gate and drain biases ([$V_{G,Q}$, $V_{D,Q}$]) of [0, 0 V] (blue dashed line), [−50, 0 V] (red solid line), and [−50, 80 V] (green dashed-dotted line) that correspond to dispersion-free, gate-lag, and drain-lag conditions, respectively [34]

mechanism instead of peak electric fields in the gate–drain depletion region, which is consistent with the off-state drain leakage current being one to two orders of magnitude higher than the gate leakage current (I_G) and suggests that the intrinsic V_{br} should have been much higher, as to be expected from simulations. By developing a high-density composite oxide stack for the field plate as well as adopting a SiO_2 gate insulator to provide a larger conduction band offset than with Al_2O_3 [35, 36], Zeng et al. have obtained a considerably improved V_{br} of 1.85 kV for a FP-MOSFET that utilizes an MBE-grown Sn-doped channel [37] and spin-on-glass source/drain doping [38], yet the accompanying specific on-resistance ($R_{on,sp}$) is almost two orders of magnitude larger than that of the 755-V device.

Ga_2O_3 FP-MOSFETs have also been fabricated with a source-connected FP [39, 40], which is often preferred to a gate-connected FP from the perspective of high-speed performance since comparable electric field profiles can be engineered without introducing a parasitic gate–drain feedback capacitance. Devices reported by Lv et al. [39] feature Si^+-implanted source/drain contacts and a highly doped channel containing at least twice as much charge as in the aforementioned FP-MOSFETs, thereby giving rise to a high I_D of >200 mA/mm. A V_{br} of 480 V and a corresponding $R_{on,sp}$ of 4.57 mΩ cm^2 translate to a large power figure of merit ($V_{br}^2/R_{on,sp}$) of 50.4 MW/cm^2. In a separate work by Mun et al. [40], field-plated devices show enhancement in V_{br} up to 2.32 kV (corresponding $R_{on,sp}$ = 959 mΩ cm^2).

32.2.5 MOSFET with Ge-Doped Channel

All the Ga_2O_3 MOSFETs discussed so far have incorporated either Sn or Si as a dopant in their conducting *n*-type channels. Ge has been proposed as a more favorable *n*-type dopant in Ga_2O_3 for its better match than Si^{4+} with the Ga^{3+} cationic size [41] and for being potentially more effective than Sn^{4+} by favoring the fourfold-coordinated tetrahedral Ga(I) site over the sixfold-coordinated octahedral Ga(II) site [42]. A prototype Ge-doped MOSFET grown by MBE on an Fe-doped semi-insulating Ga_2O_3 (010) substrate shows a maximum I_D of >75 mA/mm, an I_{on}/I_{off} of $>10^8$, an off-state V_{br} of 479 V, and stable operation at 250 °C [43]. These results mirror the performance of previously reported Sn-and Si-doped MOSFETs, thereby attesting to the promise of Ge doping for Ga_2O_3 devices though clear performance advantages are yet to be demonstrated.

32.2.6 Critical Field Strength in Ga_2O_3 MOSFETs

The experimental critical electric fields of the Ga_2O_3 MOSFETs introduced so far in this section, calculated by approximating a linear potential gradient between the gate and drain terminals, are smaller than 1 MV/cm and thus fall far short of the projected theoretical value of 8 MV/cm [23]. This is because those devices have an L_{GD} larger than the minimum required to support the V_{br} achievable at their respective channel doping concentrations. In the limit of a very large L_{GD}, the average breakdown field approaches zero while the peak field is determined by the extent of the gate–drain depletion region; on the other hand, the lateral electric field in a scaled gate–drain drift region becomes more uniform in magnitude such that the averaged field more accurately reflects the peak field. Indeed, improved average field strengths of 3.8 and 2.2 MV/cm—the former of which being the highest ever measured in a lateral transistor—have been reported in MOSFETs possessing smaller L_{GD} of 0.6 and 1.8 μm, respectively [5, 37].

32.3 Thermal Characteristics of Ga_2O_3 MOSFETs

MOSFETs in the on-state experience highly localized power dissipation and heat generation within a submicron region of the channel at the drain side of the gate where the maximum electric field occurs. Self-heating in Ga_2O_3 devices are exacerbated by the low thermal conductivity of Ga_2O_3 that ranges from 11 to 27 $Wm^{-1}K^{-1}$ at room temperature along the three principal crystallographic directions [18, 44–47]. Inefficient heat transport leads to higher temperatures in the active region of Ga_2O_3 transistors than for other technologies at comparable power densities (P_D). A large thermal resistance (R_{th}) of 48 mm K/W (i.e., a temperature rise of 48 K per 1 W/mm of P_D) is extracted for a Ga_2O_3 FP-MOSFET under room temperature operation by an electrical method that determines a spatially averaged channel temperature (T_{ch}) via isothermal pulsed current–voltage measurements with quiescent biases and sweep parameters chosen to minimize dispersion effects related to trapping (Fig. 32.10) [48]. Raman nanoparticle thermography, which enables

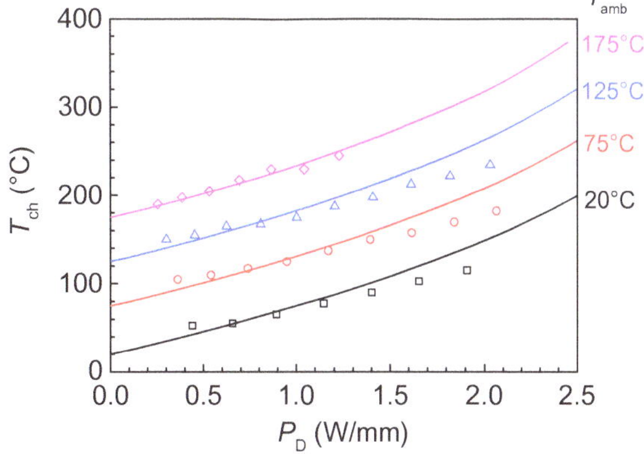

Fig. 32.10 Dependence of T_{ch} (symbol—measurement; line—simulation) in Ga_2O_3 FP-MOSFET on P_D at ambient temperatures (T_{amb}) of 20, 75, 125, and 175 °C. Reprinted from [48], with the permission of AIP Publishing

localized measurement of the peak T_{ch}, yields a larger R_{th} of 88 mm K/W for a similar device structure [49]. These measured R_{th} values exceed those of AlGaN/GaN high electron mobility transistors fabricated on sapphire substrates [50] and are a factor of three to five larger than those of GaN devices that benefit from efficient heat sinking through thermally conductive substrates like SiC [50] or diamond [51, 52]. Significantly higher junction temperatures have also been revealed in a Ga_2O_3 Schottky barrier diode than in a SiC diode at comparable P_D [53].

Performance boost in Ga_2O_3 FETs is evident when self-heating effects due to the material's low thermal conductivity are mitigated by means of pulsed-mode operation. A typical manifestation of reduced heat generation is an increase in pulsed I_D over the corresponding DC values [54]. Pulsed load-pull measurements have also yielded higher large-signal RF gain and efficiency compared with continuous-wave measurements [34]. These results indicate that proper thermal mounting and heat spreading techniques will dramatically improve the electrical performance of Ga_2O_3 devices. It has recently been demonstrated that direct attachment of an n^+-Ga_2O_3 wafer to a thermally conductive polycrystalline n^+-SiC (poly-SiC) substrate via surface activated bonding forms a void-free interface with a negligible R_{th} (Fig. 32.11) [55]. The effective out-of-plane thermal conductivity of the composite Ga_2O_3/poly-SiC structure matches that of bulk Si (156 $Wm^{-1}K^{-1}$) for a Ga_2O_3 thickness of 21 μm and is expected to be comparable to that of GaN (220 $Wm^{-1}K^{-1}$) if the Ga_2O_3 is further thinned down to about 10 μm. Furthermore, the heterointerface exhibits only a small specific electrical resistivity of 2×10^{-4} Ω cm^2, which can be further reduced with higher doping. This wafer bonding technique, in combination with top-side heat extraction strategies, thus provides a viable thermal solution for vertical Ga_2O_3 power devices.

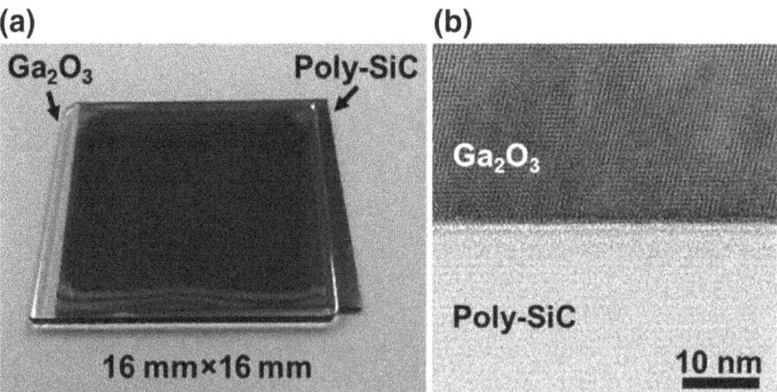

Fig. 32.11 Surface activated bonding of Ga_2O_3 and poly-SiC: **a** photograph of bonded substrate and **b** cross-sectional transmission electron micrograph of bonded heterointerface between Ga_2O_3 and poly-SiC. Reprinted from [55], with the permission of AIP Publishing

32.4 Ga_2O_3 MOSFETs for Radiation-Hard Electronics

High-energy radiation presents severe challenges to the performance and long-term reliability of semiconductor electronic devices by causing physical damage in the constituent materials and/or transient effects related to ionization. Defects created by bombardment of ions, electrons, neutrons, and/or photons form midgap states that act as traps and generation-recombination centers in semiconductors, leading to carrier scattering and charge compensation while causing degradations in device performance that typically manifest as deteriorated transport properties, aggravated gate/drain-lag effects, or compromised detector efficiency. Knowledge of the mechanisms behind radiation-induced degradation is relevant to and important for building reliable devices and circuits to be deployed in space missions and nuclear facilities, where these electronics will be exposed to intense particle and electromagnetic radiation.

It has been established that wide-bandgap semiconductors (notably SiC [56], GaN [57, 58], and diamond [59, 60]) are much less susceptible to displacement damage by particle irradiation than elemental and compound semiconductors with narrower bandgaps. A higher displacement threshold reflects stronger bond strength and serves as an indication of enhanced radiation hardness. Demonstrations of operational Ga_2O_3 solar-blind ultraviolet photodetectors [61], Schottky rectifiers [62–64], and exfoliated quasi-two-dimensional nanobelt FETs [65] that have been exposed to energetic protons or electrons at high fluences ($\geq 10^{15}$ cm^{-2}) reveal the strong prospects of Ga_2O_3 for radiation-hard applications. The resilience of Ga_2O_3 devices is also manifest in the effective recovery of Ga_2O_3 nanobelt FETs from 10-MeV proton damage by rapid thermal annealing at a modest temperature of 500 °C, despite a high proton fluence of 2×10^{15} cm^{-2} that corresponds to approximately 10^5 times the intensity of a solar proton event [65]. These pioneering studies on radiation-hard Ga_2O_3 devices have spurred a growing body of related theoretical and experimental materials research that aims at understanding fundamental defect properties and carrier dynamics in radiation-damaged Ga_2O_3 [66–75].

Effects of ionizing radiation on epitaxial Ga_2O_3 FETs have been studied by exposing MOSFETs to gamma-rays (γ-rays) emitted from a ^{60}Co source with an average photon energy of 1.25 MeV [76]. The devices show no discernible increase in on-resistance (R_{on}), reduction in I_D, or shift in V_T up to a cumulative γ-ray dose of 500 kGy(SiO_2), where a dose of 1 Gray (Gy) corresponds to an absorbed radiation energy of 1 J per kg of mass (SiO_2 being a legacy reference material). The unchanged R_{on} and I_D reflect insignificant generation of radiation-induced bulk traps or scattering centers in the Ga_2O_3 channel, while a stable V_T indicates insignificant carrier trapping or fixed charge accumulation under the gate. All tested devices remain functional at cumulative doses of 1 MGy(SiO_2) and beyond, which are more than ten times the target γ-ray tolerance for space and radiation detector

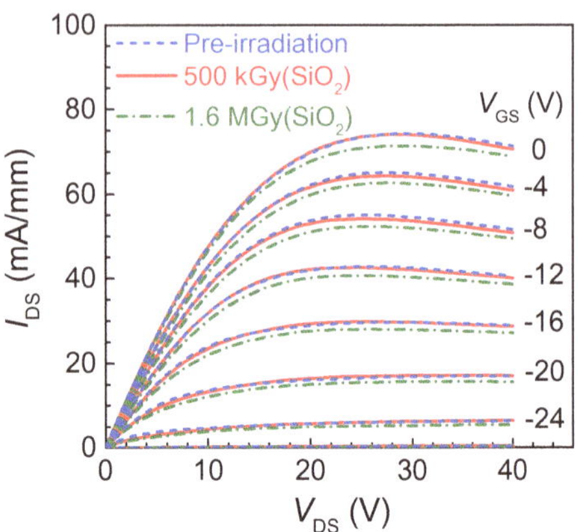

Fig. 32.12 DC I_D–V_D characteristics of Ga_2O_3 MOSFET before γ-ray irradiation (blue dashed line) and after exposure to cumulative γ-ray doses of 500 kGy(SiO_2) (red solid line) and 1.6 MGy(SiO_2) (green dashed-dotted line). Reprinted from [76], with the permission of AIP Publishing

applications [77]. At the highest cumulative dose of 1.6 MGy(SiO_2), the devices preserve the small hysteresis of the transfer characteristics from their virgin state; however, they show slight reductions of less than 5% in the maximum I_D and peak transconductance (g_m) with an associated positive V_T shift of +1 to +1.5 V that are ascribable to bulk carrier trapping (Fig. 32.12). Such heavy irradiation also causes an increase in off-state I_G and degradation in dynamic R_{on} due to gate dielectric damage and radiation-induced electron trapping in the drain access region, respectively.

On the basis of the electrical characteristics described above, it can be deduced that the total-dose γ-ray response of Ga_2O_3 MOSFETs is dominated by the deleterious effects of dielectric damage and interface charge trapping instead of the intrinsic radiation hardness of Ga_2O_3. A recent study of Ga_2O_3 MOS structures irradiated with 10-keV x-rays also affirms that radiation-induced trapping in the dielectric layer dictates the electrical properties of the irradiated devices [78], whereas dielectric-free Ga_2O_3 Schottky rectifiers and solar-blind photodetectors exemplify strong γ-ray tolerance with relatively stable device characteristics [79, 80]. In general, the radiation tolerance of devices that benefit from low susceptibility of the bulk semiconductor to radiation damage is limited by the reliability of dielectrics, dielectric/semiconductor interfaces [77, 81–89], and metal–semiconductor contacts [90, 91]. It is well known that SiC MOSFETs are much less amenable to stable operation under γ-rays than SiC MESFETs owing to radiation-induced buildup of positive gate oxide charges [92], the impact of which can be diminished with thinner oxide insulators [87, 89]. Robust processes for the gate and passivating dielectrics are therefore paramount to enhancing the overall radiation hardness of Ga_2O_3 MOSFETs.

32.5 Lateral Enhancement-Mode Ga_2O_3 FETs

32.5.1 Overview

While the rapid progress of D-mode Ga_2O_3 transistors has established the viability of Ga_2O_3 for power electronics, normally-off power switches are preferred for practical applications to ensure fail-safe operation and eliminate negative-polarity power supplies for simplified circuit designs and system architecture. The majority of enhancement-mode (E-mode) Ga_2O_3 MOSFETs has been implemented as *n*-channel accumulation-mode devices, until very recently when inversion-mode operation via acceptor doping of the Ga_2O_3 channel layer is also proposed. Both planar and non-planar topologies have been pursued.

32.5.2 Planar MOSFETs

Planar E-mode Ga_2O_3 MOSFETs can be realized by decreasing the thickness and/or doping concentration of the channel layer to enable full depletion under the gate at a gate voltage (V_G) of 0 V. The earliest device fabricated following this simple consideration shows a V_T of +3 V and a respectable V_{br} of 390 V, but large parasitic resistances in the ungated source and drain access regions render its I_{on} extremely low ($\sim$0.1 mA/mm) [93]. One potential approach to high-current operation involves eliminating the access regions by overlapping the gate with n^+ source and drain contacts formed by selective-area Si^+ implantation doping of a UID Ga_2O_3 channel grown by ozone MBE with a low background carrier concentration ($\leq 4\times10^{14}$ cm^{-3}) (Fig. 32.13) [94]. Despite a significant reduction in the total parasitic resistance to a mere 2.2 Ω mm, severe charge trapping effects at the dielectric interface limit the maximum I_D to 1.4 mA/mm (Fig. 32.14) and the corresponding I_{on}/I_{off} to 10^6 while causing large positive V_T shifts. Based on the device structure in [94], a normally-off Ga_2O_3 MOSFET aiming at inversion-mode operation to stabilize the V_T at a larger value is fabricated by employing plasma-assisted MBE to grow a Ga_2O_3 channel layer doped unintentionally with deep nitrogen (N) acceptors at a concentration of 1×10^{18} cm^{-3} (Fig. 32.15) [95]. A large V_T of over +8 V is achieved. Even though on-state device operation is restricted to the subthreshold regime owing to limited V_G swing, an I_{on}/I_{off} ratio exceeding five orders of magnitude is obtained (Fig. 32.16).

E-mode Ga_2O_3 MOSFETs that feature gate–source and gate–drain overlaps resemble prototypical Si transistors for complementary logic and are therefore unamenable to high-voltage operation. Devices with both high V_{br} and high I_D are demonstrated by using a gate-recess process to partially remove a highly doped channel under the gate while the ungated access regions remain at full thickness to minimize R_{on} (Fig. 32.17) [96]. This process produces T-shaped gates with a built-in FP. Figure 32.18 shows the DC output characteristics of a laterally scaled

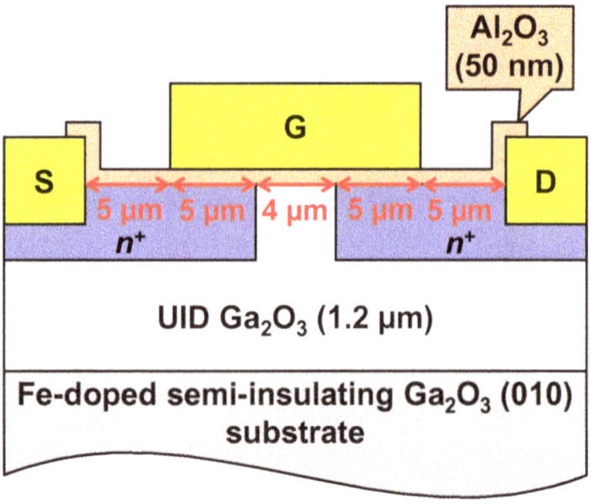

Fig. 32.13 Cross-sectional schematic of E-mode Ga_2O_3 MOSFET with UID channel and Si^+-implanted source/drain contacts [94]. Copyright 2017, the Japan Society of Applied Physics

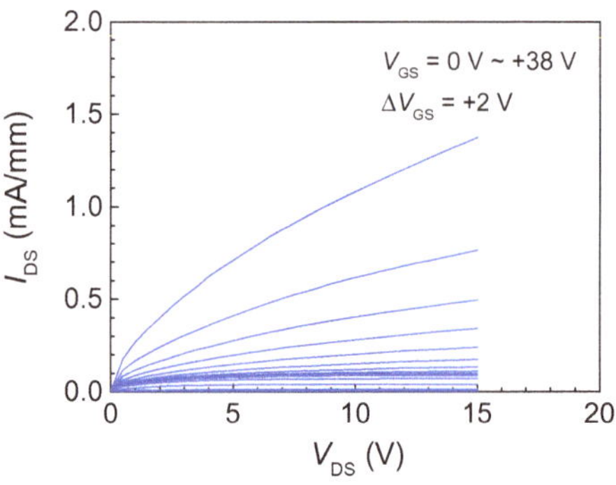

Fig. 32.14 DC I_D–V_D characteristics of E-mode Ga_2O_3 MOSFET with UID channel and Si^+-implanted source/drain contacts [94]. Copyright 2017, the Japan Society of Applied Physics

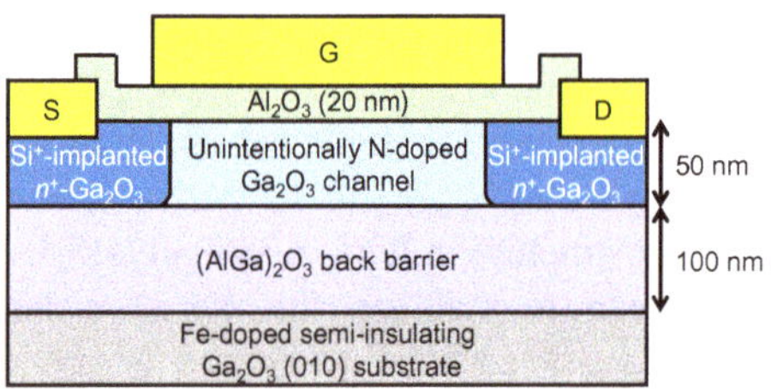

Fig. 32.15 Cross-sectional schematic of E-mode Ga_2O_3 MOSFET with N-doped channel and Si^+-implanted source/drain contacts. © 2019 IEEE. Reprinted, with permission, from [95]

Si-doped (010) device with a V_T of +2 V. The R_{on} of 215 Ω mm, I_D of 40 mA/mm, and I_{on}/I_{off} of 10^9 are the best reported for homoepitaxial E-mode Ga_2O_3 devices and have only been surpassed by non-epitaxial nanomembrane FETs [97, 98]. An off-state V_{br} of 505 V is measured at V_G = 0 V.

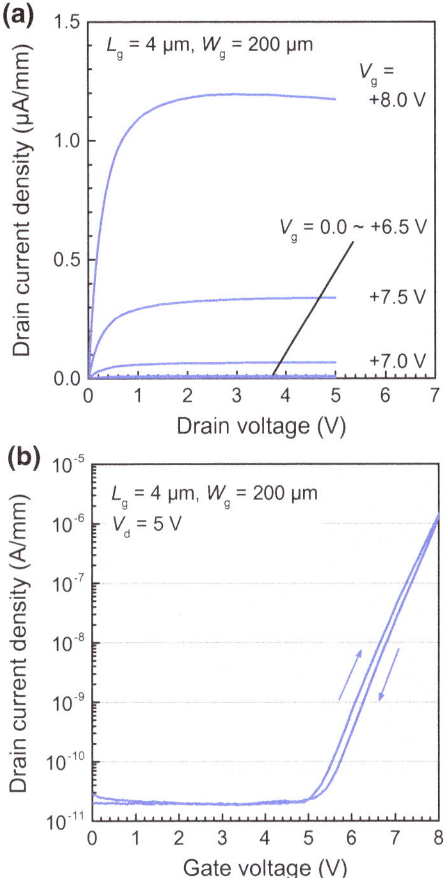

Fig. 32.16 DC characteristics of E-mode Ga_2O_3 MOSFET with N-doped channel and Si^+-implanted source/drain contacts: **a** I_D–V_D and **b** I_D–V_G. © 2019 IEEE. Reprinted, with permission, from [95]

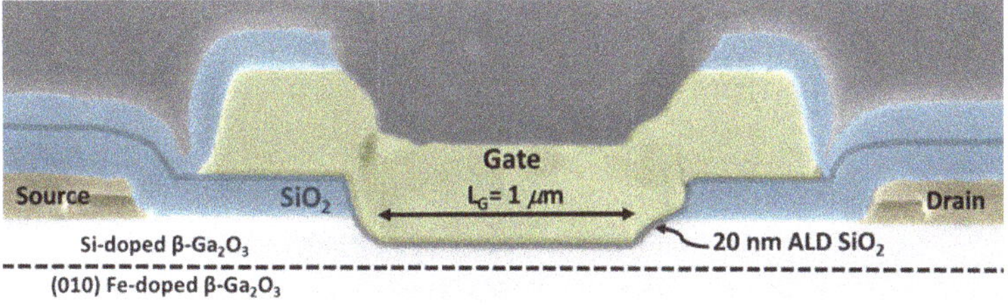

Fig. 32.17 Cross-sectional scanning electron micrograph (false-colored) of E-mode Ga_2O_3 MOSFET with gate recess and T-shaped metal gate. © 2018 IEEE. Reprinted, with permission, from [96]

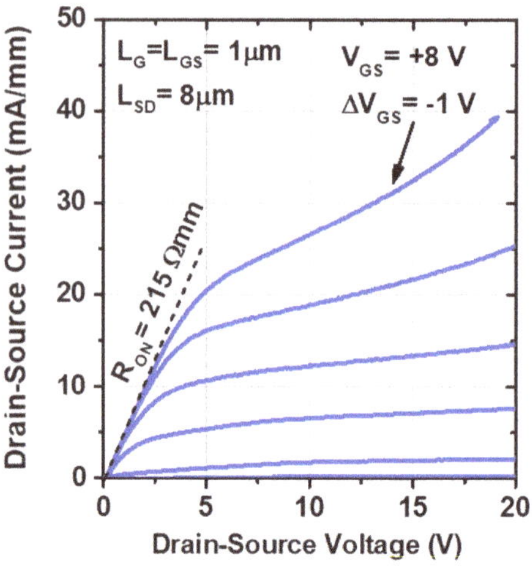

Fig. 32.18 DC I_D–V_D characteristics of recessed-gate E-mode Ga_2O_3 MOSFET. © 2018 IEEE. Reprinted, with permission, from [96]

32.5.3 *Fin-Array MOSFETs*

A non-planar conception of normally-off Ga_2O_3 FETs capitalizes on the enhanced electrostatic control offered by wrap-gate architecture to deplete non-planar fin-shaped channels from the sidewalls, thereby shifting the V_T to a positive value without sacrificing doping or volume current densities [99]. The fins are formed by top-down BCl_3 plasma etching of a planar Sn-doped Ga_2O_3 film grown by metalorganic vapor phase epitaxy on a Mg-doped native Ga_2O_3 (100) substrate. FinFETs (Fig. 32.19) with a 20-nm-thick Al_2O_3 dielectric and a 2-μm-long wrap-gate demonstrate normally-off operation with a V_T between 0 and +1 V. The I_{on}/I_{off} is greater than 10^5 and is mainly limited by a low I_{on} of less than 1 mA/mm (Fig. 32.20), but substantial improvements can be expected with R_{on} optimization.

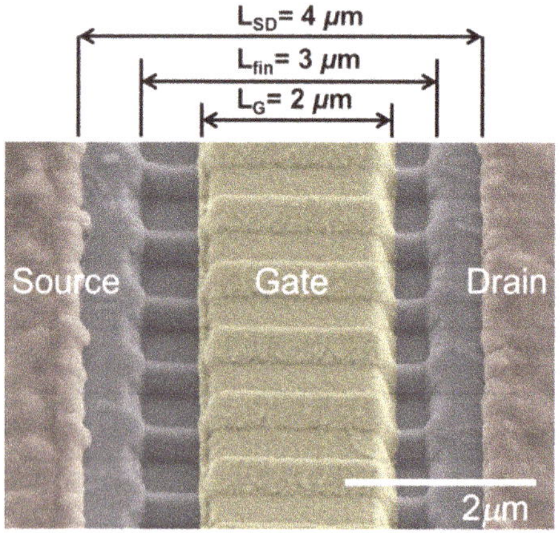

Fig. 32.19 Tilted scanning electron micrograph (false-colored) of fin-array E-mode Ga_2O_3 MOSFETs. Reprinted from [99], with the permission of AIP Publishing

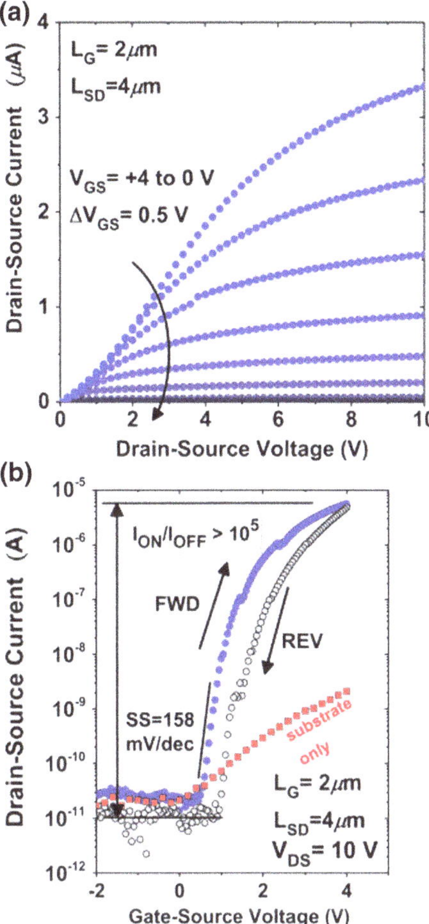

Fig. 32.20 DC characteristics of E-mode Ga_2O_3 finFETs: **a** I_D–V_D and **b** I_D–V_G. Reprinted from [99], with the permission of AIP Publishing

Due to a lack of proper sidewall surface treatment, plasma etch damage of the fin sidewalls results in trapping effects and a relatively low electron mobility of 24 cm^2/Vs, in spite of which an acceptable subthreshold slope of 158 mV/dec is obtained. At V_G = 0 V, a V_{br} of 612 V is achieved in the absence of a field plate. This performance represents the highest breakdown for any transistor technology utilizing non-planar device channels.

32.6 Current Aperture Vertical Ga_2O_3 MOSFET

The Ga_2O_3 transistors reported to date have predominantly taken on a lateral geometry. However, lateral devices are not amenable to the large currents and voltages required for many applications owing to their sizeable layout areas and potential reliability concerns arising from self-heating and surface instabilities. For applications demanding high voltage ratings and high power levels, vertical switching devices are desirable since they allow for higher current drives, simplified thermal management, and far superior field termination.

Two types of vertical Ga_2O_3 MOSFETs have been proposed. Owing to a lack of shallow acceptor dopants for Ga_2O_3 [100–102], a number of early devices have pursued a non-planar fin architecture featuring side-gate modulation to circumvent the need for *pn* junction isolation between source and drain [103–106]. Normally-off operation with an I_D exceeding 1 kA/cm^2, an I_{on}/I_{off} of 10^9, and a V_{br} of 1.6 kV has been demonstrated in these transistors [104–106]. The other implementation—known as the current aperture vertical Ga_2O_3 MOSFET [107, 108]—is inspired by the commercially successful SiC double-implanted MOSFET [109, 110] and modeled on the evolving GaN current aperture vertical electron transistor [111–113], both of which have been conceived from the Si vertical double-diffused MOSFET [114]. The planar-gate topology of the current aperture approach affords an unetched gating surface and relative ease of integration compared to the fin architecture. Among the family of ultrawide-bandgap semiconductors, Ga_2O_3 is distinctively amenable to doping with both shallow donors (Si) and deep acceptors (Mg, N) by ion implantation with efficient dopant activation at a low thermal budget [27, 115], thereby enabling the fabrication of current aperture vertical Ga_2O_3 MOSFETs based on a manufacturable all-ion-implanted process resembling that for Si and SiC MOSFETs.

Figure 32.21 illustrates the generic structure and operating principles of a D-mode current aperture vertical Ga_2O_3 MOSFET. In this implementation, the top source contact is electrically isolated from the bottom drain contact by a buried electron barrier called a current blocking layer (CBL), except at an aperture opening (L_{ap}) bounded by CBLs through which I_D is conducted. The CBL also serves as a back-barrier for a top-gated lateral channel, where the gate–CBL overlap dimension (L_{go}) defines the gate length. Initial devices have adopted a Mg-ion-implanted CBL, but they suffer a large source–drain leakage due to rapid thermal diffusion and a concomitant loss of Mg from the CBL upon implant activation annealing [107]. In contrast to Mg, N exhibits much lower thermal diffusivity in Ga_2O_3, thus enabling the use of higher annealing temperatures to maximize N activation efficiency without drastically affecting the dopant profile [115, 116]. A D-mode current aperture vertical MOSFET is subsequently accomplished through the integration of N-ion (N^{++}) and Si^+ implantation doping [108].

The current aperture vertical Ga_2O_3 MOSFET with N^{++}-implanted CBLs consists of a 5-µm-thick Si-doped (2.5×10^{16} cm^{-3}) n^--Ga_2O_3 drift layer grown by halide vapor phase epitaxy (HVPE) on a Sn-doped n^+-Ga_2O_3 (001) substrate.

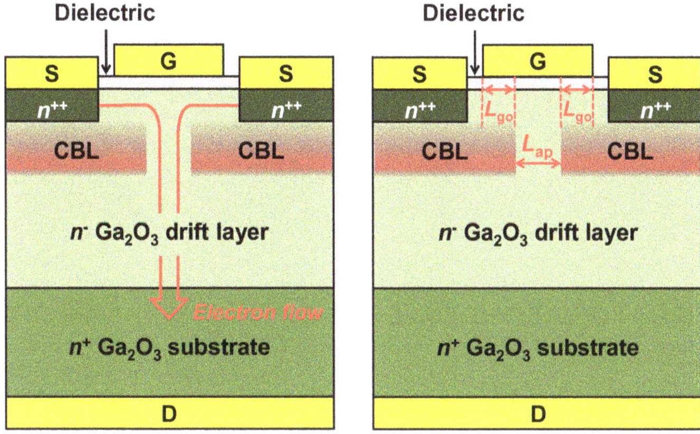

Fig. 32.21 Structure and operation of a generic D-mode current aperture vertical Ga_2O_3 MOSFET [107]. Copyright 2018, the Japan Society of Applied Physics

HVPE is the epitaxial technique of choice since its capabilities of growing single-crystal Ga_2O_3 films with a low background electron density ($<10^{13}$ cm^{-3}) and at high speeds (>10 μm/h) are desirable for obtaining thick, lightly doped drift layers required for high V_{br} [117–119]. Three ion implantation steps are employed to form the Si-doped n^{++} source contacts, Si-doped n-type channel, and N-doped CBLs. Figure 32.22 shows a cross-sectional schematic of a device with a 150-nm-thick channel, a 50-nm-thick Al_2O_3 gate dielectric, an L_{ap} of 20 μm, and an L_{go} of 2.5 μm. This MOSFET delivers a maximum I_D of 0.42 kA/cm^2 (normalized to the Si channel implant area) and an $R_{on,sp}$ of 31.5 mΩ cm^2 (Fig. 32.23a); the corresponding values are 1.37 kA/cm^2 and 9.7 mΩ cm^2 when normalized to the aperture area. A high I_{on}/I_{off} of over 10^8 is achieved (Fig. 32.23b), where the I_{off} is dominated by I_G. The V_{br} is limited by Al_2O_3 breakdown to less than 30 V owing to the high Si doping over the aperture. Under 5-μs gate-pulsed conditions with a duty

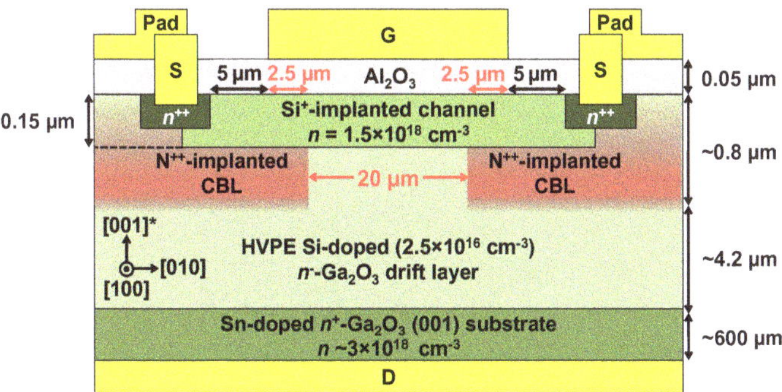

Fig. 32.22 Cross-sectional schematic of D-mode current aperture vertical Ga_2O_3 MOSFET with L_{go} = 2.5 μm and L_{ap} = 20 μm

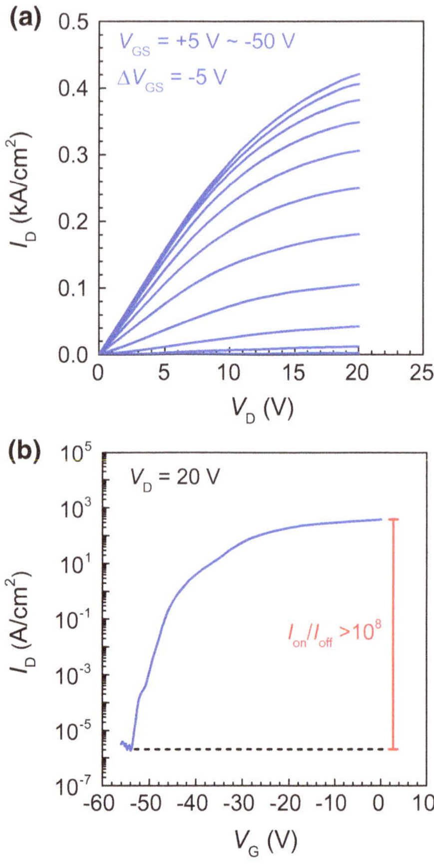

Fig. 32.23 DC characteristics of vertical Ga_2O_3 MOSFET: **a** I_D–V_D and **b** I_D–V_G. © 2019 IEEE. Reprinted, with permission, from [108]

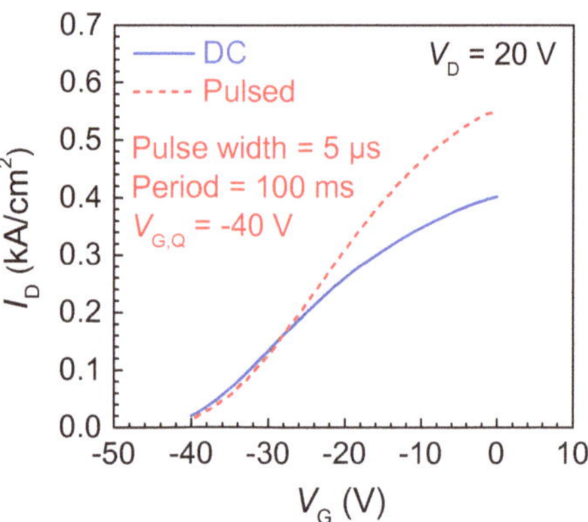

Fig. 32.24 DC and pulsed transfer characteristics of vertical Ga_2O_3 MOSFET. © 2019 IEEE. Reprinted, with permission, from [108]

cycle of 0.005% at a constant drain voltage (V_D) of 20 V, the MOSFET demonstrates potential for high-speed switching operation by virtue of a high I_D when pulsed on, yet a small positive V_T shift attributable to bulk or interface electron trapping under the gate is observed (Fig. 32.24). It is envisaged that performance limitations can be readily overcome with improved gate dielectric quality and optimized doping schemes.

32.7 Summary and Conclusions

The past few decades have witnessed technological innovations driven by physical electronics solutions embodying novel materials and innovative devices that fundamentally change our lives today. Ga_2O_3 is emerging as the next major wide-bandgap semiconductor to engender paradigm shifts in power conversion and harsh-environment applications by offering higher efficiency and enhanced functionality at an affordable cost. In a matter of years, Ga_2O_3 transistors have taken enormous strides including demonstrations of kV breakdown, normally-off operation, and vertical device concepts, all of which serve to position Ga_2O_3 as a relevant contender for practical applications. With the full effort of the community to establish a more complete understanding of the materials and devices, a continued steep growth trajectory of Ga_2O_3 will generate far-reaching technological and socioeconomic impacts.

Acknowledgements This work was partially supported by Council for Science, Technology and Innovation (CSTI), Cross-ministerial Strategic Innovation Promotion Program (SIP), "Next-generation power electronics" (funding agency: New Energy and Industrial Technology Development Organization).

References

1. M. Higashiwaki, G.H. Jessen, Appl. Phys. Lett. **112**, 060401 (2018)
2. H.H. Tippins, Phys. Rev. **140**, A316 (1965)
3. M. Orita, H. Ohta, M. Hirano, H. Hosono, Appl. Phys. Lett. **77**, 4166 (2000)
4. T. Onuma, S. Saito, K. Sasaki, T. Masui, T. Yamaguchi, T. Honda, M. Higashiwaki, Jpn. J. Appl. Phys. **54**, 112601 (2015)
5. A.J. Green, K.D. Chabak, E.R. Heller, R.C. Fitch, Jr., M. Baldini, A. Fiedler, K. Irmscher, G.Wagner, Z. Galazka, S.E. Tetlak, A. Crespo, K. Leedy, G.H. Jessen, IEEE Electron Device Lett. **37**, 902 (2016)
6. K. Konishi, K. Goto, H. Murakami, Y. Kumagai, A. Kuramata, S. Yamakoshi, M. Higashiwaki, Appl. Phys. Lett. **110**, 103506 (2017)
7. C. Joishi, S. Rafique, Z. Xia, L. Han, S. Krishnamoorthy, Y. Zhang, S. Lodha, H. Zhao, S. Rajan, Appl. Phys. Express **11**, 031101 (2018)
8. X. Yan, I.S. Esqueda, J. Ma, J. Tice, H. Wang, Appl. Phys. Lett. **112**, 032101 (2018)

9. W. Li, Z. Hu, K. Nomoto, R. Jinno, Z. Zhang, Q.T. Thieu, K. Sasaki, A. Kuramata, D. Jena, H.G. Xing, in *Proceedings of IEEE International Electron Devices Meeting*, 1–5 Dec 2018, San Francisco, CA, USA, pp. 8.5.1–8.5.4
10. J. Kim, S.J. Pearton, C. Fares, J. Yang, F. Ren, S. Kim, A.Y. Polyakov, J. Mater. Chem. C **7**, 10 (2019)
11. D. Szalkai, Z. Galazka, K. Irmscher, P. Tüttő, A. Klix, D. Gehre, IEEE Trans. Nucl. Sci. **64**, 1574 (2017)
12. X. Lu, L. Zhou, L. Chen, X. Ouyang, B. Liu, J. Xu, H. Tang, Appl. Phys. Lett. **112**, 103502 (2018)
13. N. Ueda, H. Hosono, R. Waseda, H. Kawazoe, Appl. Phys. Lett. **70**, 3561 (1997)
14. E. G. Víllora, K. Shimamura, Y. Yoshikawa, K. Aoki, N. Ichinose, J. Cryst. Growth **270**, 420 (2004)
15. N. Suzuki, S. Ohira, M. Tanaka, T. Sugawara, K. Nakajima, T. Shishido, Phys. Status Solidi C **4**, 2310 (2007)
16. Y. Tomm, P. Reiche, D. Klimm, and T. Fukuda, J. Cryst. Growth **220**, 510 (2000)
17. Z. Galazka, R. Uecker, K. Irmscher, M. Albrecht, D. Klimm, M. Pietsch, M. Brützam, R. Bertram, S. Ganschow, R. Fornari, Cryst. Res. Technol. **45**, 1229 (2010)
18. Z. Galazka, K. Irmscher, R. Uecker, R. Bertram, M. Pietsch, A. Kwasniewski, M. Naumann, T. Schulz, R. Schewski, D. Klimm, M. Bickermann, J. Cryst. Growth **404**, 184 (2014)
19. Z. Galazka, R. Uecker, D. Klimm, K. Irmscher, M. Naumann, M. Pietsch, A. Kwasniewski, R. Bertram, S. Ganschow, M. Bickermann, ECS J. Solid State Sci. Technol. **6**, Q3007 (2017)
20. H. Aida, K. Nishiguchi, H. Takeda, N. Aota, K. Sunakawa, Y. Yaguchi, Jpn. J. Appl. Phys. **47**, 8506 (2008)
21. A. Kuramata, K. Koshi, S. Watanabe, Y. Yamaoka, T. Masui, S. Yamakoshi, Jpn. J. Appl. Phys. **55**, 1202A2 (2016)
22. K. Hoshikawa, E. Ohba, T. Kobayashi, J. Yanagisawa, C. Miyagawa, Y. Nakamura, J. Cryst. Growth **447**, 36 (2016)
23. M. Higashiwaki, K. Sasaki, A. Kuramata, T. Masui, S. Yamakoshi, Appl. Phys. Lett. **100**, 013504 (2012)
24. K. Sasaki, A. Kuramata, T. Masui, E. G. Víllora, K. Shimamura, S. Yamakoshi, Appl. Phys. Express **5**, 035502 (2012)
25. K. Sasaki, M. Higashiwaki, A. Kuramata, T. Masui, S. Yamakoshi, J. Cryst. Growth **392**, 30 (2014)
26. M. Higashiwaki, K. Sasaki, T. Kamimura, M.H. Wong, D. Krishnamurthy, A. Kuramata, T. Masui, S. Yamakoshi, Appl. Phys. Lett. **103**, 123511 (2013)
27. K. Sasaki, M. Higashiwaki, A. Kuramata, T. Masui, S. Yamakoshi, Appl. Phys. Express **6**, 086502 (2013)
28. M. Higashiwaki, K. Sasaki, M.H. Wong, T. Kamimura, D. Krishnamurthy, A. Kuramata, T. Masui, S. Yamakoshi, in *Proceedings of IEEE International Electron Devices Meeting*, 9–11 Dec 2013, Washington, DC, USA, pp. 28.7.1–28.7.4
29. M.H. Wong, K. Sasaki, A. Kuramata, S. Yamakoshi, M. Higashiwaki, IEEE Electron Device Lett. **37**, 212 (2016)
30. M.H. Wong, K. Sasaki, A. Kuramata, S. Yamakoshi, M. Higashiwaki, Appl. Phys. Lett. **106**, 032105 (2015)
31. M.H. Wong, K. Sasaki, A. Kuramata, S. Yamakoshi, M. Higashiwaki, Jpn. J. Appl. Phys. **55**, 1202B9 (2016)
32. J.F. McGlone, Z. Xia, Y. Zhang, C. Joishi, S. Lodha, S. Rajan, S.A. Ringel, A. R. Arehart, IEEE Electron Device Lett. **39**, 1042 (2018)
33. C. Joishi, Z. Xia, J. McGlone, Y. Zhang, A.R. Arehart, S. Ringel, S. Lodha, S. Rajan, Appl. Phys. Lett. **113**, 123501 (2018)
34. M. Singh, M.A. Casbon, M.J. Uren, J.W. Pomeroy, S. Dalcanale, S. Karboyan, P.J. Tasker, M.H. Wong, K. Sasaki, A. Kuramata, S. Yamakoshi, M. Higashiwaki, M. Kuball, IEEE Electron Device Lett. **39**, 1572 (2018)

35. T. Kamimura, K. Sasaki, M.H. Wong, D. Krishnamurthy, A. Kuramata, T. Masui, S. Yamakoshi, M. Higashiwaki, Appl. Phys. Lett. **104**, 192104 (2014)
36. K. Konishi, T. Kamimura, M.H. Wong, K. Sasaki, A. Kuramata, S. Yamakoshi, M. Higashiwaki, Phys. Status Solidi B **253**, 623 (2016)
37. K. Zeng, A. Vaidya, U. Singisetti, IEEE Electron Device Lett. **39**, 1385 (2018)
38. K. Zeng, J.S. Wallace, C. Heimburger, K. Sasaki, A. Kuramata, T. Masui, J.A. Gardella, Jr., U. Singisetti, IEEE Electron Device Lett. **38**, 513 (2017)
39. Y. Lv, X. Zhou, S. Long, X. Song, Y. Wang, S. Liang, Z. He, T. Han, X. Tan, Z. Feng, H. Dong, X. Zhou, Y. Yu, S. Cai, M. Liu, IEEE Electron Device Lett. **40**, 83 (2019)
40. J.K. Mun, K. Cho, W. Chang, H.-W. Jung, and J. Do, ECS J. Solid State Sci. Technol. **8**, Q3079 (2019)
41. E.G. Víllora, K. Shimamura, Y. Yoshikawa, T. Ujiie, K. Aoki, Appl. Phys. Lett. **92**, 202120 (2008)
42. J.B. Varley, J.R. Weber, A. Janotti, C.G. Van de Walle, Appl. Phys. Lett. **97**, 142106 (2010)
43. N. Moser, J. McCandless, A. Crespo, K. Leedy, A. Green, A. Neal, S. Mou, E. Ahmadi, J. Speck, K. Chabak, N. Peixoto, G. Jessen, IEEE Electron Device Lett. **38**, 775 (2017)
44. E.G. Víllora, K. Shimamura, T. Ujiie, K. Aoki, Appl. Phys. Lett. **92**, 202118 (2008)
45. M. Handwerg, R. Mitdank, Z. Galazka, S. F. Fischer, Semicond. Sci. Technol. **30**, 024006 (2015)
46. Z. Guo, A. Verma, X. Wu, F. Sun, A. Hickman, T. Masui, A. Kuramata, M. Higashiwaki, D. Jena, T. Luo, Appl. Phys. Lett. **106**, 111909 (2015)
47. M.D. Santia, N. Tandon, J.D. Albrecht, Appl. Phys. Lett. **107**, 041907 (2015)
48. M.H. Wong, Y. Morikawa, K. Sasaki, A. Kuramata, S. Yamakoshi, M. Higashiwaki, Appl. Phys. Lett. **109**, 193503 (2016)
49. J.W. Pomeroy, C. Middleton, M. Singh, S. Dalcanale, M.J. Uren, M.H. Wong, K. Sasaki, A. Kuramata, S. Yamakoshi, M. Higashiwaki, M. Kuball, IEEE Electron Device Lett. **40**, 189 (2019)
50. S. Martin-Horcajo, A. Wang, M.-F. Romero, M.J. Tadjer, F. Calle, IEEE Trans. Electron Devices **60**, 4105 (2013)
51. J. Pomeroy, M. Bernardoni, A. Sarua, A. Manoi, D.C. Dumka, D.M. Fanning, M. Kuball, in *Proceedings of IEEE Compound Semiconductor Integrated Circuit Symposium*, 13–16 Oct 2013, Monterey, CA, USA, pp. 1–4
52. H. Sun, J.W. Pomeroy, R.B. Simon, D. Francis, F. Faili, D.J. Twitchen, M. Kuball, IEEE Electron Device Lett. **37**, 621 (2016)
53. B. Chatterjee, A. Jayawardena, E. Heller, D.W. Snyder, S. Dhar, S. Choi, Rev. Sci. Instrum. **89**, 114903 (2018)
54. N.A. Moser, J.P. McCandless, A. Crespo, K.D. Leedy, A.J. Green, E.R. Heller, K.D. Chabak, N. Peixoto, G.H. Jessen, Appl. Phys. Lett. **110**, 143505 (2017)
55. C.-H. Lin, N. Hatta, K. Konishi, S. Watanabe, A. Kuramata, K. Yagi, M. Higashiwaki, Appl. Phys. Lett. **114**, 032103 (2019)
56. A.L. Barry, B. Lehmann, D. Fritsch, D. Bräunig, IEEE Trans. Nucl. Sci. **38**, 1111 (1991)
57. D.C. Look, D.C. Reynolds, J.W. Hemsky, J.R. Sizelove, R.L. Jones, R.J. Molnar, Phys. Rev. Lett. **79**, 2273 (1997)
58. A. Ionascut-Nedelcescu, C. Carlone, A. Houdayer, H.J. von Bardeleben, J.-L. Cantin, S. Raymond, IEEE Trans. Nucl. Sci. **49**, 2733 (2002)
59. J.C. Bourgoin, B. Massarani, Phys. Rev. B **14**, 3690 (1976)
60. J. Koike, D.M. Parkin, T.E. Mitchell, Appl. Phys. Lett. **60**, 1450 (1992)
61. S. Ahn, Y.-H. Lin, F. Ren, S. Oh, Y. Jung, G. Yang, J. Kim, M.A. Mastro, J.K. Hite, C.R. Eddy, Jr., S.J. Pearton, J. Vac. Sci. Technol. B **34**, 041213 (2016)
62. J. Yang, F. Ren, S.J. Pearton, G. Yang, J. Kim, A. Kuramata, J. Vac. Sci. Technol. B **35**, 031208 (2017)
63. J. Yang, Z. Chen, F. Ren, S.J. Pearton, G. Yang, J. Kim, J. Lee, E. Flitsiyan, L. Chernyak, A. Kuramata, J. Vac. Sci. Technol. B **36**, 011206 (2018)

64. J. Yang, C. Fares, Y. Guan, F. Ren, S.J. Pearton, J. Bae, J. Kim, A. Kuramata, J. Vac. Sci. Technol. B **36**, 031205 (2018)
65. G. Yang, S. Jang, F. Ren, S.J. Pearton, J. Kim, ACS Appl. Mater. Interfaces **9**, 40471 (2017)
66. A.Y. Polyakov, N.B. Smirnov, I.V. Shchemerov, E.B. Yakimov, J. Yang, F. Ren, G. Yang, J. Kim, A. Kuramata, S.J. Pearton, Appl. Phys. Lett. **112**, 032107 (2018)
67. M.E. Ingebrigtsen, J.B. Varley, A.Y. Kuznetsov, B.G. Svensson, G. Alfieri, A. Mihaila, U. Badstübner, L. Vines, Appl. Phys. Lett. **112**, 042104 (2018)
68. J. Lee, E. Flitsiyan, L. Chernyak, J. Yang, F. Ren, S.J. Pearton, B. Meyler, Y. Joseph Salzman, Appl. Phys. Lett. **112**, 082104 (2018)
69. M.F. Chaiken, T.E. Blue, IEEE Trans. Nucl. Sci. **65**, 1147 (2018)
70. H. Gao, S. Muralidharan, N. Pronin, M.R. Karim, S.M. White, T. Asel, G. Foster, S. Krishnamoorthy, S. Rajan, L.R. Cao, M. Higashiwaki, H. von Wenckstern, M. Grundmann, H. Zhao, D.C. Look, L.J. Brillson, Appl. Phys. Lett. **112**, 242102 (2018)
71. A.Y. Polyakov, N.B. Smirnov, I.V. Shchemerov, E.B. Yakimov, S.J. Pearton, C. Fares, J. Yang, F. Ren, J. Kim, P.B. Lagov, V.S. Stolbunov, A. Kochkova, Appl. Phys. Lett. **113**, 092102 (2018)
72. A.Y. Polyakov, N.B. Smirnov, I.V. Shchemerov, S.J. Pearton, F. Ren, A.V. Chernykh, P.B. Lagov, T.V. Kulevoy, APL Mater. **6**, 096102 (2018)
73. E. Farzana, M.F. Chaiken, T.E. Blue, A.R. Arehart, S.A. Ringel, APL Mater. **7**, 022502 (2019)
74. M.E. Ingebrigtsen, A.Y. Kuznetsov, B.G. Svensson, G. Alfieri, A. Mihaila, U. Badstübner, A. Perron, L. Vines, J.B. Varley, APL Mater. **7**, 022510 (2019)
75. H.J. von Bardeleben, S. Zhou, U. Gerstmann, D. Skachkov, W.R.L. Lambrecht, Q.D. Ho, P. Deák, APL Mater. **7**, 022521 (2019)
76. M.H. Wong, A. Takeyama, T. Makino, T. Ohshima, K. Sasaki, A. Kuramata, S. Yamakoshi, M. Higashiwaki, Appl. Phys. Lett. **112**, 023503 (2018)
77. Y. Tanaka, S. Onoda, A. Takatsuka, T. Ohshima, T. Yatsuo, Mat. Sci. Forum **645–648**, 941 (2010)
78. M.A. Bhuiyan, H. Zhou, R. Jiang, E.X. Zhang, D.M. Fleetwood, P.D. Ye, T.-P. Ma, IEEE Electron Device Lett. **39**, 1022 (2018)
79. J. Yang, G.J. Koller, C. Fares, F. Ren, S.J. Pearton, J. Bae, J. Kim, D.J. Smith, ECS J. Solid State Sci. Technol. **8**, Q3041 (2019)
80. B.R. Tak, M. Garg, A. Kumar, V. Gupta, R. Singh, ECS J. Solid State Sci. Technol. **8**, Q3149 (2019)
81. T. Ohshima, M. Yoshikawa, H. Itoh, Y. Aoki, I. Nashiyama, Jpn. J. Appl. Phys., Part 2 **37**, L1002 (1998)
82. T. Ohshima, H. Itoh, M. Yoshikawa, J. Appl. Phys. **90**, 3038 (2001)
83. D.C. Sheridan, G. Chung, S. Clark, J.D. Cressler, IEEE Trans. Nucl. Sci. **48**, 2229 (2001)
84. T. Chen, Z. Luo, J.D. Cressler, T.F. Isaacs-Smith, J.R. Williams, G. Chung, S.D. Clark, Solid-State Electron. **46**, 2231 (2002)
85. K.K. Lee, T. Ohshima, H. Itoh, IEEE Trans. Nucl. Sci. **50**, 194 (2003)
86. M. Nawaz, C. Zaring, S. Onoda, T. Ohshima, M. Östling, in *Proceedings of the 67th Device Research Conference*, 22–24 June 2009, University Park, PA, USA, pp. 279–280
87. A. Akturk, J. M. McGarrity, S. Potbhare, and N. Goldsman, IEEE Trans. Nucl. Sci. **59**, 3258 (2012)
88. S.J. Pearton, F. Ren, E. Patrick, M.E. Law, A.Y. Polyakov, ECS J. Solid State Sci. Technol. **5**, Q35 (2016)
89. S. Mitomo, T. Matsuda, K. Murata, T. Yokoseki, T. Makino, A. Takeyama, S. Onoda, T. Ohshima, S. Okubo, Y. Tanaka, M. Kandori, T. Yoshie, Y. Hijikata, Phys. Status Solidi A **214**, 1600425 (2017)
90. G.A. Umana-Membreno, J.M. Dell, G. Parish, B.D. Nener, L. Faraone, U.K. Mishra, IEEE Trans. Electron Devices **50**, 2326 (2003)
91. J. Kim, F. Ren, G.Y. Chung, M.F. MacMillan, A.G. Baca, R.D. Briggs, D. Schoenfeld, S.J. Pearton, Appl. Phys. Lett. **84**, 371 (2004)

92. S. Onoda, N. Iwamoto, S. Ono, S. Katakami, M. Arai, K. Kawano, T. Ohshima, IEEE Trans. Nucl. Sci. **56**, 3218 (2009)
93. K. Zeng, K. Sasaki, A. Kuramata, T. Masui, U. Singisetti, in *Proceedings of the 74th Device Research Conference*, 19–22 June 2016, Newark, DE, USA, pp. 1–2
94. M.H. Wong, Y. Nakata, A. Kuramata, S. Yamakoshi, M. Higashiwaki, Appl. Phys. Express **10**, 041101 (2017)
95. T. Kamimura, Y. Nakata, M.H. Wong, M. Higashiwaki, IEEE Electron Device Lett. **40**, 1064 (2019)
96. K.D. Chabak, J.P. McCandless, N.A. Moser, A.J. Green, K. Mahalingam, A. Crespo, N. Hendricks, B.M. Howe, S.E. Tetlak, K. Leedy, R. C. Fitch, D. Wakimoto, K. Sasaki, A. Kuramata, G.H. Jessen, IEEE Electron Device Lett. **39**, 67 (2018)
97. H. Zhou, M. Si, S. Alghamdi, G. Qiu, L. Yang, P.D. Ye, IEEE Electron Device Lett. **38**, 103 (2017)
98. H. Zhou, K. Maize, G. Qiu, A. Shakouri, P.D. Ye, Appl. Phys. Lett. **111**, 092102 (2017)
99. K.D. Chabak, N. Moser, A.J. Green, D.E. Walker, Jr., S.E. Tetlak, E. Heller, A. Crespo, R. Fitch, J.P. McCandless, K. Leedy, M. Baldini, G. Wagner, Z. Galazka, X. Li, G. Jessen, Appl. Phys. Lett. **109**, 213501 (2016)
100. A. Kyrtsos, M. Matsubara, E. Bellotti, Appl. Phys. Lett. **112**, 032108 (2018)
101. J.L. Lyons, Semicond. Sci. Technol. **33**, 05LT02 (2018)
102. T. Gake, Y. Kumagai, F. Oba, Phys. Rev. Mater. **3**, 044603 (2019)
103. K. Sasaki, Q.T. Thieu, D. Wakimoto, Y. Koishikawa, A. Kuramata, S. Yamakoshi, Appl. Phys. Express **10**, 124201 (2017)
104. Z. Hu, K. Nomoto, W. Li, N. Tanen, K. Sasaki, A. Kuramata, T. Nakamura, D. Jena, H.G. Xing, IEEE Electron Device Lett. **39**, 869 (2018)
105. Z. Hu, K. Nomoto, W. Li, Z. Zhang, N. Tanen, Q.T. Thieu, K. Sasaki, A. Kuramata, T. Nakamura, D. Jena, H.G. Xing, Appl. Phys. Lett. **113**, 122103 (2018)
106. Z. Hu, K. Nomoto, W. Li, R. Jinno, T. Nakamura, D. Jena, H. G. Xing, in *Proceedings of the 31st International Symposium on Power Semiconductor Devices and ICs*, 19–23 May 2019, Shanghai, China, pp. 483–486
107. M.H. Wong, K. Goto, Y. Morikawa, A. Kuramata, S. Yamakoshi, H. Murakami, Y. Kumagai, M. Higashiwaki, Appl. Phys. Express **11**, 064102 (2018)
108. M.H. Wong, K. Goto, H. Murakami, Y. Kumagai, M. Higashiwaki, IEEE Electron Device Lett. **40**, 431 (2019)
109. J.N. Shenoy, J.A. Cooper, Jr., M.R. Melloch, IEEE Electron Device Lett. **18**, 93 (1997)
110. T. Kimoto, Jpn. J. Appl. Phys. **54**, 040103 (2015)
111. M. Kanechika, M. Sugimoto, N. Soejima, H. Ueda, O. Ishiguro, M. Kodama, E. Hayashi, K. Itoh, T. Uesugi, T. Kachi, Jpn. J. Appl. Phys., Part 2 **46**, L503 (2007)
112. S. Chowdhury, B.L. Swenson, M.H. Wong, and U.K. Mishra, Semicond. Sci. Technol. **28**, 074014 (2013)
113. H. Nie, Q. Diduck, B. Alvarez, A.P. Edwards, B.M. Kayes, M. Zhang, G. Ye, T. Prunty, D. Bour, I.C. Kizilyalli, IEEE Electron Device Lett. **35**, 939 (2014)
114. B.J. Baliga, in *Fundamentals of Power Semiconductor Devices* (Springer, New York, 2008), p. 284
115. M.H. Wong, C.-H. Lin, A. Kuramata, S. Yamakoshi, H. Murakami, Y. Kumagai, M. Higashiwaki, Appl. Phys. Lett. **113**, 102103 (2018)
116. H. Peelaers, J.L. Lyons, J.B. Varley, C.G. Van de Walle, APL Mater. **7**, 022519 (2019)
117. K. Nomura, K. Goto, R. Togashi, H. Murakami, Y. Kumagai, A. Kuramata, S. Yamakoshi, A. Koukitu, J. Cryst. Growth **405**, 19 (2014)
118. H. Murakami, K. Nomura, K. Goto, K. Sasaki, K. Kawara, Q.T. Thieu, R. Togashi, Y. Kumagai, M. Higashiwaki, A. Kuramata, S. Yamakoshi, B. Monemar, A. Koukitu, Appl. Phys. Express **8**, 015503 (2015)
119. K. Goto, K. Konishi, H. Murakami, Y. Kumagai, B. Monemar, M. Higashiwaki, A. Kuramata, S. Yamakoshi, Thin Solid Films **666**, 182 (2018)

Chapter 33
Field-Effect Transistors 3

β-$(Al_xGa_{1-x})_2O_3/Ga_2O_3$ Modulation-Doped Field-Effect Transistors

Yuewei Zhang, Sriram Krishnamoorthy and Siddharth Rajan

Abstract The availability of monoclinic aluminum gallium oxide (β-$(Al_xGa_{1-x})_2O_3$) alloys and their staggered band alignment with β-Ga_2O_3 makes it possible to realize two-dimensional electron gas (2DEG) through modulation doping. This brings unique advantages in improving the electrical transport properties in β-Ga_2O_3, and at the same time, it provides a great platform for the evaluation of a range of the fundamental materials and electrical properties of β-Ga_2O_3. In this chapter, we describe the early efforts in the design and epitaxy of the heterostructures and discuss the confirmation of a 2DEG through temperature-dependent Hall measurements as well as the first observation of the quantum oscillations in the material system. We will further discuss the realization of modulation-doped field-effect transistors (MODFETs), which have contributed to the evaluation of the saturation velocity and demonstration of high breakdown field in the material system. We will also discuss the methods to improve the 2DEG density and present the first double heterostructure MODFET (DH-MODFET). The early efforts on β-$(Al_xGa_{1-x})_2O_3/Ga_2O_3$ heterostructures confirmed that the modulation-doped structure could be a promising architecture for electronic device applications.

Y. Zhang (✉)
Materials Department, University of California, Santa Barbara, CA 93106, USA
e-mail: yueweizhang@ucsb.edu

S. Krishnamoorthy
Electrical and Computer Engineering, The University of Utah, Salt Lake City, UT 84112, USA
e-mail: sriram.krishnamoorthy@utah.edu

S. Rajan
Electrical and Computer Engineering, The Ohio State University, Columbus, OH 43210, USA
e-mail: rajan.21@osu.edu

M. Higashiwaki and S. Fujita (eds.), *Gallium Oxide*, Springer Series in Materials Science 293, https://doi.org/10.1007/978-3-030-37153-1_33

33.1 Introduction

Beta-phase gallium oxide (β-Ga_2O_3) has been recognized recently as a promising candidate for electronic device applications because of its wide bandgap energy (4.6 eV) [1] and the expected high breakdown field near 8 MV/cm [2]. This, together with good doping properties, ohmic contacts, and a relatively high estimated saturation velocity of $\sim 2 \times 10^7$ cm/s [3], leads to a high figure of merit for both power electronics and high-frequency electronics applications. Moreover, the availability of high-quality single-crystal material using melt-based techniques is beneficial for mass production of low-defect wafers [4–6], especially when compared to other wide bandgap semiconductors, such as SiC and GaN. Recent efforts have led to the demonstration of various device structures with promising device performance [7–14]. Experimental observations of high breakdown fields above 5 MV/cm have been reported for vertical Schottky diodes [15, 16], and near 4 MV/cm for lateral MOSFET transistors [17], which have already surpassed the material limit for GaN and SiC. Current modulation and the capability toward high-frequency operations were demonstrated as well. However, most of the existing research relied on bulk-doped β-Ga_2O_3 channels, leading to low transconductance (g_m). Besides, the on-resistance and the ability to operate under velocity saturation regime have also been limited by the low channel mobility below 150 cm^2/Vs at room temperature mainly due to polar optical phonon scattering. These issues greatly limit the applications of β-Ga_2O_3 in both power and high-frequency devices.

Improvements in the electron mobility are feasible in a modulation-doped two-dimensional electron gas (2DEG) channel by reducing the effect of ionized impurity scattering. Theoretical calculations predicted an enhanced screening effect for high 2DEG densities above 5×10^{12} cm^{-2}, and this could lead to an improved channel mobility above 500 cm^2/Vs at room temperature [18]. The higher channel mobility could bring significant research opportunities for β-Ga_2O_3 in a range of device applications. Additionally, the 2DEG channel provides the feasibility for charge scaling and aggressive device scaling in both lateral and vertical directions, making it especially favorable for high-frequency electronics and power-switching applications.

In this chapter, we will review the early development of $(AlGa)_2O_3/Ga_2O_3$ modulation-doped field-effect transistors (MODFETs) and discuss the design and experimental confirmation of the 2DEG channel at the heterointerface. The modulation-doped structures were further used to determine several electronic properties of β-Ga_2O_3, including low-field transport properties in a wide temperature range, quantum transport, electron saturation velocity in β-Ga_2O_3, and breakdown strength.

Monoclinic phase $(AlGa)_2O_3$ forms a staggered band alignment to β-Ga_2O_3, with a positive conduction band offset and small valence band offset [19]. When a thin sheet of dopants is placed in the $(AlGa)_2O_3$ layer, the free carriers could be transferred to the $(AlGa)_2O_3/Ga_2O_3$ interface [20], leading to the formation of a

2DEG. The equilibrium 2DEG density (n_s) in the system is determined by electrostatics and can be estimated by

$$n_s = \frac{eN_d^+ d_\delta - \varepsilon(\phi_b - \Delta E_c/e)}{eD}, \tag{33.1}$$

where N_d^+, ϕ_b, and ΔE_c represent the activated donor concentration in the delta-doped layer, the Schottky barrier height and the conduction band offset, respectively. e and ε are the unit charge and dielectric constant. The physical definitions for the thicknesses d_δ and D are schematically shown in Fig. 33.1a. For 2DEG densities above 10^{12} cm^{-2}, the contribution from the second term of (1) is negligible and the 2DEG density is mainly determined by the first term. Therefore, more charge could be transferred to the 2DEG channel when the spacer thickness is reduced. Increasing the intentional doping concentration could lead to a similar increase of the 2DEG density until the onset of a parasitic channel formed in the $(AlGa)_2O_3$ barrier layer. The low conductivity in the parasitic channel could compromise the overall conductivity of the modulation-doped structure and impact the device performance. Therefore, significant effort has been devoted to the optimization of the modulation-doped structures to avoid the formation of the parasitic channel, and this also sets an upper limit on the pure 2DEG density for a given conduction band offset.

Alloying of Al with β-Ga_2O_3 provides flexibility for heterostructure device designs. β-$(Al_xGa_{1-x})_2O_3$ has been demonstrated to be stable with Al composition up to 80%, even though Al_2O_3 is stable in the corundum polytype [21]. Epitaxial growth of β-$(Al_xGa_{1-x})_2O_3$ has been attempted by several growth techniques, including molecular beam epitaxy (MBE) [22–25], pulsed laser deposition (PLD) [26–29], mist chemical vapor deposition (CVD) [30] and metal-organic chemical vapor deposition (MOCVD) [31]. Because of the atomic layer precision, controlled n-type doping provided by MBE growth and the maturity in the growth

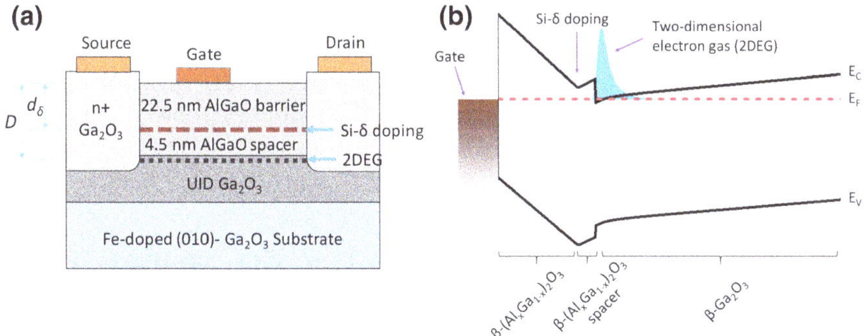

Fig. 33.1 **a** Schematic structure for the modulation-doped field-effect transistor. **b** Schematic energy band diagram of the MODFET. A 2DEG is formed at the $(AlGa)_2O_3/Ga_2O_3$ interface through modulation doping

technique, it was the first technique used for the realization of the MODFET structures [13, 14, 32].

The MBE growth condition of β-($Al_xGa_{1-x})_2O_3$ was first explored by Takayoshi Oshima et al. on (100)-oriented β-Ga_2O_3 substrates [22]. At a substrate temperature of 800 °C, an Al composition up to 61% was achieved. However, the films showed high surface roughness, and at the same time, the growth on (100)-oriented substrates suffered from significantly lower growth rate compared to the growth on (010)-oriented substrates since it is a cleavage plane and has higher desorption rate at the growth temperatures. β-($Al_xGa_{1-x})_2O_3$ growth on (010)-oriented substrates was investigated by Stephen Kaun et al. [23], who demonstrated smooth surface morphology and sharp heterointerfaces between β-($Al_xGa_{1-x})_2O_3$ and β-Ga_2O_3. However, the maximum Al composition achieved in the monoclinic phase was found to be less than 25%, which was further confirmed by other researchers. This limits the conduction band offset at the $(AlGa)_2O_3/Ga_2O_3$ interface below 0.4 eV [19]. As a result, the maximum 2DEG density that can be confined at the heterointerface before the onset of the parasitic channel in the $(AlGa)_2O_3$ barrier layer is estimated to be $\sim 2 \times 10^{12}$ cm^{-2} for the structure shown in Fig. 33.1a, which has a 4.5 nm spacer layer and 22.5 nm $(AlGa)_2O_3$ barrier above the delta-doped layer.

Preliminary demonstrations on β-($Al_xGa_{1-x})_2O_3/Ga_2O_3$ modulation-doped field-effect transistors have been reported by several groups using either Si delta doping [14, 32, 33] or Ge doping in the β-($Al_xGa_{1-x})_2O_3$ layer [13]. Carrier confinement was demonstrated based on capacitance-voltage measurements [13, 14, 34]. However, high sheet charge densities above 5×10^{12} cm^{-2} were measured for the channels, suggesting the existence of a parasitic channel in the $(AlGa)_2O_3$ barrier layer. This was further confirmed by the absence of an improvement of the channel mobility, which could have been reduced by parallel conduction through the low mobility channel in the $(Al_xGa_{1-x})_2O_3$ layer.

Development of the β-($Al_xGa_{1-x})_2O_3/Ga_2O_3$ modulation-doped structures by Y. Zhang et al. focused on the optimization of the intentional doping concentration in the $(AlGa)_2O_3$ barrier layer to avoid the formation of a parasitic channel [32–35]. The Si delta doping concentration was controlled by varying both the cell temperature and the shutter pulsing time during growth [32]. The impurity distribution within one monolayer of $(Al_xGa_{1-x})_2O_3$ was expected. Successes in the removal of the parasitic channel led to the demonstration of single 2DEG channels with densities below 2×10^{12} cm^{-2}. This allowed for the characterization of the intrinsic 2DEG transport properties in β-Ga_2O_3.

33.2 Electrical Transport in $(AlGa)_2O_3/Ga_2O_3$ MODFETs

In this section, we discuss the characterization of a 2DEG at the $(AlGa)_2O_3/Ga_2O_3$ interface through the measurement of the electrical transport characteristics of the structure. Similar to the other 2DEG systems in AlGaAs/GaAs MODFET or AlGaN/GaN HEMT structures, the 2DEG carrier density is expected to have weak temperature dependence because of the electrostatics. At the same time, the reduced impact from polar optical phonon scattering is expected to enable superior transport characteristics at cryogenic temperatures. Therefore, unique quantum transport properties of the 2DEG are anticipated in the $(AlGa)_2O_3/Ga_2O_3$ modulation-doped structure. In this section, we discuss direct evidences of quantum confinement of a 2DEG at the β-$(Al_xGa_{1-x})_2O_3/Ga_2O_3$ interface based on temperature-dependent Hall measurements and Shubnikov-de Haas (SdH) oscillations.

The modulation-doped β-$(Al_xGa_{1-x})_2O_3/Ga_2O_3$ structure is shown in Fig. 33.1. To achieve ohmic contact, a degenerately doped β-Ga_2O_3 layer was grown to enable sidewall contact to the electron channel [32, 36]. A metal stack of Ti/Au was then evaporated on the regrown Ga_2O_3 regions and annealed at 470 °C for 1 min to form ohmic contact. TLM measurement for the source/drain contacts indicated ohmic performance with contact resistivity of ~9 to ~1 Ω mm depending on the total charge density in the channel [32, 33, 36]. The degenerate doping in the β-Ga_2O_3 contact layer prevented carrier freeze-out and ensured ohmic contact at cryogenic temperatures.

The temperature dependence of Hall charge density and Hall mobility was measured using a Van der Pauw structure, as shown in Fig. 33.2. Here, the only difference between sample A and B is the UID Ga_2O_3 buffer layer thickness, which is 130 nm for sample A and 360 nm for sample B.Weak temperature dependence in the measured Hall charge density was observed for the modulation-doped structures. This is in contrast with the carrier freeze-out in bulk-doped β-Ga_2O_3 at low temperatures [37–39] and serves as a direct proof of a degenerate 2DEG at the β-$(Al_xGa_{1-x})_2O_3/Ga_2O_3$ interface. One challenge that needs to be addressed is the possible appearance of a parasitic channel in the $(Al_xGa_{1-x})_2O_3$ barrier layer because of the difficulty in the intentional doping control. It was experimentally observed that the 2DEG density that can be confined in the adopted $(AlGa)_2O_3/Ga_2O_3$ modulation-doped structure with Al composition of ~18% (Fig. 33.2) is

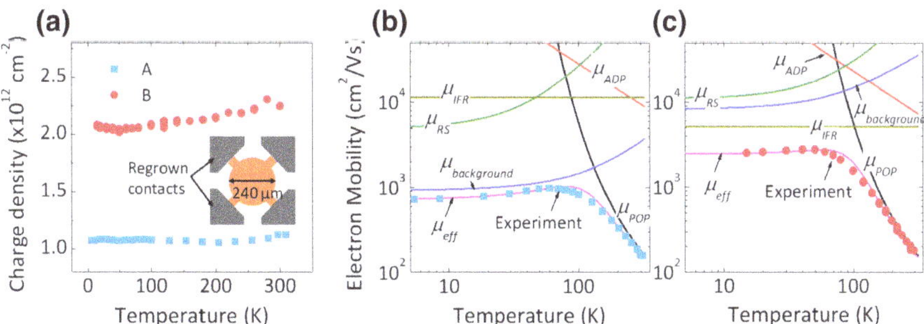

Fig. 33.2 **a** Temperature dependence of charge density measured using a Van der Pauw configuration as shown in the inset. (**b**, **c**) Experimental and calculated electron mobilities for sample A (**b**) and sample B (**c**) by considering various scattering mechanisms, including: polar optical phonon scattering (μ_{POP}), remote impurity scattering (μ_{RS}), background impurity scattering ($\mu_{background}$), interface roughness scattering (μ_{IFR}), and acoustic deformation potential scattering (μ_{ADP}). Reproduced from [17], with the permission of AIP Publishing [32]

approximately $\sim 2 \times 10^{12}$ cm^{-2} because of the small conduction band offset, which was predicted to be less than 0.4 eV based on first-principle calculations [19]. Further increase in the 2DEG charge density requires higher conduction band offset or using a relatively thinner spacer layer.

The room temperature mobility of 180 cm^2/Vs was measured in a modulation-doped structure. This is similar to the highest reported values in lightly doped bulk β-Ga_2O_3 films [39]. The channel mobility increased significantly upon lowering the measurement temperature, and a low-temperature mobility value of 2790 cm^2/Vs was obtained in the β-Ga_2O_3 material system. To understand the scattering process, the temperature dependence of the electron mobility was analyzed by considering various scattering mechanisms, including polar optical phonon scattering (μ_{POP}), remote impurity scattering (μ_{RS}), background impurity scattering ($\mu_{\mathrm{background}}$), acoustic deformation potential scattering (μ_{ADP}), interface roughness scattering (μ_{IFR}) and alloy scattering (μ_{Alloy}). The total effective mobility was obtained based on Matthiessen's rule: $1/\mu_{\mathrm{eff}} = 1/\mu_{\mathrm{RS}} + 1/\mu_{\mathrm{background}} + 1/\mu_{\mathrm{POP}} + 1/\mu_{\mathrm{ADP}} + 1/\mu_{\mathrm{IFR}} + 1/\mu_{\mathrm{Alloy}}$.

The electron scattering was found to be dominated by impurity scattering at low temperature, and by polar optical phonon scattering in the high temperature range. A significant improvement of the low-temperature mobility was observed when the UID β-Ga_2O_3 buffer layer thickness was increased from 130 to 360 nm. Mobility calculation suggested a corresponding reduction of the background impurity concentration through the growth of thicker buffer layer. At low temperatures, mobility is limited by a combination of interface roughness and background impurity scattering. Further optimization of the epitaxial material quality is expected to further increase the 2DEG mobility at low temperatures. Recently, the development of growth techniques, such as MOCVD [39] and HVPE [40], has led to high-quality epitaxial layers with low background charge concentrations below 10^{16} cm^{-3}, and therefore, low-temperature electron mobility above 5000 cm^2/Vs was demonstrated [40]. However, the difference in the background impurity concentration showed minimal impact on the room temperature mobility, which is intrinsically limited by the low optical phonon energy [41, 42]. Increasing the 2DEG density to take advantage of the enhanced screening effect provides a viable solution to overcome such mobility limits [18].

The high channel mobility at low temperature and the existence of a degenerate 2DEG allowed for the observation of quantum transport in the Ga_2O_3 material system. The dependence of the transverse magnetoresistance (R_{xx}) of the modulation-doped structure on the magnetic field perpendicular to the sample surface was measured, as shown in Fig. 33.3. Multiple periods of oscillations were observed for the transverse magnetoresistance (R_{xx}), and plateaus corresponding to Landau splitting were observed in the measured Hall resistance (R_{xy}). These results represent the first demonstration of Shubnikov-de Haas oscillations in the Ga_2O_3 material system [32].

The oscillation component of R_{xx} was extracted by subtracting the non-oscillating background and was plotted as a function of reciprocal magnetic field ($1/B$) as shown in Fig. 33.3. The oscillation period $\Delta(1/B)$ is determined by the

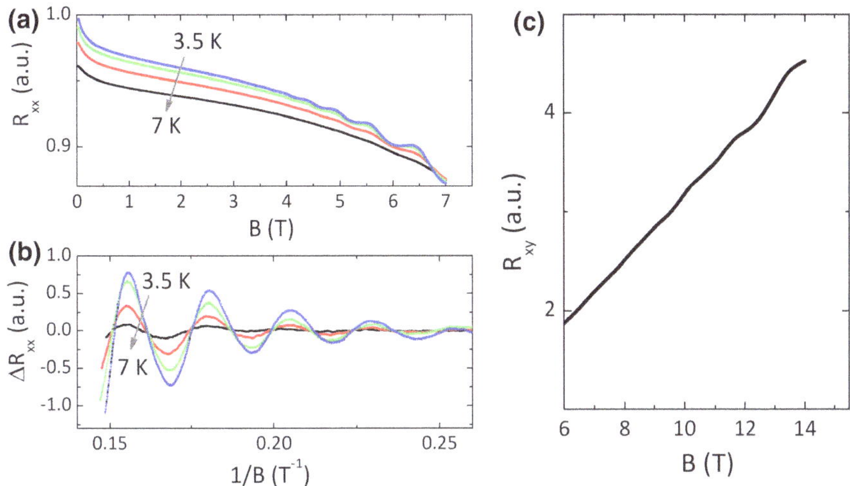

Fig. 33.3 **a** Temperature dependence of the transverse magnetoresistance of a $(AlGa)_2O_3/Ga_2O_3$ 2DEG channel as a function of the magnetic field perpendicular to the sample surface. **b** SdH oscillations of the transverse magnetoresistances. **c** The Hall resistances (R_{xy}) of the channel showing plateaus, which corresponds to the quantization of Landau levels. Reproduced from [17] with the permission of AIP Publishing [32]

2DEG carrier density through the expression: $\Delta(1/B) = e/\hbar n_{2D}$. The carrier density extracted from the SdH oscillations is 1.96×10^{12} cm^{-2}, which matches well with the low-field Hall measurements.

The first observation of quantum oscillations suggests that the β-$(Al_xGa_{1-x})_2O_3/Ga_2O_3$ modulation-doped structure could be a great platform for the studies of quantum transport in the material system. This could be used for the extraction of a range of fundamental material properties and investigate novel quantum phenomena in β-Ga_2O_3. The electron effective mass could be extracted based on the SdH oscillation amplitude as a function of the measurement temperature. The fitting to the experimental data suggested an effective mass of $m^* = 0.313 \pm 0.015\ m_0$, which is close to what was estimated from the theoretical calculations, band structure measurements and optical Hall measurements. At the same time, the quantum scattering time could be extracted to be 0.33 ps. This provides a reference for the analysis of the dominating scattering mechanism for low-field electron transport in β-Ga_2O_3 [32].

As discussed above, further increasing the 2DEG density in a single $(AlGa)_2O_3/Ga_2O_3$ heterostructure is challenging because of the limited conduction band offset that can be obtained in the present growth technologies. Two major solutions are possible to increase the 2DEG density. The first one is to develop novel growth conditions to increase Al incorporation. Higher growth temperature was found to be favorable for the stability of $(AlGa)_2O_3$ in the monoclinic phase. There are now significant research efforts toward higher temperature growth of $(AlGa)_2O_3$ using both MOCVD and MBE [25, 31]. A recent demonstration of an In-catalyzed

$(AlGa)_2O_3$ growth method pushed the MBE growth temperature to more than 200 °C higher than the typical growth temperatures, which provides a tremendous growth window to explore [25]. However, heterostructures with higher Al compositions have not been demonstrated. A second method to increase the 2DEG density is to introduce multiple heterostructures [33].

Y. Zhang et al. demonstrated experimentally a double heterostructure as shown in Fig. 33.4 [33]. Compared to the single heterostructure, the electrons are confined in a Ga_2O_3 quantum well layer sandwiched between two $(AlGa)_2O_3$ barrier layers. Since Si delta doping exists in both $(AlGa)_2O_3$ layers, electrons could be transferred from both below and above the β-Ga_2O_3 quantum well, resulting in a higher achievable 2DEG density in the quantum well. Using this structure, a confined 2DEG charge density of 3.85×10^{12} cm^{-2} was estimated from low-temperature Hall measurement, which doubled what was achievable in a single heterostructure. The experimental confirmation of a higher 2DEG density in the double heterostructure suggests that further charge scaling is feasible by increasing the number of quantum wells. This provides an elegant method to reach high charge densities required for high-current and high-power device operations.

The field effect mobility of the quantum well channel was evaluated using three-terminal measurement on a fat transistor with gate length/width of L_G/W = 100/100 μm. The field-effect mobility was found to peak at 150 cm^2/Vs in the β-Ga_2O_3 quantum well layer as shown in Fig. 33.5; however, it showed a significant reduction to 35 cm^2/Vs in the β-$(Al_xGa_{1-x})_2O_3$ layers, which could be limited by the strong ionized impurity scattering and alloy scattering in the parasitic channel in the $(AlGa)_2O_3$ layer. The field-effect mobility could be translated to Hall mobility by multiplying with the Hall factor, which was estimated to be ~1.7 by Ma et al. [41]. This leads to a Hall mobility of 255 cm^2/Vs, which is higher than the

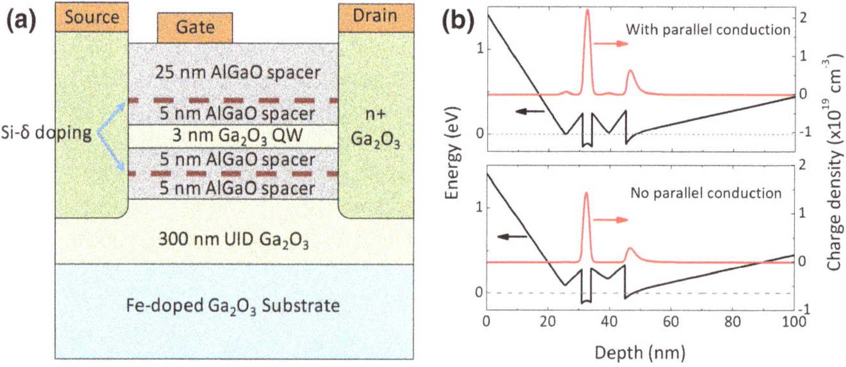

Fig. 33.4 **a** Schematic epitaxial stack of the double heterostructure MODFET. **b** Equilibrium energy band diagrams with and without parasitic channel and the corresponding 2DEG charge distributions. Reproduced from [23], with the permission of AIP Publishing [33]

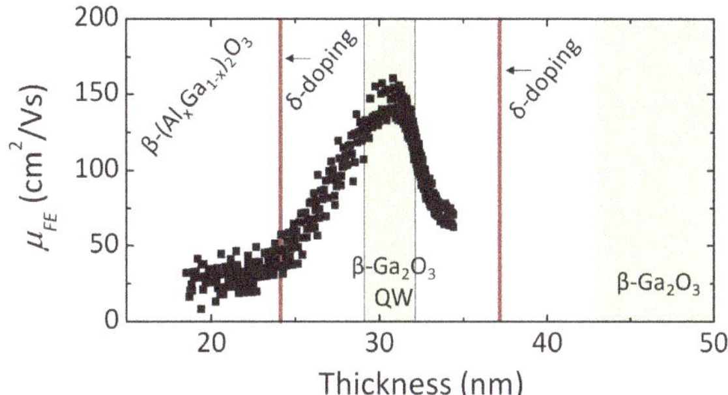

Fig. 33.5 Depth profile of the field effect mobility (μ_{FE}) extracted from a fat transistor at V_{DS} = 0.1 V. Reproduced from [23], with the permission of AIP Publishing [33]

measured Hall mobility of 180 cm^2/Vs in a single heterostructure and further confirms the enhanced screening effect on the scattering centers, especially the polar optical phonons, by increasing the 2DEG density [33].

33.3 Modulation-Doped Field-Effect Transistors

$(AlGa)_2O_3/Ga_2O_3$ modulation-doped structure holds great promise for both high-frequency and high-power device applications, mainly because of the excellent transport properties and the feasibility for aggressive device scaling in both the vertical and the lateral directions. Preliminary device demonstrations on modulation-doped field-effect transistors were reported by Ahmadi et al. [13] and Krishnamoorthy et al. [14]; however, the device performance was compromised by the poor contacts and the dominant parasitic channel in the $(AlGa)_2O_3$ barrier layer. Recently, $(AlGa)_2O_3/Ga_2O_3$ single heterostructure MODFETs with pure 2DEG channel was demonstrated by Zhang et al. [32–35]. Using Pt/Au Schottky gate contact, the device showed normally-off operation because of the depletion of the low 2DEG charge density $\sim 2 \times 10^{12}$ cm^{-2} under the gate.

As shown in Fig. 33.6, the MODFET device showed excellent drain current on/off ratio above 10^9. The subthreshold slope is $\sim$90 mV/decade over more than five decades of the subthreshold current, suggesting high material quality of the β-$(Al_xGa_{1-x})_2O_3$ barrier layer. The devices showed low output conductance after saturation, indicating good gate control of the 2DEG channel. The peak transconductance (g_m) was measured to be 39 mS/mm at (V_{DS}, V_{GS}) of (10, 1.5 V) for the device with a gate length (L_G) of 0.7 μm. While the transconductance is still limited by the low channel mobility because of the long channel length and low-charge density, the obtained transconductance is among the highest reported values for Ga_2O_3 transistors. The drain current of the devices was limited by the low-charge density of 2×10^{12} cm^{-2}, and higher drain current above 250 mA/mm was achieved for double heterostructure MODFETs by increasing the total channel charge density [33].

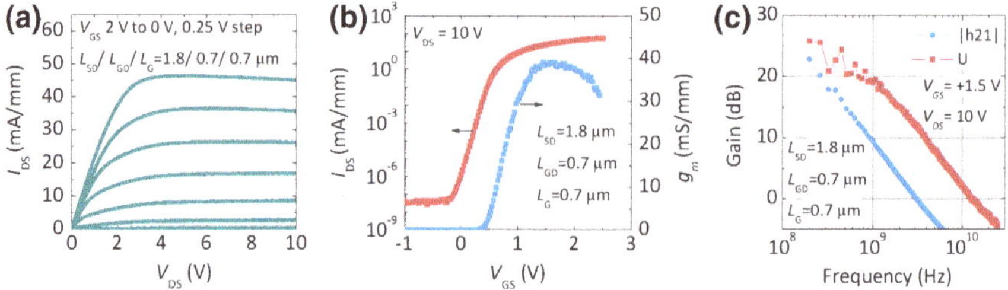

Fig. 33.6 **a** Output characteristics measured with gate bias V_{GS} from 2 to 0 V at a step of 0.25 V. **b** Transfer characteristics measured under a drain bias of V_{DS} = 10 V. **c** RF characteristics measured at V_{DS} = 10 V and V_{GS} = 1.5 V. The gate length, gate-drain spacing, source-drain spacing of the device are L_G = 0.7 μm, L_{GD} = 0.7 μm and L_{SD} = 1.8 μm, respectively. Reproduced from [17], with the permission of AIP Publishing [32]

The RF performance of the β-$(Al_xGa_{1-x})_2O_3/Ga_2O_3$ MODFETs was characterized as shown in Fig. 33.6c. The small-signal measurements on the MODFET device with L_G = 0.7 μm showed a cutoff frequency of 3.1 GHz, and maximum oscillation frequency of 13.1 GHz at V_{DS} of 10 V and V_{GS} of 1.5 V. Compared to AlGaN/GaN HEMTs or AlGaAs/GaAs MODFETs with similar gate length, the high-frequency performance of β-$(Al_xGa_{1-x})_2O_3/Ga_2O_3$ MODFETs is still limited due to the lower channel mobility, which prevents electrons from reaching saturated velocity. However, due to the significant increase in the 2DEG mobility at low temperatures (as shown in Fig. 33.2), it is in fact possible to achieve velocity saturation inside the channel at lower temperatures. The small-signal RF measurement of a β-$(Al_xGa_{1-x})_2O_3/Ga_2O_3$ MODFET device with gate length of L_G = 0.61 μm indicated a significant increase in both f_t and f_{max} from 4.0/11.8 GHz at room temperature to 17.4/40.8 GHz at 50 K. The analysis of the low-temperature f_t based on device simulations indicated a peak velocity of 1.2×10^7 cm/s [35]. The current gain cutoff frequency is comparable to the GaN devices with similar gate length, further suggesting that the saturation velocity in β-Ga_2O_3 is comparable to that in GaN.

Three-terminal off-state breakdown measurements of $(AlGa)_2O_3/Ga_2O_3$ MODFETs suggested a breakdown voltage of 122.4 V for the device with gate-drain spacing of L_{GD} = 0.38 μm. This translates into an average breakdown field of 3.22 MV/cm. Further incorporation of SiO_2 dielectric layer for field management has led to improved breakdown field near 4 MV/cm and high breakdown voltages above 1 kV for the $(AlGa)_2O_3/Ga_2O_3$ MODFETs [43].

33.4 Summary

The design and development of 2DEG channels in $(AlGa)_2O_3/Ga_2O_3$ modulation-doped structures were reviewed in this chapter. While the research on modulation doping in the β-Ga_2O_3 system is still in its early stage, tremendous progress has been achieved. The main results are summarized below.

1. Improved transport properties were observed for 2DEG channels as compared to the bulk-doped channels. The room temperature mobility was found to be 180 cm^2/Vs in $(AlGa)_2O_3/Ga_2O_3$ single heterostructures, while the estimated Hall mobility reached 255 cm^2/Vs in the quantum well channel of the double heterostructure MODFETs. The low-temperature mobility of 2790 cm^2/Vs (50 K) measured in the 2DEG channel is within the highest reported mobility values in β-Ga_2O_3 regardless of a high background charge estimated to be $\sim 10^{17}$ cm^{-3} in the channel region. Further improvement of the material quality and reduction of the impurity concentration could lead to substantial improvement in the low-temperature mobility, making it possible to investigate quantum transport characteristics in β-Ga_2O_3.
2. Due to the high channel mobility 2DEG, quantum oscillations were observed in β-Ga_2O_3 for the first time. The analysis of the oscillation amplitude as a function of the measurement temperature and magnetic field led to the extraction of the effective mass to be $m^* = 0.313 \pm 0.015\ m_0$ and quantum scattering time to be 0.33 ps.
3. The high channel mobility at low temperatures made it possible to extract the electron saturation velocity in β-Ga_2O_3. A saturation velocity of 1.2×10^7 cm/s was estimated based on both the measured velocity-field profile and the analysis on the small-signal RF measurements at 50 K, which showed a current gain cutoff frequency of 17.4 GHz for the device with gate length of 0.61 μm.
4. The three-terminal off-state breakdown measurements on the MODFETs suggested high average breakdown fields near 4 MV/cm.

The experimental confirmation of the enhanced transport properties, high saturation velocity and high breakdown field in the $(AlGa)_2O_3/Ga_2O_3$ MODFETs makes them promising candidates for future electronic device applications.

References

1. M. Orita, H. Ohta, M. Hirano, H. Hosono, Appl. Phys. Lett. **77**(25), 4166 (2000)
2. M. Higashiwaki, K. Sasaki, A. Kuramata, T. Masui, S. Yamakoshi, Appl. Phys. Lett. **100**(1), 013504 (2012)
3. K. Ghosh, U. Singisetti, J. Appl. Phys. **122**(3), 035702 (2017)
4. Y. Tomm, P. Reiche, D. Klimm, T. Fukuda, J. Cryst. Growth **220**(4), 510 (2000)
5. Y. Tomm, J. Ko, A. Yoshikawa, T. Fukuda, J. Cryst. Growth **66**(1–4), 369 (2001)

6. H. Aida, K. Nishiguchi, H. Takeda, N. Aota, K. Sunakawa, Y. Yaguchi, Jpn. J. Appl. Phys. **47**(11R), 8506 (2008)
7. M. Higashiwaki, K. Sasaki, T. Kamimura, M. Hoi Wong, D. Krishnamurthy, A. Kuramata, T. Masui, S. Yamakoshi, Appl. Phys. Lett. **103**(12), 123511 (2013)
8. W.S. Hwang, A. Verma, H. Peelaers, V. Protasenko, S. Rouvimov, H. Xing, A. Seabaugh, W. Haensch, C. Van de Walle, Z. Galazka, M. Albrecht, R. Fornari, D. Jena, Appl. Phys. Lett. **104**(20), 203111 (2014)
9. M.H. Wong, K. Sasaki, A. Kuramata, S. Yamakoshi, M. Higashiwaki, IEEE Electron Device Lett. **37**(2), 212 (2016)
10. S. Krishnamoorthy, Z. Xia, S. Bajaj, M. Brenner, S. Rajan, Appl. Phys. Express **10**(5), 051102 (2017)
11. N. Moser, J. McCandless, A. Crespo, K. Leedy, A. Green, A. Neal, S. Mou, E. Ahmadi, J. Speck, K. Chabak, IEEE Electron Device Lett. **38**(6), 775 (2017)
12. A.J. Green, K.D. Chabak, M. Baldini, N. Moser, R. Gilbert, R.C. Fitch, G. Wagner, Z. Galazka, J. McCandless, A. Crespo, IEEE Electron Device Lett. **38**(6), 790 (2017)
13. E. Ahmadi, O.S. Koksaldi, X. Zheng, T. Mates, Y. Oshima, U.K. Mishra, J.S. Speck, Appl. Phys. Express **10**(7), 071101 (2017)
14. S. Krishnamoorthy, Z. Xia, C. Joishi, Y. Zhang, J. McGlone, J. Johnson, M. Brenner, A.R. Arehart, J. Hwang, S. Lodha, S. Rajan, Appl. Phys. Lett. **111**(2), 023502 (2017)
15. K. Konishi, K. Goto, H. Murakami, Y. Kumagai, A. Kuramata, S. Yamakoshi, M. Higashiwaki, Appl. Phys. Lett. **110**(10), 103506 (2017)
16. C. Joishi, S. Rafique, Z. Xia, L. Han, S. Krishnamoorthy, Y. Zhang, S. Lodha, H. Zhao, S. Rajan, Appl. Phys. Express **11**(3), 031101 (2018)
17. A.J. Green, K.D. Chabak, E.R. Heller, R.C. Fitch, M. Baldini, A. Fiedler, K. Irmscher, G. Wagner, Z. Galazka, S.E. Tetlak, IEEE Electron Device Lett. **37**(7), 902 (2016)
18. K. Ghosh, U. Singisetti, J. Mater. Res. **32**(22), 4142 (2017)
19. H. Peelaers, J.B. Varley, J.S. Speck, C.G. Van de Walle, Appl. Phys. Lett. **112**(24), 242101 (2018)
20. T. Mimura, S. Hiyamizu, T. Fujii, K. Nanbu, Jpn. J. Appl. Phys. **19**(5), L225 (1980)
21. B.W. Krueger, C.S. Dandeneau, E.M. Nelson, S.T. Dunham, F.S. Ohuchi, M.A. Olmstead, J. Am. Ceram. Soc. **99**(7), 2467 (2016)
22. T. Oshima, T. Okuno, N. Arai, Y. Kobayashi, S. Fujita, Jpn. J. Appl. Phys. **48**(7R), 070202 (2009)
23. S.W. Kaun, F. Wu, J.S. Speck, J. Vac. Sci. Technol. A **33**(4), 041508 (2015)
24. Q. Feng, X. Li, G. Han, L. Huang, F. Li, W. Tang, J. Zhang, Y. Hao, Opt. Mater. Express **7** (4), 1240 (2017)
25. P. Vogt, A. Mauze, F. Wu, B. Bonef, J.S. Speck, Appl. Phys. Express **11**(11), 115503 (2018)
26. F. Zhang, K. Saito, T. Tanaka, M. Nishio, M. Arita, Q. Guo, Appl. Phys. Lett. **105**(16), 162107 (2014)
27. C. Kranert, M. Jenderka, J. Lenzner, M. Lorenz, H. von Wenckstern, R. Schmidt-Grund, M. Grundmann, J. Appl. Phys. **117**(12), 125703 (2015)
28. R. Schmidt-Grund, C. Kranert, H. Von Wenckstern, V. Zviagin, M. Lorenz, M. Grundmann, J. Appl. Phys. **117**(16), 165307 (2015)
29. X. Wang, Z. Chen, F. Zhang, K. Saito, T. Tanaka, M. Nishio, Q. Guo, AIP Adv. **6**(1), 015111 (2016)
30. H. Ito, K. Kaneko, S. Fujita, Jpn. J. Appl. Phys. **51**(10R), 100207 (2012)
31. R. Miller, F. Alema, A. Osinsky, I.E.E.E. Trans, Semicond. Manuf. **31**(4), 467 (2018)
32. Y. Zhang, A. Neal, Z. Xia, C. Joishi, J.M. Johnson, Y. Zheng, S. Bajaj, M. Brenner, D. Dorsey, K. Chabak, Appl. Phys. Lett. **112**(17), 173502 (2018)
33. Y. Zhang, C. Joishi, Z. Xia, M. Brenner, S. Lodha, S. Rajan, Appl. Phys. Lett. **112**(23), 233503 (2018)
34. T. Oshima, Y. Kato, N. Kawano, A. Kuramata, S. Yamakoshi, S. Fujita, T. Oishi, M. Kasu, Appl. Phys. Express **10**(3), 035701 (2017)

35. Y. Zhang, Z. Xia, J. McGlone, W. Sun, C. Joishi, A.R. Arehart, S.A. Ringel, S. Rajan, I.E.E.E. Trans, Electron Devices **66**, 1574 (2019)
36. Z. Xia, C. Joishi, S. Krishnamoorthy, S. Bajaj, Y. Zhang, M. Brenner, S. Lodha, S. Rajan, IEEE Electron Device Lett. **39**(4), 568 (2018)
37. T. Oishi, Y. Koga, K. Harada, M. Kasu, Appl. Phys. Express **8**(3), 031101 (2015)
38. S. Rafique, M.R. Karim, J.M. Johnson, J. Hwang, H. Zhao, Appl. Phys. Lett. **112**(5), 052104 (2018)
39. Y. Zhang, F. Alema, A. Mauze, O.S. Koksaldi, R. Miller, A. Osinsky, J.S. Speck, APL Mater. **7**(2), 022506 (2019)
40. K. Goto, K. Konishi, H. Murakami, Y. Kumagai, B. Monemar, M. Higashiwaki, A. Kuramata, S. Yamakoshi, Thin Solid Films **666**, 182 (2018)
41. N. Ma, N. Tanen, A. Verma, Z. Guo, T. Luo, H. Xing, D. Jena, Appl. Phys. Lett. **109**(21), 212101 (2016)
42. Y. Kang, K. Krishnaswamy, H. Peelaers, C.G. Van de Walle, J. Phys. Condens. Matter **29** (23), 234001 (2017)
43. C. Joishi, Y. Zhang, Z. Xia, W. Sun, A.R. Arehart, S. Ringel, S. Lodha, S. Rajan, IEEE Electron Device Lett. **40**, 1241 (2019)

Chapter 34
Field-Effect Transistors 4

Nano-Membrane β-Ga_2O_3 Field-Effect Transistors

Hong Zhou, Jinhyun Noh, Hagyoul Bae, Mengwei Si and Peide D. Ye

Abstract β-Ga_2O_3-based field-effect transistor (FET) is regarded as a promising candidate for the next-generation power electronics due to its ultrawide bandgap of 4.5–4.8 eV, estimated critical field of 8 MV/cm and decent intrinsic electron mobility limit of 250 cm^2/Vs, yielding a high BFOM of more than 3000, which is several times higher than GaN and SiC. Meanwhile, β-Ga_2O_3 crystal also possesses a unique property that it has a large lattice constant of 12.23 Å along [100] direction, which allows a facile cleavage into thin belts or nano-membranes. Therefore, by transferring β-Ga_2O_3 nano-membrane from its bulk substrate to a foreign substrate, we can fabricate β-Ga_2O_3 on insulator FETs and then explore material and device potentials before β-Ga_2O_3 epitaxy technology becomes mature and the cost of the epi-wafers reduces significantly. In this chapter, we will focus on the nano-membrane-based FETs and their electrical interfaces, demonstrating record high drain current density of the devices, minimize the self-heating effect by the integration of nano-membrane on high thermal conductivity substrates and expand research direction toward a low-power and wide bandgap logic application.

Recently, β-Ga_2O_3 has shown its great promise for next-generation high-power device applications due to its ultrawide bandgap of 4.6–4.9 eV [1, 2]. This allows β-Ga_2O_3 to possess a corresponding empirical estimated electrical breakdown field (E_{br}) of 8 MV/cm and decent intrinsic electron mobility limit of 250 cm^2/Vs, yielding a high BFOM of more than 3000, which is several times higher than GaN and SiC. There are five different phases for Ga_2O_3, namely α, β, γ, δ, and ε, with β phase is the most stable one and other four polymorphs are metastable and can be converted to β phase at an elevated temperature of 750–900 °C. Meanwhile, β-Ga_2O_3 also has the advantage of a low-cost native bulk substrate that can be synthesized in large size through melt-grown Czochralski, edge-defined film-fed growth, and floating zone method [3–8]. Besides those aforementioned material

H. Zhou · J. Noh · H. Bae · M. Si · P. D. Ye (✉)
School of Electrical and Computer Engineering and Birck Nanotechnology Center, Purdue University, West Lafayette, IN 47907, USA
e-mail: yep@purdue.edu

M. Higashiwaki and S. Fujita (eds.), *Gallium Oxide*, Springer Series in Materials Science 293, https://doi.org/10.1007/978-3-030-37153-1_34

characteristics, β-Ga_2O_3 crystal also has some unique properties. For instance, its (100) plane has a large lattice constant of 12.23 Å along [100] direction, which allows a facile cleavage into thin belts or nano-membranes [9, 10]. Similar to many 2D materials work, by transferring β-Ga_2O_3 nano-membrane from its bulk substrate to a foreign substrate, β-Ga_2O_3 on insulator (GOOI) FETs can be fabricated and then the fundamental properties of this emerging electronic material and its device potentials can be studied before β-Ga_2O_3 epitaxy technology becomes mature and the cost of the epi-wafers reduces significantly. Although the monoclinic structure of bulk β-Ga_2O_3 crystals would allow a facile cleavage into nano-membrane along the [100] direction, β-Ga_2O_3 is not a van der Waals 2D material so that how to passivate the cleaved surface and form device quality interface is an important research direction for the device development. Meanwhile, since β-Ga_2O_3 itself has a wide bandgap, the choice of dielectric and their interface characterization are different from the conventional semiconductors. A crucial disadvantage of β-Ga_2O_3 as the channel of a power device is its low thermal conductivity of 0.1–0.3 W/cm K depends on different crystal orientations [11, 12]. One approach to solve the low thermal conductivity issue of β-Ga_2O_3 is to introduce a high thermal conductivity substrate rather than the β-Ga_2O_3 native substrate. In general, it will take time and efforts to develop the integration of β-Ga_2O_3 on high thermal conducting substrates such as sapphire, SiC, or diamond. However, β-Ga_2O_3 membrane offers the flexibility to address some of the potential issues on the heterogeneous integration of β-Ga_2O_3. Although most of β-Ga_2O_3 research is for power device applications, we attempt to explore its potential for low-power digital applications due to its ultrawide bandgap of the channel and extremely high on/off ratio using ferroelectric gate dielectric, especially for applications in extreme environments such as high temperature and high radiation. In this chapter, we focused on β-Ga_2O_3 nano-membrane and explored its potential device performance, electrical interfaces with different dielectrics, and thermal interfaces with different substrates. The details are described in the following sections.

34.1 Back-Gated Depletion/Enhancement Modes GOOI FETs on SiO_2/Si Substrate with Record Drain Current Density of 1.5/1 A/mm

34.1.1 Back-Gate GOOI FET Fabrication

β-Ga_2O_3 metal-oxide-semiconductor field-effect transistors (MOSFETs) have demonstrated a high blocking voltage of 750 V and an E_{br} of 3.8 MV/cm in depletion-mode (D-mode) [13, 14] and a breakdown voltage (BV) of more than 600 V in enhancement-mode (E-mode) [15]. High BV of 257 V and 1000 V have also been demonstrated in β-Ga_2O_3 metal-semiconductor field-effect transistors (MESFETs) and Schottky barrier diodes (SBDs), respectively [16, 17]. Except for the high direct current (DC) breakdown characteristic, a high-pulsed drain current (I_D) of 478 mA/mm has also been achieved for β-Ga_2O_3 MOSFETs [18]. Cutoff frequency and maximum oscillation frequency (f_T and f_{max}) of 3.3 GHz and 12.9 GHz on β-Ga_2O_3 MOSFET were also demonstrated [19]. Even though β-Ga_2O_3 is not a van der Waals 2D material, the unique monoclinic structure of bulk β-Ga_2O_3 crystals allows a facile cleavage into nano-membrane along the [100] direction (lattice constant of 12.23 Å along [100]). This opens a mechanical exfoliation-based fabrication process for β-Ga_2O_3 MOSFETs. In this section, we focused on using these cleaved β-Ga_2O_3 nano-membranes and explored the potential of this emerging material for power device applications.

Figure 34.1a, b show the schematic of a GOOI FET and atomic force microscopy (AFM) image of a β-Ga_2O_3 surface after cleavage, which shows the atomically flat and uniform nano-membrane. The success exfoliation of β-Ga_2O_3 thin film, possibly due to its large lattice constant of 12.23 Å along this [100] direction, makes it possible to study its fundamental transport properties and device integration. Device fabrication started from (−201) β-Ga_2O_3 bulk substrates with Sn doping concentration of 3×10^{18} and 8×10^{18} cm^{-3}. The Sn doping concentration is determined by capacitance–voltage (C–V) measurements [20]. Thin β-Ga_2O_3 nano-membranes were transferred onto a solvent-cleaned 300 nm SiO_2/p^+ Si substrate. Ti/Al/Au (15/60/50 nm) source and drain contacts were formed by the VB6 electron beam lithography (EBL), e-beam evaporation and lift-off processes. The fabricated GOOI FETs have channel thicknesses from 50 nm to 150 nm, measured by the AFM. The successful integration of β-Ga_2O_3 on a foreign arbitrary substrate shows the potential of migrating the low thermal conductivity issue of β-Ga_2O_3 substrate through wafer bonding β-Ga_2O_3 on a high thermal conductivity sapphire or diamond substrates, which is discussed in great details in Sect. 34.2 of this chapter. In addition, the advantage of this device fabrication process can also enable the study of β-Ga_2O_3 channel thickness-dependent threshold voltage (V_T) and

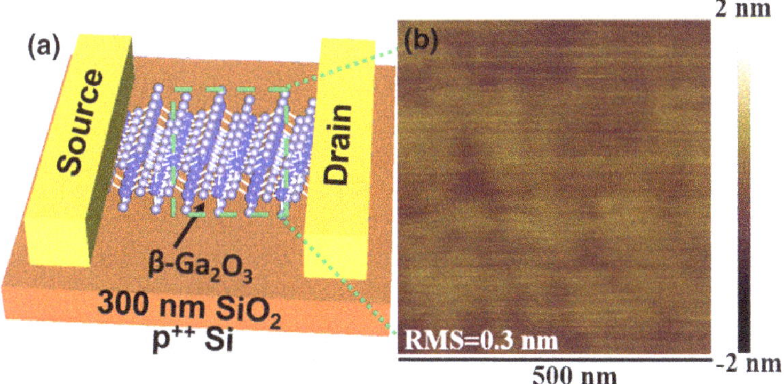

Fig. 34.1 **a** Schematic view of a GOOI FET with a 300 nm SiO_2 layer on Si substrate and **b** AFM image of the atomic flat β-Ga_2O_3 surface after cleavage. The atomic flatness of the surface after cleavage is quite impressive

provide a higher bandgap insulating material underneath the β-Ga_2O_3 channel for future BV enhancement.

34.1.2 D/E-Mode GOOI FETs with Record High Drain Current Densities

Maximum on-state drain current density is used to evaluate the new semiconductor material for the potential device applications. Figure 34.2a shows the D-mode direct-current (DC) output characteristics (I_D–V_{DS}) for devices with doping concentration of 3.0×10^{18} cm^{-3} (lower-doping channel) and 8.0×10^{18} cm^{-3} (higher-doping channel). The devices with higher-doping channel have a channel length (L_{CH}, here is the same as source to drain spacing L_{SD}) of 0.3 μm, channel width (W) of 0.15 μm, and channel thickness (t) of 70 nm. The devices with lower-doping channel have $L_{CH} = 0.3$ μm, $W = 0.6$ μm, and $t = 100$ nm. The device dimensions are accurately determined by scanning electron microscopy (SEM). A record high maximum I_D (I_{Dmax}) of 1.5 A/mm is obtained on the higher-doping depletion-mode device, which is twice more than device with lower-doping channel [21]. The device with higher-doping channel also has a much lower R_{on} (5 Ω mm), compared to that of 11 Ω mm on the device with lower-doping channel. At low V_{DS}, I_D–V_{DS} of higher-doping channel device shows better linear behavior while the lower-doping counterpart displays a Schottky-like contacts, because higher-doping concentration membrane has lower Schottky barrier width thus lower contact resistance of the device. Therefore, the R_c in device with higher-doping concentration is significantly reduced. Figure 34.2b is the I_D–V_{GS} and g_m–V_{GS} characteristics of the same D-mode GOOI FET with 8.0×10^{18} cm^{-3} doping concentration as in Fig. 34.2a. The threshold voltage (V_T) of this D-mode GOOI FET is −135 V, extracted at V_{DS} = 1 V and I_D = 0.1 mA/mm. A peak transconductance (g_{max}) of 9.2 mS/mm is achieved on device with higher-doping channel, which is twice of the g_{max} on the device with lower-doping channel, indicating a much improved R_c. Figure 34.2c, d illustrate the I_D–V_{DS}, I_D–V_{GS} and g_m–V_{GS} characteristics of an E-mode GOOI FET with $L_{CH} = 0.3$ μm, $t = 55$ nm and $W = 0.17$ μm and with 8.0×10^{18} cm^{-3} channel doping concentration. The performance of device with $t = 75$ nm, $W = 0.45$ μm and 3.0×10^{18} cm^{-3} channel doping concentration is also shown in Fig. 34.2c as black-dashed curves for comparison. A record high I_{Dmax} = 1.0 A/mm is obtained on devices with higher-doping channel, which is more than 80% higher than device with lower-doping channel. E-mode GOOI FET has a V_T of 2 V, extracted from the I_D–V_{GS} at V_{DS} = 1 V and I_D = 0.1 mA/mm. No I_D saturation is observed until the V_{DS} reaches device breakdown. The I_{Dmax} of device with $L_{CH} = 0.3$ μm and 3.0×10^{18} cm^{-3} channel doping concentration is similar compared to our previous work, slightly increased from 600/450 to 710/550 mA/mm for D/E-modes [22]. Further scaling of the L_{CH} shows minor effect on boosting the I_{Dmax}, indicating the device

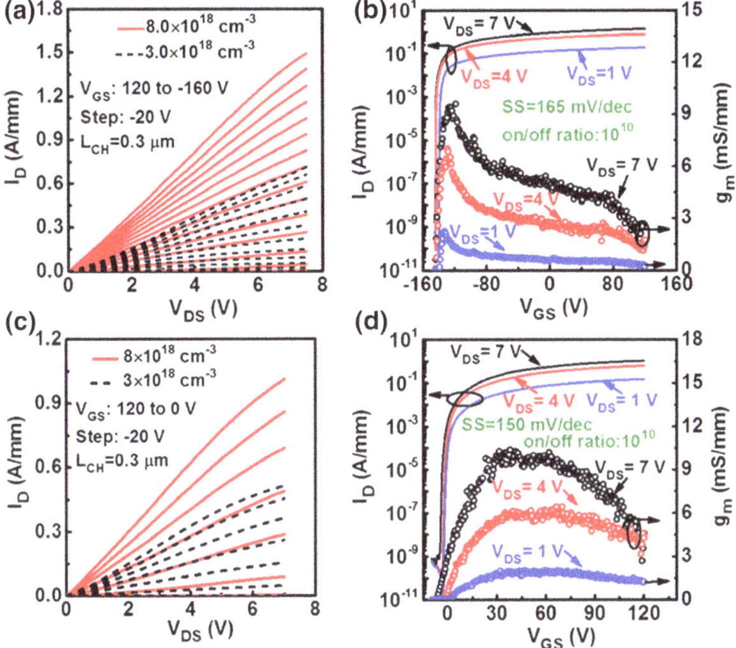

Fig. 34.2 **a**, **c** are I_D–V_{DS} output characteristics of D- and E-mode GOOI FETs with 3.0 × 10^{18} and 8.0 × 10^{18} cm^{-3} doping channel, respectively. **b**, **d** are I_D–g_m–V_{GS} transfer characteristics of D-mode and E-mode GOOI FETs with 8.0 × 10^{18} cm^{-3} doping channel, respectively. Record high I_{Dmax} of 1.5 and 1.0 A/mm are demonstrated for D- /E-mode devices. Both D- and E-mode devices have high on/off ratio of 10^{10} and low SS of 150–165 mV/dec for 300 nm SiO_2

on-state performance is primarily limited by the high R_c of low doping channel underneath the contacts. The drain-induced barrier lowering (DIBL) is extracted to be 0.73 and 0.38 V/V for D/E-modes devices with higher-doping channel. Both D/E-mode devices achieved high on/off ratio of 10^{10} and low subthreshold swing (SS) of 150–165 mV/dec even with a thick 300 nm SiO_2 gate insulator, benefiting from the wide bandgap of β-Ga_2O_3 and the low damage transfer fabrication process in this work.

34.1.3 V_T Dependence on the β-Ga_2O_3 Nano-membrane Thickness

Instead of engineering channel doping concentration or metal gate work-functions, we can directly modify and shrink β-Ga_2O_3 nano-membrane thickness to make V_T shift from negative values in D-mode devices to positive values in E-mode devices. Figure 34.3a shows the I_D–V_{GS} characteristics of β-Ga_2O_3 devices with different

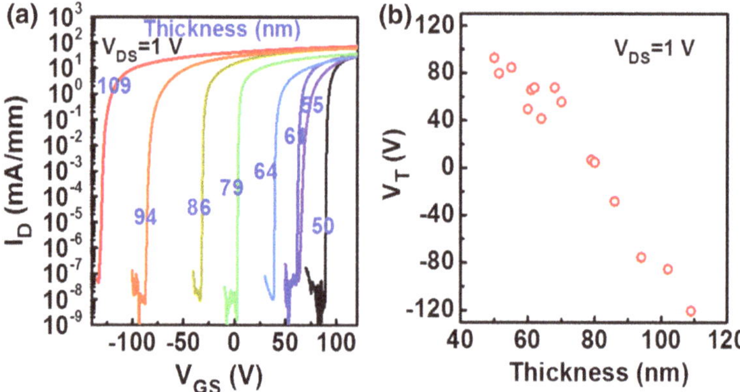

Fig. 34.3 **a** Thickness-dependent I_D–V_{GS} plots of various GOOI FETs from D-mode of thicker β-Ga_2O_3 to E-mode of thinner β-Ga_2O_3. **b** Thickness-dependent V_T extracted at V_{DS} = 1 V of 15 devices

channel thicknesses, showing V_T increases when channel thickness reduces. Figure 34.3b summarizes V_T versus channel thickness of 15 devices. The control of V_T by channel thickness is valuable to realize high performance E-mode GOOI FETs.

The significant thickness-dependent V_T shift is the result of surface depletion effect. The dangling bonds on the surface of 3D Sn-doped β-Ga_2O_3 lead to the surface trap states. The surface trap states could deplete all mobile carriers in ~80 nm thick β-Ga_2O_3 nano-membrane (doping concentration of 3.0 × 10^{18} cm^{-3}), resulting in a low current density at zero V_{GS}, as shown in Fig. 34.3a. Therefore, E-mode GOOI FETs with V_T greater than zero can be realized. The surface depletion effect also exists on the top surface of GOOI FETs. The V_T is significantly reduced by more than 70 V after the 15 nm Al_2O_3 ALD passivation, as shown in Fig. 34.4a, suggesting the effect of top surface depletion. Similar depletion effect induced by surface or interface charges was also reported by Moser et al. [5]. The surface depleted charge density (n_s) is determined and simulated by TCAD C-V simulation to be 1.2 × 10^{13} and 2.2 × 10^{13} cm^{-2} for 3.0 × 10^{18} and 8.0 × 10^{13} cm^{-3} nano-membranes with thickness of 80 and 55 nm. Higher n_s in β-Ga_2O_3 nano-membrane with higher doping may be related with more Sn^{+4} dopants induced surface states. Figure 34.4b shows the simulated C–V characteristics for E-mode GOOI FET, in good agreement with the V_T from I_D–V_{GS} characterization. The simulated band diagrams of the E-mode GOOI FET at V_{GS} = 0 V for both lower-doping and higher-doping channels are shown in Fig. 34.4c, d. Very few mobile carriers exist in the nano-membrane because the conduction band of β-Ga_2O_3 is pulled up by the surface depletion. The surface depletion led by the surface states is widely observed in other conventional semiconductors, but it is fundamentally different from 2D materials with atomic layer thin channel and dangling bond free surface.

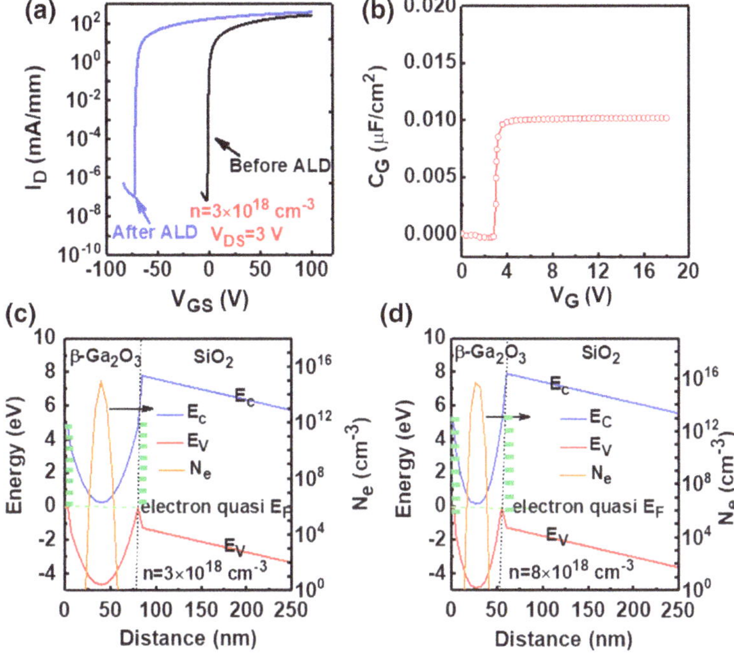

Fig. 34.4 **a** I_D–V_{GS} comparison between GOOI FETs with and without ALD passivation for β-Ga_2O_3 nano-membrane with doping concentration of 3.0 × 10^{18} cm^{-3}, **b** simulated C–V curve for E-mode GOOI FET at a β-Ga_2O_3 nano-membrane thickness of 80 nm and doping concentration of 3.0 × 10^{18} cm^{-3} after considering the top and bottom negative surface charge (n_s = 1.2 × 10^{13} cm^2) depletion effect. Band diagram and electron density distribution of E-mode GOOI FETs with surface negative charge depletion on **c** lower doping (n_s = 1.2 × 10^{13} cm^{-2}) and **d** high doping (n_s = 2.2 × 10^{13} cm^{-2}) β-Ga_2O_3 nano-membrane channels at V_{GS} = 0 V

34.1.4 Interface Characterization of β-Ga_2O_3 Nano-membrane on Conventional Dielectrics

As discussed above, the β-Ga_2O_3 nano-membrane surface is usually unpassivated due to its 3D material nature. How to passivate the surface and how to quantitatively characterize the interface with ALD dielectric or the dielectric where the nano-membrane transferred on are very important for the device development. Unlike the conventional GaN high-electron mobility transistors (HEMTs) with Schottky gate structures, metal-oxide-semiconductor (MOS)-based devices using a gate insulator effectively achieve more sufficient gate modulation and suppress gate leakage current. Therefore, the interface quality between the gate insulator and the active channel material becomes a critical issue. Hence, the full experimental characterization of interface trap density $D_{it}(E)$ over the forbidden bandgap from the valence band maximum (E_V) to the conduction band minimum (E_C) becomes very important in device research. The interface traps, that arise from the interface between the gate oxide and the channel surface could lead to short- or long-term device degradation and impede the high performance and reliability of the devices. Several techniques have been developed to characterize and analyze D_{it} in the MOS system, such as the high/low frequency, photo-assisted current–voltage (I–V), AC conductance, deep-level optical and transient spectroscopy, and Terman method [23, 24]. However, the simultaneous extraction of both $D_{it\_D}(E)$ and $D_{it\_A}(E)$ eV^{-1}cm^{-2} over a wide range of bandgap energies is essential. A novel characterization technique based on only experimental photonic I–V data in β-Ga_2O_3 FETs has been developed [25].

Here, we propose an I–V-based sub-bandgap optoelectronic characterization technique for the simultaneous extraction of $D_{it\_D}(E)$ and $D_{it\_A}(E)$ using deep UV light with sub-bandgap photon energy (E_{ph} = 3.6 eV and λ = 390 nm). By employing the measured I_{DS}–V_{GS} curves under both dark and photonic states, a consistent mapping of the surface potential (ψ_S) for $D_{it\_D}(E)$ and $D_{it\_A}(E)$ over the bandgap energy ($E_C < E < E_V$) was applied in two distinguishable subthreshold regions ($V_{ON} < V_{GS} < V_{FB}$ and $V_{FB} < V_{GS} < V_T$). Figure 34.5a shows the measurement setup and an equivalent circuit model for the photonic I–V characterization of the β-Ga_2O_3 FETs with bottom-gate structure. A schematic illustration of the energy band diagram for the extraction of $D_{it\_D}(E)$ and $D_{it\_A}(E)$ under a photonic state is shown in Fig. 34.5b, c. The UV optical source illuminates the β-Ga_2O_3 channel of the fabricated device vertically. UV light (λ = 390 nm, E_{ph} = 3.6 eV < E_{g_Ga2O3} = 4.8 eV, and P_{opt} = 2.8 mW) was used to pump the trapped electrons in the channel surface region from E_C–E_{ph} to E_C. The work principle is illustrated in Fig. 34.5a–c.

The detailed work can be found in [25]. We conclude from the experiments that the $D_{it\_D}(E)$ and $D_{it\_A}(E)$ obtained from this newly developed photonic I–V technique over the subgap energy ranges are 0.5×10^{11} eV^{-1}cm^{-2} to 5.5×10^{11} eV^{-1}cm^{-2} and of 2.1×10^{11} eV^{-1}cm^{-2} to 1.1×10^{12} eV^{-1}cm^{-2}, respectively. The extracted values obtained in this work are comparable to those of other reports [20, 26].

34.2 Minimized Self-heating Effect of Top-Gate GOOI FETs on High Thermal Conductivity Substrates

Self-heating effect induced temperature increase and non-uniform distribution of dissipated power have emerged as one of the most important concerns in the degradation of the I_D, output power density (P), as well as the gate leakage current, device variability, and reliability [27]. This effect would become more severe in high power device with a low thermal conductivity (κ) substrate such as β-Ga_2O_3 (κ is just 10–25 W/m K). This β-Ga_2O_3 thermal concern has become one of the major challenges to realize practical applications. An effective approach to mitigate low κ problem is to utilize a higher κ substrate rather than the β-Ga_2O_3 native

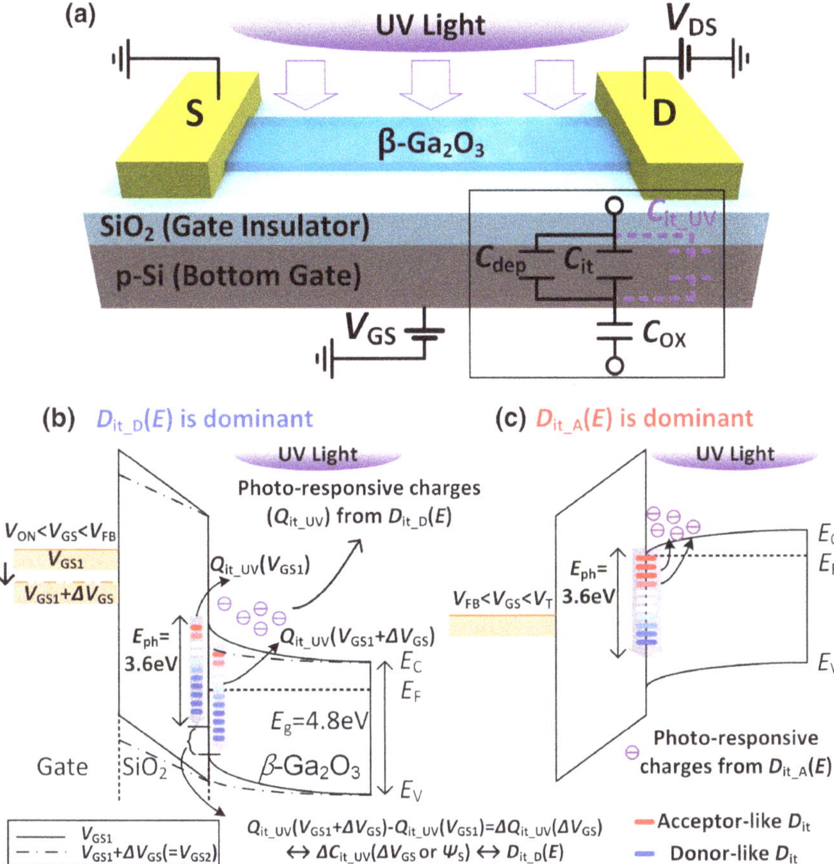

Fig. 34.5 **a** Schematic view of the measurement setup under a photonic state with the sub-bandgap UV light including the equivalent circuit model as the inset. **b**, **c** Energy band diagrams for the concept of simultaneous extractions of both $D_{it\_D}(E)$ and $D_{it\_A}(E)$

substrate through a potential wafer bonding technique. Meanwhile, high breakdown field in the substrate is also required to enhance the breakdown voltage of β-Ga_2O_3 devices.

In this section, we demonstrate β-Ga_2O_3 FETs on SiO_2/Si substrate (κ = 1.5 W/m K for 270 nm SiO_2), sapphire substrate (κ = 40 W/m K) and diamond substrate (κ = 1,000–2,200 W/m K) [28, 29]. An ultra-fast, high-resolution thermos-reflectance (TR) imaging technique is applied to examine the local surface temperature on the β-Ga_2O_3 FETs on different substrates. The high thermal conductivity substrates lead to a remarkable improvement in on-current density, with I_{Dmax} of 325, 535 and a record high 980 mA/mm on SiO_2/Si, sapphire and diamond substrates, respectively. The self-heating effect is also systematically studied by TR imaging, showing a significant reduction on surface temperature by applying the higher thermal conductivity sapphire and diamond substrates.

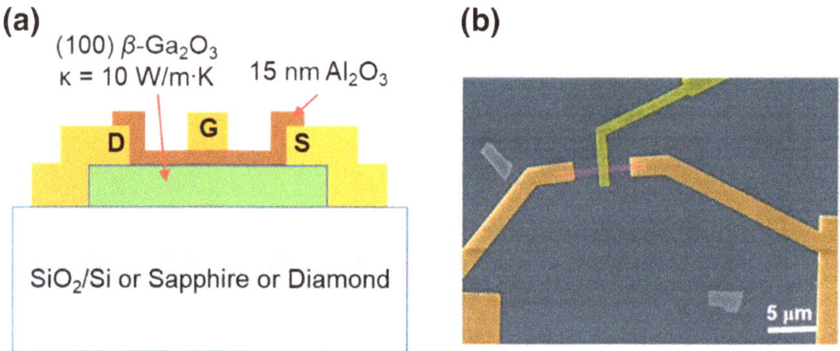

Fig. 34.6 **a** Cross-section schematic view of a top-gate GOOI FET. 15 nm of Al_2O_3 is used as the gate dielectric, Ti/Al/Au (15/60/50 nm) is used as the source/drain electrodes, and Ni/Au is used the gate electrode. **b** False-colored SEM top-view of a GOOI FET with $L_G = 1$ μm and $L_{SD} = 6$ μm

34.2.1 Top-Gate GOOI FET Fabrication

Similar to the device fabrication process described previously, source and drain regions were defined by the VB6 EBL, followed by Ti/Al/Au (15/60/50 nm) metallization and lift-off processes. 15 nm of Al_2O_3 was deposited by ASM F-120 ALD at 250 °C using tri-methyl-aluminum (TMA) and H_2O as precursors. Ni/Au is deposited as the gate electrode, followed by a lift-off process. Figure 34.6a, b show the device schematic and a SEM image of a fabricated GOOI FET.

34.2.2 I–V Electrical Characterization

Figure 34.7a–c show the well-behaved DC I_D–V_{DS} of three top-gate GOOI FETs with L_{SD} of 6/6.5/6 μm, L_G of 1 μm, and channel thickness of 73/75/80 nm on SiO_2/Si, sapphire and diamond substrates, respectively. The typical range of channel width of the nano-membrane devices is 0.5–1.5 μm, determined by SEM as shown in Fig. 34.6b. The measurements start from applying the V_{GS} to 8 or 6 V and then stepping to the device pinch-off voltage of −28 with −2 V as a step, while the V_{DS} is swept from 0 to 27 or 30 V for the SiO_2/Si, sapphire and diamond devices. I_{Dmax} of 325 mA/mm, 535 mA/mm and 980 mA/mm are obtained for devices on SiO_2/Si, sapphire and diamond substrates. I_{Dmax} of GOOI FET are significantly improved on higher thermal conductivity substrates, originating from better transport properties at a lower device temperature. A record high maximum drain current of 980 mA/mm for top-gate β-Ga_2O_3 field-effect transistors is achieved on diamond substrate.

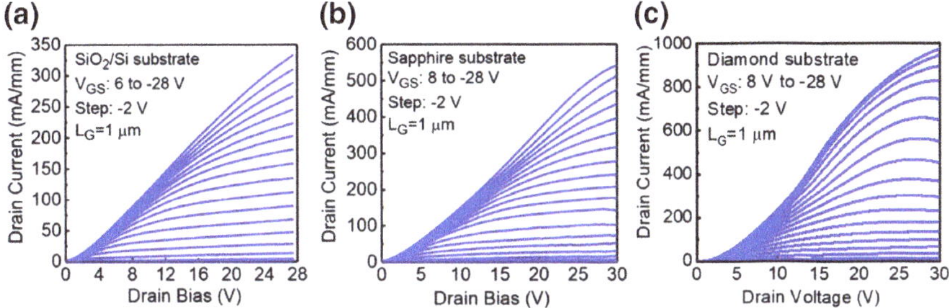

Fig. 34.7 I_D–V_{DS} output characteristics of GOOI FETs on **a** SiO_2/Si, **b** sapphire and **c** diamond substrates with L_G = 1 μm

34.2.3 Suppression of Self-heating Effects by High Thermal Conductivity Substrates

The self-heating effects (SHE) must be considered for power electronics because of the high voltage and large current density. The power consumption (P) of a device, which is proportional to the heat generation rate, can be estimated as the product of I_D and V_{DS}. In a GOOI FET, the device power consumption (on sapphire substrate as an example) can be estimated as $P = I_D \times V_{DS}$ = 30 V × 540 mA/mm × 1 μm = 1.62 × 10^{-2} W. The heat dissipation through air can be estimated by equation $F = hA\Delta T$, where A is the nano-membrane area and ΔT (43 K in this case) is the device temperature rise above room temperature. This heat flux (4.6 × 10^{-10} W) is negligible compared to power consumption of the device. Therefore, substrate acts as the heat sink for heat dissipation and high thermal conductivity substrate is important to reduce the temperature of the GOOI FET. In the TR characterization, temperature of the gate electrode metal Au is measured by laser light illumination and analyzing reflected signals. The gate electrode metal Au is directly on top of the channel region, so that the Au temperature is a direct measurement of channel temperature.

Figure 34.8a–c show merged optical and TR thermal image of GOOI FETs on SiO_2/Si, sapphire and diamond substrates under different P conditions at steady state. In this measurement, V_{GS} is biased at 0 V, while a V_{DS} pulse signal (pulse width of 1 ms and 10% duty) is swept from low V_{DS} to high V_{DS} to reach different P conditions. As reported in the transient temperature measurement in [28], 1 ms pulse is sufficient to reach the saturation temperature for steady-state measurement. The temperature probing optical pulse width was 100 μs, synchronized right prior to the falling edge (at the end of the V_{DS} pulse) to measure the TR image [28–30]. The I_D is measured simultaneously to calculate the normalized power density by area. As can be seen clearly in Fig. 34.8, the device is heated up at higher V_{DS}. At P of 717 W/mm^2 on the SiO_2/Si substrate, ΔT is measured to be 106 K, while in contrast the ΔT is 43 K at a higher P of 917 W/mm^2 on the sapphire substrate. Moreover, ΔT reduced to 21.6 K at an even higher P of 1237.6 W/mm^2 on the diamond

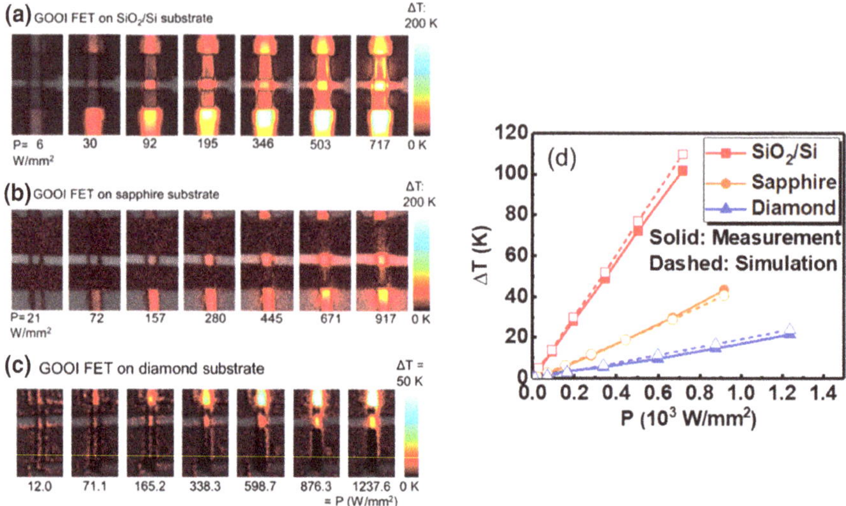

Fig. 34.8 TR and charge-coupled device (CCD) camera merged images of GOOI FET **a** SiO_2/Si, **b** sapphire and **c** diamond substrates when device P is increased by increasing V_{DS} at $V_{GS} = 0$ V. Results show the device heating at steady state, and significant SHE is observed on GOOI FET on the SiO_2/Si substrate at high P. **d** Measured and simulated local gate Au surface ΔT comparison between GOOI FET on SiO_2/Si, sapphire, and diamond substrates

substrate. Figure 34.8d shows the measured and simulated ΔT versus P for GOOI FETs on the three substrates. For both measurement and simulation results, the GOOI FET on higher thermal conductivity substrate has lower ΔT. The thermal resistance (R_T) of device on diamond, sapphire, and SiO_2/Si substrates is calculated to be 1.71×10^{-2}, 4.62×10^{-2}, and 1.47×10^{-1} mm^2 K/W, through $R_T = \Delta T/P$. The enhanced on-current (record high on diamond substrate), reduced SHE and R_T of GOOI FET on sapphire and diamond demonstrate that a higher κ substrate is more effective in heat dissipation. SHE can also be further reduced from top dielectric and metal gate thermal engineering. Novel dielectrics which have thermal match to β-Ga_2O_3 are needed to enhance the heat dissipation for power devices.

34.3 β-Ga_2O_3 Nano-membrane NC-FET with Steep SS for Wide Bandgap Logic Application

Solid-state devices and circuits operated in extreme environment are required for many applications such as high temperature and high radiation. CMOS circuits based on wide bandgap semiconductors for logic applications are promising in such applications. To employ the negative capacitance effect in ferroelectric insulators in the gate stack of a MOSFET as a negative capacitance field-effect transistor

(NC-FET) can further improve the energy efficiency of the CMOS circuits by reducing the subthreshold slope (SS) of the transistor [31–33]. In this section, β-Ga_2O_3 NC-FETs with ferroelectric hafnium zirconium oxide (HZO) as gate stack is discussed [34]. SS less than 60 mV/dec at room temperature is obtained for both forward and reverse V_{GS} sweeps with a minimum value of 34.3 mV/dec at the reverse V_{GS} sweep and 53.1 mV/ dec at the forward V_{GS} sweep at $V_{DS} = 0.5$ V. The E-mode operation is achieved by tuning the thickness of β-Ga_2O_3 with a V_T of 0.47 V for the forward V_{GS} sweep, a V_T of 0.38 V for the reverse V_{GS} sweep, and a low hysteresis less than 0.1 V.

34.3.1 β-Ga_2O_3 Nano-membrane NC-FET Structure and Fabrication

Figure 34.9a shows the schematic diagram of β-Ga_2O_3 NC-FETs with an 86 nm thick β-Ga_2O_3 nano-membrane as the channel, a 3 nm Al_2O_3/20 nm HZO as gate stack, a heavily doped silicon substrate as the gate electrode, and a Ti/Au source/ drain as the metal contacts. The device fabrication process is similar to the above GOOI FETs, except that an ALD HZO ferroelectric gate stack is used. The ALD process of ferroelectric HZO is discussed in great details in our previous work [34, 35]. Figure 34.9b shows the false-color SEM image of the fabricated β-Ga_2O_3 NC-FETs. Figure 34.9c shows the cross-sectional TEM image of the Al_2O_3/HZO gate stack, showing the polycrystalline HZO and the amorphous Al_2O_3. The ferroelectricity of the Al_2O_3/HZO gate stack is confirmed by the P–V measurement on the same structure but with a Ni top electrode, as shown in the P–V hysteresis loop in Fig. 34.9d.

34.3.2 β-Ga_2O_3 NC-FET with Bidirectional SS < 60 mV/dec and Small Hysteresis

Figure 34.10a shows the I_D–V_{GS} characteristics of a β-Ga_2O_3 NC-FET at V_{GS} from −0.4 to 2 V at V_{DS} of 0.1, 0.5 and 0.9 V. This device has a channel length of 0.5 μm and a channel thickness of 86 nm. This particular thickness is chosen to tune the V_T slightly above zero, as discussed previously in the thickness-dependent V_T in GOOI FETs in Sect. 34.1.3. The I_D–V_{GS} characteristics were measured bidirectionally, with V_{GS} from low to high as forward sweep and V_{GS} from high to low as reverse sweep. SS is extracted as a function of I_D for both forward sweep ($SS_{min,For}$) and reverse sweep ($SS_{min,Rev}$). Figure 34.10b shows the SS–I_D characteristics at V_{DS} = 0.5 V of the same device as in Fig. 34.10a. The device exhibits $SS_{min,For}$ = 53.1 mV/dec and $SS_{min,Rev}$ = 34.3 mV/dec at V_{DS} = 0.5 V. SS less than 60 mV/dec at room temperature is demonstrated for both forward and reverse V_{GS}

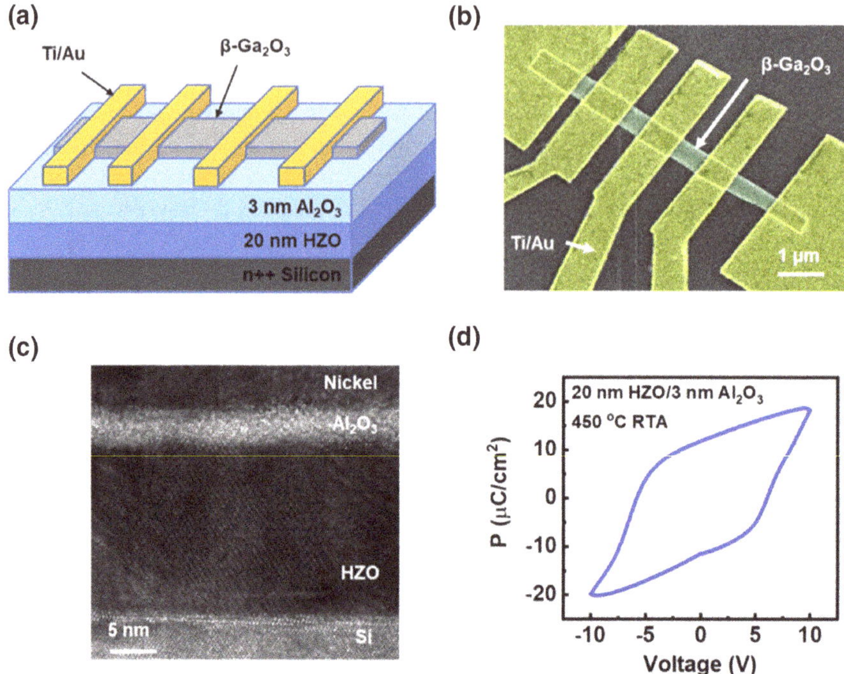

Fig. 34.9 **a** Schematic view of β-Ga_2O_3 NC-FETs. The gate stack includes a heavily n-doped Si as the gate electrode, 20 nm HZO as the ferroelectric insulator, 3 nm Al_2O_3 as the capping layer. Ti/Au (30/60 nm) is used as the source/drain electrodes. Sn-doped n-type β-Ga_2O_3 (86 nm) is used as the channel. **b** Top-view false-color SEM image of representative β-Ga_2O_3 NC-FETs on the same membrane with different channel lengths. **c** Cross-sectional view of the HZO/Al_2O_3 gate stack, capturing the polycrystalline HZO and the amorphous Al_2O_3. **d** P–V characteristics of the 20 nm HZO/3 nm Al_2O_3 gate stack, showing clear ferroelectric hysteresis loop

sweeps. The bidirectional steep slope operation of Ga_2O_3 NC-FETs here supports our DC enhancement model in Fig. S3 of [35]. In contrast, the Ga_2O_3 MOSFETs with 15 nm Al_2O_3 as a gate dielectric has a minimum SS = 118.8 mV/dec. Figure 34.10c shows the I_D–V_{DS} characteristics the same β-Ga_2O_3 NC-FET as in Fig. 34.10a. V_T is extracted to be 0.47 V in forward sweep and 0.38 V in the reverse V_{GS} sweep, indicating an E-mode device operation and a small hysteresis. After more β-Ga_2O_3 material growth development, i.e., amorphous or polycrystalline thin films grown at 450 °C or less, it can also be a very interesting oxide for back-of-the-line (BOEL) device development for logic applications.

In conclusion, the potential of β-Ga_2O_3 nano-membrane as an emerging ultra-wide bandgap semiconductor is intensively studied and its device applications are also explored. High performance both back-gated and top-gated GOOI FETs are fabricated and characterized with record high I_{Dmax} of 1.5/1 A/mm for D/E-modes, lower R_T on a high thermal conductivity substrate and hysteresis-free bidirectional

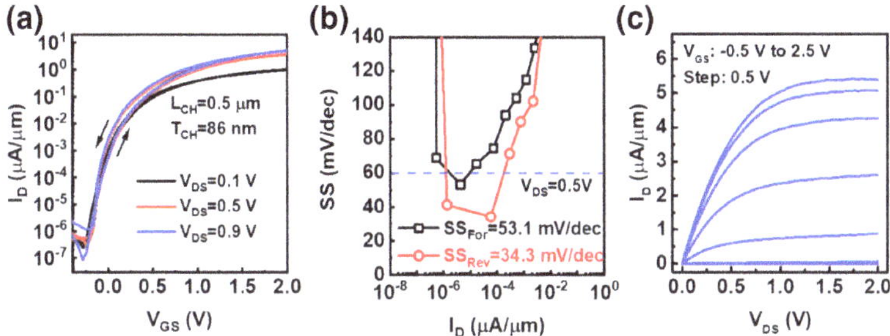

Fig. 34.10 **a** I_D–V_{GS} characteristics of a β-Ga_2O_3 NC-FET. This device has a channel length of 0.5 μm and a channel thickness of 86 nm. **b** SS versus I_D characteristics of the same device in **a** at V_{DS} = 0.5 V. SS less than 60 mV/dec at room temperature is demonstrated for both forward and reverse V_{GS} sweeps. **c** I_D−V_{DS} characteristics of the same β-Ga_2O_3 NC-FET as in (**a**)

steep SS < 60 mV/dec are demonstrated, showing the great promise of β-Ga_2O_3 as a channel material for not only high power devices but also potentially for low-power wide bandgap logic devices.

Acknowledgements This work was supported in part by ASCENT, one of the six centers in JUMP, a Semiconductor Research Corporation (SRC) program sponsored by DARPA. This work was also supported in part by the Office of Naval Research's Naval Enterprise Partnership Teaming with Universities for National Excellence (under Grant N00014-15-1-2833) and Defense Threat Reduction Agency (under Grant HDTRA1-12-1-0025). The authors also would like to thank the collaborations with Kerry Maize, Sami Alajlouni, Marko J. Tadjer, James C. Culbertson, Peter A. Bermel, Ali Shakouri on thermal studies of β-Ga_2O_3 in general.

References

1. M. Higashiwaki, K. Sasaki, H. Murakami, Y. Kumagai, A. Koukitu, A. Kuramata, T. Masui, S. Yamakoshi, Semicond. Sci. Technol. **31**, 034001 (2016)
2. M. Higashiwaki, G.H. Jessen, Appl. Phys. Lett. **112**, 060401 (2018)
3. K. Irmscher, Z. Galazka, M. Pietsch, R. Uecker, R. Fornari, J. Appl. Phys. **110**, 063720 (2011)
4. Z. Galazka, K. Irmscher, R. Uecker, R. Bertram, M. Pietsch, A. Kwasniewski, M. Naumann, T. Schulz, R. Schewski, D. Klimm, M. Bickermann, J. Cryst. Growth **404**, 184 (2014)
5. H. Aida, K. Nishighuchi, H. Takeda, N. Aota, K. Sunakawa, Y. Yaguchi, Jpn. J. Appl. Phys., Part 1 **47**, 8506 (2008)
6. A. Kuramata, K. Koshi, S. Watanabe, Y. Yamaoka, T. Masui, S. Yamakoshi, Jpn. J. Appl. Phys. **55**, 1202A2 (2016)
7. N. Ueda, H. Hosono, R. Waseda, H. Kawazoe, Appl. Phys. Lett. **70**, 3561 (1997)
8. E.G. Víllora, K. Shimamura, Y. Yoshikawa, K. Aoki, N. Ichinose, J. Cryst. Growth **270**, 420 (2004)
9. W.S. Hwang, A. Verma, H. Peelaers, V. Protasenko, S. Rouvimov, H. Xing, A. Seabaugh, W. Haensch, C. Van de Walle, Z. Galazka, M. Albrecht, R. Fornari, D. Jena, Appl. Phys. Lett. **104**, 203111 (2014)

10. S. Ahn, F. Ren, J. Kim, S. Oh, J. Kim, M.A. Mastro, S.J. Pearton, Appl. Phys. Lett. **109**, 062102 (2016)
11. M.D. Santia, N. Tandon, J.D. Albrecht, Appl. Phys. Lett. **107**, 041907 (2015)
12. Z. Guo, A. Verma, X. Wu, F. Sun, A. Hickman, T. Masui, A. Kuramata, M. Higashiwaki, D. Jena, T. Luo, Appl. Phys. Lett. **106**, 111909 (2015)
13. M.H. Wong, K. Sasaki, A. Kuramata, S. Tamakoshi, M. Higashiwaki, IEEE Electron Device Lett. **37**, 212 (2016)
14. A.J. Green, K.D. Chabak, E.R. Heller, R.C. Fitch, M. Baldini, A. Fiedler, K. Irmscher, G. Wagner, Z. Galazka, S.E. Tetlak, A. Crespo, K. Leedy, G.H. Jessen, IEEE Electron Device Lett. **37**, 902 (2016)
15. K.D. Chabak, N. Moser, A.J. Green, D.E. Walker Jr., S.E. Tetlak, E. Heller, A. Crespo, R. Fitch, J.P. McCandless, K. Leedy, M. Baldini, G. Wagner, Z. Galazka, X. Li, G. Jessen, Appl. Phys. Lett. **109**, 213501 (2016)
16. M. Higashiwaki, K. Sasaki, A. Kuramata, T. Masui, S. Yamakoshi, Appl. Phys. Lett. **100**, 013504 (2012)
17. K. Konishi, K. Goto, H. Murakami, Y. Kumagai, A. Kuramata, S. Yamakoshi, M. Higashiwaki, Appl. Phys. Lett. **110**, 103506 (2017)
18. N.A. Moser, J.P. Mccandless, A. Crespo, K.D. Leedy, A.J. Green, E.R. Heller, K.D. Chabak, N. Peixoto, G.H. Jessen, Appl. Phys. Lett. **110**, 143505 (2017)
19. A.J. Green, K.D. Chabak, M. Baldini, N. Moser, R.C. Gilbert, R. Fitch, G. Wagner, Z. Galazka, J. Mccandless, A. Crespo, K. Leedy, G.H. Jessen, IEEE Electron Device Lett. **38**, 790 (2017)
20. H. Zhou, S. Alghmadi, M. Si, G. Qiu, P.D. Ye, IEEE Electron Device Lett. **37**, 1411 (2016)
21. H. Zhou, K. Maize, G. Qiu, A. Shakouri, P.D. Ye, Appl. Phys. Lett. **111**, 092102 (2017)
22. H. Zhou, M. Si, S. Alghamdi, G. Qiu, L. Yang, P.D. Ye, IEEE Electron Device Lett. **38**, 103 (2017)
23. Z. Zhang, E. Farzana, A.R. Arehart, S.A. Ringel, Appl. Phys. Lett. **108**, 052105 (2016)
24. Y. Nakano, ECS J. Solid State Sci. Technol. **6**, 615 (2017)
25. H. Bae, J. Noh, S. Alghamdi, M. Si, P.D. Ye, IEEE Electron Device Lett. **39**, 1708 (2018)
26. K. Zeng, Y. Jia, U. Singisetti, IEEE Electron Device Lett. **37**, 906 (2016)
27. S.-H. Shin, M.A. Wahab, M. Masuduzzaman, K. Maize, J. Gu, M. Si, A. Shakouri, P.D. Ye, M.A. Alam, I.E.E.E. Trans, Electron Devices **62**, 3516 (2015)
28. H. Zhou, K. Maize, J. Noh, A. Shakouri, P.D. Ye, ACS Omega **2**, 7723 (2017)
29. J. Noh, M. Si, H. Zhou, M.J. Tadjer, P.D. Ye, IEEE J. Electron Devices Soc. **7**, 914 (2019)
30. K. Maize, A. Ziabari, W.D. French, P. Lindorfer, B. OConnell, A. Shakouri, IEEE Trans. Electron Devices **61**, 3047 (2014)
31. S. Salahuddin, S. Datta, Nano Lett. **8**, 405 (2008)
32. M. Si, C. Jiang, N.J. Conrad, H. Zhou, K. Mazie, G. Qiu, C.-T. Wu, A. Shakouri, M.A. Alam, P.D. Ye, Nat. Nanotechnol. **13**, 24 (2018)
33. M.A. Alam, M. Si, P.D. Ye, Appl. Phys. Lett. **114**, 090401 (2019)
34. M. Si, L. Yang, H. Zhou, P.D. Ye, ACS Omega **2**, 7136 (2017)
35. M. Si, X. Lyu, P.D. Ye, A.C.S. Appl, Electron. Mater. **1**, 745 (2019)

Chapter 35
Field-Effect Transistors 5

Vertical Ga_2O_3 Fin-Channel Field-Effect Transistors and Trench Schottky Barrier Diodes

Zongyang Hu, Wenshen Li and Huili Grace Xing

Abstract Recently, significant progresses have been made on the demonstration and development of vertical gallium oxide power devices. The goal of this chapter is to give a brief review on vertical gallium oxide FinFETs and Schottky barrier diodes based on fin-channel structures. In both devices, vertical fin-shaped channels are the key to achieving high breakdown voltages and low on-resistances simultaneously. The chapter starts with a short introduction of vertical transistor concepts and development in various wide band gap semiconductors, then moves on to topics including vertical gallium oxide FinFET structures, device operation, high voltage design and baseline fabrication process flow. Fundamental electrical performance will be discussed, followed by our analysis on threshold voltage control, drain-induced barrier lowering effects and breakdown mechanisms. Finally, we will briefly mention our key results on vertical trench Schottky barrier diodes.

Z. Hu (✉) · W. Li · H. G. Xing
School of Electrical and Computer Engineering, Cornell University, Ithaca, NY 14853, USA
e-mail: zh249@cornell.edu

W. Li
e-mail: wl552@cornell.edu

H. G. Xing
e-mail: grace.xing@cornell.edu

H. G. Xing
Department of Materials Science and Engineering, Cornell University, Ithaca, NY 14853, USA

H. G. Xing
Kavli Institute at Cornell for Nanoscale Science, Cornell University, Ithaca, NY 14853, USA

M. Higashiwaki and S. Fujita (eds.), *Gallium Oxide*, Springer Series in Materials Science 293, https://doi.org/10.1007/978-3-030-37153-1_35

35.1 Brief Introduction of Vertical FinFET Device Concept

In comparison with lateral structures, vertical device structures are more preferable for discrete power electronic devices. In general, vertical power devices have higher current density per chip area than the lateral counterparts. This is because the footprint of the vertical devices does not depend on the drift layer thickness, which is larger for devices with higher voltage ratings, whereas the footprint of lateral devices increases with drift layer width. In addition, vertical devices have a more uniform heat distribution than lateral devices, which tends to have localized hot spots or areas. With the availability of high-quality conductive bulk substrates, together with halide vapor phase epitaxy (HVPE) for high-rate growth of high-quality epitaxial layers, vertical device structures are particularly attractive in β-Ga_2O_3.

In the case of power transistors, normally-off (or enhancement-mode) operation is much preferred for fail-safe operations. To realize a normally-off conduction channel, the carriers inside the channel need to be well-confined and depleted at zero gate bias. Due to the lack of conductive *p*-type doping in β-Ga_2O_3, it is not viable to use inversion channels. Alternatively, lightly doped *n*-type channels etched into fins or nanowires can be utilized, thanks for the excellent electrostatic control in such FET structures. By carefully designing the fin width, the carriers in the vertical fin channel can be depleted at zero bias by the built-in work function difference between the gate metal and the β-Ga_2O_3 fin channel; thus, normally-off operation can be achieved in the vertical fin-channel power field-effect transistors (FinFETs). To minimize leakage between the gate and channel, a metal-insulator-semiconductor gate structure is preferred over a metal-semiconductor junction.

The fin-channel width needs to be well below 1 μm to achieve normally-off operations for a fin carrier concentration higher than mid-10^{15} cm^{-3}. These sub-micron channels may appear demanding on lithography, especially when compared with traditional power devices, whose lateral feature size is mostly around or larger than 1 μm. But this requirement can be readily satisfied with the advanced patterning technologies today. As shown in Fig. 35.1, the achievable lateral feature size in Si CMOS FETs is well below the wavelength of the photolithography light source. From the device physics perspective, the vertical fin transistors are similar to the charge-coupled/reduced surface field (RESURF) devices, in which the specific on-resistance can be reduced by reducing the fin width/pitch size ratio, thus breaking the one-dimensional unipolar limit of the semiconductor. In this sense, it is highly desired to take advantage of the ever-evolving lithography technologies driven by the CMOS and keep scaling down the lateral feature size for the RESURF-engineered power devices. The vertical fin-channel MISFETs without employing *p*-type pillars are particularly attractive with respect to scaling, since the MIS-gate on the fin sidewall does not pose limitation on the pitch size as does the *p*-type pillar (for its finite carrier concentration in comparison with metal).

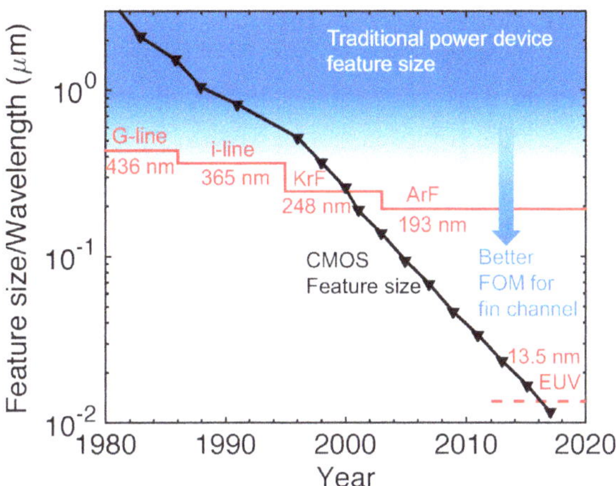

Fig. 35.1 Trend of the photolithography light source wavelength (red line) along with the lateral feature size of Si CMOS transistors (black line) and power devices (top shaded area) with time. Modern power devices can be readily designed to the sub-micron scale by taking advantage of today's patterning technologies

In the following sections, we will discuss the vertical fin-channel MISFETs in details, covering the design, fabrication, operation and characterization of these devices.

35.2 Ga_2O_3 Vertical FinFET Structure Design

35.2.1 Ga_2O_3 Vertical FinFET Device Structure

The basic device structure of our vertical Ga_2O_3 FinFETs resembles that of a static induction transistor (SIT) or junction field-effect transistor (JFET), which includes a vertically aligned fin shaped FET channel, gate junctions formed by Schottky contacts (Fig. 35.2a), p-n junctions (Fig. 35.2b) or Metal-Oxide-Semiconductor junctions (Fig. 35.2c) on both sides of the channel, source ohmic contact on top of the channel, a thick drift layer connected to bottom of the channel and a blanket drain contact on the bottom of the wafer. The concept of SIT dates back to the 1950s when it was first proposed as a solid-state device in analogy to the vacuum tube triode [1] and was experimentally demonstrated in 1972 with a non-saturation output characteristic [2, 3]. During the transistor on-state operation, the drain current flows (carried by electrons for n-channel transistors) vertically through the fin channel and can be modulated by the gate-source voltage (V_{gs}) that controls the width of the conductive channel. During the off-state, V_{gs} is biased below the channel pinch-off voltage and causes a high electron energy barrier inside the channel, therefore, prevents electron flow and turns off the transistor. The name "static induction" refers to the lowering of the energy barrier height in the channel with increased drain voltages when V_{gs} is not far below the pinch-off voltage and output current is relatively low, i.e., the drain-induced barrier lowering (DIBL) effect. Compared to planar FETs, the vertical SIT/JFET/FinFET allows very small

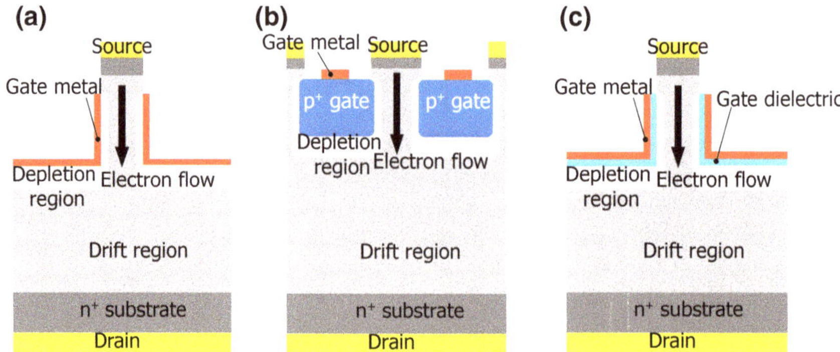

Fig. 35.2 Three typical designs of vertical fin-channel transistors: **a** with Schottky contact gates, **b** vertical SIT/JFET with p-n junction gates and **c** vertical MISFET/FinFET with MOS/MIS gates

cell pitch sizes and very densely packed vertical channels. In addition, high voltage engineering is easier and more reliable due to large separation between the drain and the source/gate electrodes.

The gate structure choices in the vertical SIT/JFET/FinFET type of devices, i.e., vertical FETs with sidewall gates, are an interesting topic to review. All three options in Fig. 35.2 have been adopted to date in various semiconductors depending on the materials properties and device fabrication technologies available.

The first Si-based SITs were fabricated with p-n junction gate structures because both diffusion and ion implantation doping were readily available for Si devices. For GaAs, the p-type GaAs in the gate structures were fabricated by either ion implantation [4] or selective epitaxial regrowth [5]. In addition, the p-n junction gates and direct band gap nature of GaAs were used to realize optical modulation of the output currents [4].

SiC has been the most popular choice of material for the SIT type of devices. Schottky contact-based gate structures were used in the early stage of development, but they were quickly replaced by p-n junction gate structures once p^+ ion implantation and activation annealing technologies became mature for SiC [6]. It was clear that p-n junction gates had numerous advantages over Schottky contact gates with the major ones listed here: (1) higher voltage operation is achieved due to improved edge termination; (2) normally-off operation is easier to realize because of higher junction built-in voltage; (3) non-destructive breakdown allows for electrostatic discharge; (4) operation is robust under elevated temperatures, radiation, etc.

More recently, GaN-based vertical FinFETs/SITs using similar structures but with MIS junctions [7–9] or Schottky gate [10] structures have been developed. The main limiting factors are the same as those for early development of SiC—the lack of a high-quality selective p-type doping technique, thus lateral and buried p-n junctions suffer from high leakage currents and low breakdown voltages. Fortunately, when the sidewalls of Ga-polar *n*-type GaN fins defined by dry etching are subject to a base (TMAH, KOH, etc.) wet etching at elevated temperatures (80 ~ 90 °C), only the m-planes of the GaN crystal are exposed. It turns out that MIS junctions made on these vertical (90°) atomically smooth sidewalls have low trap densities. Therefore, MIS-based vertical FinFETs/SITs are presently one of the best structures for GaN power transistors.

For the newly developed β-Ga_2O_3, since epitaxial p-type doping and conductivity have not been realized to date, we decided to use MIS junction gate structures in the vertical FinFET devices aiming to achieve E-mode operation with high breakdown voltages and low off-state leakage currents. The high breakdown voltage sustained in an MIS junction means that the MIS capacitor stays in deep depletion under high reverse voltages, i.e., a charge inversion layer should "never" form at the insulator/semiconductor interface. Since the vertical MIS-FinFETs under discussion are unipolar n-channel devices without an ohmic access to the valence band of the semiconductor, the only way to form an inversion layer is by thermal activation or impact ionization. We estimate the time constant to form an inversion p-channel due to thermally generated holes as $\tau = \frac{2n\tau_0}{n_i}$, where n is the majority electron concentration, τ_0 is electron/hole lifetime and n_i is the intrinsic carrier concentration, which is reduced exponentially with increased band gap. A rough calculation of this time constant is shown in Fig. 35.3. This figure shows that MIS junctions alone in Si and GaAs power devices cannot sustain high DC voltages since under a high electric field near the insulator/semiconductor interface, an inversion layer forms within milliseconds to minutes! Once an inversion layer is formed, most of the high voltages have to be sustained across the dielectric layer, thus leading to inevitable dielectric breakdown. Therefore, one can

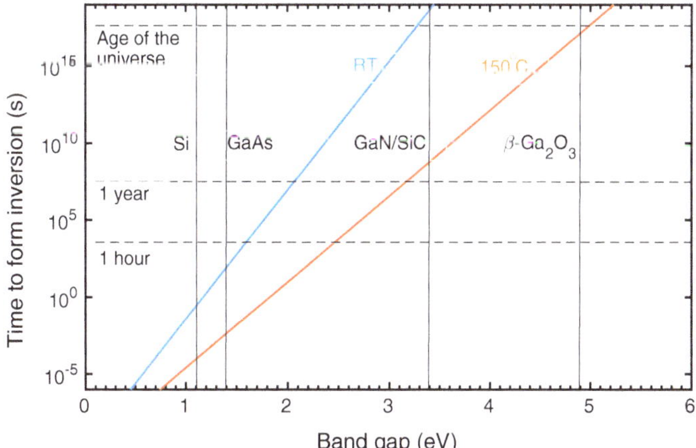

Fig. 35.3 Estimated time constant associated with thermal excitation only to form an inversion layer in various semiconductor MOS junctions under inversion bias, assuming a donor concentration of 10^{15} cm^{-3} and a constant electron/hole lifetime of 85 ns at RT. Though the carrier lifetime deviates from a constant value, this plot shows the general trend on the band gap dependence of the time constant to thermally excite minority carriers in a MOS structure

see why semiconductors with relatively narrow band gap critically rely on the p-type regions designed in power devices to take out the excess holes in order to sustain high voltage operation. Wide band gap (WBG) semiconductors, especially Ga_2O_3 with a gap > 4 eV, are suited for MIS-based power device designs because an inversion layer due to thermal excitation only can "never" form during the device' operation lifetime even at moderately elevated temperatures provided the semiconductor is pure. To the first order of considerations, p-regions are not needed in order to extract holes; unipolar leakage mechanisms such as Fowler-Nordheim tunneling and trap-assisted tunneling often dominate the WBG device breakdown behavior. However, holes generated by impact ionization should be considered separately, and this is still an open question in Ga_2O_3.

Figure 35.4 shows a schematic cross section and a SEM image of a vertical FinFET developed in our lab using β-Ga_2O_3 (001) grown by HVPE on a bulk Ga_2O_3 (001) substrate. Important device parameters are: gate length (L_g), fin/channel width (W_{ch}), gate dielectric thickness (t_{ox}), channel (N_{ch}) and drift layer (N_{dr}) doping concentrations. These normally-off transistors are able to sustain a drain–source voltage over 1 kV at off-state (V_{gs} = 0 V) and deliver ~1 kA/cm^2 output current density in the active region at on-state.

35.2.2 Ga_2O_3 Vertical FinFET Device Operation

The Ga_2O_3 vertical FinFET is essentially a double-gated MOSFET. The transistor operation can be divided into the following regions in its I_d–V_{ds} plot:

(1) Ohmic region: when V_{gs} is high (V_{gs} > the threshold voltage V_{th}), and V_{ds} is relatively low, the channel is not completely depleted near the center vertical

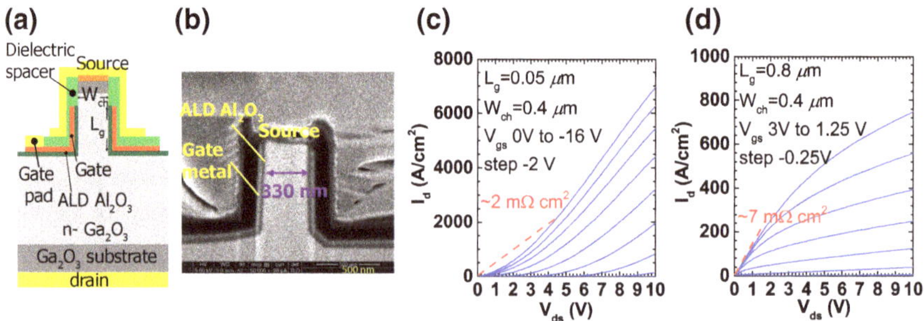

Fig. 35.4 **a** Schematic cross-section and **b** 52° SEM cross-section image of a completed Ga_2O_3 vertical power FinFET with a fin width W_{ch} = 330 nm, a gate length L_g = 795 nm and a gate dielectric thickness t_{ox} = 30 nm [11]. Measured I_d–V_{ds} family *I*–*V* curves of Ga_2O_3 vertical FinFETs with different channel geometries: (c) Triode-like *I*–*V*s with L_g = 0.05 μm, W_{ch} = 0.4 μm, Ga_2O_3 (−201) substrate with $N_{ch} = N_{dr} = 10^{17}$ cm^{-3} (Gen1) [12] and **d** pentode-like *I*–*V*s with L_g = 0.8 μm, W_{ch} = 0.44 μm, $N_{ch} = 10^{16}$ cm^{-3}, $N_{dr} = 10^{15}$ cm^{-3}, HVPE Ga_2O_3 (001) on bulk substrates (Gen2) [13]

line, and the transistor operates as a variable resistor, with its channel resistance controlled by the gate voltage. Under this condition, the total on-resistance of the transistor is estimated as the following:

$$R_{\mathrm{on,sp}} = R_{\mathrm{c}} + R_{\mathrm{ac}} + R_{\mathrm{dr}} \frac{W_{\mathrm{ch}}}{W_{\mathrm{dr}}} + R_{\mathrm{sub}} \frac{W_{\mathrm{ch}}}{W_{\mathrm{sub}}} + \frac{L_{\mathrm{g}}}{qN_{\mathrm{d}}\mu} \cdot \frac{W_{\mathrm{ch}}}{W} \tag{35.1}$$

in which the five components sequentially from left to right are: specific contact resistance R_c coming from the source and drain contact resistances; access resistance R_{ac} representing the resistance of the ungated access region between source and gate; drift region-specific resistance R_{dr}, substrate resistance R_{sub} are both evaluated after considering current spreading, where W_{ch} is the geometric width of the fin channel, W_{dr} and W_{sub} are the averaged width of vertical current flow in the drift region and the substrate; The last term channel resistance is calculated using gate length L_g, electron charge q, channel doping concentration N_d, electron mobility μ and effective width of the channel W. The ohmic region can be clearly observed in the Gen2 Ga_2O_3 vertical FinFETs (Fig. 35.4d). In Gen1 device $I-V$s shown in Fig. 35.4c, the Schottky-like ohmic contacts have an adverse impact on the ohmic region, and the R_{on} is also higher than expected because of this. Equation (35.1) predicts a linear dependence of the transistor's on-resistance with W_{ch}/W, which is consistent with our experimental observation. The exact expression for W depends on what type of junction is used as the gate. For MOS-type gate structures used in our Ga_2O_3 FinFETs, based on simplified 1-D band diagram, we get:

$$\frac{\Delta E_{\mathrm{c}}}{q} - \frac{qN_{\mathrm{d}}(W_{\mathrm{ch}} - W)}{8\varepsilon} - t_{\mathrm{ox}} \frac{qN_{\mathrm{d}}(W_{\mathrm{ch}} - W)/2 - Q_{\mathrm{it}}}{\varepsilon_{\mathrm{ox}}} = V_{\mathrm{gs}} \tag{35.2}$$

where ε and ε_{ox} are relative dielectric constants of the semiconductor and the gate dielectric, Q_{it} is the (fixed) dielectric/semiconductor interface sheet charge concentration usually found in Ga_2O_3 MOS capacitors, ΔE_c is the energy offset between conduction band of the semiconductor and work function of the gate metal, and V_{gs} is the applied gate voltage. From this equation, threshold voltage V_{th} can also be obtained by setting $W = 0$. This relationship can also be used to study V_{th} scaling and dielectric/semiconductor interface properties.

(2) Saturation region: when the transistor is at the on-state, i.e., $V_{gs} > V_{th}$, V_{ds} can be increased to reach the drain current saturation region, when a negative feedback starts to modify the channel resistance and slows down the increment of the drain current. There are three main mechanisms for the negative feedback, each happens with different types of devices and geometries, and they are briefly summarized as follows:

a. Current saturation due to channel pinch-off. This happens in a planar long channel FET as well as a vertical FinFET in which the channel length to width ratio L_g/W_{ch} is very large. Under this condition, the transistor becomes a voltage-controlled current source. At saturation, the gate–drain voltage $V_{gd} = V_{th}$, therefore the channel vanishes near the drain end. Further increase of the drain voltage causes a longer depletion region, while the electric field in the un-depleted channel remains the same because the gate screens the drain voltage. The depleted channel region has no intrinsic carriers and has a very high longitudinal field, which quickly sweeps away the injected carriers from the source side. The current is limited by the injection of carriers into the depleted channel. This mechanism is described by the long channel model of a FET.
b. Velocity saturation. In short channel FETs, the longitudinal field may reach saturation electric field before the channel is pinched off. The carrier velocity (v)–field (E) relationship considering velocity saturation can be modeled as:

$$v = \frac{\mu E}{1 + E/E_{sat}} \tag{35.3}$$

where μ is the low-field mobility and the electric field parameter $E_{sat} = 2v_{sat}/\mu$. In β-Ga_2O_3, the electron saturation velocity is calculated to be $\sim 2\times10^7$ cm/s at electric field of ~ 0.2 MV/cm [14]. In our vertical FinFET I–Vs shown in Fig. 35.4d, a "saturation" I–V behavior is seen at $V_{ds} < 5$ V with a ~ 1 μm long channel. Therefore, this phenomenon is unlikely due to velocity saturation.
Combining the field-dependent velocity and the long channel model, the drain current is described as [15]:

$$I_d = \mu C_{ox}\frac{W}{L}\left(V_g - V_{th} - \frac{V_{ds}}{2}\right)\frac{V_{ds}}{1 + V_{ds}/E_{sat}L} \tag{35.4}$$

and the saturation current is:

$$I_{d,sat} = WC_{ox}(V_{gs} - V_{th} - V_{ds,sat})v_{sat} \tag{35.5}$$

which also gives saturation drain voltage:

$$V_{ds,sat} = \frac{E_{sat}L(V_{gs} - V_{th})}{E_{sat}L + (V_{gs} - V_{th})} \tag{35.6}$$

c. Space charge limited current (SCLC). In a typical SIT, the on-state drain current beyond ohmic region saturates very weakly (or does not saturate at all). When a high gate voltage and a high drain voltage are applied, the energy barrier inside the channel is almost zero. But the conductive channel is not charged neutral anymore because it is flooded with injected carriers. The carrier concentration is the highest near the source where the electric field is the lowest, therefore creating a small energy barrier there. This operation region is referred to as SCLC. Under the assumption that the injected carrier concentration $n \gg N_d$, and electrons are traveling with saturation velocity, the SCLC is written as [16]:

$$I_d = 2\varepsilon v_{sat} \frac{V_{ds}}{\left(L_{gs} + L_g + L_{gd}\right)^2} A_{eff} \tag{35.7}$$

where A_{eff} is the effective channel area. The shorter the gate length, the higher the current.

(3) Exponential region or the thermionic emission region: This happens in the sub-threshold region where drain current depends exponentially the energy barrier height in the channel. It is the special operation mode for SIT due to its short gate length (more accurately, a small L_g/W_{ch} ratio) and in analogy to the vacuum tube triodes. With the channel already pinched off by the gate voltage $V_{gs} < V_{th}$, the drain current is determined by the thermionic emission current over the energy barrier along the least resistive path, which is the central line of the channel. A 3-d plot of potential distribution inside the channel looks like a saddle. Due to comparable L_g and W_{ch}, a deeper pinch-off from the gate can be overcome by a higher drain voltage. This is known as the static induction hence the name SIT, which is so-called drain-induced barrier lowering (DIBL) in MOSFETs. This counterbalance effect between the gate and the drain is described by the voltage blocking gain for SITs:

$$\mu = \left.\frac{\partial V_{ds}}{\partial V_{gs}}\right|_{I_d = \text{const}} \tag{35.8}$$

Meanwhile, DIBL is commonly defined as the shift of V_{th} as a function of V_{ds} at a constant I_d, i.e., DIBL = $1/\mu$.

The thick drift region in a power device has a significant impact on the voltage blocking gain or DIBL because it sustains most of the drain voltage. More details of the analytical models of device operation can be found in [13, 16]. SITs were primarily developed for power amplification applications; therefore, a high voltage blocking gain is desired, i.e., a small DIBL. Similarly, DIBL should be minimized to achieve high breakdown voltage in power FinFETs.

35.2.3 Ga_2O_3 Vertical FinFET Device Fabrication

The baseline process flow and key fabrication steps of vertical Ga_2O_3 MISFETs are described in Fig. 35.5. A typical device process flow is as follows. The epitaxial layers are grown by HVPE on *n*-type bulk Ga_2O_3 (001) substrates ($n = 2 \times 10^{18}$ cm^{-3}). The 10 μm-thick n^--Ga_2O_3 epitaxial layer is doped with Si with a target doping concentration of $<2 \times 10^{16}$ cm^{-3}. Device fabrication starts with Si ion implantation with a box profile of $\sim 10^{20}$ cm^{-3} in the top 50 ~ 100 nm of the drift layer, followed by an activation annealing in N_2 at 1000 °C and 1 atm for 30 min to reduce the contact resistance. Pt metal masks are then patterned by electron beam lithography and deposited by electron beam evaporation on the sample surface to define position and size of the FET channels. Vertical fin channels are etched in an

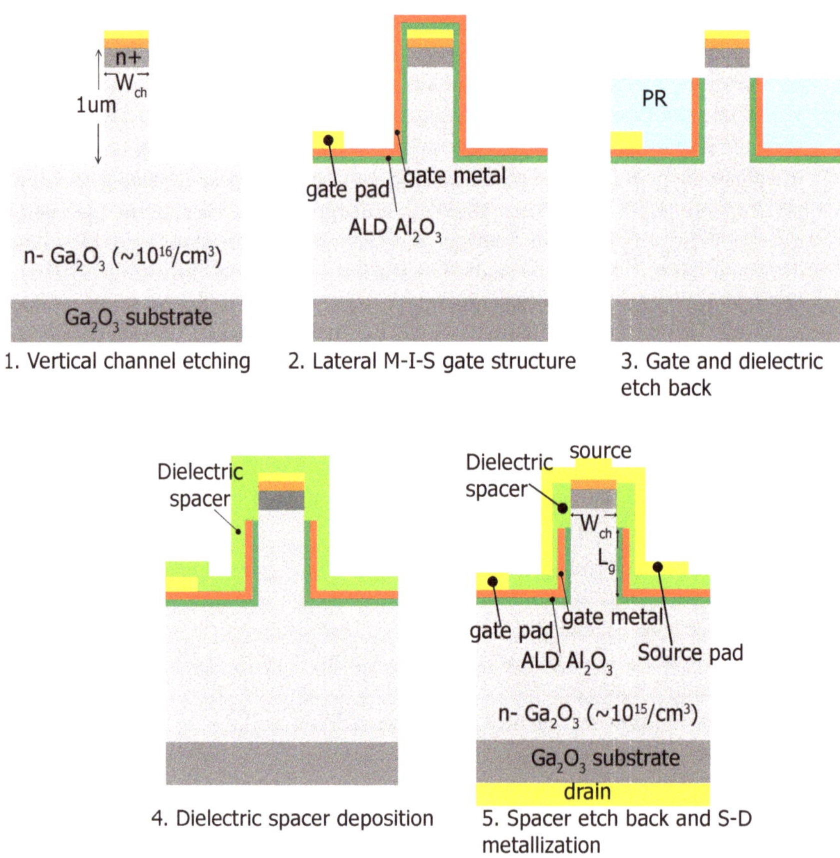

Fig. 35.5 Baseline process flow of Ga_2O_3 vertical FinFETs at Cornell, including an ion implantation and activation step, one gate dielectric deposition and metallization, two planarization/etch back and two ohmic contact formation steps

inductively coupled plasma (ICP) system with a target height of ~ 1.0 μm. Then, a 30 nm Al_2O_3 is deposited in an atomic layer deposition (ALD) system at 300 °C as the gate dielectric, immediately followed by a 50 nm thick Cr sputtered as the gate metal. The gate pads are deposited a few microns away from the channel for the convenience of electrical measurements. A photoresist planarization process is used to selectively etch away the gate metal/dielectric on top of the n^+-Ga_2O_3 source. Then, a 200 nm SiO_2 is deposited by plasma-enhanced chemical vapor deposition (PECVD) and a second planarization process is used to etch away the SiO_2 on top of the n^+ source. The remaining SiO_2 in the device structure serves as the isolation spacer between the gate metal and the source metal. Finally, source ohmic contacts are formed by depositing Ti/Al/Pt, and device isolation is realized by etching away SiO_2 and Cr between active devices. The schematic cross section of a completed device with a single-finger channel is shown in Fig. 35.4, along with a 52° cross-section image of a completed device taken in a focused ion beam (FIB) scanning electron microscopy (SEM) system. The channel width is measured to be 330 nm and the vertical gate length is 795 nm excluding the rounded corners at the bottom of the etched channel.

Unlike conventional planar device fabrication, gate lengths, source-drain distances and alignment in this process are determined by the precise control of dielectric thicknesses and etching thicknesses in both vertical and lateral directions. There are several critical steps that are worthwhile to discuss:

(1) Vertical channel formation. A low-damage dry etching is necessary for definition of the vertical channels. We have developed ICP-RIE dry etching processes that can form 90° sidewalls. Successful application of this etching in our past generations of devices has demonstrated a robust threshold voltage control. A wet etching process should be applied after the dry etching to remove the damaged materials on the sidewalls. There have been literature reports on hot (>200 °C) concentrated (>80 wt%) H_3PO_4, H_2SO_4 and HF etching of Ga_2O_3, with strong etch rate dependence on crystallographic planes [17]. Inverse Metal-Assisted Chemical Etching is reported as an effective way to form high aspect ratio vertical fins [18]. These provide various possible ways of achieving atomically smooth planes on vertical sidewalls.
(2) Gate dielectric. In our past generations of devices, we have identified dielectric breakdown as one of the main limitations of transistor breakdown voltages. To improve this, strong dielectric materials with large band gap, breakdown electric field and dielectric constants are needed to exploit the potential of Ga_2O_3. Because of high growth temperatures, MOCVD in situ or ex situ dielectrics such as AlN and SiN_x [19] may give stronger dielectrics and yield good interfaces with Ga_2O_3.
(3) Edge termination. Off-state breakdown usually happens at the edge of the gate metal pads. Novel field plate technologies include multi-step and slanted field plate structures that can be explored to better high voltage performance. Due to lack of p-type doping, high-resistive Ga_2O_3 can be used instead, which can be

achieved by N ion implantation. One concern is that the breakdown field in the high-resistive Ga_2O_3 may be lower than the as-grown material due to implantation damage.

(4) Device yield and scalability. The main challenge to maximize device yield is control of the gate length and the gate-source spacing. Variation of photoresist thickness across the sample is a challenge for the planarization process. To this end, yield is expected to improve after moving onto large area wafers due to reduced wafer edge areas.

35.2.4 Design for High Breakdown Voltages

In Ga_2O_3 vertical FinFETs, the off-state breakdown voltage (BV) is limited by the electric field peaks shown in the schematic of Fig. 35.6. Position 1 refers to the bottom corners of the fin channels, and Position 2 refers to the edge of the gate metal pads. Based on 2-D device simulations, it is found that the electric field peak at Position 1 depends strongly on the channel width. For typical doping concentrations (channel doping $N_{ch} = 10^{16}$ cm^{-3} and drift layer doping $N_{dr} = 10^{15}$ cm^{-3}), the magnitude of Peak 1 is close to the electric field under the gate pad when W_{ch} is very small (<1 μm), and it asymptotically approaches the peak magnitude at the gate pad edge with increasing W_{ch}. The fin-width dependence of Peak 1 can be confirmed under two extreme cases: (1) $W_{ch} \to 0$, the gap between two electrodes is closed, so the electric field should be equal to that underneath the gate pad. (2) $W_{ch} \gg 0$, the electric field near one of the corners is not affected by the other one; therefore, it approaches the field at the gate edge. A smaller W_{ch} also means the threshold voltage V_{th} is more positive, moving toward normally-off transistors. The electric field peaks at Position 2 is almost independent of the channel geometries; therefore, it is very critical to design edge terminations near Position 2 in order to achieve high breakdown voltages. It should be noted that the electric field peak magnitude in numerical simulations depends on the mesh size used in the device structure, but the same trend shown in Fig. 35.6 should be observed regardless of the choice of mesh sizes.

Lack of conductive p-type doping in Ga_2O_3, edge terminations such as junction termination extension cannot be easily implemented. Resistive implantation isolation is a viable edge termination technique for Ga_2O_3 devices, and it could be combined with field plates (FP) to achieve higher BV. MIS junctions and field plates (FP) are considered in our current device design (Fig. 35.6c and d). The design parameters are the field plate length (L_{FP}) and thickness (t_{FP}) of the supporting dielectric layer. Because of the addition of a field plate, there are two electric field peaks: one at the original gate edge that increases with t_{FP}, and the other one at the FP edge that decreases with t_{FP}. The optimal design is determined when the two electric field peaks reach the same value ($t_{FP} \sim 480$ nm for the design considered in Fig. 35.6c). In addition, L_{FP} needs to be long enough for effective

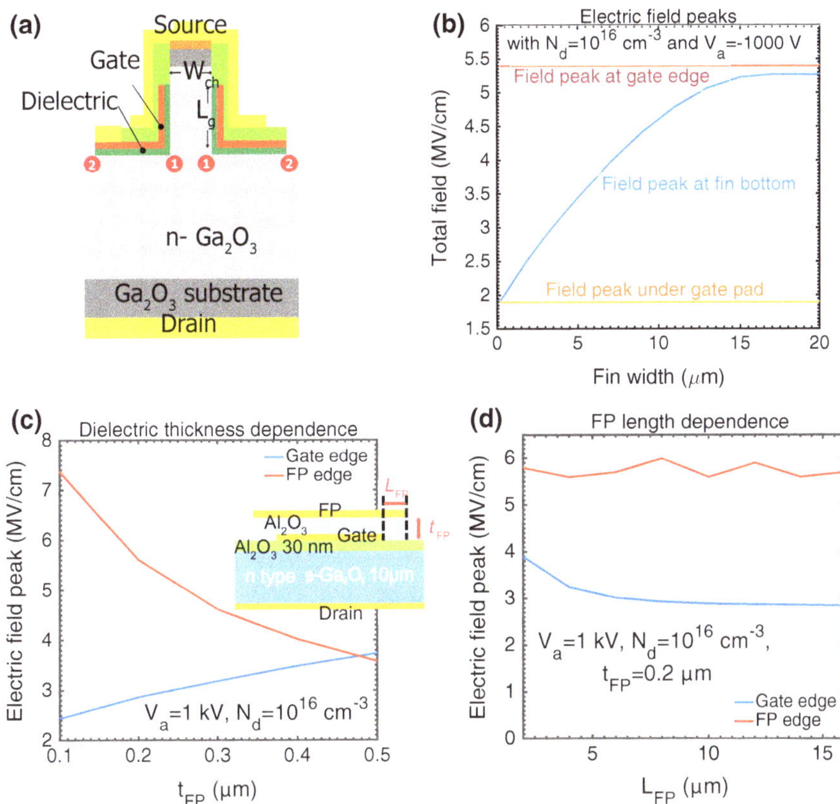

Fig. 35.6 **a**, **b** Simulated electric field values in Ga_2O_3 vertical FinFETs and their dependence on the fin-channel width. "Field peak at fin bottom" and "Field peak at gate edge" correspond to Positions 1 and 2 in the schematic, respectively. "Field peak under gate pad" means the middle area between the two positions. Simulations are for $N_{ch} = N_{dr} = 10^{16}$ cm^{-3} under a reverse bias of 1000 V. **c**, **d** Simulated electric field peaks in field plate designs for drift layer doping concentration of $N_{dr} = 10^{16}$ cm^{-3} and a reverse bias of −1000 V. Electric field values are read at a depth of 0.1 μm below the Ga_2O_3/Al_2O_3 interface

suppression of the electric field peak under the gate edge (Fig. 35.5d), while the peak near the FP edge is almost unaffected by the FP length.

In power devices, the Baliga's figure of merit (BFOM) describes a trade-off relationship between BV and unipolar R_{on} using material parameters including mobility, breakdown electric field and dielectric constant. Employing an optimal edge termination scheme, the "effective critical electric field E_c" defined as the peak electric field in the planar part of a junction approaches the ideal critical field of the semiconductor. Figure 35.7 shows the necessary drift layer thickness and doping concentration for a given "effective E_c". For a better edge termination, the effective E_c is higher. Hence, the drift layer thickness reduces and the doping level can be increased, which leads to lower drift resistances and a lower cost. For the case of

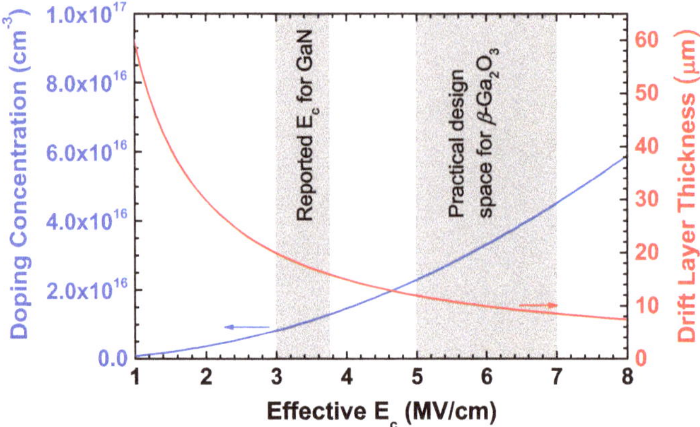

Fig. 35.7 Design space for the drift layer thickness and doping concentration under various edge termination schemes (i.e., effective E_c) for a 3 kV Ga_2O_3 device

GaN, our group, along with a few other groups, have consistently observed a highest effective field between 3 and 3.75 MV/cm. For β-Ga_2O_3, an experimental value of 5.2 MV/cm was reported on a thin film with a thickness less than 200 nm [20]; our group observed a value of 5.9 MV/cm in trench junction barrier didoes [21]. Therefore, it is reasonable to consider a practical design space between 5 and 7 MV/cm for Ga_2O_3.

35.3 Current–Voltage Characteristics of Ga_2O_3 Vertical FinFETs

35.3.1 Device Characteristics

Example device characteristics of Ga_2O_3 vertical FinFETs are shown in Fig. 35.8. A charge concentration of $\sim 10^{16}$ cm^{-3} in the 10 μm HVPE epitaxial layer is extracted from *C–V* measurements. For devices with a fin width W_{ch} of 0.44 μm, the drain current density reaches $\sim$750 A/cm^2 with a V_{gs} of 3 V and a V_{ds} of 10 V. The differential on-resistance R_{on} calculated near $V_{ds} = 0$ V is $\sim$7 mΩ cm^2. The transistor shows a noticeable output conductance since an aspect ratio L_g/W_{ch} of $\sim$1.8 is not high enough to completely suppress the DIBL effects. A high current on/off ratio of $\sim 10^9$ and an on-current of >1 kA/cm^2 (normalized to the area of the source contact) is measured. The maximum drain current I_d observed in the I_d–V_{gs} transfer curve is higher than that in the I_d–V_{ds} curves due to device self-heating and D_{it} at the gate dielectric and channel interface. The sub-threshold slope (SS) is extracted as $\sim$80 mV/dec near the drain current of 1 mA/cm^2. Since the SS of a MOSFET is written as [22]:

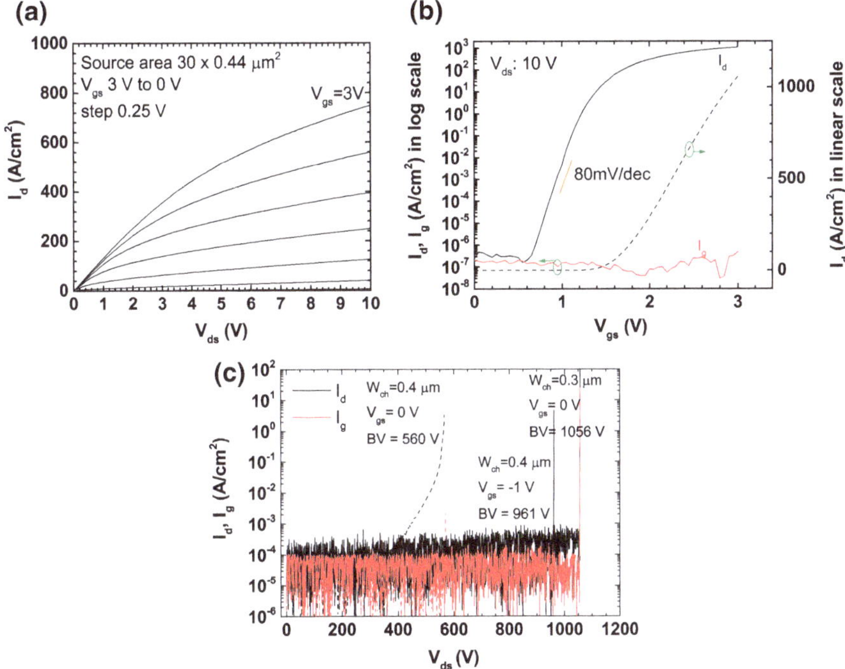

Fig. 35.8 **a**, **b** Current-voltage characteristics of Ga_2O_3 vertical FinFETs developed at Cornell (channel width = 0.44 μm) [13]. **c** Breakdown voltages of vertical FinFETs with 0.3 and 0.4 μm channel widths

$$\mathrm{SS} = \frac{kt}{q}\ln 10\left(1 + \frac{C_{\mathrm{d}} + qD_{\mathrm{it}}}{C_{\mathrm{ox}}}\right) \tag{35.9}$$

where k is the Boltzmann constant, T is the temperature, q is the elementary charge, C_{d} is the depletion capacitance of the channel (at pinch-off condition $C_{\mathrm{d}} \sim 0$ for double-gated junctionless MISFETs including FinFETs), C_{ox} is the capacitance per unit area associated with the gate dielectric, and D_{it} the interface state density. We can thus estimate D_{it} to be $>3\times10^{11}$ cm^{-2} eV^{-1}. This value is comparable to a few other reports on interface state extraction using Ga_2O_3 MOS capacitors [23, 24]. V_{th} extracted by linear extrapolation of the drain current is $\sim$1.6 V. The peak g_{m} at V_{gs} = 2.25 V and V_{ds} = 10 V is $\sim$850 S/cm^2. Off-state BVs of these transistors are measured and shown in Fig. 35.8c. For both 0.3 and 0.4 μm channel widths, the breakdown voltages are near 1000 V. For the 0.4 μm channel transistor, a blocking voltage of 560 V is measured at V_{gs} = 0 V. A clear soft breakdown behavior of the drain current appears at V_{ds} > 380 V while the gate current stays low at the instrumentation limit. This suggests that the transistor breakdown at V_{gs} = 0 V is limited by DIBL. To further probe this hypothesis, another set of devices is measured under negative gate bias. At V_{gs} = −1 V, much higher BVs of up to $\sim$960 V are observed; moreover, both gate and drain currents increase abruptly and simultaneously at breakdown, which is destructive. These observations exclude the likelihood of avalanche breakdown at $V_{gs} = 0$ V, thus confirming DIBL is the dominant mechanism. Based on the highest BV and R_{on}, a Baliga's figure of merit of 124 MW/cm^2 can be calculated from, highest in all Ga_2O_3 transistors to date.

35.3.2 Threshold Voltage Control and Normally-on Versus Normally-off FinFETs

When the fin-channel widths are varied from 0.4 to 0.1 μm, based on (35.2), the threshold voltage V_{th} is expected to be more positive, moving toward normally-off transistors. As shown in the transfer characteristics of the FinFETs in Fig. 35.9, an increase in V_{th} from ~1.6 to ~2.4 V is observed when the fin width (W_{fin}) is reduced from 0.4 μm to 0.1 μm. However, with 30 nm ALD Al_2O_3 gate dielectric and Cr gate, if we assume there is no interface-trapped charge Q_{it}, the calculated V_{th} is much lower than the measured values. This indicates the presence of negatively charged interface traps. This is further supported by the frequency-dependent *C–V* measurement, where the high-frequency (5 MHz) capacitance near 10 V saturates at a value smaller than the gate oxide capacitance (C_{ox}), indicating the presence of interface-trap density (D_{it}) near the conduction band edge (E_C) and negative Q_{it}. The D_{it} profile is extracted using the conductance method. Using the extracted $Q_{it}(\varphi_s)$ and D_{it} profile, an excellent match between the calculated and measured V_{th} is achieved. The interface-trapped charge at the Al_2O_3–Ga_2O_3 interface is around -3.5×10^{12} cm^{-2}. The charge trapping could come from: (1) Trap states in the dielectric, such as border traps. (2) D_{it} due to Ga_2O_3-dielectric interface states. (3) Deep states that are located under the Ga_2O_3 surface (into the body). Processing damages such as dry-etched damage and CMP damage are believed to create subsurface damage, and thus the source of (3). The preparation of the dielectric influences both (1) and (2). The interface states can usually be reduced by surface treatment including piranha etching, concentrated HF and other wet etching methods. Post deposition annealing is also found to reduce the charge trapping. This work demonstrated a simple method to engineer V_{th} in Ga_2O_3 fin-channel transistors by varying the fin width and important understandings of the interface between Al_2O_3 and processed Ga_2O_3.

35.3.3 Voltage Blocking Gain and Drain-Induced Barrier Lowering Effects

In Fig. 35.10a [13], the transfer *I–V* of the same device is shown at varied V_{ds} biases. In order to minimize the effect from D_{it}, V_{gs} is swept from 0 to 3 V with the

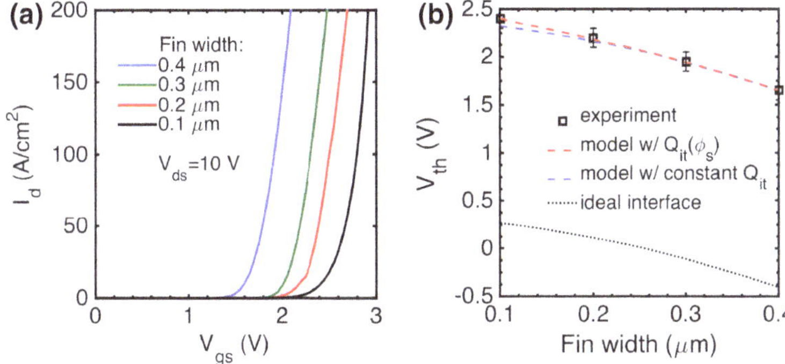

Fig. 35.9 Experimental transfer curves of Ga_2O_3 vertical FinFETs with different fin widths (left) and threshold voltage (V_{th}) versus fin width from experiment and modeling (right). The calculation using the extracted Q_{it} and D_{it} profile obtains the best match with the experiment. With a fixed Q_{it}, the slope of the curve is shallower than the experimental data

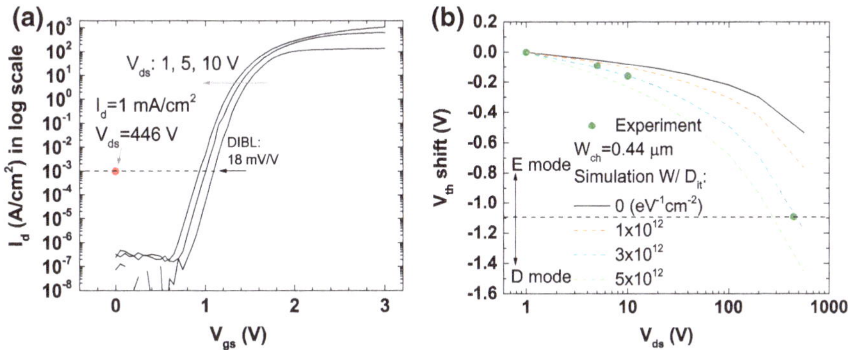

Fig. 35.10 **a** I_d–V_{gs} transfer characteristics of a vertical Ga_2O_3 power FinFET in the semi-log scale measured at 1, 5 and 10 V. **b** V_{th} shifts as a result of increased V_{ds}. Dashed lines are simulation results assuming various interface state densities. The solid line is simulation without interface states. Solid symbols are experimental data. These transistors have a V_{th} at 1 mA/cm^2 of ~1.1 V near V_{ds} ~ 0 V. The V_{th} shift is calculated by $\Delta V_{th}(V_{ds}) = V_{gs}(V_{ds}) - V_{gs}(V_{ds} = 1\text{ V})$ at $I_d = 1$ mA/cm^2

same sweeping rate in all three curves. The DIBL effect, defined as DIBL = $\Delta V_{gs}/\Delta V_{ds}$, is measured to be 18 mV/V near the current density of 1 mA/cm^2, corresponding to a voltage blocking gain of ~56.

Due to the complexity associated with the thick drift region, edge termination near the gate as well as the interface states, 2-dimensional simulations using Sentaurus are employed for the quantitative analysis of DIBL in these vertical Ga_2O_3 power FinFETs. Figure 35.10b shows the simulated V_{th} shift as a function of V_{ds} with and without the impact of interface states. Experiment data at 1, 5 and 10 V are extracted from V_{gs} = 1.093 V, 1.004 V and 0.935 V at I_d = 1 mA/cm^2;

we observe $V_{ds} \sim 446$ V at $V_{gs} = 0$ V and $I_d = 1$ mA/cm^2 thus the net V_{th} shift can be calculated to be −1.093 V at $V_{ds} \sim 446$ V. The comparison between the simulations and experimental results indicates the presence of D_{it} with a value of $\sim 3 \times 10^{12}$ cm^{-2} eV^{-1} under the bias conditions considered here. Based on the simulation, it is expected that a MISFET with the same geometry but an optimized interface treatment would have a blocking voltage beyond 1 kV at $V_{gs} = 0$ V. In addition, it is observed that the V_{th} shift curve for devices with a W_{ch} of 0.55 μm and zero interface states roughly aligns with the curve for devices with a W_{ch} of 0.44 μm and a D_{it} of 3×10^{12} cm^{-2} eV^{-1} at all voltages. This further confirms that the existence of interface states is equivalent to a wider channel in terms of their impact on DIBL. The simulation also indicates that though DIBL is reduced at higher V_{ds}, the V_{th} shift (integration of DIBL with respect to V_{ds}) shows no trend of saturation, thus remaining a challenge for high voltage transistor operations.

35.4 Summary of Vertical Power FinFETs and Brief Mention of Trench SBDs

Over the short time of development since 2017, the Ga_2O_3 vertical power FinFETs/MISFETs have shown encouraging performance compared to other Ga_2O_3 transistors types. These include >1 kV breakdown voltage and ~kA/cm^2 maximum drain currents. Figure 35.11 is a summary of transistor performance in terms of

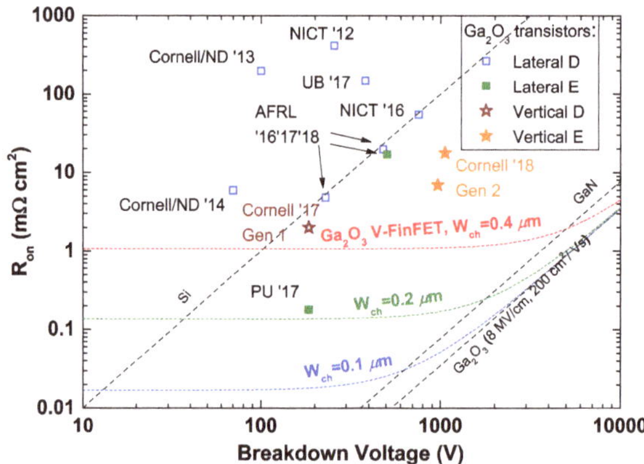

Fig. 35.11 Benchmark of on-resistance and breakdown voltage values of the state-of-the-art Ga_2O_3 lateral and vertical power transistors. To calculate R_{on}, the active device area excluding ohmic contacts and metal pads has been used, i.e., channel width source–drain distance for lateral devices and the source area on top of the fin channel for vertical devices. Channel resistances are calculated by assuming the cell pitch size W_p and channel length L_g both scales with the channel width W_{ch}, and channel doping scales with $1/W_{ch}^2$ so that V_{th} can remain the same. References used in the figure: ND '13 [25], '14 [26], NICT '12 [27], '16 [28], AFRL '16 [29], '17 [30], '18 [31], UB '17 [32], PU '17 [33], Cornell '17 [11], '18 [12, 13]

their R_{on} and BV. Compared with lateral transistors, Ga_2O_3 vertical transistors have already shown advantage in their power density performance. In addition, normally-off transistors have also been achieved for power switching applications. In order to improve device performance, one of the most important tasks is to improve edge termination so that a higher effective E_c can be realized. This includes looking for better stronger dielectric materials, i.e., higher E_c (breakdown electric field) $\times$ ε (dielectric constant), and electric field management by designing edge termination.

The vertical FinFET has several advantages compared to lateral FETs regarding high-power applications.

(1) Fully vertical channel allows very efficient use of the chip area and very small pitch sizes. Since the metal gate is placed on the sidewalls of the channel, the cell footprint is minimized. Multiple cells can be packed very densely to deliver high currents. To this end, vertical FinFETs can benefit from the advanced lithographic technologies shown in Fig. 35.1 and achieve the best power densities.
(2) Placing drain contacts on the backside of the wafer is advantageous for high BV. The breakdown electric field inside the bulk of a semiconductor is typically much higher than that on the surface, minimizing the adverse effects of surface defects. Besides, when designing devices for higher BVs, the chip area can be kept the same but to increase the drift layer thickness. This is not the case for lateral transistors because the gate–drain spacing has to increase to support higher voltages.
(3) Double-side gating of the fin channel or gate-all-around structure of a nanowire channel provides superior electrostatic control compared to the conventional top gate control of a planar power MOSFET without an electrically controlled back barrier. It has been shown in Sect. 35.2.4 that double gating can also improve BV with properly designed edge terminations.

However, vertical power transistors require precise control of doping concentration and compensation levels over tens of microns of epitaxial layers, and a high-quality low-dislocation substrate to start with. Ga_2O_3 single-crystal bulk substrates are available through melt-growth methods. This is one of the main advantages of Ga_2O_3 vertical power devices compared to those based on SiC and GaN.

It can be shown that miniaturization of the lateral feature sizes in vertical FinFET is beneficial to their high-power performance. When channel resistance has to be considered as in the case of a power transistor, the theoretical limits of the BV versus R_{on} can be calculated as the dotted lines in Fig. 35.11. One can see that the reduction of the channel width reduces the lower limit of R_{on}, especially for the low voltage range where the channel resistance is not negligible compared to the drift resistance. Moreover, it has also been shown in Fig. 35.6 that reduction of channel width improves BV through the RESURF effect, which further improves the BFOM. However, the negative impact of dielectric/semiconductor interface traps

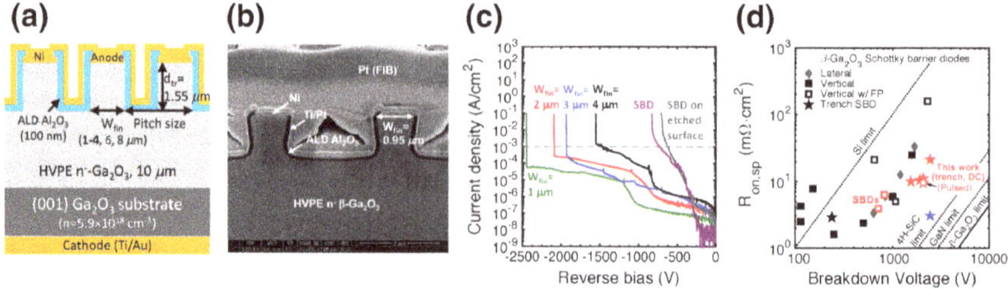

Fig. 35.12 From left to right: schematic cross section of the Ga_2O_3 trench SBDs designed at Cornell, SEM image of fabricated trench SBDs, reverse blocking voltage of trench SBDs with varied fin widths, and the benchmark plot of the demonstrated SBDs. The red symbols in the benchmark plot are results from Cornell and the blue star is the projected performance if and only if the electron mobility in the HVPE Ga_2O_3 were 150 cm²/Vs instead of 71 cm²/Vs as extracted from the experimental data adopted from Reference [21]

may become more challenging when the channel width is scaled down, and threshold instability, mobility degradation, on-resistance and dynamic switching behavior should all be evaluated to determine the optimal device design.

Trench Schottky barrier diodes (SBDs) share many similar device layout characteristics to the vertical power FinFETs. Due to space limitation, we will briefly mention some of our key findings in trench SBDs published in [21, 34] and also shown in Fig. 35.12.

The reduced surface electric field at the Schottky metal and Ga_2O_3 interface is achieved by etching trenches into a planar SBD structure, which can be alternatively described as current-carrying fins. On the sidewalls of the fins, MIS structures are formed. The space charge in the fin channel images with the metal contact on the sidewalls; as a result, the narrower the fins are, the stronger the RESURF effect. In turn, we expect to observe higher breakdown voltage in narrower fins, which was successfully observed in our devices. The extracted peak electric field in Ga_2O_3 in these devices is about 5.9 MV/cm, the highest ever reported in Ga_2O_3, most likely among all semiconductors.

In conclusion, RESURF effects are key in achieving the high-power performance Ga_2O_3 promises. One can employ combinations of the three types of basic junctions: Schottky, p-n and MIS junctions to optimize the RESURF design. Vertical power FinFETs and trench SDBs are two examples of device topologies that take advantage of the properties of a MIS junction.

References

1. Y. Watanabe, J. Nishizawa, Japanese Patent 205,068, 1950
2. J. Nishizawa, in *Technical Digest of the IEEE International Electron Devices Meeting*, 1972
3. J. Nishizawa, T. Terasaki, J. Shibata, I.E.E.E. Trans, Electron Devices **22**, 185 (1975)

4. P. Hadizad, J. Hur, H. Zhao, K. Kaviani, M. Gundersen, H. Fetterman, IEEE Electron Device Lett. **14**, 190 (1983)
5. J. Nishizawa, P. Plotka, T. Kurabayashi, IEE Proc. Circ. Devices Syst. **146**, 27 (1999)
6. G.C. DeSalvo, Int. J. High Speed Electron. Syst. **15**(4), 997 (2005)
7. M. Sun, M. Pan, X. Gao, and T. Palacios, in *Proceedings of the Device Research Conference*, 2016
8. Y. Zhang, M. Sun, D. Piedra, J. Hu, Z. Liu, Y. Lin, X. Gao, K. Shepard, T. Palacios, in *Technical Digest of the IEEE International Electron Devices Meeting*, 2017
9. Z. Hu, W. Li, K. Nomoto, M. Zhu, X. Gao, M. Pilla, D. Jena, H.G. Xing, *in Proceedings of the Device Research Conference*, 2017
10. W. Li, D. Ji, R. Tanaka, S. Mandal, M. Laurent, S. Chowdhury, IEEE J. Electron Devices Soc. **5**(6), 485 (2017)
11. Z. Hu, K. Nomoto, W. Li, N. Tanen, K. Sasaki, A. Kuramata, D. Jena, H.G. Xing, IEEE Electron Device Lett. **39**(6), 869 (2018)
12. Z. Hu, K. Nomoto, W. Li, L. Zhang, J.-H. Shin, N. Tanen, T. Nakamura, D. Jena, H.G. Xing, in *Proceedings of Device Research Conference,* 2017
13. Z. Hu, K. Nomoto, W. Li, Z. Zhang, N. Tanen, Q.T. Thieu, K. Sasaki, A. Kuramata, T. Nakamura, D. Jena, H.G. Xing, Appl. Phys. Lett. **113**, 122103 (2018)
14. R.S. Muller, T.I. Kamins, M. Chan, *Device Electronics for integrated Circuits*, 3rd edn. (Wiley, 2003), Chap. 9
15. A. Przadka, Dissertation, Purdue University, 1999
16. C. Bulucea, A. Rusu, Solid-State Electron. **30**(12), 1227 (1987)
17. S.J. Pearton, J. Yang, P.H. Cary IV, F. Ren, J. Kim, M.J. Tadjer, M.A. Mastro, Appl. Phys. Rev. **5**, 011301 (2018)
18. S.H. Kim, P.K. Mohseni, Y. Song, T. Ishihara, X. Li, Nano Lett. **15**, 641 (2015)
19. R. Li, Y. Cao, M. Chen, R. Chu, IEEE Electron Device Lett. **37**(11), 1466 (2016)
20. X. Yan, I.S. Esqueda, J. Ma, J. Tice, H. Wang, Appl. Phys. Lett. **112**(3), 032101 (2018)
21. W. Li, K. Nomoto, Z. Hu, R. Jinno, Z. Zhang, T.Q. Tu, K. Sasaki, A. Kuramata, D. Jena, H. G. Xing, in *Technical Digest of the IEEE International Electron Devices Meeting*, 2018
22. S.M. Sze, *Physics of Semiconductor Devices*, 2nd edn. (Wiley, 1981), Chap. 8.2.3, p. 447
23. H. Zhou, S. Alghmadi, M. Si, G. Qiu, P.D. Ye, IEEE Electron Device Lett. **37**(11), 1411 (2016)
24. K. Zeng, Y. Jia, U. Singisetti, IEEE Electron Device Lett. **37**(7), 906 (2016)
25. W.S. Hwang, A. Verma, V. Protasenko, S. Rouvimov, H.G. Xing, A. Seabaugh, W. Haensch, C. Van de Walle, Z. Galazka, M. Albrecht, R. Fornari, D. Jena, in *Proceedings of the Device Research Conference*, 2013
26. W.S. Hwang, A. Verma, H. Peelaers, V. Protasenko, S. Rouvimov, H.G. Xing, A. Seabaugh, W. Haensch, C. Van de Walle, Z. Galazka, M. Albrecht, R. Fornari, D. Jena, Appl. Phys. Lett. **104**, 203111 (2014)
27. M. Higashiwaki, K. Sasaki, A. Kuramata, T. Masui, S. Yamakoshi, Appl. Phys. Lett. **100**, 013504 (2012)
28. K.D. Chabak, N. Moser, A.J. Green, D.E. Walker Jr., S.E. Tetlak, E. Heller, A. Crespo, R. Fitch, J.P. McCandless, K. Leedy, M. Baldini, G. Wagner, Z. Galazka, X. Li, G. Jessen, Appl. Phys. Lett. **109**, 213501 (2016)
29. A.J. Green, K.D. Chabak, E.R. Heller, R.C. Fitch, M. Baldini, A. Fiedler, K. Irmscher, G. Wagner, Z. Galazka, S.E. Tetlak, A. Crespo, K. Leedy, G.H. Jessen, IEEE Electron Device Lett. **37**(7), 902 (2016)
30. N. Moser, J. McCandless, A. Crespo, K. Leedy, A. Green, A. Neal, S. Mou, E. Ahmadi, J. Speck, K. Chabak, N. Peixoto, G. Jessen, IEEE Electron Device Lett. **38**(6), 775 (2017)
31. K.D. Chabak, J.P. McCandless, N.A. Moser, A.J. Green, K. Mahalingam, A. Crespo, N. Hendricks, B.M. Howe, S.E. Tetlak, K. Leedy, R.C. Fitch, D. Wakimoto, K. Sasaki, A. Kuramata, G.H. Jessen, IEEE Electron Device Lett. **39**(1), 67 (2018)
32. K. Zeng, J.S. Wallace, C. Heimburger, K. Sasaki, A. Kuramata, T. Masui, J.A. Gardella Jr., U. Singisetti, IEEE Electron Device Lett. **38**(4), 513 (2017)

33. H. Zhou, M. Si, S. Alghamdi, G. Qiu, L. Yang, P.D. Ye, IEEE Electron Device Lett. **38**(1), 103 (2017)
34. W. Li, Z. Hu, K. Nomoto, Z. Zhang, J.-Y. Hsu, T.Q. Tu, K. Sasaki, A. Kuramata, D. Jena, H. G. Xing, Appl. Phys. Lett. **113**, 202101 (2018)

Chapter 36
Diodes 1

Vertical Geometry Ga_2O_3 Rectifiers

Jiancheng Yang, Minghan Xian, Randy Elhassani, Fan Ren, S. J. Pearton, Marko J. Tadjer and Akito Kuramata

Abstract There is increasing interest in Schottky rectifiers made on wide bandgap semiconductors because of their fast switching speed, which is important for improving the efficiency of inductive motor controllers and power supplies, as well as their low forward voltage drop and high-temperature operability. The advantage of simple Schottky rectifiers over p-n diodes is the shorter switching times due to the absence of minority carriers. This, however, leads to higher on-state resistance (R_{ON}) values than in p-i-n rectifiers. The availability of a wider bandgap than Si improves the rectifier performance, with lower on-state resistance at a given reverse voltage. Both GaN and SiC power Schottky diodes have demonstrated shorter turn-on delays than comparable Si devices. Recently, vertical geometry rectifiers fabricated on thick epitaxial layers of β-Ga_2O_3 on conducting substrates grown by edge-defined film-fed growth (EFG) have shown promising performance in terms of high reverse breakdown voltage (V_B > 1 kV) and low R_{ON}, leading to good power figure-of-merits (V_B^2/R_{ON}) Electrical breakdown caused by impact ionization process will preferentially occur at the contact periphery if the maximum electric field in these areas is

J. Yang (✉) · M. Xian · R. Elhassani · F. Ren · S. J. Pearton
University of Florida, Gainesville, FL 32611, USA
e-mail: yjcallen@ufl.edu

M. Xian
e-mail: mxian@ufl.edu

R. Elhassani
e-mail: randyelhassani@ufl.edu

F. Ren
e-mail: fren@che.ufl.edu

S. J. Pearton
e-mail: spear@mse.ufl.edu

M. J. Tadjer
U.S. Naval Research Laboratory, Washington, DC 20375, USA
e-mail: marko.tadjer@nrl.navy.mil

A. Kuramata
Novel Crystal Technology, Inc., Sayama, Saitama 350-1305, Japan
e-mail: kuramata@novelcrystal.co.jp

M. Higashiwaki and S. Fujita (eds.), *Gallium Oxide*, Springer Series in Materials Science 293, https://doi.org/10.1007/978-3-030-37153-1_36

not reduced by proper edge termination design. This chapter summarizes recent advances in the design and implementation of Ga_2O_3 vertical geometry rectifiers.

36.1 Introduction

There is strong interest in the development of ultra-high power inverter modules based on wide bandgap semiconductors [1–10]. These would have application in a range of power conditioning systems, including pulsed power for avionics and electric ships, in solid-state drivers for heavy electric motors and in advanced power management and control electronics [4–10]. The development of high power electronics capable of operating at elevated temperatures, without the need for extensive system cooling requirements, is attractive in many industrial and particularly, military applications for power switching devices, boosting conversion efficiency [7–10]. A large plurality (~80%) of electronic systems, from communications to industrial manufacturing and e-mobility, requires conversion of primary electricity into another form of electricity. Therefore, highly efficient energy conversion for these systems is critical, and this depends mainly on the ability of power switching transistors to provide low resistance on-state and highly resistive off-state conditions.

Schottky rectifiers are a key element of inverter modules because of their high switching speeds and low switching losses [5–9]. The saturation current in a Schottky barrier diode is 10^3–10^8 greater than the current in a p-n junction [1]. This is advantageous, for example, in low-voltage, high-current rectifiers. To produce the same current level, a much smaller forward bias voltage is necessary in a Schottky rectifier compared to that in a p-n junction. The static conduction loss (P_{cond}) of a Schottky or p-n diode is proportional to the diode series resistance, R_S, and forward voltage drop across the diode, V_D, by [7, 10]

$$P_{\mathrm{cond}} = I_{\mathrm{DC}} \cdot V_{\mathrm{D}} + I_{\mathrm{DC}}^2 \cdot R_{\mathrm{S}}.$$

where I_{DC} is the conduction current. Given the smaller forward voltage drop across the Schottky diode, a much smaller loss is expected compared to the loss across a p-n junction device. A Schottky rectifier is often the last stage in a switching power supply to convert low-voltage AC power to filtered DC [10]. The efficiency of the power converters used to transmit the energy source power to the electric machines can be a significant factor in the overall performance of the system.

The use of wide bandgap semiconductors is particularly advantageous to overcome the limitations of Si pin diodes, which suffer from high reverse recovery charge and therefore reduce the usable switching speed and exhibit high switching losses. For Si, the critical field E_c is about 20 V μm^{-1} while for wider bandgap materials, E_c is close to 300 V μm^{-1}, as shown in Fig. 36.1. The figures of merit (FOM) for wide bandgap semiconductors are much larger than for Si in these applications [2, 6]. In general, the large bandgap of Ga_2O_3 or the more established SiC or GaN allows high-temperature operation without complete carrier ionization

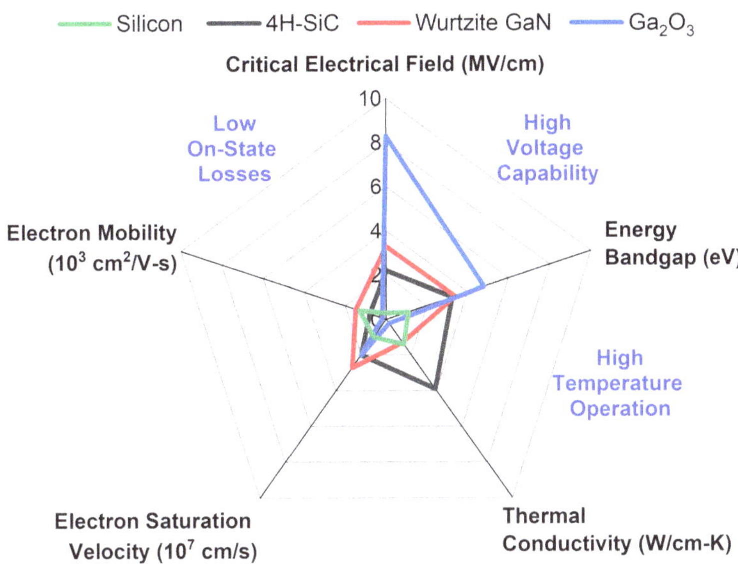

Fig. 36.1 The pentagon diagram showing the critical material properties important to power semiconductor devices. A larger pentagon is preferred. The data is taken from [6, 7, 10]

and high-voltage operation (relative to maximum breakdown). There are also advantages of specific materials systems, for example, the high electron mobility of an AlGaN/GaN high electron mobility transistor (HEMT) is advantageous as R_{on} is inversely proportional to mobility. Similarly, the large thermal conductivity of SiC minimizes self-heating effects [7]. The wide bandgap of Ga_2O_3 allows device operation at elevated temperature (>300 °C) without degradation. To evaluate the performance of power electronics, the Baliga's figure of merit (FOM) ($\varepsilon_r \mu E br^3$) where ε_r is the relative dielectric constant and μ is the mobility), is often used [2]. β-Ga_2O_3 has a 4 × larger Baliga's FOM than SiC and GaN [4, 5, 9], indicating β-Ga_2O_3 power electronics have the potential to outperform SiC and GaN devices [11–19].

One of the most important advantages of β-Ga_2O_3 is the availability of cost-effective single-crystal substrates [5, 20–22]. Edge-defined film-fed growth (EFG) is one of the most popular methods for producing large-sized β-Ga_2O_3 substrates due to its low cost and compatibility with mass production [4, 21, 22]. High-quality two-inch (201) substrates grown by the EFG method have been commercialized with controllable doping concentrations ranging from 10^{16} to 10^{19} cm^{-3} and four-inch substrates have also been demonstrated [21, 22]. Additionally, Ga_2O_3 has a high saturation electron velocity ($v_{sat} = 2 \times 10^7$ cm/s), which is advantageous for high current density, I_{max} ($I_{max} \sim qnv_{sat}$ where $q = 1.6 \times 10^{-19}$ C, n = charge density, v_s = electron saturation velocity)[(5)]. Despite the relatively low thermal conductivity of Ga_2O_3, the rapid development of high-quality native Ga_2O_3 substrates lowers the overall cost of production and avoids many of the defect-related issues that have hampered GaN and SiC devices. It is expected that Ga_2O_3-based devices will be competitive with Si-based medium-power as well as GaN and SiC-based high-power electronic devices [10] and there are aggressive plans to lower the cost of Ga_2O_3 substrates.

36.2 Applications

The initial thrust on Ga_2O_3 electronics is targeted toward high-power converters for both DC/DC and DC/AC applications [4–6, 18–20, 23–29]. Ga_2O_3 Schottky diodes could displace 600 V Si or SiC rectifiers targeted at switch mode power converters [7, 30]. For power switching applications, the operating voltage is limited by the breakdown electric field strength (E_{br}) and the background doping in epitaxial drift layers [31]. The total energy loss is determined by resistive power dissipation during on-state current conduction and the capacitive loss during dynamic switching. The power frequency product for Ga_2O_3 is comparable to that for GaN, even though the saturation electron velocity in Ga_2O_3 is lower [5, 10]. The difference is compensated by the higher E_{br} of Ga_2O_3. Figure 36.2 shows some suggested application spaces for Ga_2O_3 to take advantage of its large breakdown capability. One limitation to Ga_2O_3 material for power electronics is its low thermal conductivity (11–27 W/mK, cf. Si at 130 W/mK, SiC at 360–490 W/mK, and GaN at 150–200 W/mK). Clever packaging solutions can provide alternative thermal pathways other than through the substrate. Figure 36.1 also shows that Ga_2O_3 could have applications in very high power defense and transportation systems.

The relationship between bandgap and breakdown field in wide bandgap semiconductors is given by $\varepsilon_c = a(E_g)^n$ where a and n are fitting parameters [1–10]. For indirect semiconductors, the values for the fitting parameters are $a = 2.38 \times 10^5$ and $n = 1.995$, for direct semiconductors $a = 1.73 \times 10^5$ and

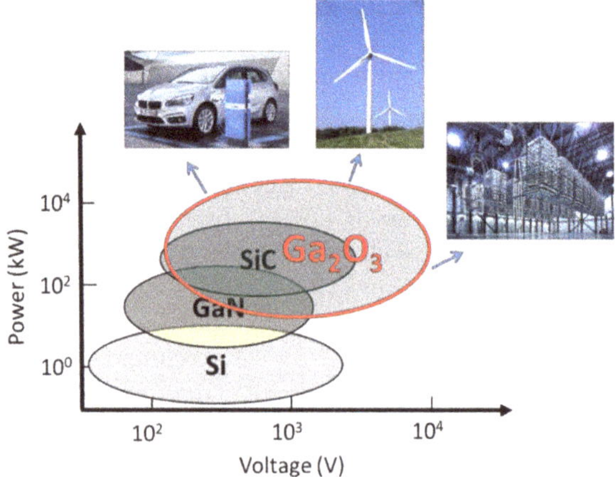

Fig. 36.2 Applications for Si, SiC, GaN and Ga_2O_3 power electronics in terms of power and voltage requirements, including electric vehicle charging infrastructure, control of renewable energy into the existing power grid and high-density data storage

n = 2.506, and for all semiconductors $a = 1.75 \times 10^5$ and n = 2.359 [1, 9]. The generally accepted value breakdown field for β-type of Ga_2O_3 is 8 MV/cm [4, 5]. Experimentally, in transistor structures, a E_b of 3.8 MV/cm was obtained for a 0.65 μm channel length device [10, 11]. The reported breakdown voltage in transistors of 755 V corresponds to 0.5 MV/cm breakdown strength, obtained for a 15 μm channel length device [31]. As material quality improves, it is expected that the experimental breakdown voltages will approach the theoretical predictions.

There are two geometries for rectifiers, vertical and lateral, with only the former being capable of practical conduction currents [32–40]. Lateral devices can achieve high breakdown voltage by using large contact separations, but cannot simultaneously achieve high forward current due to the low total conduction thickness and they also have high on-state resistance. For example, while excellent reverse blocking voltages (V_B) have been achieved in lateral GaN rectifiers (V_B up to ~9.7 kV), these devices have limited utility because of their low on-state current [32, 33] and attention quickly turned to vertical geometry rectifiers fabricated on free-standing substrates [32]. These showed excellent forward current characteristics, with total currents >1.7 A for 7 mm diameter devices and low forward turn-on voltages (V_F ~ 1.8 V) [9, 32].

The reverse breakdown voltages in these structures in all wide bandgap semiconductors are still limited by avalanche breakdown at defects and/or surfaces. The rapid progress in improving both defect density and purity of free-standing Ga_2O_3 substrates makes them the most promising approach for achieving both high V_B and on-state currents [21, 34–36], as well as structures involving lightly doped drift layers grown by techniques such as hydride vapor phase epitaxy or metal-organic chemical vapor deposition [19, 23–27]. It is important in the latter case to have a low effective doping concentration. For example, a simple model for avalanche breakdown in GaN resulting from impact ionization produces the relation [6, 8, 9, 33]

$$V_B \sim 1.98 \times 10^{15}\, N^{-0.75}$$

where N is the doping concentration in the GaN. This emphasizes that low doping density is needed for high breakdown in wide bandgap rectifiers, including Ga_2O_3. Currently, all Ga_2O_3 rectifiers show performance limited by the presence of defects and by breakdown initiated in the depletion region near the electrode corners. In SiC rectifiers, a wide variety of edge termination methods have been employed to smooth out the electric field distribution around the rectifying contact periphery [7, 30], including mesas, high resistivity layers created by ion implantation, field plates and guard rings. The situation is far less developed for GaO_3, with just a few reports of field plate termination [26, 29, 31–38].

36.3 Reverse Breakdown Voltage

The phenomenon of reverse breakdown is explained by avalanche multiplication, which involves impact ionization between host atoms and high-energy carriers. When a high-energy hole or electron under high electric field impacts an electron in the valence band, it will produce a new electron-hole pair (EHP). This newly generated EHP will cause other collisions and rapidly multiply carriers. Avalanche breakdown is defined to occur when [1]

$$\int_0^{W_D} \alpha_p \exp\left[\int_0^x (\alpha_n - \alpha_p)\mathrm{d}x\right]\mathrm{d}x > 1$$
$$\alpha_i = \alpha_0 \exp\left(\frac{-b0}{E}\right)$$

where W_D is the depletion width, α_a and α_p are the ionization rates of electrons and holes. The electron ionization rate has been calculated for Ga_2O_3 as $\alpha_n = 0.79 \times 10^6 \frac{\exp(-2.92 \times 10^7)}{E} (\mathrm{cm}^{-1})^{(41)}$. The rates for both carrier types have also been calculated for GaN and SiC and we give them here for comparison. The hole-initiated and electron-initiated ionization rate of electrons and holes, respectively, for both wurtzite and zinc blende GaN are $\alpha_{n,p} = 8.85 \times 10^6 \exp(-\frac{2.6 \times 10^7}{E})$ (cm^{-1}) while for 6H-SiC $\alpha_n = 1.66 \times 10^6 \exp\left(-\frac{1.273 \times 10^7}{E}\right) (\mathrm{cm}^{-1})$ and $\alpha_p = 5.18 \times 10^6 \exp\left(-\frac{1.4 \times 10^7}{E}\right) (\mathrm{cm}^{-1})$. To calculate the integrals of impact ionization is time-consuming. Thus, a power-law approximation can make the calculation easier in Fulop's form; $\alpha_{\mathrm{eff}} = AE^n$ where for GaN, $\alpha_{\mathrm{eff}} = AE^n = 9.1 \times 10^{-43} E^7$ and for 6H-SiC $\alpha_{\mathrm{eff}} = AE^n = 4.55 \times 10^{-35} E^6$ [7, 24].

The simplified breakdown condition is expressed by the following ionization integral.

$$\int_0^{W_C} \alpha_{\mathrm{eff}}\,\mathrm{d}x = 1$$

Therefore, the depletion layer width at breakdown for GaN, for example, is

$$W_c = \left(\frac{8}{A}\right)^{\frac{1}{8}} \left(\frac{\varepsilon\varepsilon_0}{qN_B}\right)^{\frac{7}{8}}$$

The critical electric field can be calculated from the 1-D Poisson's equation ($\mathrm{d}E/\mathrm{d}x = qN_B/\varepsilon\varepsilon_0$) and obtained through numerical substitutions.

$$\frac{\mathrm{d}^2V}{\mathrm{d}x^2} = -\frac{\mathrm{d}E}{\mathrm{d}x} = -\frac{qN_\mathrm{B}}{\varepsilon\varepsilon_0}$$

$$E_\mathrm{c} = \left(\frac{8}{A}\cdot\frac{qN_\mathrm{B}}{\varepsilon\varepsilon_0}\right)^{\frac{1}{8}} = \left(\frac{8}{AW_\mathrm{c}}\right)^{\frac{1}{7}} = 3.4\times10^4\,N_\mathrm{B}^{\frac{1}{8}}$$

If Poisson's equation is solved with the voltage and electric field relationship, the breakdown voltage for non-punch-through junction case is given by $\mathrm{BV_{pp}} = 2.87\times10^{15}\,N_\mathrm{B}^{-\frac{3}{4}}$. Similar analysis for the depletion layer width, the critical electric field, and the breakdown voltage for 6H–SiC yields:

$$W_\mathrm{c} = \left(\frac{7}{A}\right)^{\frac{1}{7}}\left(\frac{\varepsilon\varepsilon_0}{qN_\mathrm{B}}\right)^{\frac{6}{7}} = 6.2\times10^4\,N_\mathrm{B}^{-\frac{6}{7}}$$

$$E_\mathrm{c} = \left(\frac{7}{A}\cdot\frac{qN_\mathrm{B}}{\varepsilon\varepsilon_0}\right)^{\frac{1}{7}} = \left(\frac{7}{Aw_\mathrm{c}}\right)^{\frac{1}{6}} = 1.16\times10^4\,N_\mathrm{B}^{\frac{1}{7}}$$

$$\mathrm{BV_{pp}} = 3.59\times10^{14}\,N_\mathrm{B}^{-\frac{5}{7}}$$

In the case of punch-through junction diode, the breakdown voltage is given by

$$\mathrm{BV_{PT}} = E_\mathrm{c}W_\mathrm{PT} - \frac{qN_\mathrm{B}W_\mathrm{PT}^2}{2\varepsilon\varepsilon_0}$$

While the relevant relations are still being refined for Ga_2O_3, initial calculations of the same type can produce a plot of theoretical breakdown voltage of Ga_2O_3 punch-through diodes as a function of doping concentration and drift region thickness. Figure 36.3 shows relation between breakdown voltage, electric field and doping in a vertical geometry rectifier consisting of a lightly doped drift region on a more heavily doped layer on a conducting substrate. Figure 36.3 (bottom) is a plot of theoretical breakdown voltage of Ga_2O_3 punch-through diodes as a function of doping concentration and drift region thickness. It can be seen that 3 μm epi layer with doping concentration of 10^{16} cm^{-3} gives ~1800 V of breakdown voltage. The actual experimental value of breakdown voltage is far from these theoretical predictions. Materials imperfections such as the threading dislocations lead to premature breakdown [41, 42]. Therefore, the edge termination technique should be developed for Ga_2O_3 to prevent the early breakdown, and the crystal quality should be advanced to improve the device performance. Note that lateral Ga_2O_3-based devices may break down due to mechanisms other than avalanche, such as breakdown of the surface dielectric passivation.

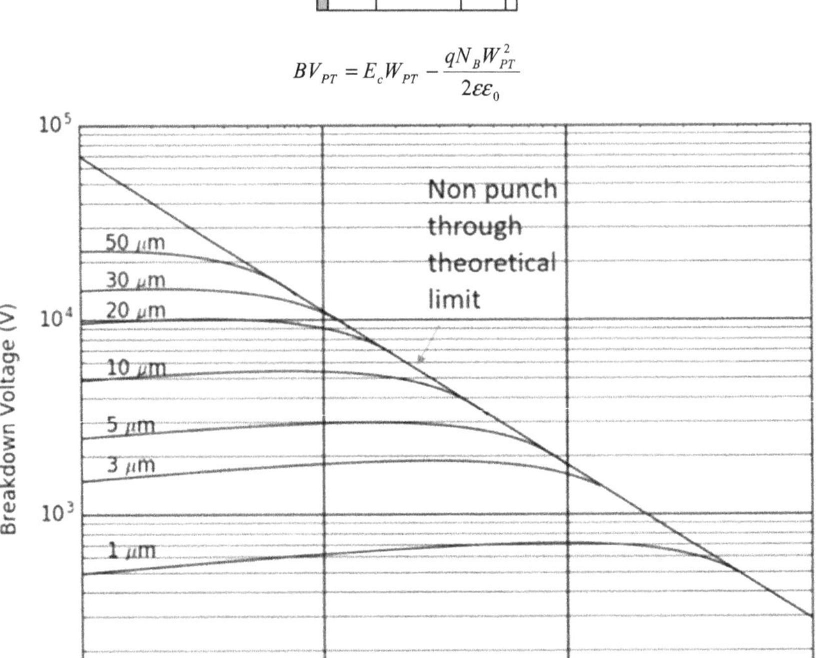

Fig. 36.3 (Top) Schematic of unterminated vertical punch-through rectifier structure and relationship between voltage, electric field, depletion depth and doping. E_C is critical electric field, W_P the drift region thickness, N_B the doping concentration and ε the permittivity. (Bottom) The reverse breakdown voltage of punch-through junction for Ga_2O_3 as a function of doping concentration and drift region thickness

36.4 On-State Resistance

The specific on-state resistance of unipolar diode is a sum of the drift region resistance, the contact resistance, and the substrate resistance [1, 6, 8].

$$R_{\text{diode}} = R_{\text{drift}} + R_{\text{sub}} + R_{\text{contact}}$$

The specific on-state resistance of drift region is given by

$$R_{\mathrm{ON}} = \int \frac{\mathrm{d}x}{q\mu N_{\mathrm{B}}} = \frac{W_{\mathrm{D}}}{q\mu N_{\mathrm{B}}}$$

where μ is the low-field mobility, N_{B} is the doping concentration of drift region and W_{D} is the drift region thickness. The on-state resistance of drift region for Ga_2O_3 can be related to the reverse breakdown voltage and given by $R_{\mathrm{ON}} = 2.4 \times 10^{-12} \cdot \mathrm{BV}^{2.5}$ (Ω cm^2), while for Si, the corresponding relation is $R_{\mathrm{ON}} = 5.91 \times 10^{-9} \cdot \mathrm{BV}^{2.5}$ (Ω cm^2).

36.5 Edge Termination Design

Edge termination is critical for obtaining high breakdown voltage and reduced on-state resistance in vertical rectifiers [7, 26, 29, 31, 38, 39]. Severe electric field crowding occurs around the metal contact periphery in these structures, as shown schematically in Fig. 36.4. There are high leakage current and breakdown at this area of highest electric field (Fig. 36.5). The use of edge termination techniques produces a lateral spread of the electric field crowding and depletion layer. This approach is widely used in power device design due to the straightforward inclusion

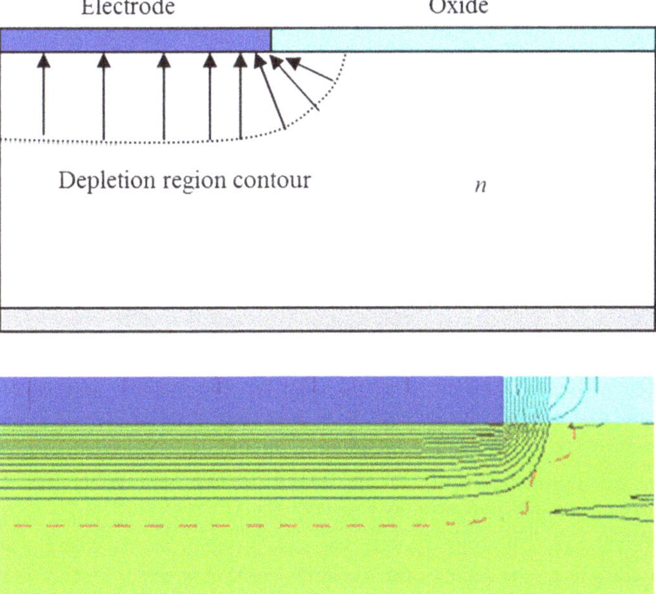

Fig. 36.4 Schematic of field crowding at the edge of the anode contact on an unterminated vertical geometry rectifier (top) and field strength contours simulated in Medici (bottom)

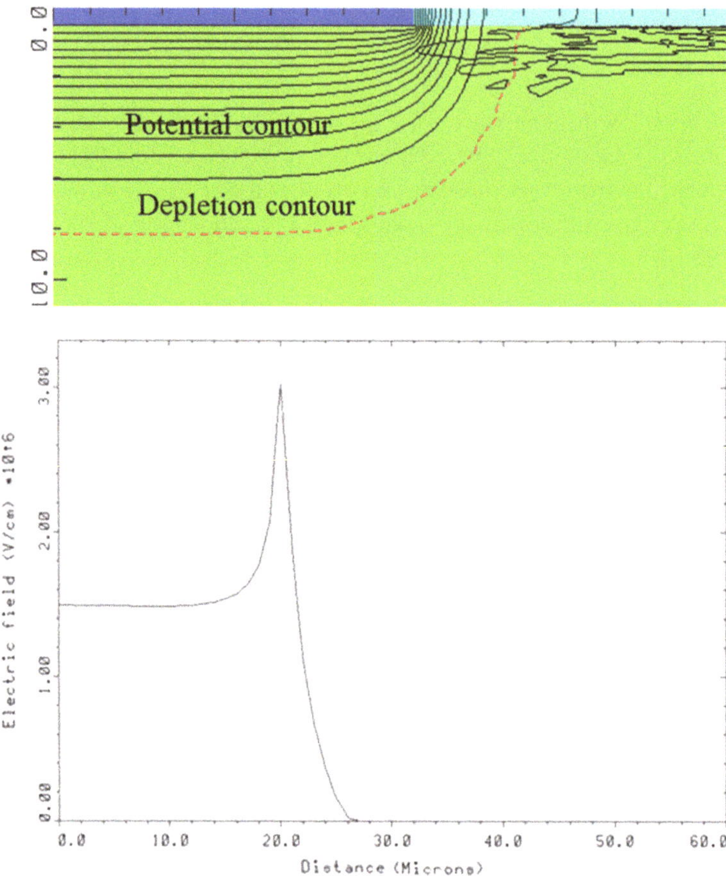

Fig. 36.5 Expanded view of field and depletion contours from Medici simulation (top) and dependence of electric field as a function of distance near the edge of anode contact (bottom)

in the fabrication process. There are a number of common edge termination techniques:

(i) Field plate edge termination—this uses extension of the Schottky metal over a dielectric layer at the edge to improve the reverse blocking capability. A schematic is shown in Fig. 36.6, together with results of a simulation to determine the dependence of breakdown voltage on extent of metal overlap onto the dielectric. Some of the experimental parameters that go into the design and must be optimized are the metal overlap distance (x_0), dielectric thickness (t_{ox}), the type of dielectric (SiO_2, SiN_x, Sc_2O_3, MgO, and AlN), the ramp angle on the edge of the dielectric and interface state charge density. To eliminate metal contact corner breakdown, a dielectric thickness of at least

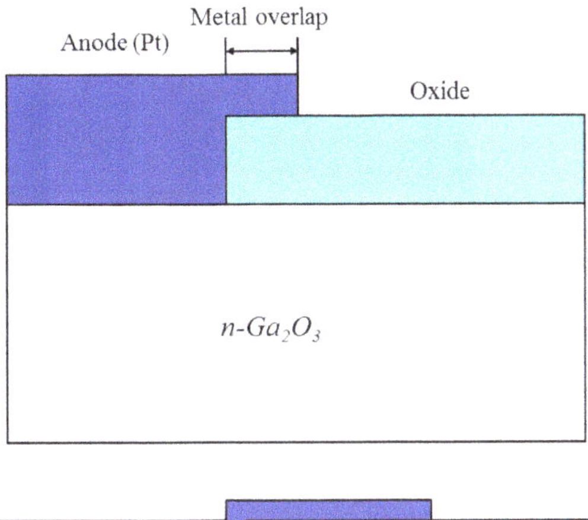

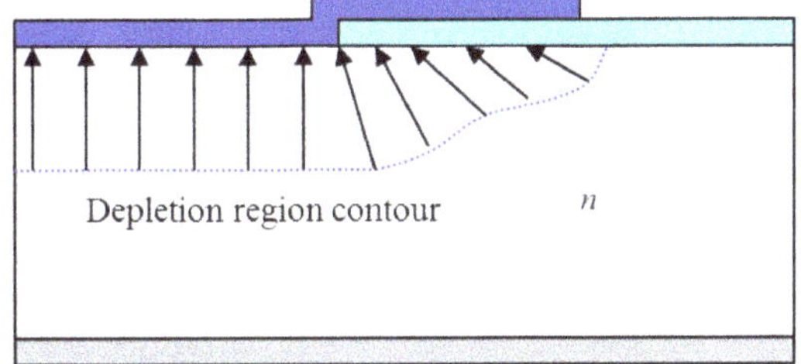

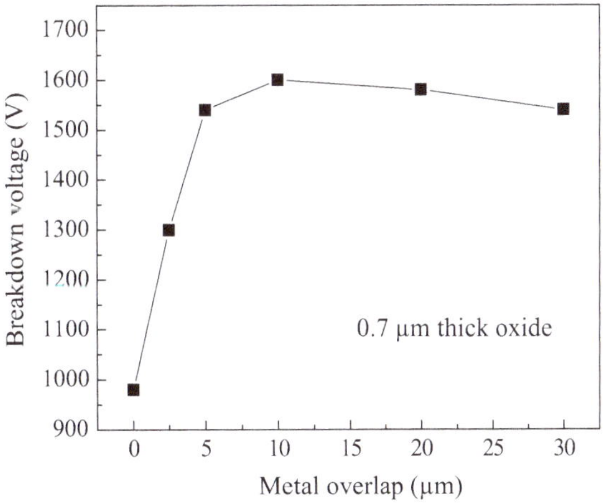

Fig. 36.6 Schematic of simulated, field-terminated Ga_2O_3 rectifier structure (top), the effect of the dielectric overlap in reducing field crowding (center) and simulated breakdown field as a function of metal overlap onto the dielectric (bottom)

0.2 μm should be used. There is usually observed to be no further improvement in V_B beyond 10 μm overlap.

(ii) P-guard ring edge termination—Another alternative for edge terminations is the use of equipotential (floating) p^+ guard rings Fig. 36.7 (top). As the bias in the main junction increases, the space charge region extends until it reaches the guard ring. The potential of the guard ring is given by

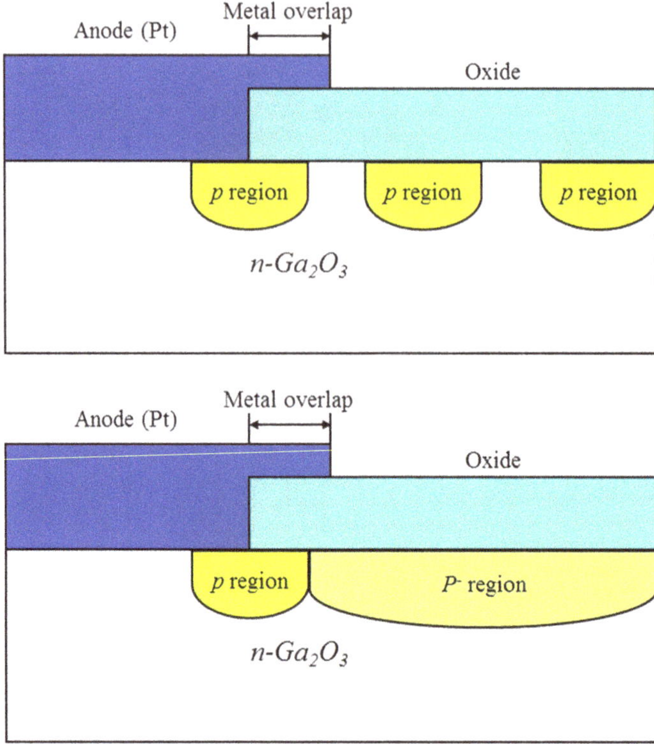

Fig. 36.7 Schematic of field-plated rectifiers with either p-guard rings (top) or with planar junction termination extension and one p[+] guard ring (bottom)

$$V_{\mathrm{FFR}} = \left(\frac{2eN_{\mathrm{A}}W_{\mathrm{S}}2V_{\mathrm{A}}}{\varepsilon}\right)^{\frac{1}{2}} + \frac{eN_{\mathrm{A}}W_{\mathrm{S}}^{2}}{2\varepsilon}$$

where N_A is the acceptor concentration, ε the Ga_2O_3 permittivity, W_S the field ring spacing and V_A the applied Schottky bias on the Schottky contact. As the bias is further increased, this potential increases, tracking the value of the equipotential line from the main junction [2]. At a given bias, the potential of the ring is equal to that of the highest potential around it and lower than the value on the main junction. The net effect is to reduce field crowding near the junction curvature. The two key parameters to optimize are the junction spacing and doping. The maximum electric field should be induced at the outside of the junction.

(iii) Junction Termination (JTE)—this consists of extending the highly doped main junction by a connected surrounding region of the same type of conductivity but with a lower doping level, to allow the spreading of the equipotential lines emerging below the junction edge curvature toward the surface. Traditional JTE designs require precise control of dopants in the JTE layer in order to completely deplete it at the desired blocking voltage. For a given doping and thickness of the substrate, and extension junction width are the main parameters affecting the blocking voltage. Junction termination extensions have been widely used, but JTEs are difficult to optimize and often require multiple zones of decreasing implant dose in order to achieve ideal breakdown for a junction. A schematic of the JTE approach is shown in Fig. 36.7 (bottom).

Other techniques include beveling the edge of the contact area and this can be combined with a field plate, resistive regions or JTE along the bevel surface [39]. The lack of available p-type doping for Ga_2O_3 means that field plates and beveling have been the only edge termination methods applied to date. Deposition of p-type NiO_2 or some other p-type oxide to create alternating p-n regions such as in the guard ring or JTE approaches may be an alternative to p-doping in the Ga_2O_3 itself. The choice of edge termination should be based on the device type, size, and the effectiveness of termination method. The edge termination designs are usually optimized using a numerical solution technique, such as the process simulators Silvaco Atlas or Medici. In the next sections, we show the results of these types of simulation for Ga_2O_3 rectifiers.

(a) **Field Plate Termination**

The structure at the top of Fig. 36.6 was used as our standard for simulation and is based on the available Ga_2O_3 epilayers on bulk substrates, which currently have a doping density of $\sim 10^{16}$ cm^{-3}. The depletion depth is limited by the background doping and we used a thickness of 30 μm in the simulations. The parameters we investigated in this study were the dielectric material, its thickness and the extent of metal overlap onto the field plate. Important design parameters influencing V_B are the length of the metal overlap and the dielectric thickness. For a given value of the plate length, the dielectric thickness must be optimized in order to balance the electric field peaks at the junction and plate edges, as reported previously for GaN. The simulations were carried out using the MEDICI™ code. The back n-ohmic contact resistance was assumed to be 10^{-5} Ω cm^2, which is consistent with our past experimental data, and an interface state density of 5×10^{11} eV^{-1} cm^{-2} was assumed for the dielectric/Ga_2O_3 interface. Once again, this is based on past experimental results. After designing the particular basic structure, a mesh of nodes is created to allow the solutions to the transport equations to be obtained. The program includes Shockley-Read-Hall and Auger recombination, an incomplete ionization model and an average of the available high-field saturation and avalanche models. We assumed a conduction band density of states of 2.6×10^{18} cm^{-3} for the n-type Ga_2O_3, a surface recombination velocity of 10^3 cm s^{-1} and Shockley-Read-Hall lifetime of 1 ns.

The maximum electric field in an unterminated rectifier occurs directly under the corner of the Schottky contact and emphasizes that avalanche breakdown is more likely to initiate at that location. The breakdown voltage was 980 V for our chosen test structure.

Figure 36.6 (bottom) shows the calculated breakdown voltages obtained for a 0.7 μm thick SiO_2 field plate on top of the rectifier, as a function of the extent of the overlap of the Schottky contact onto the SiO_2. Note that the V_B values increase rapidly for metal overlaps up to ~10 μm, with a maximum increase of ~63% in breakdown voltage relative to the unterminated device. Beyond an overlap of 10 μm, there is no further improvement in breakdown voltage from this given thickness of SiO_2 field plate. We believe this is due to the fact that the lateral spread of the depletion layer becomes comparable to the depth of this layer, so that extending the field plate into undepleted regions does not affect the breakdown behavior.

The effect of SiO_2 thickness at a given metal overlap distance of 10 μm was also simulated and showed an almost linear increase in V_B with increasing oxide thickness up to 0.7 μm. At thickness > 1 μm, the simulations showed that the electric field inside the oxide began to increase. One must therefore choose a thickness such that the field strength inside the oxide does not exceed its breakdown strength. The fact that very thick oxide layers do not lead to an improvement in V_B is an advantage from a practical viewpoint because such layers would require very long deposition times and introduce problems such as stress.

Other dielectrics that should demonstrate reasonably low interface state densities on Ga_2O_3 include AlN, MgO and Sc_2O_3. SiO_2 produces the highest breakdown voltage rectifiers for these conditions because of its large bandgap and low dielectric constant, as shown in the simulation data of Fig. 36.8. Choi et al. [43]

	Oxide	Nitride	AlN	MgO	Sc_2O_3
ε	3.9	7.5	8.5	9.8	14
Eg (eV)	9	4.7	6.2	8	6.3

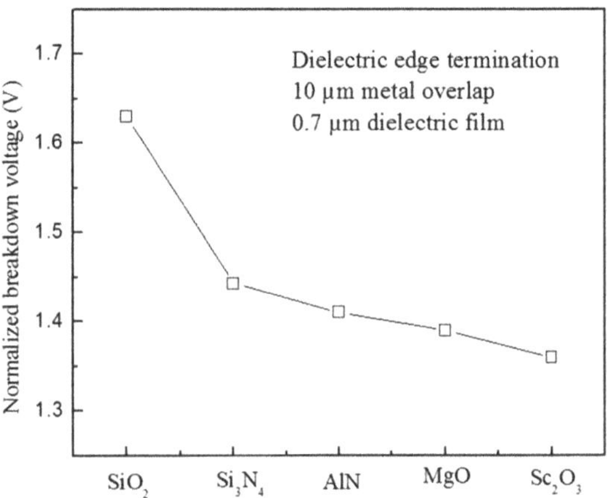

Fig. 36.8 Simulated normalized breakdown voltages for 0.7 μm thick field plates as a function of dielectric material for rectifiers with a fixed metal overlap of 10 μm

recently published simulations comparing SiO_2, HfO_2 and Al_2O_3 as potential field plate materials for Ga_2O_3 rectifiers and found that HfO_2 with an extension of at least 3 μm was a superior choice. In real devices, it should be considered that reliability and ease of deposition are of utmost importance and it is not necessarily the case that HfO_2 or SiO_2 would be the best choice with this consideration in mind. Obviously, work needs to be done to establish experimentally the relative tradeoff between V_B and long-term device stability. The main findings of our simulation study are:

(i) The use of an optimized SiO_2 field plate edge termination can increase the reverse breakdown voltage of vertical Ga_2O_3 rectifiers by up to a factor of two compared to unterminated devices.
(ii) The dielectric material, thickness (and ramp angle if using a bevel edge termination) all influence the resulting V_B of the rectifier by determining where the maximum field strength occurs in the device structure. The key aspect in designing the field plate edge termination is to shift the region of the high field region away from the periphery of the rectifying contact.

(b) **Junction Termination/Guard rings**

Referring back to Fig. 36.7 (bottom) of the schematic of a rectifier employing planar junction termination with a dielectric field plate, the breakdown point is extended beyond the contact periphery by this approach. In our case of one p-guard ring with additional planar junction termination, the simulated V_B shows a slight improvement over the planar junction termination with the field plate. Even though there is no demonstrable robust p-type doping capability for Ga_2O_3, we also simulated the effect of guard ring spacing (Fig. 36.9). In our geometry, the maximum V_B is obtained for a spacing of $\sim$3 μm. The benefits of the field spreading are lost at either very small or large spacing. The use of additional guard rings is also beneficial. For example, using two rings, separated by 2 μm and 3 μm respectively, was found to significantly increase V_B.

The JTE method extends the high-doped side of the main junction by a connected region of lower doping level. The net effect is once again to spread the field lines and avoid premature breakdown. A key design parameter is obviously the doping concentration in the JTE region. In our case, a doping concentration of 4×10^{17} cm^{-3} produced a V_B value of 2720 V. The value of V_B was strongly peaked for a JTE doping around 4×10^{17} cm^{-3}, although V_B values above 2000 V were achieved in the simulations for a reasonable range of concentrations.

Finally, Fig. 36.10 provides summary of the V_B values calculated for the different edge termination methods. The single JTE approach provides an almost fivefold increase relative to an unterminated rectifier and points out the clear necessity to employ one or more methods in order to maximize the blocking voltage. For Ga_2O_3 power rectifiers to mature into a manufacturable technology, attention must be paid to the design and implementation of edge termination methods that maximize the reverse breakdown voltage.

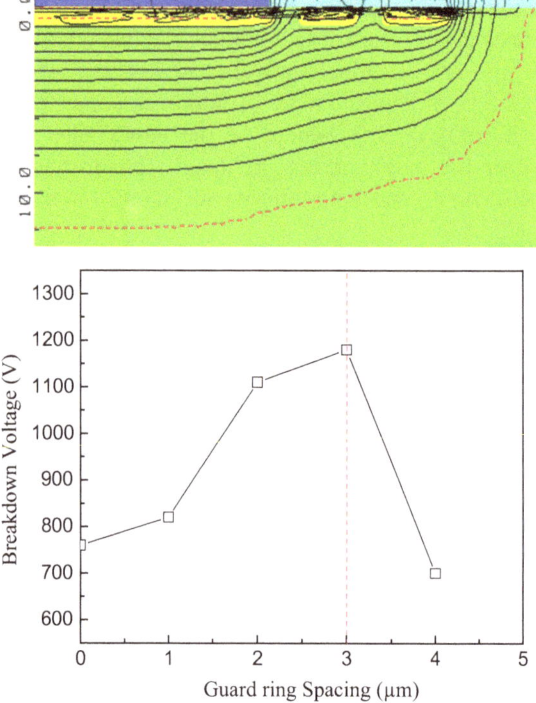

Fig. 36.9 Simulated field line contours (top) and effect of guard ring spacing on V_B (bottom)

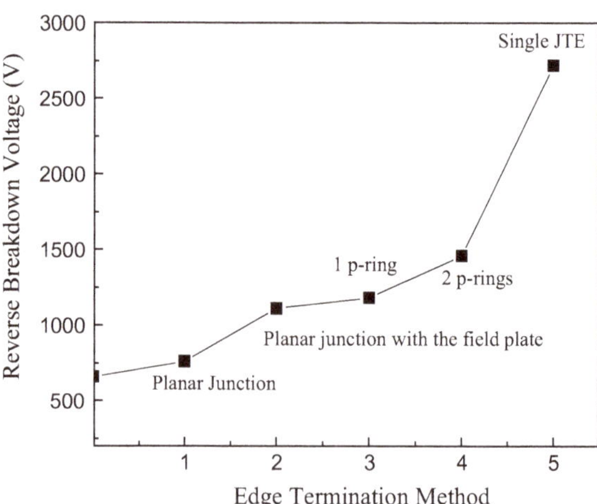

Fig. 36.10 Comparison of V_B values for Ga_2O_3 rectifiers employing different edge termination methods

36.6 Summary of Literature on Ga_2O_3 Rectifiers

Compared with lateral diodes grown on insulating substrates, vertical geometry Schottky diodes on conducting substrates can deliver higher power with full back side Ohmic electrodes and have higher current capability since they take advantage of the entire conducting area [22–27, 36, 38]. Edge termination can also enhance the performance by preventing premature breakdown due to field crowding around the contact periphery. One of the key components of a rectifier is the Schottky contact. Schottky contacts to β-Ga_2O_3 have been characterized using a number of metals, including Ni, Pt, Cu and Au. Tadjer et al. [44] carried out temperature I–V and Schottky barrier height studies on e-beam evaporated Pt/Au bilayers and compared these with ALD TiN (65 nm thick) contacts on (−201) β-Ga_2O_3. Both contacts showed thermionic emission, similar barrier heights of about 1 eV and near-unity ideality factor values, independent of temperature. The TiN contact actually had a much lower reverse current at room temperature. Richardson plots for the two types of contacts yielded barrier heights of 1.01 eV for Pt and 0.98 eV for TiN and this difference was much lower than the ~2.6 eV predicted by the difference in work functions. This may result from image force lowering. Ideality factors as a function of temperature were extracted at low bias in order to eliminate series resistance effects. Both samples had ideality factors close to unity over the whole temperature range investigated. The only drawback of the TiN was a tendency to oxidize at higher temperatures.

Vertical rectifiers with high breakdown voltage, low on-resistance, and low conduction losses due to the reduced drift layer thickness and resistance are attainable as long as the donor concentration in the drift layer is controllably low ($< 10^{16}$ cm^{-3}). The device structures reported to date in the literature for the most part have been relatively simple [20–40, 43–58]. These rectifiers show performance limited by the presence of defects and by breakdown initiated in the depletion region near the electrode corners [41–48, 59–65]. Vertical geometry, large-area planar devices can produce large total forward currents, while maintaining adequate reverse breakdown.

Reverse breakdown voltages of over 1 kV, with a maximum in the 2.4 kV range, for β-Ga_2O_3 have been reported [24–27], even without edge termination [24, 25, 27]. The highest reverse breakdown voltages have been achieved with similar layer structures, consisting of a thick epitaxial layer grown a high-quality substrate, approximately 7–20 μm thick of lightly Si-doped n-type Ga_2O_3 grown by HVPE on n^+ bulk, (−201) Sn-doped (3.6×10^{18} cm^{-3}) single crystal wafers. Diodes have full area back Ohmic contacts of Ti/Au (20/80 nm), while the Schottky contacts were Ni/Au (20/80 nm) on the epitaxial layers. There may be some pre-treatment of the back surface to enhance conductivity and lower the contact resistance. This may be plasma exposure, ozone cleaning, or ion implantation of donor dopants. Notable are reverse breakdown voltages (V_B) of 2300 V for a 150 μm diameter device (area = 1.77×10^{-4} cm^{-2}) [47] and 2440 V breakdown in trench structures [53].

Yang et al. [24, 25] reported that vertical geometry Ni/Au/β-Ga_2O_3 Schottky rectifiers displayed a reverse breakdown voltage (without edge termination) that was a function of the diode diameter. Specifically, they found a reverse breakdown voltage of 920–1016 V for a diode diameter of 105 μm and 810 V for a diode diameter of 210 μm. The authors found a Schottky barrier height of 1.1 at 25 °C with an ideality factor of 1.08. An increase in temperature to 100 °C led to a decrease in the Schottky barrier height to 0.94 with an ideality factor 1.28. They reported a figure-of-merit (V_{BR}^2/R_{on}) of 154.07 MW cm^2 for a 105 μm diameter diode, with an on-state resistance of R_{on} = 6.7 mΩ/cm^2. Tests on switching from +5 to −5 V revealed a reverse recovery time of 26 ns.

Ahn et al. [40] compared Schottky diodes based on Ni/Au contacts fabricated on a Si doped Ga_2O_3 epitaxial layer on bulk Sn-doped Ga_2O_3 versus Pt/Au contacts on bulk β-Ga_2O_3. The Ni/Au Schottky diodes displayed a barrier height of 1.07 eV and the Pt/Au Schottky diodes a barrier height of 1.04 eV at 25 °C. The on-state resistance decreased with temperature from 25 to 200 °C, and the barrier height increased with temperature over the same range. Defining a linear reverse breakdown relationship as

$$V_{RB} = V_{RB_0} + \beta(T - T_0),$$

Ahn et al. [40] extracted a temperature coefficient of reverse breakdown voltage, β, as −0.1 mV/K for Pt/Au and −4 mV/K for Ni/Au. For the Ni/Au diodes, the figure-of-merit (V_B^2/R_{ON}) was approximately 3 MW cm^{-2} at 25 °C and approximately 1 MW cm^{-2} at 200 °C.

Fu et al. [28] investigated the differences in rectifier properties between vertical (201) and (010) β-Ga_2O_3 Schottky rectifiers. The devices were fabricated on single-crystal substrates grown by EFG. The (201) and (010) exhibited on-resistances (R_{on}) of 0.56 and 0.77 mΩ cm^2, turn-on voltages (V_{on}) of 1.0 and 1.3 V, Schottky barrier heights (SBH) of 1.05 and 1.20 eV, electron mobilities of 125 and 65 cm^2 V^{-1} s^{-1}, respectively, with an on-current of ~1.3 kA cm^{-2} and on/off ratio of ~10^9. The (010) diode had a larger V_{on} and barrier height due to anisotropic surface properties. The homogeneous barrier height was also extracted: 1.33 eV for (201) and 1.53 eV for the (010). The (201) diode showed a larger leakage current due to its lower barrier height and thus a smaller breakdown voltage [28].

Another thing relevant here is that while α-Ga_2O_3 actually has a larger bandgap (~5.16 eV) than β-Ga_2O_3, the few rectifiers demonstrated on this material to date have not shown particularly large breakdown voltages, perhaps at this stage due to the need for continued optimization of material quality [36].

Konishi et al. [26] obtained high reverse breakdown voltages in excess of 1 kV for diodes employing an SiO_2 field plate with 300 nm thickness and length 20 μm. The device structure is shown in Fig. 36.11 (top), consisting of approximately 7 μm thick of lightly Si-doped n-type Ga_2O_3 grown by HVPE on n$^+$ bulk, (−201) Sn-doped (3.6 × 10^{18} cm^{-3}) Ga_2O_3 single crystal wafers. The dislocation density from etch pit observation was approximately 10^3 cm^{-2}. The reverse leakage current

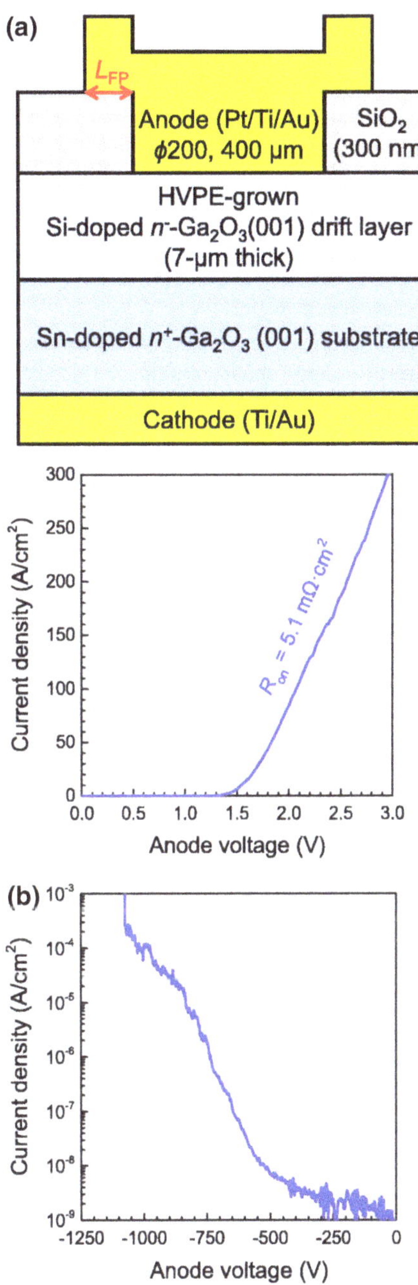

Fig. 36.11 Schematic cross section of field-plated Ga_2O_3 rectifier (top) and forward (center) and reverse (bottom) J–V characteristics at room temperature. Reprinted with permission from Konishi et al. Appl. Phys. Lett. 110, 103506 (2017), copyright AIP

has been closely correlated to the dislocation density in (0–10) oriented bulk β-Ga_2O_3 revealed upon a hot H_3PO_4 acid delineation etch for 1 h. This was designed using simulation software to optimize the breakdown voltage, which was

1 kV with low on-resistance (Fig. 36.11, center and bottom). The simulated maximum electric field under the anode edge was 5.1 MV/cm, much larger than the theoretical limits for SiC and GaN and similar to the breakdown field for lateral Ga_2O_3 MOSFETs. Even higher breakdown voltage should be possible by demonstrating a junction barrier Schottky (JBS) diode architecture, where in reverse bias the drift layer depletes away from the surface by employing a p-n junction. In the case of Ga_2O_3, p-n-type heterojunctions have been demonstrated using Li-doped NiO_2 deposited on Ga_2O_3 and Ga_2O_3 deposited on 6H–SiC, where the β-Ga_2O_3/6H–SiC anisotype heterojunction appeared to have achieved minority carrier injection [44].

To give an example of breakdown achievable on fairly resistive Ga_2O_3, current-voltage measurements were performed on O_2-annealed MBE Ga_2O_3 films, producing high (2.47 kV for the lateral geometry) reverse breakdown voltage [44], as shown in Fig. 36.12. At low bias, there was minimal current, independent of bias polarity. This high resistivity was not consistent with the carrier concentration measured from the O_2-annealed substrates and could not be explained by passivation of donor states. Secondary ion mass spectrometry (SIMS) indicated significant out diffusion of Fe from the MBE sample (Fe of about 10^{16} cm^{-3}). No trace of Fe was detected in as-grown Ga_2O_3 epilayers even when a Fe-doped substrate was used.

Joishi et al. [38] reported field plate bevel mesa Schottky diode using LPCVD-grown β-Ga_2O_3 epi layers. The devices had maximum reverse breakdown of 190 V, corresponding to a breakdown field of 4.2 MV cm^{-1}, extrinsic R_{ON} of 3.9 mΩ cm^2. The bevel was formed by dry etching.

Devices without edge termination also show the capability of the Ga_2O_3 to withstand high field strengths. The diameter of these contacts ranged from 20 μm to 0.53 mm. The V_{BR} was approximately 1600 V for 20 μm diameter smaller diode and 250 V for 0.53 mm diameter. This trend is typical of newer materials technologies still being optimized in terms of defect density [39–45, 59–65]. Kasu et al.

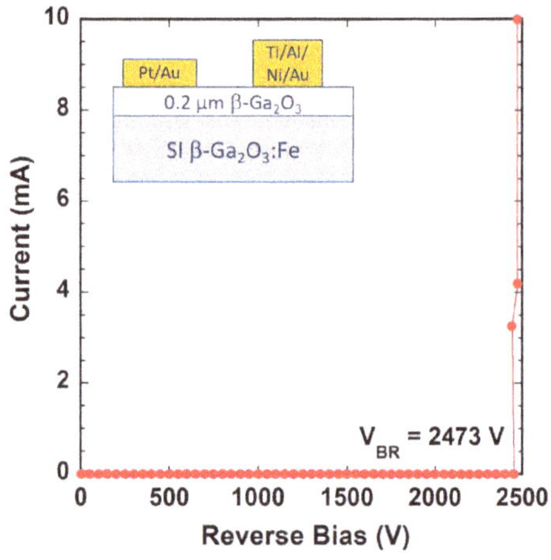

Fig. 36.12 Breakdown voltage measurement of a lateral Schottky diode fabricated on O_2-annealed MBE Ga_2O_3. The anode-cathode spacing was 7.5 μm

[42, 64] examined the effect of crystal defects revealed by etch pit delineation and found that dislocations are closely related to the reverse leakage current in the rectifier and that not all voids produce leakage current. Dislocation defects along the [010] direction were found to act as paths for leakage current, while the Si doping did not affect this dislocation-related leakage current [41, 42, 45, 60–65]. By contrast, in the [102] orientation, three types of etch pits were present, namely a line-shaped etch pattern originating from a void and extending toward the [010] direction, arrow-shaped pits in the [102] direction and gourd-shaped pits in the [102] direction. Their average densities were estimated to be 5×10^2, 7×10^4, and 9×10^4 cm^{-2}, respectively, but in this orientation there was no correlation between the leakage current in rectifiers and these crystalline defects [41, 42, 45, 60–65]. Thus, the orientation of the substrate used determines the sensitivity to defect density.

Large forward current (2 A) rectifiers were reported by Yang et al. [46–48]. These were fabricated on ~10–20 μm. Thick layers grown by HVPE on conducting substrates. The device structure is shown at the top of Fig. 36.13. A range of contact sizes were used, as shown in the optical image at the bottom of Fig. 36.13. Forward I–V characteristics of the devices were measured by applying a square wave voltage pulse (0 V − V_F) to the Schottky contact and monitoring the current using a wideband current probe connected to a 500 MHz Agilent Infiniium 50,662 oscilloscope. Both the 0.2×0.3 and 0.1×0.3 cm^2 devices could be pushed to above 2 A, with a maximum of 2.2 A for the former. Since these were single sweeps, excessive self-heating was not a significant issue and the devices showed no degradation in performance. Figure 36.14 shows the reverse I–V characteristic from a smaller (150 μm) diodes, with a breakdown voltage of ~2300 V. Schematics of 0.2×0.3 cm^2 rectifiers with front side Ni/Au

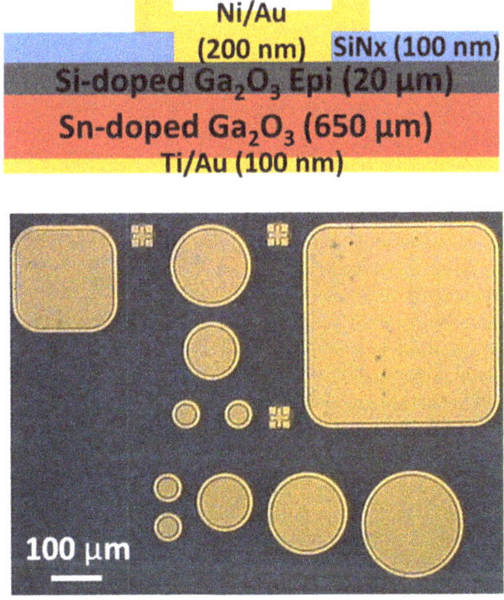

Fig. 36.13 Schematic cross section of rectifiers with front side Ni/Au rectifying contacts and full area backside Ti/Au Ohmic contacts (top) and optical microscope image of different size devices on the mask-set (bottom)

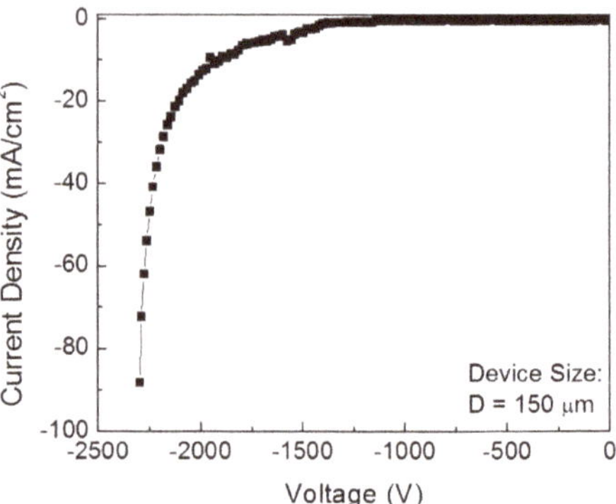

Fig. 36.14 Reverse I–V characteristic for a 150 μm diameter vertical rectifier diode

rectifying contacts and full area backside Ti/Au Ohmic contacts are shown in 15 (top) and temperature dependence of forward current for a 0.2 × 0.3 cm^2 diode. The inset shows an expanded view of the low-voltage region, yielding a barrier height 1.08 eV, with a Richardson's constant of 48 A cm^{-2} K^{-2} (Fig. 36.15).

It is worth noting that the reverse breakdown exhibits a negative temperature coefficient. This is usually the result of defects that enhance multiplication, leading to reduced breakdown and has been reported previously in the early stage development of SiC and GaN power electronics [39, 49, 61]. The impact ionization coefficients (α_p) for holes measured near defects were found to be higher than those measured at a non-defective regions [49]. Also, the values measured near defects were found to increase with increasing temperature in contrast with a

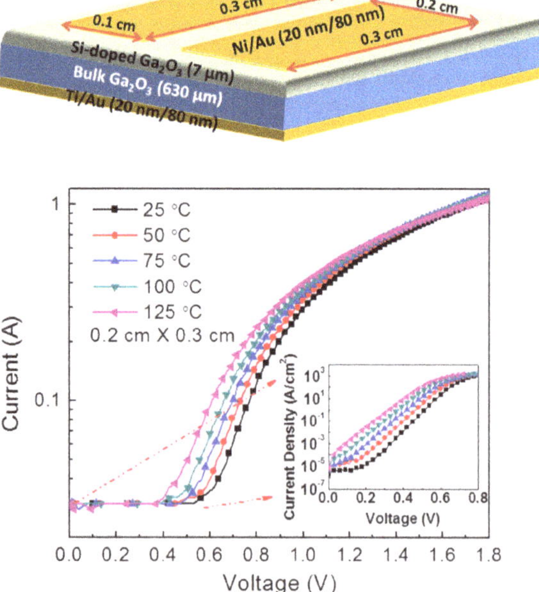

Fig. 36.15 Schematic of 0.2 × 0.3 cm^2 rectifiers with front side Ni/Au rectifying contacts and full area backside Ti/Au Ohmic contacts (top) and temperature dependence of forward current for 0.2 × 0.3 cm^2 diode. The inset shows an expanded view of the low-voltage region

defect-free diode where α_p decreased with increasing temperature, clearly indicating that the defects produce the observed negative temperature coefficient of breakdown voltage.

There have also been recent reports of high-voltage vertical Ga_2O_3 MISFETs on HVPE layers on bulk Ga_2O_3 (001) substrates [50]. The three-terminal breakdown voltage was 1057 V without field plates [50]. The devices operate in the enhancement mode with a threshold voltage of ~1.2–2.2 V, a current on/off ratio of ~10^8, and an on-resistance of ~13–18 mΩ cm^2, and an output current of >300 A/cm^2 [50].

In summary, there have been a number of 1 kV breakdown voltage rectifiers reported. Table 36.1 collects some of the important studies to date, while Fig. 36.16 shows the experimental data compared to the theoretical performance. The maximum current density reported is 3 kA cm^{-2}, with a lowest reported R_{on} of 0.1 mΩ cm^2 and a turn-on voltage of 1.7 V. The growth methods have included HVPE, MBE and Mist-CVD and both vertical and horizontal geometries have been demonstrated. It is worth noting that annealing of EFG-grown materials under oxygen ambients typically leads to a reduction in net carrier concentration of up to an order of magnitude, which can lead to higher reverse breakdown voltages. The usual Schottky contacts are Pt/Au, Au or Pt/Ti/Au and the Ohmic metallization is usually Ti/Au [51] and the most common edge termination method used has been field plates, although Ar implantation has also been employed [66].

Table 36.1 Summary of vertical geometry Ga_2O_3 rectifiers reported in literature

References	Epi thickness (μm)	Drift layer doping (cm^{-3})	Edge termination	V_B (V)	R_{ON} (Ω cm^2)
Konishi et al. [26]	10	1.8×10^{16}	Yes-field plate	1076	5.1×10^{-3}
Yang et al. [24]	10	4.02×10^{15}	No	1600	25×10^{-3}
Yang et al. [25]	10	2×10^{16}	No	1016	6.7×10^{-3}
Sasaki et al. [18]	Unintentionally doped substrate	3×10^{16}	No	150	4.3×10^{-3}
Li et al. [57]	10	2×10^{16}	Trench	1350	15×10^{-3}
Li et al. [53]	10	2×10^{16}	Trench	2440	25×10^{-3}
Oh et al. [35]	2	Undoped, <3×10^{16}	No	210	2582
He et al. [19]	Unintentionally doped substrate	2×10^{14}	No	>40	12.5×10^{-3}
Tadjer et al. [27]	~10	8×10^{12}	No	2380	n/a
Fu et al. [28]	Sn-doped EFG substrates	4×10^{18}	No	low	0.77×10^{-3}

(continued)

Table 36.1 (continued)

References	Epi thickness (μm)	Drift layer doping (cm^{-3})	Edge termination	V_B (V)	R_{ON} (Ω cm^2)
Joishi et al. [38]	2	2.5×10^{17}	Bevel	190	3.9×10^{-3}
Yang et al. [46]	10	1.33×10^{16}	Yes-field plate	650	1.58×10^{-2}
Yang et al. [47]	20	2.1×10^{15}	Yes-field plate	2300	0.25
Yang et al. [48]	7	2×10^{16}	No	466	0.26-5.9×10^{-4}
Li [51]	15, exfoliated	Sn doped	No	97	2.1×10^{-3}
Gao et al. [66]	10, exfoliated	Sn doped	Ar implantation	550	1.7×10^{-3}
This work	8	4.4×10^{15}	Yes-field plate	760	22.2×10^{-3}

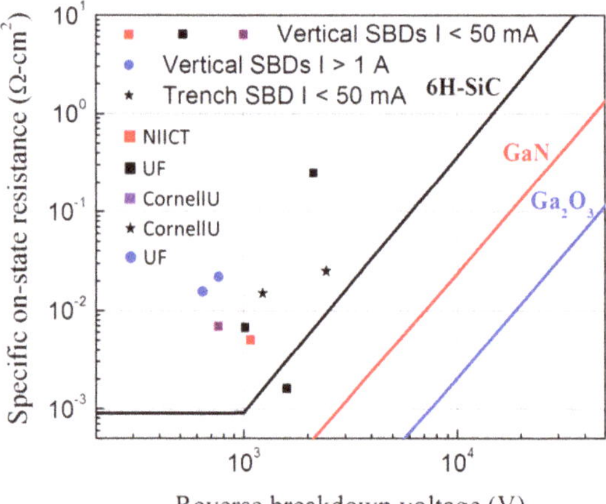

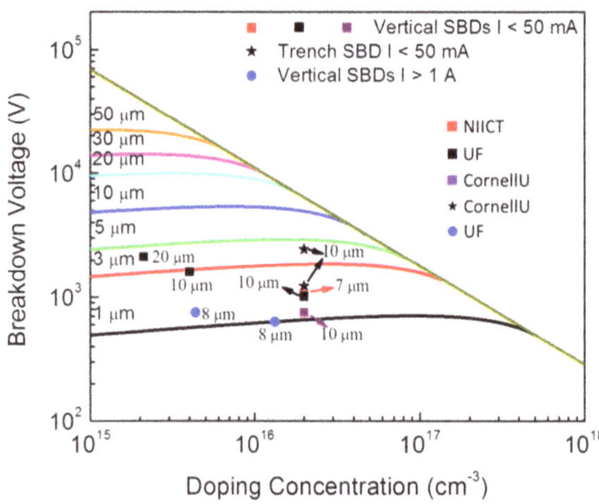

Fig. 36.16 The reverse breakdown voltage of punch-through junctions for Ga_2O_3 as a function of doping concentration and drift region thickness. (Bottom) Specific R_{ON} versus V_B of state-of-the-art vertical β-Ga_2O_3 rectifiers. Reprinted with permission from Yang et al., ECS J. Solid State Sci. Technol. 8, Q3159 (2019), copyright ECS

36.7 Summary

The performance of technologically important high-voltage rectifiers benefits from the larger critical electric field of β-Ga_2O_3 relative to either SiC or GaN. However, the absence of clear demonstrations of p-type doping in Ga_2O_3, which may be a fundamental issue resulting from the band structure, makes it very difficult to simultaneously achieve low turn-on voltages and ultra-high breakdown because of the absence of a p-i-n rectifier technology. Devices based on β-Ga_2O_3 have not yet reached the expected 8 MV/cm theoretical value for breakdown voltage. The best Schottky barrier diodes based on β-Ga_2O_3 have achieved a breakdown strength of ~ 4 MV/cm. Recent results in Schottky diodes fabricated from 0.43 μm mist grown α-Ga_2O_3 displayed a breakdown strength of 12 MV/cm [36]. Care must be taken to ensure proper and efficient termination of the junction at the edge of the die; if the junction is poorly terminated, the device breakdown voltage can be as low as 10–20% of the ideal case. Such severe degradation in breakdown voltage can seriously compromise device design and lead to reduced current rating. The purpose of the various edge termination techniques is to reduce electron-hole avalanche generation by lowering the peak electric field strength along the semiconductor surface and thereby shifting the avalanche breakdown location into the bulk of the device. There have been demonstrations of use of p-type NiO as a junction termination method [52] on Ga_2O_3 MOSFETs.

The future direction in this area is that large-area and vertical devices will mirror the designs of silicon and silicon carbide power devices, particularly with an emphasis on vertical design with defined conduction paths. The relatively low thermal conductivity of Ga_2O_3 creates self-heating effects that must be mitigated in order to utilize Ga_2O_3 for high power at even moderate switching frequencies. The question remains as to whether Ga_2O_3 will have commercial advantages over the more mature SiC and GaN technology for power switching and power amplifier applications. While the initial device performance looks very promising, challenges exist including growth maturity, thermal limits, cost, and device reliability. Lastly, the absence of p-type β-Ga_2O_3 eliminates utilization of p-n junctions to define the current path in vertical devices. The continued optimization of material device design and process technology should lead to significant advances in power performance.

Acknowledgements The project at UF was sponsored by the Department of the Defense, Defense Threat Reduction Agency, HDTRA1-17-1-011, monitored by Jacob Calkins. The content of the information does not necessarily reflect the position or the policy of the federal government, and no official endorsement should be inferred. Research at NRL was supported by the Office of Naval Research, partially under Award Number N00014-15-1-2392. Part of the work at Tamura was supported by "The research and development project for innovation technique of energy conservation" of the New Energy and Industrial Technology Development Organization (NEDO), Japan. Research at Novel Crystal Technology is partially supported by ONR Global (Grant # N62909-16-1-2217). We also thank Dr. Kohei Sasaki from Tamura Corporation for fruitful discussions.

References

1. S.M. Sze, G. Gibbons, Appl. Phys. Lett. **8**, 111 (1966)
2. B.J. Baliga, IEEE Electron Device Lett. **10**, 455 (1989)
3. B.J. Baliga, J. Appl. Phys. **53**, 1759 (1982)
4. M. Higashiwaki, A. Kuramata, H. Murakami, Y. Kumagai, J. Phys. D Appl. Phys. **50**, 333002 (2017)
5. M. Higashiwaki, G.H. Jessen, Appl. Phys. Lett. **112**, 060401 (2018)
6. J.Y. Tsao, S. Chowdhury, M. Hollis, D. Jena, N. Johnson, K. Jones, R. Kaplar, S. Rajan, C. Van de Walle, E. Bellotti, C.L. Chua, R. Collazo, M.E. Coltrin, J.A. Cooper, K.R. Evans, S. Graham, T. Grotjohn, E. Heller, M. Higashiwaki, M. Islam, P. Juodawlkis, M.A. Khan, A. Koehler, J. Leach, U. Mishra, R. Nemanich, R. Pilawa-Podgurski, J. Shealy, Z. Sitar, M. Tadjer, A. Witulski, M. Wraback, J.A. Simmons, Adv. Electron. Mater. **4**, 1600501 (2018)
7. A.Q. Huang, Proc. IEEE **105**, 2019 (2017)
8. H. Amano, Y. Baines, E. Beam, M. Borga, T. Bouchet, P. Chalker, M. Charles, K.J. Chen, N. Chowdhury, R. Chu, C. De Santi, M. De Souza, S. Decoutere, L. Di Cioccio, B. Eckardt, T. Egawa, P. Fay, J. Freedsman, L. Guido, O. Häberlen, G. Haynes, T. Heckel, D. Hemakumara, P. Houston, J. Hu, M. Hua, Q. Huang, A. Huang, S. Jiang, H. Kawai, D. Kinzer, M. Kuball, M. Kumar, K. Lee, X. Li, D. Marcon, M. März, R. McCarthy, G. Meneghesso, M. Meneghini, E. Morvan, A. Nakajima, E. Narayanan, S. Oliver, T. Palacios, D. Piedra, M. Plissonnier, R. Reddy, M. Sun, I. Thayne, A. Torres, T. Trivellin, N. Unni, M. Uren, M. Van Hove, D. Wallis, J. Wang, J. Xie, S. Yagi, S. Yang, C. Youtsey, R. Yu, E. Zanoni, S. Zeltner, Y. Zhang, J. Phys. D Appl. Phys. **51**, 163001 (2018)
9. T.J. Flack, B.N. Pushpakaran, S.B. Bayne, J. Electron. Mater. **45**, 2673 (2016)
10. B. Bayraktaroglu, *Assessment of Gallium Oxide Technology, Air Force Research Lab, Devices for Sensing Branch, Aerospace Components & Subsystems Division*, Report AFRL-RY-WP-TR-2017-0167 (2017)
11. M. Higashiwaki, K. Sasaki, A. Kuramata, T. Masui, S. Yamakoshi, Appl. Phys. Lett. **100**, 013504 (2012)
12. M. Higashiwaki, K. Sasaki, T. Kamimura, M.H. Wong, D. Krishnamurthy, A. Kuramata, T. Masui, S. Yamakoshi, Appl. Phys. Lett. **103**, 123511 (2013)
13. K.D. Chabak, J. McCandless, N. Moser, A.J. Green, K. Mahalingam, A. Crespo, N. Hendricks, B. Howe, S. Tetlak, K. Leedy, R. Fitch, D. Wakimoto, K. Sasaki, A. Kuramata, G. Jessen, IEEE Electron Device Lett. **39**, 67 (2016)
14. M.J. Tadjer, N.A. Mahadik, V.D. Wheeler, E. Glaser, L. Ruppalt, A. Koehler, K. Hobart, C.R. Eddy, F. Kub, ECS J. Solid State Sci. Technol. **5**, P468 (2016)
15. A. Green, K. Chabak, E. Heller, R. Fitch, M. Baldini, A. Fiedler, K. Irmscher, G. Wagner, Z. Galazka, S. Tetlak, A. Crespo, K. Leedy, G. Jessen, IEEE Electron Device Lett. **37**, 902 (2016)
16. K. Chabak, N. Moser, A. Green, D. Walker, S. Tetlak, E. Heller, A. Crespo, R. Fitch, J. McCandless, K. Leedy, M. Baldini, G. Wagner, Z. Galazka, X. Li, G. Jessen, Appl. Phys. Lett. **109**, 213501 (2016)
17. M.H. Wong, Y. Nakata, A. Kuramata, S. Yamakoshi, M. Higashiwaki, Appl. Phys. Express **10**, 041101 (2017)
18. K. Sasaki, M. Higashiwaki, A. Kuramata, T. Masui, S. Yamakoshi, IEEE Electron Device Lett. **34**, 493 (2013)
19. Q. He, W. Mu, H. Dong, S. Long, Z. Jia, H. Lv, Q. Liu, M. Tang, X. Tao, M. Liu, Appl. Phys. Lett. **110**, 093503 (2017)
20. T. Oishi, Y. Koga, K. Harada, M. Kasu, Appl. Phys. Express **8**, 031101 (2015)
21. A. Kuramata, K. Koshi, S. Watanabe, Y. Yamaoka, T. Masui, S. Yamakoshi, Jpn. J. Appl. Phys. **55**, 1202A2 (2016)
22. T. Oishi, K. Harada, Y. Koga, M. Kasu, Jpn. J. Appl. Phys. **55**, 030305 (2016)

23. M. Higashiwaki, K. Konishi, K. Sasaki, K. Goto, K. Nomura, Q. Thieu, R. Togashi, H. Murakami, Y. Kumagai, B. Monemar, A. Koukitu, A. Kuramata, S. Yamakoshi, Appl. Phys. Lett. **108**, 133503 (2016)
24. J. Yang, S. Ahn, F. Ren, S.J. Pearton, S. Jang, J. Kim, A. Kuramata, Appl. Phys. Lett. **110**, 192101 (2017)
25. J. Yang, S. Ahn, F. Ren, S.J. Pearton, S. Jang, A. Kuramata, IEEE Electron Device Lett. **38**, 906 (2017)
26. K. Konishi, K. Goto, H. Murakami, Y. Kumagai, A. Kuramata, S. Yamakoshi, M. Higashiwaki, Appl. Phys. Lett. **110**, 103506 (2017)
27. M. Tadjer, A. Koehler, N. Mahadik, E. Glaser, J. Freitas, B. Feigelson, V. Wheeler, K. Hobart, F. Kub and A. Kuramata, in *Proceedings of the 232nd ECS Meeting*, National Harbor, Maryland, 1–5 October 2017
28. H. Fu, H. Chen, X. Huang, I. Baranowski, J. Montes, T. Yang, Y. Zhao, I.E.E.E. Trans, Electron Devices **65**, 3507 (2018)
29. J. Bae, H. Kim, I. Kang, G. Yang, J. Kim, Appl. Phys. Lett. **112**, 122102 (2018)
30. X. She, A. Huang, O. Lucia, B. Ozpineci, IEEE Trans. Ind. Electron. **64**, 8193 (2017)
31. M.H. Wong, K. Sasaki, A. Kuramata, S. Yamakoshi, M. Higashiwaki, IEEE Electron Device Lett. **37**, 212 (2016)
32. A.P. Zhang, G. Dang, F. Ren, H. Cho, K. Lee, S.J. Pearton, J.I. Chyi, T. Nee, C. Lee, C. Chuo, IEEE Trans. Electron Devices **48**, 407 (2001)
33. A. Zhang, J. Johnson, F. Ren, J. Han, A. Polyakov, N. Smirnov, A. Govorkov, J. Redwing, K. Lee, S.J. Pearton, Appl. Phys. Lett. **78**, 823 (2001)
34. T. Oishi, Y. Koga, K. Harada, M. Kasu, Appl. Phys. Express **8**, 31101 (2015)
35. S. Oh, G. Yang, J. Kim, ECS J. Solid State Sci. Technol. **6**, Q3022 (2017)
36. M. Oda, R. Tokuda, H. Kambara, T. Tanikawa, S. Sasaki, T. Hitora, Appl. Phys. Express **9**, 021101 (2016)
37. M. Tadjer, V. Wheeler, D. Shahin, C. Eddy, F. Kub, ECS J. Solid State Sci. Technol. **6**, P165 (2017)
38. C. Joishi, S. Rafique, Z. Xia, L. Han, S. Krishnamoorthy, Y. Zhang, T. Lodha, H. Zhao, S. Rajan, Appl. Phys. Express **11**, 031101 (2018)
39. W. Sung, B. Baliga, A. Huang, I.E.E.E. Trans, Electron Devices **63**, 1630 (2016)
40. S. Ahn, F. Ren, L. Yuan, S.J. Pearton, A. Kuramata, ECS J. Solid State Sci. Technol. **6**, P68 (2017)
41. O. Ueda, N. Ikenaga, K. Koshi, A. Kuramata, T. Moribayashi, S. Yamakoshi, K. Hanada, M. Kasu, Jpn. J. Appl. Phys. **55**, 1202BD (2016)
42. M. Kasu, T. Oshima, K. Hanada, T. Moribayashi, A. Hashiguchi, T. Oishi, K. Koshi, K. Sasaki, A. Kuramata, O. Ueda, Jpn. J. Appl. Phys. **56**, 091101 (2017)
43. J.H. Choi, C.H. Cho, H.Y. Cha, Results Phys. **9**, 1170 (2018)
44. M. Tadjer, N. Mahadik, J. Freitas, E. Glaser, A. Koehler, L. Luna, B. Feigelson, K. Hobart, F. Kub, A. Kuramata, Proc. SPIE 10532, GaN Materials and Devices XIII, 1053212 (2018)
45. K. Nakai, T. Nagai, K. Noami, T. Futagi, Jpn. J. Appl. Phys. **54**, 051103 (2015)
46. J.C. Yang, F. Ren, M. Tadjer, S.J. Pearton, A. Kuramata, AIP Adv. **8**, 055026 (2018)
47. J.C. Yang, F. Ren, M.J. Tadjer, S.J. Pearton, A. Kuramata, ECS J. Solid State Sci. Technol. **7**, Q92 (2018)
48. J.C. Yang, F. Ren, S.J. Pearton, A. Kuramata, I.E.E.E. Trans, Electron Devices **65**, 2790 (2018)
49. R. Raghunathan, B.J. Baliga, Appl. Phys. Lett. **72**, 3196 (1998)
50. Z. Hu, K. Nomoto, W. Li, N. Tanen, N. Sasaki, A. Kuramata, T. Nakamura, D. Jena, H.G. Xing, IEEE Electron Device Lett. **39**, 869 (2018)
51. A. Li, Q. Feng, J. Zhang, Z. Hu, Z. Feng, K. Zhang, C. Zhang, H. Zhou, Y. Hao, Superlattices Microstruct. **119**, 212 (2018)
52. K. Sasaki, Q.T. Thieu, D. Wakimoto, Y. Koishikawa, A. Kuramata, S. Yamakoshi, Appl. Phys. Express **10**, 124201 (2017)

53. W. Li, Z. Hu, K. Nomoto, R. Jinno, Z. Zhang, T.Q. Tu, K. Sasaki, A. Kuramata, D. Jena, H.G. Xingt, in *Technical Digest of the IEEE International Electron Devices Meeting* (2018)
54. A. Takatsuka, K. Sasaki, D. Wakimoto, Q.T. Thieu, Y. Koishikawa, J. Arima, J. Hirabayashi, D. Inokuchi, Y. Fukumitsu, A. Kuramata, S. Yamakoshi, in *Proceedings of the 76th Device Research Conference* (2018)
55. Q. He, W. Mu, B. Fu, Z. Jia, S. Long, Z. Yu, Z. Yao, W. Wang, H. Don, Y. Qin, IEEE Electron Device Lett. **39**, 556 (2018)
56. G. Jian, Q. He, W. Mu, B. Fu, H. Dong, Y. Qin, Y. Zhang, H. Xue, S. Lon, Z. Jia, AIP Adv. **8**, 015316 (2018)
57. W. Li, Z. Hu, K. Nomoto, Z. Zhang, J.-Y. Hsu, Q.T. Thieu, K. Sasaki, A. Kuramata, D. Jena, H.G. Xing, Appl. Phys. Lett. **113**, 202101 (2018)
58. J. Yang, F. Ren, Y.T. Chen, Y.T. Liao, C.W. Chang, J. Lin, M.J. Tadjer, S.J. Pearton, A. Kuramata, J. Electron. Dev. Soc. **7**, 57 (2019)
59. K. Ghosh, U. Singisetti, J. Appl. Phys. **124**, 085707 (2018)
60. K. Sasaki, D. Wakimoto, Q. Thieu, Y. Koishikawa, A. Kuramata, M. Higashiwaki, S. Yamakoshi, IEEE Electron Device Lett. **38**, 783 (2017)
61. M. Oda, J. Kikawa, A. Takatsuka, R. Tokuda, T. Sasaki, K. Kaneko, S. Fujita, T. Hitora, in *Proceedings of the 73rd Device Research Conference* (2015)
62. K. Hanada, T. Moribayashi, T. Uematsu, S. Masuya, K. Koshi, K. Sasaki, A. Kuramata, O. Ueda, M. Kasu, Jpn. J. Appl. Phys. **55**, 030303 (2016)
63. K. Hanada, T. Moribayashi, K. Koshi, K. Sasaki, A. Kuramata, O. Ueda, M. Kasu, Jpn. J. Appl. Phys. **55**, 1202BG (2016)
64. M. Kasu, K. Hanada, T. Moribayashi, A. Hashiguchi, T. Oshima, T. Oishi, K. Koshi, K. Sasaki, A. Kuramata, O. Ueda, Jpn. J. Appl. Phys. **55**, 1202BB (2016)
65. T. Oshima, A. Hashiguchi, T. Moribayashi, K. Koshi, K. Sasaki, A. Kuramata, O. Ueda, T. Oishi, M. Kasu, Jpn. J. Appl. Phys. **56**, 086501 (2017)
66. Y. Gao, A. Li, Q. Feng, Z. Hu, Z. Feng, K. Zhang, X. Lu, Nanoscale Res. Lett. **14**, 8 (2019)

Chapter 37
Diodes 2

All-Oxide *pn*-Heterojunction Diodes with β-Ga_2O_3

Daniel Splith, Peter Schlupp, Holger von Wenckstern and Marius Grundmann

Abstract Due to the unipolarity and small predicted hole mobilities of β-Ga_2O_3, bipolar homojunction diodes are not of relevance for this material. A valid alternative is the realization of heterojunction diodes, especially with other, *p*-type oxide semiconductors, as they also exhibit a high chemical stability as well as typically large bandgaps. Possible candidates for such *p*-type oxide semiconductors are e. g., nickel oxide, cuprous oxide, or zinc cobalt oxide. Here, the properties of β-Ga_2O_3/*p*-type oxide semiconductor heterojunction diodes are discussed for the different *p*-type semiconductors and different device layouts. Characteristic properties like ideality factor, rectification ratio, built-in potential, and breakdown voltage are compared for the different devices.

37.1 Introduction

β-Ga_2O_3 has promising material properties to be exploited in e. g., high-power applications. However, its unipolarity is disadvantageous. Until now, no significant *p*-type conduction was reported at room temperature. Also, a flat valence band at the Γ-point is predicted from band structure calculations, leading to large effective hole masses and subsequently small hole mobilities [1–3]. This makes the realization of bipolar β-Ga_2O_3 homojunction devices unfeasible. However, several other oxide materials show *p*-type conduction, therefore allowing the fabrication of oxide *pn*-heterojunction diodes. Early reports concern a thin, nominally undoped β-Ga_2O_3

D. Splith (✉) · P. Schlupp · H. von Wenckstern · M. Grundmann
Felix Bloch Institute for Solid State Physics, Linnéstr. 5, 04103 Leipzig, Germany
e-mail: daniel.splith@physik.uni-leipzig.de

P. Schlupp
e-mail: peter.schlupp@physik.uni-leipzig.de

H. von Wenckstern
e-mail: wenckst@physik.uni-leipzig.de

M. Grundmann
e-mail: grundmann@physik.uni-leipzig.de

M. Higashiwaki and S. Fujita (eds.), *Gallium Oxide*, Springer Series in Materials Science 293, https://doi.org/10.1007/978-3-030-37153-1_37

layer in Cu_2O-based solar cells, thereby significantly increasing the open circuit voltage [4, 5]. Also, some *pn*-heterojunction diodes and other devices with non-oxide *p*-type semiconductors like SiC or GaN were demonstrated [6–9]. The first all-oxide heterostructure diode made with conducting β-Ga_2O_3 was published by Kokubun et al. in 2016, employing Li-doped nickel oxide as *p*-type material [10] deposited on a β-Ga_2O_3 single crystal. Recently, Watahiki et al. demonstrated Cu_2O/β-Ga_2O_3 heterojunction diodes on commercially available homoepitaxial thin films grown by halide vapor phase epitaxy (HVPE) [11]. Further, Schlupp et al. demonstrated nickel oxide and zinc cobalt oxide *pn*-heterojunctions with heteroepitaxial β-Ga_2O_3 thin films on a Ga-doped ZnO back contact layer and *p*-type materials grown at room temperature [12].

Herein, we discuss the properties of the mentioned *p*-type materials and compare the characteristics and properties of the published heterojunction diodes to diodes fabricated on β-Ga_2O_3 single crystals (unintentionally doped, (010)-oriented from Tamura Corporation) and thin films grown by pulsed laser deposition (PLD) on (001)-MgO and c-plane sapphire substrate from a target with 0.5 wt% SiO_2 for doping. Details on the thin film growth can be found in [13].

37.2 Suitable *p*-Type Materials

In order to fabricate all-oxide *pn*-heterojunction diodes from β-Ga_2O_3, suitable *p*-type oxide semiconductors must be used. In [14], a comprehensive overview on possible *p*-type materials for oxide heterojunction diodes is provided. Here, we focus on materials for which *pn*-heterojunction diodes with β-Ga_2O_3 were successfully demonstrated. Important material properties as reported in literature are compiled in Table 37.1.

Table 37.1 Important material properties of the *p*-type oxide semiconductors discussed here. Effective hole mass refers to the light hole, if different masses are given in the reference. m_0 is the electron rest mass

	NiO	Cu_2O	ZCO
Bandgap $E_{g,p}$ (eV)	3.4–4.3 [15–20]	2–2.4 [21–23]	2.26–2.63 [21–23]
Electron affinity χ_p (eV)	1.46–1.8 [24, 25]	3–3.2 [22, 23, 26]	
Work function W_p (eV)			5.3–5.6 [27]
Relative dielectric constant ϵ_r	11.8 [28]	7.11 [22]	
Effective hole mass m_h (m_0)	6 [29]	0.56–0.66 [30, 31]	≈18 [32]

37.2.1 Nickel(II) Oxide

The first all-oxide heterojunction diode with β-Ga_2O_3 was published by Kokubun et al. using nickel(II)-oxide (NiO) doped with Li. Further, Schlupp et al. demonstrated the usage of nominally undoped NiO fabricated at room temperature. NiO is a material with a rocksalt crystal structure. Reported values for its optical band gap range between 3.4 and 4.3 eV and depend on the growth conditions [15–20]. Further, values for its electron affinity between 1.46 and 1.8 eV were reported [24, 25]. Intrinsic *p*-type conduction most likely arises from nickel vacancies, which are particularly formed if the NiO is grown under oxygen rich conditions. Ni^{2+} ions in the vicinity of such vacancies are converted to Ni^{3+} ions, contributing holes to the semiconductor [33]. Those, however, are localized near the defect, making charge transport only possible by a hopping process [16, 34], which leads to a low mobility of the holes. Extrinsic doping is possible with group-I elements. In literature, especially Li-doped NiO is investigated [35].

Assuming an electron affinity of 4 eV for β-Ga_2O_3 [36] and neglecting the difference between Fermi energy and valence or conduction band of NiO and β-Ga_2O_3, respectively, a type-II band lineup with built-in potentials between 0.86 and 2.1 V and conduction (valence) band discontinuities larger than 2.2 eV (2.4 eV) is expected for the NiO/β-Ga_2O_3 heterojunction.

The *p*-type NiO discussed in this work was grown by pulsed laser deposition at room temperature. Details on the fabrication can be found in [37].

37.2.2 Copper(I) Oxide

Watahiki et al. demonstrated a *pn*-heterojunction diode with nitrogen doped copper(I)-oxide (Cu_2O) as *p*-type material. The fundamental band gap of Cu_2O was determined to be between 2 and 2.4 eV [21–23], with the first allowed optical transition occurring at an energy of about 2.6 eV [22]. The electron affinity is in the range between 3 and 3.2 eV [22, 23, 26]. Again, intrinsic *p*-type conduction is explained by the formation of cation vacancies. Acceptor doping can for example be achieved using nitrogen [38].

Neglecting the Fermi energy difference, again a type-II band lineup with built-in potentials between 1 and 1.6 V and conduction (valence) band discontinuities between 0.8 and 1 eV (2.9 and 3.5 eV) is expected for the Cu_2O/β-Ga_2O_3 heterojunction.

37.2.3 Zinc Cobalt Oxide

First *pn*-heterojunction diodes on β-Ga_2O_3 utilizing amorphous zinc cobalt oxide ($ZnCo_2O_4$, abbreviation: ZCO) were demonstrated by Schlupp et al. Crystalline ZCO has a spinel structure, and experimental values for the optical bandgap range between 2.26 and 2.63 eV [39–41]. To the best of our knowledge, no experimental data for the electron affinity exist. However, Zakutayev et al. determined the work function of crystalline ZCO to be between 5.3 and 5.6 eV (depending on the annealing temperature) [27]. The intrinsic *p*-type conduction originates from an antisite defect: In the normal crystal, the zinc ions are located on a lattice site with tetrahedral coordination, while the cobalt ions are located on a site with octahedral coordination. If the sites of these cations are exchanged, the zinc ion on the site with octahedral coordination will form an acceptor-like defect, while the cobalt ion on the site with tetrahedral coordination will form a donor-like defect. The later, however, lies below the valence band edge and is therefore not electrically active [32].

Using the value range for the work function of ZCO, again, a type-II band lineup with an estimated built-in potential between 1.3 and 1.6 eV and conduction (valence) band discontinuities between 0.66 and 1.33 eV (2.9 and 3.2 eV) are expected.

The *p*-type ZCO layers in this work were grown by pulsed laser deposition at room temperature. Details on the fabrication can be found in [37].

37.3 Diode Properties

In the following, we compare and discuss the properties of diodes published in literature and in this work.

37.3.1 Device Layout and Junction Symmetry

Different device layouts can be used in order to realize heterojunction diodes. In literature, most publications have a vertical device layout: The ohmic contact to the junction is realized by a layer that lies below the junction, so that the current is expected to flow perpendicular to the thin film surface. Another possibility is a horizontal layout. Here, the ohmic contact is fabricated on the top of the thin film or single crystal, and therefore, the current is expected to mostly flow in the thin film plane. Schematics of the different layouts are shown in Fig. 37.1 for single crystals (a, b), homoepitaxial (c, d) and heteroepitaxial thin films (e, f). By using the different layouts, the path of the current through the layer of the highest resistance can be minimized, leading to a decrease in series resistance. Especially for thin film samples for which the thin film layer is significantly thinner than the distance of the top contacts, a vertical layout using a back contact layer reduces the series resistance.

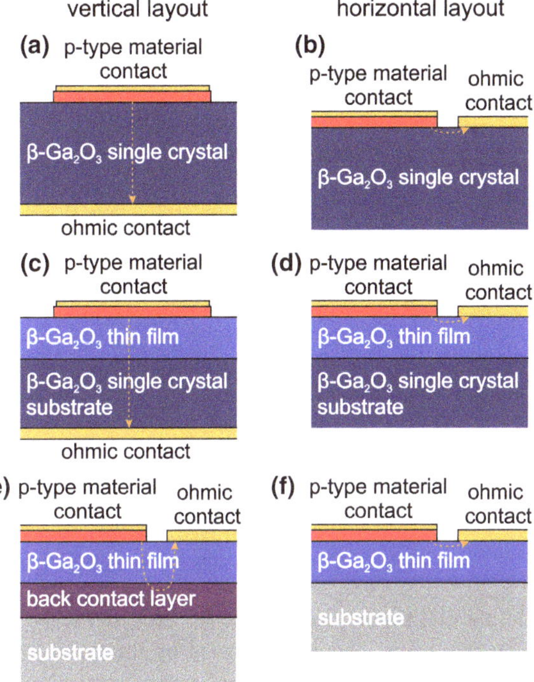

Fig. 37.1 Schematics of the horizontal (**a**, **c**, **e**) and vertical (**b**, **d**, **f**) device layouts for single crystals (**a**, **b**), homoepitaxial thin films (**c**, **d**) and heteroepitaxial thin films (**e**, **f**). The orange arrow indicates the expected current flow direction for each layout. Please note that the thicknesses of the different layers as well as the lateral and vertical dimensions are not to scale

However, lateral junctions are generally more easy to fabricate, as there is no need for a contact on the backside of the sample or an additional layer for the back contact. Note that the type of layout will also influence the direction in which the space charge region will mainly expand for increasing reverse biases.

The diodes fabricated and discussed in this work use a horizontal device layout with thin film layer thicknesses of ≈400 nm. The contacts are circular with diameters between 150 and 750 μm. The ohmic contacts surround the heterojunction diodes, and the distance between ohmic contact and diode contact is 30 μm.

Depending on the net-doping concentration of the *n*-type and the *p*-type material as well as their dielectric constants [42], a larger part of the voltage will drop across one or the other material. This also leads to different widths of the space charge regions within the respective materials. In the cases discussed here, an asymmetric junction with a small space charge region inside the *p*-type material and a large space charge region inside the *n*-type material is expected, since for all published diodes as well as the diodes shown here, the net-doping density of the *p*-type material (acceptors) is higher than the net-doping density of the *n*-type material (donors). A possible exception may be the NiO/β-Ga_2O_3 heterodiodes on heteroepitaxial thin films. Schlupp et al. estimated the free carrier concentration of similarly fabricated NiO to be larger than 6×10^{17} cm^{-3} [12], while the net-doping density of the β-Ga_2O_3 thin films in our work here is in the order of 1×10^{18} cm^{-3}. However, since the intrinsic acceptor-like defect in NiO is expected to have a rather large activation energy, the

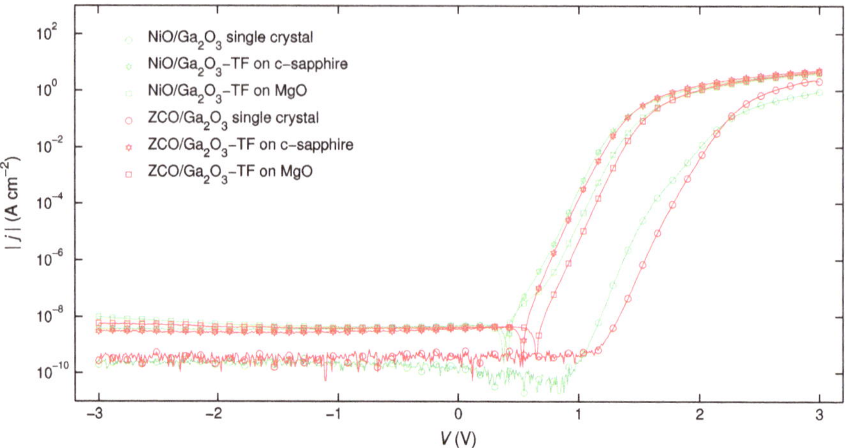

Fig. 37.2 IV-characteristics of NiO and ZCO *pn*-heterojunction diodes with β-Ga_2O_3 for thin film samples on c-plane sapphire and MgO as well as for single crystal samples. Instead of the current, the absolute value of the current density is plotted

actual net-doping density can be expected to be much higher than the free carrier concentration. This is confirmed by the quasi-static capacitance-voltage measurements. If the NiO would form a symmetric junction with the β-Ga_2O_3, a net-doping density different to the net-doping density of the ZCO/β-Ga_2O_3 junction would be expected, which was not the case.

37.3.2 Current-Voltage Characteristics

In Fig. 37.2, room temperature current-voltage (IV) characteristics of different all-oxide *pn*-heterojunction diodes fabricated on β-Ga_2O_3 single crystals and PLD-grown heteroepitaxial thin films are shown. Instead of the current, the absolute value of the current density is plotted for better comparability.

For all curves shown here, the same qualitative behavior expected for a diode can be observed: (i) Starting from negative voltages to voltages of about 1 V, the absolute value of the current density stays about constant at a value below 10^{-8}A cm^{-2} (the current is negative in this range), (ii) in the range between 1 and 1.5 V, the current increases exponentially (linearly in the semi-log plot), and (iii) starting at about 1.5 V, the series resistance begins to limit the current flow, leading to a linear dependence of the current on the voltage (logarithmic in the semi-log plot). The difference of the reverse current between the *pn*-diodes on single crystals and heteroepitaxial layers published in this work can be explained by a charging current and the difference in net-doping density: During the measurement, a voltage sweep is performed with constant time difference in the low current range. This leads to a current proportional to the capacitance of the diode [43], which is lower for the diodes on single crystals due to the lower net-doping density in comparison with the thin films. Also, the sign of this current depends on the sweep direction. In Fig. 37.2, the sweep from positive to negative voltages is shown.

IV-characteristics of the *pn*-heterodiodes comprising β-Ga_2O_3 in literature show a similar rectifying behavior [10–12]. The lowest on-resistance was achieved by Watahiki et al. on a homoepitaxial β-Ga_2O_3 layer grown by HVPE with values of 8.2 mΩ cm. Values below 0.1 Ω cm were achieved for some of the NiO/β-Ga_2O_3 diodes on single crystals in this work. The values for the other diodes shown here and in literature range between 0.1 and 1.3 Ω cm, independent of the contact layout (horizontal or vertical layout). Note, however, that for other contacts with larger diameters on the thin film samples, the on-resistance is larger, which may be due to current crowding [44].

Similar to the results of Kokubun et al., Watahiki et al., and Schlupp et al., the *pn*-heterodiodes in this work were evaluated by fitting the exponential range of the characteristics with the Shockley equation in order to determine the ideality factor η. Additionally, the rectification ratio S_R at ± 3 V, which is the ratio of the absolute currents at that voltages, was calculated. The results of both evaluations in comparison with the values published in literature are shown in Fig. 37.3. For the data of Watahiki et al., the rectification ratio was estimated from the published forward and reverse characteristics. Note that the actual value of the rectification ratio may be significantly higher, since it seems like the noise level of the measurement instrument used is dominating the reverse current. The empty markers for the diodes published in this work show mean values of multiple contacts on the same sample (> 20 diodes on the thin films, 15 (5) diodes on the single crystal samples for NiO (ZCO)). The corresponding filled markers show the values for the device with the highest rectification ratio of the ensemble. For some of the heterojunction ensembles, the differences between best rectification ratio and mean rectification ratio are rather large, hinting toward a large variability of the diode properties. This might be due to inhomogeneities in the β-Ga_2O_3 thin films or in the room temperature fabricated *p*-type material. Nevertheless, the highest rectification ratios were achieved for the *pn*-heterojunction diodes on β-Ga_2O_3 single crystal. The maximum rectification ratio is 1.8×10^{10} (3.6×10^{10}) for NiO (ZCO) diodes.

Most of the ideality factors of the diodes bunch around 2, indicating that interface recombination is the dominating current transport mechanism [42]. Exceptions to this is the diode published by Watahiki et al. and two of the five investigated ZCO/β-Ga_2O_3 diodes on single crystals, which show ideality factors below 1.5. For the diode presented by Watahiki et al., even two distinct regions with different ideality factors were observed. As stated by Watahiki et al., it is probable that here, another current transport mechanism like diffusion dominates the current flow.

Overall, although only few results exist in literature, the *pn*-heterodiodes show promising IV-characteristics with high rectification ratios and good ideality factors.

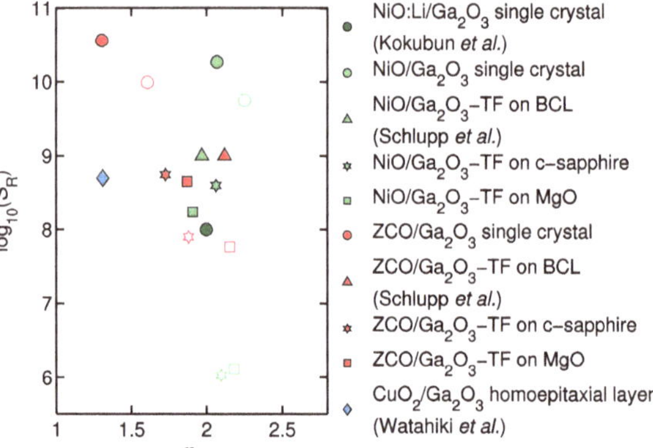

Fig. 37.3 Values for the rectification ratios (all determined at ±3 V) and ideality factors of published diodes as well as for the ones investigated here. Empty markers show the mean value of the device ensemble on each sample, and filled markers show the values for the device with the highest rectification ratio or the published values

37.3.3 Capacitance Voltage Measurements and Built-In Potential

Using capacitance voltage measurements (CV measurements), the built-in voltage and the net-doping density of the *pn*-heterojunctions can be determined experimentally. The available literature values as well as the values determined in this work are shown in Fig. 37.4. For the literature values, the given free carrier concentration was used instead of the net-doping density. In place of usual CV measurements, quasi-static CV measurements were employed for the diodes fabricated in this work. Our NiO/β-Ga_2O_3 heterojunction diodes consistently have higher built-in potential than the ZCO-based diodes fabricated on a piece of the same β-Ga_2O_3 thin film. Independent of the material used, an overall trend can be observed: With increasing doping concentration of the β-Ga_2O_3, the built-in potential decreases. This is, however, not the expected behavior: Since the distance of the Fermi energy to the conduction band decreases with increasing net-doping density, the built-in poten-

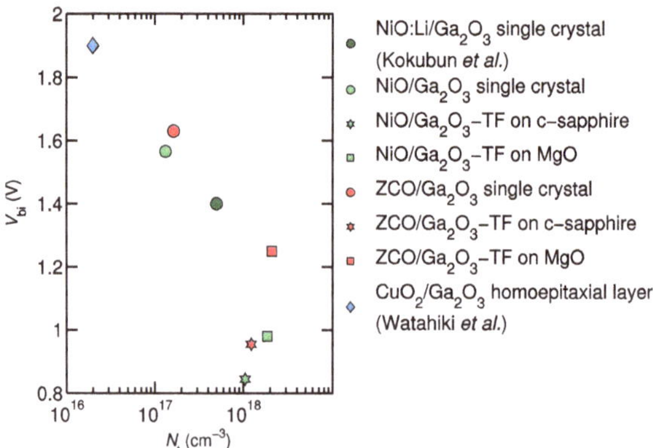

Fig. 37.4 Results of CV measurements: built-in voltage in dependence on the net-doping density. For the literature values, the free carrier concentration was used instead of the net-doping density

tial should increase. The ideality factors bunching around 2 for most of the diodes indicate the existence of interface states, and therefore, Fermi level pinning might be a reason for this unexpected behavior. Also, upward band-bending of at least 0.5 eV was reported for the surface of β-Ga_2O_3 single crystals with low doping concentrations of $5 \times 10^{16}\,cm^{-3}$ [45], while for high doping concentrations flat bands were observed. This also affects the barrier height of Schottky barrier diodes [46]. Fermi level pinning and upwards band-bending could also explain the differences between the built-in potentials of the ZCO heterojunction diodes as well as the Cu_2O heterojunction diode published by Watahiki et al. and the respective ranges for the expected built-in potentials discussed in Sect. 37.2. However, further investigations are necessary to fully understand this behavior.

37.3.4 Breakdown of the Diodes

Since one of the most promising features of β-Ga_2O_3 is its high estimated breakdown field of $8\,MV\,cm^{-1}$, especially the breakdown behavior is interesting for *pn*-heterojunction diodes on β-Ga_2O_3. In order to exploit the high breakdown field of β-Ga_2O_3, the bigger part of the voltage needs to drop across the β-Ga_2O_3 side of the junction, meaning that also the space charge region needs to expand mainly into the β-Ga_2O_3. For this, an asymmetric junction is favorable. Also, the higher the net-doping density, the smaller the width of the space charge region, the higher the maximum electric field in the space charge region, and therefore the lower the breakdown voltage (applied in reverse direction). For the samples shown here, the breakdown characteristics can be found in Fig. 37.5. From the characteristics, it is clear that the breakdown voltage of the *pn*-heterojunction diodes on single crystals is superior to those fabricated on the different thin films. For the latter, an almost expo-

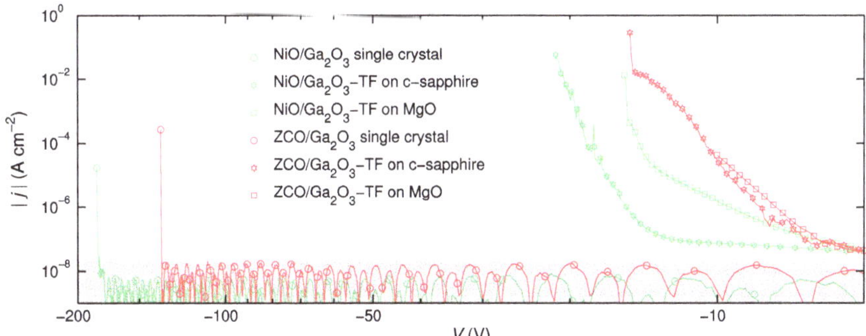

Fig. 37.5 Breakdown characteristics of the diodes investigated in this work. The gray box indicates the noise level of the used measurement system. For the ZCO/β-Ga_2O_3-diode on MgO, no hard breakdown was observed in the reverse measurement, however, the characteristics of the diode significantly deteriorate after applying −10V

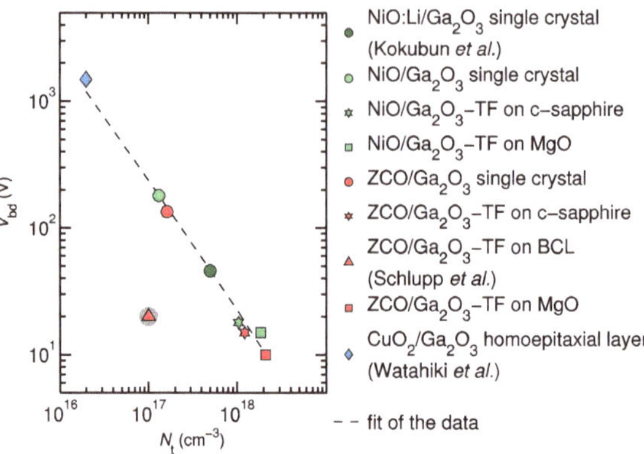

Fig. 37.6 Results of the breakdown measurements: breakdown voltage (applied in reverse direction) in dependence on the net-doping densities taken from Fig. 37.4. A linear trend in the log-log plot is observed for all diodes except the one published by Schlupp et al. By fitting the trend (black line), a similar breakdown field of 3 MV cm^{-1} can be estimated for all *pn*-heterojunction diodes

nential increase of the reverse current can be observed with increasing reverse bias. A possible reason for this current could be band-to-band tunneling [44] due to the small width of the space charge region. The highest breakdown voltage achieved here is 183 V for a lateral NiO/β-Ga_2O_3 heterojunction on a single crystal. Watahiki et al. report breakdown voltages of up 1490 V for their vertical *pn*-heterojunction diode on homoepitaxially grown β-Ga_2O_3 with doping concentrations of $2 \times 10^{16}\,cm^{-3}$. A comparison of the published breakdown voltages for *pn*-heterojunctions with β-Ga_2O_3 as well as the values determined here can be found in Fig. 37.6 in dependence on the net-doping density. Note that for the diodes investigated in this work, the net-doping density was determined by CV measurements on the *pn*-heterojunction diodes. Therefore, only under the assumption that the net-doping density is significantly higher in the *p*-type material, the net-doping density of the junction is close to the net-doping density of the β-Ga_2O_3 thin film or single crystal. Note also that for the diodes fabricated by Schlupp et al., no breakdown was observed for the NiO/β-Ga_2O_3 diode in the measured voltage range. Also, since the CV measurements were not conclusive, the net-doping densities were only estimated. All other points in the plot in Fig. 37.6 obey the same trend: An increase of the net-doping density by one order of magnitude decreases the breakdown voltage by one order of magnitude. One can easily show that, under the assumption that most of the voltage drops over the β-Ga_2O_3 side of the junction, the following equation holds for the dependence of the breakdown voltage on the net-doping density (again, bias is applied in reverse direction):

$$V_{\mathrm{bd}} = \frac{\varepsilon_{\mathrm{bd}}^2 \epsilon_{\mathrm{r}} \epsilon_0}{2 e N_{\mathrm{t}}} - V_{\mathrm{bi}} \,. \quad (37.1)$$

Here, ε_{bd} is the breakdown field, ϵ_0 is the dielectric constant, and e is the elementary charge. Therefore, if we assume that the difference of the built-in potential for the different p-materials is small in comparison with the breakdown voltage, the linear dependence seen in the log-log plot in Fig. 37.6 means that for all junctions, a similar breakdown field was achieved. By fitting, the breakdown field was determined to be $3\,\text{MV}\,\text{cm}^{-1}$ (using $\epsilon_r = 10$ for β-Ga_2O_3 [47]). Note that for all pn-heterojunction diodes discussed here, no field-plate structure was used. Therefore, the field at the edge of the junction might be significantly larger. Recently, it was demonstrated that using a field-plated structure can increase the breakdown voltage of Schottky barrier diodes on HVPE-grown homoepitaxial layers from around 500 V [48] to 1 kV [49]. The breakdown field of the field-plated structure was estimated to be $5\,\text{MV}\,\text{cm}^{-1}$.

37.4 Summary and Outlook

In Table 37.2, the properties of all published pn-heterojunction diodes are summarized. Overall, the realization of pn-heterojunction diodes with β-Ga_2O_3 is very promising. High rectification ratios, low on-resistances, and good ideality factors were already achieved, even though only a few publications exist. Further investigation is necessary to understand the dependence of the built-in potential on the net-doping density, which seems to be independent on the p-type material used. The high breakdown voltage of the device published by Watahiki et al. [11] demonstrates the performance capability of such heterojunction diodes for high-power electronics. The breakdown field of almost all published diodes is about the same. A further increase of the breakdown voltage is expected if a field-plated structure is used.

Acknowledgements We thank H. Hochmuth for PLD growth of the thin films and M. Hahn for the preparation of the PLD targets and photolithography of the samples. This work was financially supported by the European Social Fund within the Young Investigator Group "Oxide Heterostructures" (SAB 100310460) and partly by Deutsche Forschungsgemeinschaft in the Framework of Sonderforschungsbereich 762 "Functionality of Oxide Interfaces" and through GR 1011/27. We acknowledge support by Universität Leipzig within the research profile area "Complex Matter".

Table 37.2 Properties of published *pn*-heterojunction diodes. For measurements performed on multiple contacts on the same sample, the best values of the ensemble are shown in each case (highest V_{on} and S_R, lowest R_{on} and η). Note that for the diodes investigated in this work, the different values were not necessarily achieved for the same device. Values in brackets correspond to the free carrier concentration determined by Hall-effect measurements (*n* or *p*)

β-Ga_2O_3 properties					*p*-type material properties			Diode properties							References
Sample type	FM	or.	sub.	$N_{t,n}$ (or *n*) (cm^{-3})	Material	FM	$N_{t,p}$ (or *p*) (cm^{-3})	l.	V_{on} (V)	R_{on} ($\Omega\,cm^2$)	S_R	η	V_{bi} (V)	V_{bd} (V)	
SC	FZ	(1 0 0)		(5×10^{17})	NiO:Li	SG	1.1×10^{18}	h.	1.4	1	10^8	≈2.0	1.4	46	[10]
HoTF	HVPE	(0 0 1)		(2×10^{16})	Cu_2O	SP	$>10^{18}$	v.	1.7	0.0082	$\approx 5 \times 10^{8\dagger}$	1.31	1.9	1490	[11]
HeTF	PLD	$(\bar{2}\,0\,1)$	BCL	$\approx 10^{17}$	NiO	PLD	$(>6 \times 10^{17})$	v.	1.95	0.81	$\approx 10^9$	1.97		18	[12]
HeTF	PLD	$(\bar{2}\,0\,1)$	BCL	$\approx 10^{17}$	a-ZCO	PLD	$(>10^{20})$	v.	2.27	0.65	$\approx 10^9$	2.12		<18	[12]
SC	EFG	(0 1 0)		$\approx 1 \times 10^{17}$	NiO	PLD	$(>6 \times 10^{17})$	h.	2.61	0.016	1.8×10^{10}	1.7	1.56	183	t. w.
SC	EFG	(0 1 0)		$\approx 1 \times 10^{17}$	a-ZCO	PLD	$(>10^{20})$	h.	2.62	0.12	3.6×10^{10}	1.3	1.63	135	t. w.
HeTF	PLD	$(\bar{2}\,0\,1)$	c-Al_2O_3	$\approx 1 \times 10^{18}$	NiO	PLD	$(>6 \times 10^{17})$	h.	1.62	0.32	4×10^8	1.72	0.84	18	t. w.
HeTF	PLD	$(\bar{2}\,0\,1)$	c-Al_2O_3	$\approx 1 \times 10^{18}$	a-ZCO	PLD	$(>10^{20})$	h.	1.63	0.28	5×10^8	1.73	0.95	15	t. w.
HeTF	PLD	(1 0 0)	MgO	$\approx 1 \times 10^{18}$	NiO	PLD	$(>6 \times 10^{17})$	h.	1.78	0.19	1.5×10^8	1.63	0.98	15	t. w.
HeTF	PLD	(1 0 0)	MgO	$\approx 1 \times 10^{18}$	a-ZCO	PLD	$(>10^{20})$	h.	1.76	0.16	4×10^8	1.85	1.25	10	t. w.

SC: single crystal, HoTF: homoepitaxial thin film, HeTF: heteroepitaxial thin film, FM: fabrication method, FZ: floating zone method, HVPE: halide vapor phase epitaxy, PLD: pulsed laser deposition, EFG: edge-defined film-fed growth, SG: sol-gel, SP: sputtering, or.: orientation of the β-Ga_2O_3, sub.: substrate, BCL: c-ZnO:Ga back contact layer on a-Al_2O_3 (see [50] for details), MgO: (1 0 0)-MgO, l.: layout, v.: vertical, h.: horizontal, t. w.: this work

†estimated from published characteristic

References

1. H. He, R. Orlando, M. Blanco, R. Pandey, E. Amzallag, I. Baraille, M. Rérat, Phys. Rev. B **74**(19), 195123 (2006)
2. K. Yamaguchi, Solid State Commun. **131**(12), 739 (2004)
3. J.B. Varley, J.R. Weber, A. Janotti, C.G. Van de Walle, Appl. Phys. Lett. **97**(14), 142106 (2010)
4. T. Minami, Y. Nishi, T. Miyata, Appl. Phys. Express **6**(4), 044101 (2013)
5. Y.S. Lee, D. Chua, R.E. Brandt, S.C. Siah, J.V. Li, J.P. Mailoa, S.W. Lee, R.G. Gordon, T. Buonassisi, Adv. Mater. **26**(27), 4704 (2014)
6. S. Nakagomi, T. Momo, S. Takahashi, Y. Kokubun, Appl. Phys. Lett. **103**(7), 072105 (2013)
7. S. Nakagomi, K. Yokoyama, Y. Kokubun, J. Sens. Sens. Syst. **3**(2), 231 (2014)
8. S. Nakagomi, T. Sato, Y. Takahashi, Y. Kokubun, Sens. Actuator A-Phys. **232**, 208 (2015)
9. S. Nakagomi, T. Sakai, K. Kikuchi, Y. Kokubun, Phys. Status Solidi A **216**(5), 1700796 (2018)
10. Y. Kokubun, S. Kubo, S. Nakagomi, Appl. Phys. Express **9**(9), 091101 (2016)
11. T. Watahiki, Y. Yuda, A. Furukawa, M. Yamamuka, Y. Takiguchi, S. Miyajima, Appl. Phys. Lett. **111**(22), 222104 (2017)
12. P. Schlupp, D. Splith, H. von Wenckstern, M. Grundmann, Phys. Status Solidi A **216**(7), 1800729 (2019)
13. S. Müller, H. von Wenckstern, D. Splith, F. Schmidt, M. Grundmann, Phys. Status Solidi A **211**(1), 34 (2014)
14. M. Grundmann, F. Klüpfel, R. Karsthof, P. Schlupp, F.L. Schein, D. Splith, C. Yang, S. Bitter, H. von Wenckstern, J. Phys. D **49**(21), 213001 (2016)
15. R. Newman, R.M. Chrenko, Phys. Rev. **114**(6), 1507 (1959)
16. D. Adler, J. Feinleib, Phys. Rev. B **2**(8), 3112 (1970)
17. G.A. Sawatzky, J.W. Allen, Phys. Rev. Lett. **53**(24), 2339 (1984)
18. H. Sato, T. Minami, S. Takata, T. Yamada, Thin Solid Films **236**(1), 27 (1993)
19. L. Ai, G. Fang, L. Yuan, N. Liu, M. Wang, C. Li, Q. Zhang, J. Li, X. Zhao, Appl. Surf. Sci. **254**(8), 2401 (2008)
20. M. Guziewicz, J. Grochowski, M. Borysiewicz, E. Kaminska, J.Z. Domagala, W. Rzodkiewicz, B.S. Witkowski, K. Golaszewska, R. Kruszka, M. Ekielski, A. Piotrowska, Opt. Appl. **41**(2), 431 (2011)
21. J. Ghijsen, L.H. Tjeng, J. van Elp, H. Eskes, J. Westerink, G.A. Sawatzky, M.T. Czyzyk, Phys. Rev. B **38**(16), 11322 (1988)
22. B.K. Meyer, A. Polity, D. Reppin, M. Becker, P. Hering, P.J. Klar, T. Sander, C. Reindl, J. Benz, M. Eickhoff, C. Heiliger, M. Heinemann, J. Bläsing, A. Krost, S. Shokovets, C. Müller, C. Ronning, Phys. Status Solidi B **249**(8), 1487 (2012)
23. Y. Takiguchi, Y. Takei, K. Nakada, S. Miyajima, Appl. Phys. Lett. **111**(9), 093501 (2017)
24. H. Wu, L.S. Wang, J. Chem. Phys. **107**(1), 16 (1997)
25. D. Kawade, S.F. Chichibu, M. Sugiyama, J. Appl. Phys. **116**(16), 163108 (2014)
26. C.S. Tan, S.C. Hsu, W.H. Ke, L.J. Chen, M.H. Huang, Nano Lett. **15**(3), 2155 (2015)
27. A. Zakutayev, T.R. Paudel, P.F. Ndione, J.D. Perkins, S. Lany, A. Zunger, D.S. Ginley, Phys. Rev. B **85**(8), 085204 (2012)
28. P.J. Gielisse, J.N. Plendl, L.C. Mansur, R. Marshall, S.S. Mitra, R. Mykolajewycz, A. Smakula, J. Appl. Phys. **36**(8), 2446 (1965)
29. A.J. Bosman, H.J. van Daal, Adv. Phys. **19**(77), 1 (1970)
30. J.W. Hodby, T.E. Jenkins, C. Schwab, H. Tamura, D. Trivich, J. Phys. C: Solid State Phys. **9**(8), 1429 (1976)
31. A. Goltzene, C. Schwab, H.C. Wolf, Solid State Commun. **18**(11), 1565 (1976)
32. J.D. Perkins, T.R. Paudel, A. Zakutayev, P.F. Ndione, P.A. Parilla, D.L. Young, S. Lany, D.S. Ginley, A. Zunger, N.H. Perry, Y. Tang, M. Grayson, T.O. Mason, J.S. Bettinger, Y. Shi, M.F. Toney, Phys. Rev. B **84**(20), 205207 (2011)
33. J.H. de Boer, E.J.W. Verwey, Proc. Phys. Soc. **49**(4S), 59 (1937)
34. P. Lunkenheimer, A. Loidl, C.R. Ottermann, K. Bange, Phys. Rev. B **44**(11), 5927 (1991)

35. R.R. Heikes, W.D. Johnston, J. Chem. Phys. **26**(3), 582 (1957)
36. M. Mohamed, K. Irmscher, C. Janowitz, Z. Galazka, R. Manzke, R. Fornari, Appl. Phys. Lett. **101**(13), 132106 (2012)
37. H. von Wenckstern, D. Splith, S. Lanzinger, F. Schmidt, S. Müller, P. Schlupp, R. Karsthof, M. Grundmann, Adv. Electron. Mater. **1**(4), 1400026 (2015)
38. S. Ishizuka, S. Kato, T. Maruyama, K. Akimoto, Jpn. J. Appl. Phys. **40**(4S), 2765 (2001)
39. H.J. Kim, I.C. Song, J.H. Sim, H. Kim, D. Kim, Y.E. Ihm, W.K. Choo, J. Appl. Phys. **95**(11), 7387 (2004)
40. M. Dekkers, G. Rijnders, D.H.A. Blank, Appl. Phys. Lett. **90**(2), 021903 (2007)
41. S. Kim, J.A. Cianfrone, P. Sadik, K.W. Kim, M. Ivill, D.P. Norton, J. Appl. Phys. **107**(10), 103538 (2010)
42. M. Grundmann, R. Karsthof, H. von Wenckstern, A.C.S. Appl, Mater. Interfaces **6**(17), 14785 (2014)
43. D. Splith, S. Müller, H. von Wenckstern, M. Grundmann, Proc. SPIE **10533** (2018)
44. S.M. Sze, K.K. Ng, *Physics of Semiconductor Devices* (Wiley, Hoboken, NJ, 2006)
45. T.C. Lovejoy, R. Chen, X. Zheng, E.G. Villora, K. Shimamura, H. Yoshikawa, Y. Yamashita, S. Ueda, K. Kobayashi, S.T. Dunham, F.S. Ohuchi, M.A. Olmstead, Appl. Phys. Lett. **100**(18), 181602 (2012)
46. S. Müller, H. von Wenckstern, F. Schmidt, D. Splith, F.L. Schein, H. Frenzel, M. Grundmann, Appl. Phys. Express **8**(12), 121102 (2015)
47. M. Passlack, N.E.J. Hunt, E.F. Schubert, G.J. Zydzik, M. Hong, J.P. Mannaerts, R.L. Opila, R.J. Fischer, Appl. Phys. Lett. **64**(20), 2715 (1994)
48. M. Higashiwaki, H. Murakami, Y. Kumagai, A. Kuramata, Jpn. J. Appl. Phys. **55**, 1202A1 (2016)
49. K. Konishi, K. Goto, H. Murakami, Y. Kumagai, A. Kuramata, S. Yamakoshi, M. Higashiwaki, Appl. Phys. Lett. **110**(10), 103506 (2017)
50. D. Splith, S. Müller, F. Schmidt, H. von Wenckstern, J.J. van Rensburg, W.E. Meyer, M. Grundmann, Phys. Status Solidi A **211**(1), 40 (2014)

Chapter 38
Photodetectors

Overview of Ga_2O_3-Based Photodetectors

Takayoshi Oshima

Abstract This chapter provides an overview of Ga_2O_3-based photodetectors, which have been the subject of many published papers. I classify and summarize these detectors in several tables, roughly according to the device types: film-based photodetectors (photoconductors, metal-semiconductor-metal photodetectors, and alloy-film-based photodetectors), vertical Schottky photodiodes that use bulk single crystals, heterojunction photodetectors, and nanostructure-based photodetectors. In each class of detectors, some interesting devices are spotlighted.

38.1 Introduction

Solar-blind photodetectors—which are blind to photons with wavelengths longer than 280 nm and which, thus, enable the detection of deep-UV radiation without detecting infrared or visible light, particularly from the sun or room illumination (see Fig. 38.1)—have been attracting attention owing to their potential applications in civil and military areas. Such applications include chemical and biological analysis, flame detection (including fire alarms, missile-warning systems, or combustion monitoring), inter-satellite communications, emitter calibration (instrumentation and UV lithography), and astronomical studies [1–3]. Solar-blind detectors have previously been constructed using photomultiplier tubes; however, such instruments are fragile and need high-voltage sources. This makes them bulky and heavy, and they are relatively expensive, which limits their applications. Therefore, solid-state alternatives to photomultiplier tubes have also been employed as photodetectors, depending on the applications. Ultraviolet-enhanced, Si-based photodetectors are one successful alternative; however, they suffer from low quantum efficiency and require expensive optical filters due to their narrow bandgaps. This is also a drawback to operations in high-temperature and radioactive

T. Oshima (✉)
Department of Electrical and Electronic Engineering, Saga University, 1 Honjo-Machi, Saga 840-8502, Japan
e-mail: oshima@cc.saga-u.ac.jp

M. Higashiwaki and S. Fujita (eds.), *Gallium Oxide*, Springer Series in Materials Science 293, https://doi.org/10.1007/978-3-030-37153-1_38

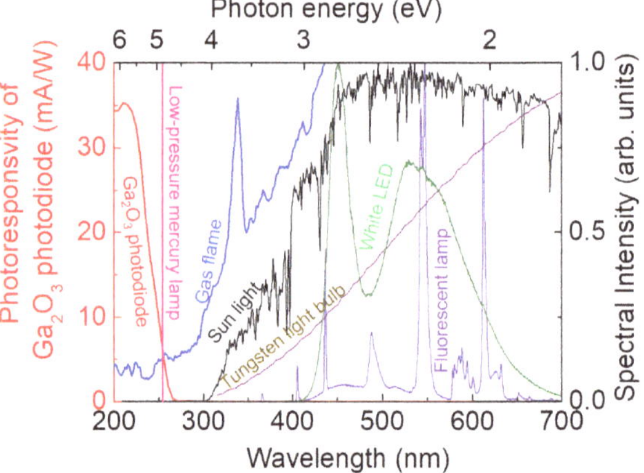

Fig. 38.1 Spectral response of a typical Ga_2O_3, solar-blind photodiode and the spectra of far-UV light sources, sunlight, and various types of room illumination

environments. Thus, wide-bandgap semiconductors with absorption edges at wavelengths smaller than 280 nm (corresponding to bandgap energies E_g wider than 4.43 eV) are required for the construction of high-performance, low-cost, and reliable solid-state, solar-blind photodetectors [1–3].

Among the possible wide-bandgap semiconductor candidates—AlGaN [4], MgZnO [5], β-Ga_2O_3 [6], and diamond [7]—β-Ga_2O_3 is the most promising for the production of cost-effective solar-blind detectors. Its absorption edge at 260–280 nm (corresponding to E_g of 4.4–4.7 eV [8, 9]) is very close to the solar-blind edge at 280 nm, which eliminates the need for optical filters and the need to tune E_g by alloying, which is essential for AlGaN- and MgZnO-based detectors. Moreover, large, bulk single crystals of β-Ga_2O_3 synthesized by melt-growth production methods (e.g., edge-defined film-fed growth [10], floating zone [11], or the Czochralski process [12]) enable the production of low-cost wafers, which are more advantageous than small and expensive diamond wafers. In addition, the crystalline quality of such bulk crystals is sufficiently high for photodetection, and thus epitaxial growth is not absolutely necessary for the fabrication of photodetectors using bulk crystals. Even if epitaxial growth is necessary, a wide variety of solid-, liquid-, and vapor-phase growth methods can be employed to obtain the films [13].

Studies of Ga_2O_3-based photodetectors date back to 1965, the year Tippins published a well-known paper about the optical absorption and photoconductivity near the band edge of β-Ga_2O_3 [9]. In that paper, he presented the spectral response of β-Ga_2O_3 for the first time, as shown in Fig. 38.2. The wavelength-selective photoresponse in the deep-UV region shows that β-Ga_2O_3 can be employed in deep-UV photodetectors. After this first demonstration, many β-Ga_2O_3-based photodetectors have been reported. In this chapter, I summarize this literature to facilitate a better understanding of the development of β-Ga_2O_3-based photodetectors.

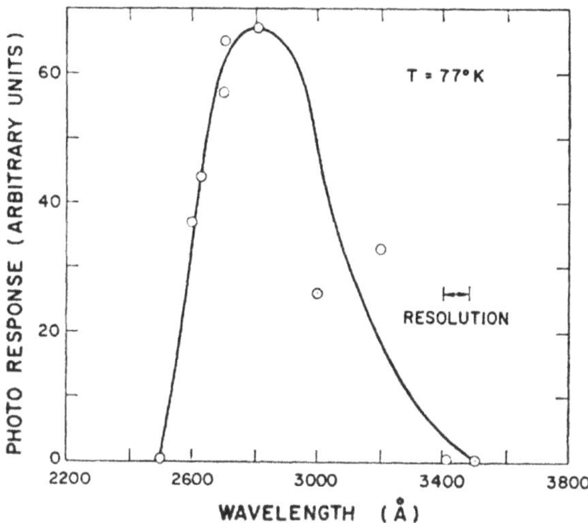

Fig. 38.2 Wavelength dependence of the photoconductivity of the first β-Ga_2O_3 photodetector [9]. Reprinted with permission from American Physical Society

38.2 Types of Photodetectors

Photodetectors based on β-Ga_2O_3 can be categorized as photoconductors, metal-semiconductor-metal (MSM) photodetectors, Schottky photodiodes, and heterojunction photodetectors, as shown in Fig. 38.3. Most reported photodetectors are photoconductors or MSM photodetectors fabricated using β-Ga_2O_3 films or nanostructures on insulating substrates. These two types of detectors are distinguished by their contact behavior (ohmic or Schottky, respectively). In both cases, interdigital electrodes are often employed to apply electric fields efficiently to the illuminated films. In contrast, Schottky photodiodes are fabricated on conductive β-Ga_2O_3 single-crystal substrates, with semi-transparent Schottky contacts. In heterojunction photodetectors, junctions with other *p*-type or narrower-band *n*-type semiconductors are used. In the following sections, I describe the details of these types of photodetectors.

38.3 Film-Based Photoconductors and MSM Photodetectors

As summarized in Table 38.1, many photodetectors have been fabricated using Ga_2O_3 films formed on insulating substrates. In most cases, these have been sapphire substrates, and various fabrication methods have been used, such as sol-gel, spray, metal-organic chemical vapor deposition (MOCVD), molecular beam epitaxy (MBE), laser MBE (LMBE), pulsed laser deposition (PLD), sputtering, or the oxidation of GaN or GaAs. The prepared, undoped films tend to be highly resistive, and thus it is more difficult to form ohmic contacts on Ga_2O_3 films rather than

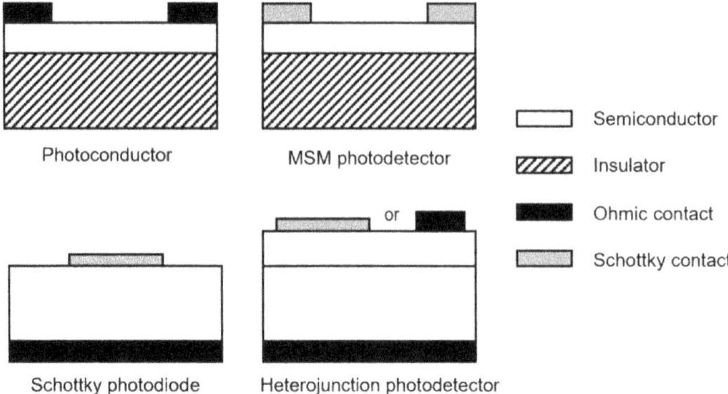

Fig. 38.3 Schematic structure of different β-Ga_2O_3-based photodetectors

Schottky contacts. Therefore, most of the literature deals with MSM-type photodetectors, and there are few reports on photoconductors.

To fabricate photoconductors, Oh et al. used ion implantation to perform Si doping, obtaining low-resistivity β-Ga_2O_3 films, thus, enabling the formation of ohmic contacts [22]. The fabricated photoconductors exhibited a relatively high responsivity of 1.45 A/W (5 V, 254 nm); however, the dark current was comparable to the photocurrent.

Photoconductors can also be fabricated using unintentionally doped (UID) β-Ga_2O_3 films grown under reducing conditions. Guo et al. grew β-Ga_2O_3 films by LMBE under an oxygen partial pressure of 5×10^{-3} Pa at 750 °C (reducing conditions), and some of the grown films were annealed in an O_2 atmosphere for comparison [21]. The Ti/Al contacts on the as-grown and the annealed films exhibited ohmic and Schottky behaviors, respectively, as shown in Fig. 38.4a, indicating that the oxidation state of the film influences the behavior of the photodetectors fabricated thereon. These investigators also found that, at 1 V, the MSM photoconductor exhibited a smaller dark current ($\sim$0.3 nA) and a faster response ($\sim$0.62 s) to 254 nm radiation than those of the ohmic photoconductor ($\sim$60 nA and 1.91 s, respectively) owing to the presence of Schottky barriers, as shown in Fig. 38.4b.

To clarify the conduction mechanism in MSM photodetectors, Ghose et al. measured the I–V curves of MSM photodetectors grown on β-Ga_2O_3 films by MBE [37]. As seen in Fig. 38.5, the photocurrents and dark currents increase with the applied voltage. By fitting transport models to the I–V curves, they showed that the carrier-injection mechanisms are dominated by ohmic conduction ($I \propto V$) under illumination by 254 nm radiation and by space-charge-limited conduction (SCLC) ($I \propto V^a$, where $a \geq 2$) under illumination by 405 nm radiation and in the dark.

The performance of an MSM photodetector is strongly dependent on the device geometry and the crystallinity of the films, including the density of defects. Wang et al. investigated the dependence of the photocurrent flowing in an MSM photoconductor on the β-Ga_2O_3 film thickness [41]. The ratio of the photocurrent to the

Table 38.1 Summary of Ga_2O_3 film-based photodetectors

Type	Film	Method	Substrate	Electrode	R (A/W)	V (V)	λ (nm)	Rejection ratio	t_r/t_d	τ_r/τ_d	D (Jones)	Remark	References
	β-Ga_2O_3	Spray	Quartz				254						[14]
MSM	β-Ga_2O_3	Sol-gel	Sapphire	Au	3E-5	10	230						[15]
MSM	β-Ga_2O_3	MBE	Sapphire	Ti/Au	0.037	10	254						[16]
MSM	β-Ga_2O_3	Oxidation	GaN on Sapphire	Ti/Al/Ti/Au	0.453	5	260						[17]
MSM	β-Ga_2O_3	MOD	Sapphire	Ti/Al	0.76	20	254		50/30 ms				[18]
MSM	β-Ga_2O_3	LMBE	Sapphire	Ti/Au		10	254			0.86/1.02 s			[19]
MSM	β-Ga_2O_3	MOCVD	Sapphire	IZO	3.2E-4	10	185	3000			2.8E12	Seethrough	[20]
PC	β-Ga_2O_3	LMBE	Sapphire	Ti/Au		1	254			1.91 s/2.16 s		As-grown	[21]
MSM	β-Ga_2O_3	LMBE	Sapphire	Ti/Au		1	254			0.62/0.83 s		Annealed in O_2	[21]
PC	Si:β-Ga_2O_3	MOCVD	Sapphire	Ti/Au	1.45	5	254	10.8 (254/365)		0.58/1.2 s		Ion implantation	[22]
MSM	β-Ga_2O_3	MOCVD	Sapphire	Au	17	20	255	8.5E6 (255/450)			7.0E12		[23]
MSM	β-Ga_2O_3	PLD	Sapphire	Ni/Au	0.903	5	250						[24]
MSM	Au/β-Ga_2O_3	LMBE	Sapphire	Ti/Au			254			1.3/1.2 s		Plasmon (Au)	[25]
MSM	α-Ga_2O_3	LMBE	Sapphire	Ti/Au	0.032	5	254					α phase	[26]
PC	Si:β-Ga_2O_3	MOCVD	Sapphire	Ti/Au	5	5	254	9.4 (254/365)				25 °C	[27]
PC	Si:β-Ga_2O_3	MOCVD	Sapphire	Ti/Au	36	5	254	216 (254/365)				350 °C	[27]
MSM	β-Ga_2O_3	LMBE	Sapphire	Ti/Au	0.009	40	244					Film	[28]

(continued)

Table 38.1 (continued)

Type	Film	Method	Substrate	Electrode	R (A/W)	V (V)	λ (nm)	Rejection ratio	t_r/t_d	τ_r/τ_d	D (Jones)	Remark	References
										0.40/ 0.18 s			
MSM	β-Ga_2O_3	Exfoliation	β-Ga_2O_3	Ti/Au	0.05	40	252			0.45/ 0.24 s		Bulk	[28]
PC	Si:β-Ga_2O_3	MOCVD	Sapphire	Ti/Au	0.3	5	254					Proton irradiation	[29]
MSM	β-Ga_2O_3	MBE	Sapphire	Ti/Al	54.9	10	254	3220 (254/ 350)	2/4 s		3.71E14		[30]
MSM	β-Ga_2O_3	LMBE	Sapphire	graphene	9.66	10	254			0.96/ 0.81 s		Vertical	[31]
MSM	β-Ga_2O_3	LMBE	Sapphire	Ti/Au	0.06	10	259			0.47/ 0.28 s			[32]
MSM	a-Ga_2O_3	Sputtering	Quartz	ITO	0.19	10	254			−/ 19.1 μs		Amorphous seethrough	[33]
MSM	a-Ga_2O_3	Sputtering	PEN	ITO		10	254					Flexible, amorphous, seethrough	[33]
MSM	Zn:β-Ga_2O_3	MOCVD	Sapphire	Ni/Au	210	20	232	5E4 (232/ 320)	3.2/ 1.4 s				[34]
MSM	Mg:β-Ga_2O_3	Sputtering	Sapphire	Ti/Au	0.0238	10	254			0.33/ 0.02 s			[35]
MSM	β-Ga_2O_3	LPCVD	Sapphire	Ti/Au	0.11	1	250	1.4E2 (250/370)		0.4/ 0.45 s		As-grown	[36]
MSM	β-Ga_2O_3	LPCVD	Sapphire	Ti/Au	0.14	1	250	8.4E4 (250/370)		0.3/ 0.2 s		Annealed in O_2	[36]
MSM	β-Ga_2O_3	MBE	Sapphire	Ti/Au			254			5/6 s			[37]
MSM	β-Ga_2O_3	Oxidation	GaAs	Al	0.292	20	270		1.4/ 0.1 s				[38]

(continued)

Table 38.1 (continued)

Type	Film	Method	Substrate	Electrode	R (A/W)	V (V)	λ (nm)	Rejection ratio	t_r/t_d	τ_r/τ_d	D (Jones)	Remark	References
MSM	β-Ga_2O_3	MBE	Sapphire	Ni/Au	>1.5	4	236	>1E5 (236/420)	3.33/ 0.4 s			Passivation	[39]
MSM	β-Ga_2O_3	MBE	Sapphire	Ti/Al	4.21	10	254	1.61E4 (250/350)		0.41/ 0.04 s			[40]
MSM	a-Ga_2O_3	Sputtering	Sapphire	Ti/Al	70.26	10	254	1.15E5 (250/350)		0.41/ 0.02 s		Amorphous	[40]
MSM	β-Ga_2O_3	sputtering	Sapphire	Ti/Au			254			210/ 16 ms		Thickness dependence	[41]
MSM	Ga/β-Ga_2O_3	Sputtering	Quartz	ITO	2.85	10	260					Plasmon (Ga)	[42]
MSM	β-Ga_2O_3	Mist CVD	Sapphire	Al	>150	20	254			1.8/ 0.3 s			[43]
MSM	β-Ga_2O_3	Sputtering	Si	Cr/Au	96.13	5	254	>1E3 (250/400)		32/ 78 ms			[44]
MSM	β-Ga_2O_3	Sol-gel	Sapphire	Au	5.86E-5	30	254			0.10/ 0.10 s			[45]
MSM	β-Ga_2O_3	Sputtering	Sapphire	Ti/Au	0.8926	10	250	194 (250/400)		305/ 251 ms		Array	[46]
MSM	β-Ga_2O_3	MBE	Sapphire	Ni/Au, Ti/Au	0.014	0	255	1E2 (255/450)	1.1/ 0.3 s		2.0E12	Self-bias	[47]
MSM	β-Ga_2O_3	MBE	Sapphire	Ni/Au, Ti/Au	1.14	5	255	1E5 (255/450)					[47]
MSM	β-Ga_2O_3	MOCVD	Sapphire	Ti/Au	26.1	10	255	3E2 (255/405)		0.48/ 0.18 s	1.25E13		[48]

R: responsivity, *V*: volt, λ = peak or measured wavelength, t_r/t_d: rise/decay times, τ_r/τ_d: rise/decay time constants, PC: photoconductor, and a-Ga_2O_3: amorphous-Ga_2O_3

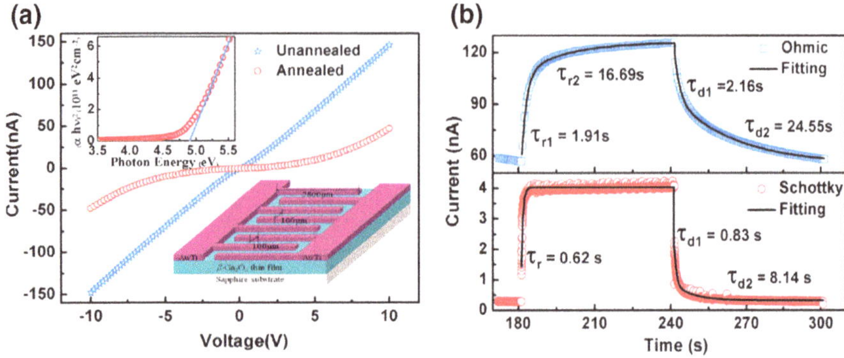

Fig. 38.4 **a** The dark I–V characteristics of photodetectors fabricated using as-grown (annealed) and oxidized (annealed) β-Ga_2O_3 films. **b** Time-dependent photoresponse of the ohmic photoconductor and the MSM photodetector [21]. Reprinted with permission from AIP Publishing

Fig. 38.5 I–V curves of the MSM photodetector in the dark, under 405 nm radiation and under 254 nm radiation, each fitted with a carrier-injection mechanism [37]. Reprinted with permission from AIP Publishing

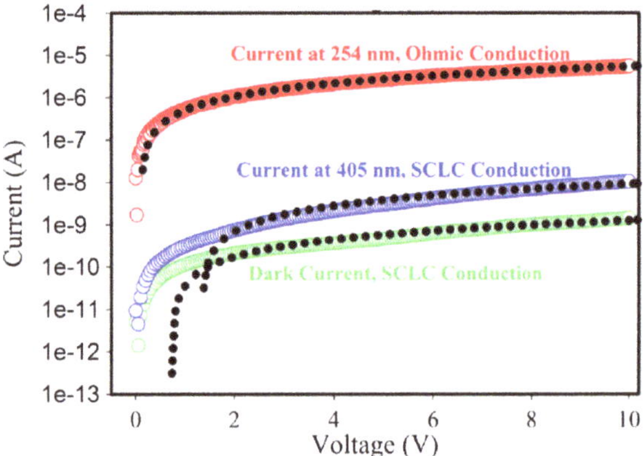

dark current peaks at a thickness of 303 nm. The investigators explained that the light absorption is the main factor in generating the photocurrent up to 303 nm, at which thickness 85% of the 254 nm radiation is absorbed. They also found that the distribution of the electric field over the thickness of the film plays a major role. Rafique et al. analyzed the oxidation state of the as-grown and oxygen-annealed β-Ga_2O_3 films grown by low-pressure CVD (LPCVD), and they evaluated the MSM photodetectors fabricated using these films [36]. By comparing the data they obtained, they found that thermal oxidation reduces the number of oxygen vacancies and improves the responsivity at 250 nm, the rejection ratio, and the time response, as shown in Table 38.1. Feng et al. compared the device characteristics of the MSM photodetectors fabricated on a β-Ga_2O_3 single-crystal substrate with those of an LMBE-grown heteroepitaxial film grown on a sapphire substrate [28]. Owing to its higher crystallinity, the detector constructed on the bulk single crystal exhibited a higher peak responsivity (0.05 A/W at 252 nm) than that grown on the epitaxial film (0.009 A/W at 244 nm).

For future imaging detectors, Peng et al. designed and fabricated MSM photodetector arrays of 32 × 32, 16 × 16, 8 × 8, and 4 × 4 pixels using a sputter-deposited β-Ga_2O_3 film on a 2-inch sapphire substrate. In these arrays, the cathode and anode signal lines are isolated by SiO_2 spacers. At 5 V, 16 devices in one of the 4 × 4 arrays exhibited dark currents in the range 0.111–0.528 nA and maximum responsivities of 0.2119–0.5851 A/W. The average values with standard deviations were 0.330 ± 0.173 nA and 0.3346 ± 0.066 A/W, respectively.

38.4 Photodetectors Based on $(Al_xGa_{1-x})_2O_3$ and $(In_xGa_{1-x})_2O_3$ Alloy Films

To shift the peak wavelength of the spectral response of a β-Ga_2O_3 film-based photodetector, $(Al_xGa_{1-x})_2O_3$ and $(In_xGa_{1-x})_2O_3$ alloy films have been used, because their E_g are wider and narrower than that of β-Ga_2O_3. Tables 38.2 and 38.3, respectively, summarize the properties of photodetectors based on $(Al_xGa_{1-x})_2O_3$ and on $(In_xGa_{1-x})_2O_3$ alloy films.

Zhang et al. demonstrated that, among the alloy-film-based detectors, $(In_xGa_{1-x})_2O_3$-based photodetectors can be employed as UV-A, B, and C detectors [53]. They prepared numerous MSM photodetectors on an $(In_xGa_{1-x})_2O_3$ composition-spread film deposited on a 2-inch sapphire substrate by PLD. The composition of the $(In_xGa_{1-x})_2O_3$ photodetectors ranged from $x = 0.013$ to $x = 0.789$, resulting in cutoff energies between 4.83 and 3.22 eV. As Fig. 38.6 shows, the corresponding peak-response wavelengths of the MSM detectors ranged from 360 to 250 nm. The wide controllability of the spectral response—covering all the UV-A, B, and C regions—is particularly interesting for UV-protection applications.

38.5 Schottky Photodiodes Using β-Ga_2O_3 Single-Crystal Substrates

Vertical Schottky photodiodes have been fabricated using conductive β-Ga_2O_3 single-crystal substrates, as listed in Table 38.4. Some photodiodes were fabricated without using epitaxial layers, owing to the high quality of the β-Ga_2O_3 single crystals, while others were fabricated with light-absorbing layers prepared by thermal oxidation of the crystals or by thin-film growth.

The maximum photoreponsivity of the β-Ga_2O_3 Schottky photodiodes under zero bias is comparable to those of commercial devices based on other wide-bandgap semiconductor photodiodes. Alema et al. grew a Ge-doped

Table 38.2 Summary of $(Al_xGa_{1-x})_2O_3$ film-based photodetectors

Type	x	Method	Substrate	Electrode	R (A/W)	V (V)	λ (nm)	D (Jones)	Remark	References
MSM	0–	Sol-gel	Sapphire	Au	<8E-5	10	215–250		Thermal diffusion	[16]
MSM	0–0.35	LMBE	Sapphire	Ti/Au	0.01–1.5	40	238–252			[49]
MSM	0–0.03	Sputtering	Sapphire	Ti/Au	0.025–1.38	5	220–250	1.2E11–1.8E12		[50]

Table 38.3 Summary of $(In_xGa_{1-x})_2O_3$ film-based photodetectors

Type	x	Method	Substrate	Electrode	R (A/W)	V (V)	λ (nm)	τ_r/τ_d	Remark	References
MSM	0–0.2	Sol-gel	Sapphire	Au	0.0001–0.2	10	245–260			[51]
MSM	0.009–0.066	PLD	Sapphire	Pt	0.02–0.5	10	250–264		CCS	[52]
MSM	0.013–0.789	PLD	Sapphire	Pt	0.03–0.5	4	250–360		CCS	[53]
PC	0.25–0.49	PLD	Sapphire	Ti/Au	0.1–50	5–10	260–280			[54]
MSM	0–0.08	PLD	Sapphire	Ti/Au	0.07–6.12	30	258–267	0.63–0.82/0.39–0.52 s		[55]

CCS: continuous composition spread

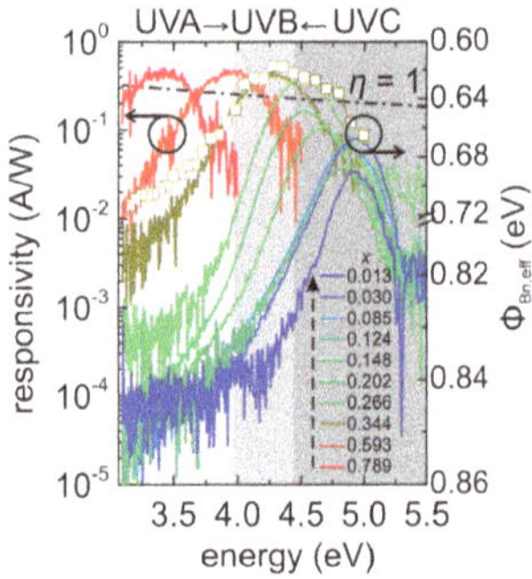

Fig. 38.6 Spectral response of 10 $(In_xGa_{1-x})_2O_3$ MSM photodetectors with various In compositions [53]. Reprinted with permission from AIP Publishing

homoepitaxial film on a Sn-doped β-Ga_2O_3 single crystal and fabricated a Schottky photodiode using a semi-transparent Pt Schottky contact [62]. When operated at zero bias, the detector exhibited solar-blind sensitivity, with a 230 to 350 nm rejection ratio as high as 10^4. The maximum peak responsivity was 0.09 A/W at 230 nm, which corresponds to an external quantum efficiency of 52%. This value is comparable to those of commercial devices based on GaN, SiC, and AlGaN.

Schottky photodiodes based on β-Ga_2O_3 can be employed for flame detection. Under self-biased operation, the photodiodes can eliminate undesired dark currents, which enables the detection of very weak radiation in the solar-blind region emitted from a flame. Oshima et al. utilized a semi-insulating (SI) layer formed by thermal oxidation of a UID β-Ga_2O_3 single crystal as a light-absorbing layer to fabricate a Schottky photodiode [58]. The band diagram of the SI-n structure can be obtained by analyzing the capacitance-voltage measurements of Schottky electrodes with different areas. These investigators also tested PEDOT:PSS, a well-known organic conductor, as a highly transparent Schottky electrode. The photodiode exhibited relatively high performance, even under zero-bias operation. The maximum photoreponsivity at 250 nm was 0.037 A/W, with a high 250–300 nm rejection ratio of 1.5×10^4. The on/off response times were as small as 9 ms. Putting the photodiode into a current–voltage conversion circuit to amplify the voltage signal, the investigators produced a detection system that successfully detected a flame as shown in Fig. 38.7. This device distinguished the 1.5 nW cm^{-2} solar-blind radiation emitted by the flame from strong fluorescent-lamp illumination without requiring any visible cutoff filters.

The mechanism of photoconductive gain in β-Ga_2O_3-based Schottky photodiodes has been discussed by Armstrong et al. [61]. As can be seen in Table 38.4, the responsivities of photodiodes operating with bias voltages exceed ~0.2 A/W, which corresponds to an external quantum efficiency of unity. This demonstrates that these detectors exhibit photoconductive gain. It is unlikely that avalanche multiplication is responsible for the gain, because in all cases the estimated electric fields are far below the critical point (~8 MV/cm). Armstrong et al. explained that the gain arises from the self-trapping of holes near the Schottky contact. As Fig. 38.8 shows, the trapped holes lower the effective Schottky barrier height under reverse bias, resulting in photoconductive gain. Note that the self-trapping of holes

Table 38.4 Summary of Schottky photodiodes based on single-crystal β-Ga_2O_3

Film	Method	Substrate	Schottky electrode	Ohmic electrode	R (A/W)	V (V)	λ (nm)	Rejection ratio	t_r/t_d	D (Jones)	Remark	References
SI-β-Ga_2O_3	Oxidation	β-Ga_2O_3	Ni/Au	Ti/Au	8.7	10	230					[56]
–		β-Ga_2O_3	Au	Ti/Al	1E3	5	240	1E6 (240/350)				[57]
SI-β-Ga_2O_3	Oxidation	β-Ga_2O_3	PEDOT:PSS	In	0.037	0	250	1.5E4 (250/300)	9/9 ms		Flame detection	[58]
SI-β-Ga_2O_3	Sol-gel	β-Ga_2O_3	Au	Ti/Al	4.3	3 (forward)	250	>1E5 (250/350)			Forward bias operation	[59]
–		β-Ga_2O_3	Graphene	Cr/Au	3.93	20	254			5.92E13		[60]
–		β-Ga_2O_3	Ni	Ti	>8	20	<220	>1E5 (220/280)				[61]
Ge: β-Ga_2O_3	MBE	β-Ga_2O_3	Pt	Ti/Al/Ni	0.09	0	230	>1E4 (230/350)				[62]
–		β-Ga_2O_3	Cu	Ti/Au				1E3 (241/400)	0.729/1.209 s			[63]

SI: semi-insulating

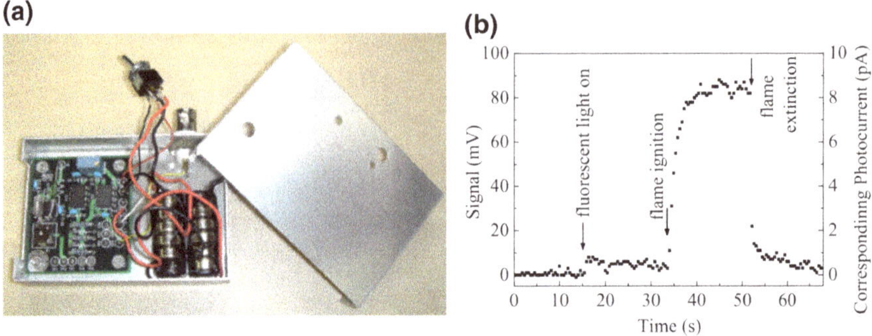

Fig. 38.7 **a** Image of the flame-detection system [58]. **b** Signal from the detection system during a demonstration of flame sensing while under illumination from a fluorescent light [58]

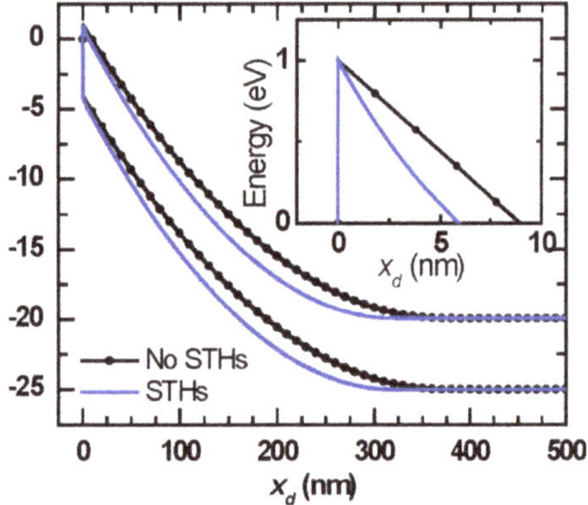

Fig. 38.8 Energy-band diagrams showing the lowering of the effective Schottky barrier height by self-trapped holes (STHs) near the surface [61]. Reprinted with permission from AIP Publishing

is an intrinsic property of β-Ga_2O_3, and thus the gain occurs for any combination of metallic Schottky contacts and surface orientations. This mechanism can also explain the gain reported for MSM photodetectors.

Oshima et al. described MOS photodiodes derived from Schottky photodiodes [64]. β-Ga_2O_3 Schottky photodiodes inevitably exhibit large leakage currents under reverse bias, due to the metal/semiconductor junction. This makes it difficult for such devices to detect very weak UV radiation except at zero bias. To decrease this undesired reverse-leakage current, they introduced HfO_2 between the metal and the β-Ga_2O_3 junction to make type-II MOS photodiodes. This increases the apparent barrier height and width for electrons without forming a barrier for holes. Figure 38.9 compares the I–V curves of the Schottky and MOS photodiodes in the dark and under illumination. A substantial decrease in the leakage current is seen for the MOS photodiodes, compared with the Schottky diodes. Interestingly, the MOS structure is also effective in suppressing the gain associated with STHs that is observed for Schottky photodiodes. These results suggest that the MOS

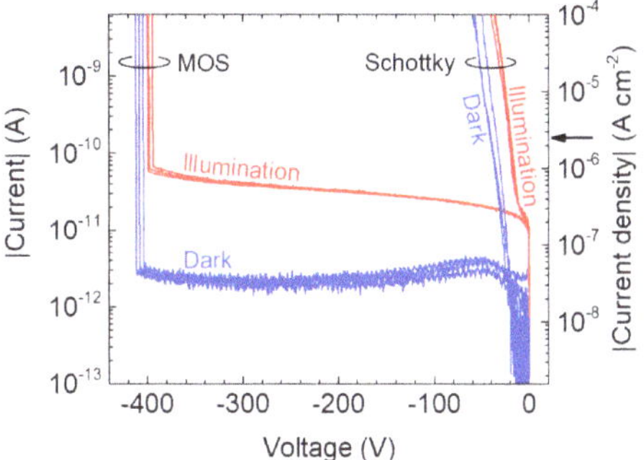

Fig. 38.9 Reverse I–V characteristics for Schottky and MOS photodiodes in the dark (blue) and under illumination (red) [64]

photodiodes behave like p-n junction photodiodes and that the MOS structure can be a basis for future developments of avalanche photodiodes without p-n junctions.

38.6 Heterojunction Photodetectors

Heterojunctions between β-Ga_2O_3 and other semiconductors have also been applied to photodetectors, as summarized in Table 38.5. Most of the heterojunctions are formed by depositing β-Ga_2O_3 films using various methods [oxidation, evaporation, LMBE, sputtering, and plasma-enhanced CVD (PECVD)] on conducting substrates or epitaxial wafers. All the heterojunction photodetectors are vertical types.

One advantage of heterojunction photodetectors is p-n junction photodiodes can suppress dark leakage currents, as compared with metal-Schottky photodiodes. Nakagomi et al. demonstrated n-β-Ga_2O_3/p-GaN and n-β-Ga_2O_3/p-SiC heterojunction photodiodes [66, 67, 77]. Up to a reverse bias of −5 V, the dark currents in these devices were as small as $\sim 10^{-10}$ A. In both cases, the carrier concentrations of p-GaN and p-SiC were higher than those of n-β-Ga_2O_3; therefore, carrier depletion occurred mainly in the β-Ga_2O_3 layer, as shown in the band diagrams of Fig. 38.10a and b. The detectors reflected these band diagrams. As Fig. 38.11 shows, the spectral responses peak at around 225 nm, indicating that the photodetection occurs mainly in the β-Ga_2O_3 layers.

Another advantage of heterojunction photodetectors is they are capable of multiband light detection, owing to the use of two semiconductors having different bandgap energies E_g. Kaira et al. fabricated UV-A and UV-C multiband photodetectors based on β-Ga_2O_3/GaN heterojunctions grown on sapphire [75]. In these cases, the Fermi levels of the *n*-type β-Ga_2O_3 and GaN layers are almost equal, and light detection occurs in both layers. The spectral response of the photodetector under forward bias (β-Ga_2O_3 biased positive) shows both UV-A and UV-C multiband photodetections.

Table 38.5 Summary of heterojunction photodetectors

Film	Method	Substrate	Electrode	R (A/W)	V (V)	λ (nm)	Rejection ratio	t_r/t_d	τ_r/τ_d	D (Jones)	Remark	References
β-Ga_2O_3	Oxidation	n-GaN on sapphire	Au, Ti/Al/Ti/Au			230−360						[65]
β-Ga_2O_3	Evaporation	p-SiC	Au, Al/Ti/Al/Pt	0.07	2	225		<1.2/1.5 ms				[66]
β-Ga_2O_3	Evaporation	p-GaN on sapphire	Au, Pt/Ti/Pt/Au	0.18	2	225		0.3/0.3 ms				[67]
β-Ga_2O_3	LMBE	p-Si	Ti/Au, Au	370	3	254			1.79/0.27 s		Forward bias	[68]
β-Ga_2O_3/i-SiC	LMBE	p-Si	Ti/Au, Au			254						[69]
β-Ga_2O_3	LMBE	n-SiC	Graphene, Au	0.18	5	254			0.65/1.73 s			[70]
β-Ga_2O_3	Sputtering	n-ZnO	In, In	0.35	5	254			0.62/0.67 s			[71]
α-Ga_2O_3	LMBE	n-ZnO	Au, In	0.5	5	255	>1E3 (230/450)		−/238 μs	9.66E12	α phase	[72]
β-Ga_2O_3	LMBE	n-ZnO	Ti/Au, Au	0.00763	0	260			0.179/0.272 s			[73]
β-Ga_2O_3	Sputtering	n-STO	In, In	43.31	10	254			1.08/0.65 s			[74]
β-Ga_2O_3	PLD	n-GaN/AlN on Si	Ni/Au, Ni/Au	3.7	5	256−365	>1E3 (365/500)			4.7E10		[75]
β-Ga_2O_3	PECVD	p-Diamond	Ti/Au, Au	0.002	0	244	140 (244/400)			6.9E9		[76]
β-Ga_2O_3	Evaporation	p-SiC	Au, Al/Ti/Al/Pt	0.053		260			<30/30 μs			[77]

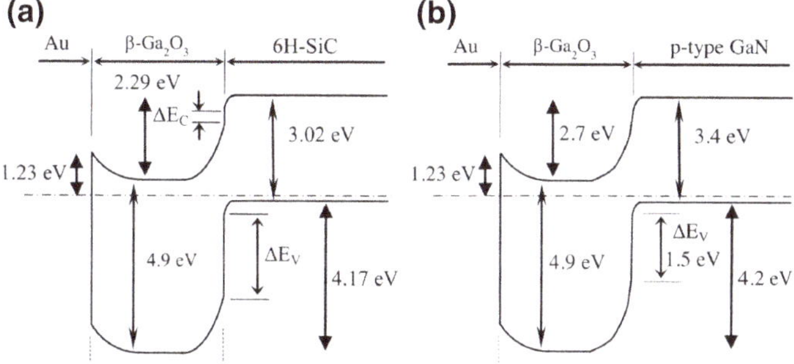

Fig. 38.10 Band diagrams of **a** n-β-Ga_2O_3/p-GaN and **b** n-β-Ga_2O_3/p-SiC heterojunctions [66, 67]. Reprinted with permission from AIP Publishing and Elsevier

Fig. 38.11 Spectral responses of photodiodes based on β-Ga_2O_3/GaN and β-Ga_2O_3/SiC heterojunction photodiodes [67]. Reprinted with permission from Elsevier

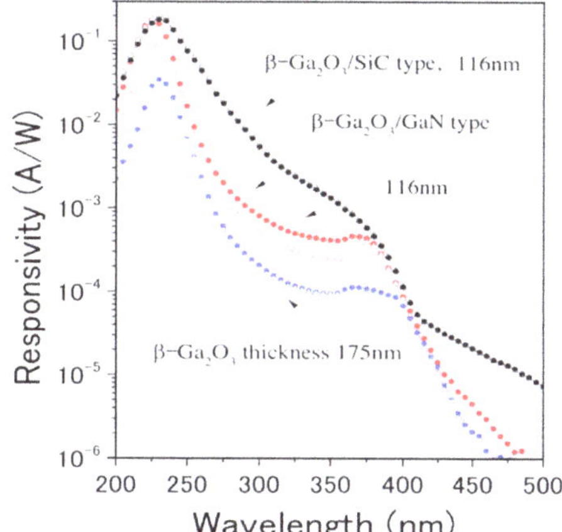

38.7 Photodetectors Based on β-Ga_2O_3 Nanostructures

Nanostructure-based photodetectors have also been studied, in addition to film- and bulk-based photodetectors. Various nanostructures consisting of nanowires, nanobelts, nanosheets, nanoflakes, islands, and core-shell structures have been fabricated by a variety of methods, including thermal evaporation, oxidation of GaSb, CVD, LMBE, and exfoliation from bulk β-Ga_2O_3. In turn, as listed in Table 38.6, these nanostructures have been used to fabricate MSM, Schottky, heterojunction, and field-effect transistor photodetectors.

Among the various types of nanostructures, nano-/micro-flakes exfoliated from β-Ga_2O_3 bulk single crystals have proven to be the most interesting, owing to their single crystallinity and small surface-to-volume ratio, both of which improve photoreponsivity. Oh et al. exfoliated a (100) β-Ga_2O_3 micro-flake from a UID β-Ga_2O_3 single crystal by using an adhesive tape, and they transferred it onto a thermally oxidized SiO_2/Si substrate to fabricate an MSM photodetector using Ni/

Table 38.6 Summary of photodetectors based on β-Ga_2O_3 nanostructures

Structure	Method	Electrode	Type	R (A/W)	V (V)	λ (nm)	Rejection ratio	t_r/t_d	τ_r/τ_d	D (Jones)	Remark	References
β-Ga_2O_3 nanowires	Thermal evaporation	Au	MSM					0.22/ 0.20 s				[78]
β-Ga_2O_3 nanowires	Thermal evaporation	Au	MSM				2E3 (250/ 280)	–/ < 20 ms				[79]
β-Ga_2O_3 nanobelt	Thermal evaporation	Cr/Au	MSM	37.6	30	250		11.8/ < 0.3 s			Individual	[80]
β-Ga_2O_3 nanowires	Thermal evaporation	Cr/Au	MSM	0.0343	5	255						[81]
Sn:β-Ga_2O_3 nanowires	Thermal evaporation	Ag	MSM	0.012	8	230					Individual	[82]
β-Ga_2O_3 nanosheet	Oxidation of GaSe nanoflakes	Cr/Au	MSM	3.3	10	254				4.0E12	Individual	[83]
ZnO-β-Ga_2O_3 core-shell microwire	CVD	In, Ti/Au	Heterojunction	1300	6	254	5E3 (254/ 400)		20/ 42 μs	9.91E14	Individual	[84]
β-Ga_2O_3 micro/ nano sheet	CVD	Au, Cr/Au	Schottky	19.31	1	254			<20/ 23 ms		Individual	[85]
Ga_2O_3/Al_2O_3 islands	LMBE	Ti/Au	MSM	1.40	10	254			1.26/ 1.99 s			[86]
β-Ga_2O_3 nanowire array film	Oxidation of Ga	Au, Ga	Schottky	0.006	10	258	2E3 (258/ 400)				Vertical	[87]
β-Ga_2O_3 nanowires	CVD	Au	MSM	377	10	231	>1E3 (231/290)		0.2/ < 0.5 s			[88]

(continued)

Table 38.6 (continued)

Structure	Method	Electrode	Type	R (A/W)	V (V)	λ (nm)	Rejection ratio	t_r/t_d	τ_r/τ_d	D (Jones)	Remark	References
β-Ga_2O_3 nano-flake	Exfoliation	Cr/Au	Transistor	1.8E5	15	254	2.93 (254/365)			7.05E11		[89]
ZnO-β-Ga_2O_3 core-shell microwire	CVD	In, Ti/Au	Heterojunction	0.097	0	251	6.9E2 (251/400)	100/900 μs			Self-bias	[90]
β-Ga_2O_3 micro-flake	Exfoliation	Ni/Au	MSM	1.68	30	254	1.92E3 (254/365)	1.76/0.53 s		3.73E10		[91]

Au Schottky electrodes [91]. The I–V characteristics of the detector exhibited the relatively high photoreponsivity of 1.68 A/W at 254 nm under 30 V bias, with a 254 to 365 nm rejection ratio of 1.92×10^3 and a detectivity of 3.73×10^{10} Jones units. This group also investigated the photodetection properties of a back-gated field-effect transistor fabricated using an exfoliated nano-flake on a Si substrate with a 300 nm thick, thermally oxidized SiO_2 insulating layer [89]. In this case, the source and drain electrodes were Cr/Au deposited on the micro-flake, and the Si substrate acted as the back gate. The advantage of this transistor is dark currents can be decreased by decreasing the gate voltage, owing to the residual carrier depletion. This detector exhibited a very high responsivity of 1.8 × 105A/W at 254 nm under a drain voltage of 15 V.

38.8 Conclusion

Ga_2O_3 is a particularly attractive material for UV photodetectors because of its solar-blind nature and the availability of high-quality single crystals and many thin-film growth methods. These facts have encouraged photodetector studies, which are reflected in numerous reports on the fabrication and characterization of various types of photodetectors, as listed in the tables above. From these investigations, we can obtain an overview of the current status of Ga_2O_3 photodetectors, while additional studies will be necessary for future applications of these detectors. In particular, it will be important to investigate the suppression of leakage currents and the control of gain using MOS, heterojunction, or other novel structures; the tuning of the peak wavelength of the spectral response by using alloy films or heterojunctions; and the integration of photodetectors into actual devices.

References

1. Z. Alaie, S. Mohammad Nejad, M.H. Yousefi, Mater. Sci. Semicond. Process. **29**, 16 (2015)
2. E. Monroy, F. Omnès, F. Calle, Semicond. Sci. Technol. **18**, R33 (2003)
3. M. Razeghi, Proc. IEEE **90**, 1006 (2002)
4. E. Muñoz, E. Monroy, J.L. Pau, F. Calle, F. Omnès, P. Gibart, J. Phys. Condens. Matter **13**, 7115 (2001)
5. Y. Hou, Z. Mei, X. Du, J. Phys. D Appl. Phys. **47**, 283001 (2014)
6. S.J. Pearton, J. Yang, P.H. Cary, F. Ren, J. Kim, M.J. Tadjer, M.A. Mastro, Appl. Phys. Rev. **5**, 011301 (2018)
7. M.Y. Liao, L.W. Sang, T. Teraji, M. Imura, J. Alvarez, Y. Koide, Jpn. J. Appl. Phys. **51**, 090115 (2012)
8. T. Onuma, S. Saito, K. Sasaki, T. Masui, T. Yamaguchi, T. Honda, M. Higashiwaki, Jpn. J. Appl. Phys. **54**, 112601 (2015)
9. H.H. Tippins, Phys. Rev. **140**, A316 (1965)
10. A. Kuramata, K. Koshi, S. Watanabe, Y. Yamaoka, T. Masui, S. Yamakoshi, Jpn. J. Appl. Phys. **55**, 1202A2 (2016)

11. E.G. Víllora, K. Shimamura, Y. Yoshikawa, K. Aoki, N. Ichinose, J. Cryst. Growth **270**, 420 (2004)
12. Z. Galazka, K. Irmscher, R. Uecker, R. Bertram, M. Pietsch, A. Kwasniewski, M. Naumann, T. Schulz, R. Schewski, D. Klimm, M. Bickermann, J. Cryst. Growth **404**, 184 (2014)
13. H. von Wenckstern, Adv. Electron. Mater. **3**, 1600350 (2017)
14. Z. Ji, J. Du, J. Fan, W. Wang, Opt. Mater. **28**, 415 (2006)
15. T. Oshima, T. Okuno, S. Fujita, Jpn. J. Appl. Phys. **46**, 7217 (2007)
16. Y. Kokubun, K. Miura, F. Endo, S. Nakagomi, Appl. Phys. Lett. **90**, 031912 (2007)
17. W.Y. Weng, T.J. Hsueh, S.J. Chang, G.J. Huang, H.T. Hsueh, IEEE Sens. J. **11**, 999 (2011)
18. P. Guo, J. Xiong, X. Zhao, T. Sheng, C. Yue, B. Tao, X. Liu, J. Mater. Sci. Mater. Electron. **25**, 3629 (2014)
19. D. Guo, Z. Wu, P. Li, Y. An, H. Liu, X. Guo, H. Yan, G. Wang, C. Sun, L. Li, W. Tang, Opt. Mater. Express **4**, 1067 (2014)
20. T.-C. Wei, D.-S. Tsai, P. Ravadgar, J.-J. Ke, M.-L. Tsai, D.-H. Lien, C.-Y. Huang, R.-H. Horng, J.-H. He, IEEE J. Sel. Top. Quantum Electron. **20**, 3802006 (2014)
21. D.Y. Guo, Z.P. Wu, Y.H. An, X.C. Guo, X.L. Chu, C.L. Sun, L.H. Li, P.G. Li, W.H. Tang, Appl. Phys. Lett. **105**, 023507 (2014)
22. S. Oh, Y. Jung, M.A. Mastro, J.K. Hite, C.R. Eddy, J. Kim, Opt. Express **23**, 28300 (2015)
23. G.C. Hu, C.X. Shan, N. Zhang, M.M. Jiang, S.P. Wang, D.Z. Shen, Opt. Express **23**, 13554 (2015)
24. F.-P. Yu, S.-L. Ou, D.-S. Wuu, Opt. Mater. Express **5**, 1240 (2015)
25. Y.H. An, D.Y. Guo, Z.M. Li, Z.P. Wu, Y.S. Zhi, W. Cui, X.L. Zhao, P.G. Li, W.H. Tang, RSC Adv. **6**, 66924 (2016)
26. D.Y. Guo, X.L. Zhao, Y.S. Zhi, W. Cui, Y.Q. Huang, Y.H. An, P.G. Li, Z.P. Wu, W.H. Tang, Mater. Lett. **164**, 364 (2016)
27. S. Ahn, F. Ren, S. Oh, Y. Jung, J. Kim, M.A. Mastro, J.K. Hite, C.R. Eddy, S.J. Pearton, J. Vac. Sci. Technol. B **34**, 041207 (2016)
28. Q. Feng, L. Huang, G. Han, F. Li, X. Li, L. Fang, X. Xing, J. Zhang, W. Mu, Z. Jia, D. Guo, W. Tang, X. Tao, Y. Hao, I.E.E.E. Trans, Electron Devices **63**, 3578 (2016)
29. S. Ahn, Y.-H. Lin, F. Ren, S. Oh, Y. Jung, G. Yang, J. Kim, M.A. Mastro, J.K. Hite, C.R. Eddy, S.J. Pearton, J. Vac. Sci. Technol. B **34**, 041213 (2016)
30. L.-X. Qian, H.-F. Zhang, P.T. Lai, Z.-H. Wu, X.-Z. Liu, Opt. Mater. Express **7**, 3643 (2017)
31. M. Ai, D. Guo, Y. Qu, W. Cui, Z. Wu, P. Li, L. Li, W. Tang, J. Alloys Compd. **692**, 634 (2017)
32. L. Huang, Q. Feng, G. Han, F. Li, X. Li, L. Fang, X. Xing, J. Zhang, Y. Hao, IEEE Photonics J. **9**, 6803708 (2017)
33. S. Cui, Z. Mci, Y. Zhang, H. Liang, X. Du, Adv. Opt. Mater. **5**, 1700454 (2017)
34. F. Alema, B. Hertog, O. Ledyaev, D. Volovik, G. Thoma, R. Miller, A. Osinsky, P. Mukhopadhyay, S. Bakhshi, H. Ali, W.V. Schoenfeld, Phys. Status Solidi A **214**, 1600688 (2017)
35. Y.P. Qian, D.Y. Guo, X.L. Chu, H.Z. Shi, W.K. Zhu, K. Wang, X.K. Huang, H. Wang, S.L. Wang, P.G. Li, X.H. Zhang, W.H. Tang, Mater. Lett. **209**, 558 (2017)
36. S. Rafique, L. Han, H. Zhao, Phys. Status Solidi A **214**, 1700063 (2017)
37. S. Ghose, S. Rahman, L. Hong, J.S. Rojas-Ramirez, H. Jin, K. Park, R. Klie, R. Droopad, J. Appl. Phys. **122**, 095302 (2017)
38. D. Patil-Chaudhari, M. Ombaba, J.Y. Oh, H. Mao, K.H. Montgomery, A. Lange, S. Mahajan, J.M. Woodall, M.S. Islam, IEEE Photonics J. **9**, 2300207 (2017)
39. A. Singh Pratiyush, S. Krishnamoorthy, S. Vishnu Solanke, Z. Xia, R. Muralidharan, S. Rajan, D.N. Nath, Appl. Phys. Lett. **110**, 221107 (2017)
40. L.-X. Qian, Z.-H. Wu, Y.-Y. Zhang, P.T. Lai, X.-Z. Liu, Y.-R. Li, ACS Photonics **4**, 2203 (2017)
41. X. Wang, Z. Chen, D. Guo, X. Zhang, Z. Wu, P. Li, W. Tang, Opt. Mater. Express **8**, 2918 (2018)

42. S.-J. Cui, Z.-X. Mei, Y.-N. Hou, Q.-S. Chen, H.-L. Liang, Y.-H. Zhang, W.-X. Huo, X.-L. Du, Chin. Phys. B **27**, 067301 (2018)
43. Y. Xu, Z. An, L. Zhang, Q. Feng, J. Zhang, C. Zhang, Y. Hao, Opt. Mater. Express **8**, 2941 (2018)
44. K. Arora, N. Goel, M. Kumar, M. Kumar, ACS Photonics **5**, 2391 (2018)
45. H. Shen, Y. Yin, K. Tian, K. Baskaran, L. Duan, X. Zhao, A. Tiwari, J. Alloys Compd. **766**, 601 (2018)
46. Y. Peng, Y. Zhang, Z. Chen, D. Guo, X. Zhang, P. Li, Z. Wu, W. Tang, IEEE Photonics Technol. Lett. **30**, 993 (2018)
47. A.S. Pratiyush, S. Krishnamoorthy, S. Kumar, Z. Xia, R. Muralidharan, S. Rajan, D.N. Nath, Jpn. J. Appl. Phys. **57**, 060313 (2018)
48. D. Zhang, W. Zheng, R.C. Lin, T.T. Li, Z.J. Zhang, F. Huang, J. Alloys Compd. **735**, 150 (2018)
49. Q. Feng, X. Li, G. Han, L. Huang, F. Li, W. Tang, J. Zhang, Y. Hao, Opt. Mater. Express **7**, 1240 (2017)
50. S.-H. Yuan, C.-C. Wang, S.-Y. Huang, D.-S. Wuu, IEEE Electron Device Lett. **39**, 220 (2018)
51. Y. Kokubun, T. Abe, S. Nakagomi, Phys. Status Solidi A **207**, 1741 (2010)
52. H. von Wenckstern, D. Splith, M. Purfürst, Z. Zhang, C. Kranert, S. Müller, M. Lorenz, M. Grundmann, Semicond. Sci. Technol. **30**, 024005 (2015)
53. Z. Zhang, H. von Wenckstern, J. Lenzner, M. Lorenz, M. Grundmann, Appl. Phys. Lett. **108**, 123503 (2016)
54. F. Zhang, H. Li, M. Arita, Q. Guo, Opt. Mater. Express **7**, 3769 (2017)
55. K. Zhang, Q. Feng, L. Huang, Z. Hu, Z. Feng, A. Li, H. Zhou, X. Lu, C. Zhang, J. Zhang, Y. Hao, IEEE Photonics J. **10**, 6802508 (2018)
56. T. Oshima, T. Okuno, N. Arai, N. Suzuki, S. Ohira, S. Fujita, Appl. Phys. Express **1**, 011202 (2008)
57. R. Suzuki, S. Nakagomi, Y. Kokubun, N. Arai, S. Ohira, Appl. Phys. Lett. **94**, 222102 (2009)
58. T. Oshima, T. Okuno, N. Arai, N. Suzuki, H. Hino, S. Fujita, Jpn. J. Appl. Phys. **48**, 011605 (2009)
59. R. Suzuki, S. Nakagomi, Y. Kokubun, Appl. Phys. Lett. **98**, 131114 (2011)
60. W.-Y. Kong, G.-A. Wu, K.-Y. Wang, T.-F. Zhang, Y.-F. Zou, D.-D. Wang, L.-B. Luo, Adv. Mater. **28**, 10725 (2016)
61. A.M. Armstrong, M.H. Crawford, A. Jayawardena, A. Ahyi, S. Dhar, J. Appl. Phys. **119**, 103102 (2016)
62. F. Alema, B. Hertog, A.V. Osinsky, P. Mukhopadhyay, M. Toporkov, W.V. Schoenfeld, E. Ahmadi, J. Speck, in *Proceedings of SPIE 10105, Oxide-Based Mater. Devices VIII*, ed. by F.H. Teherani, D.C. Look, D.J. Rogers, p. 101051M (2017)
63. C. Yang, H. Liang, Z. Zhang, X. Xia, P. Tao, Y. Chen, H. Zhang, R. Shen, Y. Luo, G. Du, RSC Adv. **8**, 6341 (2018)
64. T. Oshima, M. Hashikawa, S. Tomizawa, K. Miki, T. Oishi, K. Sasaki, A. Kuramata, Appl. Phys. Express **11**, 112202 (2018)
65. W.Y. Weng, T.J. Hsueh, S.J. Chang, G.J. Huang, H.T. Hsueh, IEEE Photonics Technol. Lett. **23**, 444 (2011)
66. S. Nakagomi, T. Momo, S. Takahashi, Y. Kokubun, Appl. Phys. Lett. **103**, 072105 (2013)
67. S. Nakagomi, T. Sato, Y. Takahashi, Y. Kokubun, Sens. Actuators A **232**, 208 (2015)
68. X.C. Guo, N.H. Hao, D.Y. Guo, Z.P. Wu, Y.H. An, X.L. Chu, L.H. Li, P.G. Li, M. Lei, W.H. Tang, J. Alloys Compd. **660**, 136 (2016)
69. Y. An, Y. Zhi, Z. Wu, W. Cui, X. Zhao, D. Guo, P. Li, W. Tang, Appl. Phys. A **122**, 1036 (2016)
70. Y. Qu, Z. Wu, M. Ai, D. Guo, Y. An, H. Yang, L. Li, W. Tang, J. Alloys Compd. **680**, 247 (2016)
71. D.Y. Guo, H.Z. Shi, Y.P. Qian, M. Lv, P.G. Li, Y.L. Su, Q. Liu, K. Chen, S.L. Wang, C. Cui, C.R. Li, W.H. Tang, Semicond. Sci. Technol. **32**, 03LT01 (2017)

72. X. Chen, Y. Xu, D. Zhou, S. Yang, F. Ren, H. Lu, K. Tang, S. Gu, R. Zhang, Y. Zheng, J. Ye, A.C.S. Appl, Mater. Interfaces **9**, 36997 (2017)
73. Z. Wu, L. Jiao, X. Wang, D. Guo, W. Li, L. Li, F. Huang, W. Tang, J. Mater. Chem. C **5**, 8688 (2017)
74. D. Guo, H. Liu, P. Li, Z. Wu, S. Wang, C. Cui, C. Li, W. Tang, A.C.S. Appl, Mater. Interfaces **9**, 1619 (2017)
75. A. Kalra, S. Vura, S. Rathkanthiwar, R. Muralidharan, S. Raghavan, D.N. Nath, Appl. Phys. Express **11**, 064101 (2018)
76. Y.-C. Chen, Y.-J. Lu, C.-N. Lin, Y.-Z. Tian, C.-J. Gao, L. Dong, C.-X. Shan, J. Mater. Chem. C **6**, 5727 (2018)
77. S. Nakagomi, T. Sakai, K. Kikuchi, Y. Kokubun, Phys. Status Solidi A **216**, 1700796 (2018)
78. P. Feng, J.Y. Zhang, Q.H. Li, T.H. Wang, Appl. Phys. Lett. **88**, 153107 (2006)
79. Y. Li, T. Tokizono, M. Liao, M. Zhong, Y. Koide, I. Yamada, J.-J. Delaunay, Adv. Funct. Mater. **20**, 3972 (2010)
80. L. Li, E. Auer, M. Liao, X. Fang, T. Zhai, U.K. Gautam, A. Lugstein, Y. Koide, Y. Bando, D. Golberg, Nanoscale **3**, 1120 (2011)
81. Y.L. Wu, S.-J. Chang, W.Y. Weng, C.H. Liu, T.Y. Tsai, C.L. Hsu, K.C. Chen, IEEE Sens. J. **13**, 2368 (2013)
82. I. López, A. Castaldini, A. Cavallini, E. Nogales, B. Méndez, J. Piqueras, J. Phys. D Appl. Phys. **47**, 415101 (2014)
83. W. Feng, X. Wang, J. Zhang, L. Wang, W. Zheng, P. Hu, W. Cao, B. Yang, J. Mater. Chem. C **2**, 3254 (2014)
84. B. Zhao, F. Wang, H. Chen, Y. Wang, M. Jiang, X. Fang, D. Zhao, Nano Lett. **15**, 3988 (2015)
85. M. Zhong, Z. Wei, X. Meng, F. Wu, J. Li, J. Alloys Compd. **619**, 572 (2015)
86. W. Cui, D. Guo, X. Zhao, Z. Wu, P. Li, L. Li, C. Cui, W. Tang, RSC Adv. **6**, 100683 (2016)
87. X. Chen, K. Liu, Z. Zhang, C. Wang, B. Li, H. Zhao, D. Zhao, D. Shen, A.C.S. Appl, Mater. Interfaces **8**, 4185 (2016)
88. J. Du, J. Xing, C. Ge, H. Liu, P. Liu, H. Hao, J. Dong, Z. Zheng, H. Gao, J. Phys. D Appl. Phys. **49**, 425105 (2016)
89. S. Oh, J. Kim, F. Ren, S.J. Pearton, J. Kim, J. Mater. Chem. C **4**, 9245 (2016)
90. B. Zhao, F. Wang, H. Chen, L. Zheng, L. Su, D. Zhao, X. Fang, Adv. Funct. Mater. **27**, 1700264 (2017)
91. S. Oh, M.A. Mastro, M.J. Tadjer, J. Kim, ECS J. Solid State Sci. Technol. **6**, Q79 (2017)

Chapter 39
Image Sensors

Ga_2O_3/Se Photodiodes for Image Sensor Applications

Keitada Mineo

Abstract The advent of next-generation broadcasting systems such as 8K Super Hi-Vision has increased the demand for high-performance cameras. However, the low sensitivity of 8K cameras is challenging. To address this issue, we have developed the stacked complementary metal-oxide semiconductor (CMOS) image sensor overlaid with a gallium oxide (Ga_2O_3)/crystalline selenium (c-Se) photodiode. Using Ga_2O_3 decreased the dark current resulting from the injection of holes from the electrode. In addition, Ga_2O_3 doped with tin (Sn), which has higher carrier concentration, effectively reduced the operating voltage because the depletion layer spread into c-Se more easily as compared to that in non-doped Ga_2O_3. Furthermore, the crystallization of Ga_2O_3 improved the crystal orientation of the Se formed on β-Ga_2O_3, thereby decreasing the dark current.

39.1 Introduction

The demand for video systems with high definition [1, 2] and high frame rate [3, 4] is increasing. The 8K Super Hi-Vision system developed by the Japan Broadcasting Corporation (NHK) as a next-generation broadcasting technology is one such system [5]. It features 33 Mp (7,680 $\times$ 4,320) resolution, which is 16 times greater than that of the current 2K Hi-Vision system and has a frame rate of 120 fps (frames per second) with progressive scanning [6, 7]. This next-generation broadcasting system provides highly realistic video. However, the low sensitivity of 8K cameras, which stems from a reduction in the amount of light received per pixel and per frame, pose a serious challenge. In this chapter, we describe a new stacked complementary metal-oxide semiconductor (CMOS) image sensor overlaid with a gallium oxide (Ga_2O_3)/crystalline selenium (c-Se) photodiode, which results in an image sensor with improved sensitivity.

K. Mineo (✉)
NHK Science and Technology Research Laboratories, Kinuta Setagaya-ku, Tokyo 157-8510, Japan
e-mail: mineo.k-ge@nhk.or.jp

M. Higashiwaki and S. Fujita (eds.), *Gallium Oxide*, Springer Series in Materials Science 293, https://doi.org/10.1007/978-3-030-37153-1_39

39.2 Theory

39.2.1 Stacked Image Sensor

Figure 39.1 shows structures of conventional and stacked CMOS image sensors. The stacked CMOS image sensor has several advantages over the conventional one. First, the incident light is not attenuated by metal layers because the photoconversion layer is on the top surface of the image sensor. Second, the materials of the photoconversion layer are selected freely instead of being limited to conventional silicon (Si). Third, charge multiplication in the photoconversion layer enables the development of a highly sensitive image sensor.

39.2.2 Ga_2O_3/Se Photodiode

Figure 39.2a shows a cross-sectional view of the photodiode selected for the photoconversion layer of the new stacked CMOS image sensor. This photodiode comprises c-Se as the *p*-type layer and Ga_2O_3 as the *n*-type layer. c-Se has a higher absorption coefficient than Si over the entire visible light range (Fig. 39.3). Therefore, incident visible light is almost fully absorbed even when using a thin c-Se film. The electric field required to induce charge multiplication can thus be obtained with a low voltage.

Ga_2O_3 has a wide bandgap of 4.9 eV [8], and the large energy barrier between the work function of an indium tin oxide (ITO) electrode and the valence band of Ga_2O_3 prevents hole injection from the electrode (Fig. 39.2b), thereby increasing the dark current. The carrier concentration of Ga_2O_3 is known to be increased by

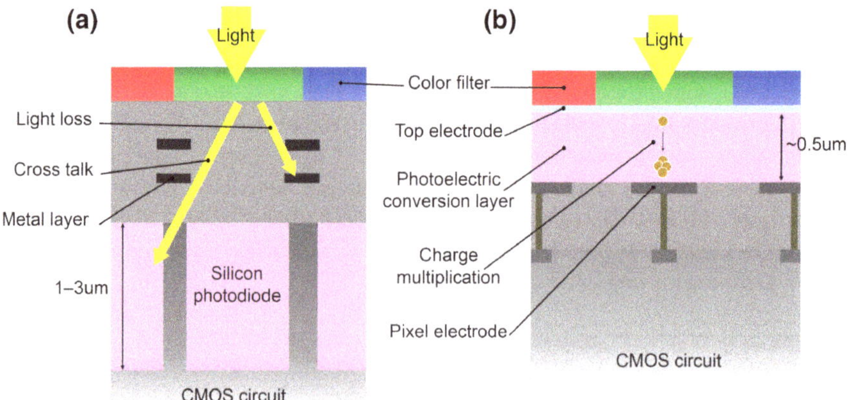

Fig. 39.1 Structures of **a** a conventional image sensor and **b** a stacked image sensor

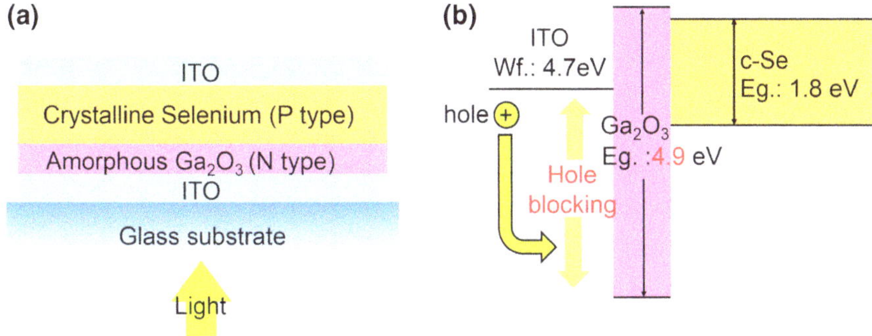

Fig. 39.2 **a** Structure and **b** band diagram of a Ga_2O_3/c-Se photodiode

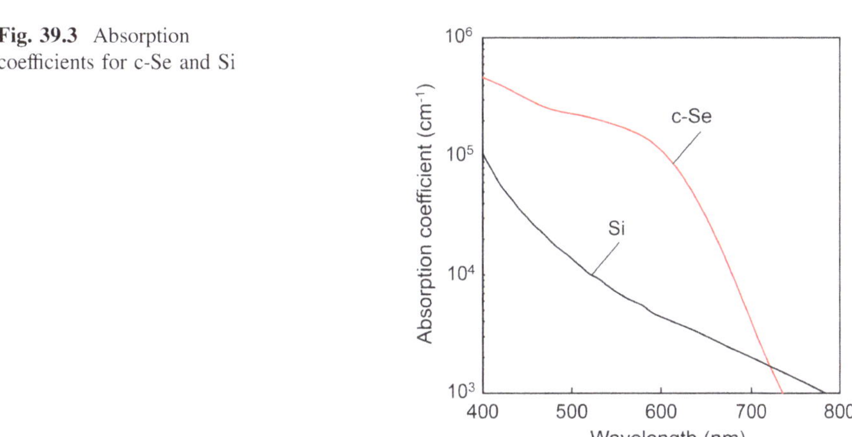

Fig. 39.3 Absorption coefficients for c-Se and Si

doping with tin (Sn) as a shallow donor [9]. Thus, the incorporation of Sn-doped Ga_2O_3 can potentially decrease the operating voltage of the photodiode through control of the depletion layer's width.

39.3 Fabrication of Photodiode

After ITO was deposited onto a glass substrate, amorphous Ga_2O_3 (thickness: 20 nm) was formed via radio-frequency (RF) magnetron sputtering with a sputtering power of 200 W under a mixture gas of argon (Ar) and oxygen (O_2) (50:1). To investigate the effect of Sn-doping, we formed Ga_2O_3 with two different Sn concentrations (5 and 5.6 mol%) via RF sputtering using Ga_2O_3 targets with 10 and 20 at.% Sn, respectively. On each Ga_2O_3 layer, a 500-nm-thick layer of amorphous Se (a-Se) was deposited via vacuum evaporation at room temperature and then completely converted into c-Se by annealing at 200 °C for 1 min in air. Next, ITO with a thickness of 30 nm was deposited as an electrode through direct-current magnetron sputtering. The current–voltage characteristics were measured for incident light irradiated from the glass-substrate side. The incident light intensity and wavelength were 2.5 μW/cm^2 and 450 nm, respectively. The illuminated area was 0.00785 cm^2.

39.4 Results and Discussion

39.4.1 Characteristics of Ga_2O_3/Se Photodiode

Figure 39.4 shows a dark current of an ITO/c-Se Schottky photodiode and a Ga_2O_3/c-Se *pn*-junction photodiode [10]. The dark current of the Ga_2O_3/c-Se photodiode is much lower than that of the ITO/c-Se Schottky photodiode. These results confirm that Ga_2O_3 functions as an effective barrier against hole injection.

Figure 39.5 shows a photocurrent of the irradiated photodiodes comprising Ga_2O_3 with various Sn concentrations (0, 5, and 5.6 mol%) [10]. With increasing Sn concentration, the photocurrent increases at a lower voltage because the increase in carrier concentration in Ga_2O_3 achieved via Sn-doping causes the depletion layer to mainly spread into c-Se. Moreover, the phenomenon of increasing photocurrent is observed at approximately 15–20 V in all photodiodes, suggesting that avalanche multiplication leads to the realization of a high-sensitivity 8K CMOS image sensor in the Ga_2O_3/Se photodiode.

Figure 39.6 shows an image captured by the stacked image sensor overlaid with a Ga_2O_3/c-Se photodiode [11]. The number of pixels and pixel pitch are 992 × 636 and

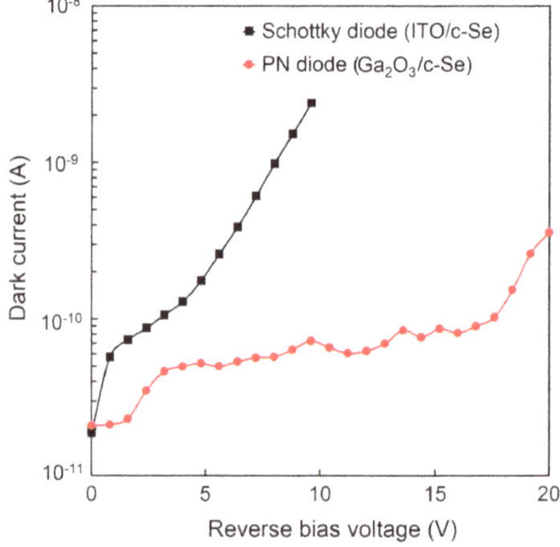

Fig. 39.4 Dark current for an ITO/c-Se Schottky photodiode and a Ga_2O_3/c-Se *pn*-junction photodiode

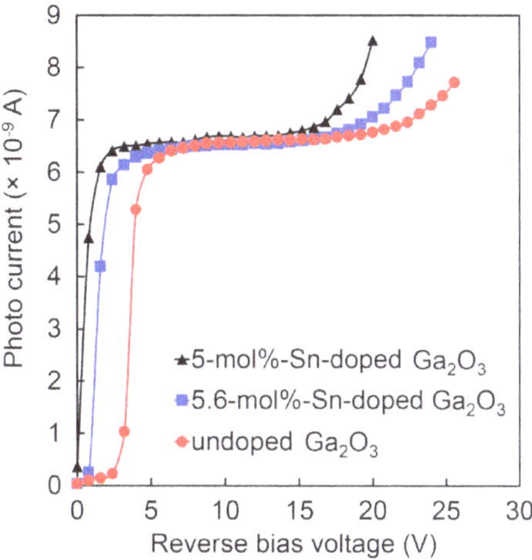

Fig. 39.5 Photocurrent for Sn-doped Ga_2O_3/c-Se photodiodes with various Sn concentrations

Fig. 39.6 Image captured with the stacked image sensor

3 μm, respectively. Clear images are observed. To the best of our knowledge, this is the first image captured by a stacked image sensor overlaid with a Ga_2O_3/c-Se photodiode.

39.4.2 *Improving Characteristics via Crystallization of Ga_2O_3*

To further reduce the dark current and operating voltage for avalanche multiplication in the Ga_2O_3/c-Se photodiode, we investigated the effect of the crystallization of Ga_2O_3 [12]. After ITO and amorphous Ga_2O_3 were deposited onto a sapphire substrate, they were annealed at 600, 800, or 1000 °C for 1 h under O_2 atmosphere. The β-Ga_2O_3 diffraction pattern was observed in the samples annealed at 800 and 1000 °C, as shown in Fig. 39.7. These results indicate that the change in crystallinity from amorphous to β-Ga_2O_3 is induced by annealing.

Figure 39.8 shows the results of X-ray diffraction (XRD) ω-scans of Se (100) conducted for c-Se deposited onto β-Ga_2O_3 and amorphous Ga_2O_3, respectively. The crystal orientation of c-Se on β-Ga_2O_3 is improved as compared with that of c-Se on amorphous Ga_2O_3. Furthermore, Fig. 39.9 shows a dark current of the photodiodes with β-Ga_2O_3 and amorphous Ga_2O_3. The dark current of the photodiode with β-Ga_2O_3 is obviously reduced owing to the decrease in defect density in c-Se resulting from the improvement of the crystal orientation.

Figure 39.10 compares secondary-ion mass spectrometry (SIMS) results between β-Ga_2O_3 and amorphous Ga_2O_3 on an ITO/*c*-plane sapphire substrate. Sn diffuses into Ga_2O_3 from the ITO electrode during annealing. Figure 39.11 shows a photocurrent of a photodiode with β-Ga_2O_3 and that of a photodiode with amorphous Ga_2O_3. Notably, the photocurrent curve for the photodiode with β-Ga_2O_3 is

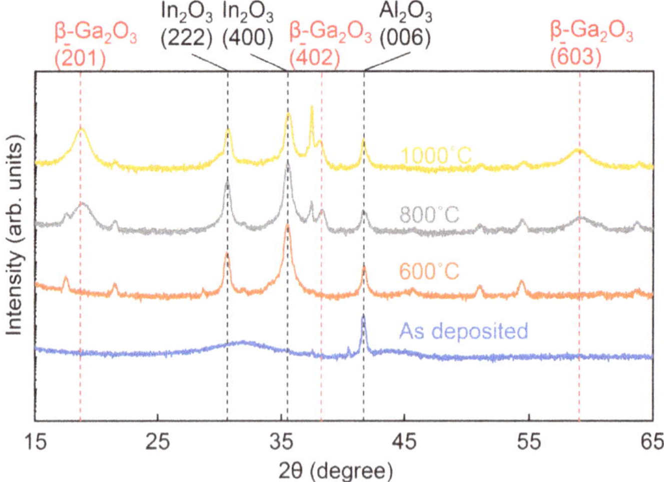

Fig. 39.7 XRD 2θ–ω scan patterns of Ga_2O_3/ITO/*c*-plane sapphire substrates annealed at various temperatures

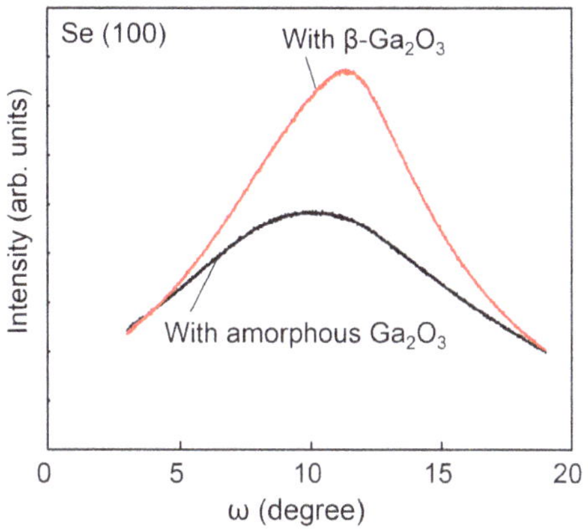

Fig. 39.8 XRD ω-scan pattern of Se on amorphous Ga_2O_3 and β-Ga_2O_3

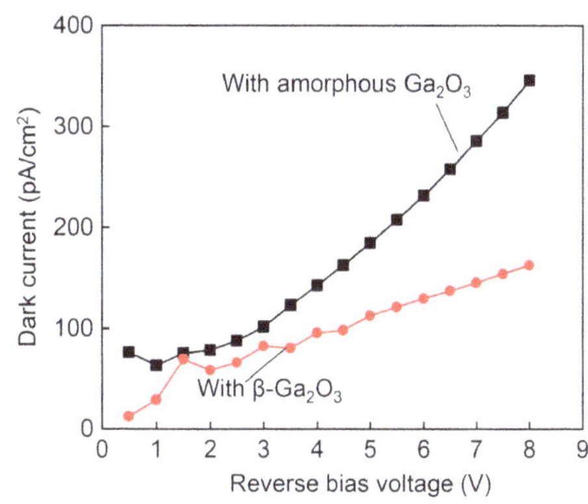

Fig. 39.9 Dark-current density of amorphous Ga_2O_3 and β-Ga_2O_3/c-Se photodiodes

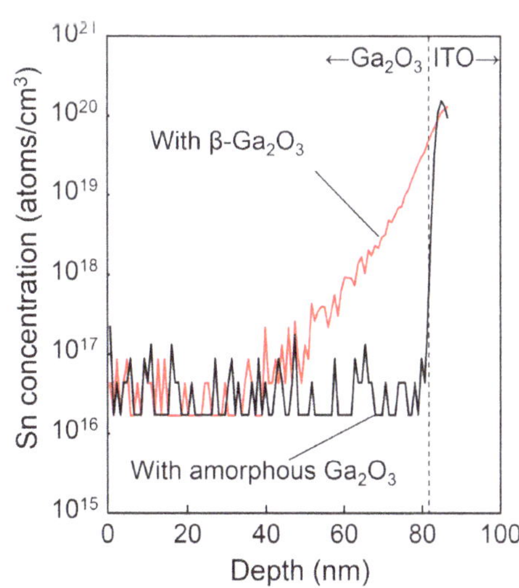

Fig. 39.10 SIMS of amorphous Ga_2O_3 and β-Ga_2O_3/ITO/*c*-plane sapphire substrate

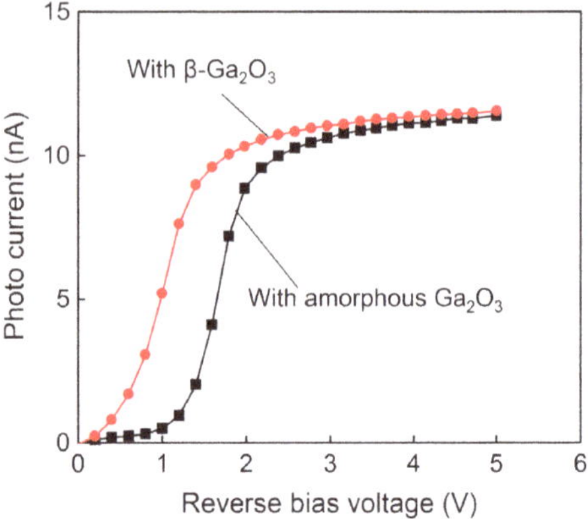

Fig. 39.11 Photocurrent of amorphous Ga_2O_3 and β-Ga_2O_3/c-Se photodiodes

shifted toward lower applied voltages. These results indicate that the depletion layer in the β-Ga_2O_3/c-Se photodiode spreads into c-Se even at a low applied voltage for the same reason as that previously mentioned in this chapter.

39.5 Conclusion

This study of a stacked CMOS image sensor overlaid with a Ga_2O_3/c-Se photodiode was conducted to realize next-generation high-sensitivity 8K cameras. The results show that Ga_2O_3 is a promising material for the aforementioned purpose owing to its wide bandgap because it enables control of carrier concentration. We speculate that the applications of Ga_2O_3 will expand into the fields of image sensors as well as power devices and high-frequency devices [13].

References

1. Y. Kim, W. Choi, D. Park, H. Jeoung, B. Kim, Y. Oh, S. Oh, B. Park, E. Kim, Y. Lee, T. Jung, Y. Kim, S. Yoon, S. Hong, J. Lee, S. Jung, C. Moon, Y. Park, D. Lee, D. Chang, in *Proceedings of the 2018 IEEE International Solid-State Circuits Conference Digest of Technical Papers (ISSCC)* (IEEE, 2018), p. 84
2. H. Totsuka, T. Tsuboi, T. Muto, D. Yoshida, Y. Matsuno, M. Ohmura, H. Takahashi, K. Sakurai, T. Ichikawa, H. Yuzurihara, S. Inoue, in *Proceedings of the 2016 IEEE International Solid-State Circuits Conference Digest of Technical Papers (ISSCC)* (IEEE, 2016), p. 116
3. T. Yasue, K. Tomioka, R. Funatsu, T. Nakamura, T. Yamasaki, H. Shimamoto, T. Kosugi, S. Jun, T. Watanabe, M. Nagase, T. Kitajima, S. Aoyama, S. Kawahito, in *Proceedings of the 2018 IEEE International Solid-State Circuits Conference Digest of Technical Papers (ISSCC)* (IEEE, 2018), p. 90

4. T. Yamazaki, H. Katayama, S. Uehara, A. Nose, M. Kobayashi, S. Shida, M. Odahara, K. Takamiya, Y. Hisamatsu, S. Matsumoto, L. Miyashita, Y. Watanabe, T. Izawa, Y. Muramatsu, M. Ishikawasa, in *Proceedings of the 2017 IEEE International Solid-State Circuits Conference Digest of Technical Papers (ISSCC)* (IEEE, 2017), p. 82
5. Recommendation ITU-R BT.2020, Parameter values for ultra-high definition television systems for production and international programme exchange (2012)
6. R. Funatsu, S. Huang, T. Yamashita, K. Stevulak, J. Rysindki, D. Estrada, S. Yan, T. Soeno, T. Nakamura, T. Hayashida, H. Shimamoto, B. Mansoorian, in *Proceedings of the 2015 IEEE International Solid-State Circuits Conference Digest of Technical Papers (ISSCC)* (IEEE, 2015), p. 112
7. T. Watabe, K. Kitamura, T. Sawamoto, T. Kosugi, T. Akahori, T. Iida, K. Isobe, T. Watanabe, H. Shimamoto, H. Ohtake, S. Aoyama, S. Kawahito, N. Egami, in *Proceedings of the 2012 IEEE International Solid-State Circuits Conference Digest of Technical Papers (ISSCC)* (IEEE, 2012), p. 388
8. H.H. Tippins, Phys. Rev. **140**, A316 (1965)
9. N. Ueda, H. Hosono, R. Waseda, H. Kawazoe, Appl. Phys. Lett. **70**, 3561 (1997)
10. S. Imura, K. Kikuchi, K. Miyakawa, H. Ohtake, M. Kubota, Appl. Phys. Lett. **104**, 242101 (2014)
11. S. Imura, K. Kikuchi, K. Miyakawa, H. Ohtake, M. Kubota, T. Okino, Y. Hirose, Y. Kato, N. Teranishi, I.E.E.E. Trans, Electron Devices **63**, 86 (2016)
12. K. Mineo, S. Imura, K. Miyakawa, K. Hagiwara, M. Namba, H. Ohtake, M. Kubota, in *Abstract of the 2nd International Workshop on* Ga_2O_3 *and Related Materials (IWGO)*, University of Parma, Italy, 12–15 Sep 2017
13. M. Higashiwaki, G. Jessen, Appl. Phys. Lett. **112**, 060401 (2018)

Special Contribution

Chapter 40
Gallium Oxide Materials and Devices

A Personal Recent History

Debdeep Jena

Abstract Gallium oxide has recently been found to be of high interest as the widest bandgap semiconductor for which single crystals bulk substrates are available and whose electronic conductivity can be controlled by n-type doping. Because wide-bandgap semiconductors lead to high breakdown voltages in small length scales, and the resistive losses over small length scales are low, the material has several attractive attributes for high-voltage electronic diodes and switches. In this chapter, I give a personal account of the materials science and physics of this exciting new semiconductor material and its initial use in device demonstrations. The intention of the personal account is to share the twisted and connected paths that lead one to a specific research direction—this is certainly true for how I ended up being interested in gallium oxide as an interesting semiconductor material.

40.1 Pre–2008

As a graduate student at the University of California (Santa Barbara) in 1998–99 (yes, last century!) I was taking a class on nanoscale fabrication taught by Professor Evelyn Hu (she is now at Harvard University). She described the many virtues of silicon dioxide as the amazing dielectric it is for Silicon microelectronics, but lamented the absence of a similar high-quality dielectric for GaAs, the king of compound semiconductors at that time. She described how efforts to oxidize GaAs resulted in the formation of a mixture of various oxides of gallium—which were typically amorphous or volatile··· before discussing the discovery of wet oxidation of Al-containing III–V semiconductors by Dalesasse and Holonyak at the University of Illinois [1]. This oxidation process was all the rage because it is very useful for achieving current confinement in lasers. Now UCSB was a powerhouse of III–V semiconductor materials and devices, and it was an unwritten rule there not to speak

D. Jena (✉)
Departments of Electrical and Computer Engineering and Materials Science and Engineering, Cornell University, Ithaca, NY 14850, USA
e-mail: djena@cornell.edu

M. Higashiwaki and S. Fujita (eds.), *Gallium Oxide*, Springer Series in Materials Science 293, https://doi.org/10.1007/978-3-030-37153-1_40

too much of Silicon. (Unbeknownst to many, the stalwarts of Silicon at the time were hard at work to replace the loved and revered SiO_2 with high-K dielectrics with metal gates—a double whammy that replaced both SiO_2 and the poly-Si gate—released as a product in 2005.)

An alternate oxide for III–V semiconductors was being investigated at that time by the Bell Labs group of Matthias Passlack [2]. I really wanted to join Bell Labs after finishing my Ph.D., so I used to read every paper from Bell Labs that I came across and could understand. I highly recommend reading Passlack et al.'s paper because it explains clearly how they deposited and investigated the properties of amorphous Ga_2O_3 by electron-beam evaporation using a source of $Gd_3Ga_5O_{12}$ over large-area single-crystal GaAs and Si wafers. They measured the stoichiometry (Ga:O=2:3), optical bandgap ($\sim$4.4 eV), low-frequency dielectric constant ($\sim$10), breakdown field ($\sim$3.6 MV/cm), and the fact that the amorphous material was resistant to etching and high electric fields. If the amorphous Ga_2O_3 was annealed to $\sim$600 °C, it would become polycrystalline and etch in 10% HF, and the breakdown field would drop to $\sim$1 MV/cm. Now if the amorphous phase of the material could be this good, imagine what a high-quality single crystal could do! But all I remembered from that time was that the natural oxide of GaAs was a mess of amorphous and volatile sub-oxides of Ga. Over the next decade, several research groups investigated better oxides for III–Vs such as GaAs and InGaAs and have succeeded in finding very good ones. Meanwhile, as a new assistant professor at Notre Dame, I was very busy not worrying about GaAs at all and spent most of my time putting together a young research group investigating electron and hole transport in the III-nitride semiconductors—and mostly in heterostructures buried at a safe distance away from the surface (the exposed surfaces definitely had oxygen!).

40.2 2008–2012

We will take a brief detour from oxides and III–V semiconductors into what may seem an unrelated distraction—but is crucial to how I ended up working on gallium oxide. Since my early graduate school days, I have been fascinated by the physics of transport phenomena: of electrons, heat, light, spin, or what have you—that has been an underlying theme for most of my research effort to date. Around 2007, researchers even remotely interested in electron transport properties could not resist the charm of 2D graphene. I had read the remarkable papers from Manchester and Columbia with great interest. I was in the organizing committee of the DRC/EMC conferences, and Professor Phillip Kim from Columbia University came over and gave a beautiful talk on how they were able to peel off flakes of monolayer graphene and observe the integer quantum Hall effect in it. Now that was at the same time amazing and depressing (to a crystal grower!). From a course project I did with Professor Arthur Gossard at UCSB, I had learnt the amazing discoveries of the integer and the fractional quantum Hall effect through the 1980s and that these effects being most robust in ultraclean samples grown by Molecular Beam Epitaxy.

In fact, Prof. Gossard had himself grown by MBE the AlGaAs/GaAs samples in which the fractional quantum Hall effect was discovered (for which the team that was awarded the 1998 Physics Nobel Prize curiously did not include Prof. Gossard in the awardee list!). So, I was really amazed that the QHE could be observed in a naturally occurring crystal isolated with scotch tape! A cohort of students working with me at the time (Aniruddha Konar, Tian Fang, Kristof Tahy, and Xiangning Luo), and I probably brainwashed each other and decided to start a small project on graphene. That led very fast to expertise on scotch tapes and exfoliation—and soon, we were making graphene FETs and investigating transport in this 2D atomic crystal and its nanostructures. In spite of the amazing transport properties of graphene, the absence of a bandgap bugged me (and all device-minded people). So, it was a relief when in 2010 Andras Kis at EPFL was able to isolate single layers of MoS_2. I had the opportunity to invite him to DRC/EMC, and a brief discussion with him convinced me to investigate this material. Samsung Labs approached me in 2011 with a MoS_2 transistor project because they really liked one of our graphene papers, and I was required to find a theoretical collaborator. I was lucky that Professor Chris van de Walle from UCSB agreed to join the team. Dr. Hartwin Peelaers, a postdoctoral scholar with him at the time, was the chief theorist, who worked out the band structure of van der Waals monolayer and multilayer materials and its effects on strain.

I had been aware of the relatively long and sustained effort in Professor James Speck's group at UC Santa Barbara on the MBE growth studies of various III-oxides in collaboration with the groups of Henning Reichert (at the Paul Drude Institute) and Zbigniew Galazka (at the Institute of Crystal Growth), both in Berlin. In particular, I had seen the very nice work on the MBE growth of In_2O_3 by Dr. Oliver Bierwagen as summarized in his recent review article [3] and wondered if the decent mobilities he was seeing ($\sim$240 cm^2/V s at room temperature) were harbingers of bigger things to come. Even amorphous films of indium gallium zinc oxide (IGZO) can obtain room temperature electron mobilities of $\sim$30 cm^2/V s (hence making them the oxides of choice for energy-efficient TFTs for displays today), meaning that this material system had finally become highly attractive for electronics. However, for high-performance electronics, in which one switches either small voltages at very fast speeds (logic and amplification), or switches very high voltages relatively fast (power electronics), these materials were not competitive to the established Silicon, GaAs, GaN, and SiC transistors. Since I was already working on GaN (and some on SiC), I did not feel the urge to start investigating the high mobility oxides. But one pain point with GaN was the lack of native substrates of high quality—though Ammono substrates were coming along around 2012 from Warsaw, they were very expensive and hard to get. So, we had to be content growing GaN on SiC, Si, or sapphire. I also had especially noticed the first transistor paper on single crystal Ga_2O_3 from Dr. Masataka Higashiwaki's group that can be considered to be the pioneering paper on device applications of this material [4]. I had known Masataka for a long time as a friend and colleague from the GaN HEMT days and met him at the DRC/EMC in 2012, where he showed with his collaborators of Tamura Corporation an ultra-bright GaN LED grown on n-type Ga_2O_3, with the n-type Ga_2O_3 serving as a transparent electron injection contact, unlike sapphire which is insulating. I thought (and still think) that this is a very clever application of an n-type wide-bandgap oxide substrate and will likely become attractive in the future.

40.3 2013–2014

Around the end of 2012 during our collaboration with Prof. van de Walle on 2D materials on the Samsung project, I learnt from him about the recent developments in IKZ Berlin on single-crystal growth and fundamental studies of bulk single-crystal β-Ga_2O_3. Chris was collaborating with them to understand the electronic structure of this crystal. I learnt of the recent exciting developments of the growth of this crystal in Germany by the Czochralski technique [5] and already knew of the work in Japan at Tamura by the Edge-Fed Gradient (EFG) technique. Chris had learnt from the group at IKZ—that when they were polishing the single-crystal β-Ga_2O_3 to create very thin films for plan-view TEM imaging, instead of forming polycrystalline residues, they observed very large single-crystal flakes of β-Ga_2O_3. He suggested to Dr. Roberto Fornari and Dr. Zbigniew Galazka from IKZ to consider sending some β-Ga_2O_3 to my group, as we were "experts" in making transistors out of flakes! Chris, more than anyone else recognized the novelty of the flakes and put 2 and 2 together—and—sure enough, within a few weeks after the introduction, I found a packet in the mail for me from Dr. Galazka as seen in Fig. 40.1. In addition to a few flakes of β-Ga_2O_3, there was one single-crystal box (in the center) which was a $1 \times 1 \times 1$ cm single crystal of β-Ga_2O_3! I remembered Zbigniew's number that these crystals had an unintentionally doped mobile electron density of $\sim 10^{17}$ cm^{-3} and found it amusing that there were $\sim 10^{17}$ conduction electrons in that cube. The reason for my amusement was more because of the fact that I had never held such a large single crystal that was perfectly transparent and conductive at the same time in my hands, as seen in Figure 40.2. Scientific motivation sometimes is not entirely scientific!

Dr. Wan-Sik Hwang at that time was a postdoctoral scholar working with me and Prof. Grace Xing, and he was an expert in E-beam lithography and making FETs out of semiconductor flakes. He used scotch tape to exfoliate thin flakes of β-Ga_2O_3 and within a few days was able to write ohmic contacts to the flake by E-Beam lithography and make an interesting looking transistor in the very first try. The transistor switched over several orders of magnitude and showed decent electron mobility ($\sim 100\,\text{cm}^2/\text{V s}$) and current saturation (=gain), and I was hooked. I remember looking at the beautiful large cube of the crystal (I still have it!) and Wan-Sik's 1st-transistor data side by side—that was likely the moment I decided that we must start investigating this material in earnest immediately. The main motivation was that to that date, I had never seen a transparent and conductive single crystal of GaN that big. A direct measurement of the optical transmission spectrum through the flakes of β-Ga_2O_3 performed by Amit Verma directly confirmed the bandgap to be ~ 4.4 to 4.6 eV, depending on the direction of light propagation, matching well with Hartwin

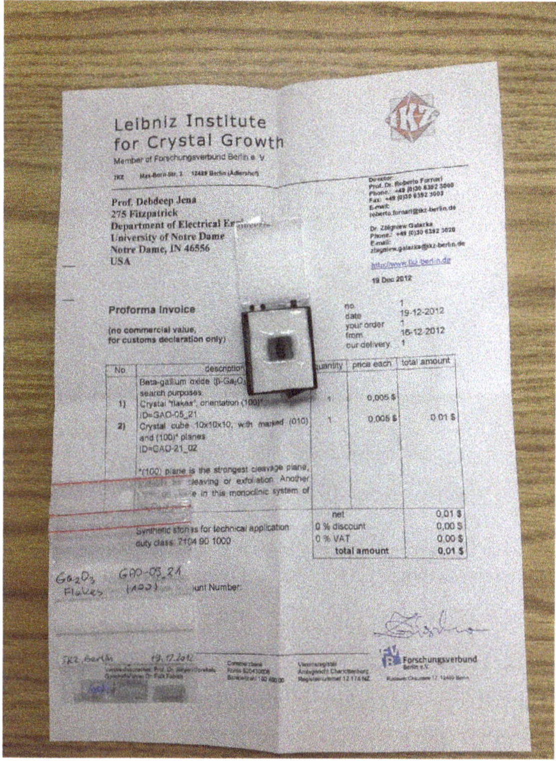

Fig. 40.1 A $1\,cm^3$ β-Ga_2O_3 crystal from IKZ Berlin (Dec. 2012)

Peelaers and van de Walle's first-principle calculations. We wrote up this report in a joint paper with the groups at IKZ Berlin and UC Santa Barbara [6], our first paper on gallium oxide and the first gallium oxide FET outside of Japan! Van de Walle's group had investigated the band structure of β-Ga_2O_3 for several years, and Hartwin wrote a definitive account in [7] indicating a predominantly Ga–4*s* orbital conduction band edge, with an electron effective mass of $\sim 0.28\,m_e$, where m_e is the rest mass of a free electron.

Publication of the Ga_2O_3 nanomembrane FET paper generated significant interest from the 2D materials community. I presented Wan-Sik's experimental results as a part of an invited talk on 2D materials at the June 2014 Graphene Week in Gothenburg. It was a veiled attempt on my part to pass a 3D quasi-layered material as a 2D material because of its tendency to form flakes! I was waiting at the Gothenburg airport lounge to catch my flight back to the USA, when the discoverer of graphene—Professor Andre Geim stopped by and requested if we could share a few flakes. We shipped him some—if we were experts in making FETs with flakes, Geim's group were (and are) the absolute masters! I was supposed to meet him two months later at the International Conference on the Physics of Semiconductors (ICPS) in August 2014

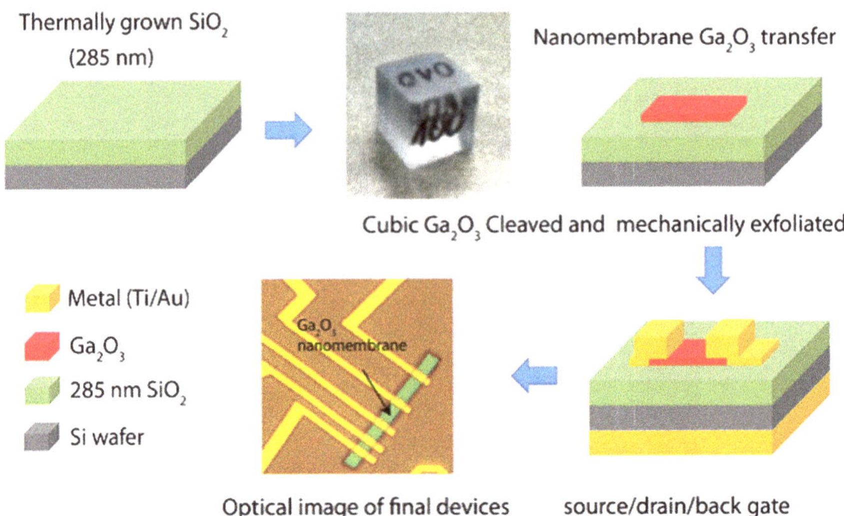

Fig. 40.2 Nanoscale Ga_2O_3 layers are found to easily exfoliate using scotch tape. This property was used to fabricate nanomembrane Ga_2O_3 channel transistors (2013/2014)

in Austin, Texas, to discuss their results. Prof. Geim was a speaker at the Nobel Prize session of the ICPS, but rather uncharacteristically did not show up. It turns out that before the conference he was taking a short vacation, and was out on a boat in Colorado. The boat capsized, and though Prof. Geim was safe, he was on the verge of hypothermia. While he was waiting on the river bank to be picked up by a rescue helicopter, to add insult to injury, he got bitten by a snake! If that is not a hell of an excuse not to show up for a meeting, I don't know what is. He recovered after a few days in the hospital, but went through quite the ordeal. Prof. Geim did email and follow up that they had managed to exfoliate flakes of β-Ga_2O_3 down to ~5 nm (down from several 10 s of nm in our group) and still could observe good transistor action with on/off switching. But the electron mobilities they observed at Manchester were not that high (at least not as high as graphene!), and it did not show any interesting magnetic behavior up to 15 Tesla—which is why they did not pursue it further—and did not publish. However, it was very interesting to know that transistor action was preserved in β-Ga_2O_3 flakes down to 5 nm, and the transistor action got better at 4 K than at room temperature, all hallmarks of robust s-orbital electrons!

The second major reason why I decided to take the plunge was the strong influence of two colleagues—Dr. Masataka Higashiwaki and Dr. Gregg Jessen—both of whom I had the fortune of knowing as good friends back from the days when they used to work on GaN transistors. Masataka had moved back to NICT from a visit to UCSB and had just published the pioneering paper on planar MESFETs with β-Ga_2O_3 as mentioned earlier, on substrates grown at Tamura by EFG, and subsequently the epitaxial layer by MBE growth [5]. Dr. Gregg Jessen following his illustrious work at AFRL on high speed GaN HEMTs, had spent a few years in Japan as an AFOSR program manager, and at more than one occasion reminded me at conferences and workshops of the developments in β-Ga_2O_3 in Dr. Higashiwaki's group which he was excited about. As he was heading back to the AFRL after his overseas appointment, he was seriously considering initiating a project on gallium oxide electronics, motivated by Masataka's group's results.

At that time, it became very clear to me that this material had a bright future from a scientific viewpoint—and it may even find some real applications! Also, MBE could be used to investigate its properties—which is something I really liked. β-Ga_2O_3 remains to this day the widest bandgap large-area single-crystal bulk substrate which can be controllably doped *n*-type. When Dr. Higashiwaki came to know I had started working on β-Ga_2O_3, he immediately invited me to stop by and visit NICT and Tamura, both near Tokyo. This trip to Tamura and NICT in May 2013 opened my eyes to the effort on the EFG large-area single crystals, and we decided to collaborate right there, including us purchasing and investigating their EFG single-crystal substrates. I also made new friends with Dr. Kuramata and Dr. Sasaki from Tamura right away, something that will be crucial for our later efforts. I also remember making a trip back to Japan later in August 2013, for a bi-annual workshop called TWHM in Hakodate in Hokkaido island, where during a memorable evening of dinner Masataka suggested the idea of potentially bringing together the small number of people working on Ga_2O_3 in a workshop—and asked if I could help shore up attendees from the USA. I was all too glad to help him and strongly encouraged him to lead the organization.

Meanwhile, it was very opportune that Prof. Xing and I together were able to obtain funds for the purchase of an oxide MBE system in 2013. The oxide system was purchased and installed at Notre Dame in Fall 2013. Oxygen was supplied through a plasma source, and effusion cells were used for the metals and dopants. MBE growths of β-Ga_2O_3 began in early 2014 in our laboratory, and very soon graduate students Amit Verma and Kasra Pourang, together with an undergraduate student Austin Hickman, were able to obtain high-quality homoepitaxial growth on single-crystal substrates from Tamura and from IKZ Berlin. The plasma-MBE growth conditions discovered by Prof. Speck's group at UCSB and those used at Tamura/NICT worked like a charm for homoepitaxy. Example transmission electron microscope images of a homoepitaxially grown single-crystal sample along the 100 direction, and a thin film ($\bar{2}01$)-oriented β-Ga_2O_3 layer heteroepitaxially grown during this period on c-plane sapphire is shown in Fig. 40.3. These images are taken by Dr. Sergei Rouvimov at Notre Dame and show that homoepitaxial growth in the right conditions produces nearly perfect β-Ga_2O_3 without extended defects. And that growth on c-plane sapphire results in ($\bar{2}01$)-oriented β-Ga_2O_3 with quite some dislocations but not that different, from say GaN grown on Silicon or sapphire. Now sapphire has a corundum structure, and the alpha-phase of Ga_2O_3 can be grown on it too—it is a metastable phase of Ga_2O_3 compared to the stable (monoclinic) beta phase. Metastable phases can potentially be stabilized by epitaxial strain for thin layers. This has been recently investigated by several groups, including Prof. Fujita's group by his student Riena Jinno by mist-CVD [8], and Prof. Oshima's group by plasma MBE [9].

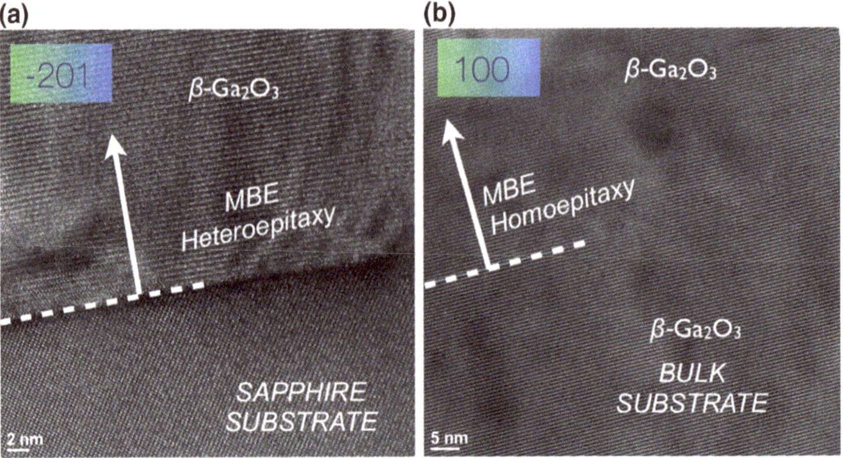

Fig. 40.3 Plasma Oxygen Molecular Beam Epitaxy (MBE) growth of β-Ga_2O_3 in 2014. **a** Heteroepitaxy on c-plane sapphire showing the [$\bar{2}01$] direction of Ga_2O_3 aligns with the c-axis of sapphire. **b** Homoepitaxy of (100)-oriented Ga_2O_3 on bulk single-crystal substrates

However, all our efforts to dope the epitaxial Ga_2O_3 layers with silicon in plasma MBE failed to make it conductive at that time. We were using a standard Si effusion cell in the plasma oxygen environment without any special (differential) pumping of the dopant source. Figure 40.4 shows example SIMS profiles of homoepitaxial β-Ga_2O_3 layers with Silicon-doping stacks at various miscut angles—these samples were grown simultaneously on co-loaded wafers. Because of the oxidation of the Silicon source material in the oxygen plasma environment, it was found that the deposited species were oxides of silicon rather than elemental silicon. A loss of control of the Silicon-doping density was observed in the oxygen plasma environment, compared to a nitrogen plasma environment. SIMS characterization of the samples indicated high Silicon incorporation, but the samples were insulating. It is possible that there are ways to protect the silicon dopant from oxidation in the plasma oxygen MBE environment (and control doping over a wide range of densities) that we are not aware of yet. A differentially pumped Silicon source, or even a gas source may have better success than a standard effusion cell as a Silicon dopant in a conventional oxygen plasma MBE.

During my visit to Tamura and NICT in Japan mentioned earlier, together with Dr. Kuramata and Dr. Higashiwaki, we had decided to investigate the rather important question of the thermal conductivity of β-Ga_2O_3. A brief discussion of potential thermoelectric applications had come up that could potentially exploit the low thermal conductivity and the high electrical conductivity of the crystal, especially at high temperatures—answering this question would require a measurement of the Seebeck coefficient. Upon my return, Prof. Tengfei Luo, at that time a young assistant professor, agreed to measure it using time-domain thermoreflectance method in his laboratory. The measurements of thermal conductivity were combined with Amit

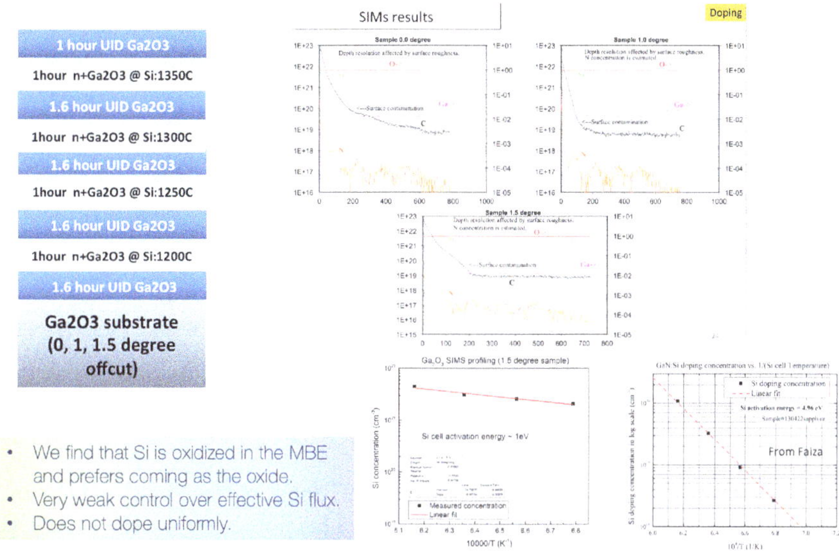

Fig. 40.4 SIMS profiles of Si-doped Ga_2O_3 layers grown by Plasma MBE in 2014. These initial layers were insulating, indicating the silicon incorporated was not able to provide free electrons in the band

Verma's measurements of the heat capacity of the crystal along three primary orientations. This paper led by Prof. Luo was completed at the end of 2014 and appeared in 2015 [10]. The low thermal conductivity of β-Ga_2O_3 (~10 to 30 W/m K) is an intrinsic property, and every device that handles a lot of energy must be engineered with the electro-thermal coupling and self-heating factored into the device design from the ground up. It is an order of magnitude lower than the thermal conductivities of SiC and GaN, and there is simply no escaping this fact! It is conceivable that the capacity of the material to flake can be put to good use to find innovative solutions to this problem.

40.4 2015–2017

Prof. Xing and I moved our research groups and laboratories to Cornell University in early 2015. We could not move the oxide MBE, resulting in a ~2.5-year hiatus in our plasma MBE growth efforts of Ga_2O_3. Since in the initial phases at Cornell we were unable to grow materials, we focused on careful studies of electron transport and optical properties in the material. During my visit to NICT, I had met Prof. Onuma and shared some of our initial optical data with him, as seen in Fig. 40.5.

In particular, Dr. Protasenko and Amit Verma in our group had observed by single-photon excitation at 157 nm (and confirmed by two-photon spectroscopy) that the

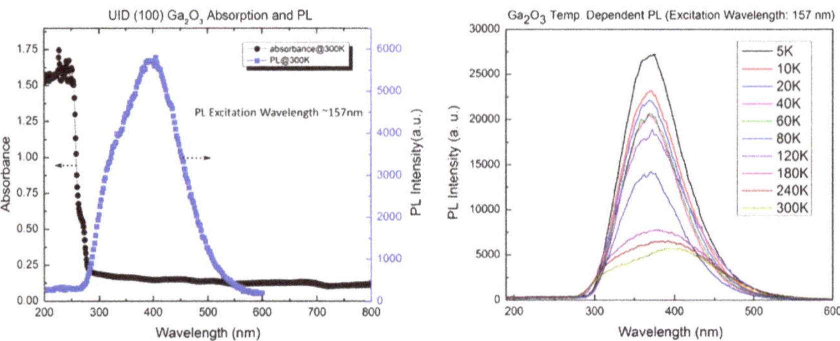

Fig. 40.5 Absorption spectra and photoluminescence of unintentionally doped Ga_2O_3 substrate (left), and a temperature-dependent photoluminescence spectra (right) measured in 2014

onset of the absorption was at ~4.5 to 4.6 eV (~270 nm), but the photoluminescence peaked in the ~300 to 400 nm window. This was rather interesting—and we had a discussion on the potential reason for what appeared to be an uncharacteristically large Stoke's shift between the absorption edge and the emission peak. The temperature-dependent photoluminescence intensity increased by 5X when the temperature was lowered from 300 to 5 K. A crude estimation indicates an internal quantum efficiency of ~20%. Now this is a rather large IQE if we consider β-Ga_2O_3 to be an indirect bandgap semiconductor. On the other hand, if the hole in the valence band is highly localized by virtue of being an oxygen $2p$ state, coupled with the large polarity of the crystal and potential defects, the hole can essentially get trapped in its own potential well and become immobile forming a polaron. This interesting aspect is still under investigation by various groups, and it is also possible that vacancies and defects play a role in the emission process. Prof. Onuma published a clear paper in 2015 on the absorption edges in β-Ga_2O_3 measured by polarized light absorption and was able to conclude the ordering of the valence bands [11]. With careful Raman spectroscopy, he had identified a large number of phonon modes as may be expected from the rather large unit cell of the crystal, and there were quite a few polar optical modes around 10 s of meV. For comparison, the polar optical phonon energy of GaN is ~90 meV.

Because at the time we did not have the plasma MBE running at Cornell, I discussed with Dr. Kuramata and Dr. Sasaki from Tamura, who had spun out a Ga_2O_3 company from Tamura called Novel Crystal Technology (NCT), about the possibility of creating a 2D electron gas in a modulation-doped $(Al_xGa_{1-x})_2O_3$ heterostructure. Dr. Sasaki agreed to try to grow it in collaboration with us. These were rather expensive samples for us, so we wanted to get it right.

Figure 40.6 shows the heterostructure we designed for the 2DEG by performing energy band diagram calculations with mostly unknown(!) band offsets and dopant activation energies of $(AlGa)_2O_3$. Dr. Sasaki put in substantial effort at NCT to grow the structure: Fig. 40.7 shows the structure grown at NCT by Dr. Sasaki by MBE with Sn modulation doping, the corresponding X-ray diffraction of the structure, and a TEM image of the structure.

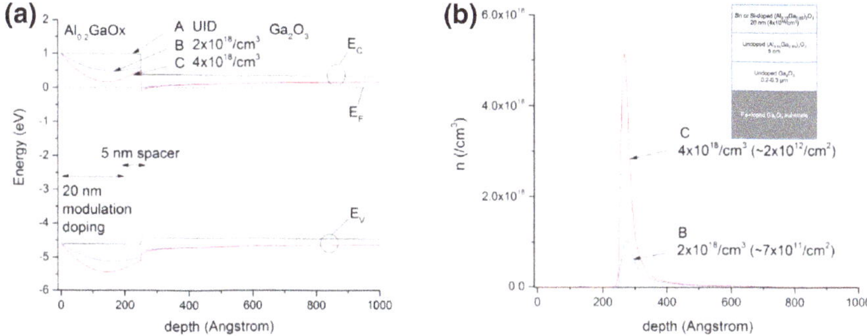

Fig. 40.6 **a** Proposed modulation-doped $(Al,Ga)_2O_3$ heterostructure for the observation of a 2D electron gas (November 2015). The calculated self-consistent Schrodinger/Poisson energy band diagram (**a**) and corresponding 2D electron gas (**b**) assumed a shallow dopant level, and a conduction band offset of ~0.5 eV, unknown at the time

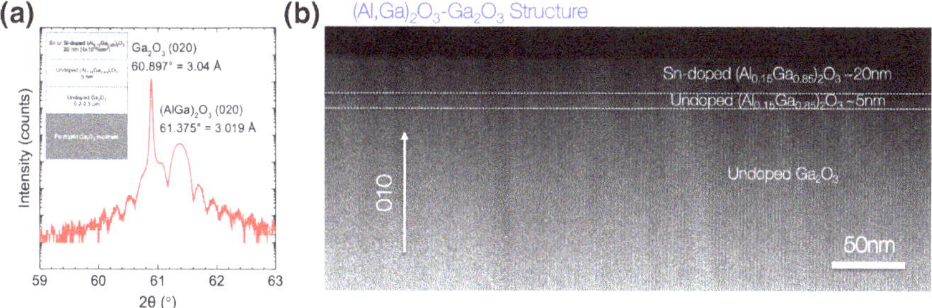

Fig. 40.7 **a** X-ray diffraction pattern, and **b** TEM image of a modulation-doped $(Al, Ga)_2O_3$ heterostructure aimed at the formation of a 2D electron gas at the heterojunction (December 2015). The sample was grown by Dr. Sasaki of NCT, and the imaging performed by Celesta Chang at Cornell University

The TEM image is taken by Celesta Chang, a Ph.D. student working with Prof. David Muller at Cornell. The $(Al_xGa_{1-x})_2O_3$ heterostructure was doped with Sn with a spacer layer and exhibited a reasonable heterointerface. The samples looked structurally what was desired for a modulation-doped 2DEG. We were however not able to measure a 2DEG in this structure by standard van der Pauw Hall effect. As discussed later, the 2DEG in similar structures was reported in 2017 by Prof. Speck's group at UCSB, and Prof. Rajan's group at Ohio State. But the Sn-doped heterostructures we had at hand in early 2016 were completely insulating—Dr. Verma tried several tricks to form ohmic contacts to a buried 2DEG, but there were no signs of free electrons in the structure. We are to this day still trying to figure out exactly why!

Tamura researchers had successfully demonstrated MOCVD growth of blue InGaN quantum well LEDs for solid state lighting already. In fact, it was an initial motivation for the growth of single-crystal β-Ga_2O_3 in the first place! To me,

Fig. 40.8 Participants of the 1st International Workshop on Gallium Oxide (IWGO) in front of the ROHM building of Kyoto University (November 2015)

an interesting question was (and remains)—if MBE could create nitride/oxide heterostructures of electronic quality. Since at the time, we could grow GaN by plasma MBE, we began the growth of heterostructures of GaN and AlN directly on β-Ga_2O_3. These growths were performed on the $(\bar{2}01)$ surface of β-Ga_2O_3 so that the hexagonal lattice symmetry necessary for wurtzite GaN is available. We could observe nice heterojunction Schottky barriers and even signatures of mobile carrier confinement at the surface of *n*-doped Ga_2O_3 when AlN was grown on it. Such structures may be of interest in the future for nitride/oxide heterojunctions that potentially offer so many unique and untapped possibilities.

The first International Workshop on Gallium Oxide (IWGO-1) was held in the ROHM building at the outskirts of Kyoto University. It was organized by Dr. Higashiwaki of NICT and Prof. S. Fujita of Kyoto University—whose group has investigated the growth of gallium oxide for several years using the mist-CVD growth method. As I had promised Dr. Higashiwaki, I served on the program committee and helped out with participants from USA. It was a memorable workshop, most of the early adopters of this material system were there (Fig. 40.8). The atmosphere was very exciting, as befits a new material system with infinite possibilities. In addition to researchers and scientists, there were several interested industrial attendees scoping out the potential applications of this new wide-bandgap semiconductor material.

At Cornell, a new Ph.D. student Nick Tanen and Dr. Amit Verma (now a postdoctoral scholar) had been measuring the electron mobility in β-Ga_2O_3 bulk crystals at various temperatures, and postdoctoral scholar Dr. Nan Ma was modeling various scattering mechanisms to explain the measured data. The connection of the low optical phonon energies, coupled with the rather large ionicity of the crystal, was important in formulating an explanation for the measured dependence of the electron mobility on temperature. In particular, in our resulting 2016 paper, the expected intrinsic electron mobilities in β-Ga_2O_3 were estimated, by comparing experimental data to a model of the various of scattering mechanisms [12]. In this work, a single "effective" phonon mode was used to explain the experimental data. Parsini and Fornari had also come to a similar conclusion, albeit finding another phonon mode to be the dominant scattering mechanism [13]. A more comprehensive story with a first-principles evaluation of the phonon modes was found by Prof. van de Walle's group [14], and by Prof. Uttam Singisetti's group at Buffalo [15].

The IKZ group, led by Dr. Klaus Irmscher, had in 2011 found the room temperature electron mobilities to be not too far from $\sim 100\,cm^2/V\,s$ and had identified several deep level defects using the excellent crystals grown by Dr. Galazka [4]. I had the opportunity to attend a gallium oxide workshop organized jointly by the Paul Drude Institute and the IKZ in Berlin in 2016 September. There I learnt from Dr. Irmscher the spin-resonance measurements they had performed to identify the nature of the defects in β-Ga_2O_3. In that workshop, we were able to witness the Czochralski oxide growth systems of Dr. Galazka, and the oxide MOCVD system operated by the group of Dr. Gunther Wagner. Dr. Wagner gave us a memorable tour of the facilities—one of the vivid images I have is their effort to produce a small "soccer-ball" size single crystal of Silicon—in which every atom of Silicon was the same isotope! Such a crystal is now used to redefine the fundamental standard of mass.

At the 2016 International Conference on the Physics of Semiconductors (ICPS) in Beijing, I was discussing the new developments of β-Ga_2O_3 with Prof. Peide Ye of Purdue University. Among the several excellent research achievements of Prof. Ye, his group is prolific in making excellent transistors with 2D materials. He became especially fascinated by the fact that nanomembranes of β-Ga_2O_3 could be exfoliated and could be made into rudimentary transistors. After returning to Purdue, he initiated an effort in his group on nanomembrane transistors of this material and soon had good success. With a sustained effort, his group has now succeeded in making β-Ga_2O_3 nanomembrane FETs that carry more than 1 mA/μm current densities [16], which shows clearly the potential of this material. Also, by integrating nanomembranes with high thermal conductivity substrates, he is exploring ways to overcome the limitations posed by the material's intrinsic low thermal conductivity. He has also been able to integrate the nanomembranes with ferroelectric gate dielectrics, and he has also observed transistors that switch below the thermal Boltzmann limit of 60 mV/decade at room temperature exploiting the proposed phenomena of effective negative capacitance [17].

At Cornell, Prof. Darrell Schlom, an expert in the MBE growth of a wide range of perovskite oxide materials (semiconductors, ferroelectrics, ferromagnets, multiferroics, superconductors, etc.), got interested in β-Ga_2O_3. He made his ozone MBE system available for epitaxial growth. Dr. Amit Verma and Ph.D. student Nick Tanen, with Dr. Hanjong Paik began the exploration of ozone MBE growth of β-Ga_2O_3 on sapphire and on single-crystal substrates. Following reports from Prof. Speck's group at UCSB on Ge-doping of β-Ga_2O_3 by plasma MBE, they were able to demonstrate the growth of Ge-doped conducting Ga_2O_3 epitaxial layers by ozone MBE too. The doping densities that were achieved were on the high 10^{18} cm^{-3} side, with electron mobilities below 100 cm^2/V s, with smooth surfaces and high homoepitaxial layer quality. Meanwhile, a master's student, Liheng Zhang at Cornell, was developing ICP-RIE etching of Ga_2O_3. Liheng was taking a course on nanofabrication from Prof. Grace Xing (rings a bell from the 1st paragraph of this article!) and got the idea to design the experiment for a physical versus chemical etch of Ga_2O_3. Together with Dr. Amit Verma, they were able to find a condition that led to nearly vertical etched sidewalls, which was an important finding for some laterally gated devices to follow [18]. Though the sidewalls are not completely devoid of electronic traps, we have the opportunity now to solve that problem using ideas from other semiconductors.

In 2017, an old GaN MBE system was reconfigured at Cornell with a plasma oxygen source, and we were back to growing β-Ga_2O_3 by plasma MBE. Nick Tanen was successful in Ge-doping of thin film Ga_2O_3 (similar to the results from Prof. Speck's group at UCSB), but success in controlling low-doping levels has remained elusive. Controlled, low-doping levels in the 10^{15} cm^{-3} range are necessary for high-voltage devices. On the brighter side, several $(Al_xGa_{1-x})_2O_3/Ga_2O_3$ heterostructures, multiple quantum wells, and heterostructures have been realized up to 15% Al composition.

In September 2017, the second edition of IWGO was held at Parma, Italy, chaired by Prof. Roberto Fornari who had moved to Parma from IKZ Berlin since our initial collaboration. It was a very well-organized conference, with a much larger attendee list than initially anticipated, showing the health of the growing field. In the conference, several excellent papers were presented, and the science of growth, electronic, thermal, and optical properties were discussed at depth. Of special interest (at least to me), the rather remarkable enhancement of the β-Ga_2O_3 growth rate in plasma MBE reported by the catalytic use of Indium by Vogt from Dr. Oliver Bierwagen's group at PDI Berlin [19], the report of experimental observation of potential ferroelectricity in epsilon-phase Ga_2O_3 by Prof. Fornari's group in Parma, and the reports of modulation-doped 2DEG structures by Prof. James Speck's [20] and Prof. Siddharth Rajan's group [21] are major steps forward for this material system. These advances in the growth science will bring a high level of control of β-Ga_2O_3 layers and heterostructures for up and coming electronic devices.

40.5 2018–Future

Several attractive high-voltage Schottky diodes and lateral MOSFETs on the β-Ga_2O_3 platform have been demonstrated by the leading research groups of Dr. Higashiwaki and Dr. Man Hoi Wong at NICT, Japan, and Dr. Gregg Jessen and Dr. Kelson Chabak at AFRL over the past few years. Some of this is reviewed in their comprehensive recent survey article of the field [22]. The growth of β-Ga_2O_3 by HVPE, pioneered by Prof. Kumagai and now commercialized by NCT has made impressive strides and makes it industrially attractive with mobilities exceeding 5000 cm^2/V s at low temperatures [23]. Using the vertical etch profiles, some of the first kilovolt-class normally-off β-Ga_2O_3 vertical FETs have now been demonstrated in Prof. Grace Xing's group at Cornell [24]. Industrial interest in this material system is thus picking up—in an opportune time, when the control of high voltages in confined geometries is of high interest for energy-efficient power electronics.

That the bandgap of the material is so large also has raised interest in the material platform for RF electronics at high voltages. These applications must wrestle with the relatively lower electron mobility and low thermal conductivity of the β-Ga_2O_3 crystal. But one must not underestimate the resourcefulness of an active community of researchers in overcoming at least some of these problems. Gallium oxide enthusiasts can find solace in the fact that GaAs has a thermal conductivity of ~50 W/m K and is a workhorse in RF electronics and lasers, both of which require efficient heat dissipation. Much of the materials science of the various phases of β-Ga_2O_3, its doping and heterostructures with In and Al and Sn remain mostly unexplored. The baton to host and cultivate this young community passes from our Japanese colleagues (IWGO-1, 2015) and European colleagues (IWGO-2, 2017) to American researchers. Prof. Speck and I made a bid at IWGO-2 for researchers from the USA to host the next IWGO, which was accepted by the committee. IWGO-3 will be held in 2019 in Columbus, and we expect it to be as full of potential as the several new and young researchers it looks forward to welcome.

Acknowledgements Much of the initial phase of the work in my research group was kindly supported by the National Science Foundation DMREF Program under Grant 1534303 monitored by Dr. J. Schlueter and in part by AFOSR under Grant FA9550-17-1-0048 monitored by K. Goretta. Because of interest in this field and the formative work in the past few years, several countries have increased financial support of this material. For example, following the formative materials and device work in Japan, the GraFOX initiative was launched in Germany to study fundamental properties of oxides. In the USA, a Multidisciplinary University Research Initiative (MURI) was launched by Dr. Ali Sayir of the AFOSR in 2018. I believe these are the first steps toward building a firm foundation for an emerging field, and there will be several more in the future.

I would like to acknowledge the many collaborators mentioned in this chapter, and beyond, for introducing me to β-Ga_2O_3, and giving me the opportunity to work on this semiconductor system. It is inevitable that I have skipped or missed several important topics, developments, and people in this personal review—for which I sincerely apologize. The purpose of this article is not so much to present a comprehensive review of the field for which we have this whole book and its excellent chapters. Rather, I have discussed candidly the various unexpected connections, unanticipated opportunities, constant struggles, and most importantly, the joy of discovery in this emerging field that I have experienced by my fortune of being an early adopter. I hope this chapter gives young researchers entering the field a sense that research in science and engineering is after all done by human beings (at least till now, till machines can learn and A.I. takes over!) and is a beautiful example of how friends across countries and continents work together to make true advance in a field and uncover new materials and phenomena. Personal friendships and connections, love for the subject, and a strong belief go a long way in uncovering new science and building new technologies from a mere few atoms in the seemingly vast abyss of combination of elements in the periodic table.

References

1. J.M. Dallesasse, N. Holonyak, J. Appl. Phys. **113**, 051101 (2013)
2. M. Passlack, E.F. Schubert, W.S. Hobson, M. Hong, N. Moriya, S.N.G. Chu, K. Konstadinidis, J.P. Mannaerts, M.L. Schnoes, G.J. Zydik, J. Appl. Phys. **77**, 686 (1995)
3. O. Bierwagen, Semicond. Sci. Technol. **30**, 024001 (2015)
4. K. Irmscher, Z. Galazka, M. Pietsch, R. Uecher, R. Fornari, J. Appl. Phys. **110**, 063720 (2011)
5. M. Higashiwaki, K. Sasaki, A. Kuramata, T. Masui, S. Yamakoshi, Appl. Phys. Lett. **100**, 013504 (2012)
6. W.S. Hwang, A. Verma, H. Peelaers, V. Protasenko, S. Rouvimov, H. Xing, A. Seabaugh, W. Haensch, C. van de Walle, Z. Galazka, M. Albrecht, R. Fornari, D. Jena, Appl. Phys. Lett. **104**, 203111 (2014)
7. H. Peelaers, C.G. Van de Walle, Phys. Status Solidi B **252**, 828 (2015)
8. R. Jinno, T. Uchida, K. Kaneko, S. Fujita, Appl. Phys. Express **9**, 071101 (2016)
9. T. Oshima, Y. Kato, M. Imura, Y. Nakayama, M. Takeguchi, Appl. Phys. Express **11**, 065501 (2018)
10. Z. Guo, A. Verma, X. Wu, F. Sun, A. Hickman, T. Masui, A. Kuramata, M. Higashiwaki, D. Jena, T. Luo, Appl. Phys. Lett. **106**, 111909 (2015)
11. T. Onuma, S. Saito, K. Sasaki, T. Masui, T. Yamaguchi, T. Honda, M. Higashiwaki, Jpn. J. Appl. Phys. **54**, 112601 (2015)
12. N. Ma, N. Tanen, A. Verma, Z. Guo, T. Luo, H. Xing, D. Jena, Appl. Phys. Lett. **109**, 212101 (2016)
13. A. Parsini, R. Fornari, Semicond. Sci. Technol. **31**, 035023 (2016)
14. Y. Kang, K. Krishnamurthy, H. Peelaers, C. van de Walle, J. Phys.: Condens. Matter **29**, 234001 (2017)
15. K. Ghosh, U. Singisetti, J. Mater. Res. **32**, 4142 (2017)
16. H. Zhou, K. Maize, G. Qiu, A. Shakouri, P. Ye, Appl. Phys. Lett. **111**, 092102 (2017)
17. M. Si, L. Yang, H. Zhou, P. Ye, ACS Omega **2**, 7136 (2017)
18. L. Zhang, A. Verma, H. Xing, D. Jena, Jpn. J. Appl. Phys. **56**, 030304 (2017)
19. P. Vogt, O. Brandt, H. Reichert, J. Lahnemann, O. Bierwagen, Phys. Rev. Lett. **119**, 196001 (2017)
20. E. Ahmadi, O.S. Koksaldi, X. Zheng, T. Mates, Y. Oshima, U.K. Mishra, J.S. Speck, Appl. Phys. Express **10**, 071101 (2017)
21. Y. Zhang, A. Neal, Z. Xia, C. Joishi, J. Johnson, Y. Zheng, S. Bajaj, M. Brenner, D. Dorsey, K. Chabak, G. Jessen, J. Hwang, S. Mou, J. Heremans, S. Rajan, Appl. Phys. Lett. **112**, 173502 (2018)
22. M. Higashiwaki, G.H. Jessen, Appl. Phys. Lett. **112**, 060401 (2018)
23. K. Goto, K. Konishi, H. Murakami, Y. Kumagai, B. Monemar, M. Higashiwaki, A. Kuramata, S. Yamakoshi, Thin Solid Films **666**, 182 (2018)
24. Z. Hu, K. Nomoto, W. Li, N. Tanen, K. Sasaki, A. Kuramata, T. Nakamura, D. Jena, H.G. Xing, IEEE Electron Device Lett. **39**, 869 (2018)

Index

M. Higashiwaki and S. Fujita (eds.), *Gallium Oxide*, Springer Series in Materials Science 293, https://doi.org/10.1007/978-3-030-37153-1

E

H

I

T

GPSR Compliance
The European Union's (EU) General Product Safety Regulation (GPSR) is a set of rules that requires consumer products to be safe and our obligations to ensure this.

If you have any concerns about our products, you can contact us on

ProductSafety@springernature.com

In case Publisher is established outside the EU, the EU authorized representative is:

Springer Nature Customer Service Center GmbH
Europaplatz 3
69115 Heidelberg, Germany

www.ingramcontent.com/pod-product-compliance
Ingram Content Group UK Ltd.
Pitfield, Milton Keynes, MK11 3LW, UK
UKHW021918270726
14059UKWH00002B/89